Bent Functions

Sihem Mesnager

Bent Functions

Fundamentals and Results

Sihem Mesnager
University of Paris VIII
Paris, France

ISBN 978-3-319-81322-6 ISBN 978-3-319-32595-8 (eBook)
DOI 10.1007/978-3-319-32595-8

Softcover reprint of the hardcover 1st edition 2016

Printed on acid-free paper

This Springer imprint is published by Springer Nature
The registered company is Springer International Publishing AG Switzerland

In everyone's life, at some time, our inner fire goes out.
It is then burst into flame by an encounter with another human being.
We should all be thankful for those people who rekindle the inner spirit.

— Albert Schweitzer

To my husband Laurent

List of Symbols and Notation

General

Symbol	Description	Page
$\#S$	Cardinality of the set S	viii
$\mathbb{C}$	Field of complex numbers	viii
$\mathbb{F}_{2^n}$	Finite field with 2^n elements	viii
$\mathbb{F}_q$	Finite field with q elements	viii
$\mathbb{N}$	Semiring of non-negative integers	viii
$\mathbb{R}$	Field of real numbers	viii
$\mathbb{Z}$	Ring of integers	viii
$\overline{K}$	An algebraic closure of K	viii
K, L	A perfect field or a number field	viii
l	A prime number different from p	viii
p	A prime number	viii
q	A prime number's power	viii
$\mathbb{F}_2^n$	The vectorspace of all binary vectors of length n	viii

Boolean Functions

Symbol	Description	Page
$\mathcal{RM}(r, n)$	Reed–Muller code of order r and length 2^n	83
$AI(f)$	Algebraic immunity of f	49
$nl(f)$	The nonlinearity of a Boolean function f	48

Tu–Deng Conjecture

Symbol	Description	Page
$a \simeq b$	Cyclotomic equivalence for modular integers	59
k	Exponent for modulus $2^k - 1$	58
$P_{t,k}$	Fraction of modular integers in $S_{t,k}$	62
$r(a, t)$	Number of carries occurring while adding a to t	61
$S_{t,k}$	Set of the Tu–Deng conjecture	62
$S_{t,v,u,k}$	Set of the generalized Tu–Deng conjecture	58
t, a, b	Modular integers	58

Elliptic Curves

Symbol	Description	Page
$\mathbb{Z}_{p^n}$	Unramified extension of degree n of $\mathbb{Z}_p$	300
E_a	Elliptic curve associated with a	266

Sets and Families

Symbol	Description	Page
$\mathbb{Q}(\xi_p)$	The cyclotomic field	42
$\mathcal{F}_n$	The set of functions $f_{a_r}(x) = \sum_{r\in R} \mathrm{Tr}_1^n a_r x^{r(2^m-1)}$	207
$\mathcal{H}_n$	The set of functions $f_{a,b}(x) = \sum_{r\in R} \mathrm{Tr}_1^n a_r x^{r(2^m-1)} + \mathrm{Tr}_1^t b x^{\frac{2^m+1}{\tau}(2^m-1)}$	281
$\mathcal{H}_n$	Family of functions of the form $f_{a,b}(x) = \sum_{r\in R} \mathrm{Tr}_1^n a_r x^{r(2^m-1)} + \mathrm{Tr}_1^t b x^{s(2^m-1)}$	230
$\mathcal{PS}$	Partial Spread class	96
$\mathcal{PS}^+$	Partial Spread class	94
$\mathcal{PS}^-$	Partial Spread class	94
$\mathcal{PS}_{ap}$	Partial Spread class	95
$\mathcal{PS}_{ap}^{\#}$	Class $\mathcal{PS}_{ap}^{\#}$	174
$\mathfrak{F}_n$	The set of functions $f_{a,b}(x) = \mathrm{Tr}_1^n(ax^{(2^m-1)}) + \mathrm{Tr}_1^2 b x^{\frac{2^n-1}{3}}$	176
$\mathfrak{G}_n$	The set of functions $g_{a,b}(x) = \mathrm{Tr}_1^n(ax^{3(2^m-1)}) + \mathrm{Tr}_1^2 b x^{\frac{2^n-1}{3}}$	193
$\mathfrak{H}_n$	The set of functions $f_{a_r,b}(x) := \sum_{r\in R} \mathrm{Tr}_1^n a_r x^{r(2^m-1)} + \mathrm{Tr}_1^2 b x^{\frac{2^n-1}{3}}$	208
$\mathfrak{W}_n$	The set of functions $f_{a,b}(x) := \sum_{r\in R} \mathrm{Tr}_1^n a_r x^{r(2^m-1)} + \mathrm{Tr}_1^4 b x^{\frac{2^n-1}{5}}$	222
$PG_2(2^n)$	The projective space of dimension 2 over finite field $\mathbb{F}_q$	161
$PG_n(q)$	The projective space of dimension n over $\mathbb{F}_{2^n}$	161
$\mathcal{M}$	Maiorana–McFarland class	93

Finite Fields and Exponential Sums

Symbol	Description	Page
χ_f	The "sign" function of a Boolean function f	22
$\mathcal{F}(f)$	The Fourier transformation of f	23
$\mathcal{T}_i$		28
$\overline{\mathcal{F}(f)}$	The normalized Fourier transformation of f	23
$\Xi(f)$	The exponential sum associated with f	28
$\widehat{\chi_f}(\omega)$	The Walsh Hadamard transform of f at ω.	24
$\widehat{\chi_f}(\omega, k)$	The extended Walsh–Hadamard transform ($\omega \in \mathbb{F}_{2^n}$, k an integer co-prime with $2^n - 1$)	25
$C_m(a, b)$	The binary cubic sums on $\mathbb{F}_{2^m}$ associate with a and b	27
$D_r(X)$	Binary Dickson polynomial of degree r	19
$K_n^{(p)}$	Kloosterman sums over $\mathbb{F}_{p^n}$	340
$K_m(a)$	The Binary Kloosterman sums on $\mathbb{F}_{2^m}$ associate with a	26
$S_f(b)$	Walsh transform of f at $b \in \mathbb{F}_{p^n}$	330
$T_i(f)$	The partial exponential sum on $\mathcal{T}_i$ associated with f	28
$TR_1^m(\cdot)$	Trace function from the Galois ring $\mathbb{R}$ to $\mathbb{Z}_4$	402
U	The cyclic group of $(2^m + 1)$-st roots of unity in $\mathbb{F}_{2^n}$	29

Contents

List of Figures

List of Tables

List of Algorithms

Overview

Boolean functions are important objects in discrete mathematics. They play a role in mathematics and almost all the domains of computer science. In this book, we are mainly interested in their relationships with error-correcting codes and private-key cryptography. Mathematically, Boolean functions are mostly considered in this book in their univariate representations over finite fields. The theory of finite fields is a branch of modern algebra that has come to the fore in the last 60 years because of its diverse applications in combinatorics, coding theory, and cryptology, among others.

The book is devoted to special families of Boolean functions which are viewed as important objects in combinatorics and the information theory framework (namely, cryptography and coding theory).

In fact, one of the most important cryptographic characteristics of a Boolean function is its nonlinearity. Most interest is attracted by the extremal nonlinear functions. *Bent functions* are maximally nonlinear Boolean functions with an even number of variables and are optimal combinatorial objects.

In the mathematical field of combinatorics, a bent function is a special type of Boolean function. Defined and named in the 1960s by Oscar Rothaus [3] in research not published until 1976, bent functions are so called because they are as different as possible from all linear and affine functions. The first paper on bent functions has been written in 1966 by O. Rothaus (as indicated by J. Dillon in his thesis), but its final version was published ten years later in [3]. The definition of bent function can be extended in several ways, leading to different classes of generalized bent functions that share many of the useful properties of the original.

Bent functions are wonderful creatures, initially studied by John Francis Dillon in his PhD thesis [2]. They have attracted a lot of research, especially in the last 20 years for their own sake as interesting combinatorial objects (e.g., difference sets), in design theory (any difference set can be used to construct a symmetric design) but also for their relations to coding theory (e.g. Reed–Muller codes, Kerdock codes, etc.) and applications in cryptography (design of stream ciphers) and sequence theory. A jubilee survey paper on bent functions giving a historical perspective and making pertinent connections to designs, codes, and cryptography is [1].

In cryptography, bent functions play a central role in the robustness of stream and block ciphers, since they are the only source of their nonlinearity, by providing confusion in these cryptosystems. The main cryptographic weakness of these functions in symmetric cryptography, forbidding to directly use them in stream ciphers, is that bentness makes it impossible for them to be balanced (i.e., to have output uniformly distributed over the smallest field of cardinality 2); this induces a statistical correlation between the plaintext and the ciphertext.

A natural generalization of Boolean functions is the multi-output Boolean functions. Such vectorial functions constitute the so-called substitution boxes (S-boxes) in symmetric cryptosystems which are fundamental parts of block ciphers. Bent vectorial functions can be involved in the substitution boxes (S-boxes) of block ciphers, whose role is also to bring some amount of nonlinearity allowing them to resist differential and linear attacks.

Bent functions are particular plateaued functions. The notion of plateaued function has been introduced in 1999 by Zheng and Zhang as good candidates for designing cryptographic functions since they possess desirable various cryptographic characteristics. They are defined in terms of the Walsh–Hadamard spectrum. Plateaued functions bring together various nonlinear characteristics and include two important classes of Boolean functions defined in even dimension: the well-known bent functions and the semi-bent functions. Very recently, the study of semi-bent functions has attracted the attention of several researchers. Many progresses in the design of such functions have been made.

Bent functions, their subclasses (e.g., hyper-bent functions), and their generalizations (e.g., plateaued functions) have many theoretical and practical applications in combinatorics, coding theory, (symmetric) cryptography, and sequence theory. Bent functions (including their constructions) have been extensively investigated since 1974. A complete classification of bent functions is elusive and looks hopeless today; therefore, not only their characterization but also their generation is a challenging problem.

The research activity on bent functions has been important for four decades and remains very intensive. However, very recently, many advances have been obtained on super-classes of bent functions (plateauted functions, partially bent functions, etc.), related classes of bent functions (semi-bent functions, near-bent functions, etc.) and subclasses (hyper-bent functions, Niho bent functions, symmetric bent functions, bent nega-bent functions, etc.). In particular, many new connections in the framework of semi-bent functions with other domains of mathematics and computer science (Dickson polynomial, Kloosterman sums, spreads, oval polynomial, finite geometry, coding, cryptography, sequences, etc.) have been exhibited. The research in this framework is relatively new and becomes very active.

This book provides a detailed survey of main results in binary and generalized bent functions, presents a systematic overview of their generalizations, their variations, and their applications, considers open problems in classification and systematization of bent functions, discusses proofs of several results, and reflects recent developments and trends in the field. Up to now, there is no analog of this book in detail and completeness of material on bent functions, their variations, and their generalizations. It is the first book in this field collecting essential material and is complementary to the existing surveys, since the emphasis is on bent functions via a univariate approach based on finite fields.

In this book we have aimed at presenting both the classical and the application-oriented aspects of the subject. The reader will find many results and several techniques that are of importance. Because of the vastness of the subject, limitations had to be imposed on the choice of the material: we are mostly dealing with binary bent functions. The book tries to be as self-contained as possible. It contains information from highly regarded sources. Wide varieties of references are listed.

The book is split into 18 chapters. In most chapters, we bring some preliminaries providing enough background for the unfamiliar reader to understand the content of the chapter in which we present advanced results, significant advancements, and the recent contributions of the researchers to the subject.

The noteworthy prerequisite for the book is a background in linear algebra and basic concepts in finite fields such as the general structure theory of finite fields, the theory of polynomials over finite fields, and the theory of Boolean functions.

Chapter 1 is basic for the rest of the book as it contains the general notions related to Boolean functions as well as notions and concepts used throughout the book. In Chap. 2, we provide several technical results and some mathematical tools that we need subsequently in several chapters. Chapter 3 presents and discusses Boolean functions as important primitives of symmetric cryptosystems playing a central role in their security. From Chap. 4, we enter in the core of the main subject of the book. After a short historical note, we present definitions, main properties, classes, and main constructions of binary bent functions as well as their relationship with error-correcting codes and private-key cryptography and combinatorics. Chapters 5, 6, and 7 are devoted to primary and secondary constructions of bent Boolean functions. In Chap. 8, we will be interested in the connection between the theory of bent functions and some important objects from finite geometry. Chapters 9, 10, and 11 concern a subclass of bent functions: the so-called hyper-bent functions. We shall show how we can treat the property of being hyperbent using tools from the theory of exponential sums and the one of hyper-elliptic curves. Chapter 12 is dealing with multi-output bent functions. In Chap. 13, we study bent functions in arbitrary characteristic. Chapter 14 deals with connections of bent functions and spreads. Chapter 15 is devoted to various cryptographic and algebraic generalizations of bent functions. We shall present partially bent functions, rotation symmetric bent functions, homogeneous bent functions, negabent functions, and several generalized bent functions. Chapters 16 and 17 are concerned with the so-called plateaued functions which are cryptographic generalizations of bent functions. In particular, we discuss near and semi-bent functions. Finally, the last chapter is devoted to recent advances related to linear error-correcting codes with few weights constructed via bent functions.

We hope that this book will be useful firstly to researchers in discrete mathematics and their applications in cryptography and coding theory and to students and professors of mathematical and theoretical computer science. It will also be useful to all interested in mathematical foundations of cryptography, engineers, and managers in security. It can be used as a material for such university courses as discrete mathematics, Boolean functions, symmetric cryptography, etc. The book will contain parts of different level: from basic (available to first-year master's students) to very advanced (for specialists in discrete mathematics, cryptography, coding theory, sequences, etc.).

This book is both a reader book on an exciting field and a reference of an extreme wealth. The *Notation* section can be found before the table of contents. An *Index* toward the end of this book gives some terms used. Readers are encouraged to send their comments to: smesnager@univ-paris8.fr

Acknowledgments I am indebted to Gérard Cohen for his reading of the whole book, always in a cheerful manner, his very constructive and valuable comments, his interesting discussions on bent functions related to combinatorics and coding theory as well as his great support. I am very grateful to Claude Carlet. My meeting with him has completely changed my life as a researcher! I never imagined finding so much energy and force in me to learn a new area just after my PhD thesis in algebra and modern algebraic geometry. I never forgot his encouragement and the interesting problems that he had suggested during my conversion from research in mathematics in algebra to discrete mathematics for cryptography and coding theory. I also thank him for his invitation to write together the jubilee survey on bent functions. Many thanks to Jean-Pierre Flori for his nice contribution and to Matthew Geoffrey Parker for his great contribution in writing Sect. 15.7 of Chap. 15. My gratitude to Joachim von zur Gathen, William M. Kantor, Henning Stichtenoth and Alev Topuzoglu for very interesting discussions and their support. I am grateful for help on various matters to Pascale Charpin, Cunsheng Ding, Tor Helleseth, Harald Niederreiter and Ferruh Özbudak. I would like to thank also Kanat Abdukhalikov, Lilya Budaghyan, Anne Canteaut, Ayca Ceşmelioğlu, Keqin Feng, Guang Gong, Alexander Kholosha, Gohar Kyureghyan, Gilles Lachaud, Philippe Langevin, Petr Lisoněk, Gary McGuire, Wilfried Meidl, Stanica Pantelimon, Enes Pasalic, Alexander Pott, François Rodier, Leo Storme, and Jacques Wolfmann for attractive discussions on bent functions during the redaction of this book.

Although I speak of its wonderful works in almost all my talks, I had never got the chance to meet John Dillon!

References

1. C. Carlet and S. Mesnager. Four decades of research on bent functions. In *Journal Designs, Codes and Cryptography, Vol. 78, No. 1*, pages 5–50, 2016.
2. J. Dillon. Elementary Hadamard difference sets. In *PhD dissertation, University of Maryland.*
3. O.S. Rothaus. On "bent" functions. In *J. Combin. Theory Ser A 20*, pages 300–305, 1976.

Chapter 1
Generalities on Boolean Functions and p-Ary Functions

1.1 Boolean Functions

1.1.1 Background on Boolean Functions

In mathematics, a Boolean function f is a map from the vectorspace $\mathbb{F}_2^n$ of all binary vectors of length n to the finite field with two elements $\mathbb{F}_2$ i.e.: $f : \mathbb{F}_2^n \to \mathbb{F}_2$. The vectorspace $\mathbb{F}_2^n$ will sometimes be also endowed with the structure of field—the field $\mathbb{F}_{2^n}$ (also denoted by $\mathbb{F}_{2^n}$); indeed, this field being an n-dimensional vectorspace over $\mathbb{F}_2$, each of its elements can be identified with the binary vector of length n of its coordinates relative to a fixed basis. The set of all Boolean functions $f : \mathbb{F}_2^n \to \mathbb{F}_2$ will be denoted by $\mathcal{B}_n$. The *Hamming weight* $\mathrm{wt}(x)$ of a binary vector $x = (x_1, \cdots, x_n)$ is the number of its nonzero coordinates (*i.e.* the size of $\{i \in N | x_i \neq 0\}$ where N denotes the set $\{1, \cdots, n\}$, called the *support*. The *Hamming weight of a Boolean function* f on $\mathbb{F}_2^n$ denoted by $\mathrm{wt}(f)$, is (also) the size of the support of the function, *i.e.* the set $\{x \in \mathbb{F}_2^n | f(x) \neq 0\}$.The *Hamming distance* $d_H(f, g)$ between two functions f and g is the size of the set $\{x \in \mathbb{F}_2^n | f(x) \neq g(x)\}$. Thus it equals $w_H(f \oplus$ [1] $g)$.

The *first derivative* of a Boolean function f in the direction of $a \in \mathbb{F}_{2^n}$ is defined as $D_a f(x) = f(x) + f(x + a)$ and the *second order derivative* of f with respect to $(a, b) \in \mathbb{F}_{2^n}^2$ is defined as $D_b D_a f(x) = f(x) + f(x + b) + f(x + a) + f(x + a + b)$. The derivative is also defined for mappings from $\mathbb{F}_{2^n}$ to $\mathbb{F}_{2^m}$. More precisely, let

[1] Some additions of bits will be considered in $\mathbb{Z}$ (in characteristic 0) and denoted then by $+$, and some will be computed modulo 2 and denoted by $\oplus$ or by $+$ if there is no ambiguity. These two different notations will be necessary because some representations of Boolean functions will live in characteristic 2 and some representations of the same functions will live in characteristic 0. But the additions of elements of the finite field $\mathbb{F}_{2^n}$ will be denoted by +, as it is usual in mathematics. So, for simplicity (since $\mathbb{F}_2^n$ will often be identified with $\mathbb{F}_{2^n}$) and because there will be no ambiguity, we shall also denote by $+$ the addition of vectors of $\mathbb{F}_2^n$ when $n > 1$.

S. Mesnager, *Bent Functions*, DOI 10.1007/978-3-319-32595-8_1

$F : \mathbb{F}_{2^n} \to \mathbb{F}_{2^m}$ be a mapping. For $a \in \mathbb{F}_{2^n}$, the function D_aF given by $D_aF(x) = F(x) + F(x + a), \forall x \in \mathbb{F}_{2^n}$ is called the *derivative of F* in the direction of a.

Next we give definitions of linear translator and linear structure.

Definition 1.1.1. Let $n = rk$, $1 \leq k \leq n$. Let f be a function from $\mathbb{F}_{2^n}$ to $\mathbb{F}_{2^k}$, $\gamma \in \mathbb{F}_{2^n}^*$ and b be a constant of $\mathbb{F}_{2^k}$. Then γ is a *b-linear translator* of f if $f(x) + f(x+u\gamma) = ub$ for all $x \in \mathbb{F}_{2^n}$ and $u \in \mathbb{F}_{2^k}$. If $f(x)+f(x+b) = b$ for all $x \in \mathbb{F}_{2^n}$, then γ is called a *b-linear structure* of f. Note that being b-linear translator is stronger than being b-linear structure if $k > 1$ and they are the same if $k = 1$.

The notion of a linear structure exists for p-ary functions (p being a prime number), that is, functions from $\mathbb{F}_{p^n}$ to $\mathbb{F}_p$. Functions with linear structures are considered as weak for some cryptographic applications.

Definition 1.1.2. Let q be a power of a prime number and $\mathbb{F}_{q^n}$ be the finite field of order q^n. A non-zero element $\alpha \in \mathbb{F}_{q^n}$ is called an a-linear translator for the mapping $f : \mathbb{F}_{q^n} \to \mathbb{F}_q$ if $f(x + u\alpha) - f(x) = ua$ holds for any $x \in \mathbb{F}_{q^n}$, $u \in \mathbb{F}_q$ and a fixed $a \in \mathbb{F}_q$. In particular, when $q = 2$, $\alpha \in \mathbb{F}_{2^n}^\star$ usually said to be an a-linear structure for the Boolean function f (where $a \in \mathbb{F}_2$), that is, $f(x + \alpha) + f(x) = a$, for any $x \in \mathbb{F}_{2^n}$.

Note that if α is a a-linear structure of f, then necessarily $a = f(\alpha) - f(0)$. The notions of linear translators and derivatives are related. The *linear kernel of f* is the linear subspace of vectors a such that D_af is a constant function. Any element of the linear kernel is in fact a linear structure of f.

Boolean functions play an important role in both cryptographic and error correcting coding activities. Indeed, cryptographic transformations (pseudo-random generators in stream ciphers, S-boxes in block ciphers) can be designed by appropriate composition of nonlinear Boolean functions. Moreover, every code of length 2^n, for some positive integer n, can be interpreted as a set of Boolean functions, since every n-variable Boolean function can be represented by its truth table (an ordering of the set of binary vectors of length n being first chosen) and thus associated with a binary word of length 2^n, and *vice versa*; important codes such as Reed–Muller and Kerdock codes can be defined this way as sets of Boolean functions.

In both frameworks, n is rarely large, in practice. The error correcting codes derived from n-variable Boolean functions have length 2^n; so, taking $n = 11$ already gives codes of length 2048. For reason of efficiency, the S-boxes used in most block ciphers are concatenations of sub S-boxes on at most eight variables. In the case of stream ciphers, n was in general at most equal to 11 until recently. The cryptographic situation has changed a little in practice since the apparition of the (fast) algebraic attacks but the number of variables is now most often limited to 20. An excellent reference for Boolean functions is the book's chapter of C. Carlet [1, Chapter 8]. Another interesting reference is the book of T. W. Cusick and P. Stănică, [4].

1.1.2 Boolean Functions: Representations

There exist several representations of a given Boolean function. We shall recall only the representations of Boolean functions that we need in this book.

1.1.2.1 Algebraic Normal Form

The algebraic normal Form (in brief the ANF) is the classical representation of Boolean functions. It is the most used in cryptography and coding. The Algebraic Normal Form of an Boolean function f on $\mathbb{F}_2^n$ is the n-variable polynomial representation over $\mathbb{F}_2$, of the form

$$f(x) = \bigoplus_{I \in \mathcal{P}(N)} a_I \left(\prod_{i \in I} x_i \right) = \bigoplus_{I \in \mathcal{P}(N)} a_I x^I, \tag{1.1}$$

where $\mathcal{P}(N)$ denotes the power set of $N = \{1, \cdots, n\}$. Every coordinate x_i appears in this polynomial with exponents at most 1, because every bit in $\mathbb{F}_2$ equals its own square. This representation belongs to $\mathbb{F}_2[x_1, \cdots, x_n]/(x_1^2 \oplus x_1, \cdots, x_n^2 \oplus x_n)$ and it is unique.

The *algebraic degree* of the function f (this makes sense thanks to the existence and uniqueness of the ANF) denoted by $\deg(f)$ equals the maximum degree of those monomials whose coefficients are nonzero in its algebraic normal form, that is, $\deg(f) = \max\{|I| / \, a_I \neq 0\}$, where $|I|$ denotes the size of I.

The algebraic degree is an affine invariant (it is invariant under the action of the general affine group).

x_1	x_2	x_3	$f(x_1, x_2, x_3)$	$g(x_1, x_2, x_3)$	$h(x_1, x_2, x_3)$
0	0	0	0	0	1
0	0	1	0	1	0
0	1	0	1	1	0
0	1	1	1	1	1
1	0	0	1	1	0
1	0	1	0	0	1
1	1	0	1	1	1
1	1	1	0	0	0

Example 1.1.3. $f(x_1, x_2, x_3) = x_1 + x_2 + x_1x_2 + x_1x_3$; $deg(f) = 2$; $g(x_1, x_2, x_3) = x_1 + x_2 + x_3 + x_1x_2 + x_2x_3 + x_1x_2x_3$; $deg(g) = 3$; $h(x_1, x_2, x_3) = 1 + x_1 + x_2 + x_3$; $deg(h) = 1$.

There exists a simple algorithm for calculating the ANF from the truth table (or vice versa):

Algorithm 1.1: Computing the ANF of a Boolean function given by its truth table

Input: A Boolean function f given by its truth table
Output: The Algebraic Normal Form (ANF) of f

1. write the truth-table of f, in which the binary vectors of length n are in reverse lexicographic order;
2. let f_0 be the restriction of f to $\mathbb{F}_2^{n-1} \times \{0\}$ and f_1 the restriction of f to $\mathbb{F}_2^{n-1} \times \{1\}$; the table of values of f_0 (resp. f_1) corresponds to the upper (resp. lower) half of the table of f; replace the values of φ_1 by those of $\varphi_0 \oplus \varphi_1$;
3. apply recursively step 2, separately to the functions now obtained in the places of φ_0 and φ_1.

Table 1.1 ANF of f is $x_2 \oplus x_1x_2x_3 \oplus x_1x_4$

x_1	x_2	x_3	x_4	$f(x)$				
0	0	0	0	0	0	0	0	0
1	0	0	0	0	0	0	0	0
0	1	0	0	1	1	1	1	1
1	1	0	0	1	1	1	1	0
0	0	1	0	0	0	0	0	0
1	0	1	0	0	0	0	0	0
0	1	1	0	1	1	0	0	0
1	1	1	0	0	0	1	1	1
0	0	0	1	0	0	0	0	0
1	0	0	1	1	1	1	1	1
0	1	0	1	1	0	0	0	0
1	1	0	1	0	1	1	1	0
0	0	1	1	0	0	0	0	0
1	0	1	1	1	1	0	0	0
0	1	1	1	1	0	0	0	0
1	1	1	1	1	1	0	0	0

When the algorithm ends (*i.e.* when it reaches functions on one variable each), the global table gives the coefficients of the ANF.

The complexity of the above algorithm is $O(n2^n)$.

Example 1.1.4. : Let $n = 4$. In Table 1.1 we give the ANF of a Boolean function in four variables given by its truth table.

1.1.2.2 Numerical Normal Form

Any n-variable Boolean function can be viewed as an integer-valued mapping taking values in the subset $\{0, 1\}$ of $\mathbb{Z}$. Now, any integer-valued mapping f can be uniquely represented as a multivariate polynomial over $\mathbb{Z}$:

$$\forall x \in \mathbb{F}_2^n, \quad f(x) = \sum_{I \in \mathcal{P}_n} \lambda_I \prod_{i \in I} x_i \tag{1.2}$$

where the λ_I's are in $\mathbb{Z}$. This representation is unique and called the *numerical normal form*.[2] The degree of the numerical normal form of an integer-valued map f is called its *numerical degree*. To ensure that f takes values in $\{0, 1\}$, that is, satisfies $f^2(x) = f(x)$ for every $x \in \mathbb{F}_2^n$, the coefficients λ_I's have to satisfy

$$\forall I \in \mathcal{P}_n, \quad \Big(\sum_{J \subset I} \lambda_J\Big)^2 - \sum_{J \subset I} \lambda_J = 0. \tag{1.3}$$

where $\sum_{J \subset I}$ denotes the summation over all the subsets J of $\{1, \ldots, n\}$ which are contained in the subset I.

Note that the numerical normal form leads to a (simple) characterization of the bent functions[3] [3].

1.1.2.3 Trace Function and the Polynomial Form

Now, we consider a Boolean function f defined on $\mathbb{F}_{2^n}$, that is, is an $\mathbb{F}_2$-valued function on the Galois field $\mathbb{F}_{2^n}$ of order 2^n. The *weight* of f, denoted by wt(f), is the Hamming weight of the image vector of f, that is, the cardinality of its support *supp*(f) $:= \{x \in \mathbb{F}_{2^n} \mid f(x) = 1\}$.

For any positive integer k, and r dividing k, the *trace function* from $\mathbb{F}_{2^k}$ to $\mathbb{F}_{2^r}$ is denoted by $\mathrm{Tr}_r^k(\cdot)$. It is defined as

$$\mathrm{Tr}_r^k(x) = \sum_{i=0}^{\frac{k}{r}-1} x^{2^{ir}} = x + x^{2^r} + x^{2^{2r}} + \cdots + x^{2^{k-r}}.$$

In particular, we denote the *absolute trace* over $\mathbb{F}_2$ of an element $x \in \mathbb{F}_{2^n}$ by $\mathrm{Tr}_1^n(x) = \sum_{i=0}^{n-1} x^{2^i}$. Recall some basic properties of trace functions:

1. the trace function Tr_k^n is surjective;
2. $\mathrm{Tr}_k^n(ax + by) = a\mathrm{Tr}_k^n(x) + b\mathrm{Tr}_k^n(y)$ for $a, b \in \mathbb{F}_{2^k}$ and $x, y \in \mathbb{F}_{2^n}$;
3. $\mathrm{Tr}_k^n(x^{2^k}) = \mathrm{Tr}_k^n(x)$ for $x \in \mathbb{F}_{2^n}$;
4. when $\mathbb{F}_{2^k} \subset \mathbb{F}_{2^r} \subset \mathbb{F}_{2^n}$, the trace function Tr_k^n satisfies the transitivity property, that is, $\mathrm{Tr}_k^n = \mathrm{Tr}_k^r \circ \mathrm{Tr}_r^n$.

There exist several kinds of possible trace (univariate) representations of Boolean functions (see for instance, [1, p. 266]) which are not necessary unique and use the identification between the vectorspace $\mathbb{F}_2^n$ and the field $\mathbb{F}_{2^n}$. In this book, we will extensively study the so-called *Bent functions*. Those functions are often better viewed in their bivariate representation can also be viewed in their univariate

[2]This representation has been introduced by Carlet and Guillot in [2] in the framework of cryptography.

[3]The definition of a bent function is given later in Chap. 4.

representation. The univariate representation of any Boolean function is defined as follows: we identify $\mathbb{F}_2^n$ with $\mathbb{F}_{2^n}$ (which is an n-dimensional vectorspace over $\mathbb{F}_2$) and consider then the input to f as an element of $\mathbb{F}_{2^n}$. An inner product in $\mathbb{F}_{2^n}$ is $x \cdot y = \mathrm{Tr}_1^n(xy)$ where $\mathrm{Tr}_1^n(\cdot))$ is the absolute trace function over $\mathbb{F}_{2^n}$ defined as $\mathrm{Tr}_1^n(x) = \sum_{i=0}^{n-1} x^{2^i}$. There exists a unique univariate polynomial $\sum_{i=0}^{2^n-1} a_i x^i$ over $\mathbb{F}_{2^n}$ such that f is the polynomial function over $\mathbb{F}_{2^n}$ associated to it (this is true for every function from $\mathbb{F}_{2^n}$ to $\mathbb{F}_{2^n}$). Moreover, f being Boolean, its univariate representation can be written as a unique trace expansion of the form

$$f(x) = \sum_{j \in \Gamma_n} \mathrm{Tr}_1^{o(j)}(a_j x^j) + \epsilon(1 + x^{2^n-1}), \quad a_j \in \mathbb{F}_{2^{o(j)}},$$

valid for all $x \in \mathbb{F}_{2^n}$ and called its *polynomial form*. In the above expression:

- Γ_n is the set of integers obtained by choosing one element in each cyclotomic coset modulo $2^n - 1$ (including the trivial coset containing 0 and only 0), the most usual choice being the smallest element in each cyclotomic coset, called the coset leader,
- $o(j)$ is the size of the cyclotomic coset containing j (that is, $o(j)$ is the smallest positive integer such that $j2^{o(j)} \equiv j \pmod{2^n - 1}$),
- and $\epsilon = wt(f) \pmod 2$.

Let us explain the existence of the polynomial form of a Boolean function which is an important representation used in almost the whole book.

Any function $f : \mathbb{F}_{2^n} \to \mathbb{F}_{2^n}$ admits a unique representation: $f(x) = \sum_{j=0}^{2^n-1} a_j x^j\ ; a_j, x \in \mathbb{F}_{2^n}$. f is Boolean if and only if $f(x)^2 = f(x), \forall x \in \mathbb{F}_{2^n}$. Using the univariate representation, we obtain:

f is Boolean if and only if

$$a_0, a_{2^n-1} \in \mathbb{F}_2 \text{ and } a_{2j \pmod{2^n-1}} = (a_{j \pmod{2^n-1}})^2; 0 < j < 2^n - 1.$$

Let us split the range $[1, 2^n - 2]$ as follows: $[1, 2^n - 2] = \cup_{r=1}^{c} \Gamma_r$; where $\Gamma_r = \{j_r mod(2^n-1), 2.j_r mod(2^n-1), \cdots, 2^{o(j_r)-1} j_r mod(2^n-1)\}$.

One can prove by induction on p that for $i \in [1, 2^n - 2]$, we have $a_{2^p i mod(2^n-1)} = a_{i \pmod{2^n-1}}^{2^p}$ for every positive integer p. Therefore, we have:

$$\begin{aligned} f(x) &= a_0 + a_{2^n-1}x^{2^n-1} + \sum_{r=1}^{c} \sum_{s=0}^{o(j_r)-1} a_{2^s j_r \pmod{2^n-1}} x^{2^s j_r} \\ &= a_0 + a_{2^n-1}x^{2^n-1} + \sum_{r=1}^{c} \sum_{s=0}^{o(j_r)-1} (a_{j_r \pmod{2^n-1}} x^{j_r})^{2^s} \\ &= a_0 + a_{2^n-1}x^{2^n-1} + \sum_{r=1}^{c} \mathrm{Tr}_1^{o(j_r)}(a_{j_r \pmod{2^n-1}} x^{j_r});\ a_0, a_{2^n-1} \in \mathbb{F}_2, \\ &\qquad a_{j_r \pmod{2^n-1}} \in \mathbb{F}_{2^{o(j_r)}}. \end{aligned}$$

The proof of the uniqueness of the polynomial form is trivial.

Example 1.1.5. Let $n = 4$. $f : \mathbb{F}_{2^4} \to \mathbb{F}_2$, $f(x) = \sum_{j\in\Gamma_4} \mathrm{Tr}_1^{o(j)}(a_j x^j) + \epsilon(1 + x^{15})$, $a_j \in \mathbb{F}_{2^{o(j)}}$.

Γ_4 is the set obtained by choosing one element in each cyclotomic class of 2 modulo $2^n - 1 = 2^4 - 1 = 15$. $C(j)$ the cyclotomic coset of 2 modulo 15 containing j.

$C(j) = \{j, j2, j2^2, j2^3, \cdots, j2^{o(j)-1}\}$ where $o(j)$ is the smallest positive integer such that $j2^{o(j)} \equiv j \pmod{15}$.

The cyclotomic cosets modulo 15 are:

$C(0) = \{0\}$;
$C(1) = \{1, 2, 4, 8\}$;
$C(3) = \{3, 6, 12, 9\}$;
$C(5) = \{5, 10\}$;
$C(7) = \{7, 14, 11, 13\}$.
One can choose $\Gamma_4 = \{0, 1, 3, 5, 7\}$.

Therefore,

$$\begin{aligned} f(x) &= \mathrm{Tr}_1^{o(1)}(a_1 x^1) + \mathrm{Tr}_1^{o(3)}(a_3 x^3) + \mathrm{Tr}_1^{o(5)}(a_5 x^5) + \mathrm{Tr}_1^{o(7)}(a_7 x^7) + a_0 + \epsilon(1 + x^{15}); \\ &= \mathrm{Tr}_1^4(a_1 x) + \mathrm{Tr}_1^4(a_3 x^3) + \mathrm{Tr}_1^2(a_5 x^5) + \mathrm{Tr}_1^4(a_7 x^7) + a_0 + \epsilon(1 + x^{15}). \end{aligned}$$

where $a_1, a_3, a_7 \in \mathbb{F}_{2^4}$, $a_5 \in \mathbb{F}_{2^2}$, $a_0, \epsilon \in \mathbb{F}_2$ with

$\mathrm{Tr}_1^4 : \mathbb{F}_{2^4} \to \mathbb{F}_2; x \mapsto x + x^2 + x^{2^2} + x^{2^3}$ and
$\mathrm{Tr}_1^2 : \mathbb{F}_{2^2} \to \mathbb{F}_2; x \mapsto x + x^2$.

Recall the univariate representation of $f : f(x) = \sum_{j=0}^{2^n-1} a_j x^j$; $a_j, x \in \mathbb{F}_{2^n}$.

Let $\alpha_0, \cdots, \alpha_{n-1}$ be a basis of the $\mathbb{F}_2$-vectorspace $\mathbb{F}_{2^n}$. Every element $x \in \mathbb{F}_{2^n}$ can be uniquely decomposed as: $x = \sum_{i=0}^{n-1} x_i \alpha_i$ with $x_i \in \mathbb{F}_2$. Therefore (by using the binary expansion of j: $j = \sum_{s=0}^{n-1} j_s 2^s$ where $j_s \in \{0, 1\}$)

$$\begin{aligned} f(x) &= \sum_{j=0}^{2^n-1} a_j \Big(\sum_{i=0}^{n-1} x_i \alpha_i\Big)^j \\ &= \sum_{j=0}^{2^n-1} a_j \Big(\sum_{i=0}^{n-1} x_i \alpha_i\Big)^{\sum_{s=0}^{n-1} j_s 2^s} ; \; j_s \in \{0, 1\} \\ &= \sum_{j=0}^{2^n-1} a_j \prod_{s=0}^{n-1} \Big(\sum_{i=0}^{n-1} x_i \alpha_i^{2^s}\Big)^{j_s} =: \sum_{j=0}^{2^n-1} a_j \prod_{s=0}^{n-1} l_s^{j_s}. \end{aligned}$$

The linear forms on $\mathbb{F}_{2^n}$: $\{l_s; s \in [0, n-1] : l_s(x_0, \cdots, x_{n-1}) = \sum_{i=0}^{n-1} x_i \alpha_i^{2^s}\}$ are independent.

The degree of $\prod_{s=0}^{n-1} l_i^{j_s}$ is thus equals $\sum_{s=0}^{n-1} j_s = w_2(j)$ where $w_2(j)$ is the Hamming weight of the binary expansion of j.

Consequently, the *algebraic degree* of f given in polynomial form is equal to the maximum 2-weight of an exponent j for which $a_j \neq 0$ if $\epsilon = 0$ and to n if $\epsilon = 1$. Recall that the 2-weight $w_2(j)$.[4] We have $deg(f) = max\{w_2(j);\, j \text{ s.t. } a_j \neq 0\}$.

The computation of the algebraic degree is not always simple in general when the function is given in its univariate representation (in contrast with the ANF).

Example 1.1.6. Let $n = 2m$, $a \in \mathbb{F}_{2^n}^\star$. Then

- for $f(x) = \mathrm{Tr}_1^n(ax^{2^m+1})$, $deg(f) = 2$ (that is, f is quadratic);
- for $f(x) = \mathrm{Tr}_1^n(ax^{2^m-1})$, $deg(f) = m$;
- for $f(x) = \mathrm{Tr}_1^n(ax^{3(2^m-1)})$, $deg(f) = m$.

Note that the above expression of f can also be written under a non-unique form $\mathrm{Tr}_1^n(P(x))$ where $P(x)$ is a polynomial over $\mathbb{F}_{2^n}$.

Going from the non-unique trace representation to the unique one basically amounts to take the traces of the coefficients from $\mathbb{F}_{2^n}$ to $\mathbb{F}_{2^{o(j)}}$. Going the other way around relies on the surjectivity of the trace map from $\mathbb{F}_{2^n}$ to $\mathbb{F}_{2^{o(j)}}$.

1.1.2.4 The Bivariate Representation

The *bivariate representation* of Boolean functions is defined only when $n = 2m$ is even as follows: we identify $\mathbb{F}_2^n$ with $\mathbb{F}_{2^m} \times \mathbb{F}_{2^m}$ and consider then the input to f as an ordered pair (x, y) of elements of $\mathbb{F}_{2^m}$. There exists a unique bivariate polynomial

$$\sum_{0 \leq i,j \leq 2^m-1} a_{i,j} x^i y^j$$

over $\mathbb{F}_{2^m}$ such that f is the bivariate polynomial function over $\mathbb{F}_{2^m}$ associated to it. Then the algebraic degree of f equals

$$\max_{(i,j)\,|\,a_{i,j} \neq 0} (w_2(i) + w_2(j)),$$

and f being Boolean, its bivariate representation can be written in the form

$$f(x, y) = \mathrm{Tr}_1^m(P(x, y))$$

where $P(x, y)$ is some polynomial in two variables over $\mathbb{F}_{2^m}$.

[4] More precisely, for $j \in \mathbb{Z}/(2^k - 1)\mathbb{Z}$, $w_2(j)$ is the Hamming (or binary) weight of the unique representative of j in $\{0, \cdots, 2^k - 2\}$

1.1.2.5 From a Representation to Another

1. In the following we explain how we convert a Boolean function given in univariate representation into a Boolean function in bivariate representation. Let $n = 2m$. As $\mathbb{F}_{2^n}$ is a vectorspace over $\mathbb{F}_{2^m}$ of dimension 2, there exists a basis $\{v, w\}$ of elements in $\mathbb{F}_{2^n}$. Without loss of generality, one can choose v and w such that $Tr_m^n(vw^{2^m}) \neq 0$. Indeed, one has necessary $Tr_m^n(vw^{2^m}) \neq 0$ otherwise

$$\begin{aligned} Tr_m^n(vw^{2^m}) = 0 &\iff vw^{2^m} + v^{2^m}(w^{2^m})^{2^m} = 0 \\ &\iff vw^{2^m} + v^{2^m}w^{2^n} = 0 \\ &\iff vw^{2^m} = v^{2^m}w^{2^n} \\ &\iff v^{2^m-1} = w^{2^m-1} \\ &\iff (\frac{v}{w})^{2^m-1} = 1 \\ &\iff \frac{v}{w} \in \mathbb{F}_{2^m} \end{aligned}$$

leading to a contradiction with the fact that $\{v, w\}$ is a basis of $\mathbb{F}_{2^n}$ as $\mathbb{F}_{2^m}$ vectorspace.

Now, given $x \in \mathbb{F}_{2^n}$, there exists a unique couple $(y, z) \in \mathbb{F}_{2^m} \times \mathbb{F}_{2^m}$ such that $x = yv + zw$ with $Tr_m^n(vw^{2^m}) \neq 0$. Therefore, $w^{2^m}x = yvw^{2^m} + zw^{2^m+1}$. Hence (using the linearity of Tr_m^n over $\mathbb{F}_{2^m}$) $Tr_m^n(w^{2^m}x) = Tr_m^n(yvw^{2^m}) + Tr_m^n(zw^{2^m+1})$. But $zw^{2^m+1} \in \mathbb{F}_{2^m}$ (since $z \in \mathbb{F}_{2^m}$ and $w^{2^m+1} \in \mathbb{F}_{2^m}$), thus $Tr_m^n(zw^{2^m+1}) = zw^{2^m+1}Tr_m^n(1) = 0$. Therefore, $Tr_m^n(yvw^{2^m}) = yTr_m^n(vw^{2^m})$. Consequently, $y = \frac{Tr_m^n(w^{2^m}x)}{Tr_m^n(vw^{2^m})}$.

Similarly, starting from $x = yv + zw$, we have $v^{2^m}x = yv^{2^m+1} + zwv^{2^m}$. Hence, $Tr_m^n(v^{2^m}x) = Tr_m^n(yv^{2^m+1}) + Tr_m^n(zwv^{2^m})$. But $y \in \mathbb{F}_{2^m}$ and $v^{2^m+1} \in \mathbb{F}_{2^m}$. Hence $yv^{2^m+1} \in \mathbb{F}_{2^m}$. We thus obtain that $Tr_m^n(v^{2^m}x) = yv^{2^m+1}Tr_m^n(1) + zTr_m^n(wv^{2^m})$, from which, we deduce $z = \frac{Tr_m^n(v^{2^m}x)}{Tr_m^n(wv^{2^m})}$. To conclude, we obtain the following decomposition of x:

$$x = \frac{Tr_m^n(w^{2^m}x)}{Tr_m^n(vw^{2^m})}v + \frac{Tr_m^n(v^{2^m}x)}{Tr_m^n(wv^{2^m})}w.$$

2. Now, we explain the relationship between the univariate representation of a Boolean function and its multivariate representation (ANF). Let $n > 1$ be a positive integer. Let $\{\gamma_1, \cdots, \gamma_n\}$ be a basis of $\mathbb{F}_{2^n}$. We identify $\mathbb{F}_2^n$ with $\mathbb{F}_{2^n}$ through the mapping $(x_1, \cdots, x_n) \mapsto \sum_{i=1}^n x_i\gamma_i$. In this way, each variable taking value from $\mathbb{F}_{2^n}$ can be written as $x = x_1\gamma_1 + \cdots x_n\gamma_n$. The univariate representation $f(x)$ over $\mathbb{F}_{2^n}$ is therefore represented in multivariate representation defined over $\mathbb{F}_2^n$ by $f(X) := f(x_1\gamma_1 + \cdots x_n\gamma_n)$, where $X := (x_1, \cdots, x_n)$.

For a positive integer j we form the corresponding cyclotomic coset $C(j)$ of 2 modulo $2^n - 1$ containing j, that is,

$$C(j) := \{j, j2, j2^2, j2^3, \cdots, j2^{o(j)-1}\},$$

where $o(j)$ is the smallest positive integer such that $j2^{o(j)} \equiv j \pmod{2^n - 1}$. The 2-coset $C(j)$ has a size $o(j)$ which divides n.

To each 2-coset $C(j)$, we associate polynomials $\left(P_{j,\beta_k}(x)\right)_{k\in[1,o(j)]}$ defined by

$$P_{j,\beta_k}(x) = \sum_{i=0}^{o(j)-1} (\beta_k x^j)^{2^i},$$

where $\{\beta_1, \beta_2, \cdots, \beta_{o(j)}\}$ is a basis of $\mathbb{F}_{2^{o(j)}}$ viewed as a vectorspace over $\mathbb{F}_2$ of dimension $o(j)$. Since $\{\beta_k\}_{k\in[1,o(j)]}$ forms a linearly independent set over $\mathbb{F}_2$, the set $\{x \mapsto P_{j,\beta_k}(x) \mid k \in [1, o(j)]\}$ is also a linearly independent set over $\mathbb{F}_2$. Moreover, the functions $x \mapsto P_{j,\beta_k}(x)$ are Boolean functions over $\mathbb{F}_{2^n}$. Indeed, for every $x \in \mathbb{F}_{2^n}^\star$ we have

$$\begin{aligned}
\left(P_{j,\beta_k}(x)\right)^2 &= \left(\sum_{i=0}^{o(j)-1} (\beta_k x^j)^{2^i}\right)^2 \\
&= \sum_{i=0}^{o(j)-1} (\beta_k x^j)^{2^{i+1}} \\
&= \sum_{i=1}^{o(j)} (\beta_k x^j)^{2^i} \\
&= \sum_{i=1}^{o(j)-1} (\beta_k x^j)^{2^i} + (\beta_k x^j)^{2^{o(j)}} \\
&= \sum_{i=1}^{o(j)-1} (\beta_k x^j)^{2^i} + \beta_k^{2^{o(j)}} x^{j2^{o(j)}}.
\end{aligned}$$

But $\beta_k^{2^{o(j)}} = \beta_k$ (since $\beta_k \in \mathbb{F}_{2^{o(j)}}$) and $x^{j2^{o(j)}} = x$ (since $j2^{o(j)} \equiv j \pmod{2^n - 1}$). Therefore, $\left(P_{j,\beta_k}(x)\right)^2 = P_{j,\beta_k}(x)$, $\forall x \in \mathbb{F}_{2^n}$.

Now, let us denote by $C(j_1), C(j_2), \cdots, C(j_t)$ all the distinct cyclotomic coset of 2 modulo $2^n - 1$ which partition the set of integers modulo $2^n - 1$. We have $\bigcup_{k=1}^{t} C(j_k) = \mathbb{Z}/(n-1)\mathbb{Z}$ and $\sum_{k=1}^{t} o(j_k) = 2^n - 1$. The set Ω composed by the multi-index polynomials $\left(P_{j_r,\beta_k}(x)\right)_{k\in[1,o(j)],r\in[1,t]}$ together with the trivial polynomial $\{x^{2^n-1}\}$ is a linearly independent set over $\mathbb{F}_2$. Moreover the $\mathbb{F}_2$-vectorspace spanned by Ω is of cardinality 2^{2^n} in which every non-zero element defines a non-zero function over $\mathbb{F}_{2^n}$. Consequently, the $\mathbb{F}_2$-vectorspace spanned by Ω coincides with the set of all the Boolean functions over $\mathbb{F}_{2^n}$, that is,

$$Span(\Omega) :=< \mathbf{1}, P_{j_r,\beta_k} \mid k \in [1, o(j)], r \in [1,t] >= \{f : \mathbb{F}_{2^n} \to \mathbb{F}_2\}.$$

In particular the set of Boolean functions of algebraic degree at most d (where $0 \leq d \leq n$) coincides with the set of functions $f(X)$ $(X = (x_1, \cdots, x_n))$ where $f(x) := f(x_1\gamma_1 + \cdots x_n\gamma_n)$ are such that f ranges through the set $Span\Big(\{P_{j,\beta_k} \mid w_2(j) \leq d, k \in [1, o(j)]\}\Big)$.

1.1.3 Boolean Functions of Low Degree

Affine functions are those of algebraic degree at most 1; they can be represented as follows: $\mathrm{Tr}_1^n(ax) + \epsilon$ where $a, x \in \mathbb{F}_{2^n}$ and $\epsilon \in \mathbb{F}_2$.

Quadratic functions are those of algebraic degree 2; they can be represented as follows: when n is even, $\epsilon + \sum_{i=0}^{\frac{n}{2}-1} \mathrm{Tr}_1^n(a_i x^{2^i+1}) + \mathrm{Tr}_1^{\frac{n}{2}}(a_{\frac{n}{2}} x^{2^{\frac{n}{2}}+1})$ where $\epsilon \in \mathbb{F}_2$, $a_i \in \mathbb{F}_{2^n}, \forall i, 0 \leq i \leq n/2$ and $a_{\frac{n}{2}} \in \mathbb{F}_{2^{n/2}}$; when n is odd, $f(x) = \epsilon + \sum_{i=0}^{\frac{n-1}{2}} \mathrm{Tr}_1^n(a_i x^{2^i+1}), a_i \in \mathbb{F}_{2^n}$. The *rank* of a quadratic function f is defined as follows: $rank(f) = n - dim_{\mathbb{F}_2} rad(B_f)$ where $rad(B_f) := \{x \in \mathbb{F}_{2^n} \mid B_f(x,y) = 0, \forall y \in \mathbb{F}_{2^n}\}$ with B_f the bilinear symplectic form defined as: $B_f(x,y) := f(x+y) + f(x) + f(y) + f(0)$. Set $k_f := dim_{\mathbb{F}_2} rad(B_f)$. Then 2 divides $(n - k_f)$. Any quadratic Boolean function on $\mathbb{F}_{2^n}$ has a rank $2t$ with $0 \leq t \leq \lfloor \frac{n}{2} \rfloor$ [5] and can be obtained as follows: the rank of f equals $2t$ if and only if the equation $B_f(x,y) = 0$ for any $y \in \mathbb{F}_{2^n}$ in x has exactly 2^{n-2t} solutions.

1.2 An Equivalent Notion on Boolean Functions

Boolean functions $f, g : \mathbb{F}_{2^n} \to \mathbb{F}_2$ are *extended-affine equivalent* (in brief, EA-equivalent) if there exist an affine permutation L of $\mathbb{F}_{2^n}$ and an affine function $\ell : \mathbb{F}_{2^n} \to \mathbb{F}_2$ such that $g(x) = (f \circ L)(x) + \ell(x)$. A class of functions is *complete* if it is a union of EA-equivalence classes. The *complete class* of a set of functions C is the smallest possible complete class that contains C.

1.3 On p-Ary Functions

1.3.1 Trace Functions

Let n and k be two positive integers. Assume k divides n. Let $\mathbb{F}_{p^n}$ and $\mathbb{F}_{p^k}$ be the finite fields of order p^n and p^k, respectively.

Definition 1.3.1. The trace function $Tr_{p^n}^{p^k}$ from the finite field $\mathbb{F}_{p^n}$ to its subfield $\mathbb{F}_{p^k}$ is defined by

$$Tr_{p^k}^{p^n}(x) = \sum_{i=0}^{\frac{n}{k}-1} x^{p^{ki}}.$$

For $k = 1$ we have the *absolute trace* and use the notation $tr_n(\cdot)$ for $Tr_p^{p^n}(\cdot)$.

Note that $(x, y) \rightarrow Tr_{p^n}^{p^k}(xy)$ defines a non-degenerate bilinear form.

Some important and useful properties of the trace mapping are provided in the next proposition.

Proposition 1.3.2. *i)* $tr_n(ax + by) = atr_n(x) + btr_n(y)$ *for any* $x, y \in \mathbb{F}_p^n$ *and* $a, b \in \mathbb{F}_p$.
ii) $tr_n(x^p) = tr_n(x)$ *for any* $x \in \mathbb{F}_p^n$.
iii) $tr_k(Tr_{p^k}^{p^n}(x)) = tr_n(x)$ *for any* $x \in \mathbb{F}_p^n$.
iv) $tr_n(ax)$ *takes on all elements on* $\mathbb{F}_p$ *equally often when* $a \neq 0$.

1.3.2 p-Ary Functions-Representations

A p-ary function is a function from $\mathbb{F}_p^n$ to $\mathbb{F}_p$. If we identify $\mathbb{F}_p^n$ with $\mathbb{F}_{p^n}$ (by considering $\mathbb{F}_{p^n}$ as an n-dimensional vectorspace over $\mathbb{F}_p$), all p-ary functions can be described in the so-called *univariate form*, which can be put in the (non unique) *trace form* $tr_n(F(x))$ for some function F from $\mathbb{F}_{p^n}$ to $\mathbb{F}_{p^n}$ of degree at most $p^n - 1$. The univariate representation is not unique. However, a unique univariate form of a Boolean function, called *trace representation* is given by

$$f(x) = \sum_{j\in\Gamma_n} tr_{o(j)}(A_j x^j) + A_{p^n-1}x^{p^n-1},$$

where

- Γ_n is the set of integers obtained by choosing the smallest element in each cyclotomic coset modulo $p^n - 1$ (with respect to p);
- $o(j)$ is the size of the cyclotomic coset containing j;

- $A_j \in \mathbb{F}_{p^{o(j)}}$;
- $A_{p^n-1} \in \mathbb{F}_p$.

The algebraic degree of f is equal to $max_{\{j|A_j \neq 0\}}(w_p(j))$, where $w_p(j)$ is the weight of the p-ary expansion of j. In particular, p-ary linear functions are exactly all functions of the form $tr_n(ax)$ for some $a \in \mathbb{F}_{p^n}$.

Example 1.3.3. Let $p = 3$. Denote by $C(j)$ the cyclotomic coset of 3 modulo $3^n - 1$. We have:

$$C(j) = \{j, 3j, 3^2j, \cdots 3^{o(j)-1}j\},$$

where $o(j)$ is the smallest positive integer such that $j \equiv j3^{o(j)} \pmod{3^n - 1}$. Note that $o(j)$ divides n. The subscript j is selected as the smallest integer in $C(j)$, and j is said to be the coset leader of $C(j)$. For example, for $n = 3$, the cyclotomic cosets modulo 26 are: $C(0) = \{0\}$, $C(1) = \{1, 3, 9\}$, $C(2) = \{2, 6, 18\}$, $C(4) = \{4, 10, 12\}$, $C(5) = \{5, 15, 19\}$, $C(7) = \{7, 11, 21\}$, $C(8) = \{8, 20, 24\}$, $C(13) = \{13\}$, $C(14) = \{14, 16, 22\}$ and $C(17) = \{17, 23, 25\}$. The coset leaders modulo 26 are: $0, 1, 2, 4, 5, 7, 8, 13, 14$ and 17.

Any nonzero function f from $\mathbb{F}_{3^n} \rightarrow \mathbb{F}_3$ can be represented as $f(x) = \sum_{j\in\Gamma_n} tr_{o(j)}(A_j x^j) + A_{3^n-1}x^{3^n-1}$ with $A_j \in \mathbb{F}_{3^{o(j)}}$, $A_{3^n-1} \in \mathbb{F}_3$ where Γ_n is the set consisting of all coset leaders modulo $3^n - 1$, $o(j)$ is the size of the cyclotomic coset of 3 modulo $3^n - 1$ containing j (we have $o(j)|n$), and $tr_{o(j)}$ is the trace function from $\mathbb{F}_{3^{o(j)}}$ to $\mathbb{F}_3$.

If we do not identify $\mathbb{F}_p^n$ with $\mathbb{F}_{p^n}$, the p-ary function has a representation as a unique multinomial in $x_1, \cdots, x_n$, where the variables x_i occur with exponent at most $p-1$. This is called the *multivariate representation* or algebraic normal form (ANF). The algebraic degree of a p-ary function is the global degree of its multivariate representation.

The *bivariate* representation of a p-ary function of even dimension $n = 2m$ is defined as follows: we identify $\mathbb{F}_p^n$ with $\mathbb{F}_{p^m} \times \mathbb{F}_{p^m}$ and consider the argument of f as an ordered pair (x, y) of elements in $\mathbb{F}_{p^m}$. There exists a unique bivariate polynomial

$$\sum_{0\le i,j\le p^m-1} A_{i,j}x^iy^j$$

over $\mathbb{F}_{p^m}$ that represents f. The algebraic degree of f is equal to

$$max_{\{(i,j)|A_{i,j}\neq 0\}}(w_p(i) + w_p(j)).$$

Since f takes on its values in $\mathbb{F}_p$, the bivariate representation can be written in the form $f(x, y) = tr_m(P(x, y))$ where $P(x, y)$ is some polynomial of two variables over $\mathbb{F}_{p^m}$.

More generally, the reader notices that every function from $\mathbb{F}_q$ to $\mathbb{F}_q$ can be expressed as a polynomial over $\mathbb{F}_q$.

1.3.3 Quadratic Form Over Finite Fields in Characteristic p

A function $Q(x)$ from $\mathbb{F}_{p^n}$ to $\mathbb{F}_p$ is called quadratic over $\mathbb{F}_p$ if Q is a homogeneous polynomial of degree two in the form

$$Q(x_1, x_2, \cdots, x_n) = \sum_{1 \leq i \leq j n} a_{ij} x_i x_j,$$

where $a_{ij} \in \mathbb{F}_p$, and we use a basis $\{\alpha_1, \alpha_2, \cdots, \alpha_n\}$ of $\mathbb{F}_{p^n}$ over $\mathbb{F}_p$ and identify $x = \sum_{i=1}^n x_i \alpha_i$ with the vector $\bar{x} = (x_1, x_2, \cdots, x_n) \in \mathbb{F}_p^n$. We shall write $\bar{x}$ when an element is to thought of as a vector in $\mathbb{F}_p^n$ and write x when the same vector is to be thought of as an element of $\mathbb{F}_p^n$. The rank of the quadratic form $Q(x)$ is defined as the codimension of the $\mathbb{F}_p$-vectorspace

$$V = \{y \in \mathbb{F}_{p^n} \mid Q(x+y) - Q(x) - Q(y) = 0, \forall x \in \mathbb{F}_{p^n}\}.$$

That is, $\#V = p^{m-r}$ where r is the rank of $Q(x)$. The reader notices that the rank of a quadratic from over $\mathbb{F}_p$ is the smallest number of variables required to represent the quadratic form, up to nonsingular coordination transformations. It is known that any quadratic form of rank r can be transformed into three canonical forms. Let $\tilde{Q}_{2s}(\bar{x}) = x_1x_2 + x_3x_4 + \cdots x_{2s-1}x_{2s}$ where $s \geq 0$ is a positive integer (we assume that $\tilde{Q}_0 = 0$ when $s = 0$). Let υ be a function over $\mathbb{F}_p$ defined by $\upsilon(0) = p - 1$ and $\upsilon(\alpha) = -1$ for any $\alpha \in \mathbb{F}_p^*$. Then any quadratic form $Q(x)$ over $\mathbb{F}_p$ of rank r, in n variables is equivalent (under a change of coordinates) to one of the following three standard types:

1. **Type I**: $\tilde{Q}_r(\bar{x})$, r even;
2. **Type II**: $\tilde{Q}_{r-1}(\bar{x}) + \mu x_n^2$, r odd;
3. **Type III**: $\tilde{Q}_{r-2}(\bar{x}) + x_{r-1}^2 - \eta x_r^2$, r even;

where $\mu \in \{1, \eta\}$ and η is a fixed nonsquare in $\mathbb{F}_p$. Moreover, for any $\alpha \in \mathbb{F}_p$, the number of solutions $\bar{x} \in \mathbb{F}_p^n$ of the equation $Q(\bar{x}) = \alpha$ is

1. **Type I**: $p^{n-1} + \upsilon(\alpha)p^{n-r/2-1}$;
2. **Type II**: $p^{n-1} + \eta(\mu\alpha)p^{n-(r+1)/2}$;
3. **Type III**: $p^{n-1} - \upsilon(\alpha)p^{n-r/2-1}$;

where η is the quadratic (multiplicative) character of $\mathbb{F}_p$ with the assumption that $\eta(0) = 0$.

An interesting class of quadratic forms is the quadratic form with full rank since in this case the corresponding functions are bent functions.

References

1. C. Carlet. Boolean functions for cryptography and error correcting codes. In Yves Crama and Peter L. Hammer, editors, *Boolean Models and Methods in Mathematics, Computer Science, and Engineering*, pages 257–397. Cambridge University Press, June 2010.
2. C. Carlet and P. Guillot. A new representation of Boolean functions. In *AAECC'13, Lecture Notes in Computer Science 1719*, pages 94–103, 1999.
3. C. Carlet and P. Guillot. Bent, resilient Functions and the Numerical Normal Form. In *DIMACS, Discrete Mathematics and Theoretical Computer Science*, pages 87–96, 2001.
4. T. W. Cusick and CusickStanica.: *Cryptographic Boolean Functions and Applications*, Academic Press, San Diego, CA, 2009.
5. F. J. MacWilliams and N. J. Sloane. The theory of error-correcting codes. In *Amsterdam, North Holland*, 1977.

Chapter 2
Mathematical Foundations

This chapter contains some mathematical tools, results and algebraic techniques that will be employed throughout the book. We shall provide the proofs of some results that are of importance mainly because of their use in applications and state some standard results without proof. For basic algebraic concepts, we send the reader to the excellent book of R. Lidl and H. Niederreiter [14].

2.1 Special Polynomials Over Finite Fields

In elementary algebra, a polynomial is an expression consisting of variables (or indeterminates) and coefficients, that involves only the operations of addition, multiplication, and non-negative integer exponents. The arithmetic of polynomials is governed by familiars rules. The concept of polynomial and the associate operations can be generalized to a formal algebraic setting in a straightforward manner. Polynomials appear in a wide variety of areas of mathematics and science. They are used to define polynomial functions, which appear in several contexts ranging from basic theory in algebra to application fields and used in calculus to approximate other functions. In advanced mathematics, polynomials are used to construct polynomial rings and algebraic varieties, central concepts in algebra and algebraic geometry. In this book, we are interested in polynomials over finite fields $\mathbb{F}_q$. We have seen in Chap. 1 that every function from $\mathbb{F}_q$ to $\mathbb{F}_q$ can be expressed as a polynomial over $\mathbb{F}_q$. Since in this book we are predominantly dealing with bent Boolean functions, emphasis here is laid upon binary polynomials.

S. Mesnager, *Bent Functions*, DOI 10.1007/978-3-319-32595-8_2

2.1.1 Permutations Polynomials

Permutation polynomials over finite fields have been an interesting subject of study for many years, and have applications in coding theory, cryptography, combinatorial design theory, and other areas of mathematics and engineering. Information about properties, constructions, and applications of permutation polynomials can be found in [14] and [20]. A polynomial $f \in \mathbb{F}_q[x]$ is called a *permutation polynomial* if the associated polynomial function $f : a \mapsto f(a)$ from $\mathbb{F}_q$ to $\mathbb{F}_q$ is a permutation of $\mathbb{F}_q$. We shall see in the next chapters that some permutations polynomials are very useful to construct bent functions.

2.1.2 Polynomials *e*-*to*-1

A polynomial $f \in \mathbb{F}_q[x]$ is said to be e-to 1 if for every $b \in \mathbb{F}_q$, the equation $f(x) = b$ over $\mathbb{F}_q$ has either e solutions $x \in \mathbb{F}_q$ or no solution, where $e \geq 1$ is an integer, and e divides q. By definition, permutation polynomials are 1-to-1. We shall see in the next chapters that some special binary polynomials 2-to-1 are very useful to construct families of bent functions as well as super-families of bent functions.

2.1.3 Involutions

In many situations, both the permutation polynomial and its compositional inverse are required. For instance, in block ciphers, a permutation is used as an S-box to build the confusion layer during the encryption process. While decrypting the cipher, the compositional inverse of the S-box comes into the picture. Therefore, if both the permutation and its compositional inverse are efficient in terms of implementation, it is advantageous to the designer. This motivates the use of an *involution*. An involution is a permutation whose compositional inverse is itself, *i.e.*, the permutation P is such that $P \circ P$ is the identity.

Involutions form an important class of permutations in particular for their applications in cryptography. They have been used frequently in block cipher designs (as S-Boxes) and coding theory. The reader notices that an involution has a practical advantage since the implementation of the inverse does not require additional resources, which is particularly useful (as part of a block cipher) in devices with limited resources.

In [4], the authors have provided a detailed mathematical study of involutions over finite fields of characteristic 2 for cryptographic purposes. To this end, a systematic study of binary involutions as well as characterizations of the involution property of several classes of polynomials have been given including several constructions of binary involutions. The corpus of binary involutions has been fully described.

Further, some recent works [18–20] have highlighted the use of involutions to design bent Boolean functions.

2.1.4 Binary Dickson Polynomial

In this subsection, we recall an important well-known family of polynomials: the so-called Dickson polynomials. Great emphasis is laid upon binary Dickson polynomials for their applications in the context of bent functions.

We send the reader to the very nice book on Dickson polynomials of Lidl, Mullen and Turnwald [13].

Dickson polynomials of the first kind over $\mathbb{F}_q$ are defined by

$$D_r(X, a) = \sum_{i=0}^{\lfloor \frac{r}{2} \rfloor} \frac{r}{r-i} \binom{r-i}{i} (-a)^i X^{r-2i},$$

where $a \in \mathbb{F}_q$ and r is called the *order* of the polynomial.

The family of binary *Dickson polynomials* $D_r(X) \in \mathbb{F}_2[X]$ of degree r is defined by

$$D_r(X) = \sum_{i=0}^{\lfloor \frac{r}{2} \rfloor} \frac{r}{r-i} \binom{r-i}{i} X^{r-2i}, \quad r = 2, 3, \cdots$$

Dickson polynomials can also be defined by the following recurrence relation:

$$D_{i+2}(X) = XD_{i+1}(X) + D_i(X)$$

with initial values

$$D_0(X) = 0, \quad D_1(X) = X.$$

For any non-zero positive integers r and p, Dickson polynomials satisfy:

1. $\deg(D_r(X)) = r$,
2. $D_{rp}(X) = D_r(D_p(X))$,
3. $D_r(x + x^{-1}) = x^r + x^{-r}$.

The reader can refer to [13] for many useful properties and applications of Dickson polynomials. We give the list of the first eleven Dickson polynomials:

$D_0(X) = 0$
$D_1(X) = X$
$D_2(X) = X^2$
$D_3(X) = X + X^3$

$D_4(X) = X^4$
$D_5(X) = X + X^3 + X^5$
$D_6(X) = X^2 + X^6$
$D_7(X) = X + X^5 + X^7$
$D_8(X) = X^8$
$D_9(X) = X + X^5 + X^7 + X^9$
$D_{10}(X) = X^2 + X^6 + X^{10}$

Dillon and Dobbertin [7, pp. 355–356] remarked that a more careful analysis shows that Dickson polynomials leave fixed the elements whose inverses have a given absolute trace fixed[1].

Lemma 2.1.1 ([7], [9, Lemma 1]). *Let $r \geq 0$ be an integer and $x \in \mathbb{F}_{2^m}$. Then*

$$\mathrm{Tr}_1^m\left(\frac{1}{D_r(x)}\right) = \mathrm{Tr}_1^m\left(\frac{1}{x}\right).$$

This property was recently used and reproved in an elementary way by Charpin, Helleseth and Zinoviev [5, Proof of Lemma 14] for D_3, as well as Wang et al. [22, Proof of Proposition 5] for the case D_5, who remarked that

$$\frac{1}{D_3(x)} = \frac{1}{x} + \frac{1}{x+1} + \frac{1}{x^2+1}, \quad \frac{1}{D_5(x)} = \frac{1}{x} + \frac{x}{x^2+x+1} + \frac{x}{x^4+x^2+1}.$$

A much more general fact is actually true as we now demonstrate in an alternative manner. To this end auxiliary polynomials are needed.

Definition 2.1.2. Let $r \geq 0$ be an integer. Define the polynomial $f_r(x)$ as

$$D_r(x) = \begin{cases} x f_r(x)^2 & \text{if } r \text{ is odd,} \\ x^2 f_r(x)^2 & \text{if } r \text{ is even.} \end{cases}$$

The following relation between D_r and f_r is then verified.

Lemma 2.1.3 ([9]). *Let $r \geq 0$ be an integer. Then*

$$x + D_r(x) + x^2 f_r(x) f_{r+1}(x) + D_{r+1}(x) = 0.$$

Proof. We equivalently show that

$$x^2 + D_r(x)^2 + x^4 f_r(x)^2 f_{r+1}(x)^2 + D_{r+1}(x)^2 = 0,$$

[1] A weaker statement is also proved by Ranto [21, Lemma 4] who assumes that $k = \gcd(r, 2^m - 1) = 1$.

which can be rewritten as

$$x^2 + D_r(x)^2 + xD_r(x)D_{r+1}(x) + D_{r+1}(x)^2 = 0.$$

For $r = 0$, this is trivially verified. For $r \geq 1$, write down $D_{r+1}(x)$ as $D_{r+1}(x) = xD_r(x) + D_{r-1}(x)$ and the result follows by induction. □

As a corollary we get a general expression for $\frac{1}{D_r(x)}$ involving $f_r(x)$.

Corollary 2.1.4 ([9]). *Let $r \geq 1$ be an integer. Then*

$$\begin{aligned}\frac{1}{D_r(x)} &= \frac{1}{x} + \frac{f_{r-1}(x)}{f_r(x)} + \frac{f_{r-1}(x)^2}{f_r(x)^2},\\ &= \frac{1}{x} + \frac{f_{r+1}(x)}{f_r(x)} + \frac{f_{r+1}(x)^2}{f_r(x)^2}.\end{aligned}$$

Proof. Since $D_{2r}(x) = D_r(x)$, we can assume that r is odd without loss of generality. Then

$$\begin{aligned}\frac{1}{D_r(x)} &= \frac{1}{xf_r(x)^2} = \frac{x}{x^2 f_r(x)^2}\\ &= \frac{D_r(x) + x^2 f_r(x) f_{r+1}(x) + D_{r+1}(x)}{x^2 f_r(x)^2}\\ &= \frac{xf_r(x)^2 + x^2 f_r(x) f_{r+1}(x) + x^2 f_{r+1}(x)^2}{x^2 f_r(x)^2}\\ &= \frac{1}{x} + \frac{f_{r+1}(x)}{f_r(x)} + \frac{f_{r+1}(x)^2}{f_r(x)^2};\end{aligned}$$

the other equality being deduced in a similar way. □

Lemma 2.1.1 directly follows from Corollary 2.1.4, thus yielding an alternative and more concrete proof of it.

A well-known result by Chou, Gomez-Calderon and Mullen [6] describes the cardinality of the preimage of an arbitrary element.

Theorem 2.1.5 ([6, Theorem 9], [16, Theorem 3.26]). *Let $\mathbb{F}_{2^m}$ be the finite field with 2^m elements and $1 \leq r \leq 2^n - 1$ be an integer. Let*

$$k = \gcd(r, 2^m - 1), \quad l = \gcd(r, 2^m + 1).$$

Let $x, y \in \mathbb{F}_{2^m}$ be two elements such that $D_r(x) = y$. Then

$$\#D_r^{-1}(y) = \begin{cases} \frac{k+l}{2} & \text{if } y = 0\ ,\\ k & \text{if } y \neq 0 \text{ and } \mathrm{Tr}_1^m(1/x) = 0,\\ l & \text{if } y \neq 0 \text{ and } \mathrm{Tr}_1^m(1/x) = 1.\end{cases}$$

As a corollary, they obtain the cardinalities of the value sets of Dickson polynomials [6, Theorems 10 and 10′], [16, Theorems 3.27 and 3.30], and in particular a proof of the characterizations of Dickson polynomials as permutation polynomials [6, Corollary 11], [16, Corollary 3.28].

The proof heavily relies on the study of the map

$$\mathbb{F}_{2^n} \to \mathbb{F}_{2^m}$$
$$x \mapsto x + x^{-1}$$

and Waring's formula [16, Theorem 1.1], [15, Theorem 1.76] which ensures that [16, Equation 2.2], [15, Equation 7.8]

$$D_r(x + x^{-1}) = x^r + x^{-r}.$$

The following property is then a corollary to the above results.

Corollary 2.1.6 ([9]). *Let* $1 \leq r \leq 2^n - 1$ *be an integer. Then the map* $x \mapsto D_r(x)$ *induces a permutation of*

- $\mathcal{T}_0$ *if and only if* $k = \gcd(r, 2^m - 1) = 1$;
- $\mathcal{T}_1$ *if and only if* $l = \gcd(r, 2^m + 1) = 1$.

Proof. Lemma 2.1.1 shows that D_r maps $\mathcal{T}_i$ into $\mathcal{T}_i$ for $i \in \mathbb{F}_2$. One then concludes using Theorem 2.1.5 which gives the size of the preimage of $x \in \mathcal{T}_i$. □

We define the corresponding exponential sums as follows. Let $\chi : \mathbb{F}_2 \mapsto \mathbb{Z}$ denote the nontrivial additive character of $\mathbb{F}_2$. Given a Boolean function f on $\mathbb{F}_2^n$, the *"sign" function* of f is the integer-valued function $\chi_f = \chi(f)$ that is, $\chi_f = (-1)^f$.

Definition 2.1.7. Let $f : \mathbb{F}_{2^m} \to \mathbb{F}_2$ be a Boolean function. We denote by $T_i^r(f)$ the exponential sum on $\mathcal{T}_i$ for $i \in \mathbb{F}_2$ for $f \circ D_r$, that is

$$T_i^r(f) = \sum_{x \in \mathcal{T}_i} \chi_{f \circ D_r}(x).$$

Moreover, let $T_i(f) = T_i^1(f)$.

The following lemma is easily deduced from the equality $(-1)^{\mathrm{Tr}_1^m(x)} = 1 - 2\mathrm{Tr}_1^m(x)$ where the values of the trace are understood as the integers 0 and 1.

Lemma 2.1.8. *Let* $f : \mathbb{F}_{2^m} \to \mathbb{F}_2$ *be a Boolean function. Then*

$$T_i(f) = \frac{1}{2}\left(\sum_{x \in \mathbb{F}_{2^m}} \chi_f(x) + (-1)^i \sum_{x \in \mathbb{F}_{2^m}} \chi\left(\mathrm{Tr}_1^m(1/x) + f(x)\right)\right).$$

And we finally record the following corollary.

Corollary 2.1.9. *Let* $1 \leq r \leq 2^n - 1$ *be an integer and* $f : \mathbb{F}_{2^m} \to \mathbb{F}_2$ *be a Boolean function. Assume moreover that* $\gcd(r, 2^m - 1) = 1$. *Then*

$$T_0^r(f) = T_0(f),$$
$$T_1^r(f) = \sum_{x \in \mathbb{F}_{2^m}} \chi_{f \circ D_r}(x) - T_0(f).$$

2.2 Fourier Transform and Walsh Hadamard Transform

Let f be any complex valued function on $\mathbb{F}_2^n$. *The Fourier transform* of f denoted by $\mathcal{F}(f)$ is by definition the complex-valued mapping $\mathbb{F}_2^n \to \mathbb{C}$ defined for $x \in \mathbb{F}_2^n$ by

$$\mathcal{F}(f)(x) := \sum_{y \in \mathbb{F}_2^n} f(y)(-1)^{x \cdot y}, x \in \mathbb{F}_2^n$$

where "$\cdot$" is a scalar product on $\mathbb{F}_2^n$. Its normalized Fourier transformation is given by

$$\overline{\mathcal{F}}(f)(x) := \frac{1}{2^{n/2}} \sum_{y \in \mathbb{F}_2^n} f(y)(-1)^{x \cdot y}, x \in \mathbb{F}_2^n. \tag{2.1}$$

The Fourier transform is invertible and the inverse is given by $\mathcal{F}(\mathcal{F}(f)) = 2^n f$.

When f is represented by its truth table vector, the Fourier transform (2.1) also admits a matrix representation $\mathcal{F}(f) = Hf$ where H is the so-called *Hadamard matrix* whose coefficient at row $x \in \mathbb{F}_2^n$ and column $y \in \mathbb{F}_2^n$ is $H_{x,y} = (-1)^{x \cdot y}$. The Hadamard matrix is invertible and its inverse is given by $H^{-1} = \frac{1}{2^n} H$. As a result, it holds that $f = \frac{1}{2^n} H \mathcal{F}(f)$.

Let f be a Boolean function defined on $\mathbb{F}_2^n$. Then the *Walsh Hadamard transform* of f is the discrete Fourier transform of χ_f, whose value at $\omega \in \mathbb{F}_2^n$ is defined as follows:

$$\forall \omega \in \mathbb{F}_2^n, \quad \widehat{\chi_f}(\omega) = \sum_{x \in \mathbb{F}_2^n} (-1)^{f(x) + \omega \cdot x}.$$

(that is, $\widehat{\chi_f}(\omega) = \sum_{x \in \mathbb{F}_2^n} \chi\,(f(x) + \omega \cdot x))$ where "$\cdot$" is the scalar product in $\mathbb{F}_2^n$ defined as $x \cdot y = \sum_{i=1}^n x_i y_i$.

When dealing with Boolean functions, we rather resort to the Walsh transform that has nicer properties than the Fourier transform in most cases and is an easier tool to handle Boolean functions especially when we are interested in their cryptographic criteria.

Table 2.1 Truth table and Walsh spectrum of $f(x) = x_1 \oplus x_2 \oplus x_3 \oplus x_1x_4$

x_1	x_2	x_3	x_4	$f(x)$	$(-1)^{f(x)}$	$\widehat{\chi_f}(x)$
0	0	0	0	0	1	0
0	0	0	1	0	1	0
0	0	1	0	1	−1	0
0	0	1	1	1	−1	0
0	1	0	0	1	−1	0
0	1	0	1	1	−1	0
0	1	1	0	0	1	8
0	1	1	1	0	1	−8
1	0	0	0	1	−1	0
1	0	0	1	0	1	0
1	0	1	0	0	1	0
1	0	1	1	1	−1	0
1	1	0	0	0	1	0
1	1	0	1	1	−1	0
1	1	1	0	1	−1	8
1	1	1	1	0	1	8

There exists an algorithm *Fast Fourier Transform* (FFA) to compute the Walsh transform (resp. Fourier Hadamard transform) of Boolean function given by its truth table. Below, we present such an algorithm (to compute the Fourier Hadamard transform of f, on should omit the first step of the algorithm).

When the algorithm ends (*i.e.* when it arrives to functions on one variable each), the global table gives the values of $\widehat{\chi_f}$.

The complexity of the above algorithm is $O(n2^n)$.

Example 2.2.1. Let $n = 4$. In Table 2.1 we give the truth table (given by the column "f(x)") and Walsh spectrum (given by the last column) of the Boolean function defined over $\mathbb{F}_2^4$ by its ANF: $f(x_1, x_2, x_3, x_4) = x_1 \oplus x_2 \oplus x_3 \oplus x_1x_4$.

The notion of Walsh transform refers to a scalar product.[2] When $\mathbb{F}_2^n$ is identified with the field $\mathbb{F}_{2^n}$ by an isomorphism between these two n-dimensional vector spaces over $\mathbb{F}_2$, it is convenient to choose the isomorphism such that the canonical scalar product "·" in $\mathbb{F}_2^n$ coincides with the canonical scalar product in $\mathbb{F}_{2^n}$, which is the trace of the product : $x \cdot y = \sum_{i=1}^n x_iy_i = \mathrm{Tr}_1^n(xy)$ for $x, y \in \mathbb{F}_{2^n}$. Thus if f is a Boolean function defined on $\mathbb{F}_{2^n}$ then, the Walsh Hadamard transform of f is the discrete Fourier transform of χ_f, whose value at $\omega \in \mathbb{F}_{2^n}$ is defined as follows:

$$\forall \omega \in \mathbb{F}_{2^n}, \quad \widehat{\chi_f}(\omega) = \sum_{x \in \mathbb{F}_{2^n}} (-1)^{f(x)+\mathrm{Tr}_1^n(\omega x)}.$$

(that is, $\widehat{\chi_f}(\omega) = \sum_{x \in \mathbb{F}_{2^n}} \chi\left(f(x) + \mathrm{Tr}_1^n(\omega x)\right)$).

[2]Note that in the definition of the Walsh transform, we can take any inner product; the cryptographic properties are not related to a particular choice of the inner product therefore, the issue of the choice of the isomorphism does not arise.

The Walsh transform satisfies the well-known Parseval's relation

$$\sum_{\omega \in \mathbb{F}_{2^n}} \widehat{\chi_f}^2(\omega) = 2^{2n}$$

and also the inverse Fourier formula

$$\sum_{\omega \in \mathbb{F}_{2^n}} \widehat{\chi_f}(\omega) = 2^n(-1)^{f(0)}.$$

Note that not all values of the Walsh Hadamard transform can have the same sign, except when the function is affine. This comes from the fact that we then have $\left(\sum_{\omega \in \mathbb{F}_{2^n}} \widehat{\chi_f}(\omega)\right)^2 = \sum_{\omega \in \mathbb{F}_{2^n}} \widehat{\chi_f}^2(\omega)$ which implies that all these values are null except one.

The Walsh transform satisfies also the *Poisson summation formula*, valid for every $\mathbb{F}_2$-linear space E:

$$\sum_{u \in a+E} (-1)^{b \cdot u} \widehat{\chi_f}(u) = |E| \, (-1)^{a \cdot b} \sum_{x \in b+E^{\perp}} (-1)^{f(x)+a \cdot x}.$$

where $|E| := \{x \in \mathbb{F}_{2^n} \mid \forall y \in E, x\dot{y} = 0\}$ is the orthogonal of E

The weights and the Walsh spectra of affine functions are peculiar: the Walsh transform of the function $\ell(x) = a \cdot x \oplus \varepsilon$ takes null value at every vector $u \neq a$ and takes value $2^n \, (-1)^{\varepsilon}$ at a.

The rank of a quadratic Boolean function is connected with the distribution of its Walsh–Hadamard transform values. The following is due to Helleseth and Kumar.

Theorem 2.2.2 ([10]). *Let f be a quadratic Boolean function on $\mathbb{F}_{2^n}$ with rank $2t$, $0 \leq t \leq \lfloor \frac{n}{2} \rfloor$. Then the distribution of its Walsh transform is given by Table 2.2.*

Now, let f be a Boolean function on $\mathbb{F}_{2^n}$. Then the *extended Walsh–Hadamard transform* of f is defined as

$$\widehat{\chi_f}(\omega, k) = \sum_{x \in \mathbb{F}_{2^n}} (-1)^{f(x)+\mathrm{Tr}_1^n(\omega x^k)},$$

for $\omega \in \mathbb{F}_{2^n}$ and k an integer co-prime with $2^n - 1$.

Table 2.2 Walsh spectrum of quadratic function with rank $2t$

Value of $\widehat{\chi_f}(\omega)$, $\omega \in \mathbb{F}_{2^n}$	Number of occurrences
0	2^{n-2t}
2^{n-t}	$2^{2t-1} + 2^{t-1}$
-2^{n-t}	$2^{2t-1} - 2^{t-1}$

2.3 Some Classical Binary Exponential Sums

Exponential sums are important tools in number theory for solving problems involving integer and real numbers in general. Analogous sums can be considered in the framework of finite fields and turn out to be useful in various applications of finite fields; in particular to study the bentness property of functions or other related objects such as the computation of the weight distributions of linear codes from bent functions. An *exponential sum* is defined as a sum of the form

$$S(P) = \sum_{x} e^{2\pi i f(x)},$$

where x runs over all the integers (or some of them) from a certain interval, P is the number of summands and f(x) is an arbitrary function taking on real values under integer x. Several problems of number theory and its applications can be reduced to the study of such sums. In secret key cryptography, we are usually dealing with binary exponential sums. In this section, we present the most important binary exponential sums involved naturally when we proving the bentness property of some binary Boolean functions.

2.3.1 Binary Kloosterman Sums

Binary *Kloosterman sums* have been widely studied for a long time for their own sake as interesting mathematical objects and have recently become the focus of much research, most notably due to their applications in cryptography and for their connection to coding theory. The classical binary Kloosterman sums on $\mathbb{F}_{2^m}$ (where m is an arbitrary positive integer) are defined as follows.

Definition 2.3.1. Let $a \in \mathbb{F}_{2^m}$. The binary Kloosterman sums on $\mathbb{F}_{2^m}$ associated with a is

$$K_m(a) := \sum_{x \in \mathbb{F}_{2^m}} \chi\big(\mathrm{Tr}_1^m(ax + \frac{1}{x})\big), \quad a \in \mathbb{F}_{2^m}.$$

Kloosterman sums are generally defined on the multiplicative group $\mathbb{F}_{2^m}^{\star}$ of $\mathbb{F}_{2^m}$. In the document we extend to 0 assuming that $\chi(\mathrm{Tr}_1^m(\frac{1}{x})) = 1$ for $x = 0$ (in fact, $\mathrm{Tr}_1^m(\frac{1}{x}) = \mathrm{Tr}_1^m(x^{2^{m-1}-1})$).

In particular, such an exponential sum can be seen as the Walsh–Hadamard transform of a simple function. Indeed, the function $a \mapsto K_n(a)$ is the Walsh–Hadamard transform of the inverse function (we define $1/0 = 0$ or $1/x$ as x^{2^n-2} for all $x \in \mathbb{F}_{2^n}$).

It is an elementary fact that $K_n(a) = K_n(a^2)$. Indeed,

$$K_n(a) = 1 + \sum_{x\in\mathbb{F}_{2^n}^*} (-1)^{\mathrm{Tr}_1^n(ax+\frac{1}{x})} = 1 + \sum_{x\in\mathbb{F}_{2^n}^*} (-1)^{\mathrm{Tr}_1^n\left(a^2x^2+\frac{1}{x^2}\right)}$$

$$= 1 + \sum_{x\in\mathbb{F}_{2^n}^*} (-1)^{\mathrm{Tr}_1^n(a^2x+\frac{1}{x})} = K_n(a^2) .$$

The following proposition is directly obtained from the result of Lachaud and Wolfmann in [11] which is suitable for any positive integer m.

Proposition 2.3.2. *Let m be a positive integer. The set $\{K_m(a), a \in \mathbb{F}_{2^m}\}$, is the set of all the integers multiple of 4 in the range $[-2^{(m+2)/2} + 1, 2^{(m+2)/2} + 1]$.*

Divisibility properties of the Kloosterman sums have been studied in several recent papers. In particular, the following result is given by Charpin, Helleseth and Zinoviev in [5] on the divisibility by 3 of $K_m(a) - 1$.

Proposition 2.3.3 ([5]). *Let $m \geq 3$ be an odd integer, and let $a \in \mathbb{F}_{2^m}^\star$. Then,*

$$K_m(a) - 1 \equiv 0 \pmod 3 \iff \mathrm{Tr}_1^m(a^{1/3}) = 0.$$

2.3.2 Binary Cubic Sums

The binary cubic sums on $\mathbb{F}_{2^m}$ (where m is an arbitrary positive integer) are defined as follows.

Definition 2.3.4. The cubic sums on $\mathbb{F}_{2^m}$ are:

$$C_m(a,b) := \sum_{x\in\mathbb{F}_{2^m}} \chi\big(\mathrm{Tr}_1^m(ax^3 + bx)\big), \quad a \in \mathbb{F}_{2^m}^\star, b \in \mathbb{F}_{2^m}.$$

Of course, such exponential sums can be seen as the Walsh–Hadamard transform of a simple function. Indeed, the function $b \mapsto C_m(a,b)$ is the Walsh–Hadamard transform of $x \mapsto ax^3$.

The exact values of the cubic sums $C_m(a,a)$ on $\mathbb{F}_{2^m}$ can be computed thanks to Carlitz's result [1] by means of the Jacobi symbol (see below). Recall that the Jacobi symbol $\left(\frac{2}{m}\right)$ is a generalization of the Legendre symbol (which is defined when m is an odd prime). For m odd, $\left(\frac{2}{m}\right) = (-1)^{\frac{(m^2-1)}{8}}$.

Proposition 2.3.5 ([1]). *Let m be an odd integer. Then we have:*

1. $C_m(1,1) = \left(\frac{2}{m}\right) 2^{(m+1)/2}$,
2. *If* $\mathrm{Tr}_1^m(c) = 0$, *then* $C_m(1,c) = 0$,
3. *If* $\mathrm{Tr}_1^m(c) = 1$ *(with $c \neq 1$), then* $C_m(1,c) = \chi(\mathrm{Tr}_1^m(\gamma^3 + \gamma)) \left(\frac{2}{m}\right) 2^{(m+1)/2}$ *where* $c = \gamma^4 + \gamma + 1$ *for some* $\gamma \in \mathbb{F}_{2^m}$.

Remark 2.3.6. Note that when $\mathrm{Tr}_1^m(c) = 1$ and $c \neq 1$, then the cubic sums $C_m(1, c)$ can be computed thanks to a recent result of Charpin et al. in [2]. More precisely, if $\mathrm{Tr}_1^m(c) = 1$ (with $c \neq 1$), then $C_m(1, c) = (-1)^{\mathrm{Tr}_1^m(\gamma^3+\gamma)} \left(\frac{2}{m}\right) 2^{(m+1)/2}$ where γ is the unique element of $\mathbb{F}_{2^m}$ satisfying $c = \gamma^4 + \gamma + 1$ and $\mathrm{Tr}_1^m(\gamma) = 0$.

2.3.3 Partial Exponential Sums

Let $f : \mathbb{F}_{2^m} \to \mathbb{F}_2$ be a Boolean function. We denote the exponential sum associated with f by $\Xi(f)$, that is

$$\Xi(f) = \sum_{x \in \mathbb{F}_{2^m}} \chi_f(x).$$

With this notation, the classical binary Kloosterman sum associated with a on $\mathbb{F}_{2^m}$ is defined as

$$K_m(a) = \Xi\left(\mathrm{Tr}_1^m\left(ax + \frac{1}{x}\right)\right).$$

The following partial exponential sums are a classical tool to study hyper-bentness. Beware that the Boolean function is defined on $\mathbb{F}_{2^n}$ in the first definition and $\mathbb{F}_{2^m}$ in the second one.

Definition 2.3.7. Let $f : \mathbb{F}_{2^n} \to \mathbb{F}_2$ be a Boolean function and U be the set of $(2^m + 1)$-st roots of unity in $\mathbb{F}_{2^n}$. We define $\Lambda(f)$ as

$$\Lambda(f) = \sum_{u \in U} \chi_f(u).$$

We define $\mathcal{T}_0$ and $\mathcal{T}_1$ as follows.

Definition 2.3.8. For $i \in \mathbb{F}_2$, let $\mathcal{T}_i$ The exponential sum $\mathcal{T}_i = \{x \in \mathbb{F}_{2^m} \mid \mathrm{Tr}_1^m(1/x) = i\}$ denote the set

$$\mathcal{T}_i = \{x \in \mathbb{F}_{2^m} \mid \mathrm{Tr}_1^m(1/x) = i\}.$$

Now, we define the partial exponential sum on $\mathcal{T}_i$ associated with f as follows.

Definition 2.3.9. Let $f : \mathbb{F}_{2^m} \to \mathbb{F}_2$ be a Boolean function and, for $i \in \mathbb{F}_2$, denote by $T_i(f)$ the partial exponential sum on $\mathcal{T}_i$ associated with f, that is

$$T_i(f) = \sum_{x \in \mathcal{T}_i} \chi_f(x).$$

The following lemma is easily deduced from the equality $\chi\left(\mathrm{Tr}_1^m(x)\right) = 1 - 2\mathrm{Tr}_1^m(x)$ where the values of the trace are understood as the integers 0 and 1.

Lemma 2.3.10. *Let $f : \mathbb{F}_{2^m} \to \mathbb{F}_2$ be a Boolean function. Then*

$$T_i(f) = \frac{1}{2}\left(\Xi(f) + \chi(i)\,\Xi\left(\mathrm{Tr}_1^m(1/x) + f(x)\right)\right) .$$

Finally, we have the following relation between Kloosterman sums and the above partial exponential sums.

Corollary 2.3.11. *Let $a \in \mathbb{F}_{2^m}^*$. Then $K_m(a) = -2T_1(\mathrm{Tr}_1^m(ax)) = 2T_0(\mathrm{Tr}_1^m(ax))$.*

Proof. We have

$$T_0(\mathrm{Tr}_1^m(ax)) - T_1(\mathrm{Tr}_1^m(ax)) = K_m(a) .$$

Moreover,

$$T_0(\mathrm{Tr}_1^m(ax)) + T_1(\mathrm{Tr}_1^m(ax)) = \Xi\left(\mathrm{Tr}_1^m(ax)\right) = 0 .$$

□

2.4 Some Results on the Sum Over the Cyclic Group U of Characters

From now, $n = 2m$ is an (even) integer. Recall the well known *polar decomposition*. Let x be an element of $\mathbb{F}_{2^n}$. The *conjugate* of x over a subfield $\mathbb{F}_{2^m}$ of $\mathbb{F}_{2^n}$ will be denoted by $\bar{x} = x^{2^m}$ and the relative norm with respect to the quadratic field extension $\mathbb{F}_{2^n}/\mathbb{F}_{2^m}$ by $norm(x) = x\bar{x}$. Also, we denote by U the set $\{u \in \mathbb{F}_{2^n} \mid norm(u) = 1\} = \{u \in \mathbb{F}_{2^n} \mid u^{2^m+1} = 1\}$, which is the group of $(2^m + 1)$-st roots of unity. Since the multiplicative group of the field $\mathbb{F}_{2^n}$ is cyclic and $2^m + 1$ divides $2^n - 1$, the order of U is $2^m + 1$. Finally, the unit 1 is the single element in $\mathbb{F}_{2^m}$ of norm one and every non-zero element x of $\mathbb{F}_{2^n}$ has a unique decomposition as: $x = yu$ with $y \in \mathbb{F}_{2^m}^\star$ and $u \in U$.

In this book, U will always denote the cyclic group of $(2^m + 1)$-st roots of unity that is $\{u \in \mathbb{F}_{2^n} \mid u^{2^m+1} = 1\}$.

Remark 2.4.1. Consider the polar deposition of $x \in \mathbb{F}_{2^n}^\star$: $x = yu$ with $y \in \mathbb{F}_{2^m}^\star$ and $u \in U$. Note that $x + x^{2^m} = (u + u^{2^m})y$. If $u \neq 1$, that is, $x \notin \mathbb{F}_{2^m}$, one has $y = \frac{x+x^{2^m}}{u+u^{2^m}}$. Hence, if $x \notin \mathbb{F}_{2^m}$, $x = \lambda z$ with $\lambda = \frac{u}{u+u^{2^m}}$ and $z = x + x^{2^m} \in \mathbb{F}_{2^m}$. Note that $\lambda + \lambda^{2^m} = 1$. Set $\tilde{U} := \{\lambda \in \mathbb{F}_{2^n} \mid \lambda^{2^m} + \lambda = 1\}$. We have

- if $x \in \mathbb{F}_{2^n}^\star \setminus \mathbb{F}_{2^m}$, there is a unique pair $(z, \lambda) \in \mathbb{F}_{2^m} \times \tilde{U}$ such that $x = \lambda z$.
- if $x \in \mathbb{F}_{2^m}$ then write $x = z$.

In the following we provide several technical results and some mathematical tools that we need subsequently in the other chapters (especially, Chaps. 10 and 17). More precisely, we are interested to express some particular exponential sums over the cyclic group U by means of Kloosterman sums and cubic sums. Such expressions will be useful to exhibit conditions of bentness and semi bentness (of some Boolean functions in polynomial forms) involving Kloosterman sums and cubic sums.

First we state a well-known result. We give a proof because the result is important.

Proposition 2.4.2 ([2, 8, 11, 12]). *Let* $n = 2m$, *r a positive integer such that* $\gcd(r, 2^m + 1) = 1$ *and* $a \in \mathbb{F}_{2^m}$. *Then,*

$$\sum_{u \in U} \chi\left(\mathrm{Tr}_1^n(au^r)\right) = 1 - K_m(a).$$

Proof. Let $a \in \mathbb{F}_{2^m}^\star$. We have

$$K_m(a) = \sum_{x \in \mathbb{F}_{2^m}, \mathrm{Tr}_1^m(1/x)=0} \chi(\mathrm{Tr}_1^m(ax)) - \sum_{x \in \mathbb{F}_{2^m}, \mathrm{Tr}_1^m(1/x)=1} \chi(\mathrm{Tr}_1^m(ax)).$$

But

$$\sum_{x \in \mathbb{F}_{2^m}, \mathrm{Tr}_1^m(1/x)=0} \chi(\mathrm{Tr}_1^m(ax)) + \sum_{x \in \mathbb{F}_{2^m}, \mathrm{Tr}_1^m(1/x)=1} \chi(\mathrm{Tr}_1^m(ax)) = \sum_{x \in \mathbb{F}_{2^m}} \chi(\mathrm{Tr}_1^m(ax)) = 0.$$

Hence

$$K_m(a) = -2 \sum_{x \in \mathbb{F}_{2^m}, \mathrm{Tr}_1^m(1/x)=1} \chi(\mathrm{Tr}_1^m(ax)).$$

The mapping $u \mapsto u + u^{-1}$ is onto and 2-to-1 from $U \setminus \{1\}$ to $\{x \in \mathbb{F}_2m, \mathrm{Tr}_1^m(1/x) = 1\}$:

$$K_m(a) = - \sum_{u \in U, u \neq 1} \chi(\mathrm{Tr}_1^m(a(u + u^{-1})))$$

But $u + u^{-1} = Tr_m^n(u)$. Thus

$$K_m(a) = - \sum_{u \in U, u \neq 1} \chi(\mathrm{Tr}_1^n(au) = - \sum_{u \in U} \chi(\mathrm{Tr}_1^n(au) + 1$$

Now, it is clear that for r coprime with 2^m+1 we have, $\sum_{u\in U}\chi\left(\mathrm{Tr}_1^n(au^r)\right) = \sum_{u\in U}\chi\left(\mathrm{Tr}_1^n(au)\right)$. □

The following result extends Proposition 2.4.2.

Proposition 2.4.3 ([17]). *Let $n = 2m$ with m odd. Let $a \in \mathbb{F}_{2^m}^\star$, $b \in \mathbb{F}_4^\star$ and r a positive integer such that* $\gcd(r, 2^m+1) = 1$. *Then,*

$$\sum_{u\in U}\chi\left(\mathrm{Tr}_1^n(au^r) + \mathrm{Tr}_1^2(bu^{\frac{2^m+1}{3}})\right) = \frac{K_m(a) - 1 + \lambda C_m(a,a)}{3}$$

where $\lambda = 4$ if $b = 1$ and $\lambda = -2$ otherwise.

Proof. Set $S(a,b) := \sum_{u\in U}\chi\left(\mathrm{Tr}_1^n(au^r) + \mathrm{Tr}_1^2(bu^{\frac{2^m+1}{3}})\right)$. The mapping $u \mapsto u^r$ is a permutation of U (we denote by $x \mapsto x^{\frac{1}{r}}$ its inverse map) since $gcd(r, 2^m+1) = 1$. Hence, $S(a,b) = \sum_{u\in U}\chi\left(\mathrm{Tr}_1^n(au) + \mathrm{Tr}_1^2(bu^{\frac{1}{r}\frac{2^m+1}{3}})\right)$. Now, m being odd, we have the following decomposition of U:

$$U = V \cup \zeta V \cup \zeta^2 V$$

where $V := \{u^3 \mid u \in U\}$ and ζ is a generator of the cyclic group U. Therefore,

$$\begin{aligned} S(a,b) &= \sum_{j=0}^{2}\sum_{v\in V}\chi\left(\mathrm{Tr}_1^n(a\zeta^j v) + \mathrm{Tr}_1^2(b\zeta^{\frac{j}{r}\frac{2^m+1}{3}})\right) \\ &= \sum_{j=0}^{2}\sum_{v\in V}\chi\left(\mathrm{Tr}_1^n(a\zeta^j v)\right)\chi\left(\mathrm{Tr}_1^2(b\zeta^{\frac{j}{r}\frac{2^m+1}{3}})\right) \\ &= \sum_{j=0}^{2}\chi\left(\mathrm{Tr}_1^2(b\zeta^{\frac{j}{r}\frac{2^m+1}{3}})\right)\sum_{v\in V}\chi\left(\mathrm{Tr}_1^n(a\zeta^j v)\right). \end{aligned}$$

For $j \in \{0,1,2\}$, set

$$\sigma_j(a) := \sum_{v\in V}\chi(\mathrm{Tr}_1^n(a\zeta^j v)),$$

then we have

$$S(a,b) = \sum_{j=0}^{2}\chi(\mathrm{Tr}_1^2(b\zeta^{\frac{j}{r}\cdot\frac{2^m+1}{3}}))\sigma_j(a).$$

Remark that, for every $a \in \mathbb{F}_{2^m}$, $\sigma_1(a) = \sigma_2(a)$. Indeed, ζ^{2^m-2} is an element of V because 3 divides $(2^m + 1)$ (since m is odd) and the mapping $v \mapsto \zeta^{2^m-2} v^{2^m}$ is a permutation on V. Thus,

$$\begin{aligned}\sigma_1(a) &= \sum_{v\in V} \chi(\mathrm{Tr}_1^n(a\zeta v)) = \sum_{v\in V} \chi(\mathrm{Tr}_1^n(a\zeta^{2^m} v^{2^m}))\\ &= \sum_{v\in V} \chi(\mathrm{Tr}_1^n(a\zeta^2(\zeta^{2^m-2} v^{2^m}))) = \sigma_2(a).\end{aligned}$$

Hence,

$$S(a,b) = \chi(\mathrm{Tr}_1^2(b))\sigma_0(a) + \left(\chi(\mathrm{Tr}_1^2(b\zeta^{\frac{1}{r}\cdot\frac{2^m+1}{3}})) + \chi(\mathrm{Tr}_1^2(b\zeta^{\frac{2}{r}\cdot\frac{2^m+1}{3}}))\right)\sigma_1(a).$$

Now, note that since ζ is a generator of U, $\mathrm{Tr}_1^2(\zeta^{\frac{1}{r}\cdot\frac{2^m+1}{3}}) = \mathrm{Tr}_1^2(\zeta^{\frac{2}{r}\cdot\frac{2^m+1}{3}}) = 1$. Moreover, if $b \in \mathbb{F}_4^\star \setminus \{1\}$ then, $\chi(\mathrm{Tr}_1^2(b\zeta^{\frac{1}{r}\cdot\frac{2^m+1}{3}})) + \chi(\mathrm{Tr}_1^2(b\zeta^{\frac{2}{r}\cdot\frac{2^m+1}{3}})) = 0$.

Therefore,

$$S(a,b) = \begin{cases} \sigma_0(a) - 2\sigma_1(a) & \text{if } b = 1\\ -\sigma_0(a) & \text{if } b \neq 1.\end{cases}$$

Now, one can express the sum $\sigma_1(a)$ by means of $\sigma_0(a)$. For that, we compute in two ways the sum $\sum_{b\in\mathbb{F}_4^\star} S(a,b)$, for every $a \in \mathbb{F}_{2^m}^\star$.

Firstly,

$$\begin{aligned}\sum_{b\in\mathbb{F}_4^\star} S(a,b) &= \sum_{b\in\mathbb{F}_4^\star}\left(\sum_{j=0}^{2}\chi\left(\mathrm{Tr}_1^2(b\zeta^{\frac{j}{r}\frac{2^m+1}{3}})\right)\sigma_j(a)\right)\\ &= \sum_{j=0}^{2}\left(\sum_{b\in\mathbb{F}_4}\chi\left(\mathrm{Tr}_1^2(b\zeta^{\frac{j}{r}\frac{2^m+1}{3}})\right) - 1\right)\sigma_j(a).\end{aligned}$$

We get (since $\sigma_2(a) = \sigma_1(a)$)

$$\sum_{b\in\mathbb{F}_4^\star} S(a,b) = -\sum_{j=0}^{2}\sigma_j(a) = -\sigma_0(a) - 2\sigma_1(a).$$

Secondly, for $a \in \mathbb{F}_{2^m}^\star$, we have

$$\sum_{b\in\mathbb{F}_4^\star} S(a,b) = \sum_{u\in U} \chi(\mathrm{Tr}_1^n(au)) \sum_{b\in\mathbb{F}_4^\star} \chi(\mathrm{Tr}_1^2(bu^{\frac{1}{r}\frac{2^m+1}{3}}))$$

$$= \sum_{u\in U} \chi(\mathrm{Tr}_1^n(au)) \left(\sum_{b\in\mathbb{F}_4} \chi(\mathrm{Tr}_1^2(bu^{\frac{1}{r}\frac{2^m+1}{3}})) - 1 \right).$$

Hence,

$$\sum_{b\in\mathbb{F}_4^\star} S(a,b) = -\sum_{u\in U} \chi(\mathrm{Tr}_1^n(au)).$$

Thanks to Proposition 2.4.2, we obtain

$$\sum_{b\in\mathbb{F}_4^\star} S(a,b) = K_m(a) - 1.$$

Collecting the two expressions of $\sum_{b\in\mathbb{F}_4^\star} S(a,b)$, we finally obtain:

$$\sigma_0(a) + 2\sigma_1(a) = 1 - K_m(a)$$

That is, $\sigma_1(a) = (1 - K_m(a) - \sigma_0(a))/2$.
Thus,

$$S(a,b) = \begin{cases} 2\sigma_0(a) + K_m(a) - 1 & \text{if } b = 1 \\ -\sigma_0(a) & \text{if } b \neq 1. \end{cases}$$

To conclude, we have to express the sum $\sigma_0(a)$ by means of Kloosterman sums and cubic sums. First, recall that, since n is even, the mapping $x \mapsto x^3$ is 3-to-1 from $\mathbb{F}_{2^n}$ to $\mathbb{F}_{2^n}$. Moreover, the mapping $x \mapsto x^3$ is also 3-to-1 from U to itself (since m is odd, 3 divides $2^m + 1$, the group $\mathbb{F}_4^\star$ is then contained in U). That implies in particular that

$$\sigma_0(a) := \sum_{v\in V} \chi(\mathrm{Tr}_1^n(av)) = \frac{1}{3} \sum_{u\in U} \chi(\mathrm{Tr}_1^n(au^3)).$$

Thanks to the transitivity of trace function, we have $\mathrm{Tr}_1^n(au^3) = \mathrm{Tr}_1^m(\mathrm{Tr}_m^n(au^3)) = \mathrm{Tr}_1^m(au^3 + (au^3)^{2^m})$. Thus

$$\sigma_0(a) = \frac{1}{3} \sum_{u\in U} \chi(\mathrm{Tr}_1^m(a(u^3 + u^{-3}))).$$

Moreover, we make use of the fact, that every element $1/c$ where $c \in \mathbb{F}_{2^m}^\star$ with $\mathrm{Tr}_1^m(c) = 1$ can be uniquely represented as $u + u^{2^m}$ with $u \in U$. Note now that

$1/c^3 + 1/c = u^3 + u^{-3}$. Therefore, using the fact that the mapping $c \mapsto 1/c$ is a permutation on $\mathbb{F}_{2^m}$, we obtain

$$\begin{aligned}\sigma_0(a) &= (1 + \sum_{u\in U\setminus\{1\}} \chi(\mathrm{Tr}_1^m(a(u^3 + u^{-3}))))/3 \\ &= (1 + 2 \sum_{\substack{c\in\mathbb{F}_{2^m}\\ \mathrm{Tr}_1^m(c)=1}} \chi(\mathrm{Tr}_1^m(a/c^3 + a/c)))/3 \\ &= (1 + 2 \sum_{\substack{c\in\mathbb{F}_{2^m}\\ \mathrm{Tr}_1^m(1/c)=1}} \chi(\mathrm{Tr}_1^m(ac^3 + ac)))/3.\end{aligned}$$

Now, Charpin et al. have proved in [3] that

$$2 \sum_{c\in\mathbb{F}_{2^m}, \mathrm{Tr}_1^m(1/c)=1} \chi(\mathrm{Tr}_1^m(ac^3 + ac)) = 2C_m(a,a) - K_m(a)$$

from which we deduce that

$$\sigma_0(a) = (2C_m(a,a) + 1 - K_m(a))/3,$$

and then

$$S(a,b) = \begin{cases} (K_m(a) - 1 + 4C_m(a,a))/3 \text{ if } b = 1 \\ (K_m(a) - 1 - 2C_m(a,a))/3 \text{ if } b \neq 1 \end{cases}.$$

□

In Chaps. 9 and 17, we will need to produce necessary and sufficient conditions so that the sum involved in the previous proposition takes the value 1. To this end, we need the following statement.

Corollary 2.4.4 ([17]). *Let* $n = 2m$ *with* $m > 3$ *odd. Let* $a \in \mathbb{F}_{2^m}$, $b \in \mathbb{F}_4$ *and* r *be a positive integer such that* $\gcd(r, 2^m + 1) = 1$. *Then*

$$\sum_{u\in U} \chi\left(\mathrm{Tr}_1^n(au^r) + \mathrm{Tr}_1^2(bu^{\frac{2^m+1}{3}})\right) = 1 \tag{2.2}$$

if and only if

- $b = 0$ *and* $K_m(a) = 0$,
- *or,* $b \neq 0$ *and* $K_m(a) = 4$.

Proof. Assume that (2.2) holds. Assume $b = 0$. According to Proposition 2.4.2, (2.2) is equivalent to $1 - K_m(a) = 1$, that is, $K_m(a) = 0$.

Assume $b \neq 0$. If $\mathrm{Tr}_1^m(a^{1/3}) = 0$ then $C_m(a,a) = 0$ according to Proposition 2.3.5. Thus, according to Proposition 2.4.3, (2.2) is reduced to $\frac{K_m(a)-1}{3} = 1$, that is, (2.2) is equivalent to $K_m(a) = 4$. If $\mathrm{Tr}_1^m(a^{1/3}) = 1$. According to Proposition 2.3.5, $C_m(a,a) = C_m(1, a^{2/3}) = \epsilon_a \left(\frac{2}{m}\right) 2^{(m+1)/2}$ with $\epsilon_a = \pm 1$. Thus, according to Proposition 2.4.3, (2.2) is equivalent to $K_m(a) = 4 \pm \left(\frac{2}{m}\right) 2^{(m+3)/2}$ if $b \neq 1$ or $K_m(a) = 4 \pm \left(\frac{2}{m}\right) 2^{(m+5)/2}$ if $b = 1$. Now, Proposition 2.3.2 says that the Kloosterman sum $K_m(a)$ takes integer values in the range $[-2^{(m+2)/2} + 1, 2^{(m+2)/2} + 1]$. But, the values $4 \pm \left(\frac{2}{m}\right) 2^{(m+3)/2}$ and $4 \pm \left(\frac{2}{m}\right) 2^{(m+5)/2}$ do not belong to $[-2^{(m+2)/2} + 1, 2^{(m+2)/2} + 1]$ for every $m > 3$ proving that (2.2) is never satisfied if $\mathrm{Tr}_1^m(a^{1/3}) = 1$.

Conversely, assume that $b = 0$ and $K_m(a) = 0$. According to Proposition 2.4.2, $\sum_{u \in U} \chi(\mathrm{Tr}_1^n(au)) = 1 - K_m(a) = 1$. Assume that $b \neq 0$ and $K_m(a) = 4$. According to Proposition 2.3.3, $K_m(a) = 4$ implies that $\mathrm{Tr}_1^m(a^{1/3}) = 0$ and thus that $C_m(a,a) = 0$, thanks to Proposition 2.3.5. Equality (2.2) follows then from Proposition 2.4.3. □

The following technical result will also be useful in Chap. 17.

Proposition 2.4.5 ([17]). *Let $n = 2m$ with m odd. Let $b \in \mathbb{F}_4^\star$, $a \in \mathbb{F}_{2^m}^\star$ and ζ be a generator of the cyclic group U. Let $i \in \{0, 1\}$. Then,*

If $m \not\equiv 3 \pmod 6$, we have

$$\sum_{u \in U} \chi\left(\mathrm{Tr}_1^n(a\zeta^i u^3) + \mathrm{Tr}_1^2(bu^{\frac{2^m+1}{3}})\right) = (K_m(a) - 1 + \mu_i C_m(a,a))/3.$$

where $\mu_0 = -2$ and $\mu_1 = 1$.

Proof. Set

$$S'(a,b) := \sum_{u \in U} \chi\left(\mathrm{Tr}_1^n(au^3) + \mathrm{Tr}_1^2(bu^{\frac{2^m+1}{3}})\right);$$

$$S''(a,b) := \sum_{u \in U} \chi\left(\mathrm{Tr}_1^n(a\zeta u^3) + \mathrm{Tr}_1^2(bu^{\frac{2^m+1}{3}})\right).$$

Keeping the same notation as in the proof of Proposition 2.4.3:
for $j \in \{0, 1, 2\}$, we set

$$\sigma_j(a) := \sum_{v \in V} \chi(\mathrm{Tr}_1^n(a\zeta^j v)).$$

Now, recall that m being odd, one can decompose U as follows

$$U = V \cup \zeta V \cup \zeta^2 V$$

where $V = \{u^3 \mid u \in U\}$. Thus, every element $u \in U$ can be uniquely decomposed as $u = \zeta^j v$ with $j \in \{0, 1, 2\}$ and $v \in V$.

Thus (in the second equality, we use the fact that v is a cube of an element of U which is a cyclic group of order $2^m + 1$)

$$S'(a,b) = \sum_{j=0}^{2}\sum_{v\in V} \chi(\mathrm{Tr}_1^n(a\zeta^{3j}v^3) + \mathrm{Tr}_1^2(b\zeta^{j\frac{2^m+1}{3}} v^{\frac{2^m+1}{3}}))$$

$$= \sum_{j=0}^{2}\sum_{v\in V} \chi(\mathrm{Tr}_1^n(a\zeta^{3j}v^3) + \mathrm{Tr}_1^2(b\zeta^{j\frac{2^m+1}{3}})).$$

Similarly,

$$S''(a,b) = \sum_{j=0}^{2}\sum_{v\in V} \chi(\mathrm{Tr}_1^n(a\zeta^{3j+1}v^3) + \mathrm{Tr}_1^2(b\zeta^{j\frac{2^m+1}{3}} v^{\frac{2^m+1}{3}}))$$

$$= \sum_{j=0}^{2}\sum_{v\in V} \chi(\mathrm{Tr}_1^n(a\zeta^{3j+1}v^3) + \mathrm{Tr}_1^2(b\zeta^{j\frac{2^m+1}{3}})).$$

Now, since $m \not\equiv 3 \pmod 6$, integers 3 and $\frac{2^m+1}{3}$ are co-prime. The mapping $v \mapsto v^3$ is a permutation of V and thus for $(i,j) \in \{0,1,2\}^2$, we have (in the second equality, we use the fact that the mapping $v \mapsto \zeta^{3j}v$ is a permutation of V)

$$S'(a,b) = \sum_{j=0}^{2}\sum_{v\in V} \chi(\mathrm{Tr}_1^n(a\zeta^{3j}v) + \mathrm{Tr}_1^2(b\zeta^{j\frac{2^m+1}{3}}))$$

$$= \sum_{j=0}^{2}\sum_{v\in V} \chi(\mathrm{Tr}_1^n(av) + \mathrm{Tr}_1^2(b\zeta^{j\frac{2^m+1}{3}})).$$

Similarly,

$$S''(a,b) = \sum_{j=0}^{2}\sum_{v\in V} \chi(\mathrm{Tr}_1^n(a\zeta^{3j+1}v) + \mathrm{Tr}_1^2(b\zeta^{j\frac{2^m+1}{3}}))$$

$$= \sum_{j=0}^{2}\sum_{v\in V} \chi(\mathrm{Tr}_1^n(a\zeta v) + \mathrm{Tr}_1^2(b\zeta^{j\frac{2^m+1}{3}})).$$

Now, the set $\{b, b\zeta^{\frac{2^m+1}{3}}, b\zeta^{2\frac{2^m+1}{3}}\}$ is equal to $\mathbb{F}_4^\star$ (which contains two elements of absolute trace 1 on $\mathbb{F}_4$ and one element of absolute trace 0 on $\mathbb{F}_4$). We deduce that

$$S'(a,b) = \sum_{v\in V} \chi(\mathrm{Tr}_1^n(av)) \sum_{j=0}^{2} \chi(\mathrm{Tr}_1^2(b\zeta^{j\frac{2^m+1}{3}}))$$
$$= -\sum_{v\in V} \chi(\mathrm{Tr}_1^n(av)) = -\sigma_0(a).$$

Similarly,

$$S''(a,b) = \sum_{v\in V} \chi(\mathrm{Tr}_1^n(a\zeta v)) \sum_{j=0}^{2} \chi(\mathrm{Tr}_1^2(b\zeta^{j\frac{2^m+1}{3}}))$$
$$= -\sum_{v\in V} \chi(\mathrm{Tr}_1^n(a\zeta v)) = -\sigma_1(a).$$

We conclude thanks to the relations between $\sigma_0(a)$, $\sigma_1(a)$, $K_m(a)$ and $C_m(a,a)$ (obtained in the proof of Proposition 2.4.3) that is,

$$\sigma_0(a) = (2C_m(a,a) + 1 - K_m(a))/3$$

and

$$\sigma_1(a) = (1 - K_m(a) - \sigma_0(a))/2.$$

□

In Chap. 17, we will need to exhibit necessary and sufficient conditions so that the sum involved in the previous proposition takes the value 1. To this end, we need the following statement.

Corollary 2.4.6 ([17]). *Let $n = 2m$ with m odd. Let $a \in \mathbb{F}_{2^m}$, $b \in \mathbb{F}_4^\star$, ζ a generator of the cyclic group U and $i \in \{0,1\}$.*

a) If $m \not\equiv 3 \pmod 6$ then,

$$\sum_{u\in U} \chi\left(\mathrm{Tr}_1^n(a\zeta^i u^3) + \mathrm{Tr}_1^2(bu^{\frac{2^m+1}{3}})\right) = 1 \qquad (2.3)$$

if and only if

- *$i = 0$ and $K_m(a) = 4$,*
- *or, $i = 1$ and $K_m(a) + C_m(a,a) = 4$.*

b) If $m \equiv 3 \pmod 6$ then

$$\sum_{u\in U} \chi\left(\mathrm{Tr}_1^n(a\zeta^i u^3) + \mathrm{Tr}_1^2(bu^{\frac{2^m+1}{3}})\right) \neq 1. \qquad (2.4)$$

Proof. a) Assume $m \not\equiv 3 \pmod 6$. Assume that (2.3) holds. If $i = 0$, Eq. (2.3) is equivalent to $K_m(a) = 4 + 2C_m(a, a)$ by Proposition 2.4.5. If $\mathrm{Tr}_1^m(a^{1/3}) = 0$ then, according to Proposition 2.3.5, $C_m(a, a) = 0$. Therefore, Eq. (2.3) is equivalent to $K_m(a) = 4$. If $\mathrm{Tr}_1^m(a^{1/3}) = 1$ then, according to Proposition 2.3.5, $C_m(a, a) = C_m(1, a^{2/3}) = \epsilon_a \left(\frac{2}{m}\right) 2^{(m+1)/2}$ with $\epsilon_a = \pm 1$. But, $4 \pm \left(\frac{2}{m}\right) 2^{(m+3)/2} \notin [-2^{(m+2)/2} + 1, 2^{(m+2)/2} + 1]$.

Now, if $i = 1$, Eq. (2.3) is equivalent to $K_m(a) + C_m(a, a) = 4$, by Proposition 2.4.5.

Conversely, let us prove (2.3). Assume $i = 0$ and $K_m(a) = 4$. That implies that $\mathrm{Tr}_1^m(a^{1/3}) = 0$ (by Proposition 2.3.3) and thus that $C_m(a, a) = 0$ (by Proposition 2.3.5). Equation (2.3) follows then from the first identity of Proposition 2.4.5. Likewise, in the case where $i = 1$ and $K_m(a) + C_m(a, a) = 4$, Eq. (2.3) is deduced from the second identity of Proposition 2.4.5.

b) Assume $m \equiv 3 \pmod 6$. Then $2^m + 1$ is a multiple of 9. Therefore, the mapping $u \mapsto u^3$, being 3-to-1 from U to itself, we have

$$\sum_{u \in U} \chi\left(\mathrm{Tr}_1^n(a\zeta^i u^3) + \mathrm{Tr}_1^2(bu^{\frac{2^m+1}{3}})\right) = 3 \sum_{v \in V} \chi\left(\mathrm{Tr}_1^n(a\zeta^i v) + \mathrm{Tr}_1^2(bv^{\frac{2^m+1}{9}})\right)$$

where $V = \{u^3, u \in U\}$, which implies (2.4). □

Finally, we also have to express another type of exponential sums over U in terms of Kloosterman sums and cubic sums. This result will be useful in Chap. 9.

Proposition 2.4.7 ([17]). *Let m be an odd integer. Let $a \in \mathbb{F}_{2^m}^\star$. Then, we have*

$$\sum_{u \in U} \chi(\mathrm{Tr}_1^n(au^3)) = 1 - K_m(a) + 2C_m(a, a).$$

Proof. Using the transitivity rule of trace function, we have

$$\mathrm{Tr}_1^n(au^3) = \mathrm{Tr}_1^m(\mathrm{Tr}_m^n(au^3)) = \mathrm{Tr}_1^m(au^3 + (au^3)^{2^m}).$$

Hence

$$\sum_{u \in U} \chi(\mathrm{Tr}_1^n(au^3)) = \sum_{u \in U} \chi(\mathrm{Tr}_1^m(a(u^3 + u^{-3}))).$$

Now, recall that every element $1/c$ where $c \in \mathbb{F}_{2^m}^\star$ with $\mathrm{Tr}_1^m(c) = 1$ can be uniquely represented as $u + u^{2^m}$ with $u \in U$. Therefore, since $1/c^3 + 1/c = u^3 + u^{-3}$ (indeed, $1/c^3 = (u + u^{2^m})^3 = (u + u^{-1})^3 = u^3 + u^{-3} + uu^{-1}(u + u^{-1}) = u^3 + u^{-3} + 1/c$), we have

$$\sum_{u \in U} \chi(\mathrm{Tr}_1^n(au^3)) = 1 + \sum_{u \in U \setminus \{1\}} \chi(\mathrm{Tr}_1^m(a(u^3 + u^{-3})))$$

$$= 1 + 2 \sum_{\substack{c \in \mathbb{F}_{2^m} \\ \mathrm{Tr}_1^m(c)=1}} \chi(\mathrm{Tr}_1^m(a/c^3 + a/c))$$

$$= 1 + 2 \sum_{\substack{c \in \mathbb{F}_{2^m} \\ \mathrm{Tr}_1^m(1/c)=1}} \chi(\mathrm{Tr}_1^m(ac^3 + ac)).$$

In the last equality, we use the fact that the map $c \mapsto 1/c$ is a permutation on $\mathbb{F}_{2^m}$. Now, Charpin, Helleseth and Zinoviev have proved in [3] that when m is odd, we have

$$2 \sum_{c \in \mathbb{F}_{2^m}, \mathrm{Tr}_1^m(1/c)=1} \chi(\mathrm{Tr}_1^m(ac^3 + ac)) = 2C_m(a, a) - K_m(a)$$

from which we deduce the result. □

2.5 Some Basic Notions in Number Theory

In this section we present some basic background in number theory. In treating a number of arithmetical questions, the notion of congruence is extremely useful. This notion led us to consider the ring $\mathbb{Z}/m\mathbb{Z}$ (where m is a positive integer) and its group of units $U(\mathbb{Z}/m\mathbb{Z})$. The fact that $U(\mathbb{Z}/p\mathbb{Z})$ is a cyclic group when p is a prime is equivalent to the existence of primitive roots. The notion of primitive root can be generalized somewhat as follows.

Definition 2.5.1. Let $a, n \in \mathbb{Z}$. Then a is said to be a primitive root modulo n, the residue class of $a \pmod n$ generates $U(\mathbb{Z}/p\mathbb{Z})$. It is equivalent to require that a and n be relatively prime and that $\phi(n)$ (where ϕ denotes the Euler's function) be the smallest positive integer such that $a^{\phi(n)} \equiv 1 \pmod n$.

The following result is well-known.

Theorem 2.5.2. *Let p an odd prime and let $s \in \mathbb{N}$, then $U(\mathbb{Z}/p^s\mathbb{Z})$ is cyclic, i.e. there exist primitive roots modulo p^s.*

2.5.1 Quadaratic Residues

Definition 2.5.3. If $gcd(a, m) = 1$, a is called a *quadratic residue* modulo m if the congruence $x^2 \equiv a \pmod m$ has a solution. Otherwise a is called a quadratic nonresidue modulo m.

The following statement gives a less tedious way of deciding when a given integer is a quadratic residue modulo m.

Proposition 2.5.4. *Let $m = 2^e p_1^{e_1} \cdots p_r^{e_r}$ be the prime decomposition of m, and assume that $gcd(a, m) = 1$. Then $x^2 \equiv a \pmod{m}$ is solvable if and only if, the following conditions hold.*

1. *if $e = 2$, then $a \equiv 1 \pmod{4}$. If $e \geq 3$, then $a \equiv 1 \pmod{8}$.*
2. *For each i we have $a^{(p_i-1)/2} \equiv 1 \pmod{p_i}$.*

The above result reduces questions about quadratic residues to the corresponding questions for prime moduli. In the following p will denote an odd prime. We recall the notion of Legendre symbol which is an extremely convenient device for discussing quadratic residues.

Definition 2.5.5. The symbol $(\frac{a}{p})$ will have the value 1 if a is a quadratic residue modulo p, the value -1 if a is a quadratic nonresidue modulo p, and the value 0 if p divides a. The symbol $(\frac{a}{p})$ is called the *Legendre symbol.*

Below is a list of some properties of the Legendre symbol.

Proposition 2.5.6. *With the notation above we have*

1. *$a^{p-1)/2} \equiv (\frac{a}{p}) \pmod{p}$.*
2. $\frac{ab}{p} = (\frac{a}{p})(\frac{b}{p})$.
3. *If $a \equiv b \pmod{p}$, then $(\frac{a}{p}) = (\frac{b}{p})$.*

Particularly interesting results are given by the following well-know theorem

Theorem 2.5.7. *With the notation above, we have*

1. *$(\frac{-1}{p}) = (-1)^{(p-1)/2}$.*
2. $(\frac{2}{p}) = (-1)^{(p^2-1)/8}$.

In the following, we recall briefly another symbol: the so-called *Jacobi symbol* defined as follows.

Definition 2.5.8. Let b be an odd, positive integer and a any integer. Let $b = p_1 p_2 \cdots p_m$, where the p_i are (not necessarily distinct) primes. The symbol $(\frac{a}{b})$ defined by

$$(\frac{a}{b}) = (\frac{a}{p_1})(\frac{a}{p_2}) \cdots (\frac{a}{p_m})$$

is called the Jacobi symbol.

The Jacobi symbol has properties that are remarkably similar to the Legendre symbol, which it generalizes. A word of caution is useful. The Jacobi symbol $(\frac{a}{b})$ may equal 1 without a being a quadratic residue modulo b.

Proposition 2.5.9. *With the notation above we have*

1. *If* $a \equiv c \pmod{b}$, *then* $(\frac{a}{b}) = (\frac{c}{b})$.
2. $(\frac{ac}{b}) = (\frac{a}{b})(\frac{c}{b})$.
3. $(\frac{a}{b_1 b_2}) = (\frac{a}{b_1})(\frac{a}{b_1})$.

2.5.2 Group Characters and Gauss Sums

A basic role in setting exponential sums for finite fields is played by particular homomorphisms called *characters*. In this section, we place ourselves in a general context and present some background on characters of the finite field $\mathbb{F}_q$ where q is a power of a prime $q = p^m$. It is necessary to distinguish between two types of character-namely, additive and multiplicative characters. Throughout this subsection ξ_p will denote $e^{\frac{2\pi\sqrt{-1}}{p}}$, a primitive root th of unity. An *additive character* of $\mathbb{F}_q$ is a nonzero function χ from $\mathbb{F}_q$ to the set $\mathbb{C}$ of complex numbers such that $\chi(x+y) = \chi(x)\chi(y)$ for any pair $(x, y) \in {\mathbb{F}_q}^2$. For every $\lambda \in \mathbb{F}_q$, the mapping

$$\omega \mapsto \chi_\lambda(\omega) = \xi_p^{Tr_1^m(\lambda\omega)}$$

defines an additive character of $\mathbb{F}_q$. If $\lambda = 0$ then $\chi_0(\omega) = 1$ is called the *trivial character* of $\mathbb{F}_q$. Every additive character of $\mathbb{F}_q$ can be written as $\chi_\lambda(\omega) = \chi_1(\lambda\omega)$. Moreover, since the multiplicative group $\mathbb{F}_q^\star$ is cyclic, all the characters of the multiplicative group $\mathbb{F}_q^\star$ are given by

$$\Psi_j(\alpha^k) = e^{\frac{2\pi\sqrt{-1}jk}{q-1}}, k = 0, 1, \cdots, q-2,$$

where $0 \leq j \leq q-2$ and α is a generator of $\mathbb{F}_q^\star$. The characters Ψ_j are called *multiplicative characters* of $\mathbb{F}_q$, and form a group of order $q-1$ with identity element Ψ_0. The character $\Psi_{(q-1)/2}$ is called the *quadratic character* of $\mathbb{F}_q$ and is denoted simply by ν which can be extended at 0 by setting $\nu(0) = 0$.

We have the following well-known result

Proposition 2.5.10.

$\sum_{r=0}^{p-1} \xi_p^{ar} = p$ *if* $a \equiv 0 \pmod{p}$ *and* 0 *otherwise.*
$\sum_{r=0}^{p-1} (\frac{r}{p}) = 0$, *where* $(\frac{r}{p})$ *is the Legendre symbol.*

Gaussian sums are the most important kinds of exponential sums for finite fields. They play an important role to make the transition from the additive to the multiplicative structure and vice versa.

The *Gauss sum* $G(\nu, \chi_1)$ over $\mathbb{F}_q$ is defined by

$$G(\nu, \chi_1) := \sum_{c \in \mathbb{F}_q^*} \nu(c)\chi_1(c) = \sum_{c \in \mathbb{F}_q} \nu(c)\chi_1(c)$$

and the Gauss sum $G(\bar{\nu}, \bar{\chi}_1)$ over $\mathbb{F}_p$ is defined by

$$G(\bar{\nu}, \bar{\chi}_1) := \sum_{c \in \mathbb{F}_p^*} \bar{\nu}(c)\bar{\chi}_1(c) = \sum_{c \in \mathbb{F}_p} \bar{\nu}(c)\bar{\chi}_1(c),$$

where $\bar{\nu}$ and $\bar{\chi}_1$ are the quadratic and canonical additive characters of $\mathbb{F}_p$, respectively. The Gauss sums $G(\nu, \bar{\chi}_1)$ and $G(\nu, \bar{\chi}_1)$ can be computed as follows.

$$G(\nu, \chi_1) = (-1)^{m-1} \sqrt{-1}^{\left(\frac{p-1}{2}\right)^2 m} \sqrt{q},$$

$$G(\bar{\nu}, \bar{\chi}_1) = \sqrt{-1}^{\left(\frac{p-1}{2}\right)^2} \sqrt{p} = \sqrt{p^*},$$

where $p^* = (-1)^{(p-1)/2} p$.

The *quadratic Gauss sum* is defined by

$$g_a := \sum_{r=0}^{p-1} \left(\frac{r}{p}\right) \xi_p^{ar}.$$

We have :

$$g_a = \left(\frac{a}{p}\right) g_1,$$

and

$$g_1^2 = (-1)^{(p-1)/2} p.$$

For more details, we send the reader to Chapter 5 in [14].

2.6 Cyclotomic Field $\mathbb{Q}(\xi_p)$

In this short section, we recall some basic results on cyclotomic fields. $\mathbb{Q}$ denotes the rational field. Let p be a prime integer and $\xi_p = e^{\frac{2\pi\sqrt{-1}}{p}}$ be the primitive p-th root of unity. The ring of integers in the cyclotomic field $\mathbb{Q}(\xi_p)$ is $\mathcal{O}_K = \mathbb{Z}(\xi_p)$. An integral basis of $\mathcal{O}_{\mathbb{Q}(\xi_p)}$ is $\{\xi_p^i \mid 1 \leq i \leq p-1\}$. The field extension $\mathbb{Q}(\xi_p)/\mathbb{Q}$ is Galois of degree $p-1$ and the Galois group $Gal(\mathbb{Q}(\xi_p)/\mathbb{Q}) = \{\sigma_a \mid a \in (\mathbb{Z}/p\mathbb{Z})^*\}$, where the automorphism σ_a of $\mathbb{Q}(\xi_p)$ is defined by $\sigma_a(\xi_p) = \xi_p^a$. The field $\mathbb{Q}(\xi_p)$ has a unique quadratic subfield $\mathbb{Q}(\sqrt{p^*})$ with $p^* = (\frac{-1}{p})p = (-1)^{(p-1)/2}p$ where $(\frac{a}{p})$ denotes the Legendre symbol for $1 \leq a \leq p-1$. Note that $p^m = (\frac{-1}{p})^m \sqrt{p^*}^{2m}$. For $1 \leq a \leq p-1$, $\sigma_a(\sqrt{p^*}) = (\frac{a}{p})\sqrt{p^*}$. Hence, the Galois group $Gal(\mathbb{Q}(\sqrt{p^*})/\mathbb{Q}$ is $\{1, \sigma_\gamma\}$, where γ is any quadratic nonresidue in the prime field $\mathbb{F}_p$.

References

1. L. Carlitz. Explicit evaluation of certain exponential sums. In *Math. Scand. Vol. 44*, pages 5–16, 1979.
2. P. Charpin, T. Helleseth, and V. Zinoviev. The divisibility modulo 24 of Kloosterman sums of $GF(2^m)$, m odd. *Journal of Combinatorial Theory, Series A*, 114:322–338, 2007.
3. P. Charpin, T. Helleseth, and V. Zinoviev. Divisibility properties of Kloosterman sums over finite fields of characteristic two. In *ISIT 2008, Toronto, Canada, July 6–11*, pages 2608–2612, 2008.
4. P. Charpin, S. Mesnager, and S. Sarkar. On involutions of finite fields. In *Proceedings of 2015 IEEE International Symposium on Information Theory, (ISIT)*, 2015.
5. P. Charpin, T. Helleseth, and V. Zinoviev. Divisibility properties of classical binary Kloosterman sums. *Discrete Mathematics*, 309(12):3975–3984, 2009.
6. Wun Seng Chou, Javier Gomez-Calderon, and Gary Lee Mullen. Value sets of Dickson polynomials over finite fields. *J. Number Theory*, 30(3):334–344, 1988.
7. J. Dillon and H. Dobbertin. New cyclic difference sets with Singer parameters. *Finite Fields and Their Applications*, 10(3):342–389, 2004.
8. J. F. Dillon and H. Dobbertin. New cyclic difference sets with Singer parameters. In *Finite Fields and Their Applications Volume 10, Issue 3*, pages 342–389, 2004.
9. J-P Flori and S. Mesnager. Dickson polynomials, hyperelliptic curves and hyper-bent functions. In *7th International conference SETA 2012, LNCS 7280, Springer*, pages 40–52, 2012.
10. T. Helleseth and P. V. Kumar. Sequences with low correlation. In *Handbook of Coding Theory, Part 3: Applications, V. S. Pless, W. C. Huffman, and R. A. Brualdi, Eds. Amsterdam, The Netherlands: Elsevier, chapter. 21*, pages 1765–1853, 1998.
11. G. Lachaud and J. Wolfmann. The weights of the orthogonals of the extended quadratic binary Goppa codes. In *IEEE Trans. Inform. Theory 36 (3)*, pages 686–692, 1990.
12. G. Leander. Monomial Bent Functions. In *IEEE Trans. Inform. Theory (52) 2*, pages 738–743, 2006.
13. R. Lidl, G. L. Mullen, and G. Turnwald. Dickson Polynomials. In *ser.Pitman Monographs in Pure and Applied Mathematics. Reading, MA: Addison-Wesley, vol. 65*, 1993.
14. R. Lidl and H. Niederreiter. Finite Fields, Encyclopedia of Mathematics and its Applications. In *vol. 20, Addison-Wesley, Reading, Massachusetts*, 1983.
15. R. Lidl and H. Niederreiter. *Finite fields*, volume 20 of *Encyclopedia of Mathematics and its Applications*. Cambridge University Press, Cambridge, second edition, 1997. With a foreword by P. M. Cohn.
16. Rudolf Lidl, Gary Lee Mullen, and Gerhard Turnwald. *Dickson polynomials*, volume 65 of *Pitman Monographs and Surveys in Pure and Applied Mathematics*. Longman Scientific & Technical, Harlow, 1993.
17. S. Mesnager. Semi-bent functions from Dillon and Niho exponents, Kloosterman sums and Dickson polynomials. In *IEEE Transactions on Information Theory-IT, Vol 57, No 11*, pages 7443–7458, 2011.

18. S. Mesnager. A note on constructions of bent functions from involutions. In *IACR Cryptology ePrint Archive 2015: 982*, 2015.
19. S. Mesnager, G. Cohen, and D. Madore. On existence (based on an arithmetical problem) and constructions of bent functions. In *Proceedings of the fifteenth International Conference on Cryptography and Coding, Oxford, United Kingdom, IMACC 2015, LNCS, Springer, Heidelberg*, pages 3–19, 2015.
20. G. L. Mullen and D. Panario. Handbook of finite fields. In *CRC, Taylor and Francis Group; Series: Discrete Mathematics and Its Applications*, 2013.
21. Kalle Ranto. On algebraic decoding of the z4-linear goethals-like codes. *IEEE Transactions on Information Theory*, 46(6):2193–2197, 2000.
22. Baocheng Wang, Chunming Tang, Yanfeng Qi, Yixian Yang, and Maozhi Xu. A new class of hyper-bent Boolean functions in binomial forms. *CoRR*, abs/1112.0062, 2011.

Chapter 3
Boolean Functions and Cryptography

3.1 Cryptographic Framework for Boolean Functions

Stream ciphers are commonly used for encrypting and decrypting messages. Stream ciphers have several advantages which make them suitable for some applications. Most notably, they are usually faster and have a lower hardware complexity than block ciphers. They are for instance appropriate when buffering is limited, since the binary digits are individually encrypted and decrypted. In stream cipher the encryption and the decryption consist in adding bitwise the input stream and a pseudo-random sequence generated by a pseudo-random generator taking as input a secret information, the secret key. Classical tools to produce such pseudo-random sequences, that are called keystream, are Linear Feedback Registers (LFSR). Stream ciphers can use several LFSR or a single LFSR. As indicated by its name, LFSR are linear and their linear systems are governed by linear relationships between their inputs and outputs. Since linear dependencies can relatively easily be analyzed, stream ciphers designed only with LFSR would be highly insecure.

To produce more secure encryption schemes, Boolean functions are used to produce the keystream from LSFR entries. Those functions allow to make the relationship between the plaintext and the ciphertext as complex as possible. More precisely, a bit of the ciphertext is obtained from a bit of the plaintext by adding bitwise a key digit (the output of the Boolean function) whose dependence upon the LFSR entries (the secret information) is nonlinear. Thus, the security of such cryptosystems deeply relies on the choice of the Boolean function because the complexity of the relationship between the plaintext and the ciphertext depends entirely on the Boolean function. Indeed, some properties of the Boolean function can be exploited to gain access to the contents of encrypted messages, even if the key is unknown. Therefore Boolean functions need to have some important characteristics that are called *security criteria* to resist several types of attacks.

Classical models for such cryptosystems are stream ciphers: this design is loosely based on the one-time pad [46, 6.1.1], or Vernam cipher, for which a random keystream is used to encrypt the plaintext one bit at a time. Hence, to build a stream

S. Mesnager, *Bent Functions*, DOI 10.1007/978-3-319-32595-8_3

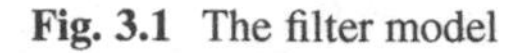

Fig. 3.1 The filter model

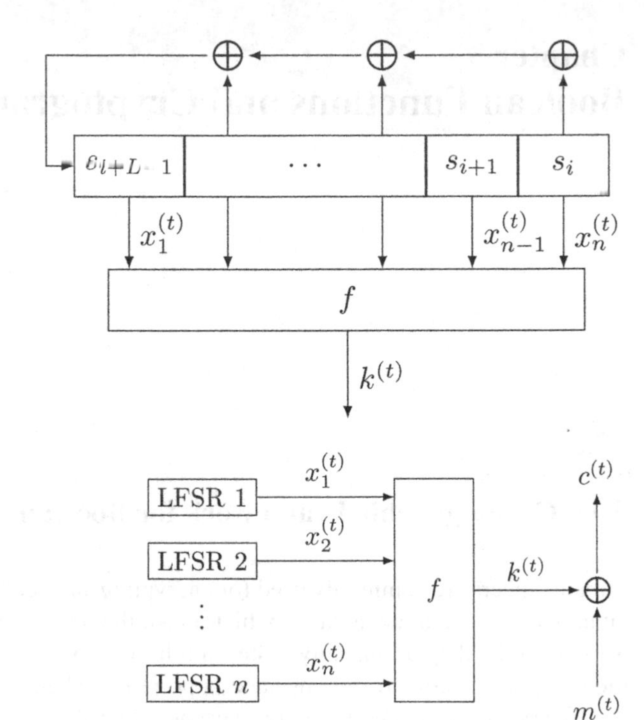

Fig. 3.2 The combiner model

cipher, a suitable pseudo-random keystream generator must be designed. A common construction is to use one or several linear feedback shift registers [46, 6.2.1] (LFSR) filtered or combined by a Boolean function. The filtered model is usually composed of one or several LFSR's, and of a nonlinear combining or filtering functions f which produces the output, given the state of the linear part. In the combiner generator model, the outputs to several Linear Feedback Shift Registers are combined by a Boolean function giving, at each clock cycle, one bit of the pseudo-random sequence. Both models are depicted in Figs. 3.1 and 3.2.

3.2 Main Cryptographic Criteria for Boolean Functions

The design of conventional cryptographic systems relies on two fundamental principles introduced by Claude Shannon in his paper [50] Communication Theory of Secrecy Systems, published in 1949: *confusion* and *diffusion*. In cryptography, confusion and diffusion are two properties of the operation of a secure cipher. Confusion aims at concealing any algebraic structure in the system. It is closely related to the complexity (that is, the cryptographic complexity, which is different from circuit complexity, for instance) of the involved Boolean functions. In Shannon's original definitions, confusion refers to making the relationship between the plaintext and the ciphertext as complex and involved as possible; diffusion refers to the property that the redundancy in the statistics of the plaintext is "dissipated" in the statistics of the ciphertext. In other words, the non-uniformity in the distribution of the individual letters (and pairs of neighbouring letters) in the plaintext should be redistributed into the non-uniformity in the distribution of much larger structures of the ciphertext, which is much harder to detect. Diffusion means that the output bits should depend on the input bits in a very complex way. In a cipher with good diffusion, if one bit of the plaintext is changed, then the ciphertext should change completely, in an unpredictable or pseudo-random manner. Diffusion consists then in spreading out the influence of any minor modification of the input data or of the key over all outputs.

These two principles were stated more than half a century ago. Since then, many attacks have been found against the diverse known cryptosystems, and the relevance of these two principles has always been confirmed. The known attacks on each cryptosystem lead to criteria that the implemented cryptographic functions must satisfy. More precisely, the resistance of the cryptosystems to the known attacks can be quantified through some fundamental characteristics (some, more related to confusion, and some, more related to diffusion) of the Boolean functions and the design of these cryptographic functions needs to consider various characteristics simultaneously.

3.2.1 *Algebraic Degree*

The linear complexity of the pseudo-random generator depends on the algebraic degree of its filtering or combining function, whence the importance for it to have a high algebraic degree in order to avoid the Berlekamp–Massey attacks [41], [46, 6.2.3], [3, 4.1.1] and, for the filter model, the more recent Rønjom–Helleseth attack [48]. It is obviously verified from the definition of the algebraic normal form that the algebraic degree of a Boolean function in n variables is at most n.

3.2.2 *Balancedness*

To avoid statistical dependence between the input (the plaintext) and the output (the ciphertext) of the stream cipher and to prevent distinguishing attacks [4, 4.1.3], cryptographic functions must be *balanced*. A balanced Boolean function is whose output yields as many 0s as 1s over its input set. This means that for a uniformly random input string of bits, the probability of getting a 1 is $\frac{1}{2}$. Equivalently, a Boolean function in n variables is said to be balanced if it has Hamming weight 2^{n-1}. Note that the algebraic degree of an n-variable Boolean balanced function is at most $n-1$.

3.2.3 Nonlinearity

The cryptographic criterion of interest in this book (in particular in Chap. 4) is that of nonlinearity and the related notion of bentness. Nonlinearity characterizes the distance between a Boolean function and the set of affine functions (i.e. those of algebraic at most 1) and is naturally defined using the Hamming distance. More precisely, the nonlinearity of f, denoted by $nl(f)$, is the minimum distance to affine functions (in terms of Reed–Muller codes, it is equal to the minimum distance of the linear code Reed–Muller code $\mathcal{RM}(1,n) \cup (f + \mathcal{RM}(1,n))$ where $\mathcal{RM}(1,n)$ denote the Reed–Muller code of order 1 and length 2^n). It can be shown that the nonlinearity of a Boolean function in n variables is upper bounded by $2^{n-1} - 2^{n/2-1}$. In order to provide confusion, cryptographic functions must lie at large Hamming distance (in the sense, close to the maximum value $2^{n-1} - 2^{n/2-1}$) to all affine functions, equivalently must be of a large nonlinearity (in the sense, close to the upper bound $2^{n-1} - 2^{n/2-1}$). Boolean functions achieving maximal nonlinearity are called *bent* functions but such functions cannot be directly used in the filter and combiner models; in particular, they are not balanced.

Nonlinearity criteria for Boolean functions are classified in view of their suitability for cryptographic design. The classification is set up in terms of the largest transformation group leaving a criterion invariant. In this respect two criteria turn out to be of special interest, the distance to linear structures and the distance to affine functions, which are shown to be invariant under all affine transformations. A high nonlinearity is surely one of the most important cryptographic criteria. In the case of stream ciphers, high nonlinearity is important to prevent fast correlation attacks [45] and best affine approximation attacks [21]. The larger is the nonlinearity, the less efficient are fast correlation attacks [1, 10, 28, 33–35, 44] and linear attack.[1]

3.2.4 Correlation Immune and Resiliency

To avoid a divide and conquer attack, called *correlation attack* (see e.g. [1, 11, 43, 52]) on the combiner model, the combining function must avoid low-order correlation . This is the reason why such a combining function is often chosen with a rather high correlation immunity order. There are two equivalent ways for characterizing the correlation immunity: either by means of the Walsh transform or by means of the sub-functions. Originally, an n-variable Boolean function f is said to be *correlation immune* of order t (or t-th order correlation immune) if

[1]We shall say that there is a correlation between a Boolean function f and a linear function ℓ if $d_H(f,\ell)$ is different from 2^{n-1}. Any Boolean function has correlation with some linear functions of its input. But this correlation should be small: the existence of affine approximations of the Boolean functions involved in a cryptosystem allows in various situations (block ciphers, stream ciphers) to build attacks on this system [29, 42].

any sub-function deduced from f by fixing at most t inputs has the same output distribution as f. On the other hand, correlation immunity can be characterized by means of the Walsh transform of f. A Boolean function f is said correlation immune of order t if and only if the Walsh transform of f vanishes at all non zero vectors of Hamming weight at most t [61]. If f is moreover balanced, then f is said to be t-*resilient*. This definition of resiliency was introduced by Siegenthaler in [51]. If f is not m-resilient, then there exists a correlation between the output to the function and (at most) m coordinates of its input; if m is small, the divide-and-conquer attack uses this weakness for attacking a system using f as combining function. To conclude, briefly, a Boolean function used in the combiner model should be resilient in order to resist correlation attacks [52]. This is not mandatory for functions used in the filter model. In the latter model, 1-resiliency is commonly considered to be sufficient and can be obtained by choosing another function in the same affine equivalence class.

3.2.5 Algebraic Immunity

Standard algebraic attacks were introduced in 2003 by Courtois and Meier [15]. Algebraic attacks recover the secret key, or at least the initialization of the system, by solving a system of multivariate algebraic equations. The idea that the key bits can be characterized as the solutions of a system of multivariate equations comes from C. Shannon [50]. In practice, for cryptosystems which are robust against the usual attacks such as the Berlekamp–Massey attack, this system is too complex to be solved[2] (its equations being highly nonlinear). However, in the case of stream ciphers, we can get a very over-defined system (i.e. a system with a number of linearly independent equations much greater than the number of unknowns). In view of these attacks, the study of the set of annihilators of a Boolean function has become very important and a Boolean function should have a high *algebraic immunity*. We define these notions below (see [43]).

Definition 3.2.1. Let f be a Boolean function in n variables. A nonzero Boolean function g is called an annihilator of f if $fg = 0$.

Definition 3.2.2. The algebraic immunity of f, denoted by $AI(f)$, is the minimum value of d such that f or its complement $1 + f$ admits an annihilator of algebraic degree d.

Clearly, the algebraic immunity of a Boolean function f is less than or equal to its algebraic degree since $1 \oplus f$ is an annihilator of f. As shown in [15], the

[2]The number of equations can then be much larger than the number of unknowns. This makes less complex the resolution of the system by using Groebner basis (see [23]), and even allows linearizing the system (i.e. obtaining a system of linear equations by replacing every monomial of degree greater than 1 by a new unknown); the resulting linear system has however too many unknowns and cannot be solved.

algebraic immunity of any n-variable function is bounded by $\lceil n/2 \rceil$. Moreover, it was shown in [17] that the Hamming weight of a Boolean function f with given algebraic immunity satisfies : $\sum_{i=0}^{AI(f)-1} \binom{n}{i} \leq \mathrm{wt}(f) \leq \sum_{i=0}^{n-AI(f)} \binom{n}{i}$. In particular, if n is odd and f has optimum algebraic immunity then f is balanced.

A high value of algebraic immunity is now an absolutely necessary cryptographic criterion for a resistance to algebraic attacks but is not sufficient, because of a more general kind of attacks indeed introduced by Courtois [14] in 2003 as well, called *fast algebraic attacks* (which work if one can find g of low degree and $h \neq 0$ of reasonable degree such that $fg = h$, see [14, 30]).

3.3 Trade-Offs Between the Different Criteria

Cryptographic functions having maximum nonlinearity (that is, bent functions) are never balanced. Moreover, Siegenthaler's bound [51] states that the algebraic degree of an n-variable t-th order correlation immune Boolean function is necessarily less than or equal to $n - t$ [51]. On the other hand, the nonlinearity of a t-th order correlation immune Boolean function is necessarily less than or equal to $2^{n-1} - 2^t$ if $t > \frac{n}{2} - 1$ and $2^{n-1} - 2^{\frac{n}{2}-1} - 2^t$ otherwise [8]. When the Boolean function is moreover balanced, the upper bounds on its algebraic degree and its nonlinearity are lower. Indeed, the algebraic degree is less than or equal to $n - t - 1$ and the nonlinearity is upper bounded by $2^{n-1} - 2^{t+1}$ if $\frac{n}{2} - 1 < t < n - 1$ and $2^{n-1} - 2^{\frac{n}{2}-1} - 2^{t+1}$ if $t \leq \frac{n}{2} - 1$. Therefore, the correlation immunity criterion is not compatible with a high algebraic degree (necessary to withstand Berlekamp–Massey attack) and a high nonlinearity (necessary for avoiding attacks using linear approximation of the function). Moreover, the recent algebraic attacks, e.g. [13, 16], highlighted the need for having a high algebraic degree as well as a high algebraic immunity so that stream ciphers can resist these attacks. Now, there seems to be some kind of contradiction for Boolean functions between having high correlation immunity and optimum or nearly optimum algebraic immunity; also, much attention having been given to algebraic immunity recently, several examples of functions having optimum algebraic immunity could be found but no example of correlation immune Boolean function with optimum algebraic immunity.

As we have seen, there are numerous cryptographic requirements for Boolean functions (in fact there exist other criteria such as Strict Avalanche criterion and propagation criterion (see [3, Section 8.4.1, pp. 303–305, 308–311]). Cryptographic functions must necessarily satisfy some of them bearing on balancedness, algebraic degree, nonlinearity, algebraic immunity and must have a good resistance to fast algebraic attacks. Such properties allow the system designer to quantify the level of resistance of the system to attacks. It is often impossible to satisfy simultaneously several criteria at once, so that compromises have to be made, and trade-offs need to be quantified . Indeed, the difficulty precisely lies in finding the best trade-offs between all criteria and proposing concrete constructions of functions achieving them. An additional important motivation is the fact that the current situation of symmetric cryptography is rather fragile because of recent progress in cryptanalysis. As explained in [3], it is difficult but not impossible to find functions satisfying good trade-offs between all these criteria. It is not clear whether it is possible to achieve additionally resiliency of a sufficient order,[3] which is necessary for the combiner model. Hence, the filter model may be more appropriate (future research will determine this).

3.4 Recent Constructions of Boolean Functions Satisfying the Main Cryptographic Criteria

Building a Boolean function meeting as many criteria as possible is a difficult task. Trade-offs must usually be made between them. Since the introduction of algebraic immunity, several constructions of Boolean functions with high algebraic immunity have been suggested, but very few of them are of optimal algebraic immunity. More importantly, those having other good cryptographic properties, as balancedness or high nonlinearity for instance, are even rarer. Among those having optimal algebraic immunity $AI(f) = \lceil n/2 \rceil$, most have a poor nonlinearity [6, 9, 18, 36, 37], close to the lower bound of Lobanov [38]:

$$\mathrm{nl}(f) \geq 2^{n-1} - \binom{n}{\lfloor \frac{n}{2} \rfloor}.$$

We now present different *good* families, i.e. meeting most of the criteria mentioned in Sect. 3.2 in a satisfactory way.

In 2008, Carlet and Feng [7] studied a family of Boolean functions introduced by Feng, Liao and Yang [24] and devised the first infinite class of functions which seems able to satisfy all of the main criteria for being used as a filtering function in a stream cipher.

Definition 3.4.1 (Construction of Carlet and Feng [7, Section 3]). Let $n \geq 2$ be a positive integer and α a primitive element of $\mathbb{F}_{2^n}$. Let f be the Boolean function in n variables defined by

$$\mathrm{supp}(f) = \left\{0, 1, \alpha, \ldots, \alpha^{2^{n-1}-2}\right\}.$$

They proved that these functions are

1. balanced,
2. of optimal algebraic degree $n-1$ for a balanced function,

[3] First-order resiliency is useful for resisting some distinguishing (less dreadful) attacks.

3. of optimal algebraic immunity $\lceil n/2 \rceil$,
4. with good immunity to fast algebraic attacks,
5. and with good nonlinearity.

$$\mathrm{nl}(f) \geq 2^{n-1} + \frac{2^{n/2+1}}{\pi} \ln\left(\frac{\pi}{2^n - 1}\right) - 1 \approx 2^{n-1} - \frac{2 \ln 2}{\pi} n 2^{n/2}.$$

Moreover, it was checked for small values of n that the functions had far better nonlinearity than the proved lower bound.

Afterwards, the same family was reintroduced in a different way by Wang et al. [5, 60] who proved a better lower bound:

$$\mathrm{nl}(f) \geq \max\left(6\lfloor\frac{2^{n-1}}{2n}\rfloor - 2, 2^{n-1} - \left(\frac{\ln 2}{3}(n-1) + \frac{3}{2}\right) 2^{n/2}\right).$$

Finally, Tang, Carlet and Tang [54] proved in 2011 that the following better lower bound is valid:

$$\mathrm{nl}(f) \geq 2^{n-1} - \left(\frac{n \ln 2}{2\pi} + 0.74\right) 2^{n/2} - 1.$$

In 2010, Tu and Deng [57] discovered that there may be Boolean functions of optimal algebraic immunity in a classical class of Partial Spread functions due to Dillon [20] provided that the following combinatorial conjecture is correct.

Conjecture 3.4.2 (Tu–Deng Conjecture). For all $k \geq 2$ and all $t \in \left(\mathbb{Z}/(2^k - 1)\mathbb{Z}\right)^*$,

$$\#\left\{(a, b) \in \left(\mathbb{Z}/(2^k - 1)\mathbb{Z}\right)^2 | a + b = t \text{ and } w_2(a) + w_2(b) \leq k - 1\right\} \leq 2^{k-1}.$$

Tu and Deng checked the validity of the conjecture for $k \leq 29$. They also proved that, if the conjecture is true, then one can get in even dimension balanced Boolean functions of optimal algebraic immunity and of high nonlinearity (better than that of the functions described above proposed by Carlet and Feng).

More explicitly, their idea was to apply the idea of Carlet and Feng to the classical construction of Dillon [19]; more precisely, functions form the so-called *partial spread* class $\mathcal{PS}_{ap}$ (see in Sect. 5.1) whose elements are defined in an explicit form: $f(x, y) = g(\frac{x}{y})$ (*i.e.* $g(xy^{2^k-2})$) with $\frac{x}{y} = 0$ if $y = 0$ where f is a Boolean function defined on $\mathbb{F}_{2^k} \times \mathbb{F}_{2^k}$ and g be a balanced Boolean function defined over $\mathbb{F}_{2^k}$ such that $g(0) = 0$. Functions in the class $\mathcal{PS}_{ap}$ are bent and have algebraic degree $n/2 = k$ [49].

Definition 3.4.3 (First Construction of Tu and Deng [57]). Let $n = 2k \geq 4$ be an even integer, α a primitive element of $\mathbb{F}_{2^n}$, $A = \left\{1, \alpha, \ldots, \alpha^{2^{k-1}-1}\right\}$ and $g : \mathbb{F}_{2^k} \to \mathbb{F}_2$ a Boolean function in k variables defined by

$$\begin{aligned}\text{supp}(g) &= \left\{\alpha^s, \alpha^{s+1}, \ldots, \alpha^{s+2^{k-1}-1}\right\}\\ &= \alpha^s A\ ,\end{aligned}$$

for any $0 \leq s \leq 2^k - 2$. Let $f : \mathbb{F}_{2^k} \times \mathbb{F}_{2^k} \to \mathbb{F}_2$ be the Boolean function in n variables defined by

$$f(x,y) = \begin{cases} g\left(\frac{x}{y}\right) & \text{if } x \neq 0, \\ 0 & \text{otherwise.}\end{cases}$$

They proved that these functions are

1. bent (because they belong to $\mathcal{PS}_{ap}$),
2. of algebraic degree $n/2 = k$ [49],
3. and of optimal algebraic immunity $n/2 = k$ if Conjecture 3.4.2 is verified.

To prove the optimal algebraic immunity, Tu and Deng have adapted to their function the approach of Carlet and Feng which consists in identifying annihilators of the Boolean function with codewords of BCH codes [39, 40, 58]. The role of the conjecture is then to deduce from the BCH bound [39, 40, 58] that those codewords are zero if the algebraic degrees of the corresponding annihilators are less than $n/2 = k$.

These functions can then be modified to give rise to functions with different good cryptographic properties as follows.

Definition 3.4.4 (Second Construction of Tu and Deng [57]). Let $n = 2k \geq 4$ be an even integer, α a primitive element of $\mathbb{F}_{2^n}$, $A = \left\{1, \alpha, \ldots, \alpha^{2^{k-1}-1}\right\}$ and $g : \mathbb{F}_{2^k} \to \mathbb{F}_2$ a Boolean function in k variables defined by

$$\text{supp}(g) = \alpha^s A,$$

for any $0 \leq s \leq 2^k - 2$. Let $f : \mathbb{F}_{2^k} \times \mathbb{F}_{2^k} \to \mathbb{F}_2$ be the Boolean function in n variables defined by

$$f(x,y) = \begin{cases} g\left(\frac{x}{y}\right) & \text{if } xy \neq 0\ , \\ 1 & \text{if } x = 0 \text{ and } y \in (\alpha A)^{-1}, \\ 0 & \text{otherwise.}\end{cases}$$

Our definition slightly differs from the original one [57], but, in the end, is equivalent because

$$\begin{aligned}(\alpha A)^{-1} &= \left\{\alpha^{-1}, \ldots, \alpha^{-(2^{k-1}-1)}, \alpha^{-2^{k-1}}\right\}\\ &= \left\{\alpha^{2^{k-1}-1}, \alpha^{2^{k-1}}, \ldots, \alpha^{2^k-2}\right\}.\end{aligned}$$

The cryptographic parameters of the function f are as follows:

1. f is balanced;
2. its algebraic degree is optimal for a balanced function, that is equal to $n-1$;
3. up to Conjecture 3.4.2, f has optimal algebraic immunity that is, $AI(f) = n$;
4. its nonlinearity satisfies

$$\mathrm{nl}(f) \geq 2^{n-1} - 2^{n/2-1} - \frac{n}{2}2^{n/4}\ln 2 - 1.$$

Afterwards, Tu and Deng [56, 62] modified their original functions to obtain a class of 1-resilient functions with high nonlinearity and high algebraic immunity.

Definition 3.4.5 (Third Construction of Tu and Deng [56, 62]). Let $n = 2k \geq 4$ be an even integer, α a primitive element of $\mathbb{F}_{2^n}$, $A = \left\{\alpha, \alpha^2, \ldots, \alpha^{2^{k-1}-1}\right\}$, $0 \leq s \leq 2^k - 2$ an integer and $B = \{0,1\} \cup A^{-1}$. Let $f : \mathbb{F}_{2^k} \times \mathbb{F}_{2^k} \to \mathbb{F}_2$ be the Boolean function in n variables defined by

$$\mathrm{supp}(f) = \bigcup \begin{cases} \{(x,y) \mid x/y \in \alpha^s A\}, \\ \{(x,y) \mid y = \alpha^{-s}y, x \in B\}, \\ \{(x,0) \mid x \in \mathbb{F}_{2^k} \setminus B\}, \\ \{(0,y) \mid y \in \mathbb{F}_{2^k} \setminus \alpha^{-s}B\}. \end{cases}$$

They proved that f satisfies the following properties:

1. f is 1-resilient;
2. f is of optimal algebraic degree $\deg(f) = n - 2$;
3. up to Conjecture 3.4.2, f has algebraic immunity $AI(f) \geq n/2 - 1$;
4. f has nonlinearity

$$\mathrm{nl}(f) \geq 2^{n-1} - 2^{n/2-1} - \frac{3}{2}n2^{n/4}\ln 2 - 7.$$

It is in fact proved that f has optimal algebraic immunity depending only on Conjecture 3.4.2 when $n/2$ is odd and on an additional assumption when $n/2$ is even [56].

Finally, Tang et al. [55] applied a degree-optimized version of an iterative construction of balanced Boolean functions with very high nonlinearity by Dobbertin [22] to the functions constructed by Tu and Deng [57, 62] and obtained functions with better nonlinearity. For $n = 2k = 2^t m \geq 4$, m odd, their first family is

1. balanced,
2. of optimal algebraic degree $n - 1$,
3. of optimal algebraic immunity $n/2$ if Conjecture 3.4.2 is verified,
4. of nonlinearity at least

$$2^{n-1} - \sum_{i=0}^{t-1} 2^{n/(2^{i+1})-1} - 2^{(m-1)/2};$$

and their second family is

1. 1-resilient,
2. of optimal algebraic degree $n-2$,
3. of algebraic immunity at least $n/2-1$ if Conjecture 3.4.2 is verified,
4. of nonlinearity at least

$$\begin{cases} 2^{n-1} - 2^{n/2-1} - 3\left(\sum_{i=1}^{t-1} 2^{n/(2^{i+1})-1} - 2^{(m-1)/2}\right) & \text{if } m = 1, \\ 2^{n-1} - 2^{n/2-1} - 3\sum_{i=1}^{t-1} 2^{n/(2^{i+1})-1} - 2^{(m+1)/2} - 6 & \text{if } m \geq 2. \end{cases}$$

Unfortunately, Carlet [2] observed that the functions introduced by Tu and Deng are weak against fast algebraic attacks and unsuccessfully tried to repair their weakness. It was subsequently shown by Wang and Johansson [59] that this family cannot be easily repaired.

Nonetheless, more recent developments have shown that the construction of Tu and Deng and the associated conjecture are not of purely aesthetic interest, but are interesting tools in a cryptographic context.

In 2011, inspired by the previous work of Tu and Deng [57], Tang, Carlet and Tang [54] constructed an infinite family of Boolean functions with many good cryptographic properties. The main idea of their construction is to replace the division in the construction of Tu and Deng by a multiplication. The associated combinatorial conjecture is then modified as follows.

Conjecture 3.4.6 (Tang–Carlet–Tang Conjecture). For all $k \geq 2$ and all $t \in (\mathbb{Z}/(2^k-1)\mathbb{Z})^*$,

$$\#\left\{(a,b) \in \left(\mathbb{Z}/(2^k-1)\mathbb{Z}\right)^2 \mid a - b = t; w_2(a) + w_2(b) \leq k-1\right\} \leq 2^{k-1}.$$

They verified it experimentally for $k \leq 29$, as well as the following generalized property for $k \leq 15$ where $u \in \mathbb{Z}/(2^k-1)\mathbb{Z}$ is such that $\gcd(u, 2^k-1) = 1$ and $\epsilon = \pm 1$.

Conjecture 3.4.7 (Tang–Carlet–Tang Conjecture). Let $k \geq 2$ be an integer, $t \in (\mathbb{Z}/(2^k-1)\mathbb{Z})^*$, $u \in \mathbb{Z}/(2^k-1)\mathbb{Z}$ such that $\gcd(u, 2^k-1) = 1$ and $\epsilon \in \{-1, 1\}$. Then

$$\#\left\{(a,b) \in \left(\mathbb{Z}/(2^k-1)\mathbb{Z}\right)^2 \mid ua + \epsilon b = t; w_2(a) + w_2(b) \leq k-1\right\} \leq 2^{k-1}.$$

This generalized conjecture includes the original conjecture proposed by Tu and Deng (Conjecture 3.4.2) for $u = 1$ and $\epsilon = +1$.

The construction of their functions is as follows.

Definition 3.4.8 (Construction of Tang, Carlet and Tang [54]). Let $n = 2k \geq 4$ be an even integer, α a primitive element of $\mathbb{F}_{2^n}$, $A = \left\{1, \alpha, \ldots, \alpha^{2^{k-1}-1}\right\}$ and $g : \mathbb{F}_{2^k} \to \mathbb{F}_2$ the Boolean function in k variables defined by

$$\mathrm{supp}(g) = \alpha^s A,$$

for any $0 \leq s \leq 2^k - 2$. Let $f : \mathbb{F}_{2^k} \times \mathbb{F}_{2^k} \to \mathbb{F}_2$ be the Boolean function in n variables defined by

$$f(x, y) = g(xy).$$

They proved that such a function f is

1. of algebraic degree $n - 2$,
2. of optimal algebraic immunity $n/2$ if Conjecture 3.4.6 is true,
3. of good immunity against fast algebraic attacks,
4. of nonlinearity at least

$$2^{n-1} - \left(\frac{\ln 2}{2\pi} n + 0.42\right) 2^{n/2} - 1.$$

The proof of the optimality of the algebraic immunity is similar to the proof of Tu and Deng [57].

These functions can then be modified using the same procedure as Tang et al. [55] to obtain balanced functions with high algebraic degree and nonlinearity. They proved that, for $n = 2k = 2^t m \geq 4$ and m odd, these modified functions are

1. balanced,
2. of optimal algebraic degree $n - 1$,
3. of optimal algebraic immunity $n/2$ if Conjecture 3.4.6 is true,
4. of good immunity against fast algebraic attacks,
5. of nonlinearity at least

$$\begin{cases} 2^{n-1} - \left(\frac{\ln 2}{2\pi} n + 0.42\right) 2^{n/2} - 2^{\frac{n/2-1}{2}} - 1 & \text{if } t = 1\ , \\ 2^{n-1} - \left(\frac{\ln 2}{2\pi} n + 0.42\right) 2^{n/2} - \sum_{i=1}^{t-1} 2^{n/(2^i+1)-1} - 2^{(m-1)/2} - 1 & \text{if } t \geq 2\ . \end{cases}$$

It should finally be mentioned that Jin et al. [32] generalized the construction of Tang, Carlet and Tang [54] in a way that included back the construction of Tu and Deng [57]. In their section, the main idea is to replace y by y^{2^k-1-u} in the construction of the function. Hence, the family of Tu and Deng is included for $u = 1$, and the family of Tang et al. for $u = 2^k - 2$. The associated combinatorial conjecture is then modified as follows.

Conjecture 3.4.9 (Jin et al. Conjecture). Let $k \geq 2$ be an integer, $t, u, v \in (\mathbb{Z}/(2^k-1)\mathbb{Z})^*$ such that $\gcd(u, 2^k - 1) = \gcd(v, 2^k - 1) = 1$. Then

$$\#\left\{(a,b) \in \left(\mathbb{Z}/(2^k-1)\mathbb{Z}\right)^2 \mid ua+vb=t; w_2(a)+w_2(b) \le k-1\right\} \le 2^{k-1}.$$

This generalized conjecture obviously includes all the previous ones.

The construction of their functions is as follows.

Definition 3.4.10 (Construction of Jin et al. [32]). Let $n = 2k \ge 4$ be an even integer, α a primitive element of $\mathbb{F}_{2^n}$, $A = \left\{1, \alpha, \ldots, \alpha^{2^{k-1}-1}\right\}$ and $g : \mathbb{F}_{2^k} \to \mathbb{F}_2$ Boolean function in k variables defined by

$$\mathrm{supp}(g) = \alpha^s A,$$

for any $0 \le s \le 2^k - 2$. Let $f : \mathbb{F}_{2^k} \times \mathbb{F}_{2^k} \to \mathbb{F}_2$ be the Boolean function in n variables defined by

$$f(x,y) = g\left(xy^{2^k-1-u}\right).$$

They proved that such a function f is

1. of algebraic degree between $n/2$ and $n-2$ depending on the value of u,
2. of optimal algebraic immunity $n/2$ if Conjecture 3.4.9 is true,
3. of nonlinearity at least

$$2^{n-1} - \frac{2}{\pi}\ln\frac{4(2^{n/2}-1)}{\pi}2^{n/2} - 1 \approx 2^{n-1} - \frac{\ln 2}{\pi}n2^{n/2}.$$

The proof of the optimality of the algebraic immunity is once again similar to the previous ones. It should be noted that resistance to fast algebraic attacks is not studied by Jin et al. [32].

Modifying these functions as before, Jin et al. obtained balanced functions with high algebraic degree and nonlinearity. They proved that for $n = 2k \ge 4$, these modified functions are

1. balanced,
2. of optimal algebraic degree $n-1$,
3. of optimal algebraic immunity $n/2$ if Conjecture 3.4.9 is true,
4. of nonlinearity at least

$$2^{n-1} - \frac{2}{\pi}\ln\frac{4(2^{n/2}-1)}{\pi}2^{n/2} - \frac{2}{\pi}\ln\frac{4(2^{n/2}-1)}{\pi}2^{n/4} - 2 \approx 2^{n-1} - \frac{\ln 2}{\pi}n2^{n/2} - \frac{\ln 2}{\pi}n2^{n/4}.$$

Jin et al. [31] applied a similar generalization to the 1-resilient Boolean function of Tu and Deng [56] and obtained a family functions which are

1. 1-resilient,
2. of optimal algebraic degree $n-2$,

3. of optimal algebraic immunity $n/2$ up to Conjecture 3.4.9 and an additional assumption,
4. of nonlinearity at least

$$2^{n-1} - \frac{2}{\pi} \ln \frac{4(2^{n/2}-1)}{\pi} 2^{n/2} - 2^{n/2-1} - \frac{4}{\pi} \ln \frac{4(2^{n/2}-1)}{\pi} 2^{n/4} - 3$$
$$\approx 2^{n-1} - \frac{\ln 2}{\pi}(n+1)2^{n/2} - \frac{2\ln 2}{\pi} n 2^{n/4} .$$

3.5 Some Results on a Conjecture About Binary Strings Distribution

As was underlined in the previous section , the good cryptographic properties of the Boolean functions of the Jin et al. family [31] and more precisely the optimality of their algebraic immunity, depend on the validity of a combinatorial conjecture. The purpose of this section, if not to prove that conjecture in its full generality, is at least to give a good insight into its expected validity not only through a thorough theoretical study.

Unless stated otherwise, we use the following notation throughout this section:

- $k \in \mathbb{N}$ is the number of bits (or length of binary strings) we are currently working on;
- $t \in \mathbb{Z}/(2^k-1)\mathbb{Z}$ is a fixed modular integer.

We use the following function of natural (or modular) integers (or binary strings). Let us denote by $S_{t,v,u,k}$ the set of interest:

$$S_{t,v,u,k} = \left\{(a,b) \in \left(\mathbb{Z}/(2^k-1)\mathbb{Z}\right)^2 \mid ua + vb = t; w_2(a) + w_2(b) \leq k-1\right\},$$

where $k \geq 2$, $t \in \left(\mathbb{Z}/(2^k-1)\mathbb{Z}\right)^*$ and $u, v \in \left(\mathbb{Z}/(2^k-1)\mathbb{Z}\right)^\times$, i.e. u and v are invertible modulo 2^k-1.

We now recall the different flavors of the conjecture already mentioned in the previous section.

Conjecture 3.4.2 (Tu–Deng Conjecture). *With the above notation,*

$$\#S_{t,+1,1,k} \leq 2^{k-1}.$$

Conjecture 3.4.6 (Tang–Carlet–Tang Conjecture). *With the above notation,*

$$\#S_{t,-1,1,k} \leq 2^{k-1}.$$

Conjecture 3.4.9 (Jin et al. Conjecture). *With the above notation,*

$$\#S_{t,v,u,k} \leq 2^{k-1}.$$

In the following, we present first general basic properties of the set $S_{t,v,u,k}$ obtained by studying the behavior of the Hamming weight under various basic transformations: *binary not* and *rotation*. In fact, in 2010, we have studied these properties in the particular case of $S_{t,+1,1,k}$ (that we denoted simply by $S_{t,k}$) since the other conjectures were formulated only in 2011. To make the book self-contained, we include the results formulated by Flori and Randriam [26] in the general case.

Definition 3.5.1. We define $\overline{a}^k$ as the modular integer whose binary expansion is the binary not on k bits of the binary expansion of the representative of a in $\{0, \dots, 2^k - 2\}$. We denote it by $\overline{a}$ when there is no ambiguity about the value of k.

Lemma 3.5.2 ([25]). *Let $a \in (\mathbb{Z}/(2^k-1)\mathbb{Z})^*$ be a non-zero modular integer, then $-a = \overline{a}$ and $w_2(-a) = k - w_2(a)$.*

Proof. Indeed $a + \overline{a} = \sum_{i=0}^{k-1} 2^i = 2^k - 1 = 0$. □

Lemma 3.5.3 ([25]). *For all $i \in \mathbb{Z}$ and $a \in \mathbb{Z}/(2^k-1)\mathbb{Z}$, we have*

$$w_2(2^i a) = w_2(a).$$

Proof. We are working in $\mathbb{Z}/(2^k-1)\mathbb{Z}$ so that $2^k = 1$ and multiplying a modular integer in $\mathbb{Z}/(2^k-1)\mathbb{Z}$ by 2 is just rotating its representation as a binary string on k bits by one bit to the left, whence the equality of the Hamming weights. □

Therefore, we say that, for any $i \in \mathbb{Z}$, $2^i a$ and a are equivalent, or that they are in the same *cyclotomic class* modulo $2^k - 1$, and we write $a \simeq 2^i a$. Remark that, for a given $a \in \mathbb{Z}/(2^k-1)\mathbb{Z}$, b must be equal to $v^{-1}(t - ua)$, whence the following lemma.

Lemma 3.5.4 ([25]). *For $k \geq 2$,*

$$\#S_{t,v,u,k} = \#\left\{a \in \mathbb{Z}/(2^k-1)\mathbb{Z} \mid w_2(a) + w_2(v^{-1}(t-ua)) \leq k-1\right\}.$$

Using the previous lemmas, we can now show that it is enough to study the conjecture for one t, but also one u and one v, in each cyclotomic class.

Lemma 3.5.5 ([25]). *For $k \geq 2$,*

$$\#S_{t,v,u,k} = \#S_{2t,v,u,k}.$$

Proof. Indeed $a \mapsto 2a$ is a permutation of $\mathbb{Z}/(2^k-1)\mathbb{Z}$ so that

$$\begin{aligned}
\#S_{2t,v,u,k} &= \#\left\{a \in \mathbb{Z}/(2^k-1)\mathbb{Z} \mid w_2(a) + w_2(v^{-1}(2t-ua)) \leq k-1\right\} \\
&= \#\left\{a \in \mathbb{Z}/(2^k-1)\mathbb{Z} \mid w_2(2a) + w_2(2v^{-1}(t-ua)) \leq k-1\right\} \\
&= \#\left\{a \in \mathbb{Z}/(2^k-1)\mathbb{Z} \mid w_2(a) + w_2(v^{-1}(t-ua)) \leq k-1\right\} \\
&= \#S_{t,v,u,k}.
\end{aligned}$$

□

Lemma 3.5.6 ([25]). *For $k \geq 2$,*

$$\#S_{t,v,u,k} = \#S_{t,v,2u,k}.$$

Proof. Using the previous lemma,

$$\begin{aligned}\#S_{t,v,2u,k} &= \#S_{2t,v,2u,k} \\ &= \#\left\{a \in \mathbb{Z}/(2^k-1)\mathbb{Z} \mid w_2(a) + w_2(v^{-1}(2t-2ua)) \leq k-1\right\} \\ &= \#\left\{a \in \mathbb{Z}/(2^k-1)\mathbb{Z} \mid w_2(a) + w_2(v^{-1}(t-ua)) \leq k-1\right\} \\ &= \#S_{t,v,u,k}.\end{aligned}$$

□

Lemma 3.5.7 ([25]). *For $k \geq 2$,*

$$\#S_{t,v,u,k} = \#S_{t,2v,u,k}.$$

Proof. Using the previous lemmas,

$$\begin{aligned}\#S_{t,2v,u,k} &= \#S_{2t,2v,2u,k} \\ &= \#\left\{a \in \mathbb{Z}/(2^k-1)\mathbb{Z} \mid w_2(a) + w_2((2v)^{-1}(2t-2ua)) \leq k-1\right\} \\ &= \#\left\{a \in \mathbb{Z}/(2^k-1)\mathbb{Z} \mid w_2(a) + w_2(v^{-1}(t-ua)) \leq k-1\right\} \\ &= \#S_{t,v,u,k}.\end{aligned}$$

□

It was shown a more elaborate relation for different values of u, v and t.

Lemma 3.5.8 ([25]). *For $k \geq 2$,*

$$\#S_{t,v,u,k} = \#S_{(uv)^{-1}t,v^{-1},u^{-1},k}.$$

Proof. We use the fact that $a \mapsto u^{-1}(-va+t)$ is a permutation of $\mathbb{Z}/(2^k-1)\mathbb{Z}$ and deduce

$$\begin{aligned}\#S_{t,v,u,k} &= \#\left\{a \in \mathbb{Z}/(2^k-1)\mathbb{Z} \mid w_2(a) + w_2(v^{-1}(t-ua)) \leq k-1\right\} \\ &= \#\left\{a \in \mathbb{Z}/(2^k-1)\mathbb{Z} \mid w_2(u^{-1}(-va+t)) + w_2(a) \leq k-1\right\} \\ &= \#\left\{a \in \mathbb{Z}/(2^k-1)\mathbb{Z} \mid w_2(v((uv)^{-1}t - u^{-1}a)) + w_2(a) \leq k-1\right\} \\ &= \#S_{(uv)^{-1}t,v^{-1},u^{-1},k}.\end{aligned}$$

□

Now, we state the following observation of Jin et al. [32].

Lemma 3.5.9. *For* $k \geq 2$ *and* $c \in \left(\mathbb{Z}/(2^k-1)\mathbb{Z}\right)^{\times}$,

$$\#S_{t,v,u,k} = \#S_{ct,cv,cu,k}.$$

Proof. Indeed, we have $ua + vb = t$ if and only if $cua + cvb = ct$ when c is invertible, whence a bijection between the sets $S_{t,v,u,k}$ and $S_{ct,cv,cu,k}$. □

Finally, as noted by Jin et al. [32], their generalized conjecture is then equivalent to the generalized conjecture of Tang et al. [54].

Now, we concentrate our efforts on the original conjecture of Tu and Deng [57] which is a natural candidate to extend the study of the other conjectures but also because we have been interested in this conjecture since 2010 while the other conjectures have been reformulated only in 2011.

In the following, we just give some results appeared in [27].[4] Readers interested in the development of this study will refer to the work of Flori and Randriam [26].

Our main approach used in this section is that of reformulating the conjecture in terms of *carries* occurring in an addition modulo $2^k - 1$. Although such an approach may at first seem quite naive to the reader, what makes the study of the conjecture seemingly so difficult is precisely that a suitable algebraic structure to cast upon the problem has yet to be found, so that only a purely combinatorial point of view is possible as of today.

Let us define the main tool we will use to study the conjecture of Tu and Deng. Note that Cohen and Flori [12] have used this tool to prove the Conjecture of Tang et al.

Definition 3.5.10. For $a \in \left(\mathbb{Z}/(2^k-1)\mathbb{Z}\right)^*$, we set

$$r(a,t) = w_2(a) + w_2(t) - w_2(a+t) \ ,$$

i.e. $r(a,t)$ is the number of carries occurring while performing the addition. By convention, we set

$$r(0,t) = k \ ,$$

i.e. 0 behaves like the $\underbrace{1\ldots1}_{k}$ binary string. We also remark that $r(-t,t) = k$.

The following statement is fundamental. It brings to light the importance of the number of carries occurring during the addition.

[4] Note that Flori had continued to study more deeply this conjecture in his PhD thesis [25] and obtained different interesting results, but unfortunately without reaching a complete proof of this conjecture.

Proposition 3.5.11 ([27]). *For $k \geq 2$ and $t \in \left(\mathbb{Z}/(2^k - 1)\mathbb{Z}\right)^*$,*

$$\#S_{t,k} = \#\{a \mid r(a,t) > w_2(t)\} \ .$$

Now, we often compute $P_{t,k} = 2^{-k}\#S_{t,k}$ rather than $\#S_{t,k}$. Therefore we use the words *proportion* or *probability* in place of *cardinality*. Moreover we often computed cardinalities considering all the binary strings on k bits, i.e. including `1...1` and `0...0`. The modular integer 0 is considered to act as the binary string `1...1`, but the binary string `0...0` should be discarded when doing final computation of $P_{t,k}$. However it ensures that variables are truly *independent*.

The original conjecture proposed by Tu and Deng [57] can be reformulated as follows.

Conjecture 3.4.2. *For $k \geq 2$ and $t \in \mathbb{Z}/(2^k - 1)\mathbb{Z}$, let $S_{t,k}$ be the following set.*[5] *:*

$$S_{t,k} = \left\{a \in \mathbb{Z}/(2^k - 1)\mathbb{Z} \mid r(a,t) > w_2(t)\right\} \ ,$$

and $P_{t,k}$ the fraction[6] *of modular integers in $S_{t,k}$:*

$$P_{t,k} = \#S_{t,k}/2^k \ .$$

Then

$$P_{t,k} \leq \frac{1}{2} \ .$$

Tu and Deng verified computationally the validity of this assumption for $k \leq 29$ in about fifteen days on a quite recent computer [57]. We also implemented their algorithm and were able to check the conjecture for $k = 39$ in about twelve hours and fifteen minutes on a pool of about four hundred quite recent cores, and $k = 40$ on a subset of these computers. The algorithm of Tu and Deng [57, Appendix] as well as the implementation have been described by Flori.

Note that conjecture is not only interesting in a cryptographic context, but also for purely arithmetical reasons. For a fixed modular integer $t \in \mathbb{Z}/(2^k - 1)\mathbb{Z}$, it is indeed natural to expect the number of carries occurring when adding a random modular integer $a \in \mathbb{Z}/(2^k - 1)\mathbb{Z}$ to t to be roughly the Hamming weight of t. Note that following this idea, Flori and Randriam [26] have studied the distribution of the number of carries around this value and proved that quite unexpectedly, the conjecture seems to indicate a kind of regularity.

---

[5]It is easy to see that this formulation is equivalent to the original one. A formal proof will be given in Corollary 3.5.11.

[6]We are fully aware that there are only $2^k - 1$ elements in $\mathbb{Z}/(2^k - 1)\mathbb{Z}$, but we will often use the abuse of terminology we make here and speak of *fraction*, *probability* or *proportion* for $P_{t,k}$.

Proposition 3.5.11 allows us to prove the conjecture in the specific case where $t \simeq -t$:

Theorem 3.5.12 ([27]). *If* $t \simeq -t$, *then* $\#S_{t,k} \leq 2^{k-1}$.

Next, we split $t(\neq 0)$ (once correctly rotated, i.e. we multiply it by a correct power of 2 so that its binary expansion on k bits begins with a `1` and ends with a `0`) in blocks of the form [`1`* `0`*] (i.e. as many `1`s as possible followed by as many `0`s as possible).

Definition 3.5.13. We denote the number of blocks by d and the numbers of `1`s and `0`s of the ith block t_i by α_i and β_i.

We have defined corresponding variables for a (a number to be added to t): γ_i the number of `0`s in front of the end of the `1`s subblock of t_i, δ_i the number of `1`s in front of the end of the `0`s subblock of t_i.

Those definitions are depicted below:

$$t = \overbrace{\texttt{1---1}}^{\alpha_1}\overbrace{\texttt{0---0}}^{\beta_1}\texttt{...}\overbrace{\texttt{1---1}}^{\alpha_i}\overbrace{\texttt{0---0}}^{\beta_i}\texttt{...}\overbrace{\texttt{1---1}}^{\alpha_d}\overbrace{\texttt{0---0}}^{\beta_d}\ ,$$

$$a = \texttt{?1}\underbrace{\texttt{0-0}}_{\gamma_1}\texttt{?0}\underbrace{\texttt{1-1}}_{\delta_1}\texttt{...?1}\underbrace{\texttt{0-0}}_{\gamma_i}\texttt{?0}\underbrace{\texttt{1-1}}_{\delta_i}\texttt{...?1}\underbrace{\texttt{0-0}}_{\gamma_d}\texttt{?0}\underbrace{\texttt{1-1}}_{\delta_d}\ ,$$

One should be aware that γ_is and δ_is depend on a and are considered as variables.

We first "approximate" the number of carries $r(a, t)$ by $\sum_{i=0}^{d} \alpha_i - \gamma_i + \delta_i$ ignoring the two following facts:

- if a carry goes out of the $i-1$st block (we say that it *overflows*) and $\delta_i = \beta_i$, the `1`s subblock produces α_i carries, whatever value γ_i takes,
- and if no carry goes out of the $i-1$st block (we say that it is *inert*), the `0`s subblock produces no carry, whatever value β_i takes.

Our first result is the exact formulas of $\#S_{t,k}$ for numbers made of only one block (i.e. for $d = 1$). More precisely, we have proved the following theorem.

Theorem 3.5.14 ([27]).

$$P_{t,k} = \begin{cases} 2^{-\alpha-\beta}\frac{1-2^{-2\alpha}}{3} & \text{if } 1 \leq \alpha \leq \frac{k-1}{2} \\ \frac{1+2^{-2\beta+1}}{3} & \text{if } \frac{k-1}{2} \leq \alpha \leq k-1 \end{cases} .$$

For $\alpha = 1$, it reads $S_{1,k} = 2^{k-2} + 1$ and for $\alpha = k - 1$, it reads $S_{-1,k} = 2^{k-1}$.

Next, we introduce the following constraint which greatly simplifies calculations:

$$\min_i(\alpha_i) \geq \sum_{i=1}^{d} \beta_i - 1 = k - w_2(t) - 1 \ .$$

That condition tells us that, if a is in $S_{t,k}$, a carry has to go through each subblock of 1s Moreover, it leads us to a proof that the conjecture is *asymptotically* (that is, when $\beta_i \to \infty$) true. More precisely, we have proved the following theorem.

Theorem 3.5.15 ([27]). *Let d be a strictly positive integer. There exists a constant K_d such that if t verifies the two following constraints:*

$$\forall i, \beta_i \geq K_d \text{ and } \min_i \alpha_i \geq k - w_2(t) - 1 \ ,$$

then $\#S_{t,k} < 2^{k-1}$.

When the number of blocks, d, goes as well to infinity, we remark that $P_{t,k}$ converges toward $1/2$.

It is possible to compute the exact value of $\#S_{t,k}$ for a given d and a corresponding set of β_is. It is worth noting that the order of the β_is does not matter because each subblock behaves the same when a is in $S_{t,k}$, i.e. it overflows. We did the computation for $d = 2$ where the symmetry of the problem leads to only one situation and gives a quite general result.

Definition 3.5.16.

$$f(x,y) = \frac{11}{27} + 4^{-x}\left(\frac{2}{9}x - \frac{2}{27}\right) + 4^{-y}\left(\frac{2}{9}y - \frac{2}{27}\right) + 4^{-x-y}\left(\frac{20}{27} - \frac{2}{9}(x+y)\right).$$

Proposition 3.5.17 ([27]).

$$P_{t,k} = f(\beta_1, \beta_2) \leq 1/2 \ .$$

Proof. An easy but quite lengthy and error-prone calculation, which can be checked with a symbolic calculus software, leads to the desired expression. The graph of f, computed with Maple™ [47], is given in Fig. 3.3. □

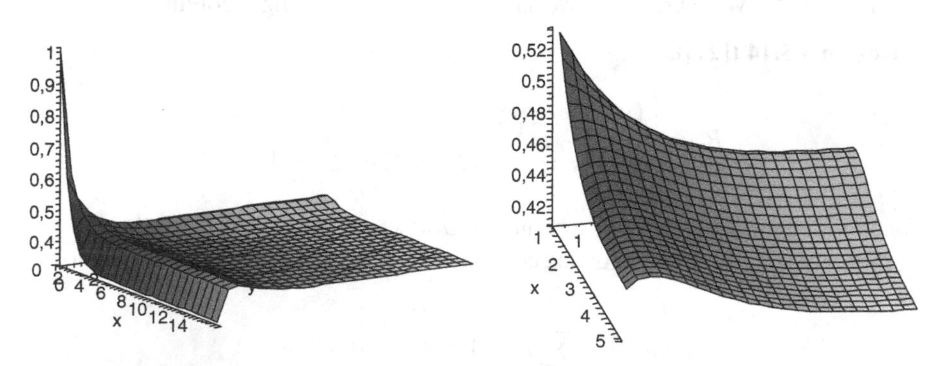

Fig. 3.3 Graph of $f(x, y)$

We have proved:

Theorem 3.5.18 ([27]). *If t verifies the following constraints:*

$$d = 2 \text{ and } \alpha_1, \alpha_2 \geq k - w_2(t) - 1 \ ,$$

then $\#S_{t,k} \leq 2^{k-1}$.

Finally, we have proved that a family of numbers reaches the bound (we believe they are the only ones to do so): In fact, we have added another constraint: $\forall i, \beta_i = 1$. The previous one becomes: $\min_i(\alpha_i) \geq k - w_2(t) - 1 = d - 1$.

Theorem 3.5.19 ([27]). *Let t verify the two following constraints:*

$$\forall i, \beta_i = 1 \text{ and } \min_i(\alpha_i) \geq k - w_2(t) - 1 = d - 1 \ ,$$

then $\#S_{t,k} = 2^{k-1}$.

Next we proved the conjecture in the following case:

Corollary 3.5.20. *Let t verify the two following constraints:*

$$\forall i, \alpha_i = 1 \text{ and } \min_i(\beta_i) \geq w_2(t) - 1 = d - 1 \ ,$$

then $\#S_{t,k} \leq 2^{k-1}$.

The theoretical study of the conjecture, together with experimental results obtained with Sage [53] made by Jean-Pierre Flori, lead to conjecture that the converse of Theorem 3.5.19 is also true, i.e. the numbers of Theorem 3.5.19 are the only ones reaching the bound of the conjecture Tu–Deng, which is obviously stronger than the original conjecture.

Conjecture 3.5.21. Let $k \geq 2$ and $t \in (\mathbb{Z}/(2^k - 1)\mathbb{Z})^*$. Then $S_{t,k} = 2^{k-1}$ if and only if t verifies the two following constraints:

- $\forall i, \ \beta_i = 1$,
- $\min_i(\alpha_i) \geq B - 1 = d - 1$.

As far we know, the conjectures presented in this section are still open. Only the variation proposed by Tang, Carlet and Tang has been proved recently by Cohen and Flori [12]. For more details, we send the reader to the PhD's thesis of Flori [25].

References

1. A. Canteaut and M. Trabbia. Improved fast correlation attacks using parity-check equations of weight 4 and 5. In *Advanced in Cryptology-EUROCRYPT 2000. Lecture notes in computer science, 1807*, pages 573–588, 2000.

2. C. Carlet. On a weakness of the Tu–Deng function and its repair. Cryptology ePrint Archive, Report 2009/606, 2009. http://eprint.iacr.org/.
3. C. Carlet. Boolean functions for cryptography and error correcting codes. In Yves Crama and Peter L. Hammer, editors, *Boolean Models and Methods in Mathematics, Computer Science, and Engineering*, pages 257–397. Cambridge University Press, June 2010.
4. C. Carlet. Boolean Functions for Cryptography and Error Correcting Codes. In *Chapter of the monography "Boolean Models and Methods in Mathematics, Computer Science, and Engineering" published by Cambridge University Press, Yves Crama and Peter L. Hammer (eds.)*, pages 257–397, 2010.
5. C. Carlet. Comments on "Constructions of cryptographically significant Boolean functions using primitive polynomials". *Information Theory, IEEE Transactions on*, 57(7):4852–4853, July 2011.
6. C. Carlet, D. Kumar Dalai, K. C. Gupta, and S. Maitra. Algebraic immunity for cryptographically significant Boolean functions: Analysis and construction. *IEEE Transactions on Information Theory*, 52(7):3105–3121, 2006.
7. C. Carlet and K. Feng. An infinite class of balanced functions with optimal algebraic immunity, good immunity to fast algebraic attacks and good nonlinearity. In Josef Pieprzyk, editor, *ASIACRYPT*, volume 5350 of *Lecture Notes in Computer Science*, pages 425–440. Springer, 2008.
8. C. Carlet and P. Sarkar. Spectral domain analysis of correlation immune and resilient Boolean functions. In *Finite Fields and their Applications*, pages 120–130, 2002.
9. C. Carlet, X. Zeng, C. Li, and L. Hu. Further properties of several classes of Boolean functions with optimum algebraic immunity. *Des. Codes Cryptography*, 52(3):303–338, 2009.
10. V. Chepyzhov and B. Smeets. On a fast correlation attack on certain stream ciphers. In *Proceedings of EUROCRYPT'91, Lecture Notes in Computer Science, 547*, pages 176–185, 1992.
11. V. V. Chepyzhov, T. Johansson, and B. Smeets. A simple algorithm for fast correlation attacks on stream ciphers. In *B. Schneier, editor, FSE 2000, volume 1978 of Lecture Notes in Computer Science, Springer-Verlag, April 10–12*, pages 181–195, 2001.
12. G. Cohen and J-P. Flori. On a generalized combinatorial conjecture involving addition mod $2^k - 1$. Cryptology ePrint Archive, Report 2011/400, 2011. http://eprint.iacr.org/.
13. N. Courtois. Higher Order Correlation Attacks, XL algorithm, and Cryptanalysis of Toyocrypts.
14. N. Courtois. Fast algebraic attacks on stream ciphers with linear feedback. In Dan Boneh, editor, *CRYPTO*, volume 2729 of *Lecture Notes in Computer Science*, pages 176–194. Springer, 2003.
15. N. Courtois and W. Meier. Algebraic attacks on stream ciphers with linear feedback. In Eli Biham, editor, *EUROCRYPT*, volume 2656 of *Lecture Notes in Computer Science*, pages 345–359. Springer, 2003.
16. N. Courtois and W. Meier. Algebraic attacks on stream ciphers with linear feedback. In *Eurocrypt 03, volume 2656 of Lecture Notes in Computer Science*, pages 345–349, 2003.
17. D. K. Dalai, K. C. Gupta, and S. Maitra. Notion of algebraic immunity and its evaluation related to fast algebraic attacks. In *International Workshop on Boolean Functions : Cryptography and Applications*, pages 13–15, 2006.
18. D. Kumar Dalai, S. Maitra, and S. Sarkar. Basic theory in construction of Boolean functions with maximum possible annihilator immunity. *Des. Codes Cryptography*, 40(1):41–58, 2006.
19. J. Dillon. Elementary Hadamard difference sets. In *PhD dissertation, University of Maryland.*
20. J.F. Dillon. *Elementary Hadamard Difference Sets*. ProQuest LLC, Ann Arbor, MI, 1974. Thesis (Ph.D.)–University of Maryland, College Park.
21. C. Ding, G. Xiao, and W. Shan. *The Stability Theory of Stream Ciphers*, volume 561 of *Lecture Notes in Computer Science*. Springer, 1991.
22. H. Dobbertin. Construction of bent functions and balanced Boolean functions with high nonlinearity. In Bart Preneel, editor, *FSE*, volume 1008 of *Lecture Notes in Computer Science*, pages 61–74. Springer, 1994.

23. J.-C. Faugère and G. Ars. An Algebraic Cryptanalysis of Nonlinear Filter Generators using Gröbner bases. In *Rapport de Recherche INRIA*, pages 4739–2003, 2003.
24. K. Feng, Q. Liao, and J. Yang. Maximal values of generalized algebraic immunity. *Des. Codes Cryptography*, 50(2):243–252, 2009.
25. J-P. Flori. Boolean functions, algebraic curves and complex multiplication. In *PhD Thesis, Telecom Paris Tech, France*, 2012.
26. J-P. Flori and H. Randriam. On the number of carries occurring in an addition mod $2^k - 1$. In *Journal Integers, vol 12*, 2012.
27. J-P. Flori, H. Randriam, G. Cohen, and S. Mesnager. On a conjecture about binary strings distribution. In C. Carlet and A. Pott, editors, *SETA*, volume 6338 of *Lecture Notes in Computer Science*, pages 346–358. Springer, 2010.
28. R. Forré. A fast correlation attack on nonlinearly feedforward filtered shift register sequences. In *Proceedings of EUROCRYPT '89, Lecture Notes in Computer Science, 434*, pages 586–595, 1990.
29. X. Guo-Zhen, C. Ding, and W. Shan. The stability theory of stream ciphers. In *Lecture Notes in Computer Science 561*, 1991.
30. P. Hawkes and G. Rose. Rewriting Variables: The Complexity of Fast Algebraic Attacks on Stream Ciphers. In *Proceedings of CRYPTO 2004, Lecture Notes in Computer Science 3152*, pages 390–406, 2004.
31. Q. Jin, Z. Liu, and B. Wu. 1-resilient Boolean function with optimal algebraic immunity. Cryptology ePrint Archive, Report 2011/549, 2011. http://eprint.iacr.org/.
32. Q. Jin, Z. Liu, B. Wu, and X. Zhang. A general conjecture similar to T-D conjecture and its applications in constructing Boolean functions with optimal algebraic immunity. Cryptology ePrint Archive, Report 2011/515, 2011. http://eprint.iacr.org/.
33. T. Johansson and F. Jönsson. Fast correlation attacks based on turbo code techniques. In *Advances in Cryptology - CRYPTO'99, no. 1666 in Lecture Notes in Computer Science*, pages 181–197, 1999.
34. T. Johansson and F. Jönsson. Improved fast correlation attack on stream ciphers via convolutional codes. In *Proceedings of EUROCRYPT'99, Lecture Notes in Computer Science, 1592*, pages 347–362, 1999.
35. T. Johansson and F. Jönsson. Fast correlation attacks through reconstruction of linear polynomials. In *Advances in Cryptology - CRYPTO 2000, no. 1880 in Lecture Notes in Computer Science*, pages 300–315, 2000.
36. Na Li and Wen-Feng Qi. Construction and analysis of Boolean functions of $2t + 1$ variables with maximum algebraic immunity. In Xuejia Lai and Kefei Chen, editors, *ASIACRYPT*, volume 4284 of *Lecture Notes in Computer Science*, pages 84–98. Springer, 2006.
37. Na Li, Longjiang Qu, Wen-Feng Qi, GuoZhu Feng, Chao Li, and DuanQiang Xie. On the construction of Boolean functions with optimal algebraic immunity. *IEEE Transactions on Information Theory*, 54(3):1330–1334, 2008.
38. Mikhail Sergeevich Lobanov. Exact relations between nonlinearity and algebraic immunity. *Diskretn. Anal. Issled. Oper.*, 15(6):34–47, 95, 2008.
39. F. J. MacWilliams and N. J. A. Sloane. *The theory of error-correcting codes. I.* North-Holland Publishing Co., Amsterdam, 1977. North-Holland Mathematical Library, Vol. 16.
40. F. J. MacWilliams and N. J. A. Sloane. *The theory of error-correcting codes. II.* North-Holland Publishing Co., Amsterdam, 1977. North-Holland Mathematical Library, Vol. 16.
41. J. L. Massey. Shift-register synthesis and BCH decoding. *Information Theory, IEEE Transactions on*, 15(1):122–127, Jan 1969.
42. M. Matsui. Linear cryptanalysis method for DES cipher. In *Proceedings of EUROCRYPT'93, Lecture Notes in Computer Science 765*, pages 386–397, 1994.
43. W. Meier, E. Pasalic, and C. Carlet. Algebraic attacks and decomposition of Boolean functions. In *Eurocrypt 2004, ser. Lecture notes in Computer Science, vol. 3027. Springer-Verlag*, pages 474–491, 2004.
44. W. Meier and O. Staffelbach. Fast correlation attacks on stream ciphers. In *Advances in Cryptology, EUROCRYPT'88, Lecture Notes in Computer Science 330*, pages 301–314, 1988.

45. W. Meier and O. Staffelbach. Fast correlation attacks on stream ciphers (extended abstract). In *EUROCRYPT'88*, pages 301–314, 1988.
46. A. J. Menezes, P. C. van Oorschot, and S. A. Vanstone. *Handbook of Applied Cryptography*. CRC Press, 1996.
47. Michael Burnett Monagan, Keith Oliver Geddes, K. Michael Heal, George Labahn, Stefan M. Vorkoetter, James McCarron, and Paul DeMarco. *Maple 10 Programming Guide*. Maplesoft, Waterloo ON, Canada, 2005.
48. S. Rønjom and T. Helleseth. A new attack on the filter generator. In *IEEE Transactions on Information Theory, vol. 53, no. 5*, pages 1752–1758, 2007.
49. O.S. Rothaus. On "bent" functions. In *J. Combin. Theory Ser A 20*, pages 300–305, 1976.
50. C.E. Shannon. Communication theory of secrecy systems. In *Bell system technical journal, 28*, pages 656–715, 1949.
51. T. Siegenthaler. Correlation-immunity of nonlinear combining Boolean functions for cryptographic applications. In *IEEE Transactions on Information Theory, Vol 30, no. 5*, pages 776–779, 1984.
52. T. Siegenthaler. Decrypting a Class of Stream Ciphers Using Ciphertext. In *IEEE Transactions on Computer, vol. C-34, no 1*, pages 81–85, 1985.
53. William Arthur Stein et al. *Sage Mathematics Software (Version 4.7)*. The Sage Development Team, 2011. http://www.sagemath.org.
54. D. Tang, C. Carlet, and X. Tang. Highly nonlinear Boolean functions with optimal algebraic immunity and good behavior against fast algebraic attacks. Cryptology ePrint Archive, Report 2011/366, 2011. http://eprint.iacr.org/. To appear in IEEE- IT.
55. X. Tang, D. Tang, X. Zeng, and L. Hu. Balanced Boolean functions with (almost) optimal algebraic immunity and very high nonlinearity. Cryptology ePrint Archive, Report 2010/443, 2010. http://eprint.iacr.org/.
56. Z. Tu and Y. Deng. Boolean functions with all main cryptographic properties. Cryptology ePrint Archive, Report 2010/518, 2010. http://eprint.iacr.org/.
57. Z. Tu and Y. Deng. A conjecture about binary strings and its applications on constructing Boolean functions with optimal algebraic immunity. *Des. Codes Cryptography*, 60(1):1–14, 2011.
58. J. H. van Lint. *Introduction to coding theory*, volume 86 of *Graduate Texts in Mathematics*. Springer-Verlag, Berlin, third edition, 1999.
59. Qichun Wang and Thomas Johansson. A note on fast algebraic attacks and higher order nonlinearities. In Xuejia Lai, Moti Yung, and Dongdai Lin, editors, *Information Security and Cryptology*, volume 6584 of *Lecture Notes in Computer Science*, pages 404–414. Springer Berlin / Heidelberg, 2011. 10.1007/978-3-642-21518-6-28.
60. Qichun Wang, Jie Peng, Haibin Kan, and Xiangyang Xue. Constructions of cryptographically significant Boolean functions using primitive polynomials. *IEEE Transactions on Information Theory*, 56(6):3048–3053, 2010.
61. G-Z Xiao and J.L. Massey. A spectral characterization of correlation-immune combining functions. In *IEEE Transactions on Information Theory, Vol 34 no. 3*, pages 569–571, 1988.
62. Z.Tu and Y. Deng. A class of 1-resilient function with high nonlinearity and algebraic immunity. Cryptology ePrint Archive, Report 2010/179, 2010. http://eprint.iacr.org/.

Chapter 4
Bent Functions-Generalities

4.1 Bent Functions: Introduction-Historical Notes

Bent functions were invented and named in 1966 by Oscar Rothaus[1] (1927–2003) in research not published until May 1976. So the final version was published ten years later in [67] in which Rothaus presented the basic properties and a large general family of bent functions. Dillon considers bent functions as wonderful creatures. Between 1960 and 1976, two documents on bent functions were written by J. Dillon, precisely in 1972 and 1974, but they had a limited distribution; these are the first papers he wrote on this subject [32] (where he mentions however an earlier paper, [26]) and the chapter he devoted to them in his nice PhD thesis [31]. A paper also appeared in 1975, based on Dillon's thesis [33]. In this preliminary period, several people (including the authors of [26]) mentioned by Dillon in [32] were interested in bent functions. In [67], two names are also cited by Rothaus: Lloyd Welch, the well-known specialist of codes and sequences, and Gerry Mitchell who contributed to a computer investigation.

[1]Oscar Rothaus was graduated from Princeton University (PhD in 1958, Salomon Bochner as advisor). He served in the US Army Signal Corps during the Korean War, then as a mathematician at the National Security Agency. Rothaus joined the Cornell faculty as a professor in 1966 and served as the mathematics department chair from 1973 to 1976. In 1995 he became acting department chair. Rothaus was a visiting professor at Hebrew University in Jerusalem, the University of Strasbourg in France and King's College London. Today the mathematical tool is used in speech recognition systems and for analyzing DNA. However, during the Cold War, Rothaus helped develop a vital military mathematical tool that simulates physical processes called the Hidden Markov Model, or HMM, for military purposes as a member of the Communication Research Division at the Defense Department's Institute for Defense Analyses (IDA) in Princeton, N.J., in the early 1960s. The model was declassified in the early 1970s. Rothaus' other mathematical research included combinatorics and coding theory, Lie and Jordan algebras, and Sobolev and logarithmic Sobolev inequalities. From 1960 to 1966 he worked at the Defense Department's Institute for Defense Analyses (IDA) and had the recognized authority.

S. Mesnager, *Bent Functions*, DOI 10.1007/978-3-319-32595-8_4

In a note written by Tokareva [70], it is mentioned that bent functions have been in fact studied in the Soviet Union since 1920 under the name of *minimal functions*. In particular, it seems that V.A. Eliseev and O.P. Stepchenkov had proved in 1962 that the degree of a bent function is not more than $\frac{n}{2}$ if $n \geq 4$ and had proposed an analog of the McFarland construction eleven years before the publication by R.L. McFarland. Their results were published as technical reports but never declassified.

In the early seventies bent functions were studied by American mathematicians J.F. Dillon [32], P.J. Chase, K.D. Lerche [26] in connection to differential sets and R.L. McFarland [62] who constructed a large class of bent functions.

Since the eighties, bent functions were widely studied in the international level. They have been extensively studied for their applications in cryptography, but have also been applied to cryptography, spread spectrum, coding theory, and combinatorial design. The definition can be extended in several ways, leading to different classes of generalized bent functions that share many of the useful properties of the original. Up to now, there exist more than two hundreds papers about bent functions and related topics. Several constructions of bent functions are obtained but they are not classified and the general structure is still not clear.

In Chap. 3 we emphasized the fact that a cryptographic Boolean function should verify several (contradictory) criteria. Constructing satisfying functions is therefore a difficult task, and trade-offs between the different criteria have to be made. In the present chapter, our approach will be slightly different: we solely focus on one criterion—nonlinearity—and more precisely on functions achieving maximum nonlinearity: *bent* functions. Recall that the significance of this aspect has again been demonstrated by the recent development of linear cryptanalysis initiated by Matsui [60, 61]. It is therefore especially important when Boolean functions are used as part of S-boxes in symmetric cryptosystems.

4.2 Bent Boolean Functions: Definition and Properties

Recall that the *nonlinearity* of a Boolean function f, denoted by $nl(f)$, is the minimum Hamming distance between f and all affine functions. In Fig. 4.1, we give the distribution of all 4-variable Boolean functions with respect to its nonlinearity.

A powerful mathematical tool to measure the nonlinearity of a Boolean function is the *Walsh transform*. The nonlinearity of an n-variable Boolean function can be expressed by means of the Walsh transform as follows:

$$nl(f) = 2^{n-1} - \frac{1}{2} max_{b \in \mathbb{F}_2^n} \left| \widehat{\chi_f}(b) \right|.$$

Because of the well-known Parseval's relation $\sum_{b \in \mathbb{F}_2^n} \widehat{\chi_f}(b)^2 = 2^{2n}$, $nl(f)$ is upper bounded by $2^{n-1} - 2^{n/2-1}$. This bound is tight for n even.

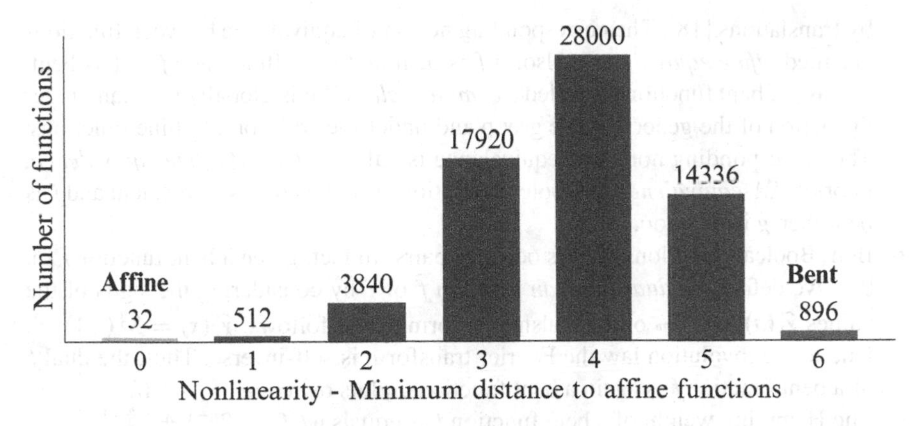

Fig. 4.1 Distribution of all 4-variable to nonlinearity

Definition 4.2.1. Let n be an even integer. An n-variable Boolean function is called bent if the upper bound $2^{n-1} - 2^{n/2-1}$ on its nonlinearity $nl(f)$ is achieved with equality.

Consequently, we have the following main characterization (which is independent of the choice of the inner product on $\mathbb{F}_2^n$) of the bentness for Boolean functions:

Proposition 4.2.2. *Let n be an even integer. An n-variable Boolean function f is bent if and only if its Walsh (Hadamard) transform satisfies* $\widehat{\chi_f}(a) = \pm 2^{\frac{n}{2}}$ *for all* $a \in \mathbb{F}_2^n$.

Hence, the Walsh transform provides a basic characterization of bentness. However, it can definitely not be used in practice to test efficiently bentness of a given function, especially if all its values are computed naively one at a time as exponential sums. Nevertheless, it should be noted that all the values of the Walsh–Hadamard transform can be computed *at once* using the so-called *fast Walsh–Hadamard transform*, a kind of Fast Fourier Transform. The complexity of the *fast Walsh–Hadamard transform* is $O(2^n n^2)$ bit operations and $O(2^n n)$ memory [4] which limits the calculations at the most to $n = 40$.

In the following, we present the main properties of bent functions:

- The algebraic degree of any bent Boolean function on $\mathbb{F}_{2^n}$ is at most m (in the case that $n = 2$, the bent functions have degree 2).
- The set of n-variable bent Boolean functions is invariant under the action of the general affine group of $\mathbb{F}_{2^n}$ and the addition of n-variable affine Boolean functions. In particular, if f and f' are two n-variable Boolean functions such that f' is linearly equivalent to f (that is, there exists an $\mathbb{F}_2$-linear automorphism L of $\mathbb{F}_{2^n}$ such that $f' = f \circ L$) then, f is bent if and only if f' is bent.
- The automorphism group of the set of bent functions (i.e., the group of permutations π on $\mathbb{F}_2^n$ or $\mathbb{F}_{2^n}$ such that $f \circ \pi$ is bent for every bent function f) is the general affine group, that is, the group of linear automorphisms composed

by translations [18]. The corresponding notion of equivalence between functions is called *affine equivalence*. Also, if f is bent and ℓ is affine, then $f + \ell$ is bent. A class of bent functions is called a *complete class* if it is globally invariant under the action of the general affine group and under the addition of affine functions. The corresponding notion of equivalence is called *extended affine equivalence*, in brief, *EA-equivalence*. If Boolean functions f and g are EA-equivalent and f is bent then g is bent too.

- Bent Boolean functions always occur in pairs. In fact, given a bent function f on $\mathbb{F}_{2^n}$, we define the *dual Boolean function* $\tilde{f}$ of f by considering the signs of the values $\widehat{\chi_f}(a), a \in \mathbb{F}_{2^n}$ of the Walsh transform of f as follows: $\widehat{\chi_f}(x) = 2^{\frac{n}{2}}(-1)^{\tilde{f}(x)}$. Due to the involution law the Fourier transform is self-inverse. Thus, the dual $\tilde{f}$ of a bent function f is again a bent function and its own dual is f itself.
- The Hamming weight of a bent function f is equals $wt(f) = 2^{n-1} \pm 2^{\frac{n}{2}-1}$.
- Any function f is bent if and only if, for any nonzero vector a, the Boolean function, called the *derivative* at a $D_af(x) = f(x) + f(x + a)$ is balanced (i.e. has Hamming weight 2^{n-1}). For this reason, bent functions are also called *perfect nonlinear functions*. Bent functions have also the property that, for every even positive integer w, the sum $\sum_{a\in\mathbb{F}_2^n} \widehat{\chi_f}^{\,w}(a)$ is minimum.
- Bent functions are the indicators of difference sets in elementary Abelian 2-groups.

4.3 Equivalent Characterizations of Bent Boolean Functions

Using algebraic and combinatorial tools, several attempts have been made to generate bent functions. To this end, several algebraic and combinatorial characterizations of bent functions have been introduced. All of them are equivalent and they differ from which point of view we look at bent functions. In the following we shall present them concisely.

- By definition, bent functions are maximally nonlinear Boolean functions and exist only with even number of inputs. More precisely, $f : \mathbb{F}_2^n \to \mathbb{F}_2$ (n even) is said to be a bent function if $nl(f) = 2^{n-1} - 2^{\frac{n}{2}-1}$. Various characterizations of bentness have been introduced in the literature. We present the most important ones this section.
- In his thesis [31], Dillon has introduced some characterizations of bentness. A main one is expressed in terms of the Walsh transform. Such notion is an important tool for research in cryptography. It plays an important role to characterize many cryptographic criteria for Boolean functions but also to define some significant cryptographic Boolean functions used in various type of symmetric cryptosystems. A Boolean function f on $\mathbb{F}_2^n$ is bent if and only if $\widehat{\chi_f}(\omega) = \pm 2^{\frac{n}{2}}, \quad \forall \omega \in \mathbb{F}_{2^n}$.

The Walsh transform as well as the bentness have been extended later to p-ary functions (that is, functions from $\mathbb{F}_p^n$ to $\mathbb{F}_p$ where $\mathbb{F}_p^n$ is the finite field of size p^n). In fact, the Walsh transform of f at $w \in \mathbb{F}_{p^n}$ is defined as: $\widehat{\chi_f}(w) = \sum_{x\in\mathbb{F}_{p^n}} \chi_p\Big(f(x) - \mathrm{Tr}_1^n(wx)\Big)$ where $\chi_p(a) = \xi_p^a$ and ξ_p is a primitive pth-root of unity. A p-ary function f is said to be bent if $|\widehat{\chi_f}(w)|^2 = p^n$ for every $w \in \mathbb{F}_{p^n}$ and said to be *regular bent* if there exists $\tilde{f} : \mathbb{F}_{p^n} \to \mathbb{F}_p$ such that $\widehat{\chi_f}(w) = \chi_p(\tilde{f}(w))p^{\frac{n}{2}}$ for all $w \in \mathbb{F}_{p^n}$. In characteristic 2, all bent functions are regular bent. Moreover, when p is odd, regular bent functions can exist only if $p \equiv 1 \mod 4$.

- An equivalent characterization of bentness related to the notion of *difference sets* has also been introduced. Dillon [32] has in fact firstly studied bent functions in connection with difference sets.

Let G be a finite (abelian) group of order μ. A subset D of G of cardinality k is called (μ, k, λ)-difference set in G if every element $g \in G$, different from the identity, can be written as $d_1 - d_2$, $d_1, d_2 \in D$, in exactly λ different ways. It is observed in Rothaus' paper [67] and developed in Dillon's thesis [31] that a Boolean function $f : \mathbb{F}_2^n \mapsto \mathbb{F}_2$ is bent if and only if its support *supp(f)* is a difference set. It is known from Mann that the parameters of such difference set must then be $(\mu, k, \lambda) = (2^n, 2^{n-1} \pm 2^{\frac{n}{2}-1}, 2^{n-2} \pm 2^{\frac{n}{2}-1})$. Such a difference set is called *Hadamard difference set*. We have the following result due to Dillon [31] which proves an equivalence between bent functions and certain difference sets.

Theorem 4.3.1. *There exists a bent function $f : \mathbb{F}_{2^n} \to \mathbb{F}_2$ if and only if there exist a $(2^n, 2^{n-1} \pm 2^{\frac{n}{2}-1}, 2^{n-2} \pm 2^{\frac{n}{2}-1})$-difference set in $\mathbb{F}_{2^n}$.*

Example 4.3.2. Let f a bent Boolean function defined on $\mathbb{F}_2^4$ $(n = 4)$ by $f(x_1, x_2, x_3, x_4) = x_1x_4 + x_2x_3$. The support of f is

$$supp(f) = \{(1,0,0,1), (1,0,1,1), (1,1,0,1), (0,1,1,0), (0,1,1,1), (1,1,1,0)\}.$$

According to the table below, the support of f is a Hadamard $(16, 6, 2)$-difference set of $\mathbb{F}_2^4$.

Note that since "designs" and "differences sets" are closely related, bent functions can be directly characterized in terms of block designs (see below in Sect. 4.9).

- A *Hadamard matrix* of order n is an $n \times n$ matrix H in which every entry is ± 1 such that $HH^t = nI_n$ (i.e. has mutually orthogonal rows) where I_n denoted the identity matrix. Observe that we can multiply all the entries in any row (or column) of a Hadamard matrix by -1 and the result is again a Hadamard matrix. By a sequence of multiplications of this type, we can transform any Hadamard matrix into a Hadamard matrix in which every entry in the first row or column is a 1. Such a Hadamard matrix is called *standardized*.

d_1	d_2	$d_1 + d_2$
1001	1011	0010
1001	1101	0100
1001	0110	1111
1001	0111	1110
1001	1110	0111
1011	1101	0110
1011	0110	1101
1011	0111	1100
1011	1110	0101
1101	0110	1011
1101	0111	1010
1101	1110	0011
0110	0111	0001
0110	1110	1000
0111	1110	1001

Example 4.3.3. It is trivial to see that (1) and (-1) are both Hadamard matrices of order 1. The following matrices H_1 and H_2 are of order 2 and 4, respectively.

$$H_1 = \begin{pmatrix} 1 & 1 \\ 1 & -1 \end{pmatrix}$$

$$H_2 = \begin{pmatrix} 1 & 1 & 1 & 1 \\ 1 & -1 & 1 & -1 \\ 1 & 1 & -1 & -1 \\ 1 & -1 & -1 & 1 \end{pmatrix}$$

We can define the square $2^n \times 2^n$ matrix whose term at row indexed by $x \in \mathbb{F}_2^n$ and column indexed by $y \in \mathbb{F}_2^n$ equals $(-1)^{f(x+y)}$; then, f is bent if and only if this matrix is a Hadamard matrix; this is observed in Rothaus' paper, where is indicated that this was also noted by Lloyd Welch, and in Dillon's thesis as well. Below we state the main characterization of bent functions in terms of Hadamard matrices.

Theorem 4.3.4. *Let f be a Boolean function on $\mathbb{F}_2^n$ and $\chi_f = (-1)^f$. Define the matrix $H_f = (h_{x,y})$, where $h_{x,y} = \chi_f(x+y)$ for all $x, y \in \mathbb{F}_2^n$. Then f is a bent function if and only if H_f is a Hadamard matrix.*

Example 4.3.5. Let $n = 2$ and $f(x_1, x_2) = x_1x_2$, where $x_1, x_2 \in \mathbb{F}_2$. The truth table of f is $[0, 0, 0, 1]$ and the corresponding truth table of χ_f is $[1, 1, 1, -1]$. The matrix H_f is

$$H_f = \begin{pmatrix} 1 & 1 & 1 & -1 \\ 1 & 1 & -1 & 1 \\ 1 & -1 & 1 & 1 \\ -1 & 1 & 1 & 1 \end{pmatrix}$$

which is easily seen to bc a Hadamard matrix of order 4.

- The previous characterization is equivalent to the fact that the Hamming distance of a bent function to the set of all affine functions takes optimal value $2^{n-1} - 2^{\frac{n}{2}-1}$ (n even); this has a direct relationship with the fast correlation attack [63] on stream ciphers and the linear attack [59] on block ciphers.
- There exists another characterization of bentness in terms of the Walsh transform. More precisely, A Boolean function f on $\mathbb{F}_{2^n}$ is bent if and only if $\sum_{a\in\mathbb{F}_{2^n}} \widehat{\chi_f}^4(a) = 2^{3n}$ (which, by the Gauchy-Schwarz inequality and the Parseval relation, is the minimum possible value).
- Another equivalent characterization of bentness is related to the notion of *derivatives*. Let f be a Boolean function over $\mathbb{F}_{2^n}$ and $a \in \mathbb{F}_{2^n}$. The derivative of f with respect to a is defined as: $D_af(x) = f(x) + f(x+a), \forall x \in \mathbb{F}_{2^n}$. For $(a,b) \in \mathbb{F}_{2^n} \times \mathbb{F}_{2^n}$, the *second order derivative* of f with respect (a,b) is defined as: $D_bD_af(x) = f(x) + f(x+b) + f(x+a) + f(x+a+b), \forall x \in \mathbb{F}_{2^n}$. A bent function can be characterized in terms of its first derivatives. More precisely, f is bent if and only if all the derivatives D_af, $a \in \mathbb{F}_{2^n}^\star$, are balanced, i.e. their values are uniformly distributed.
- There exists also a characterization [23] of bent functions by second-order covering sequences due to Parseval identity: a Boolean function f on $\mathbb{F}_2^n$ is bent if and only if $\forall x \in \mathbb{F}_2^n, \sum_{a,b\in\mathbb{F}_2^n}(-1)^{D_aD_bf(x)} = 2^n$.
- Bent functions can also be characterized in terms of autocorrelations. The autocorrelation coefficient of a Boolean function f is an element $\lambda \in \mathbb{F}_{2^n}$ defined by $C_f(\lambda) = \sum_{x\in\mathbb{F}_{2^n}} (-1)^{f(x)+f(x+\lambda)}$. The *autocorrelation spectrum* of a Boolean function is the multiset $\{C_f(\lambda), \lambda \in \mathbb{F}_{2^n}\}$. For any bent function all the entries of the autocorrelation spectrum equal to zero. More precisely, f is bent if and only if $max_{\lambda\in\mathbb{F}_{2^n}^\star}|C_f(\lambda)| = 0$.
- Bernasconi, Codenotti and VanderKam [5] have proposed a characterization of bentness from a combinatorial approach related to *strongly regular* graphs. Let G_f be a graph Cayley: $G_f = G(\mathbb{F}_2^n, supp(f))$ of a Boolean function f. All vectors of length n re vertices of the graph. We say that there is an edge between two vertices x and y if the vector $x \oplus y$ belongs to *supp*(f). A regular graph G is said to be *strongly regular* (see also the definition below) if there exist two positives integers λ and μ such that for any vertices x, y the number of vertices incident to x and y both is equal to λ or μ and it depends on the presence or absence of the edge between x and y. Then, a Boolean function f is bent if and only if the graph Cayley G_f is strongly regular and $\lambda = \mu$.
- We mention that there exists also a characterization of bent functions through the Numerical Normal Form (NNF) (see [18, p. 100]).

- Hou and Langevin in [49] have provided a characterization of bentness. Denote by $\mathcal{A}_m$ the set of all Boolean functions g over $\mathbb{F}_2^m$ such that both g and its Fourier transformation normalized $\overline{\mathcal{F}}(g)$ are integer valued (recall that given a Boolean function g on $\mathbb{F}_2^m$, then the normalized Fourier transformation of g is given by $\overline{\mathcal{F}}(g)(x) := \frac{1}{2^{m/2}} \sum_{y \in \mathbb{F}_2^m} g(y)(-1)^{x \cdot y}$ where "$\cdot$" is a scalar product on $\mathbb{F}_2^m$). Moreover, for every $S \subset \mathbb{F}_2^m$, the characteristic function of S is denoted by 1_S. Then we have the following characterization of bent functions:

  A Boolean function f on $\mathbb{F}_2^{2m}$ is bent if and only if $f + 2^{m-1}1_{\{0\}} \in \mathcal{A}_{2m}$.

  The above characterization describes in fact bent functions as certain solutions of a system of quadratic equations. Interesting new properties of bent functions are obtained using the characterization (see [49]).
- Agievich explained in [1] that there is a one-to-one relationship between the set of bent functions and the so-called bent triangles which are special matrices with restrictions on rows and columns. He characterized the bentness of bent functions in terms of bent rectangles. Let us firstly introduce the bent rectangles. Let $v_1 = 0, v_2, \cdots, v_{2^n}$ be the lexicographically ordered vectors of $\mathbb{F}_2^n$. Let f be a Boolean function over $\mathbb{F}_2^n$. The spectral sequence is defined as the row vectors $\mathcal{F}(f) := (\mathcal{F}(f)(v_1), \cdots, \mathcal{F}(f)(v_{2^n}))$ (that is, Walsh–Hadamard coefficients). Now, a $2^m \times 2^k$ matrix with rows $a_{(i)}$ and transposed columns $a^{(j)}$ is called an (m, k) bent rectangle if each of the vectors $a_{(i)}$ and $2^{m-n}a^{(j)}$ is a spectral sequence. Let m and k be two positive integers with $m + k = 2n$. Let $\overline{\mathcal{B}}_{m,k}$ be the set of all bent rectangles. Then Agievich showed that there is a bijection between the sets of all the bent Boolean functions in $2n$ variables and $\overline{\mathcal{B}}_{m,k}$ from which he deduced the following characterization of bent functions: we denote by A_f the $2^m \times 2^k$-matrix with spectral vectors $(\mathcal{F}(f)(1), \cdots, \mathcal{F}(f)(2^m))$. Then f is bent over $\mathbb{F}_2^{2n}$ of and only if the matrix A_f is a bent rectangle. Unfortunately, until now the concept to represent a bent functions via bent rectangles was not really exploited in literature.

4.4 Enumeration of Bent Functions and Bounds

4.4.1 *An Overview on the State of the Art*

The bent functions are a small set of Boolean functions and they are very valuable in particular for cryptography. Bent functions are all known for $n \leq 8$ only (their determination for eight variables has been achieved only recently by Langevin and Leander [54][2] as well as their classification under the action of the general affine group. We give in Table 4.1 the number of n-variable bent functions for $2 \leq n \leq 8$.

For $n \geq 10$, the exact number of bent functions in n variables is still unknown. In fact, for $n \geq 10$, only classes of bent functions are known, which do not cover a large part of them, apparently. Determining all bent functions (or more practically,

[2]The number of 8-variable bent functions equals 99270589265934370305785861242880.

Table 4.1 Number of n-variable bent functions for $2 \leq n \leq 8$

n	2	4	6	8
# of bent functions	$8 = 2^3$	$896 = 2^{9.8}$	$5,425,430,528$	
$\approx$			$2^{32.3}$	$2^{106.3}$

classifying them under the action of the general affine group) seems elusive. As we will see below, some infinite classes of bent functions have been obtained, thanks to the identification between the vectorspace $\mathbb{F}_2^n$ and the Galois field $\mathbb{F}_{2^n}$.

Concerning bounds on the number of bent functions firstly, from a trivial upper bound follows the fact that the algebraic degree of a bent function cannot exceed $n/2$. This fact implies that the number of bent functions over $\mathbb{F}_2^n$ is at most $2^{1+\binom{n}{1}+\binom{n}{2}+\cdots\binom{n}{n/2}} = 2^{2^{n-1}+1/2\binom{n}{n/2}}$. In [22], Carlet and Klapper have improved this upper bound by proving that the number of bent functions over $\mathbb{F}_2^n$ (for $n \geq 6$) is upper bounded by $2^{2^{n-1}+1/2\binom{n}{n/2}-2^{n/2}+n/2+1}(1+\epsilon) + 2^{2^{n-1}-1/2\binom{n}{n/2}}$ where $\epsilon = 2^{1+\binom{n-1-d}{d-1}-\binom{n-1}{d-1}}$ with d positive less or equals n. Unfortunately, the previous upper bound is closed to the trivial one. A computational upper bound for the number of bent functions has been discussed by Wang and Zhang in [71] by introducing a bent matrix connected to a bent function. Based on its properties, the authors have obtained a computational upper bound that was claimed "the best possible" after considerations on examples. But there are no sufficient details in the paper to check it.

Concerning the lower bounds on the number of bent functions over $\mathbb{F}_2^n$, a first direct lower bound, following from the Maiorana–McFarland's construction (see Sect. 5.1 in Chap. 5), is $2^{2^{n/2}}2^{n/2}!$. In [2], Agievich has provided an algorithm to construct bent functions as well as lower bound which is probably the best known direct lower bound for the number of bent functions. However, it seems difficult to study the asymptotical behavior of such a lower bound, but for small values of n, Agievich's bound is greater than the one obtained from the Maiorana–McFarland's construction.

Finally, in spite of studies dating back from the 1960s, the cardinality of the set of bent functions is still unknown in dimension greater than 8. Lower bounds or upper bounds on this number have been established by some authors but there is quite a gap between the lower $2^{2(n/2)+log_2(n-2)-1}$ and the upper $2^{2^{n-1}+1/2\binom{n}{n/2}}$ bounds. Several approaches have been adopted but, for the most part, the bounds have been obtained by counting or estimating the solutions of linear systems. Finding an asymptotic value for the number of all bent functions has been a hard problem for a long period. Such a problem is closely connected to the problem of enumeration of Hadamard matrices which remained unsolved for more than 120 years.

Over the recent years, a novel strategy has been developed, mainly for coding-theoretic purposes, to obtain upper bounds for counting problems of some combinatorial objects. Roughly speaking, the approach is to identify constraints on the parameters of the problem and to deduce a relaxation of the counting problem in terms of a linear program or a semi-definite positive program. An additional and crucial step is to reduce the size of the optimization program by using its symmetries (group representation theory to be precise) and to get a program of manageable size. Applying this strategy to obtain numerical as well as theoretical bounds on the number of bent functions could be interesting.

4.4.2 A Conjecture Related to the Number of Bent Functions

A recent conjecture related to the number of bent functions and to a decomposition problem has been given by Tokareva in [69]. More precisely, she conjectured that all Boolean functions in n-variables of degree not more than $\frac{n}{2}$ can be represented as the sum of two bent functions in n variables with $n \geq 2$ even. It would be very surprising if the conjecture was true, but there is no proof of the contrary. This hypothesis seems to be an analog of Goldbach's conjecture in number theory unsolved since 1742: any even number $n > 4$ can be represented as the sum of two prime numbers. If one can prove the above hypothesis on bent functions then the asymptotic value of the number of all bent functions will be found. In [69] Tokareva checked the hypothesis for n-variable Boolean functions for small values of $n = 2, 4, 6$. Unfortunately, Tokareva could not check even the case $n = 8$ by explaining that there is no complete affine classification of Boolean functions of degree 4 in 8 variables. Nevertheless, she showed that any cubic Boolean function in eight variables is the sum of not more than four bent functions in eight variables. Very recently, L. Qu and C. Li [65] confirmed the hypothesis in some particular cases. Namely, they proved that all quadratic Boolean functions, Maiorana–McFarland bent functions and Partial Spread functions (see Sect. 5.1 in Chap. 5) can be represented as the sums of two bent functions. Further, in a report, Tokareva has presented and studied a weak version of her conjecture. It is proved that every n-variable Boolean function of degree d (where $d \leq \frac{n}{2}$), can be represented as the sum of a constant number of bent n-variable Boolean functions.

4.5 Classification and Equivalence

Despite their simple and natural definition, bent functions turned out to admit a very complicated structure in general. Currently, the general structure of bent functions on $\mathbb{F}_{2^n}$ is not yet clear. In particular a complete classification of bent functions looks hopeless. Bent functions which are characterized are very rare, they are a vanishingly small fraction of the total number of functions when the number of variables increases. It is stated in the literature that there is no formal method of constructing all bent functions.

Since we are not able to classify the set of bent functions, an important topic in research is then to find constructions of bent functions leading to infinite classes of bent functions, or to find directly such infinite classes, after computer investigations. As well explained in [19], when a class is obtained, it remains to see if at least some of its elements are really new. Indeed, given a bent function, some simple transformations allow to obtain other bent functions; we say then that the known function and the functions we can obtain from it are equivalent (when the correspondence results in an equivalence relation); an infinite class is new if some of its elements are inequivalent to all previously known bent functions. In the following, we describe the relevant notions of equivalence for bent functions. The automorphism group of the set of bent functions $\{\sigma$ permutation of $\mathbb{F}_2^n$ such that $f \circ \sigma$ is bent of all f bent $\}$ is the *general affine group*: $\sigma(x) = x \times A + a$ where A is an invertible matrix over $\mathbb{F}_2$. Two functions f and $f \circ \sigma$ are then said to be *affinely equivalent* (and $f(x)$ and $f(x \times A)$ are said to be linearly equivalent).

Moreover, if f is a bent function and ℓ is an affine function then $f \oplus \ell$ is bent. Two functions f and $f \circ \sigma + \ell$ are called AE-equivalent, and a class of bent functions is called *complete* if it is globally invariant under AE-equivalence. The completed version of a class is the set of all functions EA-equivalent to the functions in the class.

There exists another notion of equivalence called CCZ[3]-equivalence [7, 20] is more general than EA-equivalence for vectorial functions, but it has been shown in [6] that for Boolean functions and for bent (Boolean or vectorial) functions, CCZ equivalence coincides with EA-equivalence. Note that it is mentioned in [18] that X.-D. Hou and P. Langevin have showed that under some condition on a permutation σ, composing a bent function with σ may give another bent function. But since the condition on σ is hard to achieve, this cannot be viewed as an equivalence.

We mention that in 1998, Hou [48] has introduced the notion of a sequence of ranks $\{r_i(f), i = 1, 2, \cdots, m\}$ related to a given Boolean function f defined over $\mathbb{F}_2^m$ which are invariant under the action of the general linear group $GL(m, 2)$. He used that notion to show that if f is a cubic bent function in $2k$ variables, then when $r_s(f) < k$, f is either obtained from a cubic bent function in $2k-2$ variables, or is in a well-known family of bent functions. He also determined all cubic bent functions in eight variables which was possible since the classification of $\mathcal{RM}(3, 8)/\mathcal{RM}(2, 8)$ (where $\mathcal{RM}(s, m)$ denotes the sth-order Reed–Muller code of length 2^m) was known thanks to [47].

4.6 Geometric Properties of Bent Functions and Their Representation Over the Integers

From the Poisson formula, Dillon has deduced that the intersection between the support D of an n-variable bent function and a k-dimensional subspace E of $\mathbb{F}_2^n$ where $k \geq m$ is in between $2^{k-1} - 2^{m-1}$ and $2^{k-1} + 2^{m-1}$, that D can then contain E only if $k = m$, that if D contains E (resp. is disjoint from E), then D has balanced

[3] "CCZ" stands for the three authors "Carlet-Charpin-Zinoviev" of the paper [20].

intersection with any proper coset of E and $D \setminus E$ (resp. $D \cup E$) is also a difference set. This study has been later extended in [14] to vectorspaces of any dimensions.

It is proved in [45] that bent functions are those functions $f(x+b)$, $b \in \mathbb{F}_2^n$, where f belongs to the so-called GPS class, introduced in [16], as the set of those Boolean functions satisfying: $f(x) = \sum_{i=1}^{k} m_i 1_{E_i}(x) - 2^{n/2-1}\delta_0(x)$, where $E_1, \ldots, E_k$ are m-dimensional linear subspaces of $\mathbb{F}_2^n$ (with no constraint on the number k), 1_{E_i} is the indicator of E_i, and $m_1, \ldots, m_k$ are integers (positive or negative). The dual of such f equals $\tilde{f}(x) = \sum_{i=1}^{k} m_i 1_{E_i^\perp}(x) - 2^{n/2-1}\delta_0(x)$.

Recall that it is possible to represent Boolean functions as multi-variate polynomials in $\mathbb{Z}[x_1, \cdots, x_n]/(x_1^2 - x_1, \cdots, x_n^2 - x_n)$. Such a representation is called the *Numerical Normal Form* (NNF). The NNF gives a convenient characterization of bent functions: let $f(x) = \sum_{I \in \mathcal{P}(N)} \lambda_I x^I$ be the NNF of a Boolean function f on $\mathbb{F}_2^n$. Then f is bent if and only if:

1. for every I such that $n/2 < |I| < n$, the coefficient λ_I is divisible by $2^{|I|-n/2}$;
2. λ_N (with $N = \{1, \cdots, n\}$) is congruent to $2^{n/2-1}$ modulo $2^{n/2}$.

4.7 Decompositions of Bent Functions

In [11], it is shown that the restrictions of bent functions to subspaces of codimension 1 and 2 are highly nonlinear. The following theorem is proved in [11].

Theorem 4.7.1 ([11]). *Let n be an even integer, $n \geq 4$, and let f be an n-variable Boolean function. Then the following properties are equivalent.*

1. *f is bent.*
2. *For every (resp. for some) linear hyperplane E of $\mathbb{F}_2^n$, the Walsh transforms of the restrictions h_1, h_2 of f to E and to its complement (viewed as Boolean functions on $\mathbb{F}_2^{n-1}$) take values $\pm 2^{n/2}$ and 0 only, and the disjoint union of their supports equals the whole space $\mathbb{F}_2^{n-1}$.*

Hence, by taking the restriction of a bent function to an affine hyperplane we simply obtain a plateaued function (see the definition in Chap. 16) in an odd number of variables and with optimal nonlinearity. Note that we have also (see [11]) that, if a function in an odd number of variables is such that, for some nonzero $a \in \mathbb{F}_2^n$, every derivative $D_u f$, $u \neq 0$, $u \in a^\perp$, is balanced, then its restriction to the linear hyperplane $a^\perp$ or to its complement is bent.

It is also proved in [11] that the Walsh transforms of the four restrictions of a bent function to an $(n-2)$-dimensional vector subspace E of $\mathbb{F}_2^n$ and to its cosets have the same sets of magnitudes. It is a simple matter to see that, denoting by a and b two vectors such that $E^\perp$ is the linear space spanned by a and b, these four restrictions are bent if and only if $D_a D_b \tilde{f}$ takes on constant value 1.

In [25], Charpin has presented an extensive study of the restrictions of bent functions to affine subspaces. She proposed several methods which are mainly based on properties of the derivatives and of the dual of a given bent function. She especially described the connection, for a bent function, between the Fourier spectra of its restrictions and the decompositions of its dual. Most notably, Charpin showed that the Fourier spectra of the restrictions of a bent function to the subspaces of codimension 2 can be explicitly derived from the Hamming weights of the second derivatives of the dual function. Moreover, Chapin provided some infinite classes of bent functions which cannot be decomposed into four bent functions. The study in [25] points out that the bent functions may differ on the properties of their 4-decompositions. For instance, any bent function whose dual has degree 3 admits a decomposition into four bent functions, whereas both families Maiorana–McFarland and $\mathcal{PS}^-$ (see Chap. 5, Sect. 5.1) contain some bent functions which do not satisfy this property. In this context, it appears that the structure of a bent function highly depends on some properties of its dual, such as its degree and the Hamming weights of its second derivatives. Moreover, we have proved that the bent functions whose duals have a constant second derivative present some specificities. From any such bent function f, it is possible to derive some other bent functions. The reader can find more details on decomposing bent functions in [11, 12, 25].

4.8 Bent Functions and Normality

While considering constructions of balanced Boolean functions with high nonlinearity, Dobbertin [43], pointed out that all known bent functions, till then, were constant (or affine) over a flat of dimension $\frac{n}{2}$ (i.e., a subspace of dimension $\frac{n}{2}$ or a coset of such a subspace). This led to the notion of *normality* which was generalized to k-normality by Charpin [25]. A function is called *k-weakly-normal* (resp. *k-normal*) if its restriction to some k-dimensional flat is affine (resp. constant). A bent function is said to be *normal* if it is $\frac{n}{2}$-normal. Carlet has proved in [17] that, for every $\alpha > 1$, when n tends to infinity, random Boolean functions are almost surely $[\alpha \log_2 n]$-non-normal. This means that almost all Boolean functions have high complexity with respect to this criterion. The proof of existence of non-normal functions does not give unfortunately examples of such functions. Moreover in [15] Carlet observed that if a bent function f is normal (resp. weakly-normal), that is, constant (resp. affine) on an $n/2$-dimensional flat $b + E$ (where E is a subspace of $\mathbb{F}_2^n$), then its dual $\tilde{f}$ is such that $\tilde{f}(u) \oplus b \cdot u$ is constant on $E^\perp$ (resp. on $a + E^\perp$, where a is a vector such that $f(x) \oplus a \cdot x$ is constant on E). Thus, $\tilde{f}$ is weakly-normal (in this case, f (resp. $f(x) \oplus a \cdot x$) is balanced on each of the other cosets of the flat). Dobbertin used this idea to construct balanced functions with high nonlinearities from normal bent functions.

The question whether there are non-normal (and even non-weakly-normal) bent functions (that is, bent functions which are non-constant (resp. non-affine) on every $n/2$-dimensional flat) was answered in the affirmative by Canteaut, Daum, Dobbertin and Leander [13] almost a decade later, by demonstrating the existence of non-normal bent functions on 10 and 14 variables. More precisely, Dillon and

Dobbertin [34] have proved that the so-called *Kasami function* defined over $\mathbb{F}_{2^n}$ by $f(x) = \mathrm{Tr}_1^n\left(ax^{2^{2k}-2^k+1}\right)$, with $gcd(k, n) = 1$, is bent if n is not divisible by 3 and if $a \in \mathbb{F}_{2^n}$ is not a cube. Canteaut et al. [13] have showed that if $a \in \mathbb{F}_4 \setminus \mathbb{F}_2$ and $k = 3$, then for $n = 10$, the function $f(x) \oplus \mathrm{Tr}_1^n(b)$ is non-normal for some b, and for $n = 14$, the function f is not weakly normal. On the other hand, it has been also shown by Charpin [25] that cubic bent functions on eight variables are all normal. More recently, Leander and McGuire [55] have identified for the first time non-weakly- normal Boolean functions on 10 and 12 variables. A greedy algorithm for checking normality of cryptographic Boolean functions has been provided recently in [53].

The direct sum of two normal functions is obviously a normal function, while the direct sum of two non-normal functions can be normal. The question on the sum of a normal bent function and of a non-normal bent function has been posed. This question gave rise to more general notion of normality: the so-called normal extension introduced by Carlet, Dobbertin and Leander in [8]. Given two vectorspaces U and V over $\mathbb{F}_2$ such that $U \subseteq V$ and two bent functions $\beta : U \rightarrow \mathbb{F}_2$ and $f : V \rightarrow \mathbb{F}_2$, then f is said to a *normal extension* of β, in symbols $\beta \preceq f$, if there is a direct decomposition $V = U \oplus W_1 \oplus W_2$ such that $\beta(u) = f(u + w_1)$ for all $u \in U$, $w_1 \in W_1$ and $\dim W_1 = \dim W_2$. The authors have applied the concept of normal extension to characterize when the direct sum of bent functions is normal, and proved that the direct sum of a normal bent function and a nonnormal bent function is always nonnormal. The relation $\preceq$ is transitive and if $\beta \preceq f$ then the same relation exists between the duals: $\tilde{\beta} \preceq \tilde{f}$. Furthermore, a bent function is normal if and only if $\epsilon \preceq f$, where $\epsilon \in \mathbb{F}_2$ is viewed as a Boolean functions over the vectorspace $\mathbb{F}_2^0 = \{0\}$. In [8], the authors have provided examples of normal extensions and clarified the question on the sum of a normal bent function and of a non-normal bent function. It has also been shown that normal extension leads to a secondary construction of bent functions (see [8]).

Very recently, the notion of normality has been extended to p-ary functions. The normality of bent functions in odd characteristic has been analyzed in [24]. It turns out that in contrast to Boolean bent functions, many quadratic bent-functions in odd characteristic and even dimension are not normal. It was also shown that regular Coulter-Matthews bent functions are normal.

4.9 Bent Functions: Applications

Bent functions have been extensively studied for their applications in cryptography, but have also been applied to spread spectrum[4] (it was discovered in early 1982 that maximum-length sequences based on bent functions have cross-correlation

[4]In telecommunications and radio communication, spread-spectrum techniques are methods by which a signal (e.g. an electrical, electromagnetic, or acoustic signal) generated in a particular bandwidth is deliberately spread in the frequency domain, resulting in a signal with a wider bandwidth. These techniques are used for a variety of reasons, including the establishment of

and autocorrelation properties rivalling those of the Gold codes and Kasami codes for use in CDMA; these sequences have several applications in spread spectrum techniques), coding theory, and combinatorial design.[5] The definition can be extended in several ways, leading to different classes of generalized bent functions that share many of the useful properties of the original. In the following, we present some applications of bent functions.

4.9.1 Bent Functions in Coding Theory

For every $0 \leq r \leq n$, the *Reed–Muller code* $\mathcal{RM}(r,n)$ of order r, is a linear code of length 2^n, dimension $\sum_{i=0}^{r}\binom{n}{i}$ and minimum distance 2^{n-r}. The Reed–Muller code can be defined in terms of Boolean functions or as extended cyclic code. In terms of Boolean functions, $\mathcal{RM}(r,n)$ is the set of all n-variable Boolean functions of algebraic degrees at most r. More precisely, it is the linear code of all binary words of length 2^n corresponding to the last columns of the truth-tables of these functions. The Reed–Muller codes are nested : $\mathcal{RM}(1,n) \subset \mathcal{RM}(2,n) \subset \cdots \subset \mathcal{RM}(n-1,n)$. The Reed–Muller code $\mathcal{RM}(r,n)$ can be viewed as an extended cyclic code for every $r < n$: the zeroes of the corresponding cyclic code ($RM^\star(r,n)$, the *punctured Reed–Muller code* of order r) are the elements α^j (where α is a primitive element of $\mathbb{F}_{2^n}$) such that $1 \leq j \leq 2^n-2$ and $1 \leq w_2(j) \leq n-r-1$, where $w_2(j)$ is the number of ones in the binary expansion of j. Recall that given two Boolean functions f and g of $\mathcal{B}_n$, the *Hamming distance* $d_H(f,g)$ between f and g equals the size of the set $\{x \in \mathbb{F}_2^n \,|\, f(x) \neq g(x)\}$. Moreover, recall that $nl_r(f)$ denote the minimum Hamming distance between a given Boolean function f and all Boolean functions g of degrees at most r (that is, $g \in \mathcal{RM}(r,n)$). The *covering radius* of $\mathcal{RM}(r,n)$ denoted by $\rho(r,n)$ plays an important role in error correcting codes. An excellent reference on this topic is the famous book of Cohen, Honkala, Litsyn and Lobstein [28]). The covering radius $\rho(r,n)$ is defined as the maximum value of $nl_r(f)$ when f ranges over the set $\mathcal{B}_n$ of Boolean functions in n variables, that is,

$$\rho(r,n) = \max_{f\in\mathcal{B}_n} \min_{g\in\mathcal{RM}(r,n)} d_H(f,g).$$

The covering radius $\rho(1,n)$ of the first-order Reed–Muller codes $\mathcal{RM}(1,n)$ coincides with the maximum nonlinearity $nl_1(f)$ (that we denoted simply by $nl(f)$) of n-variable Boolean functions f, that is, the maximum distance from all affine

secure communications, increasing resistance to natural interference, noise and jamming, to prevent detection, and to limit power flux density (e.g. in satellite downlinks).

[5]Combinatorial design theory is the part of combinatorial mathematics that deals with the existence and construction of systems of finite sets whose intersections have specified numerical properties.

functions. When n is even, it is known that $\rho(1,n) = 2^{n-1} - 2^{\frac{n}{2}-1}$ and the associated n-variable Boolean functions are the *bent functions*.

On the other hand, special sets of quadratic bent functions allow to construct Kerdock codes [52] (see also [58]) that are optimal and have large code distances that grow with the code lengths. This very optimality of Kerdock codes is due to extremal properties of bent functions. For every even n, the Kerdock code $\mathcal{K}_n$ is a supercode of $R(1,n)$ (*i.e.* contains $R(1,n)$ as a subset) and is a subcode of $R(2,n)$. More precisely $\mathcal{K}_n$ is a union of cosets $f_u \oplus R(1,n)$ of $R(1,n)$, where the functions f_u are quadratic (one of them is null and all the others have algebraic degree 2). The difference $f_u \oplus f_v$ between two distinct functions f_u and f_v being bent, $\mathcal{K}_n$ has minimum distance $2^{n-1} - 2^{n/2-1}$ (n even), which is the best possible minimum distance for a code equal to a union of cosets of $R(1,n)$, according to the covering radius bound. The size of $\mathcal{K}_n$ equals 2^{2n}. This is the best possible size for such minimum distance (see [30]).

Bent functions play a role even in very practical issues through the so-called *robust error detecting codes* introduced by Karpovsky, Kulikowski and Wang. In [50] the authors have emphasized that bent functions play a role in memories with self error detection, transmission and storage of multimedia data.

An interesting survey on bent functions with the emphasis on connections with coding theory and related was given by J. Wolfmann [75] in 1999. In 2015, Xiang, Ding and the author [73] have used bent functions to construct optimal codebooks from binary codes meeting Levenshtein bounds. A codebook $\mathcal{B}$ is a collection of unit norm vectors in a finite dimensional vectorspace. In applications of codebooks such as code-division multiple-access (CDMA), those vectors in a codebook should have a small maximum magnitude of inner products, denoted by $I_{max}(\mathcal{B})$, between any pair of distinct code vectors. Some lower bounds on $I_{max}(\mathcal{B})$ exist in literature namely, the famous Welch bound and Levenshtein bounds. Codebooks meeting the Welch bound or the Levenshtein bounds are much preferred in many practical applications, for example, unitary space-time modulations, multiple description coding over erasure channels, CDMA systems, and coding theory. In the following, we give some details about the use of bent functions in the construction of codebooks meeting the Levenshtein bounds. Let $\mathcal{B} = \{\mathbf{b}_0, \ldots, \mathbf{b}_{N-1}\}$, where each $\mathbf{b}_\ell$ is a unit norm $1 \times n$ complex vector over an alphabet A. Such a set $\mathcal{B}$ is called an (N,n)-*codebook* (also called signal set). The size of A is called the alphabet size of $\mathcal{B}$. As a performance measure of a codebook in practical applications, the maximum crosscorrelation amplitude of an (N,n) codebook $\mathcal{B}$ is defined by

$$I_{\max}(\mathcal{B}) = \max_{0 \le i < j \le N-1} \left|\mathbf{b}_i \mathbf{b}_j^H\right|$$

where $\mathbf{b}^H$ stands for the conjugate transpose of the complex vector $\mathbf{b}$.

In the case when N is large, we have the following Levenshtein bounds.

Proposition 4.9.1 ([51,56]). *For any real-valued codebook $\mathcal{B}$ with $N > n(n+1)/2$, we have*

$$I_{\max}(\mathcal{B}) \geq \sqrt{\frac{3N - n^2 - 2n}{(n+2)(N-n)}}. \tag{4.1}$$

For any complex-valued codebook $\mathcal{B}$ with $N > n^2$, we have

$$I_{\max}(\mathcal{B}) \geq \sqrt{\frac{2N - n^2 - n}{(n+1)(N-n)}}. \tag{4.2}$$

In [73], the authors introduced a generic construction of codebooks based on binary codes. Doing this, a few previous constructions of optimal codebooks were extended, and a new class of codebooks almost meeting the Levenshtein bounds was presented. To this end, the authors used the fact that we are able to construct a set $\{f_a(x)\}$ of $2^{m-1}-1$ bent functions on $\mathbb{F}_{2^m}$ such that the difference $f_a(x)-f_b(x)$ of any two distinct bent functions f_a and f_b in the set $\{f_a(x)\}$ is again a bent function. In fact, we have only few examples of constructions of such a set (most of the known bent functions do not satisfy this requirement). We state the construction of such sets as an open problem.

Also, it has been shown in several recent papers [35, 64, 68, 78] that bent functions lead to the construction of interesting linear codes with few weights. Linear codes with few weights have applications in secret sharing [3, 21, 40, 41, 74] authentication codes [38], association schemes [9], and strongly regular graphs [10]. A non-exhaustive list dealing with codes with few weights is [27, 29, 36, 37, 39–41, 44, 57, 68, 72, 76, 77].

4.9.2 Bent Functions in Cryptography

The properties of bent functions are naturally of interest in modern digital cryptography. By 1988 Forré recognized that the Walsh transform of a function can be used to show that it satisfies the Strict Avalanche Criterion (SAC) and higher-order generalizations, and recommended this tool to select candidates for good S-boxes achieving near-perfect diffusion. Indeed, functions satisfying the SAC to the highest possible order are always bent. Moreover, the bent functions are as far as possible from having what are called linear structures, *i.e.* nonzero vectors a such that $f(x+a)+f(x)$ is a constant. In the language of differential cryptanalysis (introduced after this property was discovered) the derivative of a bent function f at every nonzero point a (that is, $D_a(x) = f(x+a)+f(x)$) is a balanced Boolean function, taking on each value exactly half of the time. This property is called *perfect nonlinearity*. Given such good diffusion properties, apparently perfect resistance to differential cryptanalysis, and resistance by definition to linear cryptanalysis, bent functions might at first seem the ideal choice for secure cryptographic functions such as S-boxes. Their fatal flaw is that they fail to be balanced. In particular, an invertible S-box cannot be constructed directly from bent functions. Instead, one might start with a bent function and randomly complement appropriate values until the result is balanced. The modified function still has high nonlinearity, and as such functions are very rare the process should be much faster than a brute-force search. But functions produced in this way may lose other desirable properties, even failing to satisfy the SAC -so careful testing is necessary. A number of cryptographers have worked on techniques for generating balanced functions that preserve as many of the good cryptographic qualities of bent functions as possible. Some of this theoretical research has been incorporated into real cryptographic algorithms. The CAST design procedure, used by Carlisle Adams and Stafford Tavares to construct the S-boxes for the block ciphers CAST-128 and CAST-256, makes use of bent functions. The cryptographic hash function HAVAL uses Boolean functions built from representatives of all four of the equivalence classes of bent functions on six variables. The stream cipher Grain uses an NLFSR whose nonlinear feedback polynomial is, by design, the sum of a bent function and a linear function. It is important to note that a stream cipher using a bent function is vulnerable to correlation attacks in the combiner model and is also vulnerable to fast algebraic attacks and to Rønjom-Helleseth's attack [66] for the two models (the filter model and the combiner model).

Also, note that bent functions have been used in the stream cipher Grain. Such a stream cipher was designed by M. Hell, T. Johansson and W. Meier in [46] primarily for restricted hardware environments. The cipher consists of a nonlinear feedback shift register (NFSR) and a linear feedback shift register (LFSR) both of 80 bits in size. The nonlinear feedback polynomial of the NFSR, $f(x)$, was constructed as the sum of a linear and a bent function. Such a combination was suggested by the authors of [46] in order to provide high resiliency and nonlinearity.

4.9.2.1 Weakness of the Cryptographic Bent Functions

Bent functions being not balanced (*i.e.* their values being not uniformly distributed, see above), they are improper for use in cryptosystems. For this reason, even when they exist (for n even), it is also necessary to study those functions which have large but not optimal nonlinearities, say between $2^{n-1} - 2^{\frac{n-1}{2}}$ and $2^{n-1} - 2^{n/2-1}$, among which some balanced functions exist. In [42], Dobbertin has described a general explicit construction of bent functions, which unifies well known constructions due to Maiorana–McFarland and Dillon as two opposite extremal cases. In this framework Dobbertin has showed how the constructed bent functions can be modified in order to obtain highly nonlinear balanced Boolean functions. So, bent functions cannot be directly used in symmetric cryptography but should be modified into highly nonlinear balanced Boolean functions.

To conclude, from a cryptographic viewpoint, bent functions have two main interests:

1. Their *derivatives* $D_a f : x \mapsto f(x) + f(x + a)$ are balanced, therefore any addition of a nonzero vector to the input to f induces 2^{n-1} changes among the 2^n outputs; this has an important relationship with the differential attack on block ciphers, which was already known at the NSA in the seventies.
2. The Hamming distance between f and the set of affine Boolean functions takes optimal value $2^{n-1} - 2^{\frac{n}{2}-1}$ (n even); this has a direct relationship with the fast correlation attack on stream ciphers and the linear attack on block ciphers.

Nevertheless, bent functions have two main drawbacks:

1. Bent functions are not balanced and then can hardly be used for instance in stream ciphers.
2. A pseudo-random generator using a bent function as combiner or filter is weak against some attacks, like the fast algebraic attack, even if the bent function has been modified to make it balanced, as Dobbertin described.

4.9.3 Bent Functions and Their Connections to Combinatorics

- Bent functions and graphs: bent functions are closely connected to the strongly regular graphs. Such graphs form an important class of graphs which lie somewhere between the highly structured and the apparently random. A graph is called *strongly regular* with parameters (n, k, λ, μ) is a graph on n vertices which is regular with valency k and has the following properties (1) any two adjacent vertices have exactly λ common neighbours and (2) any two nonadjacent vertices have exactly μ common neighbours. The complete and null graphs are vacuously strongly regular, the parameters μ and λ respectively being undefined for them. Often these trivial cases are excluded. The four parameters are not independent. In fact, The parameters (n, k, λ, μ) of a strongly regular graph satisfy the equation $k(k - \lambda - 1) = (n - k - 1)\mu$. The parameters also satisfy various algebraic conditions and inequalities. In [5] Bernasconi et al. have proved that bent functions can be precisely characterized in terms of a special class of strongly regular graphs.
- Bent functions and block designs: in combinatorial mathematics, a *block design* is a set together with a family of subsets (repeated subsets are allowed at times) whose elements are chosen to satisfy some set of properties that are deemed useful for a particular application. These applications come from many areas, including finite geometry, cryptography). More precisely, a (block) (μ, k, λ)-design or a block design with parameters (μ, k, λ) (the parameters are not all independent) is a system of k-element subsets (usually called "blocks") of a μ-element set such that any pair of distinct elements is contained in exactly λ blocks. A symmetric design is a block design with the same number of

blocks as elements. An example of a symmetric block design is a projective plane. The incidence matrix $M = (a_{i,j})$ of the symmetric design is a $\mu \times \mu$ binary matrix whose rows and columns are indexed by blocks and elements, respectively, and whose entry $a_{i,j}$ is 1 if element j belongs to the ith block and is 0 otherwise. Now, recall that a Boolean function over $\mathbb{F}_2^n$ is bent if and only if the set $D = \{(x,f(x)) \mid x \in \mathbb{F}_2^n\}$ is a Hadamard difference set with parameters $(2^{n+1}, 2^n, 2^{n-1})$. The latter characterization can be translated in terms of block designs as follows: a Boolean function over $\mathbb{F}_2^n$ is bent if and only if the system of sets $\{D \oplus \alpha \mid \alpha \in \mathbb{F}_2^{n+1}\}$ is a symmetric $(2^{n+1}, 2^n, 2^{n-1})$-design.

References

1. S. Agievich. On the representation of bent functions by bent rectangles. In *Proceedings of the fifth international Petrozavodsk conference on probabilistic methods in discrete mathematics (Petrozavodsk, Russia, June 1-6, 2000). Available at: URL*: http://arxiv.org/abs/math/0502087, 2000.
2. S. Agievich. On the representation of bent functions by bent rectangles. In *Probabilistic Methods in Discrete Mathematics: Proceedings of the Fifth International Petrozavodsk Conference (Petrozavodsk, 2000). Utrecht, Boston: VSP, pages 121–135*, 2002.
3. R. Anderson, C. Ding, T. Helleseth, and T. Kløve. How to build robust shared control systems. In *Designs, Codes Cryptography, vol. 15, No. 2*, pages 111–124, 1998.
4. J. Arndt. *Matters Computational: Ideas, Algorithms, Source Code*. Springer, 2010.
5. A. Bernasconi, B. Codenotti, and J.M. VanderKam. Characterization of bent functions in terms of strongly regular graphs. In *IEEE Trans. Computers, Vol, 50. No. 9. pages 984–985*, 2001.
6. L. Budaghyan and C. Carlet. CCZ-equivalence of single and multi output Boolean functions. In *AMS Contemporary Math. 518, Post-proceedings of the conference Fq 9*, pages 43–54, 2010.
7. L. Budaghyan, C. Carlet, and A. Pott. New classes of almost bent and almost perfect nonlinear functions. In *IEEE Trans. Inf. Theory 52 (3), pages 1141–1152*, 2006.
8. H. Dobbertin C. Carlet and G. Leander. Normal extensions of bent functions. In *EEE Transactions on Information Theory, vol. 50, no. 11*, pages 2880–2885, 2004.
9. A. R. Calderbank and J. M. Goethals. Three-weight codes and association schemes. In *Philips J. Res., vol. 39*, pages 143–152, 1984.
10. A. R. Calderbank and W. M. Kantor. The geometry of two-weight codes. In *Bull. London Math. Soc., vol. 18*, pages 97–122, 1986.
11. A. Canteaut, C. Carlet, P. Charpin, and C. Fontaine. On cryptographic properties of the cosets of R(1,m). In *IEEE Transactions on Information Theory, vol. 47*, pages 1494–1513, 2001.
12. A. Canteaut and P. Charpin. Decomposing bent functions. In *IEEE Transactions on Information Theory, Vol 49*, pages 2004–2019, 2003.
13. A. Canteaut, M. Daum, H. Dobbertin, and G. Leander. Normal and Non-Normal Bent Functions. In *Proceedings of the Workshop on Coding and Cryptography, pages 91–100*, 2003.
14. C. Carlet. Two new classes of bent functions. In *Proceedings of EUROCRYPT'93, Lecture Notes in Computer Science 765*, pages 77–101, 1994.
15. C. Carlet. Two new classes of bent functions. In *Proceedings of EUROCRYPT'93, Lecture Notes in Computer Science, 765*, pages 77–101, 1994.
16. C. Carlet. Generalized partial spreads. In *IEEE Trans. Inform. Theory, vol. 41, no. 5*, pages 1482–1487, 1995.
17. C. Carlet. On the degree, nonlinearity, algebraic thickness and non-normality of Boolean functions, with developments on symmetric functions. In *IEEE Transactions on Information Theory, vol. 50, pages 2178–2185*, 2004.

18. C. Carlet. Boolean functions for cryptography and error correcting codes. In Yves Crama and Peter L. Hammer, editors, *Boolean Models and Methods in Mathematics, Computer Science, and Engineering*, pages 257–397. Cambridge University Press, June 2010.
19. C. Carlet. Open problems on binary bent functions. In *Proceeding of the conference "Open problems in mathematical and computational sciences", Springer, pages 203–241*, 2014.
20. C. Carlet, P. Charpin, , and V. Zinoviev. Codes, bent functions and permutations suitable for DES-like cryptosystems. In *Designs, Codes and Cryptography, 15(2), pp. 125–156*, 1998.
21. C. Carlet, C. Ding, and J. Yuan. Linear codes from perfect nonlinear mappings and their secret sharing schemes. In *IEEE Trans. Inform. Theory, vol. 51, No. 6*, pages 2089–2102, 2005.
22. C. Carlet and A. Klapper. Upper bounds on the numbers of resilient functions and of bent functions. In *23rd Symposium on Information Theory in the Benelux, Louvain-La-Neuve, Belgique, Mays*, 2002.
23. C. Carlet and E. Prouff. On plateaued functions and their constructions. In *Proceedings of Fast Software Encryption 2003, Lecture notes in computer science 2887, pages 54–73*, 2003.
24. A. Cesmelioglu, W. Meidl, and A. Pott. Generalized Maiorana–McFarland Class and normality of p-ary bent functions. In *Finite Fields and Their Applications, Vol. 24, pages 105–117*, 2013.
25. P. Charpin. Normal Boolean functions. In *Special Issue "Complexity Issues in Coding and Cryptography", dedicated to Prof. Harald Niederreiter on the occasion of his 60th birthday, Journal of Complexity, 20, pages 245–265*, 2004.
26. P.J. Chase, J.F. Dillon, and K.D. Lerche. Bent functions and difference sets. In *R41 Technical Paper, April*, 1971.
27. S.-T. Choi, J.-Y. Kim, J.-S. No, and H. Chung. Weight distribution of some cyclic codes. In *IEEE Int. Symp. Inf. Theory*, pages 2901–2903, 2012.
28. G. Cohen, I. Honkala, S. Litsyn, and A. Lobstein. Covering codes. In *North Holland*, 1997.
29. B. Courteau and J. Wolfmann. On triple-sum-sets and two or three weights codes. In *Discrete Math. vol. 50*, pages 179–191, 1984.
30. P. Delsarte. An algebraic approach to the association schemes of coding theory. In *PhD thesis. Université Catholique de Louvain*, 1973.
31. J. Dillon. Elementary Hadamard difference sets. In *PhD dissertation, University of Maryland.*
32. J. Dillon. A survey of bent functions. In *NSA Technical Journal Special Issue*, pages 191–215, 1972.
33. J. F. Dillon. Elementary Hadamard Difference sets. In *Proceedings of the Sixth Southeastern Conference on Combinatorics, Graph Theory and Computing, F. Hoffman et al., Eds., Congressus Numerantium XIV, Winnipeg Utilitas Math, pages 237–249*, 1975.
34. J. F. Dillon and H. Dobbertin. New cyclic difference sets with Singer parameters. In *Finite Fields and Their Applications Volume 10, Issue 3*, pages 342–389, 2004.
35. C. Ding. Linear codes from some 2-Designs. In *IEEE Transactions on Information Theory 61(6)*, pages 3265–3275, 2015.
36. C. Ding, C. Li, N. Li, and Z. Zhou. Three-weight cyclic codes and their weight distributions. In *Preprint*.
37. C. Ding, J. Luo, and H. Niederreiter. Two weight codes punctured from irreducible cyclic codes. In *Proc. 1st Int. Workshop Coding Theory Cryptography, Y. Li, S. Ling, H. Niederreiter, H. Wang, C. Xing, and S. Zhang, Eds., Singapore*, pages 119–124, 2008.
38. C. Ding and X. Wang. A coding theory construction of new systematic authentication codes. In *Theoretical Comput. Sci. vol. 330, no. 1*, pages 81–99, 2005.
39. K. Ding and C. Ding. Binary linear codes with three Weights. In *IEEE Communications Letters 18(11)*, pages 1879–1882, 2014.
40. K. Ding and C. Ding. A Class of two-weight and three-weight codes and their applications in secret sharing. In *IEEE Transactions on Information Theory 61(11)*, pages 5835–5842, 2015.
41. K. Ding and C. Ding. A Class of two-weight and three-weight codes and their applications in secret sharing. In *CoRR abs/1503.06512*, 2015.
42. H. Dobbertin. Construction of bent functions and balanced Boolean functions with high nonlinearity. In *Proceedings of Fast Software Encryption, Second International Workshop, Lecture Notes in computer Science 1008*, pages 61–74, 1995.

43. H. Dobbertin, T. Helleseth, P. V. Kumar, and H. M. Martinsen. Ternary m-sequences with three-valued cross-correlation function: new decimations of Welch and Niho type. In *IEEE Transactions on Information Theory, 47(4), pages 1473–1481*, 2001.
44. K. Feng and J. Luo. Value distribution of exponential sums from perfect nonlinear functions and their applications. In *IEEE Trans. Inf. Theory, vol. 53, no. 9*, pages 3035–3041, 2007.
45. P. Guillot. Completed GPS covers all bent functions. In *Journal of Combinatorial Theory, Series A 93, pages 242–260*, 2001.
46. M. Hell, T. Johansson, and W. Meier. In *A stream cipher proposal: Grain-128. eSTREAM, ECRYPT Stream Cipher Project, 2006* http://www.ecrypt.eu.org/stream/grainpf.html.
47. X. D. Hou. $GL(m, 2) acting on R(r, m)/R(r - l, m)$. In *Discrete Mathematics, Vol 149, pages 99–122*, 1996.
48. X. D. Hou. Cubic bent functions. In *Discrete Mathematics, Vol 189, pages 149–161*, 1998.
49. X. D. Hou and P. Langevin. Results on bent functions. In *J. Comb. Theory, Series A. Vol 80, pages 232–246*, 1997.
50. M. G. Karpovsky, K. J. Kulikowski and Z. Wang.: On-line Self Error Detection with Equal Protection Against All Errors. *Int. J. High. Reliab. Electron. Syst. Des.*, (2008).
51. G. A. Kabatyanskii and V. I. Levenshtein. Bounds for packing on a sphere and in space. In *Probl. Inf. Transmission, Vol. 14, pages 1–17*, 1978.
52. A.M. Kerdock. A class of low-rate non-linear binary codes. In *Inform. Control. Vol. 20, No. 2., pages 182–187*, 1972.
53. N. Kolokotronis and K. Limniotis. A greedy algorithm for checking normality of cryptographic Boolean functions. In *Proceedings ISITA 2012, Honolulu, Hawaii, USA, pages 28–31*, 2012.
54. P. Langevin, G. Leander, P. Rabizzoni, P. Veron, and J.-P. Zanotti. Counting all bent functions in dimension eight 99270589265934370305785861242880. In *Des. Codes Cryptography 59 (1–3)*, pages 193–205, 2011.
55. G. Leander and G. McGuire. Construction of bent functions from near-bent functions. In *Journal of Combinatorial Theory, Series A 116, pages 960–970*, 2009.
56. V. I. Levenshtein. Bounds for packings of metric spaces and some of their applications. In *Probl. Cybern. (in Russian), Vol. 40, pages 43–110*, 1983.
57. C. Li, Q. Yue, and F. Li. Hamming weights of the duals of cyclic codes with two zeros. In *IEEE Trans. Inf. Theory, vol. 60, no. 7*, pages 3895–3902, 2014.
58. MacWilliams F. J., Sloane N. J.: *The theory of error-correcting codes*, Amsterdam, North Holland, 1977.
59. M. Matsui. Linear cryptanalysis method for DES cipher. In *Proceedings of EUROCRYPT'93, Lecture Notes in Computer Science 765*, pages 386–397, 1994.
60. M. Matsui. Linear cryptoanalysis method for DES cipher. In *EUROCRYPT*, pages 386–397, 1993.
61. M. Matsui and A. Yamagishi. A new method for known plaintext attack of FEAL cipher. In *EUROCRYPT*, pages 81–91, 1992.
62. R. L. McFarland. A family of noncyclic difference sets. In *Journal of Comb. Theory, Series A, No. 15, pages 1–10*, 1973.
63. W. Meier and O. Staffelbach. Fast correlation attacks on stream ciphers. In *Advances in Cryptology, EUROCRYPT'88, Lecture Notes in Computer Science 330*, pages 301–314, 1988.
64. S. Mesnager. Linear codes with few weights from weakly regular bent functions based on a generic construction. In Journal cryptography and communications (CCDS), Springer, to appear.
65. L. Qu, S. Fu, and Q. Dai C. Li. When a Boolean Function can be Expressed as the Sum of two Bent Functions. In *Cryptology ePrint Archive. 2014/048.*
66. S. Rønjom and T. Helleseth. A new attack on the filter generator. In *IEEE Transactions on Information Theory, vol. 53, no. 5*, pages 1752–1758, 2007.
67. O.S. Rothaus. On "bent" functions. In *J. Combin. Theory Ser A 20*, pages 300–305, 1976.
68. C. Tang, N. Li, Y. Qi, Z. Zhou, and T. Helleseth. Linear codes with two or three weights from weakly regular bent functions. In *ArXiv: 1507.06148v3*, 2015.

69. N. Tokareva. On the number of bent functions from iterative constructions: lower bounds and hypotheses. In *Advances in Mathematics of Communications (AMC), Vol. 5, No 4. pages 609–621*, 2011.
70. N. Tokareva. Bent Functions, results and applications to cryptography. In *Elsevier*, 2015.
71. L. Wang and J. Zhang. A best possible computable upper bound on bent functions. In *J. West of China. Vol 33. No. 2. pages 2013–2015 (in Chinese)*, 2004.
72. Y. Xia, T. Helleseth, and C. Li. Some new classes of cyclic codes with three or six weights. In *Adv. in Math. of Comm., 9(1)*, pages 23–36, 2015.
73. C. Xiang, C. Ding, and S. Mesnager. Optimal codebooks from binary codes meeting the Levenshtein bound. In *Preprint 2015.*
74. J. Yuan and C. Ding. Secret sharing schemes from three classes of linear codes. In *IEEE Trans. Inf. Theory, vol. 52, no. 1*, pages 206–212, 2006.
75. J. Wolfmann.: Bent functions and coding theory. *Difference Sets, Sequences and their Correlation Properties,* A. Pott, P. V. Kumar, T. Helleseth and D. Jungnickel, eds., pp. 393-417. Amsterdam: Kluwer (1999).
76. X. Zeng, L. Hu, W. Jiang, Q. Yue, and X. Cao. The weight distribution of a class of p-ary cyclic codes. In *Finite Fields Appl., vol. 16, no. 1*, pages 56–73, 2010.
77. Z. Zhou and C. Ding. A class of three-weight codes. In *Finite Fields Appl., vol. 25*, pages 79–93, 2014.
78. Z. Zhou, N. Li, C. Fan, and T. Helleseth. Linear Codes with Two or Three Weights From Quadratic Bent Functions. In *ArXiv: 1506.06830v1, 2015.*

Chapter 5
Bent Functions: Primary Constructions (Part I)

In this chapter we focus on constructions. There are two categories of constructions of bent functions: those which build the functions from scratch (Rothaus [36] writes that such functions are "given explicitly")—these constructions are nowadays called primary—and those, called secondary (Rothaus [36] writes "having implicit features"), which need bent functions as initial functions for building new ones (in the same or larger number of variables).

This chapter is dealing with primary constructions. Dillon has introduced a general framework in which we can construct bent functions. Such frameworks are called *models or classes*. There exist at least three classes: Maiorana–McFarland's class, the $\mathcal{PS}$ class and the class H. In this chapter we introduce the two first ones since class H has been developed 35 years after in [9] (see Chap. 8).

5.1 Two Main General Classes of Bent Functions

Several classes of bent functions have been introduced in [13, 36]. Some (like the $\mathcal{PS}$ class, recalled below) need conditions whose realizations are difficult to achieve, and so are more principles of constructions rather than explicit. Others lead to explicit bent functions (given by their ANF or their polynomial representation, univariate or bivariate). The two main ones of this last kind are the Maiorana–McFarland constructions and the Partial Spread class $\mathcal{PS}_{ap}$.

5.1.1 Maiorana–McFarland's Class

The Maiorana–McFarland constructions are the best known primary constructions of bent functions [13, 30]. The *Maiorana–McFarland class* (denoted by $\mathcal{M}$) is the set of all the n-variable Boolean functions of the form:

S. Mesnager, *Bent Functions*, DOI 10.1007/978-3-319-32595-8_5

$$f(x,y) = x \cdot \pi(y) + g(y);\ x, y \in \mathbb{F}_2^m \tag{5.1}$$

where "$\cdot$" denotes an inner product in $\mathbb{F}_2^m$, π any permutation on $\mathbb{F}_2^m$ and g any Boolean function on $\mathbb{F}_2^m$. Any such function is bent (the bijectivity of π is a necessary and sufficient condition for f being bent). The dual function $\tilde{f}(x,y)$ equals: $y \cdot \pi^{-1}(x) + g(\pi^{-1}(x))$, where π^{-1} is the inverse of π. The completed class of $\mathcal{M}$ (that is, the smallest possible complete class including $\mathcal{M}$) contains all the quadratic bent functions (that is, bent functions of algebraic degree 2) which are simple and best understood.

Proposition 5.1.1 ([13]). *A bent function $f : \mathbb{F}_{2^n} \to \mathbb{F}_2$ belongs to the completed class of $\mathcal{M}$ if and only if there exists an $n/2$-dimensional vector subspace V in $\mathbb{F}_{2^n}$ such that the second-order derivatives*

$$D_{a,b}f(t) = f(t+a+b) + f(t+a) + f(t+b) + f(t)$$

vanish for any $a, b \in V$.

5.1.2 *The Partial Spread Class $\mathcal{PS}$*

Other important classes of bent functions are the so-called $\mathcal{PS}^-$ class and $\mathcal{PS}^+$ introduced by Dillon in [14]. These classes are defined relatively to classical geometric objects that are called *partial spreads* for which we give below a definition in $\mathbb{F}_{2^n}$.

Definition 5.1.2. Let $n = 2m$ be an even integer. A partial spread of $\mathbb{F}_{2^n}$ is a set of pairwise supplementary m-dimensional subspaces of $\mathbb{F}_{2^n}$. A partial spread is a spread if the union of its elements equals $\mathbb{F}_{2^n}$

Let us now describe the elements of the $\mathcal{PS}^-$ class. A Boolean function f over $\mathbb{F}_{2^n}$, $n = 2m$, is said to be in the $\mathcal{PS}^-$ class if $f(0) = 0$ and if its support together with 0 is the union of 2^{m-1} elements of a partial spread of $\mathbb{F}_{2^m}$:

$$supp(f) \cup \{0\} = \bigcup_{i=1}^{2^{m-1}} E_i \tag{5.2}$$

where $\{E_1, \cdots, E_{2^{m-1}}\}$ is a partial spread of $\mathbb{F}_{2^n}$, that is,

1. $E_i \cap E_j = \{0\}$ for $i \neq j$;
2. $dim_{\mathbb{F}_2} E_i = m$, $\forall i \in \{1, \cdots, 2^{m-1}\}$;

In fact, Dillon shows more precisely that, a Boolean function over $\mathbb{F}_{2^n}$, such that $f(0) = 0$ and whose support is the union of elements of a partial spread, is bent if and only if the cardinality of the union is exactly 2^{m-1}. He shows that a

Boolean function defined by (5.2) is always of algebraic degree m which is the optimal algebraic degree for a bent function over $\mathbb{F}_{2^m}$. Furthermore, Dillon exhibits a subclass of $\mathcal{PS}^-$, denoted by $\mathcal{PS}_{ap}$ ("ap" for "affine plane") whose elements are defined in an explicit form as follows: let $n = 2m$ and let $\mathbb{F}_{2^n}$ be identified, as a vectorspace, with $\mathbb{F}_{2^m} \times \mathbb{F}_{2^m}$ (thanks to the choice of a basis of the two-dimensional vectorspace $\mathbb{F}_{2^n}$ over $\mathbb{F}_{2^m}$); the partial spread class $\mathcal{PS}_{ap}$ consists of all the functions f on $\mathbb{F}_{2^m} \times \mathbb{F}_{2^m}$ defined by $f(x, y) = g(\frac{x}{y})$ where g is a balanced Boolean function over $\mathbb{F}_{2^m}$ such that $g(0) = 0$ (but, in fact, this last condition is not necessary for f to be bent) with the convention that $\frac{x}{y} = 0$ if $y = 0$.

The functions from class $\mathcal{PS}_{ap}$ are in fact those whose supports are the unions of 2^{m-1} multiplicative cosets of $\mathbb{F}_{2^m}^\star$. These supports can be uniquely written as $\bigcup_{u\in S} u\mathbb{F}_{2^m}^\star$ where U is the set $\{u \in \mathbb{F}_{2^n}; u^{2^m+1} = 1\}$ and S is a subset of U of size 2^{m-1}. We shall also include in $\mathcal{PS}_{ap}$ the complements of these functions. The class $\mathcal{PS}_{ap}$ is much smaller than the Maiorana–McFarland class. It contains, up to EA-equivalence, the generalized Dillon functions and the so-called *Mesnager functions*. The functions in this class are, when viewed in univariate form, those bent functions whose restrictions to the cosets $u\mathbb{F}_{2^m}^\star$ are constant where u ranges over the cyclic group of $(2^m + 1)$st of unity of $\mathbb{F}_{2^n}^\star$. They are also hyperbent (see Chap. 9).

The dual of bent function f in $\mathcal{PS}_{ap}$ is given by $\tilde{f}(x, y) = g(\frac{y}{x})$. In fact, the subclass $\mathcal{PS}_{ap}$ is obtained by choosing partial spreads embeddable in a Desarguesian spread. In $\mathbb{F}_{2^n}$, the *Desarguesian spread* is the collection $\{u\mathbb{F}_{2^m}, u \in U\}$ ("univariate form"). Identifying $\mathbb{F}_{2^n}$ with $\mathbb{F}_{2^m} \times \mathbb{F}_{2^m}$, it can also be viewed as the collection $\{E_a, a \in \mathbb{F}_{2^m}\} \cup \{E_\infty\}$ where $E_a := \{(x, ax) ; x \in \mathbb{F}_{2^m}\}$ and $E_\infty := \{(0, y) ; y \in \mathbb{F}_{2^m}\} = \{0\} \times \mathbb{F}_{2^m}$ ("bivariate form").

The other class $\mathcal{PS}^+$ can be defined in a similar way. Indeed, a Boolean function f over $\mathbb{F}_{2^n}$, $n = 2m$, is said to be in the $\mathcal{PS}^-$ class if $f(0) = 1$ and if its support is the union of $2^{m-1} + 1$ elements of a partial spread of $\mathbb{F}_{2^m}$:

$$supp(f) = \bigcup_{i=1}^{2^{m-1}+1} E_i \tag{5.3}$$

where $\{E_1, \cdots, E_{2^{m-1}+1}\}$ is a partial spread of $\mathbb{F}_{2^n}$. Likewise the preceding class, Dillon shows that a Boolean function over $\mathbb{F}_{2^n}$ such that $f(0) = 1$ and whose support is the union of elements of a partial spread, is bent if and only if the cardinality of that union is exactly $2^{m-1} + 1$. Dillon has also seen that all those functions in $\mathcal{PS}^+$ corresponding to a partial spread which is extendable, that is, which is embeddable in a spread, has algebraic degree m. Dillon however shows that for m even, $\mathcal{PS}^+$ contains all quadratic bent functions. He deduced the interesting result that functions in $\mathcal{PS}^+$ are not necessarily the complements of functions in $\mathcal{PS}^-$. The union of $\mathcal{PS}^+$ and $\mathcal{PS}^-$ forms the partial spread class $\mathcal{PS}$. Since then, several papers have dealt $\mathcal{PS}$-like constructions (see for instance [7, 20, 34]).

Dillon [13] has also introduced bent functions obtained using, more generally, sets of subgroups of a group. This extension to subgroups has been pushed further in [22] and extended to odd characteristic (see [28, 33]).

For a long time, partial spreads considered in the literature were mostly subsets of Desarguesian spreads. Very recently, Carlet has considered André spreads and has studied in [6] the $\mathcal{PS}$ functions associated to them. But above, a particular family of partial spreads has recently arisen in the literature: partial spreads coming from finite pre-quasifields [6, 37], more precisely from the Dempwolff–Müller pre-quasifield, the Knuth pre-semifield and the Kantor pre-semifield. The first constructions of $\mathcal{PS}$ bent functions from those partial spreads have been provided by Wu [37].

5.2 Basic Primary Constructions of Bent Functions in Multivariate Representation

Bent functions over the $\mathbb{F}_2$-vectorspace $\mathbb{F}_{2^n}$ from the Maiorana–McFarland class are given explicitly in multivariate representation. Further generalizations of the Maiorana–McFarland construction have been obtained (for more details, see [4]). Firstly the Maiorana–McFarland's class has been modified by the addition of indicators of flats. Next, it has been generalized is various manners via concatenations of quadratic functions, of indicators of flats and also via more complex concatenations. Moreover, the Maiorana–McFarland's class has also been generalized into a secondary construction (see Chap. 6). Finally, in [17], Dobbertin has described a general explicit construction of bent functions, which unifies well known constructions due to Maiorana–McFarland and Dillon as two opposite extremal cases. Within this framework Dobbertin also find new ways to generate bent functions. More precisely, let $f_{g,\phi,\psi}$ be a function defined by $f_{g,\phi,\psi}(x, \phi(y)) = g(\frac{x+\psi(y)}{y})$ if $y \neq 0$ and 0 otherwise; where ϕ and $\psi\colon \mathbb{F}_{2^{n/2}} \to \mathbb{F}_{2^{n/2}}$ are two mapping such ϕ is one-to-one and g is a balanced Boolean function on $\mathbb{F}_{2^{n/2}}$. With these notation, $f_{g,\phi,\psi}$ is bent if and only if the functions g, ψ and ϕ are bent. Note that the class of these bent functions $f_{g,\phi,\psi}$ contains both classes $\mathcal{PS}_{ap}$ and Maiorana–McFarland.

5.3 Primary Constructions of Bent Functions in Bivariate Representation

Any function of the form $f(x, y) = g(x^{2^m-2}y)$ is bent if and only if g is balanced. These functions belong to $\mathcal{PS}_{ap}$ and cover this class. A particular case is $f(x, y) = tr_m(x^{2^m-1-d}y^d)$, $gcd(d, 2^m - 1) = 1$.

Bent functions in bivariate representation are much more numerous than in univariate form. In the following, we list the known primary families of bent function in bivariate trace form. Most of them are developed in detail in next chapters.

1. Bent functions of Maiorana–McFarland class (defined above) can be viewed in bivariate form: $f(x,y) = Tr_1^m(x\pi(y)) + g(y)$ where π is a permutation on $\mathbb{F}_{2^m}$;
2. Bent functions of the Dillon class $\mathcal{PS}_{ap}$ (defined above): $f(x,y) = g(xy^{2^m-2})$ where g is any balanced Boolean function over $\mathbb{F}_{2^m}$). The dual functions of elements of $\mathcal{PS}_{ap}$ class are very simple: $\tilde{f}(x,y) = g(yx^{2^m-2})$;
3. Bent functions related to Dillon's H class and *o-polynomials* (see Chap. 8);
4. Bent functions associated to the so-called *almost bent* (AB) functions. In fact, if $n/2$ is odd, then it is possible to deduce a bent Boolean function on $\mathbb{F}_2^n$ from any almost bent function from $\mathbb{F}_2^{n/2}$ to $\mathbb{F}_2^{n/2}$. A vectorial Boolean function $F : \mathbb{F}_2^m \to \mathbb{F}_2^m$ is called *almost bent* if all of the component functions $v \cdot F$, $v \neq 0$ in $\mathbb{F}_2^m$, are plateaued with amplitude $2^{\frac{m+1}{2}}$ (see in Chap. 16 the definition of these terms). The function $\gamma_F(a,b)$, $a,b \in \mathbb{F}_2^m$, equal to 1 if the equation $F(x) + F(x+a) = b$ admits solutions, with $a \neq 0$ in $\mathbb{F}_2^m$, and equal to 0 otherwise is then bent (see for instance [3]) This gives new bent functions related to the almost bent functions. A list of the known almost bent functions can be found, for instance in [3]. However, note that determining the ANF or the univariate representation of γ_F is an open problem when F is a Kasami, Welch or Niho almost bent function.
5. An isolated class: $f(x,y) = Tr_1^m(x^{2^i+1} + y^{2^i+1} + xy)$, $x,y \in \mathbb{F}_{2^n}$ where n is co-prime with 3 and i is co-prime with m (see [4]);
6. Several new infinite families of bent functions and their duals (see [32]);
7. Several new infinite families of bent functions from new permutations and their duals (see [31]).

5.4 Primary Constructions and Characterization of Bent Functions in Polynomial Form

A number of researchers have been interested in providing constructions of bent functions in polynomial form, that is, functions f whose expression is of the form (see Sect. 1.1.2.3, Chap. 1):

$$f(x) = \sum_{j\in\Gamma_n} Tr_1^{o(j)}\left(a_jx^j\right) + \epsilon(1 + x^{2^n-1}), \quad a_j \in \mathbb{F}_{2^{o(j)}}.$$

Since the algebraic degree of any bent function over $\mathbb{F}_{2^n}$ is at most $\frac{n}{2}$, the Hamming weight of f is then even, that is, ϵ equals 0. Consequently, the polynomial form of any bent function f is of type:

$$f(x) = \sum_{j\in\Gamma_n} Tr_1^{o(j)}\left(a_jx^j\right), \quad a_j \in \mathbb{F}_{2^{o(j)}}.$$

The *monomial functions* and *binomial functions* are particular cases of functions in polynomial form. Monomial functions are functions which are the traces of a single power function, that is, functions f defined on $\mathbb{F}_{2^n}$ whose expression is of the form $f(x) = Tr_1^n(ax^s)$ for given positive integer s and for some $a \in \mathbb{F}_{2^n}$. Binomial functions are functions f defined on $\mathbb{F}_{2^n}$ whose expression is of the form $f(x) = Tr_1^n(a_1x^{s_1} + a_2x^{s_2})$, $(a_1, a_2) \in (\mathbb{F}_{2^n}^\star)^2$ for a given positive integers s_1 and s_2 and for some coefficients a_1, a_2 in $\mathbb{F}_{2^n}$.

5.4.1 Monomial Bent Functions

As a first step towards a characterization of the trace forms of bent functions, many authors focus on monomial functions, that is, functions of the form $Tr_1^n(ax^s)$ for a given positive integer s and for some $a \in \mathbb{F}_{2^n}$. A *bent exponent* (always understood modulo $2^n - 1$) is an integer s such that there exists $a \in \mathbb{F}_{2^n}^\star$ for which $x \mapsto Tr_1^n(ax^s)$ is bent. The current list of known bent exponents is given in Table 5.1. We send the reader to [26] where known cases of monomial bent functions are presented. Canteaut et al. [2] have carried out an exhaustive search and shown that there is no other exponent s for $n \leq 20$. The complete classification of monomial bent functions is not yet achieved.

Note that the cyclotomic cosets modulo $2^n - 1$ of all the bent exponents presented in Table 5.1 are of maximal size n that is, $o(s) = n$. The corresponding monomial functions are then in polynomial forms.

A monomial function $f(x) = Tr_1^n(ax^s)$ cannot be bent for every non-zero a (see e.g. [26]). Moreover, there are some necessary conditions for s to be a bent exponent (see for instance [26]):

- the 2-weight of a bent exponent s is at most $\frac{n}{2}$.
- $gcd(s, 2^n - 1) > 1$; moreover, $gcd(s, 2^{\frac{n}{2}} - 1) = 1$ or $gcd(s, 2^{\frac{n}{2}} + 1) = 1$.

Remark 5.4.1. Note that bent functions have been also obtained by Dillon and McGuire [16] as the restrictions of functions on $\mathbb{F}_{2^{n+1}}$, with $n + 1$ odd, to a hyperplane of this field: these functions are the Kasami functions $Tr_1^n(x^{2^{2k}-2^k+1})$ and the hyperplane has equation $Tr_1^n(x) = 0$. The restriction is bent under the condition that $n + 1 = 3k \pm 1$.

Table 5.1 Bent exponent

Exponent	Condition	Family	References
$2^i + 1$	$\frac{n}{gcd(n,i)}$ even	$\mathcal{M}$	[19]
$a(2^{\frac{n}{2}} - 1)$	$gcd(a, 2^{\frac{n}{2}} + 1) = 1$	PS_{ap}	[10, 13, 25, 26]
$2^{2i} - 2^i + 1$	$gcd(i, n) = 1$		[15]
$(2^{\frac{n}{4}} + 1)^2$	$n = 4r$, r odd	$\mathcal{M}$	[11, 26]
$2^{\frac{n}{3}} + 2^{\frac{n}{6}} + 1$	$n \equiv 0 \mod 6$	$\mathcal{M}$	[2]

We have looked for the exponents s such that $o(s) < n$ for which there exists at least one coefficient $a \in \mathbb{F}_{2^{o(s)}}$ such that, the Boolean function $Tr_1^{o(s)}(ax^s)$ is bent. After an exhaustive search up to $n \leq 14$, we have found that, the only bent Boolean functions belonging to the set of monomial functions with exponent s such that $o(s) < n$ are of the form $Tr_1^{\frac{n}{2}}(ax^{2^{\frac{n}{2}}+1})$, for some $a \in \mathbb{F}_{2^n}$. Such functions have been given by Yu and Gong in [38]. Note that a class of quadratic functions (*i.e.* of algebraic degree 2) defined on $\mathbb{F}_{2^n}$ whose expression has the form: $f(x) = \sum_{i=1}^{\frac{n}{2}-1} a_i Tr_1^n(x^{1+2^i}) + a_{\frac{n}{2}} Tr_1^{\frac{n}{2}}(x^{2^{\frac{n}{2}}+1})$ with $a_i \in \mathbb{F}_2$, for $i \in \{1, \ldots, \frac{n}{2}\}$, was considered in several papers, in which the authors investigate the conditions on the choice of the coefficients a_i for explicit definition of an infinite class of quadratic bent functions. A non-exhaustive list of references which deals with the characterization of the bentness of this class is [12, 21, 23, 24, 29, 38].

Finally, the reader can find in Table 5.2 the known monomial bent functions of the form $Tr_1^n(ax^d)$, $a \in \mathbb{F}_{2^n}^\star$.

5.4.2 *Binomial Bent Functions with Niho Exponents*

Some constructions of binomial bent functions via Niho power functions have been given in [18]. A positive integer s (always understood modulo $2^n - 1$) is said to be a *Niho exponent*, and x^s a *Niho power function*, if the restriction of x^s to $\mathbb{F}_{2^m}$ is linear or in other words $s \equiv 2^j \pmod{2^m - 1}$ for some $j < n$. As we consider $Tr_1^n(x^d)$, without loss of generality, we can assume that s is in the normalized form, with $j = 0$, and then we have a unique representation $s = (2^m - 1)d + 1$ with $2 \leq d \leq 2^m$. The name of Niho exponent comes from a theorem dealing with power functions by Niho [35], which has been later extended to linear combinations of such power functions in [18] (see also [27]), and which relates the value of the Walsh transform of such sums to the number of solutions in U of some equation. According to Dobbertin et al. [18], three subfamilies containing bent functions can be identified in the set of binomial functions (the fractions are interpreted modulo $2^m + 1$, for instance $\frac{1}{2} = 2^{m-1} + 1$):

- $s_1 = (2^m - 1)\frac{1}{2} + 1$ and $s_2 = (2^m - 1)3 + 1$;
- $s_1 = (2^m - 1)\frac{1}{2} + 1$ and $s_2 = (2^m - 1)\frac{1}{4} + 1$ (m odd);
- $s_1 = (2^m - 1)\frac{1}{2} + 1$ and $s_2 = (2^m - 1)\frac{1}{6} + 1$ (m even).

The following statement summarizes the results given in [18].

Theorem 5.4.2 ([18]). *Let $n = 2m$. Let f be a function defined on $\mathbb{F}_{2^n}$ of the form $f(x) = Tr_1^n\left(a_1 x^{(2^m-1)\frac{1}{2}+1} + a_2 x^{s_2}\right)$, where $a_1, a_2 \in \mathbb{F}_{2^n}^\star$. Assume that $a_2^{\frac{2^m+1}{2}} = a_1 + a_1^{2^m}$.*

Table 5.2 The known monomial bent functions of the form $\mathrm{Tr}_1^n(ax^d)$, $a \in \mathbb{F}_{2^n}^\star$

Case	Exponent d	Condition-1	CNS for bentness of $\mathrm{Tr}_1^n(ax^d)$
Gold	2^s+1	$s \in \mathbb{N}$	$a \notin < \alpha^{gcd(d,2^n-1)} >$
Dillon	$s(2^{n/2}-1)$	$gcd(s, 2^{n/2}+1) = 1$	$K_n(a^{2^{n/2}+1}) = 0$
Kassami	$2^{2s}-2^s+1$	$s \in \mathbb{N}$, $gcd(3,n) = 1$, $gcd(s,n) = 1$	$a \notin < \alpha^3 >$
Leander	$(2^s+1)^2$	$s \in \mathbb{N}^\star$, $n = 4s$	s odd, $a \in \mathbb{F}_4 \setminus \mathbb{F}_2 \cdot < \alpha^{2^s+1} >$
Canteaut–Charpin–Kyureghyan	$2^{2s}+2^s+1$	$s \in \mathbb{N}^\star$, $n = 6s$	$a \in \{r \in \mathbb{F}_{2^{n/2}}^\star \mid \mathrm{Tr}_s^{n/2}(r) = 0\} \cdot < \alpha^d >$

1. *Let* $s_2 = (2^m - 1)3 + 1$. *If* $a_2 = \gamma^5$ *for some* $\gamma \in \mathbb{F}_{2^n}^\star$ *then,* f *is a bent function of degree* m *(note that if* $m \not\equiv 2 \pmod 4$ *then, the map* $x \mapsto x^5$ *is a permutation of* $\mathbb{F}_{2^n}$*).*
2. *Suppose* m *is odd. Let* $s_2 = (2^m - 1)\frac{1}{4} + 1$. *Then* f *is a bent function of degree* 3.
3. *Suppose* m *is even. Let* $s_2 = (2^m - 1)\frac{1}{6} + 1$. *Then* f *is a bent function of degree* m.

Note that, as observed in [8], there is a mistake made in [18] (Theorem 3) while computing the algebraic degree of the third Niho bent function. Indeed the degree calculated in [18] being equal to $\frac{m}{2} + 1$ is not correct. The correct degree is m and this comes from the following lemma.

Lemma 5.4.3 ([8]). *Take even* $m > 2$ *and interpret* $\frac{1}{3}$ *as an inverse of* 3 *modulo* $2^m + 1$. *Then the exponent* $2s_2 = (2^m - 1)\frac{1}{3} + 2$ *has binary weight* m.

Proof. First, note that $1/3$ modulo $2^m + 1$ is equal to $(2^m + 2)/3$. Then

$$\begin{aligned} 2s_2 &= \frac{2^n - 1}{3} + \frac{2^m - 1}{3} + 2 \\ &= \sum_{i=0}^{m-1} 2^{2i} + \sum_{i=0}^{m/2-1} 2^{2i} + 2 \\ &= \sum_{i=0}^{m/2-1} 2^{2i+1} + \sum_{i=m/2}^{m-1} 2^{2i} + 2 \end{aligned}$$

whose binary weight equals m if $m > 2$. □

Remark 5.4.4. Note that $o(s_1) = o((2^m - 1)\frac{1}{2} + 1) = \frac{n}{2}$ and $o(s_2) = n$ for $s_2 \in \{(2^m - 1)3 + 1, (2^m - 1)\frac{1}{4} + 1, (2^m - 1)\frac{1}{6} + 1\}$. The polynomial forms of the three binomial functions given in Theorem 5.4.2 are then

$$Tr_1^m(a_1{}'x^{2^m+1}) + Tr_1^n(a_2x^{s_2}), a_1{}' \in \mathbb{F}_{2^m}^\star \text{ and } a_2 \in \mathbb{F}_{2^n}^\star$$

The problem of knowing whether the duals of the binomial functions given in [18] are affine equivalent to these Niho bent functions was left open in [18]. Very recently, the bivariate representation (obtained by identifying $\mathbb{F}_{2^n}$ with $\mathbb{F}_{2^m} \times \mathbb{F}_{2^m}$ and considering the input to the Boolean function as an ordered pair $(x; y)$ of elements of $\mathbb{F}_{2^m}$) of the second Dobbertin-et-al function and the bivariate expression of its dual have been computed in [9]. We also observed in [9] that the dual is not a Niho bent function, which allows answering negatively to the open question in [18]. We will discuss on the dual of those Niho bent functions in Sect. 8.2.2.1.

5.4.3 *Binomial Bent Functions with Dillon (Like) Exponents*

Chapter 10 deals with a strong property than bentness, more precisely, hyper-bentness (since hyper bent functions are in particular bent). The known constructions of bent functions via Dillon (like) exponents are also hyper-bent. So for the constructions of binomial bent functions we refer the reader to Sect. 9.5.2 in Chap. 10 in which we will present the known constructions of binomial hyper-bent functions.

5.4.3.1 Bent Functions via Several Niho Exponents

The second class in [18] of binomial bent functions (that is obtained with the exponent $(2^m - 1)\,3 + 1$) has been extended by Leander and Kholosha [27] to the functions:

$$\mathrm{Tr}_1^n\big(at^{2^m+1} + \sum_{i=1}^{2^{r-1}-1} t^{s_i}\big),$$

where $r > 1$ such that $gcd(r, m) = 1$, $a \in \mathbb{F}_{2^n}$ such that $a + a^{2^m} = 1$, $s_i = (2^m - 1)\frac{i}{2^r} + 1 \pmod{2^m + 1}$, $i \in \{1, \ldots, 2^{r-1} - 1\}$.

Very recently, the algebraic degree of any bent function in the Leander–Kholosha class[27] has been has been computed in [1].

5.4.3.2 Bent Functions with Multiple Trace Terms via Dillon (Like) Exponents

The known bent functions whose expressions are multiple trace terms via Dillon (like) exponents are also hyper-bent. We send the reader to Chap. 10, Sects. 10.1, 10.2 and 10.4.

5.4.4 *Bent Functions in Hybrid Form and Kerdock Codes*

In this section we refer to [5]. The following infinite class f_δ is a well-known family of quadratic bent functions defined over $\mathbb{F}_{2^{n-1}} \times \mathbb{F}_2$:

$$f_\delta(z, x_n) = \sum_{i=1}^{\frac{n}{2}-1} \mathrm{Tr}_1^{n-1}((\delta z)^{2^i+1} + x_n \mathrm{Tr}_1^{n-1}(\delta z),$$

where $\delta, z \in \mathbb{F}_{2^{n-1}}$ with $\delta \neq 0$. The difference (*i.e.* the sum) of two functions of such form corresponding to two distinct values of δ is bent as well. It is easily shown that any code of length 2^n (i.e. any set of Boolean functions) is equal to the union of at least two cosets of the first-order Reed–Muller code $\mathcal{RM}(1,n)$ (i.e. the set of affine n-variable Boolean functions has minimum distance bounded above by $2^{n-1} - 2^{\frac{n}{2}-1}$) with equality if and only if all the differences between the elements of two distinct cosets are bent functions. The Kerdock code of length 2^n ($n \geq 4$, n even) equals the union of all the cosets $f_\delta + \mathcal{RM}(1,n)$ where z ranges through $\mathbb{F}_{2^{n-1}}$. Such a code is optimal (it was shown by Delsarte that no code exists with better parameters, e.g. with smaller length, same size and same minimum distance, or larger size, same length and same minimum distance, or larger minimum distance, same length and same size).

References

1. L. Budaghyan, A. Kholosha, C. Carlet, and T. Helleseth. Niho bent functions from quadratic o-monomials. In *Proceedings of ISIT 2014, pages 1827–1831*, 2014.
2. A. Canteaut, P. Charpin, and G. Kyureghyan. A New Class of Monomial Bent Functions. In *Finite Fields and Their Applications, Vol 14, no. 1*, pages 221–241, 2008.
3. C. Carlet. Vectorial Boolean Functions for Cryptography. In *Chapter of the monography Boolean Methods and Models, Y. Crama and P. Hammer eds, Cambridge University Press. Preliminary version available at* http://www-rocq.inria.fr/codes/Claude.Carlet/pubs.html.
4. C. Carlet. Boolean functions for cryptography and error correcting codes. In Yves Crama and Peter L. Hammer, editors, *Boolean Models and Methods in Mathematics, Computer Science, and Engineering*, pages 257–397. Cambridge University Press, June 2010.
5. C. Carlet. Open problems on binary bent functions. In *Proceeding of the conference "Open problems in mathematical and computational sciences", Springer, pages 203–241*, 2014.
6. C. Carlet. More PS and H-like bent functions. In *Cryptology ePrint Archive, Report 2015/168*, 2015.
7. C. Carlet and C. Ding. Highly Nonlinear Mappings. In *Special Issue: "Complexity Issues in Coding and Cryptography" of the Journal of Complexity, 20(2–3), pages 205–244*, 2004.
8. C. Carlet, T. Helleseth, A. Kholosha, and S. Mesnager. On the duals of bent functions with 2^r Niho exponents. In *IEEE International Symposium on Information Theory, ISIT 2011*, pages 703–707, 2011.
9. C. Carlet and S. Mesnager. On Dillon's class H of bent functions, Niho bent functions and o-polynomials. In *Journal of Combinatorial Theory, Series A, Vol 118, no. 8*, pages 2392–2410, 2011.
10. P. Charpin and G. Gong. Hyperbent functions, Kloosterman sums and Dickson polynomials. In *IEEE Trans. Inform. Theory (54) 9*, pages 4230–4238, 2008.
11. P. Charpin and G. Kyureghyan. Cubic monomial bent functions: A subclass of $\mathcal{M}$. In *SIAM, J. Discr. Math., Vol. 22, no. 2*, pages 650–665, 2008.
12. P. Charpin, E. Pasalic, and C. Tavernier. On bent and semi-bent quadratic Boolean functions. In *IEEE Transactions on Information Theory, vol. 51, no. 12*, pages 4286–4298, 2005.
13. J. Dillon. Elementary Hadamard difference sets. In *PhD dissertation, University of Maryland.*
14. J. F. Dillon. Elementary Hadamard Difference sets. In *Proceedings of the Sixth Southeastern Conference on Combinatorics, Graph Theory and Computing, F. Hoffman et al., Eds., Congressus Numerantium XIV, Winnipeg Utilitas Math, pages 237–249*, 1975.

15. J. F. Dillon and H. Dobbertin. New cyclic difference sets with Singer parameters. In *Finite Fields and Their Applications Volume 10, Issue 3*, pages 342–389, 2004.
16. J. F. Dillon and G. McGuire. Near bent functions on a hyperplane. In *Finite Fields and Their Applications Vol. 14, Issue 3*, pages 715–720, 2008.
17. H. Dobbertin. Construction of bent functions and balanced Boolean functions with high nonlinearity. In *Proceedings of Fast Software Encryption, Second International Workshop, Lecture Notes in computer Science 1008*, pages 61–74, 1995.
18. H. Dobbertin, G. Leander, A. Canteaut, C. Carlet, P. Felke, and P. Gaborit. Construction of bent functions via Niho Power Functions. In *Journal of Combinatorial theory, Series A 113*, pages 779–798, 2006.
19. R. Gold. Maximal recursive sequences with 3-valued recursive crosscorrelation functions. In *IEEE Trans. Inform. Theory 14 (1)*, pages 154–156, 1968.
20. X. D. Hou. q-ary bent functions constructed from chain rings. In *J. Finite Fields Appl., 4, pages 55–61*, 1998.
21. H. Hu and D. Feng. On quadratic bent functions in polynomial forms. In *IEEE Trans. Inform. Theory 53 (7)*, pages 2610–2615, 2007.
22. W. M. Kantor. Bent functions generalizing Dillon's partial spread functions. In *arXiv 1211.2600*, 2012.
23. T. Kasami. Weight enumerators for several classes of subcodes of the 2nd-order Reed–Muller codes. In *Information control, Vol 18*, pages 369–394, 1971.
24. S. H. Kim and J. S. No. New families of binary sequences with low correlation. In *IEEE Trans. Inform. Theory, vol. 49, no. 11*, pages 3059–3065, 2003.
25. G. Lachaud and J. Wolfmann. The weights of the orthogonals of the extended quadratic binary Goppa codes. In *IEEE Trans. Inform. Theory 36 (3)*, pages 686–692, 1990.
26. G. Leander. Monomial Bent Functions. In *IEEE Trans. Inform. Theory (52) 2*, pages 738–743, 2006.
27. G. Leander and A. Kholosha. Bent functions with 2^r Niho exponents. In *IEEE Trans. Inform. Theory 52 (12)*, pages 5529–5532, 2006.
28. P. Lisoněk and H. Y. Lu. Bent functions on partial spreads. In *Designs, Codes and Cryptography, Vol 73, Issue 1, pages 209–216*, 2014.
29. W. Ma, M. Lee, and F. Zhang. A new class of bent functions. In *IEICE Trans. Fundamentals, Vol E88-A , Issue 7*, pages 2039–2040, 2005.
30. R. L. McFarland. A family of noncyclic difference sets. In *Journal of Comb. Theory, Series A, No. 15, pages 1–10*, 1973.
31. S. Mesnager. Further constructions of infinite families of bent functions from new permutations and their duals. In *Journal Cryptography and Communications (CCDS), Springer. To appear.*
32. S. Mesnager. Several new infinite families of bent functions and their duals. In *IEEE Transactions on Information Theory-IT, Vol. 60, No. 7, Pages 4397–4407*, 2014.
33. S. Mesnager. On p-ary bent functions from (maximal) partial spreads. In *International conference Finite field and their Applications Fq12, New York, July*, 2015.
34. K. Nyberg. Perfect non-linear S-boxes. In *Proceedings of EUROCRYPT'91, Lecture Notes in Computer Science 547*, pages 378–386, 1992.
35. S. E. Payne. Multivalued cross-correlation functions between two maximal linear recursive sequences. In *PhD thesis, Univ. of Southern California*, 1972.
36. O.S. Rothaus. On "bent" functions. In *J. Combin. Theory Ser A 20*, pages 300–305, 1976.
37. B. Wu. PS bent functions constructed from finite pre-quasifield spreads. In *ArXiv e-prints*, 2013.
38. N. Y. Yu and G. Gong. Construction of quadratic Bent functions in polynomial forms. In *IEEE Trans. Inform. Theory (52) 7*, pages 3291–3299, 2006.

Chapter 6
Bent Functions: Secondary Constructions

We call secondary a construction of bent functions from already known bent functions, in the same number of variables or not (while primary constructions, like Maiorana–McFarland construction, build bent functions from scratch).

There exist several secondary constructions of bent functions but the best-known are those of Rothaus ("direct sum") and of Carlet ("indirect sum"). We shall not present all the constructions in detail but just mention them briefly. We send the reader interested in this research direction to Carlet's chapter [7] or the survey [8].

1. **Direct sum**: the first secondary construction was given by Dillon and Rothaus [10, 13]. Such a construction is called the *direct sum* and is defined as follows: let f be a bent function on $\mathbb{F}_2^n$ (n even) and g a bent function on $\mathbb{F}_2^m$ (m even) then the function h defined on $\mathbb{F}_2^{n+m}$ by $h(x,y) = f(x) \oplus g(y)$ is bent. Unfortunately this construction[1] has no great interest from a cryptographic point of view.
2. **Rothaus's construction:** Rothaus has proved a more interesting construction (cited by Dillon in his thesis under the name of family O'), which uses three initial n-variable bent functions to build an $n+2$-variable bent function. More precisely, it is defined as follows: if g, h, k and $g \oplus h \oplus k$ are bent on $\mathbb{F}_2^n$ (n even), then the function defined at every element (x_1, x_2, x) of $\mathbb{F}_2^{n+2}$ ($x_1, x_2 \in \mathbb{F}_2$, $x \in \mathbb{F}_2^n$) by:

$$f(x_1, x_2, x) =$$

$$g(x)h(x) \oplus g(x)k(x) \oplus h(x)k(x) \oplus [g(x) \oplus h(x)]x_1 \oplus [g(x) \oplus k(x)]x_2 \oplus x_1x_2$$

is bent. Unfortunately no general class of bent functions has been deduced from this construction.

[1] In fact, this construction produces decomposable functions (a Boolean function is called decomposable if it is equivalent to the sum of two functions that depend on two disjoint subsets of coordinates; such peculiarity is easy to detect and can be used for designing divide-and-conquer attacks, as pointed out by Dillon in [10]).

S. Mesnager, *Bent Functions*, DOI 10.1007/978-3-319-32595-8_6

3. Carlet has proposed in [3] the two classes of bent functions which are derived from Maiorana–McFarland's class, by adding to some functions of this class the indicators of some vector subspaces. His result is the following.

Theorem 6.0.1 ([3]). *Let $b+E$ be any flat in $\mathbb{F}_2^n$ (E being a linear subspace of $\mathbb{F}_2^n$). Let f be any bent function on $\mathbb{F}_2^n$. The function $f^\star = f \oplus 1_{b+E}$ is bent if and only if one of the following equivalent conditions is satisfied:*

(a) For any a in $\mathbb{F}_2^n \setminus E$, the function $D_a f$ is balanced on $b+E$;
(b) The restriction of the function $\tilde{f}(x) \oplus b\cdot x$ to any coset of $E^\perp$ is either constant or balanced.

If f and $f^\star$ are bent, then E has dimension greater than or equal to $n/2$ and the algebraic degree of the restriction of f to $b+E$ is at most $\dim(E) - n/2 + 1$.

If f is bent, if E has dimension $n/2$, and if the restriction of f to $b+E$ has algebraic degree at most $\dim(E) - n/2 + 1 = 1$, i.e. *is affine, then conversely $f^\star$ is bent too.*

4. **The indirect sum and its generalizations:** other classes of bent functions have been deduced from a construction given in [4], which generalizes the secondary constructions given in 1 and 2 above:

Theorem 6.0.2 ([4]). *Let n and m be two even positive integers. Let f be a Boolean function on $\mathbb{F}_2^{n+m} = \mathbb{F}_2^n \times \mathbb{F}_2^m$ such that, for any element y of $\mathbb{F}_2^m$, the function on $\mathbb{F}_2^n$:*

$$f_y : x \mapsto f(x,y)$$

is bent. Then f is bent if and only if, for any element s of $\mathbb{F}_2^n$, the function

$$\varphi_s : y \mapsto \widetilde{f_y}(s)$$

is bent on $\mathbb{F}_2^m$. If this condition is satisfied, then the dual of f is the function $\tilde{f}(s,t) = \widetilde{\varphi_s}(t)$ (taking as inner product in $\mathbb{F}_2^n \times \mathbb{F}_2^m$: $(x,y)\cdot(s,t) = x\cdot s \oplus y \cdot t$).

The previous result gives rise to a secondary construction due to [5] called the *indirect sum*:

Corollary 6.0.3 ([5])). *Let f_1 and f_2 be two r-variable bent functions (r even) and let g_1 and g_2 be two s-variable bent functions (s even). Define*[2]

$$h(x,y) = f_1(x) + g_1(y) + (f_1 + f_2)(x)\,(g_1 + g_2)(y);\ x \in \mathbb{F}_2^r,\ y \in \mathbb{F}_2^s. \tag{6.1}$$

Then h is bent and its dual is obtained by the same formula as h is obtained from f_1, f_2, g_1 and g_2:

[2] h is the concatenation of the four functions $f_1, f_1 \oplus 1, f_2$ and $f_2 \oplus 1$, in an order controlled by $g_1(y)$ and $g_2(y)$. This construction $(f_1, f_2, g_1, g_2) \mapsto h$ leads to construct resilient functions (see [7]).

$$\tilde{h}(x,y) = \tilde{f}_1(x) + \tilde{g}_1(y) + (\tilde{f}_1 + \tilde{f}_2)(x)\,(\tilde{g}_1 + \tilde{g}_2)(y);\ x \in \mathbb{F}_2^r,\ y \in \mathbb{F}_2^s.$$

The Rothaus construction, which uses four bent functions whose sum is null, is a particular case of this construction. In fact, the indirect sum is another particular case, using also four bent functions, but without any initial condition [4]:

Two generalizations of the indirect sum needing initial conditions are given in [9], for functions of the forms

$$f(x,y) = f_1(x) + g_1(y) + (f_1 + f_2)(x)(g_1 + g_2)(y) + (f_2 + f_3)(x)(g_2 + g_3)(y)$$

and $f(x,y) = f_0(x) + g_0(y) +$

$$(f_0 + f_1)(x)(g_0 + g_1)(y) + (f_1 + f_2)(x)(g_1 + g_2)(y) + (f_2 + f_3)(x)(g_2 + g_3)(y).$$

A modified indirect sum is also introduced in [14] in which the functions have the same form as in (6.1) but where f_1 and f_2 (resp. g_1 and g_2) are the restrictions of a bent function f (resp. g) to two hyperplanes, complementary of each other.

It is also shown in [8] that if f and g are two n-variable Boolean functions with g bent and ϕ is a mapping from $\mathbb{F}_2^n$ to itself, then the $2n$-variable function $f(x) + \tilde{g}(y) + \phi(x) \cdot y$ is bent if and only if $f(x) + g(\phi(x) + b)$ is bent for every b. It is deduced that if g and h are both bent functions then:

- if g and h differ by a quadratic function then the $2n$-variable function $(g + h)(x) + \tilde{g}(y) + x \cdot y$ is bent,
- if g is a quadratic and ϕ is an affine permutation, then the $2n$-variable function $g(\phi(x)) + h(x) + \tilde{g}(y) + \phi(x) \cdot y$ is bent,
- if $Im(\phi) = \{\phi(x);\ x \in \mathbb{F}_2^n\}$ is either included in or disjoint from any translate of $supp(g)$, then the $2n$-variable function $f(x) + \tilde{g}(y) + \phi(x) \cdot y$ is bent.

5. A very simple observation of Hou and Langevin made in [11] leads to a potentially new construction of bent functions (which does not increase the number of variables, in contrast to most other secondary constructions).
6. **A construction without extension of the number of variables:** another secondary construction not increasing in the number of variables has been provided in [6]:

Theorem 6.0.4 ([6]). *Let f_1, f_2 and f_3 be three Boolean functions on $\mathbb{F}_2^n$. Let $s_1 = f_1 + f_2 + f_3$ and $s_2 = f_1 f_2 + f_1 f_3 + f_2 f_3$. Then*

$$\widehat{\chi_{f_1}} + \widehat{\chi_{f_2}} + \widehat{\chi_{f_3}} = \widehat{\chi_{s_1}} + 2\,\widehat{\chi_{s_2}}. \tag{6.2}$$

If f_1, f_2 and f_3 are bent then:

- *if s_1 is bent and if $\widetilde{s_1} = \widetilde{f_1} + \widetilde{f_2} + \widetilde{f_3}$, then s_2 is bent, and $\widetilde{s_2} = \widetilde{f_1}\widetilde{f_2} + \widetilde{f_1}\widetilde{f_3} + \widetilde{f_2}\widetilde{f_3}$;*
- *if $\widehat{\chi_{s_2}}(a)$ is divisible by 2^m for every a (e.g. if s_2 is bent), then s_1 is bent.*

It has been observed in [12] that the converse of Carlet's construction is also true: if f_1, f_2, f_3 and s_1 are bent, then s_2 is bent if and only if $\tilde{f_1} + \tilde{f_2} + \tilde{f_3} + \tilde{s_1} = 0$. To conclude, we have the following result given in [12].

Theorem 6.0.5. *Let n be an even integer. Let f_1, f_2 and f_3 be three pairwise distinct bent functions over $\mathbb{F}_{2^n}$ such that $\psi = f_1 + f_2 + f_3$ is bent. Let g be a Boolean function defined by $f_1f_2 + f_2f_3 + f_1f_3$. Then g is bent if and only if $\tilde{f_1} + \tilde{f_2} + \tilde{f_3} + \tilde{\psi} = 0$[3] for the definition of dual functions of bent functions). Furthermore, if g is bent then its dual function $\tilde{g}$ is given by*

$$\tilde{g}(x) = \tilde{f_1}(x)\tilde{f_2}(x) + \tilde{f_2}(x)\tilde{f_3}(x) + \tilde{f_3}(x)\tilde{f_1}(x), \forall x \in \mathbb{F}_{2^n}.$$

From the above result, the author deduced several new primary constructions of bent functions (see Chap. 7).

7. Using the notion of normal extension of bent function, Dobbertin et al. [1] have proposed another secondary construction of bent functions.
8. Bent functions associated to Almost Bent functions (AB functions): there exist also bent functions associated with some vectorial (n,n)-functions called *almost bent* (AB) [2]. Almost bent functions are those vectorial (n,n)-functions having maximal nonlinearity $2^{n-1} - 2^{\frac{n-1}{2}}$ (n odd). Bent functions from AB functions are developed in Chap. 12 (Sect. 12.4).

References

1. H. Dobbertin C. Carlet and G. Leander. Normal extensions of bent functions. In *EEE Transactions on Information Theory, vol. 50, no. 11*, pages 2880–2885, 2004.
2. C. Carlet. Vectorial Boolean Functions for Cryptography. In *Chapter of the monography Boolean Methods and Models, Y. Crama and P. Hammer eds, Cambridge University Press. Preliminary version available at* http://www-rocq.inria.fr/codes/Claude.Carlet/pubs.html.
3. C. Carlet. Two new classes of bent functions. In *Proceedings of EUROCRYPT'93, Lecture Notes in Computer Science 765*, pages 77–101, 1994.
4. C. Carlet. A construction of bent functions. In *Finite Fields and Applications, London Mathematical Society, Lecture Series 233, Cambridge University Press, pages 47–58*, 1996.
5. C. Carlet. On the secondary constructions of resilient and bent functions. In *Proceedings of the Workshop on Coding, Cryptography and Combinatorics 2003, published by Birkhäuser Verlag*, pages 3–28, 2004.
6. C. Carlet. On bent and highly nonlinear balanced/resilient functions and their algebraic immunities. In *AAECC 16, Las Vegas, February 2006, volume 3857 of Lecture Notes in Computer Science, Springer*, pages 1–28, 2006.
7. C. Carlet. Boolean functions for cryptography and error correcting codes. In Yves Crama and Peter L. Hammer, editors, *Boolean Models and Methods in Mathematics, Computer Science, and Engineering*, pages 257–397. Cambridge University Press, June 2010.

[3] Recall that $\tilde{h}$ stands for the dual of a bent function h.

8. C. Carlet. Open problems on binary bent functions. In *Proceedings of the conference "Open problems in mathematical and computational sciences", September 18–20, 2013, in Istanbul, Turkey, pages 203–241*, 2014.
9. C. Carlet, F. Zhang, and Y. Hu. Secondary constructions of bent functions and their enforcement. In *Advances in Mathematics of Communications (AMC) Volume 6, Number 3, pages 305–314*, 2012.
10. J. Dillon. Elementary Hadamard difference sets. In *PhD dissertation, University of Maryland.*
11. X.-D. Hou and P. Langevin. Results on bent functions. In *Journal of Combinatorial Theory, Series A, 80*, pages 232–246, 1997.
12. S. Mesnager. Several new infinite families of bent functions and their duals. In *IEEE Transactions on Information Theory-IT, Vol. 60, No. 7, Pages 4397–4407*, 2014.
13. O.S. Rothaus. On "bent" functions. In *J. Combin. Theory Ser A 20*, pages 300–305, 1976.
14. F. Zhang, C. Carlet, Y. Hu, and W. Zhang. New Secondary Constructions of Bent Functions. In *Preprint*, 2013.

8. C. Gardner. Open problems on locally best functions. In *Proceedings of the* [illegible] *Open Problems in* [illegible] *and* [illegible] *Sciences*, [illegible], [illegible] pages [illegible].
9. [illegible]
10. [illegible]
11. [illegible]
12. [illegible]
13. [illegible]
14. [illegible]

Chapter 7
Bent Functions: Primary Constructions (Part II)

In this chapter, we present large primary constructions of bent functions derived from a secondary construction due to Carlet. We'll see different possibilities of constructions (using permutations, involutions, the linear structures etc.) whose existence is based on algebraic problems in finite fields.

7.1 Primary Bent Functions with Products of Trace Functions from a Secondary Construction and Their Duals

Let us introduce the following construction of a function g based on three Boolean functions f_i ($i \in \{1, 2, 3\}$) given in [3] for which the Walsh transform of g can be expressed as a combination of those of the f_i's.

Proposition 7.1.1 ([3, Lemma 1]). *Let n be a positive integer. Let f_1, f_2 and f_3 be three pairwise distinct Boolean functions. Define then $g : \mathbb{F}_{2^n} \to \mathbb{F}_2$:*

$$g(x) = f_1(x)f_2(x) + f_1(x)f_3(x) + f_2(x)f_3(x). \tag{7.1}$$

Then, for every $\omega \in \mathbb{F}_{2^n}$

$$\widehat{\chi_g}(\omega) = \frac{\widehat{\chi_{f_1}}(\omega) + \widehat{\chi_{f_2}}(\omega) + \widehat{\chi_{f_3}}(\omega) - \widehat{\chi_\psi}(\omega)}{2}.$$

where $\psi := f_1 + f_2 + f_3$.

Recall that the *nonlinearity* of a Boolean function f denoted by $\mathrm{nl}(f)$ is the minimum Hamming distance between f and the set of affine functions (that is, functions of degree at most 1). It well known that $\mathrm{nl}(f)$ is upper bounded by $2^{n-1} - 2^{\frac{n}{2}-1}$ for any Boolean f defined over $\mathbb{F}_{2^n}$. Bent functions attain the maximum

S. Mesnager, *Bent Functions*, DOI 10.1007/978-3-319-32595-8_7

value $2^{n-1} - 2^{\frac{n}{2}-1}$ of nonlinearity. However, when n is even, semi-bent functions f defined over $\mathbb{F}_{2^n}$ (that is, functions such that $\widehat{\chi_f}(\omega) \in \{0, \pm 2^{\frac{n+2}{2}}\}, \quad \forall \omega \in \mathbb{F}_{2^n}$) have nonlinearity equal to $2^{n-1} - 2^{\frac{n}{2}}$ (this comes from the relation: $\mathrm{nl}(f) = 2^{n-1} - \frac{1}{2} \max_{a \in \mathbb{F}_{2^n}} |\widehat{\chi_f}(a)|$). Now, note that if the four functions f_1, f_2, f_3 and $f_1+f_2+f_3$ are bent, then the Walsh spectrum of the so-constructed g in Proposition 7.1.1 equals $\{-2^{m+1}, -2^m, 0, 2^m, 2^{m+1}\}$. Its nonlinearity therefore equals $\mathrm{nl}(g) = 2^{n-1} - 2^m$, that is, the same as that of a semi-bent function in even dimension. Highly nonlinear Boolean functions can therefore be constructed from construction (7.1). The question then arises of knowing whether bent functions, that have the highest possible nonlinearity, can be built from construction (7.1). The answer is stated in the theorems below. Recall that $\tilde{f}$ stands for the dual of a Boolean bent function f.

Theorem 7.1.2 ([3, Theorem 3]). *Let n be any positive integer. Let f_1, f_2 and f_3 be three bent functions. Denote by ψ the function $f_1 + f_2 + f_3$ and by g the function $f_1f_2 + f_1f_3 + f_2f_3$. Then*

1. *If ψ is bent and if $\tilde{\psi} = \tilde{f}_1 + \tilde{f}_2 + \tilde{f}_3$, then g is bent and $\tilde{g} = \tilde{f}_1\tilde{f}_2 + \tilde{f}_1\tilde{f}_3 + \tilde{f}_2\tilde{f}_3$.*
2. *If g is bent, or if more generally $\widehat{\chi_g}(w)$ is divisible by $2^{\frac{n}{2}}$ for every $w \in \mathbb{F}_{2^n}$, then ψ is bent.*

One can complete the second assertion of Theorem 7.1.2 to get the theorem below. We shall use extensively in the sequel this theorem to construct several new families of bent functions and compute their duals.

Theorem 7.1.3 ([11]). *Let n be an even integer. Let f_1, f_2 and f_3 be three pairwise distinct bent functions over $\mathbb{F}_{2^n}$ such that $\psi = f_3 + f_2 + f_1$ is bent. Let g be a Boolean function defined by (7.1). Then g is bent if and only if $\tilde{f}_1 + \tilde{f}_2 + \tilde{f}_3 + \tilde{\psi} = 0$. Furthermore, if g is bent then its dual function $\tilde{g}$ is given by*

$$\tilde{g}(x) = \tilde{f}_1(x)\tilde{f}_2(x) + \tilde{f}_2(x)\tilde{f}_3(x) + \tilde{f}_3(x)\tilde{f}_1(x), \forall x \in \mathbb{F}_{2^n}.$$

Proof. The proof uses some arguments of the proof of [3, Theorem 3]. Recall firstly that if h is a bent Boolean function on $\mathbb{F}_{2^n}$ whose dual function is $\tilde{h}$, then its Walsh transform at ω is $\widehat{\chi_h}(\omega) = 2^m \chi(\tilde{h}(\omega))$. Proposition 7.1.1 says that the Walsh transform of g at $\omega \in \mathbb{F}_{2^n}$ is

$$\begin{aligned}
&\widehat{\chi_g}(\omega) \\
&= 2^{m-1} \left(\chi(\tilde{f}_1(\omega)) + \chi(\tilde{f}_2(\omega)) + \chi(\tilde{f}_3(\omega)) - \chi(\tilde{\psi}(\omega))\right). \\
&= 2^m (\tilde{f}_1(\omega) + \tilde{f}_2(\omega) + \tilde{f}_3(\omega) + \tilde{\psi}(\omega) + 1) \pmod{2^{m+1}}.
\end{aligned}$$

Now, recall that g is bent if and only if $\widehat{\chi_g}(\omega) \equiv 2^m \pmod{2^{m+1}}$ for every $\omega \in \mathbb{F}_{2^n}$, that is, if and only if

$$\tilde{f}_1(\omega) + \tilde{f}_2(\omega) + \tilde{f}_3(\omega) + \tilde{\psi}(\omega) \equiv 0 \pmod 2, \forall \omega \in \mathbb{F}_{2^n}$$

proving that g is bent if and only if $\tilde{f}_1 + \tilde{f}_2 + \tilde{f}_3 = \tilde{\psi}$ (addition modulo 2). □

Let us now rewrite Theorem 7.1.3 in a particular case:

$$f_i : x \mapsto f_i(x) := h(x) + Tr_1^n(\lambda_i x), \lambda_i \in \mathbb{F}_{2^n}^\star, i \in \{1, 2, 3\}$$

where h is any Boolean function defined on $\mathbb{F}_{2^n}$, λ_1, λ_2, and λ_3 are pairwise distinct elements of $\mathbb{F}_{2^n}^\star$. In this case, if we develop each term in its definition (Eq. (7.1)), some terms cancel out and one can rewrite the function g of Theorem 7.1.3 as

$$\begin{aligned} g(x) =& h(x) + Tr_1^n(\lambda_1 x)Tr_1^n(\lambda_2 x) + Tr_1^n(\lambda_1 x)Tr_1^n(\lambda_3 x) \\ &+ Tr_1^n(\lambda_2 x)Tr_1^n(\lambda_3 x). \end{aligned} \tag{7.2}$$

Assume that h is bent. Denote by $\tilde{h}$ its dual function. Note then (since the bentness is affine invariant) that the functions f_i ($i \in \{1, 2, 3\}$) are also bent. Their corresponding dual functions $\widetilde{f_i}$ ($i \in \{1, 2, 3\}$) are given by $\widetilde{f_i}(x) = \tilde{h}(x+\lambda_i)$, $\forall x \in \mathbb{F}_{2^n}$ (e.g. see [2]).

Now, recall that the first derivative of a Boolean function f in the direction of $a \in \mathbb{F}_{2^n}$ is defined as $D_a f(x) = f(x) + f(x + a)$ and the second order derivative of f with respect to $(a, b) \in \mathbb{F}_{2^n}^2$ is defined as $D_b D_a f(x) = f(x) + f(x + b) + f(x + a) + f(x + a + b)$. Note then that $f_1(x) + f_2(x) + f_3(x) = h(x) + Tr_1^n((\lambda_1 + \lambda_2 + \lambda_3)x)$. Therefore the condition on bentness of g of Theorem 7.1.3 becomes:

$$\begin{aligned} &\forall x \in \mathbb{F}_{2^n}, \sum_{i=1}^{3} \tilde{h}(x + \lambda_i) + \tilde{h}(x + \lambda_1 + \lambda_2 + \lambda_3) = 0 \\ \iff& \forall x \in \mathbb{F}_{2^n}, D_{\lambda_1+\lambda_3} D_{\lambda_1+\lambda_2} \tilde{h}(x + \lambda_1 + \lambda_2 + \lambda_3) = 0 \\ \iff& \forall x \in \mathbb{F}_{2^n}, D_{\lambda_1+\lambda_3} D_{\lambda_1+\lambda_2} \tilde{h}(x) = 0. \end{aligned}$$

Thus, one can prove the following consequence of Theorem 7.1.3.

Corollary 7.1.4 ([11]). *Let h be a bent function defined on $\mathbb{F}_{2^n}$ whose dual function $\tilde{h}$ has a null second order derivative with respect to $(a, b) \in \mathbb{F}_{2^n}^2$ with $a \neq b$. Then the function g' defined by*

$$\forall x \in \mathbb{F}_{2^n},\ g'(x) := h(x) + Tr_1^n(ax)Tr_1^n(bx)$$

is bent and its dual $\tilde{g}'$ is given by

$$\begin{aligned} \forall x \in \mathbb{F}_{2^n}, \tilde{g}'(x) =& \tilde{h}(x)\tilde{h}(x + a) + \tilde{h}(x)\tilde{h}(x + b) \\ &+ \tilde{h}(x + a)\tilde{h}(x + b). \end{aligned}$$

Proof. Let f_i be defined as above by $f_i(x) := h(x) + Tr_1^n(\lambda_i x)$ for $i \in \{1, 2, 3\}$, $f_4 := f_1 + f_2 + f_3$ and g defined by (7.1). Taking $\lambda_1 \in \mathbb{F}_{2^n}^\star$, $\lambda_2 = \lambda_1 + b$, $\lambda_3 = \lambda_1 + a$ and $\lambda_4 = \lambda_1 + a + b$. Then Eq. (7.2) becomes

$$\begin{aligned}
g(x) &= h(x) + \mathrm{Tr}_1^n(\lambda_1 x)\Big(\mathrm{Tr}_1^n(\lambda_1 x) + \mathrm{Tr}_1^n(bx)\Big) \\
&\quad + \mathrm{Tr}_1^n(\lambda_1 x)\Big(\mathrm{Tr}_1^n(\lambda_1 x) + \mathrm{Tr}_1^n(ax)\Big) \\
&\quad + \Big(\mathrm{Tr}_1^n(\lambda_1 x) + \mathrm{Tr}_1^n(bx)\Big)\Big(\mathrm{Tr}_1^n(\lambda_1 x) + \mathrm{Tr}_1^n(ax)\Big) \\
&= h(x) + \mathrm{Tr}_1^n(bx)\mathrm{Tr}_1^n(ax) + \mathrm{Tr}_1^n(\lambda_1 x) \\
&= g'(x) + \mathrm{Tr}_1^n(\lambda_1 x).
\end{aligned}$$

Note that since h is bent, the functions f_i ($i \in \{1,2,3,4\}$) are bent ($f_4(x) = \sum_{i=1}^{3} f_i(x) = h(x) + \mathrm{Tr}_1^n(\sum_{i=1}^{3} \lambda_i x)$). On the other hand, the condition on the derivatives: $D_a D_b \tilde{h}(x) = 0, \forall x \in \mathbb{F}_{2^n}$ is equivalent to

$$D_{\lambda_1+\lambda_3} D_{\lambda_1+\lambda_2} \tilde{h}(x) = 0, \forall x \in \mathbb{F}_{2^n}$$

that is, (see above)

$$\sum_{i=1}^{4} \widetilde{f_i}(x) = 0, \forall x \in \mathbb{F}_{2^n}$$

where $\widetilde{f_i}$ denotes the dual function of the bent function f_i. The conditions of Theorem 7.1.3 are thus fulfilled (note that since $a \neq b$, the four functions f_i are pairwise distinct). Therefore, g is bent which proves that g' is bent (since g and g' are EA-equivalent).

According to Theorem 7.1.3, the dual function $\tilde{g}$ of g is given by

$$\begin{aligned}
\tilde{g}(x) &= \tilde{h}(x+\lambda_1)\tilde{h}(x+\lambda_2) + \tilde{h}(x+\lambda_2)\tilde{h}(x+\lambda_3) \\
&\quad + \tilde{h}(x+\lambda_1)\tilde{h}(x+\lambda_3) \\
&= \tilde{h}(x+\lambda_1)\tilde{h}(x+\lambda_1+b) + \tilde{h}(x+\lambda_1+b)\tilde{h}(x+\lambda_1+a) \\
&\quad + \tilde{h}(x+\lambda_1+a)\tilde{h}(x+\lambda_1).
\end{aligned}$$

Now, the dual functions of g and g' are linked by the relation $\tilde{g}'(x) = \tilde{g}(x+\lambda_1)$, yielding the result. □

One can generalize Corollary 7.1.4.

Corollary 7.1.5 ([11]). *Let h_1 and h_2 be two bent functions over $\mathbb{F}_{2^n}$. Assume that $D_a\widetilde{h_1} = D_a\tilde{h}_2$ with respect to some $a \in \mathbb{F}_{2^n}^{\star}$ where $\tilde{h}_1$ and $\tilde{h}_2$ stand for the dual functions of h_1 and h_2, respectively. Let g be the Boolean function defined on $\mathbb{F}_{2^n}$ by*

$$g(x) = h_1(x) + \mathrm{Tr}_1^n(ax)(h_1(x) + h_2(x) + 1), \forall x \in \mathbb{F}_{2^n}.$$

Then g is bent and its dual function $\tilde{g}$ is given by

$$\tilde{g}(x) = \tilde{h}_1(x) + \tilde{h}_1(x)\tilde{h}_2(x) + \tilde{h}_1(x+a)\tilde{h}_2(x+a), \forall x \in \mathbb{F}_{2^n}.$$

Proof. Note that $g(x)$ can be rewritten as

$$g(x) = h_1(x)\Big(h_1(x) + h_2(x)\Big) + \Big(h_2(x) + \mathrm{Tr}_1^n(ax)\Big)\Big(h_1(x) + h_2(x) + 1\Big).$$

Set $f_1(x) := h_1(x), f_2(x) := h_2(x) + \mathrm{Tr}_1^n(ax), f_3(x) := h_1(x) + \mathrm{Tr}_1^n(ax)$ and $f_4 := f_1 + f_2 + f_3$. It is clear that the functions f_i, $i \in \{1,2,3,4\}$ are bent (since h_1 and h_2 are bent and the bentness is invariant by addition of linear functions). Denote by $\widetilde{f_i}$ its dual functions. One has:

$$\begin{aligned}
&\widetilde{f_1}(x) + \widetilde{f_2}(x) + \widetilde{f_3}(x) + \widetilde{f_4}(x) \\
&\quad = \tilde{h}_1(x) + \tilde{h}_2(x+a) + \tilde{h}_1(x+a) + \tilde{h}_2(x) \\
&\quad = D_a\tilde{h}_1(x) + D_a\tilde{h}_2(x) = 0.
\end{aligned}$$

The conditions of Theorem 7.1.3 are fulfilled. The bentness of g and the computation of its dual $\tilde{g}$ follow then from Theorem 7.1.3. The expression of the dual $\tilde{g}$ is given by applying Theorem 7.1.3:

$$\begin{aligned}
\tilde{g}(x) &= \tilde{h}_1(x)\tilde{h}_2(x+a) + \tilde{h}_1(x+a)\tilde{h}_1(x) \\
&\quad + \tilde{h}_2(x+a)\tilde{h}_1(x+a) \\
&= \tilde{h}_1(x)(\tilde{h}_2(x+a) + \tilde{h}_1(x+a)) + \tilde{h}_1(x+a)\tilde{h}_2(x+a) \\
&= \tilde{h}_1(x)(\tilde{h}_2(x) + \tilde{h}_1(x)) + \tilde{h}_1(x+a)\tilde{h}_2(x+a) \\
&\quad (\text{since } D_a\tilde{h}_1(x) + D_a\tilde{h}_2(x) = 0) \\
&= \tilde{h}_1(x) + \tilde{h}_1(x)\tilde{h}_2(x) + \tilde{h}_1(x+a)\tilde{h}_2(x+a).
\end{aligned}$$

□

Now, there are two particular classes of bent Boolean functions when considering their dual: the so-called *self-dual* and *anti-self-dual* bent functions.

Definition 7.1.6. A bent function f is said to be *self-dual* if $\tilde{f} = f$ while it is said to be *anti-self-dual* if $\tilde{f} = 1 + f$.

Another known result about dual functions is that the dual of a bent function of algebraic degree 2 is also of algebraic degree 2.

Now, one can show straightforwardly obtain from Theorem 7.1.3 the following statement providing new secondary constructions of self-dual bent functions.

Corollary 7.1.7 ([11]). *Let n be an even integer. Let f_1, f_2, f_3 be three self-dual bent functions over $\mathbb{F}_{2^n}$ such that $f_3 + f_2 + f_1$ is self-dual bent. Let g be defined as*

$$g(x) = f_1(x)f_2(x) + f_1(x)f_3(x) + f_2(x)f_3(x), \forall x \in \mathbb{F}_{2^n}.$$

Then g is self-dual bent.

Remark 7.1.8. We shall show in Sect. 7.1.4 an application of Corollary 7.1.7. Indeed, we shall construct an infinite family of cubic Boolean function from quadratic Boolean functions of the form $Tr_1^{4k}(ax^{2^k+1})$ (Theorem 7.1.27). The question then arises of knowing whether such a statement can hold for anti-self-dual bent functions. However the existence of three anti-self-dual bent functions whose sum is also anti-self-dual seems possible under some modifications of the function.

A vectorial function $F : \mathbb{F}_{2^n} \to \mathbb{F}_{2^r}$ (or a (n, r)-function) is bent if and only if all its components (Boolean) functions $f_a : x \in \mathbb{F}_{2^n} \mapsto Tr_1^r(aF(x)), a \in \mathbb{F}_{2^r}^\star$ are bent. It is well known [16] that bent (n, r)-functions exist when n is even and $r \leq \frac{n}{2}$. Note that if one can find three pairwise distinct components f_{a_1}, f_{a_2} and f_{a_3} such that $\tilde{f}_{a_1} + \tilde{f}_{a_2} + \tilde{f}_{a_3} + \tilde{f}_{a_1+a_2+a_3} = 0$, where $\tilde{f}_a$ stands for the dual function of f_a, we can deduce from Theorem 7.1.3 that

$$g(x) = f_{a_1}(x)f_{a_2}(x) + f_{a_1}(x)f_{a_3}(x) + f_{a_2}(x)f_{a_3}(x)$$

is bent and that its dual function is $\tilde{f}_{a_1}\tilde{f}_{a_2} + \tilde{f}_{a_1}\tilde{f}_{a_3} + \tilde{f}_{a_2}\tilde{f}_{a_3}$.

Problem 7.1.9. Let n be an even positive integer and r be a positive integer with $r \leq \frac{n}{2}$. Let F be a bent vectorial map from $\mathbb{F}_{2^n}$ to $\mathbb{F}_{2^r}$. For $b \in \mathbb{F}_{2^r}^\star$, denote by f_b a component of F and by $\tilde{f}_b$ its dual. Find (a_1, a_2, a_3) a 3-tuple of pairwise distinct elements of $\mathbb{F}_{2^r}^\star$ with $a_3 \neq a_1 + a_2$ such that $\tilde{f}_{a_1} + \tilde{f}_{a_2} + \tilde{f}_{a_3} + \tilde{f}_{a_1+a_2+a_3} = 0$.

In the following subsections, we show how to apply the results from the previous section to obtain several infinite classes of new primary constructions of bent functions. Note that we provide the dual functions for all the new families of bent functions that we exhibit.

7.1.1 Infinite Families of Bent Functions via Kasami Function and Niho Exponents, and Their Duals

Let $n = 2m$ be a positive even integer. Let h be the monomial Niho quadratic function

$$h : x \in \mathbb{F}_{2^n} \mapsto Tr_1^m(\lambda x^{2^m+1}) \tag{7.3}$$

where $\lambda \in \mathbb{F}_{2^m}^\star$. It is well known that h is bent. The dual function $\tilde{h}$ is given (see e.g. [14]) by

$$\tilde{h}(x) = \mathrm{Tr}_1^m(\lambda^{-1}x^{2^m+1}) + 1. \tag{7.4}$$

In the following statement, we provide a construction for a new family of bent functions from the Kasami function h and compute its dual.

Theorem 7.1.10 ([11]). *Let $n = 2m$. Let $\lambda \in \mathbb{F}_{2^m}^\star$. Let $(a, b) \in \mathbb{F}_{2^n}^\star \times \mathbb{F}_{2^n}^\star$ such that $a \neq b$ and $\mathrm{Tr}_1^n(\lambda^{-1}b^{2^m}a) = 0$. Then the Boolean function f defined on $\mathbb{F}_{2^n}$ as $f(x) = \mathrm{Tr}_1^m(\lambda x^{2^m+1}) + \mathrm{Tr}_1^n(ax)\mathrm{Tr}_1^n(bx)$ is a bent function of algebraic degree 2 and its dual function $\tilde{f}$ is given by*

$$\begin{aligned}\tilde{f}(x) &= 1 + \mathrm{Tr}_1^m(\lambda^{-1}x^{2^m+1})\\ &+ \left(\mathrm{Tr}_1^m(\lambda^{-1}a^{2^m+1}) + \mathrm{Tr}_1^n(\lambda^{-1}a^{2^m}x)\right)\\ &\times \left(\mathrm{Tr}_1^m(\lambda^{-1}b^{2^m+1}) + \mathrm{Tr}_1^n(\lambda^{-1}b^{2^m}x)\right).\end{aligned}$$

Proof. Let h and $\tilde{h}$ be defined, respectively, by (7.3) and (7.4). Compute the first derivative of $\tilde{h}$ in the direction of $b \in \mathbb{F}_{2^n}$:

$$\begin{aligned}&D_b\tilde{h}(x)\\ &= \mathrm{Tr}_1^m(\lambda^{-1}x^{2^m+1}) + \mathrm{Tr}_1^m(\lambda^{-1}(x+b)^{2^m+1})\\ &= \mathrm{Tr}_1^m(\lambda^{-1}x^{2^m+1})\\ &\quad + \mathrm{Tr}_1^m(\lambda^{-1}x^{2^m+1} + \lambda^{-1}b^{2^m+1} + \lambda^{-1}x^{2^m}b + \lambda^{-1}b^{2^m}x)\\ &= \mathrm{Tr}_1^m(\lambda^{-1}(x^{2^m}b + b^{2^m}x)) + \mathrm{Tr}_1^m(\lambda^{-1}b^{2^m+1})\\ &= \mathrm{Tr}_1^m(\lambda^{-1}\mathrm{Tr}_m^n(b^{2^m}x)) + \mathrm{Tr}_1^m(\lambda^{-1}b^{2^m+1})\\ &= \mathrm{Tr}_1^m(\mathrm{Tr}_m^n(\lambda^{-1}b^{2^m}x)) + \mathrm{Tr}_1^m(\lambda^{-1}b^{2^m+1})\\ &= \mathrm{Tr}_1^n(\lambda^{-1}b^{2^m}x) + \mathrm{Tr}_1^m(\lambda^{-1}b^{2^m+1}).\end{aligned}$$

The second order derivative of $\tilde{h}$ in the direction of $(a, b) \in \mathbb{F}_{2^n}^\star \times \mathbb{F}_{2^n}^\star$ equals:

$$\begin{aligned}D_aD_b\tilde{h}(x) &= \mathrm{Tr}_1^n(\lambda^{-1}b^{2^m}x) + \mathrm{Tr}_1^m(\lambda^{-1}b^{2^m+1})\\ &+ \mathrm{Tr}_1^n(\lambda^{-1}b^{2^m}(x+a)) + \mathrm{Tr}_1^m(\lambda^{-1}b^{2^m+1})\\ &= \mathrm{Tr}_1^n(\lambda^{-1}b^{2^m}a).\end{aligned}$$

Since a and b are such that $Tr_1^n(\lambda^{-1}b^{2^m}a) = 0$, then the function h fulfills the condition of Corollary 7.1.4, that is, $f(x) = h(x) + Tr_1^n(ax)Tr_1^n(bx)$ is bent whose dual function is

$$\tilde{f}(x) = \left(Tr_1^m(\lambda^{-1}x^{2^m+1}) + 1\right)\left(Tr_1^m(\lambda^{-1}(x+a)^{2^m+1}) + 1\right)$$
$$+ \left(Tr_1^m(\lambda^{-1}x^{2^m+1}) + 1\right)\left(Tr_1^m(\lambda^{-1}(x+b)^{2^m+1}) + 1\right)$$
$$+ \left(Tr_1^m(\lambda^{-1}(x+a)^{2^m+1}) + 1\right)\left(Tr_1^m(\lambda^{-1}(x+b)^{2^m+1}) + 1\right).$$

Now, denote by h' the Boolean function defined on $\mathbb{F}_{2^n}$ by $h'(x) = Tr_1^m(\lambda^{-1}x^{2^m+1})$. We have noted above that for every c and x elements of $\mathbb{F}_{2^m}$, we have

$$\begin{aligned} &Tr_1^m(\lambda^{-1}(x+c)^{2^m+1}) \\ &= h'(x) + h'(c) + Tr_1^m(\lambda^{-1}(c^{2^m}x + cx^{2^m})) \\ &= h'(x) + h'(c) + Tr_1^m(Tr_m^n(\lambda^{-1}c^{2^m}x)) \\ &= h'(x) + h'(c) + Tr_1^n(\lambda^{-1}c^{2^m}x). \end{aligned}$$

Hence $\tilde{f}$ can be expressed by means of h' as follows.

$$\begin{aligned} &\tilde{f}(x) \\ &= (h'(x)+1)\left(h'(x) + h'(a) + Tr_1^n(\lambda^{-1}a^{2^m}x) + 1\right) \\ &+ (h'(x)+1)\left(h'(x) + h'(b) + Tr_1^n(\lambda^{-1}b^{2^m}x) + 1\right) \\ &+ \left(h'(x) + h'(a) + Tr_1^n(\lambda^{-1}a^{2^m}x) + 1\right) \\ &\quad \times \left(h'(x) + h'(b) + Tr_1^n(\lambda^{-1}b^{2^m}x) + 1\right) \\ &= h'(a)h'(b) \\ &+ h'(a)Tr_1^n(\lambda^{-1}b^{2^m}x) + h'(b)Tr_1^n(\lambda^{-1}a^{2^m}x) + h'(x) \\ &+ Tr_1^n(\lambda^{-1}a^{2^m}x)Tr_1^n(\lambda^{-1}b^{2^m}x) + 1 \\ &= h'(x) + 1 \\ &+ \left(h'(a) + Tr_1^n(\lambda^{-1}a^{2^m}x)\right)\left(h'(b) + Tr_1^n(\lambda^{-1}b^{2^m}x)\right). \end{aligned}$$

The result follows. □

Example 7.1.11 ($n = 6$). Let ζ be a primitive element of $\mathbb{F}_{64}$ such that $\zeta^6+\zeta^4+\zeta^3+\zeta+1 = 0$. One has $Tr_1^6(\zeta) = 0$. In that case, the expression of f in Theorem 7.1.10 is $f(x) = Tr_1^3(\lambda x^9) + Tr_1^6(ax)Tr_1^6(bx) = Tr_1^3(\lambda x^9) + Tr_1^6(ax(bx + (bx)^2 + (bx)^4 + (bx)^8 + (bx)^{16} + (bx)^{32}) = Tr_1^3(\lambda x^9) + Tr_1^6((ab)^{32}x + ((ab^2) + a^2b)x^3 + ((ab^4 + a^4b)x^5 + (ab^8)x^9)$.

Taking $\lambda = \zeta$, $b = 1$ and $a = \zeta^2 + \zeta^3 + \zeta^5 \in \mathbb{F}_8$. The two conditions $a \neq b$ and $Tr_1^6(\lambda^{-1}b^8a) = Tr_1^6(\zeta + 1) = 0$ are fulfilled. Then, according to Theorem 7.1.10, the following function of algebraic degree 2

$$\begin{aligned} f(x) &= Tr_1^6((\zeta^3 + \zeta^4)x) + Tr_1^6((\zeta + \zeta^2 + \zeta^4)x^3) \\ &+ Tr_1^6((1 + \zeta^2 + \zeta^4 + \zeta^5)x^5) \\ &+ Tr_1^3((\zeta^2 + \zeta^3 + \zeta^5)x^9) \end{aligned}$$

is bent and its dual function is

$$\begin{aligned} \tilde{f}(x) &= Tr_1^6((\zeta^5 + 1)x) + Tr_1^6((\zeta + \zeta^2 + \zeta^3 + \zeta^5)x^3) \\ &+ Tr_1^6((\zeta + \zeta^4)x^5) + Tr_1^3((\zeta + \zeta^3)x^9). \end{aligned}$$

Consider now the bent function h (found by Leander and Kholosha [9]) defined on $\mathbb{F}_{2^n}$ via 2^r Niho exponents by $h(x) = Tr_1^n\left(ax^{2^m+1} + \sum_{i=1}^{2^{r-1}-1} x^{(2^m-1)\frac{i}{2^r}+1}\right)$ with $r > 1$ satisfying $\gcd(r, m) = 1$ and $a \in \mathbb{F}_{2^n}$ such that $a + a^{2^m} = 1$. Note that, for $a \in \mathbb{F}_{2^n}$ such that $a + a^{2^m} = 1$, $Tr_1^n(ax^{2^m+1}) = Tr_1^m(x^{2^m+1}Tr_m^n(a)) = Tr_1^m(x^{2^m+1})$. Therefore, the expression of h can be rewritten as

$$Tr_1^m\left(x^{2^m+1}\right) + Tr_1^n\left(\sum_{i=1}^{2^{r-1}-1} x^{(2^m-1)\frac{i}{2^r}+1}\right). \tag{7.5}$$

Take any $u \in \mathbb{F}_{2^n}$ with $u + u^{2^m} = 1$. It has been shown in [6] (see also [1]) that the dual function $\tilde{h}$ of h is given by

$$\begin{aligned} \tilde{h}(x) &= Tr_1^m\Big(\big(u(1 + x + x^{2^m}) + u^{2^{n-r}} + x^{2^m}\big) \\ &\times (1 + x + x^{2^m})^{1/(2^r-1)}\Big). \end{aligned}$$

Note that the latter expression does not depend on the choice of u as has been shown in [6]. Moreover, if $d < m$ is a positive integer defined uniquely by $dr \equiv 1 \pmod m$ then the algebraic degree of $\tilde{h}$ is equal to $d + 1$. Now, from this result and Corollary 7.1.4, one can derive the following construction of bent functions via 2^r Niho exponents.

Theorem 7.1.12 ([11]). *Let (λ, μ) be a pair of distinct elements of $\mathbb{F}_{2^m}^\star$. Define a Boolean function h over $\mathbb{F}_{2^n}$ by*

$$\begin{aligned} h(x) &= Tr_1^m(x^{2^m+1}) + Tr_1^n\left(\sum_{i=1}^{2^{r-1}-1} x^{(2^m-1)\frac{i}{2^r}+1}\right) \\ &+ Tr_1^n(\lambda x)Tr_1^n(\mu x). \end{aligned}$$

Then h is bent and its dual function $\tilde{h}$ is given by

$$
\begin{aligned}
&\tilde{h}(x) \\
&= Tr_1^m\Big(\big(u(1+x+x^{2^m}) + u^{2^{n-r}} + x^{2^m}\big) \\
&\times(1+x+x^{2^m})^{\frac{1}{2^r-1}}\Big)Tr_1^m\Big((\lambda+\mu)(1+x+x^{2^m})^{\frac{1}{2^r-1}}\Big) \\
&+Tr_1^m\Big(\big(u(1+x+x^{2^m}) + u^{2^{n-r}} + x^{2^m} + \lambda\big) \\
&\quad\times(1+x+x^{2^m})^{\frac{1}{2^r-1}}\Big) \\
&\times Tr_1^m\Big(\big(u(1+x+x^{2^m}) + u^{2^{n-r}} + x^{2^m} + \mu\big) \\
&\quad\times(1+x+x^{2^m})^{\frac{1}{2^r-1}}\Big)
\end{aligned}
$$

where u is any element in $\mathbb{F}_{2^n}$ satisfying $u + u^{2^m} = 1$.

Proof. First note that for $\lambda \in \mathbb{F}_{2^m}^\star$, $1 + (x+\lambda) + (x+\lambda)^{2^m} = 1 + x + x^{2^m}$. Set $h'(x) := Tr_1^m(x^{2^m+1}) + Tr_1^n\Big(\sum_{i=1}^{2^{r-1}-1} x^{(2^m-1)\frac{i}{2^r}+1}\Big)$. Then,

$$
\begin{aligned}
D_\lambda\widetilde{h'}(x) &= Tr_1^m((u(1+x+x^{2^m}) + u^{2^{n-r}} + x^{2^m}) \\
&\quad\times(1+x+x^{2^m})^{\frac{1}{2^r-1}} \\
&\quad+ Tr_1^m(u(1+x+x^{2^m}) + u^{2^{n-r}} + x^{2^m} + \lambda) \\
&\quad\times(1+x+x^{2^m})^{\frac{1}{2^r-1}} \\
&= Tr_1^m(\lambda(1+x+x^{2^m})^{\frac{1}{2^r-1}}).
\end{aligned}
$$

Now note that, for every μ and x, it holds that $D_\lambda\widetilde{h'}(x) = D_\lambda\widetilde{h'}(x+\mu)$, that is, $D_\mu D_\lambda\widetilde{h'}(x) = 0$. Therefore, one can apply Corollary 7.1.4. The expression of $\tilde{h}$ is then obtained by straightforward calculation. □

Example 7.1.13 ($n = 8$, $r = 2$). Let ζ be a primitive element of $\mathbb{F}_{256}$ such that $\zeta^8 + \zeta^5 + \zeta^4 + \zeta^3 + \zeta^2 + \zeta + 1 = 0$. Let $\lambda = 1$ and $\mu = \zeta^5 + \zeta^3 + 1 \in \mathbb{F}_{16}$. The following function, of algebraic degree 4, is bent:

$$
\begin{aligned}
f(x) &= Tr_1^8((\zeta^7 + \zeta^5 + \zeta^4 + \zeta^3 + \zeta^2)x) \\
&\quad+ Tr_1^8((\zeta^2 + \zeta^4 + \zeta^7)x^3) + Tr_1^8(x^5) \\
&\quad+ Tr_1^8((\zeta^2 + \zeta^4 + \zeta^7 + 1)x^9) + Tr_1^8(x^{17}) \\
&\quad+ Tr_1^8(x^{23}) + Tr_1^8(x^{38}) + Tr_1^8(x^{53}).
\end{aligned}
$$

Note that the inverse of 7 modulo $2^m - 1$ is 13. Hence, $(1+x+x^{16})^{1/7} = (1+x+x^{16})^{13}$ for every $x \in \mathbb{F}_{256}$. The dual of f is therefore equal to

$$\begin{aligned}
&\tilde{f}(x)\\
&= \mathrm{Tr}_1^4\Big(\big(u\,(1+x+x^{16})\big) + u^2 + x^{16}\big)(1+x+x^{16})^{13}\Big)\\
&+\mathrm{Tr}_1^4\left((1+x+x^{16})^{13}\right)\\
&\mathrm{Tr}_1^4\left((\zeta^5+\zeta^3+1)(1+x+x^{16})^{13}\right)
\end{aligned}$$

for some $u \in \mathbb{F}_{256}$ such that $u + u^{16} = 1$.

7.1.2 *Bent Functions from the Class of Maiorana–McFarland and Their Duals*

We consider in this subsection Boolean functions defined over $\mathbb{F}_{2^m} \times \mathbb{F}_{2^m}$ by

$$f(x,y) = \mathrm{Tr}_1^m(\phi(y)x) + g(y), \quad (x,y) \in \mathbb{F}_{2^m} \times \mathbb{F}_{2^m} \tag{7.6}$$

where m is some positive integer, ϕ is a function from $\mathbb{F}_{2^m}$ to itself and g stands for a Boolean function over $\mathbb{F}_{2^m}$. Recall that the class of bent functions given by (7.6) is the so- called *Maiorana–McFarland* 's class. That class has been widely studied because its Walsh transform can be easily computed and its elements completely characterized (e.g. see [4]).

Proposition 7.1.14 ([11]). *Let m be a positive integer. Let g be a Boolean function defined over $\mathbb{F}_{2^m}$. Define f over $\mathbb{F}_{2^m} \times \mathbb{F}_{2^m}$ by (7.6). Then f is bent if and only if ϕ is a permutation of $\mathbb{F}_{2^m}$. Furthermore, its dual function $\tilde{f}$ is*

$$\tilde{f}(x,y) = \mathrm{Tr}_1^m(y\phi^{-1}(x)) + g(\phi^{-1}(x)). \tag{7.7}$$

where ϕ^{-1} denotes the inverse mapping of the permutation ϕ.

Now, in order to apply Theorem 7.1.3, one has to find three permutations ϕ_1, ϕ_2, ϕ_3 of $\mathbb{F}_{2^m}$ and three Boolean functions g_1, g_2 and g_3 on $\mathbb{F}_{2^m}$ such that

1.

$$\psi = \phi_1 + \phi_2 + \phi_3 \tag{7.8}$$

is a permutation whose inverse function is $\psi^{-1} = \phi_1^{-1} + \phi_2^{-1} + \phi_3^{-1}$.

2.

$$g_1 \circ \phi_1^{-1} + g_2 \circ \phi_2^{-1} + g_3 \circ \phi_3^{-1} + h \circ \psi^{-1} = 0 \tag{7.9}$$

where $h = g_1 + g_2 + g_3$.

We shall from now on investigate which permutations could satisfy conditions (7.8) and (7.9). In a first step, we consider functions of the form (7.12) such that $g = 0$.

Let us consider firstly monomial permutations of $\mathbb{F}_{2^m} : \phi(y) = ay^d$, $a \in \mathbb{F}_{2^m}$, $\gcd(d, 2^m - 1) = 1$. Denote by e the inverse of d modulo $2^m - 1$ i.e. $de = 1 \pmod{2^m - 1}$. The inverse map of ϕ is defined as $\phi^{-1}(y) = a^{-e}y^e$. Condition (7.8) is fulfilled for every 3-tuple (a_1, a_2, a_3) of $\mathbb{F}_{2^m}$ such that $b := a_1 + a_2 + a_3 \neq 0$ and $a_1^{-e} + a_2^{-e} + a_3^{-e} = b^{-e}$. Therefore, we deduce the following result.

Theorem 7.1.15 ([11]). *Let d be a positive integer not a power of 2 and coprime with $2^m - 1$. For $i \in \{1, 2, 3\}$, let $f_i(x, y) = Tr_1^m(a_i y^d x)$ for some $a_i \in \mathbb{F}_{2^m}$. Assume the a_i's are pairwise distinct such that $b := a_1 + a_2 + a_3 \neq 0$ and $a_1^{-e} + a_2^{-e} + a_3^{-e} = b^{-e}$ where e stands for the inverse of d modulo $(2^m - 1)$. Let f be the Boolean function defined in bivariate form over $\mathbb{F}_{2^m} \times \mathbb{F}_{2^m}$ as*

$$\begin{aligned} f(x, y) &= Tr_1^m(a_1 y^d x) Tr_1^m(a_2 y^d x) \\ &\quad + Tr_1^m(a_1 y^d x) Tr_1^m(a_3 y^d x) \\ &\quad + Tr_1^m(a_2 y^d x) Tr_1^m(a_3 y^d x). \end{aligned}$$

Then f is bent and its dual function is given by

$$\begin{aligned} \tilde{f}(x, y) &= Tr_1^m(a_1^{-e} x^e y) Tr_1^m(a_2^{-e} x^e y) \\ &\quad + Tr_1^m(a_1^{-e} x^e y) Tr_1^m(a_3^{-e} x^e y) \\ &\quad + Tr_1^m(a_2^{-e} x^e y) Tr_1^m(a_3^{-e} x^e y). \end{aligned}$$

Remark 7.1.16. Let us indicate how the algebraic degree of the function f in Theorem 7.1.15 depends on d. Each term in the expression of f is of the form $Tr_1^m(\alpha_1 y^d x) Tr_1^m(\alpha_2 y^d x)$. If we replace the second trace term by its definition, we get

$$\begin{aligned} &Tr_1^m(\alpha_1 y^d x) Tr_1^m(\alpha_2 y^d x) \\ &\quad = Tr_1^m\Big(\alpha_1 y^d x \Big(\sum_{i=0}^{m-1} \alpha_2^{2^i} y^{2^i d} x^{2^i} \Big)\Big) \\ &\quad = \sum_{i=0}^{m-1} Tr_1^m\Big(\alpha_1 \alpha_2^{2^i} y^{(2^i+1)d} x^{2^i+1}\Big). \end{aligned}$$

Now, the algebraic degree of $y^{2d}x^2$ is $w_2((2d) \mod 2^m - 1) + 1$ while $y^{(2^i+1)d}x^{2^i+1}$ is of algebraic degree $w_2(((2^i + 1)d) \mod 2^m - 1) + 2$ if $i \neq 0$. Next, note that the algebraic degrees of $y^{(2^i+1)d}x^{2^i+1}$ and $y^{(2^{m-i}+1)d}x^{2^{m-i}+1} = \left(y^{(2^i+1)d}x^{2^i+1}\right)^{2^{m-i}}$ are equal. Therefore, the algebraic degree of f in Theorem 7.1.15 is equal to

$$\max\Big(w_2((2d) \mod 2^m - 1) + 1,$$

$$\max_{1 \le i \le \frac{m}{2}} \big(w_2(((2^i + 1)d) \mod 2^m - 1) + 2\big)\Big).$$

Example 7.1.17 ($m = 4$, $d = 7$, $e = 13$). Let ζ be a primitive element of $\mathbb{F}_{16}$ such that $\zeta^4 + \zeta^3 + \zeta^2 + \zeta + 1 = 0$. Let $a_1 = 1$, $a_2 = (\zeta + \zeta^3 + 1)$ and $a_3 = (\zeta + \zeta^2 + 1)$ satisfy $a_3 + a_2 = \zeta^2 + \zeta^3 \neq 1$ and $a_1^{-13} + a_2^{-13} + a_3^{-13} = \zeta^2 + \zeta^3 = (a_1 + a_2 + a_3)^{-13}$. Then,

$$f(x, y) = \mathrm{Tr}_1^4(y^7x)\mathrm{Tr}_1^4((\zeta^2 + \zeta^3)y^7x) + \mathrm{Tr}_1^4((\zeta + \zeta^3 + 1)y^7x)\mathrm{Tr}_1^4((\zeta + \zeta^2 + 1)y^7x)$$

is bent and its dual function is

$$\tilde{f}(x, y) = \mathrm{Tr}_1^4(yx^{13})\mathrm{Tr}_1^4((1 + \zeta^3 + \zeta^2)yx^{13}) + \mathrm{Tr}_1^4((\zeta + \zeta^2 + 1)yx^{13})\mathrm{Tr}_1^4((\zeta + \zeta^3)yx^{13}).$$

According to Remark 7.1.16, the algebraic degree of f is $\max(w_2(7) + 1, w_2(6) + 2, w_2(5) + 2) = 4$, that is, f is of optimal degree for a bent function in dimension 8. We prove it by working on the expression of f. Indeed, since $\mathrm{Tr}_1^4(a) = a + a^2 + a^4 + a^8$, $\mathrm{Tr}_1^4(\alpha a)\mathrm{Tr}_1^4(\beta a) = \mathrm{Tr}_1^4(\alpha\beta a^2 + \alpha^2\beta a^3 + \alpha^4\beta a^5 + \alpha^8\beta a^9))$. Thus

$$\begin{aligned} f(x, y) = {} & \mathrm{Tr}_1^4\Big((\zeta^2 + \zeta^3)y^{14}x^2 + (\zeta^2 + \zeta^3)y^{21}x^3 \\ & + (\zeta^2 + \zeta^3)y^{35}x^5 + (\zeta^2 + \zeta^3)y^{63}x^9\Big) \\ & + \mathrm{Tr}_1^4\Big((1 + \zeta + \zeta^2 + \zeta^3)y^{14}x^2 + (\zeta + \zeta^3)y^{21}x^3 \\ & + \zeta^3y^{35}x^5 + (\zeta^3 + \zeta^2 + 1)y^{63}x^9\Big) \\ = {} & \mathrm{Tr}_1^4((\zeta^3 + 1)y^7x + (\zeta^2 + \zeta)y^6x^3 + \zeta^2y^5x^5 + y^3x^9). \end{aligned}$$

The dual function $\tilde{f}$ is of algebraic degree $\max(w_2(11) + 1, w_2(9) + 2, w_2(5) + 2) = 4$, also optimal for a bent function in dimension 8.

We now recall the definition of PS_{ap} class. The elements $f : \mathbb{F}_{2^n} \to \mathbb{F}_2$, $n = 2m$, of PS_{ap} are the functions of the form $f(x, y) = g(xy^{2^m-2})$ where g is a balanced Boolean function. All the elements of PS_{ap} are bent and their dual functions are known : the dual function of $g(xy^{2^m-2})$ is $g(yx^{2^m-2})$. Consider now the particular case where $d = e = 2^m - 2$. In this case, the inverse of ϕ is ϕ itself and the condition $a_1^{-e} + a_2^{-e} + a_3^{-e} = b^{-e}$ rewrites as $a_1 + a_2 + a_3 = b$. Note that the expression of the function f in Theorem 7.1.15 is of the form $f(x, y) = g(xy^{2^m-2})$ with g the Boolean function defined over $\mathbb{F}_{2^m}$ as $g(z) = \mathrm{Tr}_1^m(a_1z)\mathrm{Tr}_1^m(a_2z) + \mathrm{Tr}_1^m(a_1z)\mathrm{Tr}_1^m(a_3z) + \mathrm{Tr}_1^m(a_2z)\mathrm{Tr}_1^m(a_3z)$. We can therefore deduce the following statement.

Corollary 7.1.18 ([11]). *For $i \in \{1,2,3\}$, let $f_i(x,y) = Tr_1^m(a_i y^{2^m-2} x)$ for some $a_i \in \mathbb{F}_{2^m}$ (where $0^{-1} = 0$). Assume the a_i's are pairwise distinct such that $a_1 + a_2 + a_3 \neq 0$. Let f be the Boolean function defined in bivariate form as*

$$\begin{aligned} f(x,y) &= Tr_1^m(a_1 y^{2^m-2} x) Tr_1^m(a_2 y^{2^m-2} x) \\ &+ Tr_1^m(a_1 y^{2^m-2} x) Tr_1^m(a_3 y^{2^m-2} x) \\ &+ Tr_1^m(a_2 y^{2^m-2} x) Tr_1^m(a_3 y^{2^m-2} x). \end{aligned}$$

Then f belongs to the class PS_{ap} and is of degree m.

Remark 7.1.19. We have searched for $n \in \{4,5,6,8\}$ by a computer program all the integers e, coprime with $2^n - 1$, for which the following assertion holds:

(P) There exists a 3-tuple (a_1, a_2, a_3) of pairwise distinct elements of $\mathbb{F}_{2^n}^\star$ such that $a_1^{-e} + a_2^{-e} + a_3^{-e} = (a_1 + a_2 + a_3)^{-e}$ and $a_3 \neq a_2 + a_1$.

We give in Table 7.1 the found exponents,[1] in the hope to convince the reader that condition (P) is not so restrictive and that, in any dimension, it can be fulfilled by many exponents e.

Remark 7.1.20. In Theorem 7.1.15, we have excluded that d is a power of 2, which is equivalent to exclude that $e = 1$. The reader should then notice that 1 does not appear in any list of exponents. In fact, we can prove that (P) does not hold in any dimension for $e = 1$.

Table 7.1 List of exponents e for which (P) is fulfilled for $4 \leq n \leq 8$

n	$2^n - 1$	Cyclotomic class
4	15	7, 11, 13
5	31	15, 23, 27, 29
6	63	5, 11, 13, 17, 19, 23, 25, 29, 31, 37, 41, 43, 47, 53, 55, 59, 61
7	127	3, 5, 9, 15, 17, 27, 33, 43, 45, 51, 53, 63, 65, 71, 77, 85, 89, 95, 99, 111, 113, 119, 123, 125
8	255	7, 11, 13, 19, 23, 29, 31, 37, 41, 43, 47, 49, 53, 59, 61, 67, 71, 73, 77, 79, 83, 89, 91, 97, 101, 103, 107, 109, 113, 121, 127, 131, 133, 137, 139, 143, 149, 151, 157, 161, 163, 167, 169, 173, 179, 181, 191, 193, 197, 199, 203, 209, 211, 217, 223, 227, 229, 233, 239, 241, 247, 251, 253

[1] If (P) holds for an exponent e, then it holds for every exponent e' lying in the same cyclotomic class modulo $2^n - 1$ than e. Therefore, in Table 7.1, we only write the coset leaders of the cyclotomic classes.

Let a_1, a_2 and a_3 be three non zero elements of $\mathbb{F}_{2^n}$, $n \geq 2$ such that $a_1^{-1} + a_2^{-1} + a_3^{-1} = (a_1 + a_2 + a_3)^{-1}$. Set $a_2 = xa_1$ and $a_3 = ya_1$. Then the preceding equality can be rewritten as $1 + x^{-1} + y^{-1} = (1 + x + y)^{-1}$, or equivalently, as $(1 + x^{-1} + y^{-1})(1 + x + y) = 1 = 1 + x^{-1} + y^{-1} + x + y + xy^{-1} + yx^{-1}$. That becomes $x + y + x^2y + xy^2 + x^2 + y^2 = 0 = (x + y)(1 + x)(1 + y)$. Therefore, either $x = 1$ and $a_2 = a_1$, or $y = 1$ and $a_3 = a_1$, or $x = y$ and $a_3 = a_2$. Hence, Condition (P) is false in any dimension for $e = 1$.

By the above, one can show that

Proposition 7.1.21 ([11]). *Condition* (P) *is true for any even positive integer* n *which is a multiple of* 4 *when* $e = 11$.

Proof. If we set $x = \frac{a_2}{a_1}$ and $y = \frac{a_3}{a_1}$, then equation $a_1^{-e} + a_2^{-e} + a_3^{-e} = (a_1 + a_2 + a_3)^{-e}$ can be rewritten as $1 + x^{-e} + y^{-e} = (1 + x + y)^{-e}$, or equivalently, as $(1 + x^{-e} + y^{-e})(1 + x + y)^e = 1$. Let us now specialize our calculation to $e = 11$ and multiply the preceding equality by $x^{11}y^{11}$:

$$(x^{11}y^{11} + x^{11} + y^{11})(1 + x + y)^{11} = x^{11}y^{11}.$$

Suppose from now on that $y = \frac{1}{x}$,

$$\begin{aligned}&(1 + x^{11} + x^{-11})(1 + x + x^{-1})^{11} = 1\\ \iff\ & p(x) = (x^{22} + x^{11} + 1)(x^2 + x + 1)^{11} + x^{22} = 0.\end{aligned}$$

The polynomial p can be factored as follows:

$$\begin{aligned}p(x) = &(x^4 + x + 1)(x^4 + x^3 + 1)\\ &\times(x^{11} + x^8 + x^7 + x^6 + x^4 + x^3 + 1)\\ &\times(x^{11} + x^8 + x^7 + x^5 + x^4 + x^3 + 1)\\ &\times(x^4 + x^3 + x^2 + x + 1)^2(x + 1)^6.\end{aligned}$$

Now, consider the equation $x^4 + x + 1 = 0$ over $\mathbb{F}_{2^n}$. Let us rewrite it as $(x^2 + x)^2 + (x^2 + x) + 1 = 0$. Setting $z = x^2 + x$, the equation rewrites as $z^2 + z + 1 = 0$, that is, $z^3 = 1$, or equivalently, $z \in \mathbb{F}_4 \setminus \mathbb{F}_2$ (clearly, $z \neq 1$). Now, let us mention that $Tr_1^n(z) = 0$ if and only if n is a multiple of 4 (if β stands for a primitive element of $\mathbb{F}_4$, $Tr_1^n(\beta) = 0$ if $n \equiv 0 \pmod 4$ and $Tr_1^n(\beta) = 1$ if $n \equiv 2 \pmod 4$. That latter condition is required because $x^2 + x = z$ admits solutions in $\mathbb{F}_{2^n}$ if and only if $Tr_1^n(z) = 0$. If $n \equiv 0 \pmod 4$), any x such that $x^2 + x = \beta$ or $x^2 + x = \beta^2$ is a root of $x^4 + x + 1$.

To conclude, it suffices to note that, necessarily, $x \neq 1$, and thus, $\frac{1}{x} \neq x$. Therefore, $a_2 \neq a_1$, $a_3 \neq a_1$ and $a_3 \neq a_2$ if we take $a_1 \in \mathbb{F}_{2^n}^\star$, $a_2 = xa_1$ and $a_3 = x^{-1}a_1$ with $x^4 + x + 1 = 0$. □

Inspired by the proof of Proposition 7.1.21, we give in the next theorem an explicit infinite family of bent Boolean functions (we send the reader to Remark 7.1.16 concerning their degrees).

Theorem 7.1.22 ([11]). *Let $n = 2m$ be an even positive integer which is a multiple of 4 but not of 10 (in this case $2^n - 1$ and 11 are co-prime). Let d be the inverse of 11 modulo $2^n - 1$. Let c be such that $c^4 + c + 1 = 0$ and $a \in \mathbb{F}_{2^n}^\star$. Then*

$$\begin{aligned} f(x,y) &= \mathrm{Tr}_1^m(a^{-11}x^{11}y)\mathrm{Tr}_1^m(a^{-11}c^{-11}x^{11}y) \\ &\quad + \mathrm{Tr}_1^m(a^{-11}x^{11}y)\mathrm{Tr}_1^m(c^{11}a^{-11}x^{11}y) \\ &\quad + \mathrm{Tr}_1^m(a^{-11}c^{-11}x^{11}y)\mathrm{Tr}_1^m(c^{11}a^{-11}x^{11}y) \end{aligned}$$

is bent and its dual function is given by

$$\begin{aligned} \tilde{f}(x,y) &= \mathrm{Tr}_1^m(ay^dx)\mathrm{Tr}_1^m(acy^dx) \\ &\quad + \mathrm{Tr}_1^m(ay^dx)\mathrm{Tr}_1^m(ac^{-1}y^dx) \\ &\quad + \mathrm{Tr}_1^m(acy^dx)\mathrm{Tr}_1^m(ac^{-1}y^dx). \end{aligned}$$

Now, we show that condition (7.9) can be satisfied for some functions. Indeed, assume m even. Set $m = 2r$. Denote by $\mathcal{D}_m$ the set of Boolean functions g defined on $\mathbb{F}_{2^m}$ satisfying $g(ax) = g(x)$ for every $(a,x) \in \mathbb{F}_{2^r} \times \mathbb{F}_{2^m}$. It has been shown in [14] that this set is non empty and that the univariate form of a function having that property is of the form $g(x) = \sum_{\alpha=0}^{2^r} \mathrm{Tr}_1^m(\lambda_\alpha x^{\alpha(2^r-1)})$. Another important property of an element of $\mathcal{D}_m$ is that it is of algebraic degree r. Keeping the notation above, note then that, for $i \in \{1,2,3\}$, if the ϕ_i are the same permutation monomials on $\mathbb{F}_{2^m}$ as above (that is, $\phi(x) = a_ix^d$), the a_i's in $\mathbb{F}_{2^r}$ and the g_i's in the set $\mathcal{D}_m$, then Condition (7.9) is fulfilled since $g_1(a_1^{-e}x^e) + g_2(a_2^{-e}x^e) + g_3(a_3^{-e}x^e) + g_1((a_1+a_2+a_3)^{-e}x^e) + g_2((a_1+a_2+a_3)^{-e}x^e) + g_3((a_1+a_2+a_3)^{-e}x^e) = g_1(x^e) + g_2(x^e) + g_3(x^e) + g_1(x^e) + g_2(x^e) + g_3(x^e) = 0$. Hence, we deduce the following result from the discussion above and Theorem 7.1.3.

Theorem 7.1.23 ([11]). *Let d be a positive integer co-prime with $2^m - 1$. For $i \in \{1,2,3\}$, let $f_i(x,y) = \mathrm{Tr}_1^m(a_iy^dx)$ for some $a_i \in \mathbb{F}_{2^m}$. Assume the a_i's are pairwise distinct such that $b := a_1 + a_2 + a_3 \neq 0$ and $a_1^{-e} + a_2^{-e} + a_3^{-e} = b^{-e}$. Let g_1, g_2 and g_3 be three Boolean functions on $\mathcal{D}_m := \{g : \mathbb{F}_{2^m} \to \mathbb{F}_2 \mid g(ax) = g(x), \forall (a,x) \in \mathbb{F}_{2^r} \times \mathbb{F}_{2^m}\}$. Let h be the Boolean function defined in bivariate form as*

$$\begin{aligned} h(x,y) &= (\mathrm{Tr}_1^m(a_1y^dx) + g_1(y))(\mathrm{Tr}_1^m(a_2y^dx) + g_2(y)) \\ &\quad + (\mathrm{Tr}_1^m(a_1y^dx) + g_1(y))(\mathrm{Tr}_1^m(a_3y^dx) + g_3(y)) \\ &\quad + (\mathrm{Tr}_1^m(a_2y^dx) + g_2(y))(\mathrm{Tr}_1^m(a_3y^dx) + g_3(y)). \end{aligned}$$

Then h is bent and its dual function is

$$\begin{aligned}\tilde{h}(x,y) &= (\mathrm{Tr}_1^m(a_1^{-e}x^ey) + g_1(x^e))(\mathrm{Tr}_1^m(a_2^{-e}x^ey) + g_2(x^e))\\ &+ (\mathrm{Tr}_1^m(a_1^{-e}x^ey) + g_1(x^e))(\mathrm{Tr}_1^m(a_3^{-e}x^ey) + g_3(x^e))\\ &+ (\mathrm{Tr}_1^m(a_2^{-e}x^ey) + g_2(x^e))(\mathrm{Tr}_1^m(a_3^{-e}x^ey) + g_3(x^e))\end{aligned}$$

where e stands for the inverse of d modulo $2^m - 1$.

7.1.3 An Infinite Family of Bent Functions from the Maiorana–McFarland Completed Class

In this subsection we provide a construction of a family of bent functions from the Maiorana–McFarland completed class (that is, the smallest possible complete class containing the class of Maiorana–McFarland which is globally invariant under the action of the general affine group and under the addition of affine functions).

In [5], the authors have shown that the quadratic Boolean function f defined on $\mathbb{F}_{2^{4k}}$ (where k is at least 2) by $f(x) = \mathrm{Tr}_1^{4k}(\lambda x^{2^k+1})$ for every $x \in \mathbb{F}_{2^{4k}}$ where $\lambda \in \mathbb{F}_{2^{4k}}^\star$, is self-dual bent or anti-self-dual bent when $\lambda^2 + \lambda^{2^{3k+1}} = 1$ or $\lambda^{2^k+1} + \lambda^{2^{2k}+2^{3k}} = 0$. Unfortunately, the authors in [5] do not say for which λ, f is self-dual bent or anti-self-dual bent. In the following lemma, we clarify this point.

Lemma 7.1.24 ([11]). *Let* $\lambda \in \mathbb{F}_{2^{4k}}^\star$ *such that* $\lambda + \lambda^{2^{3k}} = 1$ *or* $\lambda^{(2^k+1)^2(2^k-1)} = 1$. *Set* $q(x) := \mathrm{Tr}_1^{4k}(\lambda x^{2^k+1})$ *for every* $x \in \mathbb{F}_{2^{4k}}$. *Then q is self-dual bent.*

Proof. Firstly, note that the conditions put on λ in the lemma are equivalent to the constraints given above [5]. Indeed, the relations $\lambda^2 + \lambda^{2^{3k+1}} = 1$ and $\lambda + \lambda^{2^{3k}} = 1$ are linked by the Frobenius mapping; while the second condition comes from the fact that

$$\begin{aligned}\lambda^{2^k+1} + \lambda^{2^{2k}+2^{3k}} = 0 &\iff \lambda^{2^{2k}+2^{3k}-2^k-1} = 1\\ &\iff \lambda^{(2^{2k}-1)(2^k+1)} = 1.\end{aligned}$$

The bentness of q has been established in [5]. Therefore it remains to identify explicitly the values of λ for which q is self-dual bent and those for which q is anti-self-dual bent.

To this end, note that if we know that a Boolean function f defined on $\mathbb{F}_{2^n}$ (where n is an even integer) vanishing at 0 is self-dual bent or anti-self-dual bent, then one can decide which case holds by studying the Hamming weight of f. Indeed, one can prove that $wt(f) = 2^{n-1} - 2^{\frac{n}{2}-1}(-1)^{\tilde{f}(0)}$. Therefore, if the Hamming weight of f equals $2^{n-1} - 2^{\frac{n}{2}-1}$, then f is self-dual bent while, if it is equal to $2^{n-1} + 2^{\frac{n}{2}-1}$, then f is anti-self-dual bent. Note first that $q(0) = 0$. Let us now compute the Hamming weight of $q(x)$ for $x \in \mathbb{F}_{2^{4k}}^\star$. Recall that every element x of $\mathbb{F}_{2^{4k}}^\star$ can be uniquely decomposed as $x = uy$ with $u^{2^{2k}+1} = 1$ and $y \in \mathbb{F}_{2^{2k}}^\star$. Now, every $y \in \mathbb{F}_{2^{2k}}^\star$ can be uniquely decomposed as $y = vz$ with $v^{2^k+1} = 1$ and $z \in \mathbb{F}_{2^k}^\star$. Hence, every $x \in \mathbb{F}_{2^{4k}}^\star$ can be uniquely decomposed as $x = uvz$ with $u^{2^{2k}+1} = v^{2^k+1} = 1$ and $z \in \mathbb{F}_{2^k}^\star$. Now, $q(x) = Tr_1^{4k}(\lambda x^{2^k+1}) = Tr_1^{4k}(\lambda u^{2^k+1}z^2)$. The linear map $z \in \mathbb{F}_{2^k} \mapsto Tr_1^{4k}(\lambda u^{2^k+1}z^2)$ is of Hamming weight 2^{k-1} when $Tr_1^{4k}(\lambda u^{2^k+1}) \neq 0$ and identically equals 0 when $Tr_1^{4k}(\lambda u^{2^k+1}) = 0$. Hence, since the cardinality of $\{v \in \mathbb{F}_{2^{2k}} \mid v^{2^k+1} = 1\}$ is $2^k + 1$, one has $wt(q) = (2^k+1)2^{k-1}\#\{u \in \mathbb{F}_{2^{2k}} \mid u^{2^{2k}+1} = 1, Tr_1^{4k}(\lambda u^{2^k+1}) \neq 0\}$. We know that, since q is bent, then $wt(q) \in \{2^{4k-1} - 2^{2k-1}, 2^{4k-1} + 2^{2k-1}\}$. Clearly, $2^k + 1$ divides $2^{4k-1} - 2^{2k-1} = 2^{2k-1}(2^k + 1)(2^k - 1)$ and is coprime with $2^{4k-1} - 2^{2k-1} = 2^{2k-1}(2^{2k} + 1)$.

Therefore, one has $wt(q) = 2^{4k-1} - 2^{2k-1}$, proving that q is self-dual bent. □

Let us focus our attention on the second equation $\lambda^{2^{3k+1}} + \lambda^2 = 1$. Note that the map $\lambda \in \mathbb{F}_{2^{4k}} \mapsto \lambda^{2^{3k+1}} + \lambda^2$ is linear. Moreover, its kernel is equal to $\mathbb{F}_{2^k}$. Indeed, $\lambda^{2^{3k+1}} + \lambda^2 = 0$ if and only if $\lambda = 0$ or $\lambda \neq 0$ and $\lambda^{2^{3k}} = \lambda$, that is, λ belongs to $\mathbb{F}_{2^s}$ where $s = \gcd(4k, 3k) = k$.

The mapping $\lambda \in \mathbb{F}_{2^{4k}} \mapsto \lambda^2 + \lambda^{2^{3k+1}}$ is therefore a 2^k-to-one map and Corollary 7.1.7 yields the following result.

Theorem 7.1.25 ([11]). *Let k be a positive integer such that $k \geq 2$. Let a_1, a_2, a_3 be three pairwise distinct nonzero solutions in $\mathbb{F}_{2^{4k}}$ of the equation $\lambda^{2^{3k}} + \lambda = 1$ such that $a_1 + a_2 + a_3 \neq 0$. Let g be the Boolean function over $\mathbb{F}_{2^{4k}}$ defined as*

$$\begin{aligned} g(x) &= Tr_1^{4k}(a_1x^{2^k+1})Tr_1^{4k}(a_2x^{2^k+1}) \\ &+ Tr_1^{4k}(a_1x^{2^k+1})Tr_1^{4k}(a_3x^{2^k+1}) \\ &+ Tr_1^{4k}(a_2x^{2^k+1})Tr_1^{4k}(a_3x^{2^k+1}), \forall x \in \mathbb{F}_{2^{4k}}. \end{aligned}$$

Then g is self-dual bent of algebraic degree 4.

Example 7.1.26 (k=3). Let ζ be a primitive element of $\mathbb{F}_{4096}$ such that $\zeta^{12} + \zeta^8 + \zeta^7 + \zeta^5 + \zeta^4 + \zeta + 1 = 0$. Then $\lambda_1 = \zeta^3 + \zeta^4 + \zeta^5 + \zeta^8 + \zeta^9 + \zeta^{11}$, $\lambda_2 = \zeta + \zeta^3 + \zeta^4 + \zeta^6 + \zeta^8 + \zeta^9 + \zeta^{10} + \zeta^{11} + 1$ and $\lambda_3 = (\zeta + \zeta^3 + \zeta^4 + \zeta^6 + \zeta^8 + \zeta^9 + \zeta^{10} + \zeta^{11} + 1$ are three distinct solutions of $\lambda^{2048} + \lambda = 1$ whose sum is different from 0. The Boolean function

$$\begin{aligned} g(x) &= Tr_1^{12}(\lambda_1x^9)Tr_1^{12}(\lambda_2x^9) + Tr_1^{4k}(\lambda_1x^9)Tr_1^{12}(\lambda_3x^9) \\ &+ Tr_1^{12}(\lambda_2x^9)Tr_1^{12}(\lambda_3x^9) \end{aligned}$$

is a self-dual bent function of algebraic degree 4 over $\mathbb{F}_{4096}$.

7.1.4 An Infinite Family of Cubic Bent Functions

Let k be a positive integer such that $k \geq 2$. Let $\lambda_1 \in \mathbb{F}_{2^{2k}}$ and $\lambda_2 \in \mathbb{F}_{2^{4k}}$ such that $\lambda_2 + {\lambda_2}^{2^{3k}} = 1$. Let h_1 and h_2 be two Boolean functions defined over $\mathbb{F}_{2^{4k}}$ as:

$$h_1(x) = Tr_1^{2k}(\lambda_1 x^{2^{2k}+1}), \forall x \in \mathbb{F}_{2^{4k}}$$

and

$$h_2(x) = Tr_1^{4k}(\lambda_2 (x+\beta)^{2^k+1}), \forall x \in \mathbb{F}_{2^{4k}}$$

for some $\beta \in \mathbb{F}_{2^{4k}}$. Note that h_1 is a Niho bent and its dual function $\tilde{h}_1$ is given by $\tilde{h}_1(x) = Tr_1^{2k}({\lambda_1}^{-1} x^{2^{2k}+1}) + 1$ for every $x \in \mathbb{F}_{2^{4k}}$ (see e.g. [14]). Moreover, under the condition $\lambda_2 + {\lambda_2}^{2^{3k}} = 1$, we deduce from Lemma 7.1.24 that the function $x \in \mathbb{F}_{2^{4k}} \mapsto Tr_1^{4k}(\lambda_2 x^{2^k+1})$ is a self-dual bent function. Hence, one can deduce that the dual $\tilde{h}_2$ of (the bent function) h_2 equals: $\tilde{h}_2(x) = Tr_1^{4k}(\lambda_2 x^{2^k+1}) + Tr_1^{4k}(\beta x)$, $\forall x \in \mathbb{F}_{2^{4k}}$.

Now, in order to apply Corollary 7.1.5, one has to determine the values of β, λ_1 and λ_2 for which there exists a non zero element a of $\mathbb{F}_{2^{4k}}$ such that $D_a\tilde{h}_1 = D_a\tilde{h}_2$. Let us first compute the derivatives of $\tilde{h}_1$ and $\tilde{h}_2$ with respect $a \in \mathbb{F}_{2^{4k}}$. We have, for every $x \in \mathbb{F}_{2^{4k}}$,

$$\begin{aligned}
&D_a\tilde{h}_1(x)\\
&= Tr_1^{2k}({\lambda_1}^{-1}(x+a)^{2^{2k}+1}) + Tr_1^{2k}({\lambda_1}^{-1}x^{2^{2k}+1})\\
&= Tr_1^{2k}({\lambda_1}^{-1}(x^{2^{2k}+1} + xa^{2^{2k}} + ax^{2^{2k}} + a^{2^{2k}+1}))\\
&\quad + Tr_1^{2k}({\lambda_1}^{-1}x^{2^{2k}+1})\\
&= Tr_1^{2k}({\lambda_1}^{-1}a^{2^{2k}+1} + {\lambda_1}^{-1}xa^{2^{2k}} + {\lambda_1}^{-1}ax^{2^{2k}})\\
&= Tr_1^{2k}({\lambda_1}^{-1}a^{2^{2k}+1}) + Tr_1^{2k}({\lambda_1}^{-1}(xa^{2^{2k}} + a^{2^{4k}}x^{2^{2k}}))\\
&\qquad (\text{since } a \in \mathbb{F}_{2^{4k}})\\
&= Tr_1^{2k}({\lambda_1}^{-1}a^{2^{2k}+1}) + Tr_1^{2k}({\lambda_1}^{-1}Tr_{2k}^{4k}(a^{2^{2k}}x))\\
&= Tr_1^{2k}({\lambda_1}^{-1}a^{2^{2k}+1}) + Tr_1^{2k}(Tr_{2k}^{4k}({\lambda_1}^{-1}a^{2^{2k}}x))\\
&\qquad (\text{since } \lambda_1 \in \mathbb{F}_{2^{2k}})\\
&= Tr_1^{2k}({\lambda_1}^{-1}a^{2^{2k}+1}) + Tr_1^{4k}({\lambda_1}^{-1}a^{2^{2k}}x).
\end{aligned}$$

Moreover, for every $x \in \mathbb{F}_{2^{4k}}$,

$$
\begin{aligned}
&D_a\tilde{h}_2(x)\\
&= \mathrm{Tr}_1^{4k}(\lambda_2(x+a)^{2^k+1}) + \mathrm{Tr}_1^{4k}(\beta(x+a))\\
&\quad + \mathrm{Tr}_1^{4k}(\lambda_2 x^{2^k+1}) + \mathrm{Tr}_1^{4k}(\beta x)\\
&= \mathrm{Tr}_1^{4k}(\lambda_2 a^{2^k+1}) + \mathrm{Tr}_1^{4k}(\lambda_2 x^{2^k} a) + \mathrm{Tr}_1^{4k}(\lambda_2 a^{2^k} x)\\
&\quad + \mathrm{Tr}_1^{4k}(\beta a)\\
&= \mathrm{Tr}_1^{4k}(\lambda_2 a^{2^k+1}) + \mathrm{Tr}_1^{4k}(\lambda_2{}^{2^{-k}} a^{2^{-k}} x) + \mathrm{Tr}_1^{4k}(\lambda_2 a^{2^k} x)\\
&\quad + \mathrm{Tr}_1^{4k}(\beta a).\\
&= \mathrm{Tr}_1^{4k}(\lambda_2 a^{2^k+1}) + \mathrm{Tr}_1^{4k}((\lambda_2{}^{2^{-k}} a^{2^{-k}} + \lambda_2 a^{2^k})x)\\
&\quad + \mathrm{Tr}_1^{4k}(\beta a).
\end{aligned}
$$

Now, one has $D_a\tilde{h}_2 = D_a\tilde{h}_2$ if and only if the following Conditions (7.10) and (7.11) are satisfied.

$$
\mathrm{Tr}_1^{2k}(\lambda_1{}^{-1} a^{2^{2k}+1}) + \mathrm{Tr}_1^{4k}(\lambda_2 a^{2^k+1}) = \mathrm{Tr}_1^{4k}(\beta a) \tag{7.10}
$$

$$
\lambda_1{}^{-1} a^{2^{2k}} + \lambda_2{}^{2^{-k}} a^{2^{-k}} + \lambda_2 a^{2^k} = 0. \tag{7.11}
$$

Clearly, for every $a \in \mathbb{F}_{2^{4k}}^\star$, $\lambda_1 \in \mathbb{F}_{2^{2k}}$ and $\lambda_2 \in \mathbb{F}_{2^{4k}}$, there exist 2^{4k-1} choices for β (recall that a linear Boolean function is balanced). Let us now focus our attention on the second Condition (7.11) which is more restrictive than Condition (7.10). Indeed, Condition (7.11) is equivalent to saying that the linear map from $\mathbb{F}_{2^{4k}}$ to itself defined by $a \mapsto \lambda_1{}^{-1} a^{2^{2k}} + \lambda_2{}^{2^{-k}} a^{2^{-k}} + \lambda_2 a^{2^k}$ is not a permutation.

Assume that $\lambda_1 = 1$. Note then that for $a \in \mathbb{F}_{2^k}$, the left-hand side of the Condition (7.11) becomes $a(1 + \lambda_2 + \lambda_2{}^{2^{-k}}) = 0$. To conclude, we have the following result.

Theorem 7.1.27 ([11]). *Let $k \geq 2$ be a positive integer. Let $\lambda_2 \in \mathbb{F}_{2^{4k}}$ such that $\lambda_2 + \lambda_2{}^{2^{3k}} = 1$. Let $a \in \mathbb{F}_{2^{4k}}^\star$ be a solution of $a^{2^{2k}} + \lambda_2{}^{2^{-k}} a^{2^{-k}} + \lambda_2 a^{2^k} = 0$ and $\beta \in \mathbb{F}_{2^{4k}}$ such that $\mathrm{Tr}_1^{4k}(\beta a) = \mathrm{Tr}_1^{2k}(a^{2^{2k}+1}) + \mathrm{Tr}_1^{4k}(\lambda_2 a^{2^k+1})$. Let g be the Boolean function over $\mathbb{F}_{2^{4k}}$ defined by*

$$
\begin{aligned}
g(x) &= \mathrm{Tr}_1^{2k}(x^{2^{2k}+1}) + \mathrm{Tr}_1^{4k}(ax)\mathrm{Tr}_1^{2k}(x^{2^{2k}+1})\\
&+\mathrm{Tr}_1^{4k}(ax)\mathrm{Tr}_1^{4k}(\lambda_2(x+\beta)^{2^k+1}) + \mathrm{Tr}_1^{4k}(ax), \forall x \in \mathbb{F}_{2^{4k}}.
\end{aligned}
$$

Then the cubic function g is bent and its dual function $\tilde{g}$ is given by

$$
\begin{aligned}
\tilde{g}(x) &= \mathrm{Tr}_1^{2k}(x^{2^{2k}+1})\\
&+\Big(\mathrm{Tr}_1^{2k}(x^{2^{2k}+1}) + \mathrm{Tr}_1^{4k}(\lambda_2 x^{2^k+1}) + \mathrm{Tr}_1^{4k}(\beta x)\Big)\\
&\times\Big(\mathrm{Tr}_1^{4k}(a^{2^k} x) + \mathrm{Tr}_1^{2k}(a^{2^{2k}+1})\Big), \forall x \in \mathbb{F}_{2^{4k}}.
\end{aligned}
$$

Proof. The bentness of g is the consequence of the discussion and the calculation above. We only give here the calculation yielding the expression of its dual $\tilde{g}$. According to Corollary 7.1.5 the expression of the dual $\tilde{g}$ is $\tilde{g}(x) = \tilde{h}_1(x) + \tilde{h}_1(x)\tilde{h}_2(x)+\tilde{h}_1(x+a)\tilde{h}_2(x+a)$ where $\tilde{h}_1$ and $\tilde{h}_2$ are given by $\tilde{h}_1(x) = Tr_1^{2k}(x^{2^{2k}+1})+1$ and $\tilde{h}_2(x) = Tr_1^{4k}(\lambda_2 x^{2^k+1}) + Tr_1^{4k}(\beta x)$. By substituting in the above expression of $\tilde{g}$, we obtain

$$\begin{aligned}\tilde{g}(x) &= Tr_1^{2k}(x^{2^{2k}+1}) + 1 + \left(Tr_1^{2k}(x^{2^{2k}+1}) + 1\right)\\ &\quad\times \left(Tr_1^{4k}(\lambda_2 x^{2^k+1}) + Tr_1^{4k}(\beta x)\right)\\ &\quad+\left(Tr_1^{2k}((x+a)^{2^{2k}+1}) + 1\right)\\ &\quad\times \left(Tr_1^{4k}(\lambda_2 (x+a)^{2^k+1}) + Tr_1^{4k}(\beta(x+a))\right).\end{aligned}$$

Now,

$$\begin{aligned}Tr_1^{2k}((x+a)^{2^{2k}+1}) &= Tr_1^{2k}(x^{2^{2k}+1}) + Tr_1^{2k}(a^{2^{2k}+1})\\ &\quad+ Tr_1^{4k}(a^{2^{2k}}x);\end{aligned}$$

and

$$\begin{aligned}&Tr_1^{4k}(\lambda_2(x+a)^{2^k+1}) + Tr_1^{4k}(\beta(x+a))\\ &= Tr_1^{4k}(\lambda_2 x^{2^k+1}) + Tr_1^{4k}((\lambda_2 a^{2^k} + \lambda_2^{2^{-k}} a^{2^{-k}})x)\\ &\quad+ Tr_1^{4k}(\beta x) + Tr_1^{4k}(\beta a) + Tr_1^{4k}(\lambda_2 a^{2^k+1})\\ &= Tr_1^{4k}(\lambda_2 x^{2^k+1}) + Tr_1^{4k}(a^{2^k}x)\\ &\quad+ Tr_1^{4k}(\beta x) + Tr_1^{2k}(a^{2^k+1})\end{aligned}$$

since $a^{2^{2k}} + \lambda_2^{2^{-k}} a^{2^{-k}} + \lambda_2 a^{2^k} = 0$ and $Tr_1^{4k}(\beta a) = Tr_1^{2k}(a^{2^{2k}+1}) + Tr_1^{4k}(\lambda_2 a^{2^k+1})$.
Hence

$$\begin{aligned}\tilde{g}(x) &= Tr_1^{2k}(x^{2^{2k}+1}) + 1 + (Tr_1^{2k}(x^{2^{2k}+1}) + 1)\\ &\quad\times \left(Tr_1^{4k}(\lambda_2 x^{2^k+1}) + Tr_1^{4k}(\beta x)\right)\\ &\quad+ \left((Tr_1^{2k}(x^{2^{2k}+1}) + 1) + (Tr_1^{4k}(a^{2^k}x) + Tr_1^{2k}(a^{2^{2k}+1}))\right)\\ &\quad\times \Big((Tr_1^{4k}(\lambda_2 x^{2^k+1}) + Tr_1^{4k}(\beta x))\\ &\qquad\qquad+ (Tr_1^{4k}(a^{2^k}x) + Tr_1^{2k}(a^{2^{2k}+1}))\Big).\end{aligned}$$

The result follows. □

Example 7.1.28 ($k = 3$). Let ζ be a primitive element of $\mathbb{F}_{4096}$ such that $\zeta^{12} + \zeta^8 + \zeta^7 + \zeta^5 + \zeta^4 + \zeta + 1 = 0$. Then, $\lambda_2 = \zeta^3 + \zeta^4 + \zeta^5 + \zeta^8 + \zeta^9 + \zeta^{11}$, $a = \zeta^2 + \zeta^3 + \zeta^4 + \zeta^5 + \zeta^8 + \zeta^9 + 1$ and $\beta = \zeta + \zeta^2 + \zeta^6 + \zeta^7 + \zeta^10 + \zeta^{11}$ satisfy the conditions of Theorem 7.1.27.

Remark 7.1.29. We can weaken a little bit more the condition ${\lambda_1}^{2^{2k}+1} = 1$ in Theorem 7.1.27 by taking λ_1 such that the mapping $a \mapsto {\lambda_1}^{-1} a^{2^{2k}} + {\lambda_2}^{2^{-k}} a^{2^{-k}} + \lambda_2 a^{2^k}$ is not a permutation of $\mathbb{F}_{2^{4k}}$.

To summarize this section, the study in [11] based on Theorem 7.1.3 (which completes [3, Theorem 3]) gives rise to (at least) seven variants of the construction in [11, Theorem 4], which yield infinite families of bent functions as well as their duals. All the constructed functions are given in univariate or bivariate representations and are expressed as the sum of the product of trace functions. In the following, we provide in a synthetic way the list of the constructions obtained in [11] (recall that $\tilde{f}$ stands for the dual of a bent function f):

1. Bent functions obtained by selecting Niho bent functions:

- $f(x) = Tr_1^m(\lambda x^{2^m+1}) + Tr_1^n(ax)Tr_1^n(bx)$; $x \in \mathbb{F}_{2^n}$, $n = 2m$, $\lambda \in \mathbb{F}_{2^m}^\star$ and $(a, b) \in \mathbb{F}_{2^n}^\star \times \mathbb{F}_{2^n}^\star$ such that $a \neq b$ and $Tr_1^n(\lambda^{-1} b^{2^m} a) = 0$.

$$\tilde{f}(x) = Tr_1^m(\lambda^{-1} x^{2^m+1}) + \left(Tr_1^m(\lambda^{-1} a^{2^m+1}) + Tr_1^n(\lambda^{-1} a^{2^m} x)\right) \times \left(Tr_1^m(\lambda^{-1} b^{2^m+1}) + Tr_1^n(\lambda^{-1} b^{2^m} x)\right) + 1.$$

- $g(x) = Tr_1^m(x^{2^m+1}) + Tr_1^n\left(\sum_{i=1}^{2^{r-1}-1} x^{(2^m-1)\frac{i}{2^r}+1}\right) + Tr_1^n(\lambda x)Tr_1^n(\mu x)$; $x \in \mathbb{F}_{2^n}$, $n = 2m$, $(\lambda, \mu) \in \mathbb{F}_{2^m}^\star \times \mathbb{F}_{2^m}^\star$ $(\lambda \neq \mu)$.

$$\begin{aligned}\tilde{g}(x) = {} & Tr_1^m\left(\left(u(1+x+x^{2^m}) + u^{2^{n-r}} + x^{2^m}\right)(1+x+x^{2^m})^{\frac{1}{2^r-1}}\right) \\ & \times Tr_1^m\left((\lambda+\mu)(1+x+x^{2^m})^{\frac{1}{2^r-1}}\right) \\ & + Tr_1^m\left(\left(u(1+x+x^{2^m}) + u^{2^{n-r}} + x^{2^m} + \lambda\right)(1+x+x^{2^m})^{\frac{1}{2^r-1}}\right) \\ & \times Tr_1^m\left(\left(u(1+x+x^{2^m}) + u^{2^{n-r}} + x^{2^m} + \mu\right)(1+x+x^{2^m})^{\frac{1}{2^r-1}}\right);\end{aligned}$$

where $u \in \mathbb{F}_{2^n}$ satisfying $u + u^{2^m} = 1$.

2. Bent functions obtained by selecting bent Boolean functions of Maiorana–McFarland's class:

- $f(x, y) = Tr_1^m(a_1 y^d x)Tr_1^m(a_2 y^d x) + Tr_1^m(a_1 y^d x)Tr_1^m(a_3 y^d x) + Tr_1^m(a_2 y^d x)Tr_1^m(a_3 y^d x)$; where $(x, y) \in \mathbb{F}_{2^m} \times \mathbb{F}_{2^m}$, d is a positive integer which is not a power

of 2 and $gcd(d, 2^m - 1) = 1$, a_i's are pairwise distinct such that $b := a_1 + a_2 + a_3 \neq 0$ and $a_1^{-e} + a_2^{-e} + a_3^{-e} = b^{-e}$ where $e = d^{-1} \pmod{2^m - 1}$.
$\tilde{f}(x, y) = \mathrm{Tr}_1^m(a_1^{-e}x^e y)\mathrm{Tr}_1^m(a_2^{-e}x^e y) + \mathrm{Tr}_1^m(a_1^{-e}x^e y)\mathrm{Tr}_1^m(a_3^{-e}x^e y) + \mathrm{Tr}_1^m(a_2^{-e}x^e y)\ \mathrm{Tr}_1^m(a_3^{-e}x^e y)$.

- $g(x, y) = \mathrm{Tr}_1^m(a^{-11}x^{11}y)\mathrm{Tr}_1^m(a^{-11}c^{-11}x^{11}y)$ $\mathrm{Tr}_1^m(a^{-11}x^{11}y)\mathrm{Tr}_1^m(c^{11}a^{-11}x^{11}y) + \mathrm{Tr}_1^m(a^{-11}c^{-11}x^{11}y)\mathrm{Tr}_1^m(c^{11}a^{-11}x^{11}y)$; where $(x, y) \in \mathbb{F}_{2^m} \times \mathbb{F}_{2^m}$, $a \in \mathbb{F}_{2^n}^\star$ with $n = 2m$ is a multiple of 4 but not of 10, $c \in \mathbb{F}_{2^m}$ is such that $c^4 + c + 1 = 0$.
  $\tilde{g}(x, y) = \mathrm{Tr}_1^m(ay^d x)\mathrm{Tr}_1^m(acy^d x) + \mathrm{Tr}_1^m(ay^d x)\mathrm{Tr}_1^m(ac^{-1}y^d x) + \mathrm{Tr}_1^m(acy^d x)$ $\mathrm{Tr}_1^m(ac^{-1}y^d x)$; with $d = 11^{-1} \pmod{2^n - 1}$.
- $h(x, y) = (\mathrm{Tr}_1^m(a_1 y^d x) + g_1(y))(\mathrm{Tr}_1^m(a_2 y^d x) + g_2(y)) + (\mathrm{Tr}_1^m(a_1 y^d x) + g_1(y))(\mathrm{Tr}_1^m(a_3 y^d x) + g_3(y)) + (\mathrm{Tr}_1^m(a_2 y^d x) + g_2(y))(\mathrm{Tr}_1^m(a_3 y^d x) + g_3(y))$; where $m = 2r$, $gcd(d, 2^m - 1) = 1$, a_1, a_2 and a_3 are three pairwise distinct elements of $\mathbb{F}_{2^m}$ such that $b := a_1 + a_2 + a_3 \neq 0$ and $a_1^{-e} + a_2^{-e} + a_3^{-e} = b^{-e}$ and for $i \in \{1, 2, 3\}$, $g_i \in \mathcal{D}_m := \{g : \mathbb{F}_{2^m} \to \mathbb{F}_2 \mid g(ax) = g(x), \forall (a, x) \in \mathbb{F}_{2^r} \times \mathbb{F}_{2^m}\}$.
  $\tilde{h}(x, y) = (\mathrm{Tr}_1^m(a_1^{-e}x^e y) + g_1(x^e))(\mathrm{Tr}_1^m(a_2^{-e}x^e y) + g_2(x^e)) + (\mathrm{Tr}_1^m(a_1^{-e}x^e y) + g_1(x^e))(\mathrm{Tr}_1^m(a_3^{-e}x^e y) + g_3(x^e)) + (\mathrm{Tr}_1^m(a_2^{-e}x^e y) + g_2(x^e))(\mathrm{Tr}_1^m(a_3^{-e}x^e y) + g_3(x^e))$ where $e = d^{-1} \pmod{2^m - 1}$.

3. Self-dual bent functions obtained by selecting functions from Maiorana–McFarland completed class:
   - $g(x) = \mathrm{Tr}_1^{4k}(a_1 x^{2^k+1})\mathrm{Tr}_1^{4k}(a_2 x^{2^k+1}) + \mathrm{Tr}_1^{4k}(a_1 x^{2^k+1})\mathrm{Tr}_1^{4k}(a_3 x^{2^k+1}) + \mathrm{Tr}_1^{4k}(a_2 x^{2^k+1})\mathrm{Tr}_1^{4k}(a_3 x^{2^k+1})$; where $x \in \mathbb{F}_{2^{4k}}$, $k \geq 2$, a_1, a_2, a_3 be three pairwise distinct nonzero solutions in $\mathbb{F}_{2^{4k}}$ of the equation $\lambda^{2^{3k}} + \lambda = 1$ such that $a_1 + a_2 + a_3 \neq 0$.
4. Bent functions obtained by selecting functions from PS_{ap}:
   - $f(x, y) = \mathrm{Tr}_1^m(a_1 y^{2^m-2}x)\mathrm{Tr}_1^m(a_2 y^{2^m-2}x) + \mathrm{Tr}_1^m(a_1 y^{2^m-2}x)\mathrm{Tr}_1^m(a_3 y^{2^m-2}x) + \mathrm{Tr}_1^m(a_2 y^{2^m-2}x)\mathrm{Tr}_1^m(a_3 y^{2^m-2}x)$; where $(x, y) \in \mathbb{F}_{2^m} \times \mathbb{F}_{2^m}$, the a_i's are pairwise distinct in $\mathbb{F}_{2^m}$ such that $a_1 + a_2 + a_3 \neq 0$.
     $\tilde{f}(x, y) = f(y, x)$.
5. Bent functions obtained by combining Niho bent functions and self-dual bent functions:
   - $f(x) = \mathrm{Tr}_1^{2k}(x^{2^{2k}+1}) + \mathrm{Tr}_1^{4k}(ax)\mathrm{Tr}_1^{2k}(x^{2^{2k}+1}) + \mathrm{Tr}_1^{4k}(ax)\mathrm{Tr}_1^{4k}(\lambda_2(x+\beta)^{2^k+1}) + \mathrm{Tr}_1^{4k}(ax)$; where $x \in \mathbb{F}_{2^{4k}}$ $(k \geq 2)$, $\lambda_2 \in \mathbb{F}_{2^{4k}}$ such that $\lambda_2 + \lambda_2^{2^{3k}} = 1$, $a \in \mathbb{F}_{2^{4k}}^\star$ is a solution of $a^{2^{2k}} + \lambda_2^{2^{-k}} a^{2^{-k}} + \lambda_2 a^{2^k} = 0$ and $\beta \in \mathbb{F}_{2^{4k}}$ such that $\mathrm{Tr}_1^{4k}(\beta a) = \mathrm{Tr}_1^{2k}(a^{2^{2k}+1}) + \mathrm{Tr}_1^{4k}(\lambda_2 a^{2^k+1})$.
     $\tilde{f}(x) = \mathrm{Tr}_1^{2k}(x^{2^{2k}+1}) + \left(\mathrm{Tr}_1^{2k}(x^{2^{2k}+1}) + \mathrm{Tr}_1^{4k}(\lambda_2 x^{2^k+1}) + \mathrm{Tr}_1^{4k}(\beta x)\right) \times \left(\mathrm{Tr}_1^{4k}(a^{2^k}x) + \mathrm{Tr}_1^{2k}(a^{2^{2k}+1})\right)$.

Quite fascinatingly, despite its simplicity and the conditions that seem restrictive at first, we have derived from construction (7.1) several infinite families of bent functions. Furthermore, a very important point is that we can provide the dual functions of all their elements which is rarely achieved. It remains obviously to check whether the bent functions presented above are affinely inequivalent to known constructions or not. Another aspect would be to study their other cryptographic properties.

7.2 Further Constructions of Infinite Families of Bent Functions from New Permutations and Their Duals

In the previous section, it was shown that several new infinite classes of bent functions defined over the finite field $\mathbb{F}_{2^n}$ with their duals, have been derived recently by the author in [11]. All these families are obtained by selecting three pairwise distinct bent functions suitably chosen from general classes of bent functions. This section is devoted to constructions of bent functions defined over the finite field $\mathbb{F}_{2^{2m}} \simeq \mathbb{F}_{2^m} \times \mathbb{F}_{2^m}$ in bivariate representation in the continuation of [11]. We provide two new infinite families of bent functions (Construction 7.2.10 and Construction 7.2.14) by selecting appropriately three pairwise distinct bent functions chosen from the so-called Maiorana–McFarland's class. To this end, selecting a set of three permutations of $\mathbb{F}_{2^m}$ satisfying some conditions is sufficient (see Corollary 7.2.1). A priori such permutations seem hard to obtain. But, in [11] and in this section, we prove that this problem has many solutions and allows us to construct many infinite families of bent functions whose dual functions can be computed explicitly. In [11], we have considered in particular only the case of monomial permutations. In this section, we investigate other families of permutations. One of them is constructed using the notion of linear structure and involving Boolean functions and linear permutations. The other one is a new construction of an infinite family of permutations involving Boolean functions arbitrarily chosen (Theorem 7.2.13). The compositional inverses of such families of permutations are given. Moreover, we introduce in Sect. 7.2.2 a method to construct permutations satisfying the conditions mentioned above. We indeed show that one can construct such permutations similar to permutations in lower dimension. We then deduce from that construction other families of permutations and many other infinite families of bent functions (Construction 7.2.17, Construction 7.2.19, Construction 7.2.20 and Construction 7.2.23).

7.2.1 Further Constructions of Bent Functions from Permutations and Their Duals

In this section, we start with bent functions from the Maiorana–McFarland construction which are defined over $\mathbb{F}_{2^m} \times \mathbb{F}_{2^m}$ by (7.12):

$$f(x,y) = Tr_1^m(\phi(y)x) + g(y), \quad (x,y) \in \mathbb{F}_{2^m} \times \mathbb{F}_{2^m} \tag{7.12}$$

where m is some positive integer, ϕ is a permutation from $\mathbb{F}_{2^m}$ to itself and g stands for a Boolean function over $\mathbb{F}_{2^m}$. Recall that the duals of bent functions of the form (7.12) are expressed by means of the compositional inverse mapping of the permutation ϕ as follows:

$$\tilde{f}(x,y) = Tr_1^m(y\phi^{-1}(x)) + g(\phi^{-1}(x)) \tag{7.13}$$

where ϕ^{-1} denotes the inverse mapping of the permutation ϕ.

To apply Theorem 7.1.3 to a 3-tuple of functions of the form (7.12) with $g = 0$, one has to choose appropriately the maps ϕ involved in their expressions. More precisely, one can prove the following result [11].

Corollary 7.2.1 ([10]). *Let m be a positive integer. Let ϕ_1, ϕ_2 and ϕ_3 be three permutations of $\mathbb{F}_{2^m}$. Then,*

$$g(x,y) = Tr_1^m(x\phi_1(y))Tr_1^m(x\phi_2(y)) + Tr_1^m(x\phi_1(y))Tr_1^m(x\phi_3(y)) + Tr_1^m(x\phi_2(y))Tr_1^m(x\phi_3(y))$$

is bent if and only if

1. *$\psi = \phi_1 + \phi_2 + \phi_3$ is a permutation,*
2. $\psi^{-1} = \phi_1^{-1} + \phi_2^{-1} + \phi_3^{-1}$.

Furthermore, its dual function $\tilde{g}$ is given by

$$\begin{aligned}\tilde{g}(x,y) &= Tr_1^m(\phi_1^{-1}(x)y)Tr_1^m(\phi_2^{-1}(x)y) + Tr_1^m(\phi_1^{-1}(x)y)Tr_1^m(\phi_3^{-1}(x)y) \\ &+ Tr_1^m(\phi_2^{-1}(x)y)Tr_1^m(\phi_3^{-1}(x)y).\end{aligned} \tag{7.14}$$

Remark 7.2.2. Note that for $(i,j) \in \{1,2,3\}^2$, $i \neq j$,

$$\begin{aligned}Tr_1^m(x\phi_i(y)))Tr_1^m(x\phi_j(y)) &= Tr_1^m\left(x\phi_i(y)\sum_{l=0}^{m-1} x^{2^l}\phi_j^{2^l}(y)\right) \\ &= \sum_{l=0}^{m-1} Tr_1^m\left(x^{2^l+1}\phi_i(y)\phi_j^{2^l}(y)\right).\end{aligned}$$

Hence, one can rewrite $g(x,y)$ as

$$g(x,y) = \sum_{l=0}^{m-1} \mathrm{Tr}_1^m \left(x^{2^l+1}\left(\phi_1(y)\phi_2^{2^l}(y) + \phi_1(y)\phi_3^{2^l}(y) + \phi_2(y)\phi_3^{2^l}(y)\right)\right). \quad (7.15)$$

Remark 7.2.3. It is difficult to compute the algebraic degree of g in Corollary 7.2.1. Nevertheless, note that, for every $i \in \{1,2,3\}$, the algebraic degree of $\mathrm{Tr}_1^m(x\phi_i(y))$ is equal to $d_i + 1$ where d_i denotes the algebraic degree of ϕ_i. Obviously, multiple of a Boolean function can be of low algebraic degree, that is, the algebraic degree of $\mathrm{Tr}_1^m(x\phi_i(y))\mathrm{Tr}_1^m(x\phi_j(y))$ could be of lower degree than $\min(d_i + 1, d_j + 1)$ in some exceptional situations. However, we claim that Corollary 7.2.1 would mostly lead to nonquadratic Boolean functions.

Definition 7.2.4. Let m be a positive integer. Three permutations ϕ_1, ϕ_2 and ϕ_3 of $\mathbb{F}_{2^m}$ are said to satisfy $(\mathcal{A}_m)$ if the two following conditions hold

1. Their sum $\zeta = \phi_1 + \phi_2 + \phi_3$ is a permutation of $\mathbb{F}_{2^m}$.
2. $\zeta^{-1} = \phi_1^{-1} + \phi_2^{-1} + \phi_3^{-1}$.

Remark 7.2.5. Let us note that if ϕ_1, ϕ_2, ϕ_3 are three permutations of $\mathbb{F}_{2^m}$ which satisfy $(\mathcal{A}_m)$ then

- ϕ_1^{-1}, ϕ_2^{-1} and ϕ_3^{-1} satisfy $(\mathcal{A}_m)$.
- $A_1 \circ \phi_1 \circ A_2$, $A_1 \circ \phi_2 \circ A_2$ and $A_1 \circ \phi_3 \circ A_2$ satisfy $(\mathcal{A}_m)$ for all affine automorphisms A_1 and A_2 of $\mathbb{F}_{2^m}$ (if A is an affine automorphism of $\mathbb{F}_{2^m}$ then $A(u + v + w) = A(u) + A(v) + A(w)$ for any 3-tuple (u, v, w) of $\mathbb{F}_{2^m}$).

A priori, it seems hard to determine explicitly three permutations of $\mathbb{F}_{2^m}$ satisfying the strong condition $(\mathcal{A}_m)$. Nevertheless, we will show in this section that we are able to exhibit new classes of permutations of $\mathbb{F}_{2^m}$ for which condition $(\mathcal{A}_m)$ holds.

In [11], the case where the permutations ϕ_i ($i \in \{1,2,3\}$) are monomial has been investigated. Some potentially new bent functions have been exhibited. In this subsection, we investigate other families of permutations than those considered in [11]. We begin with families of the form

$$\phi(x) = L(x) + L(\alpha)f(x),\ x \in \mathbb{F}_{2^m} \quad (7.16)$$

where L is a linear permutation of $\mathbb{F}_{2^m}$, f is a Boolean function on $\mathbb{F}_{2^m}$ and $\alpha \in \mathbb{F}_{2^m}^\star$.

But before, recall the notion of a linear structure which exists for p-ary functions (p prime). Functions with linear structures are considered as weak for some cryptographic applications.

Definition 7.2.6. Let q be a power of a prime number. A non-zero element $\alpha \in \mathbb{F}_{q^n}$ is called an a-linear translator for the mapping $f : \mathbb{F}_{q^n} \to \mathbb{F}_q$ if $f(x+u\alpha)-f(x) = ua$ holds for any $x \in \mathbb{F}_{q^n}$, $u \in \mathbb{F}_q$ and a fixed $a \in \mathbb{F}_q$. In particular, when $q = 2$, $\alpha \in \mathbb{F}_{2^n}^\star$ usually said an a-linear structure for the Boolean function f (where $a \in \mathbb{F}_2$), that is, $f(x + \alpha) + f(x) = a$, for any $x \in \mathbb{F}_{2^n}$.

Note that if α is a a-linear structure of f, then necessarily $a = f(\alpha) - f(0)$.
The following result has been shown in [8].

Theorem 7.2.7 ([8]). *Let q be a power of a prime number. Let L be a linear permutation of $\mathbb{F}_{q^n}$. Let $b \in \mathbb{F}_q$, $h \in \mathbb{F}_q \to \mathbb{F}_q$ and $\alpha \in \mathbb{F}_{q^n}^\star$ be a b-linear structure of $f : \mathbb{F}_{q^n} \to \mathbb{F}_q$, that is, $f(x + u\alpha) - f(x) = ub$ holds for any $x \in \mathbb{F}_{q^n}$ and $u \in \mathbb{F}_q$. Then the mapping $\phi(x) = L(x) + L(\alpha)h(f(x))$ permutes $\mathbb{F}_{q^n}$ if and only if $\kappa(u) := u + bh(u)$ permutes $\mathbb{F}_q$.*

Let us specialize Theorem 7.2.7 to our case : $q = 2$, $h = Id_{\mathbb{F}_{2^n}}$ (i.e. the identity mapping over $\mathbb{F}_{2^n}$) and $b = 0$.

Proposition 7.2.8 ([10]). *Let $L : \mathbb{F}_{2^m} \to \mathbb{F}_{2^m}$ be a $\mathbb{F}_2$-linear permutation of $\mathbb{F}_{2^m}$. Let f be a Boolean function over $\mathbb{F}_{2^m}$ and α be a non zero 0-linear structure of f. Then the mapping ϕ defined by (7.16) is a permutation of $\mathbb{F}_{2^m}$ and*

$$\phi^{-1}(x) = L^{-1}(x) + \alpha f(L^{-1}(x)). \tag{7.17}$$

Proof. The fact that ϕ is a permutation is a straightforward application of Theorem 7.2.7. Note next that

$$\phi(L^{-1}(x) + \alpha f(L^{-1})(x)) = L(L^{-1}(x) + \alpha f(L^{-1})(x)) + L(\alpha)f(L^{-1}(x) + \alpha f(L^{-1}(x))).$$

Now, L is a linear permutation and α is a 0-linear structure of f, therefore

$$\phi(L^{-1}(x) + \alpha f(L^{-1})(x)) = L(L^{-1}(x)) + L(\alpha)f(L^{-1})(x)) + L(\alpha)f(L^{-1}(x)) = x$$

proving (7.32). □

Remark 7.2.9. Keeping the previous notation, if we choose for α a 1-linear structure of f (that is, $f(x+\alpha)+f(x) = 1$ for every x in $\mathbb{F}_{2^m}$) then $\phi(x) = L(x)+L(\alpha)f(x)$ is 2-to-1. Indeed, $\phi(x) = y$ is equivalent to $x+\alpha f(x) = L^{-1}(y)$ since L is a permutation, that is, $x = L^{-1}(y) + \alpha f(x) \in \{L^{-1}(y), L^{-1}(y) + \alpha\}$. Hence, every element of $\mathbb{F}_{2^n}$ has at most 2 preimages under ϕ. Now, $\phi(x+\alpha) = L(x)+L(\alpha)+L(\alpha)(f(x+\alpha)) = L(x) + L(\alpha) + L(\alpha)(f(x) + 1) = L(x) + L(\alpha)f(x) = \phi(x)$ proving that the number of preimages of every element of $\phi(\mathbb{F}_{2^m})$ is even.

Recall now that the set of 0-linear structures of f is a vector space over $\mathbb{F}_2$ (if α_1 and α_2 are two 0-linear structures of f, then $f(x + \alpha_1 + \alpha_2) = f(x + \alpha_1) = f(x)$ for every x proving that $\alpha_1 + \alpha_2$ is a 0-linear structure of f). Observe next that 0-linear structures for a Boolean function f are the points for which the derivative of f vanishes : $f(x + \alpha) = f(x)$ for every x is equivalent to say that $D_\alpha f(x) := f(x+\alpha)+f(x) = 0$ for every x. Set $\mathcal{L}_f^0 = \{\alpha \in \mathbb{F}_{2^m} \mid D_\alpha f = 0\}$ for a fixed Boolean function $f : \mathbb{F}_{2^m} \to \mathbb{F}_2$. Combining Corollary 7.2.1 with Proposition 7.2.8, we get infinite classes of bent functions.

Construction 7.2.10 ([10]) *Let m be a positive integer. Let L be a linear permutation on $\mathbb{F}_{2^m}$. Let f be a Boolean function over $\mathbb{F}_{2^m}$ such that $\mathcal{L}_f^0$ is of dimension at least two over $\mathbb{F}_2$. Let $(\alpha_1, \alpha_2, \alpha_3)$ be any 3-tuple of pairwise distinct elements of $\mathcal{L}_f^0$ such that $\alpha_1 + \alpha_2 + \alpha_3 \neq 0$. Then the Boolean function g defined in bivariate representation on $\mathbb{F}_{2^m} \times \mathbb{F}_{2^m}$ by*

$$\begin{aligned} g(x,y) = \mathrm{Tr}_1^m(xL(y)) + f(y)\Big(&\mathrm{Tr}_1^m(L(\alpha_1)x)\mathrm{Tr}_1^m(L(\alpha_2)x) \\ &+ \mathrm{Tr}_1^m(L(\alpha_1)x)\mathrm{Tr}_1^m(L(\alpha_3)x) + \mathrm{Tr}_1^m(L(\alpha_2)x)\mathrm{Tr}_1^m(L(\alpha_3)x)\Big) \end{aligned} \tag{7.18}$$

is bent and its dual function $\tilde{g}$ is given by

$$\begin{aligned} \tilde{g}(x,y) &= \mathrm{Tr}_1^m(L^{-1}(x)y) \\ &+ f(L^{-1}(x))\Big(\mathrm{Tr}_1^m(\alpha_1 y)\mathrm{Tr}_1^m(\alpha_2 y) + \mathrm{Tr}_1^m(\alpha_1 y)\mathrm{Tr}_1^m(\alpha_3 y) + \mathrm{Tr}_1^m(\alpha_2 y)\mathrm{Tr}_1^m(\alpha_3 y)\Big). \end{aligned} \tag{7.19}$$

Remark 7.2.11. To apply Theorem 7.2.10, one has to find a Boolean function f such that $\mathcal{L}_f^0$ is of dimension at least 2. If $m = rk$ with r even and $k \geq 2$, candidates are functions of the form $f(x) = h(\mathrm{Tr}_k^m(x))$ where h is a Boolean function over $\mathbb{F}_{2^k}$. Indeed note that, for every $\alpha \in \mathbb{F}_{2^k}, f(x+\alpha) = g(\mathrm{Tr}_k^m(x) + \mathrm{Tr}_k^m(\alpha)) = h(\mathrm{Tr}_k^m(x) + \alpha\mathrm{Tr}_k^m(1)) = h(\mathrm{Tr}_k^m(x))$ since $\mathrm{Tr}_k^m(1) = 0$.

Remark 7.2.12. Note that $(x,y) \in \mathbb{F}_{2^m} \times \mathbb{F}_{2^m} \mapsto \mathrm{Tr}_1^m(xL(y))$, $x \in \mathbb{F}_{2^m} \mapsto \mathrm{Tr}_1^m(L(\alpha_1)x)\mathrm{Tr}_1^m(L(\alpha_2)x) + \mathrm{Tr}_1^m(L(\alpha_1)x)\mathrm{Tr}_1^m(L(\alpha_3)x) + \mathrm{Tr}_1^m(L(\alpha_2)x)\mathrm{Tr}_1^m(L(\alpha_3)x)$ and $y \in \mathbb{F}_{2^m} \mapsto (\mathrm{Tr}_1^m(\alpha_1 y)\mathrm{Tr}_1^m(\alpha_2 y) + \mathrm{Tr}_1^m(\alpha_1 y)\mathrm{Tr}_1^m(\alpha_3 y) + \mathrm{Tr}_1^m(\alpha_2 y)\mathrm{Tr}_1^m(\alpha_3 y)$ are quadratic Boolean functions. Let f and g be two Boolean functions defined as in Construction 7.2.10. Therefore, if f is of algebraic degree d, g is of algebraic degree $d+2$. Now, it is well known that the algebraic degree of a Boolean function f over $\mathbb{F}_{2^m}$ such that $\mathcal{L}_f^0$ is of dimension k is at most $m-k$. Therefore the algebraic degrees of g and its dual function $\tilde{g}$ are at most $m-k+2$ if $\mathcal{L}_f^0$ is of dimension k.

Suppose now that m is even. Let us introduce a new infinite family of permutations involving any Boolean function as well as their compositional inverses.

Theorem 7.2.13 ([10]). *Suppose that $m = 2k$. Let $a \in \mathbb{F}_{2^k}$ and $b \in \mathbb{F}_{2^m}$ such that $b^{2^k+1} \neq a^2$. Let g be a Boolean function over $\mathbb{F}_{2^k}$. Set $\rho = a + b^{2^k}$ and*

$$\phi(x) = ax + bx^{2^k} + g(\mathrm{Tr}_k^m(\rho x)),\ x \in \mathbb{F}_{2^m}. \tag{7.20}$$

Then ϕ is a permutation of $\mathbb{F}_{2^m}$ and its compositional inverse is

$$\phi^{-1}(x) = \alpha^{-1}\left(ax + bx^{2^k} + (a+b)g(\mathrm{Tr}_k^m(x))\right) \tag{7.21}$$

where $\alpha = b^{2^k+1} + a^2 \neq 0$.

Proof. It suffices to prove that $\phi^{-1} \circ \phi = Id_{\mathbb{F}_{2^m}}$ where $Id_{\mathbb{F}_{2^m}}$ stands for the identity mapping of $\mathbb{F}_{2^m}$. Firstly, one has

$$\phi^{-1}(\phi(x)) = \alpha^{-1}\left(a\phi(x) + b\phi^{2^k}(x) + (a+b)g\left(\mathrm{Tr}_k^m(\phi(x))\right)\right).$$

Now,

$$\begin{aligned}\mathrm{Tr}_k^m(\phi(x)) &= \mathrm{Tr}_k^m(ax) + \mathrm{Tr}_k^m(bx^{2^k}) + \mathrm{Tr}_k^m(g(\mathrm{Tr}_k^m(\rho x)))\\ &= \mathrm{Tr}_k^m(ax) + \mathrm{Tr}_k^m(b^{2^k}x) + g(\mathrm{Tr}_k^m(\rho x))\mathrm{Tr}_k^m(1) = \mathrm{Tr}_k^m(\rho x)\end{aligned}$$

using the $\mathbb{F}_{2^k}$-linearity of the trace function $x \mapsto \mathrm{Tr}_k^m$, the fact that $\forall z \in \mathbb{F}_{2^m}, \mathrm{Tr}_k^m(z) = \mathrm{Tr}_k^m(z^{2^k})$ and $\mathrm{Tr}_k^m(1) = 0$. Hence

$$\begin{aligned}\phi^{-1}(\phi(x)) &= \alpha^{-1}\Big(a(ax + bx^{2^k} + g(\mathrm{Tr}_k^m(\rho x))) + b(a^{2^k}x^{2^k} + b^{2^k}x^{2^m}\\ &\quad +(g(\mathrm{Tr}_k^m(\rho x))^{2^k})) + (a+b)g(\mathrm{Tr}_k^m(\rho x))\Big)\\ &= \alpha^{-1}\Big(a(ax + bx^{2^k} + g(\mathrm{Tr}_k^m(\rho x))) + b(ax^{2^k} + b^{2^k}x\\ &\quad +g(\mathrm{Tr}_k^m(\rho x))) + (a+b)g(\mathrm{Tr}_k^m(\rho x))\Big)\\ &= \alpha^{-1}(a^2 + b^{2^k+1})x = x.\end{aligned}$$

□

Let us choose ϕ_1, ϕ_2 and ϕ_3 of the form (7.20) by fixing the coefficients a and b:

$$a \in \mathbb{F}_{2^k},\ b^{2^k+1} \neq a^2,\ \phi_i(x) = ax + bx^{2^k} + g_i(\mathrm{Tr}_k^m(\rho x)),\ x \in \mathbb{F}_{2^m}$$

for some Boolean functions g_1, g_2 and g_3. In that case,

$$\psi(x) = \phi_1(x)+\phi_2(x)+\phi_3(x) = ax+bx^{2^k}+h(\mathrm{Tr}_k^m(\rho x)),\ x \in \mathbb{F}_{2^m},\ h := g_1+g_2+g_3.$$

Hence, one can apply Theorem 7.2.13 to deduce that ψ is a permutation of $\mathbb{F}_{2^m}$. Moreover, one can observe that $\psi^{-1} = \phi_1^{-1} + \phi_2^{-1} + \phi_3^{-1}$ (since $\sum_{i=1}^{3} \phi_i^{-1}(x) = \alpha^{-1}(ax+bx^{2^k}+(a+b)h(\mathrm{Tr}_k^m(x))) = \psi^{-1}$). Hence the conditions of Corollary 7.2.1 hold. Next, observe that

$$f_i(x,y) = \mathrm{Tr}_1^m(x\phi_i(y)) = \mathrm{Tr}_1^m(axy + bxy^{2^k}) + \mathrm{Tr}_1^m(xg_i(\mathrm{Tr}_k^m(\rho y)))$$

is of the form $w_i = u + v_i$. Straightforward calculations show that, for such functions, $w_1w_2 + w_1w_3 + w_2w_3 = u + v_1v_2 + v_1v_3 + v_2v_3$. The dual functions $\tilde{f}_i$ of f_i is also of the same shape:

$$\tilde{f}_i(x) = \mathrm{Tr}_1^m\left(\phi_i^{-1}(x)y\right) = \mathrm{Tr}_1^m\left(\alpha^{-1}(axy + bx^{2^k}y)\right) + \mathrm{Tr}_1^m\left(\alpha^{-1}(a+b)yg\left(\mathrm{Tr}_k^m(x)\right)\right).$$

Therefore, one can deduce from Corollary 7.2.1 the below infinite classes of Boolean functions.

Construction 7.2.14 ([10]) *Let $m = 2k$. Let $a \in \mathbb{F}_{2^k}$ and $b \in \mathbb{F}_{2^m}$ such that $b^{2^k+1} \neq a^2$. Set $\alpha = b^{2^k+1} + a^2$ and $\rho = a + b^{2^k}$. Let g_1, g_2 and g_3 be three Boolean functions over $\mathbb{F}_{2^k}$. Then the Boolean function h defined in bivariate representation on $\mathbb{F}_{2^m} \times \mathbb{F}_{2^m}$ by*

$$\begin{aligned} h(x,y) = \mathrm{Tr}_1^m(axy + bxy^{2^k}) &+ \mathrm{Tr}_1^m(xg_1(\mathrm{Tr}_k^m(\rho y)))\mathrm{Tr}_1^m(xg_2(\mathrm{Tr}_k^m(\rho y))) \\ &+ \mathrm{Tr}_1^m(xg_1(\mathrm{Tr}_k^m(\rho y)))\mathrm{Tr}_1^m(xg_3(\mathrm{Tr}_k^m(\rho y))) \\ &+ \mathrm{Tr}_1^m(xg_2(\mathrm{Tr}_k^m(\rho y)))\mathrm{Tr}_1^m(xg_3(\mathrm{Tr}_k^m(\rho y))) \end{aligned}$$

is bent and its dual function $\tilde{h}$ is given by

$$\begin{aligned} \tilde{h}(x,y) = &\mathrm{Tr}_1^m\Big(\alpha^{-1}(axy + bx^{2^k}y)\Big) \\ &+\mathrm{Tr}_1^m\Big(\alpha^{-1}(a+b)yg_1\left(\mathrm{Tr}_k^m(x)\right)\Big)\mathrm{Tr}_1^m\Big(\alpha^{-1}(a+b)yg_2\left(\mathrm{Tr}_k^m(x)\right)\Big) \\ &+\mathrm{Tr}_1^m\Big(\alpha^{-1}(a+b)yg_1\left(\mathrm{Tr}_k^m(x)\right)\Big)\mathrm{Tr}_1^m\Big(\alpha^{-1}(a+b)yg_3\left(\mathrm{Tr}_k^m(x)\right)\Big) \\ &+\mathrm{Tr}_1^m\Big(\alpha^{-1}(a+b)yg_2\left(\mathrm{Tr}_k^m(x)\right)\Big)\mathrm{Tr}_1^m\Big(\alpha^{-1}(a+b)yg_3\left(\mathrm{Tr}_k^m(x)\right)\Big). \end{aligned}$$

7.2.2 Further Constructions of Bent Functions from Secondary-Like Constructions of Permutations

Now, we introduce a method to generate further constructions of bent functions via families of permutations.

Let m be a positive integer. Let ϕ_1, ϕ_2, ϕ_3 be three permutations satisfying $(\mathcal{A}_m)$ (see Definition 7.2.4). Let n be any non-zero multiple of m (different from m). Let ρ_1, ρ_2 and ρ_3 be three permutations of $\mathbb{F}_{2^n} \setminus \mathbb{F}_{2^m}$. Define ψ_1, ψ_2, ψ_3 on $\mathbb{F}_{2^n}$ by

$$i \in \{1,2,3\},\ \psi_i(x) = \begin{cases} \phi_i(x) \text{ if } x \in \mathbb{F}_{2^m} \\ \rho_i(x) \text{ if } x \in \mathbb{F}_{2^n} \setminus \mathbb{F}_{2^m}. \end{cases} \tag{7.22}$$

Clearly, ψ_i, $i \in \{1,2,3\}$ is a permutation of $\mathbb{F}_{2^n}$ if and only if ϕ_i and ρ_i are permutations of, respectively, $\mathbb{F}_{2^m}$ and $\mathbb{F}_{2^n} \setminus \mathbb{F}_{2^m}$. Furthermore, one has

$$i \in \{1,2,3\},\ \psi_i^{-1}(x) = \begin{cases} \phi_i^{-1}(x) \text{ if } x \in \mathbb{F}_{2^m} \\ \rho_i^{-1}(x) \text{ if } x \in \mathbb{F}_{2^n} \setminus \mathbb{F}_{2^m}. \end{cases}$$

Using the above notation, one can prove the following result which shows basically that one can construct mappings ψ_1, ψ_2 and ψ_3 satisfying the above conditions 1 and 2 from mappings satisfying identical conditions in lower dimension.

Proposition 7.2.15 ([10]). *The set of permutations $\{\psi_1, \psi_2, \psi_3\}$ satisfies $(\mathcal{A}_n)$ if and only if the following two conditions hold*

1. *the set of permutations $\{\phi_1, \phi_2, \phi_3\}$ satisfies $(\mathcal{A}_m)$.*
2. *$\nu = \rho_1 + \rho_2 + \rho_3$ is a permutation of $\mathbb{F}_{2^n} \setminus \mathbb{F}_{2^m}$ whose inverse $\nu^{-1} = \rho_1^{-1} + \rho_2^{-1} + \rho_2^{-1}$.*

Proof. Let $\lambda := \psi_1 + \psi_2 + \psi_3$ and $\tau := \phi_1 + \phi_2 + \phi_3$. One has

$$\lambda(x) = \begin{cases} \tau(x) \text{ if } x \in \mathbb{F}_{2^m} \\ \rho(x) \text{ if } x \in \mathbb{F}_{2^n} \setminus \mathbb{F}_{2^m}. \end{cases}$$

Hence, if $\{\psi_1, \psi_2, \psi_3\}$ satisfies $(\mathcal{A}_n)$ then $\{\phi_1, \phi_2, \phi_3\}$ satisfies $(\mathcal{A}_m)$. Conversely, assume that $\{\phi_1, \phi_2, \phi_3\}$ satisfies $(\mathcal{A}_m)$. Then

$$\psi_1(x) + \psi_2(x) + \psi_3(x) = \begin{cases} \phi_1(x) + \phi_2(x) + \phi_3(x) \text{ if } x \in \mathbb{F}_{2^m} \\ \rho(x) \quad \text{if } x \in \mathbb{F}_{2^n} \setminus \mathbb{F}_{2^m}. \end{cases}$$

The mapping λ is a permutation of $\mathbb{F}_{2^n}$ since τ is a permutation of $\mathbb{F}_{2^m}$ and ρ is a permutation of $\mathbb{F}_{2^n} \setminus \mathbb{F}_{2^m}$. Furthermore,

$$\lambda^{-1}(x) = \begin{cases} \tau^{-1}(x) \text{ if } x \in \mathbb{F}_{2^m} \\ \rho^{-1}(x) \text{ if } x \in \mathbb{F}_{2^n} \setminus \mathbb{F}_{2^m} \end{cases} = \begin{cases} \phi_1^{-1}(x) + \phi_2^{-1}(x) + \phi_3^{-1}(x) \text{ if } x \in \mathbb{F}_{2^m} \\ \rho^{-1}(x) + \rho^{-1}(x) + \rho^{-1}(x) \text{ if } x \in \mathbb{F}_{2^n} \setminus \mathbb{F}_{2^m}. \end{cases}$$

Hence $\lambda^{-1} = \psi_1^{-1}(x) + \psi_2^{-1}(x) + \psi_3^{-1}(x)$, which completes the proof. □

Example 7.2.16. Let $n = km$, $k \geq 2$. Set $\rho_i(x) = a_i x^{2^n-2}$, $a_i \in \mathbb{F}_{2^m}$, $i \in \{1, 2, 3\}$. We have that ρ_i, $i \in \{1, 2, 3\}$, is a permutation over $\mathbb{F}_{2^n} \setminus \mathbb{F}_{2^m}$ whose inverse mapping $\rho_i^{-1}(x) = a_i x^{2^n-2}$ (ρ_i^{-1} is the inverse function). Let $\{\phi_1, \phi_2, \phi_3\}$ be three permutations of $\mathbb{F}_{2^m}$ satisfying $(\mathcal{A}_m)$. Set

$$i \in \{1, 2, 3\},\ \psi_i(x) = \begin{cases} \phi_i(x) \quad \text{if } x \in \mathbb{F}_{2^m} \\ a_i x^{2^n-2} \ \text{if } x \in \mathbb{F}_{2^n} \setminus \mathbb{F}_{2^m}. \end{cases}$$

One notes that ψ_i is a permutation of $\mathbb{F}_{2^n}$ whose inverse is given by mapping is

$$i \in \{1, 2, 3\},\ \psi_i^{-1}(x) = \begin{cases} \phi_i^{-1}(x) \text{ if } x \in \mathbb{F}_{2^m} \\ a_i x^{2^n-2} \ \text{if } x \in \mathbb{F}_{2^n} \setminus \mathbb{F}_{2^m}. \end{cases}$$

Furthermore, if $a_1 + a_2 + a_3 \neq 0$, $\{\psi_1, \psi_2, \psi_3\}$ satisfies $(\mathcal{A}_n)$.

Set, for $i \in \{1, 2, 3\}$,

$$(x, y) \in \mathbb{F}_{2^n} \times \mathbb{F}_{2^n},\ f_i(x, y) = Tr_1^n(x\psi_i(y)) = \begin{cases} Tr_1^n(x\phi_i(y)) \text{ if } x \in \mathbb{F}_{2^m} \\ Tr_1^n(x\rho_i(y)) \text{ if } x \in \mathbb{F}_{2^n} \setminus \mathbb{F}_{2^m}. \end{cases}$$

If f_i is bent, then its dual function $\tilde{f}_i$ is

$$(x,y)\in\mathbb{F}_{2^n}\times\mathbb{F}_{2^n},\ \tilde{f}_i(x,y)=\mathrm{Tr}_1^n(\psi_i^{-1}(x)y)=\begin{cases}\mathrm{Tr}_1^n(\phi_i^{-1}(x)y) & \text{if } y\in\mathbb{F}_{2^m}\\ \mathrm{Tr}_1^n(\rho_u^{-1}(x)y) & \text{if } y\in\mathbb{F}_{2^n}\setminus\mathbb{F}_{2^m}.\end{cases}$$

On the other hand, note that, if $x\in\mathbb{F}_{2^m}$,

$$\begin{aligned} f_1(x,y)f_2(x,y)+f_1(x,y)f_3(x,y)+f_2(x,y)f_3(x,y)\\ =\mathrm{Tr}_1^n(x\phi_1(y))\mathrm{Tr}_1^n(x\phi_2(y))+\mathrm{Tr}_1^n(x\phi_1(y))\mathrm{Tr}_1^n(x\phi_3(y))\\ +\mathrm{Tr}_1^n(x\phi_2(y))\mathrm{Tr}_1^n(x\phi_3(y))\end{aligned}\tag{7.23}$$

and, if $x\in\mathbb{F}_{2^n}\setminus\mathbb{F}_{2^m}$,

$$\begin{aligned} f_1(x,y)f_2(x,y)+f_1(x,y)f_3(x,y)+f_2(x,y)f_3(x,y)\\ =\mathrm{Tr}_1^n(x\rho_1(y))\mathrm{Tr}_1^n(x\rho_2(y))+\mathrm{Tr}_1^n(x\rho_1(y))\mathrm{Tr}_1^n(x\rho_3(y))\\ +\mathrm{Tr}_1^n(x\rho_2(y))\mathrm{Tr}_1^n(x\rho_3(y))\end{aligned}\tag{7.24}$$

Combining Corollary 7.2.1, the preceding arguments and Example 7.2.16, one gets

Construction 7.2.17 ([10]) *Let n be a multiple of m where m is a positive integer and $n\neq m$. Let ϕ_1, ϕ_2 and ϕ_3 be three permutations over $\mathbb{F}_{2^m}$ satisfying $(\mathcal{A}_m)$. Let (a_1,a_2,a_3) be a 3-tuple of $\mathbb{F}_{2^m}^{\star}$ such that $a_1+a_2+a_3\neq 0$. Set*

$$\begin{aligned} g(x,y)=\mathrm{Tr}_1^n(x\phi_1(y))\mathrm{Tr}_1^n(x\phi_2(y))+\mathrm{Tr}_1^n(x\phi_1(y))\mathrm{Tr}_1^n(x\phi_3(y))\\ +\mathrm{Tr}_1^n(x\phi_2(y))\mathrm{Tr}_1^n(x\phi_3(y))\end{aligned}$$

if $(x,y)\in\mathbb{F}_{2^n}\times\mathbb{F}_{2^m}$ and

$$\begin{aligned} g(x,y)=\mathrm{Tr}_1^n(a_1xy^{2^n-2})\mathrm{Tr}_1^n(a_2xy^{2^n-2})+\mathrm{Tr}_1^n(a_1xy^{2^n-2})\mathrm{Tr}_1^n(a_3xy^{2^n-2})\\ +\mathrm{Tr}_1^n(a_2xy^{2^n-2})\mathrm{Tr}_1^n(a_3xy^{2^n-2})\end{aligned}$$

if $(x,y)\in\mathbb{F}_{2^n}\times\mathbb{F}_{2^n}\setminus\mathbb{F}_{2^m}$. Then g is bent and its dual function $\tilde{g}$ is defined by

$$\begin{aligned} \tilde{g}(x,y)=\mathrm{Tr}_1^n(\phi_1^{-1}(x)y)\mathrm{Tr}_1^n(\phi_2^{-1}(x)y)+\mathrm{Tr}_1^n(\phi_1^{-1}(x)y)\mathrm{Tr}_1^n(\phi_3^{-1}(x)y)\\ +\mathrm{Tr}_1^n(\phi_2^{-1}(x)y)\mathrm{Tr}_1^n(\phi_3^{-1}(x)y)\end{aligned}$$

if $(x,y)\in\mathbb{F}_{2^m}\times\mathbb{F}_{2^n}$ and

$$\begin{aligned} \tilde{g}(x,y)=\mathrm{Tr}_1^n(a_1x^{2^n-2}y)\mathrm{Tr}_1^n(a_2x^{2^n-2}y)+\mathrm{Tr}_1^n(a_1x^{2^n-2}y)\mathrm{Tr}_1^n(a_3x^{2^n-2}y)\\ +\mathrm{Tr}_1^n(a_2x^{2^n-2}y)\mathrm{Tr}_1^n(a_3x^{2^n-2}y)\end{aligned}$$

if $(x,y)\in\mathbb{F}_{2^n}\setminus\mathbb{F}_{2^m}\times\mathbb{F}_{2^n}$.

Remark 7.2.18. In Theorem 7.2.20, we have considered permutations of $\mathbb{F}_{2^m}$ of the shape: $\rho(x) = ax^d$, $a \in \mathbb{F}_{2^m}^\star$, whose inverse is $\rho^{-1}(x) = a^{-e}x^e$ where e stands for the inverse of d modulo $2^n - 1$. Let ρ_1, ρ_2 and ρ_3 be of that shape with $a_1 + a_2 + a_3 \neq 0$. Then, these three permutations fulfill the Condition 2 of Proposition 7.2.15 if and only if

$$(a_1 + a_2 + a_3)^{-e} = a_1^{-e} + a_2^{-e} + a_3^{-e}. \tag{7.25}$$

In [11, Proposition 20], we have investigated the case when n is a multiple of 4 and when $e = 11$. We have shown that if $a_1 \neq 0$, $a_2 = ca_1$ and $a_3 = c^{-1}a_1$ with $c^4 + c + 1 = 0$, then (a_1, a_2, a_3) is a 3-tuple of elements of $\mathbb{F}_{2^n}$ satisfying (7.25). Furthermore, since c cannot be equal to 0, note that $a_1 + a_2 + a_3 = a_1c^{-1}(1 + c + c^2) - 0$ is equivalent to $1 + c + c^2 = 0$, that is, $(1 + c + c^2)^2 = 1 + c^2 + c^4 = c + c^2 = 0$. Now, $c + c^2 = 0$ is equivalent to $c \in \mathbb{F}_2$. Clearly, $c^4 + c + 1$ has no roots in $\mathbb{F}_2$. That proves that, if we set $\rho_1(x) = ax^d$, $\rho_2(x) = acx^d$ and $\rho_3(x) = ac^{-1}x^d$ where $c^4 + c + 1 = 0$, $a \in \mathbb{F}_{2^m}^\star$ and d is the inverse of 11 modulo $2^n - 1$, then $\{\rho_1, \rho_2, \rho_3\}$ satisfies the Condition 2 of Proposition 7.2.15.

One can therefore deduce from the above remark the following family of bent functions.

Construction 7.2.19 ([10]) *Let n be a multiple of m where m is a positive integer and $n \neq m$. Let ϕ_1, ϕ_2 and ϕ_3 be three permutations over $\mathbb{F}_{2^m}$ satisfying $(\mathcal{A}_m)$. Let $a \in \mathbb{F}_{2^m}^\star$ and $c \in \mathbb{F}_{2^n}$ such that $c^4 + c + 1 = 0$. Let d be the inverse of* 11 *modulo $2^n - 1$. Set*

$$g(x, y) = Tr_1^n(x\phi_1(y))Tr_1^n(x\phi_2(y)) + Tr_1^n(x\phi_1(y))Tr_1^n(x\phi_3(y)) \\ + Tr_1^n(x\phi_2(y))Tr_1^n(x\phi_3(y))$$

if $(x, y) \in \mathbb{F}_{2^n} \times \mathbb{F}_{2^m}$ and

$$g(x, y) = Tr_1^n(axy^d)Tr_1^n(acxy^d) + Tr_1^n(axy^d)Tr_1^n(ac^{-1}xy^d) \\ + Tr_1^n(acxy^d)Tr_1^n(ac^{-1}xy^d)$$

if $(x, y) \in \mathbb{F}_{2^n} \times \mathbb{F}_{2^n} \setminus \mathbb{F}_{2^m}$. Then g is bent and its dual function $\tilde{g}$ is defined by

$$\tilde{g}(x, y) = Tr_1^n(\phi_1^{-1}(x)y)Tr_1^n(\phi_2^{-1}(x)y) + Tr_1^n(\phi_1^{-1}(x)y)Tr_1^n(\phi_3^{-1}(x)y) \\ + Tr_1^n(\phi_2^{-1}(x)y)Tr_1^n(\phi_3^{-1}(x)y)$$

if $(x, y) \in \mathbb{F}_{2^m} \times \mathbb{F}_{2^n}$ and

$$\tilde{g}(x, y) = Tr_1^n(a^{-11}x^{11}y)Tr_1^n(a^{-11}c^{-11}x^{11}y) + Tr_1^n(a^{-11}x^{11}y)Tr_1^n(a^{-11}c^{11}x^{11}y) \\ + Tr_1^n(a^{-11}c^{-11}x^{11}y)Tr_1^n(a^{-11}c^{11}x^{11}y)$$

if $(x, y) \in \mathbb{F}_{2^n} \setminus \mathbb{F}_{2^m} \times \mathbb{F}_{2^n}$.

Note that, in the particular case where $\rho_1 = \rho_2 = \rho_3$, Condition 2 of Proposition 7.2.15 is trivially fulfilled. Based on this remark, we now describe a first family of bent functions that can be derived from Proposition 7.2.15. Let $n = km$ where m is a positive integer and k is a positive integer greater than 1. Let ϕ_1, ϕ_2 and ϕ_2 be three permutations of $\mathbb{F}_{2^m}$ satisfying $(\mathcal{A}_m)$. Set $\rho(x) = \alpha x^d$ where $\gcd(d, 2^n - 1) = 1$ and $\alpha \in \mathbb{F}_{2^m}^\star$. Note that then $x \mapsto \alpha x^d$ is a permutation of $\mathbb{F}_{2^m}$ (d and $2^m - 1$ are coprime since $2^n - 1 = (2^m - 1)\sum_{j=0}^{k-1} 2^{jm}$), which proves that ρ is a permutation of $\mathbb{F}_{2^n} \setminus \mathbb{F}_{2^m}$. Set, for $i \in \{1, 2, 3\}$,

$$x \in \mathbb{F}_{2^n},\ \psi_i(x) = \begin{cases} \phi_i(x) & \text{if } x \in \mathbb{F}_{2^m} \\ \alpha x^d & \text{if } x \in \mathbb{F}_{2^n} \setminus \mathbb{F}_{2^m}. \end{cases} \tag{7.26}$$

Then ψ_i is a permutation of $\mathbb{F}_{2^n}$ whose inverse is given by

$$x \in \mathbb{F}_{2^n},\ \psi_i^{-1}(x) = \begin{cases} \phi_i^{-1}(x) & \text{if } x \in \mathbb{F}_{2^m} \\ \alpha^{-e} x^e & \text{if } x \in \mathbb{F}_{2^n} \setminus \mathbb{F}_{2^m} \end{cases} \tag{7.27}$$

where e is the inverse of d modulo $2^n - 1$. Define, for each $i \in \{1, 2, 3\}$, a Boolean function f_i over $\mathbb{F}_{2^n} \times \mathbb{F}_{2^n}$ as

$$(x, y) \in \mathbb{F}_{2^n} \times \mathbb{F}_{2^n},\ f_i(x, y) = Tr_1^n(x\psi_i(y)) = \begin{cases} Tr_1^n(x\phi_i(y)) & \text{if } x \in \mathbb{F}_{2^m} \\ Tr_1^n(\alpha x y^d) & \text{if } x \in \mathbb{F}_{2^n} \setminus \mathbb{F}_{2^m}. \end{cases}$$

According to Proposition 7.1.14, f_i, $i \in \{1, 2, 3\}$ is bent and its dual function $\tilde{f}_i$ is

$$(x, y) \in \mathbb{F}_{2^n} \times \mathbb{F}_{2^n},\ \tilde{f}_i(x, y) = Tr_1^n(\psi_i^{-1}(x)y) = \begin{cases} Tr_1^n(\phi_i^{-1}(x)y) & \text{if } y \in \mathbb{F}_{2^m} \\ Tr_1^n(\alpha^{-e}x^e y) & \text{if } y \in \mathbb{F}_{2^n} \setminus \mathbb{F}_{2^m}. \end{cases} \tag{7.28}$$

On the other hand, note that, if $x \in \mathbb{F}_{2^m}$,

$$\begin{aligned} &f_1(x, y)f_2(x, y) + f_1(x, y)f_3(x, y) + f_2(x, y)f_3(x, y) \\ &\quad = Tr_1^n(x\phi_1(y))Tr_1^n(x\phi_2(y)) + Tr_1^n(x\phi_1(y))Tr_1^n(x\phi_3(y)) \\ &\quad\quad + Tr_1^n(x\phi_2(y))Tr_1^n(x\phi_3(y)) \end{aligned} \tag{7.29}$$

and, if $x \in \mathbb{F}_{2^n} \setminus \mathbb{F}_{2^m}$, $f_1(x, y)f_2(x, y) + f_1(x, y)f_3(x, y) + f_2(x, y)f_3(x, y) = Tr_1^n(\alpha x y^d)$. Collecting all the above arguments, thanks to Theorem 7.1.3 we get the following construction.

Construction 7.2.20 ([10]) *Let n be a multiple of m where m is a positive integer and $n \neq m$. Let ϕ_1, ϕ_2 and ϕ_3 be three permutations over $\mathbb{F}_{2^m}$ satisfying $(\mathcal{A}_m)$. Let $\alpha \in \mathbb{F}_{2^m}^\star$. Let d be a positive integer such that d and $2^n - 1$ are coprime. Denote by e the inverse of d modulo $2^n - 1$. Set*

$$g(x,y) = \mathrm{Tr}_1^n(x\phi_1(y))\mathrm{Tr}_1^n(x\phi_2(y)) + \mathrm{Tr}_1^n(x\phi_1(y))\mathrm{Tr}_1^n(x\phi_3(y)) + \mathrm{Tr}_1^n(x\phi_2(y))\mathrm{Tr}_1^n(x\phi_3(y))$$

if $(x,y) \in \mathbb{F}_{2^n} \times \mathbb{F}_{2^m}$ *and* $g(x,y) = \mathrm{Tr}_1^n(\alpha xy^d)$ *if* $(x,y) \in \mathbb{F}_{2^n} \times \mathbb{F}_{2^n} \setminus \mathbb{F}_{2^m}$. *Then* g *is bent and its dual function* $\tilde{g}$ *is defined by*

$$\tilde{g}(x,y) = \mathrm{Tr}_1^n(\phi_1^{-1}(x)y)\mathrm{Tr}_1^n(\phi_2^{-1}(x)y) + \mathrm{Tr}_1^n(\phi_1^{-1}(x)y)\mathrm{Tr}_1^n(\phi_3^{-1}(x)y) + \mathrm{Tr}_1^n(\phi_2^{-1}(x)y)\mathrm{Tr}_1^n(\phi_3^{-1}(x)y)$$

if $(x,y) \in \mathbb{F}_{2^m} \times \mathbb{F}_{2^n}$ *and* $\tilde{g}(x,y) = \mathrm{Tr}_1^n(\alpha^{-e}x^e y)$ *if* $(x,y) \in \mathbb{F}_{2^n} \setminus \mathbb{F}_{2^m} \times \mathbb{F}_{2^n}$.

Example 7.2.21. If we choose $\phi_i(x) = a_i x^{2^m-2}$ in Construction 7.2.20, we get the following infinite class.

Let n be a multiple of m where m is a positive integer and $n \neq m$. Let (a_1, a_2, a_3) be a 3-tuple of $\mathbb{F}_{2^m}$ such that $a_1 + a_2 + a_3 \neq 0$. Let $\alpha \in \mathbb{F}_{2^m}^\star$ and d be a positive integer such that d and $2^n - 1$ are coprime. Denote by e the inverse of d modulo $2^n - 1$. Set

$$g(x,y) = \mathrm{Tr}_1^n(a_1xy^{2^m-2})\mathrm{Tr}_1^n(a_2xy^{2^m-2}) + \mathrm{Tr}_1^n(a_1xy^{2^m-2})\mathrm{Tr}_1^n(a_3xy^{2^m-2}) + \mathrm{Tr}_1^n(a_2xy^{2^m-2})\mathrm{Tr}_1^n(a_3xy^{2^m-2})$$

if $(x,y) \in \mathbb{F}_{2^n} \times \mathbb{F}_{2^m}$ and $g(x,y) = \mathrm{Tr}_1^n(\alpha xy^d)$ if $(x,y) \in \mathbb{F}_{2^n} \times \mathbb{F}_{2^n} \setminus \mathbb{F}_{2^m}$. Then g is bent and its dual function $\tilde{g}$ is defined by

$$\tilde{g}(x,y) = \mathrm{Tr}_1^n(a_1x^{2^m-2}y)\mathrm{Tr}_1^n(a_2x^{2^m-2}y) + \mathrm{Tr}_1^n(a_1x^{2^m-2}y)\mathrm{Tr}_1^n(a_3x^{2^m-2}y) + \mathrm{Tr}_1^n(a_2x^{2^m-2}y)\mathrm{Tr}_1^n(a_3x^{2^m-2}y)$$

if $(x,y) \in \mathbb{F}_{2^m} \times \mathbb{F}_{2^n}$ and $\tilde{g}(x,y) = \mathrm{Tr}_1^n(\alpha^{-e}x^e y)$ if $(x,y) \in \mathbb{F}_{2^n} \setminus \mathbb{F}_{2^m} \times \mathbb{F}_{2^n}$. Furthermore, if $d^2 = 1 \mod 2^n - 1$, that is $e = d$, and $\alpha^{-d} = \alpha$, then g belongs to $\mathcal{PS}_{ap}$.

In the particular case where $n = 2m$, we now introduce the following map from $\mathbb{F}_{2^n}$ to itself:

$$\rho(x) = \lambda x \left(\mathrm{Tr}_m^n(x)\right)^d. \tag{7.30}$$

where d is a positive integer and $\lambda \in \mathbb{F}_{2^m}^\star$. Then one can prove the following result.

Proposition 7.2.22 ([10]). *Suppose that* $n = 2m$. *Let* ρ *be defined by (7.30). Then* ρ *is a permutation of* $\mathbb{F}_{2^n} \setminus \mathbb{F}_{2^m}$ *if and only if* $\gcd(d+1, 2^m-1) = 1$. *Furthermore, its inverse map is:*

$$x \in \mathbb{F}_{2^n} \setminus \mathbb{F}_{2^m},\ \rho^{-1}(x) = \lambda^{-\frac{1}{d+1}} x \left(\mathrm{Tr}_m^n(x)\right)^{-\frac{d}{d+1}}.$$

Proof. Clearly, ρ is a map from $\mathbb{F}_{2^n} \setminus \mathbb{F}_{2^m}$ to itself since $\rho^{2^m}(x) = x^{2^m-1}\rho(x) \neq \rho(x)$ for every $x \in \mathbb{F}_{2^n} \setminus \mathbb{F}_{2^m}$. Let us now investigate under which conditions ρ is one-to-one. Let x and y be two elements of $\mathbb{F}_{2^n} \setminus \mathbb{F}_{2^m}$ such that $\rho(x) = \rho(y)$:

$$\lambda x \left(\mathrm{Tr}_m^n(x)\right)^d = \lambda y \left(\mathrm{Tr}_m^n(y)\right)^d \iff y = \left(\mathrm{Tr}_m^n(x)\right)^d \left(\mathrm{Tr}_m^n(y)\right)^{-d} x$$

that is, $y = \beta x$ with $\beta \in \mathbb{F}_{2^m}^\star$ ($\mathrm{Tr}_m^n(z) = 0$ if and only if $z \in \mathbb{F}_{2^m}$). If we replace y by βx in the above equation, we get $x = \beta^{1+d}x$ that is equivalent to $\beta^{1+d} = 1$. If $\gcd(d+1, 2^m - 1) = 1$ then it implies that $\beta = 1$, that is, $y = x$ proving that ρ is one-to-one. If $\delta = \gcd(d+1, 2^m - 1) > 1$, ρ is not one-to-one since $\rho(\beta x) = \rho(x)$ for every $\beta \in \mathbb{F}_{2^m}$ of order δ ($\beta^\delta = 1$). In order to prove the last statement, set

$$\mu(x) = \lambda^{-\frac{1}{d+1}} x \left(\mathrm{Tr}_m^n(x)\right)^{-\frac{d}{d+1}}.$$

Let us compute $\rho \circ \mu$:

$$\begin{aligned}
\rho(\mu(x)) &= \rho\left(\lambda^{-\frac{1}{d+1}} x \left(\mathrm{Tr}_m^n(x)\right)^{-\frac{d}{d+1}}\right) \\
&= \lambda^{1-\frac{1}{d+1}-\frac{d}{d+1}} x \left(\mathrm{Tr}_m^n(x)\right)^{-\frac{d}{d+1}} \left(\mathrm{Tr}_m^n\left(x\left(\mathrm{Tr}_m^n(x)\right)^{-\frac{d}{d+1}}\right)\right)^d \\
&= x\left(\mathrm{Tr}_m^n(x)\right)^{-\frac{d}{d+1}+d-\frac{d^2}{d+1}}. \\
&= x.
\end{aligned}$$

□

One can then straightforwardly transpose Construction 7.2.20 to the family of permutations of Proposition 7.2.22.

Construction 7.2.23 ([10]) *Let $n = 2m$ where m is a positive integer. Let ϕ_1, ϕ_2 and ϕ_3 be three permutations over $\mathbb{F}_{2^m}$ satisfying $(\mathcal{A}_m)$. Let d be a positive integer such that $d+1$ and $2^n - 1$ are coprime. Let $\lambda \in \mathbb{F}_{2^m}^\star$. Set*

$$\begin{aligned}
g(x,y) = {} & \mathrm{Tr}_1^n(x\phi_1(y))\mathrm{Tr}_1^n(x\phi_2(y)) + \mathrm{Tr}_1^n(x\phi_1(y))\mathrm{Tr}_1^n(x\phi_3(y)) \\
& + \mathrm{Tr}_1^n(x\phi_2(y))\mathrm{Tr}_1^n(x\phi_3(y))
\end{aligned}$$

if $(x,y) \in \mathbb{F}_{2^n} \times \mathbb{F}_{2^m}$ and $g(x,y) = \mathrm{Tr}_1^n\left(\lambda xy\left(\mathrm{Tr}_m^n(y)\right)^d\right)$ if $(x,y) \in \mathbb{F}_{2^n} \times \mathbb{F}_{2^n} \setminus \mathbb{F}_{2^m}$. Then g is bent and its dual function $\tilde{g}$ is defined by

$$\begin{aligned}
\tilde{g}(x,y) = {} & \mathrm{Tr}_1^n(\phi_1^{-1}(x)y)\mathrm{Tr}_1^n(\phi_2^{-1}(x)y) + \mathrm{Tr}_1^n(\phi_1^{-1}(x)y)\mathrm{Tr}_1^n(\phi_3^{-1}(x)y) \\
& + \mathrm{Tr}_1^n(\phi_2^{-1}(x)y)\mathrm{Tr}_1^n(\phi_3^{-1}(x)y)
\end{aligned}$$

if $(x,y) \in \mathbb{F}_{2^m} \times \mathbb{F}_{2^n}$ and $\tilde{g}(x,y) = \mathrm{Tr}_1^n\left(\lambda^{-\frac{1}{d+1}} x\left(\mathrm{Tr}_m^n(x)\right)^{-\frac{d}{d+1}} y\right)$ if $(x,y) \in \mathbb{F}_{2^n} \setminus \mathbb{F}_{2^m} \times \mathbb{F}_{2^n}$.

In [10], we have pushed further the study of the construction proposed in [11]. That construction gives rise to many infinite families of bent functions. A question that arises is to study the EA-equivalence of the functions constructed in [10] to the class of Maiorana–McFarland or other known classes. Although this question is left open, we emphasize that the main feature of those bent functions is that their dual functions can be computed explicitly. Computing explicitly dual functions is often a hard task and for many classes of bent functions, that question is still challenging.

7.3 Infinite Families of Bent Functions from Involutions and Their Duals

We have seen in the previous section that the problem of constructing some bent functions is linked to finding 3-tuples of permutations satisfying two conditions. The latter conditions can be simplified when the permutations are involutions.

The following statement is a straightforward consequence of Theorem 7.1.3 showing that one can derive bent functions in bivariate representation from involutions.

Corollary 7.3.1. *Let m be a positive integer. Let ϕ_1, ϕ_2 and ϕ_3 be three involutions of $\mathbb{F}_{2^m}$. Then,*

$$g(x,y) = Tr_1^m(x\phi_1(y))Tr_1^m(x\phi_2(y)) + Tr_1^m(x\phi_1(y))Tr_1^m(x\phi_3(y)) + Tr_1^m(x\phi_2(y))Tr_1^m(x\phi_3(y))$$

is bent if and only if $\psi = \phi_1 + \phi_2 + \phi_3$ is an involution.

Furthermore, its dual function $\tilde{g}$ is given by $\tilde{g}(x,y) = g(y,x)$.

Notice that this gives a very handy way to compute the dual (namely, transpose the two arguments), in stark contrast with the univariate case.

In [15], the authors have investigated bent functions from monomial involutions.

Using a monomial involution (see [7]), a first construction is given by the following statement.

Theorem 7.3.2 ([15]). *Let n be an integer. Let d be a positive integer such that $d^2 \equiv 1 \pmod{2^n-1}$. Let Φ_1, Φ_2 and Φ_3 be three mappings from $\mathbb{F}_{2^n}$ to $\mathbb{F}_{2^n}$ defined by $\Phi_i(x) = \lambda_i x^d$ for all $i \in \{1,2,3\}$, where the $\lambda_i \in \mathbb{F}_{2^n}^\star$ are pairwise distinct such that $\lambda_i^{d+1} = 1$ and ${\lambda_0}^{d+1} = 1$, where $\lambda_0 := \lambda_1 + \lambda_2 + \lambda_3$. Let g be the Boolean function defined over $\mathbb{F}_{2^n} \times \mathbb{F}_{2^n}$ by*

$$g(x,y) = Tr_1^n(\Phi_1(y)x)Tr_1^n(\Phi_2(y)x) + Tr_1^n(\Phi_2(y)x)Tr_1^n(\Phi_3(y)x) + Tr_1^n(\Phi_1(y)x)Tr_1^n(\Phi_3(y)x). \tag{7.31}$$

Then the Boolean function g defined over $\mathbb{F}_{2^n} \times \mathbb{F}_{2^n}$ by (7.31) is bent and its dual is given by $\tilde{g}(x,y) = g(y,x)$.

The existence of bent functions given in Theorem 7.3.2 is a non-trivial arithmetical problem and is discussed in [15] in which the authors have partially solved the problem from algebraic and geometric point of view using Fermat hypersurface and Lang-Weil estimates.

Using similar arguments as previously, we derive in Proposition 7.3.3 and Proposition 7.3.4 more constructions of bent functions based on some involutions of $\mathbb{F}_{2^n}$ (see [7]) as application of Corollary 7.2.1.

Proposition 7.3.3 ([15]). *Let $n = rk$ be an integer with $k > 1$ and $r > 1$. For $i \in \{1,2,3\}$, let γ_i be an element of $\mathbb{F}_{2^n}^\star$ such that $\mathrm{Tr}_k^n(\gamma_i) = 0$ and Φ_i be a mapping defined over $\mathbb{F}_{2^n}$ by*

$$\Phi_i(x) = x + \gamma_i \mathrm{Tr}_k^n(x).$$

Then the Boolean function g defined over $\mathbb{F}_{2^n} \times \mathbb{F}_{2^n}$ by (7.31) is bent and its dual function is given by $\tilde{g}(x,y) = g(y,x)$.

Proposition 7.3.4 ([15]). *Let $n = 2m$ be an even integer. Let h_1, h_2, h_3 be three linear mappings from $\mathbb{F}_{2^m}$ to itself. For $i \in \{1,2,3\}$, let Φ_i be a mapping from $\mathbb{F}_{2^n}$ to itself defined by*

$$\Phi_i(x) = h_i(\mathrm{Tr}_m^n(x)) + x.$$

Then the Boolean function g defined over $\mathbb{F}_{2^n} \times \mathbb{F}_{2^n}$ by (7.31) is bent and its dual function is given by $\tilde{g}(x,y) = g(y,x)$.

Remark 7.3.5. Set $\Phi_i'(x) = h_i(\mathrm{Tr}_m^n(x)) + x^{2^m}$. Let g' be the Boolean function derived from (7.31) using the Φ_i''s. Then g' is bent and its dual is given by $\widetilde{g'}(x,y) = g'(y,x)$. Clearly the functions g (given by the previous theorem) and g' are affinely equivalent.

In the following we show that further bent functions involving linear structures can be simply obtained from general linear involutions.

Proposition 7.3.6 ([12]). *Let $L : \mathbb{F}_{2^m} \to \mathbb{F}_{2^m}$ be a $\mathbb{F}_2$-linear involution of $\mathbb{F}_{2^m}$. Let f be a Boolean function over $\mathbb{F}_{2^m}$ and α be a non zero 0-linear structure of f. Then the mapping ϕ defined by $\phi(x) = L(x) + L(\alpha)f(x)$, $x \in \mathbb{F}_{2^m}$ is a permutation of $\mathbb{F}_{2^m}$ and*

$$\phi^{-1}(x) = L(x) + \alpha f(L(x)). \tag{7.32}$$

Let $\phi(x) = \lambda x^{2^i}$ be a linear monomial mapping where $0 < i < n$ and $\lambda \in \mathbb{F}_2^\star$. In [7], the authors have characterized linear monomials that are involutions. More precisely, $\phi(x)$ is an involution if and only if $m = \frac{n}{2}$ with n even and $\lambda^{2^m+1} = 1$. It has been shown that there is no linear monomial involution when n is odd.

A natural question is whether linear monomial involutions give rise to bent functions g of the form (7.31) or not. The next lemma gives a negative answer.

Lemma 7.3.7. *Let $n = 2m$ be an even integer and λ_i $(1 \le i \le 3)$ three pairwise distinct elements of $\mathbb{F}_{2^n}^\star$. Set $\lambda_0 := \lambda_1+\lambda_2+\lambda_3$. Then there is no 3-tuple $(\lambda_1, \lambda_2, \lambda_3)$ satisfying $\lambda_i^{2^m+1} = 1$, for $0 \le i \le 3$.*

Consequently, according to Corollary 7.3.1 and Lemma 7.3.7, there is no bent function of the form (7.31) with ϕ_i's linear monomial involutions.

We focus on some binomial involutions. The following result given in [7] characterizes linear binomials that are involutions.

Proposition 7.3.8 (Proposition 5, [7]).
Let $Q(x) = ax^{2^i} + bx^{2^j}$, $a \in \mathbb{F}_2^*$ and $b \in \mathbb{F}_2^*$, where $i < j < n$. Then we have:

- *For odd n, Q can never be an involution.*
- *For even n, $n = 2m$, Q is an involution if and only if $j = i + m$ and either*

$$i = 0,\ a^2 + b^{2^m+1} = 1;$$

or m is even,

$$i = m/2,\ ab^{2^i} + a^{2^j}b = 1 \ \text{ and } \ a^{2^i+1} + b^{2^j+1} = 0.$$

Using Corollary 7.3.1 and the first part of Proposition 7.3.8 one deduces the following construction of bent functions (see also [13]).

Theorem 7.3.9 ([12]). *Let $n = 2m$ be an even integer. Let Φ_1, Φ_2 and Φ_3 be three linear mappings from $\mathbb{F}_{2^n}$ to $\mathbb{F}_{2^n}$ defined by*

$$\Phi_i(x) = \alpha_i x + \beta_i x^{2^m}$$

for all $i \in \{0, 1, 2, 3\}$ where $(\alpha_i, \beta_i) \in (\mathbb{F}_{2^n}^\star)^2$ satisfy the following condition (C)

$$\alpha_i{}^2 + \beta_i{}^{2^m+1} = 1$$

where $\alpha_0 := \alpha_1 + \alpha_2 + \alpha_3$ and $\beta_0 := \beta_1 + \beta_2 + \beta_3$. Then the Boolean function g defined over $\mathbb{F}_{2^n} \times \mathbb{F}_{2^n}$ by (7.31) is bent and its dual is given by $\tilde{g}(x, y) = g(y, x)$.

Lemma 7.3.10 ([12]). *If β_1, β_2, β_3, α_1, α_2, α_3 satisfy condition (C) of Theorem 7.3.9, then $\beta_{\sigma(1)}$, $\beta_{\sigma(2)}$, $\beta_{\sigma(3)}$, $\alpha_{\sigma(1)}$, $\alpha_{\sigma(2)}$, $\alpha_{\sigma(3)}$ is again a solution for any permutation σ of the set $\{1, 2, 3\}$. Up to a permutation of the indices, the only solutions of condition (C) of Theorem 7.3.9 are:*

- *either $\beta_1 = a$, $\beta_2 = b$ and $\beta_3 = \frac{ab^{2^m}+c}{(a+b)^{2^m}}$ where a, b are two distinct elements of $\mathbb{F}_{2^n}^\star$ and c is an element of $\mathbb{F}_{2^m}$ such that $c \neq ab^{2^m}$;*
- *or $\beta_1 = \beta_2 = a$ and $\beta_3 = b$ where a, b are two elements of $\mathbb{F}_{2^n}^\star$.*

Furthermore, $\alpha_i := \lambda_i + 1$, for $i = 1, 2, 3$, where the λ_i's are defined by : $\beta_i = \lambda_i u_1$, with the λ_i's in $\mathbb{F}_{2^m}$ and the u_i's in the cyclic group $U := \{u \in \mathbb{F}_{2^n} \mid u^{2^m+1} = 1\}$.

Using Corollary 7.3.1 and the second part of Proposition 7.3.8 one deduces the following construction of bent functions.

Theorem 7.3.11 ([12]). *Let $n = 4k$ be an integer with $k \in \mathbb{N}^\star$. Let Φ_1, Φ_2 and Φ_3 be three linear mappings from $\mathbb{F}_{2^n}$ to $\mathbb{F}_{2^n}$ defined by*

$$\Phi_i(x) = \alpha_i x^{2^k} + \beta_i x^{2^{3k}}$$

for all $i \in \{0, 1, 2, 3\}$ where $(\alpha_i, \beta_i) \in (\mathbb{F}_{2^n}^\star)^2$ satisfy the following conditions

1. $\alpha_i \beta_i^{2^k} + \alpha_i^{2^{3k}} \beta_i = 1$;
2. $\alpha_i^{2^k+1} + \beta_i^{2^{3k}+1} = 0$;

where $\alpha_0 := \alpha_1 + \alpha_2 + \alpha_3$ and $\beta_0 := \beta_1 + \beta_2 + \beta_3$. Then the Boolean function g defined over $\mathbb{F}_{2^n} \times \mathbb{F}_{2^n}$ by (7.31) is bent and its dual is given by $\tilde{g}(x, y) = g(y, x)$.

The main question remaining in the construction of bent functions derived from Theorem 7.3.11 is the existence of $(\alpha_i, \beta_i) \in (\mathbb{F}_{2^n}^\star)^2$ satisfying the conditions 1 and 2. The next lemma gives an answer of the existence's problem.

Lemma 7.3.12 ([12]). *Consider the following system (S) of Eqs. (7.33) and (7.34) in $\mathbb{F}_{2^n}^\star$ where $n = 4k$ with $k \in \mathbb{N}^\star$ and whose unknowns are x and y:*

$$\begin{cases} xy^{2^k} + x^{2^{3k}} y = 1 & (7.33) \\ x^{2^k+1} + y^{2^{3k}+1} = 0 & (7.34) \end{cases}$$

Then (x, y) is a solution of the system (S) if and only if $x = Auv^{2^{3k}}$ and $y = (A+1)uv^{2^k}$ where $A \in \mathbb{F}_{2^n}$ is such that $A^{2^k} = A + 1$, $u \in U_k := \{u \in \mathbb{F}_{2^{4k}} \mid u^{2^k+1} = 1\}$ and $v \in U_{2k} := \{u \in \mathbb{F}_{2^{4k}} \mid u^{2^{2k}+1} = 1\}$.

References

1. L. Budaghyan, C. Carlet, T. Helleseth, A. Kholosha, and S. Mesnager. Further results on Niho bent functions. In *IEEE Transactions on Information Theory-IT, Vol 58, no.11*, pages 6979–6985, 2012.
2. A. Canteaut and P. Charpin. Decomposing bent functions. In *IEEE Transactions on Information Theory, Vol 49*, pages 2004–2019, 2003.
3. C. Carlet. On bent and highly nonlinear balanced/resilient functions and their algebraic immunities. In *AAECC 16, Las Vegas, february 2006, volume 3857 of Lecture Notes in Computer Science, Springer*, pages 1–28, 2006.
4. C. Carlet. Boolean Functions for Cryptography and Error Correcting Codes. In *Chapter of the monography "Boolean Models and Methods in Mathematics, Computer Science, and Engineering" published by Cambridge University Press, Yves Crama and Peter L. Hammer (eds.)*, pages 257–397, 2010.
5. C. Carlet, L. E. Danielsen, M. G. Parker, and P. Solé. Self-dual bent functions. In *Journal IJICoT, Vol. 1, No. 4, pages 384–399*, 2010.

6. C. Carlet, T. Helleseth, A. Kholosha, and S. Mesnager. On the duals of bent functions with 2^r Niho exponents. In *IEEE International Symposium on Information Theory, ISIT 2011*, pages 703–707, 2011.
7. P. Charpin, S. Mesnager, and S. Sarkar. On involutions of finite fields. In *Proceedings of 2015 IEEE International Symposium on Information Theory, (ISIT)*, 2015.
8. G. M. Kyureghyan. Constructing permutations of finite fields via linear translators. In *Journal of Combinatorial Theory Series A. 118 (3)*, pages 1052–1061, 2011.
9. G. Leander and A. Kholosha. Bent functions with 2^r Niho exponents. In *IEEE Trans. Inform. Theory 52 (12)*, pages 5529–5532, 2006.
10. S. Mesnager. Further constructions of infinite families of bent functions from new permutations and their duals. In *Journal Cryptography and Communications (CCDS), Springer. To appear.*
11. S. Mesnager. Several new infinite families of bent functions and their duals. In *IEEE Transactions on Information Theory-IT, Vol. 60, No. 7, Pages 4397–4407*, 2014.
12. S. Mesnager. A note on constructions of bent functions from involutions. In *IACR Cryptology ePrint Archive 2015: 982*, 2015.
13. S. Mesnager. On constructions of bent functions from involutions. *Proceedings of 2016 IEEE International Symposium on Information Theory*, ISIT 2016, 2016.
14. S. Mesnager. Bent functions from spreads. In *Journal of the American Mathematical Society (AMS), Contemporary Mathematics (Proceedings the 11th International conference on Finite Fields and their Applications Fq11), Volume 632, page 295–316*, 2015.
15. S. Mesnager, G. Cohen, and D. Madore. On existence (based on an arithmetical problem) and constructions of bent functions. In *Proceedings of the fifteenth International Conference on Cryptography and Coding, Oxford, United Kingdom, IMACC 2015, LNCS, Springer, Heidelberg*, pages 3–19, 2015.
16. K. Nyberg. On the construction of highly nonlinear permutations. In *Proceedings of EUROCRYPT'92, Lecture Notes in Computer Science 658*, pages 92–98, 1993.

Chapter 8
Class $\mathcal{H}$, Niho Bent Functions and o-Polynomials

This chapter deals with an important class of bent function introduced by Dillon in 1974, which was somehow abandoned for nearly 35 years before being extended by Carlet and the author. We will see that such an extension has been very fruitful in the field of bent functions.

8.1 Classes H and $\mathcal{H}$ in Bivariate Form

8.1.1 Class H of Dillon

In his thesis [13], Dillon introduces a family of bent functions whose expression

$$f(x,y) = Tr_1^m(y + xG(yx^{2^m-2})),$$

with $x, y \in \mathbb{F}_{2^m}$ where G is a permutation of $\mathbb{F}_{2^m}$ such that $G(x) + x$ does not vanish and, for every $\beta \in \mathbb{F}_{2^m}^\star$, the function $G(x) + \beta x$ is two-to-one (i.e. the pre-image by this function of any element of $\mathbb{F}_{2^m}$ is either a pair or the empty set). He denotes this family of bent functions by H.

The condition that $G(x) + x$ does not vanish is required only for H to be an extension of $\mathcal{PS}$ but is not necessary for f to be bent. Similarly, the linear term $Tr_1^m(y)$ can be taken off if we are only interested in the bentness of the function. We then have $f(x,y) = \begin{cases} Tr_1^m\left(xG\left(\frac{y}{x}\right)\right) & \text{if } x \neq 0 \\ 0 & \text{if } x = 0. \end{cases}$

Note that the restriction of f to the vectorspaces $\{(x, ax)\,;\, x \in \mathbb{F}_{2^m}\}$ where $a \in \mathbb{F}_{2^m}$ linear.

The reader notices that functions in the class H are defined in bivariate representation but we will see in the next section that they can also be defined in univariate representation. Moreover, the bentness of elements of H is achieved under some

S. Mesnager, *Bent Functions*, DOI 10.1007/978-3-319-32595-8_8

non-obvious condition (so the class is less explicit than class $\mathcal{M}$ or class $\mathcal{PS}_{ap}$, but it happens to be more explicit than class $\mathcal{PS}$, the condition for H being easier to satisfy than for $\mathcal{PS}$).

8.1.2 Class $\mathcal{H}$

In the following, we show how to extend the class H of Dillon. Firstly, any function whose restrictions to these vectorspaces are linear has the form:

$$g(x,y) = \begin{cases} Tr_1^m\left(x\psi\left(\frac{y}{x}\right)\right) & \text{if } x \neq 0 \\ Tr_1^m(\mu y) & \text{if } x = 0 \end{cases} \tag{8.1}$$

where $\mu \in \mathbb{F}_{2^m}$ and ψ is a mapping from $\mathbb{F}_{2^m}$ to itself. The following proposition provides a necessary and sufficient condition on ψ and μ for g being bent.

Proposition 8.1.1 ([8]). *Let g be a Boolean function over $\mathbb{F}_{2^m} \times \mathbb{F}_{2^m}$ defined by* (8.1). *Then g is bent if and only if, denoting $G(z) = \psi(z) + \mu z$, we have:*

$$G \text{ is a permutation on } \mathbb{F}_{2^m} \tag{8.2}$$

$$\text{For every } \beta \in \mathbb{F}_{2^m}^{\star}, \text{ the function } z \mapsto G(z) + \beta z \text{ is 2-to-1 on } \mathbb{F}_{2^m}. \tag{8.3}$$

Proof. For every $\alpha, \beta \in \mathbb{F}_{2^m}$, we have:

$$\begin{aligned}
\widehat{\chi_g}(\alpha,\beta) &= \sum_{x,y\in\mathbb{F}_{2^m}} (-1)^{g(x,y)+Tr_1^m(\alpha x+\beta y)} \\
&= \sum_{x\in\mathbb{F}_{2^m}^{\star},z\in\mathbb{F}_{2^m}} (-1)^{Tr_1^m(x\psi(z)+\alpha x+\beta xz)} + \sum_{y\in\mathbb{F}_{2^m}} (-1)^{Tr_1^m((\beta+\mu)y)} \\
&= \sum_{x\in\mathbb{F}_{2^m},z\in\mathbb{F}_{2^m}} (-1)^{Tr_1^m(x\psi(z)+\alpha x+\beta xz)} - 2^m + 2^m\delta_\mu(\beta) \\
&= 2^m \,\#\{z \in \mathbb{F}_{2^m} \,/\, \psi(z) + \alpha + \beta z = 0\} - 2^m + 2^m\delta_\mu(\beta).
\end{aligned}$$

We denote by $N_{\alpha,\beta}$ the cardinality of the set $\{z \in \mathbb{F}_{2^m} \,/\, \psi(z) + \alpha + \beta z = 0\}$.

Then we have $\widehat{\chi_g}(\alpha,\beta) = \begin{cases} 2^m N_{\alpha,\mu} & \text{if } \beta = \mu \\ 2^m N_{\alpha,\beta} - 2^m & \text{if } \beta \neq \mu \end{cases}$, and Conditions (8.2) and (8.3) are necessary and sufficient for g being bent. □

Note that Condition (8.3) is equivalent to saying that for every $\beta \in \mathbb{F}_{2^m}^{\star}$, the function $z \mapsto \beta G(z) + z$ is 2-to-1.

In the following we present the study given in [8] concerning the stability of functions G satisfying Conditions (8.2) and (8.3).

Let G be a function satisfying Conditions (8.2) and (8.3). Then

1. the function $z \mapsto G^{-1}(z)$ satisfies Conditions (8.2) and (8.3), since denoting $G^{-1}(z)$ by z', the equation $G^{-1}(z) + \beta z = \alpha$ is equivalent to $G(z') + \frac{1}{\beta}z' = \frac{\alpha}{\beta}$.
2. the function $z \mapsto G'(z) := (L^{-1} \circ G \circ L)(z)$ where $L(z) = z^{2^j}$ is a field automorphism of $\mathbb{F}_{2^m}$, that is $G'(z) = (G(z^{2^j}))^{2^{m-j}}$, satisfies Conditions (8.2) and (8.3).
3. the function $z \mapsto G'(z) := \lambda G(z) + \lambda'$ with $\lambda \neq 0$ satisfies Conditions (8.2) and (8.3).
4. the function $z \mapsto G'(z) := G(\lambda z + \lambda')$ with $\lambda \neq 0$ satisfies Conditions (8.2) and (8.3).
5. the function $z \mapsto G'(z) := zG(z^{2^m-2})$ if $G(0) = 0$ and more generally the function $z \mapsto G'(z) := zG(z^{2^m-2}) + zG(0)$ for any value of $G(0)$ satisfies Conditions (8.2) and (8.3). Indeed (restricting ourself without loss of generality to the case $G(0) = 0$—by replacing G by $G + G(0)$—and still assuming that $\beta \neq 0$), if $\alpha \neq 0$ then $zG(z^{2^m-2}) = \alpha$ is equivalent to $G(z^{2^m-2}) = \alpha z^{2^m-2}$ which has one solution since $G(z) + \alpha z = 0$ has two solutions and $z = 0$ is one of them, and the equation $zG(z^{2^m-2}) + \beta z = \alpha$ is equivalent to $G(z^{2^m-2}) + \alpha z^{2^m-2} = \beta$ and has therefore 0 or 2 solutions; and if $\alpha = 0$ then $zG(z^{2^m-2}) = \alpha = 0$ is equivalent to $z = 0$ and the equation $zG(z^{2^m-2}) + \beta z = \alpha = 0$ is equivalent to $z = 0$ or $G(z^{2^m-2}) = \beta$ which has one (nonzero) solution.

Note that transformations (2)–(5) translated in terms of the associated bent functions $g(x,y) = Tr_1^m\left(xG\left(\frac{y}{x}\right)\right)$ (with the convention $\frac{1}{0} = 0$) result in particular cases of EA-equivalence, since transformation (2) corresponds to applying the same field automorphism to x and y; transformations (3) and (4) correspond to multiplying x and/or y by constants in $g(x,y)$ and to adding linear functions to g; and transformation (5) corresponds when $G(0) = 0$ to swapping x and y in $g(x,y)$. On the contrary, the bent functions related by transformation (1) are not EA-equivalent, in general.

Finally, it is not difficult to see that Condition (8.3) implies Condition (8.2).

Now, one can extend the class H by introducing a more general class that we denote by $\mathcal{H}$ defined more precisely as follows.

Definition 8.1.2 ([8]). We call $\mathcal{H}$ the set of functions g defined by (8.1) and satisfying Condition (8.3).

Note that the function g defined by (8.1) satisfies

$$g(x,y) + Tr_1^m(\mu y) = \begin{cases} Tr_1^m\left(xG\left(\frac{y}{x}\right)\right) & \text{if } x \neq 0 \\ 0 & \text{if } x = 0 \end{cases}$$

and that changing $G(x)$ into $G(x) + \nu$ changes $g(x,y)$ into $g(x,y) + Tr_1^m(\nu x)$. Hence, we can assume without loss of generality (up to the addition of a linear function) that $\mu = 0$ and $G(0) = 0$.

The following proposition characterizes the duals of the functions in class $\mathcal{H}$.

Proposition 8.1.3 ([8]). *Let g be a bent function of the form (8.1) Then the dual function of g is defined on $\mathbb{F}_{2^m} \times \mathbb{F}_{2^m}$ as:*

$$\tilde{g}(\alpha, \beta) = \begin{cases} 1 & \text{if the equation } \psi(z) + \beta z = G(z) + (\beta + \mu)z = \alpha \text{ has no solution in } \mathbb{F}_{2^m} \\ 0 & \text{otherwise.} \end{cases}$$

In the following we show that is it easy to exhibit a first infinite class of functions in $\mathcal{H}$. Indeed, the Frobenius map $z \mapsto G(z) = z^2$ gives an example of functions G, which leads to a function in the class $\mathcal{H}$: $g(x, y) = Tr_1^m(y^2 x^{2^m-2})$. More generally, one can get functions in the class $\mathcal{H}$ by considering the maps $z \mapsto G(z) = z^{2^i}$ where i is co-prime with m, since the equation $z^{2^i} + \beta z = \alpha$ is equivalent, denoting $\gamma = \beta^{\frac{1}{2^i-1}}$, to $\left(\frac{z}{\gamma}\right)^{2^i} + \frac{z}{\gamma} = \frac{\alpha}{\gamma^{2^i}}$. As observed by Dillon, the related bent functions are in the completed Maiorana-MacFarland class; indeed, denoting $j = m - i$, we have then $g(x, y) = Tr_1^m(x\,(yx^{2^m-2})^{2^i}) = Tr_1^m(x^{2^j} yx^{2^m-2}) = Tr_1^m(yx^{2^j-1})$.

8.2 Class $\mathcal{H}$ in Univariate Form: Niho Bent Functions

We identify now $\mathbb{F}_{2^m} \times \mathbb{F}_{2^m}$ with $\mathbb{F}_{2^n}$ by considering a basis (u, v) of the $\mathbb{F}_{2^m}$-vectorspace $\mathbb{F}_{2^n}$ and identifying $(x, y) \in \mathbb{F}_{2^m} \times \mathbb{F}_{2^m}$ with:

$$t = xu + yv.$$

Then the vectorspaces $\{(x, ax)\,;\, x \in \mathbb{F}_{2^m}\}$ where $a \in \mathbb{F}_{2^m}$ and $\{(0, y)\,;\, y \in \mathbb{F}_{2^m}\}$ become the $2^m + 1$ multiplicative cosets of $\mathbb{F}_{2^m}^\star$ in $\mathbb{F}_{2^n}^\star$, added with 0. These cosets can be written $\omega\mathbb{F}_{2^m}^\star$ where ω ranges over the multiplicative subgroup U of $\mathbb{F}_{2^n}^\star$ of order $2^m + 1$, if we want to have a unique representation of each of them. And if we allow repetition, they are the cosets $\omega\mathbb{F}_{2^m}^\star$ where $\omega \in \mathbb{F}_{2^n}^\star$. The necessary and sufficient condition for a bent function to belong to class $\mathcal{H}$ is then that its restriction to each vectorspace $\omega\mathbb{F}_{2^m}$, $\omega \in \mathbb{F}_{2^n}^\star$, is linear.

Lemma 8.2.1 ([8]). *Let f be a Boolean function over $\mathbb{F}_{2^n}$ and $f(t) = \sum_{i=0}^{2^n-1} a_i t^i$ its univariate representation. Then the restrictions of f to the vectorspaces $\omega\mathbb{F}_{2^m}$, $\omega \in \mathbb{F}_{2^n}^\star$, are all linear if and only if the only exponents i such that $a_i \neq 0$ are congruent with powers of 2 modulo $2^m - 1$.*

Proof. The condition is clearly sufficient. Let us show that it is also necessary. Clearly, we must have $a_0 = 0$. Moreover, for every $\omega \in \mathbb{F}_{2^n}^\star$, the restriction of f to $\omega\mathbb{F}_{2^m}$ being linear, there exists $\lambda_\omega \in \mathbb{F}_{2^m}$ such that $f(\omega x) = \sum_{i=1}^{2^n-1} a_i \omega^i x^{i[\mathrm{mod}\ 2^m-1]}) = Tr_1^m(\lambda_\omega x)$ for every $x \in \mathbb{F}_{2^m}^\star$. By uniqueness of the univariate representation of a Boolean function over $\mathbb{F}_{2^m}$ (here, a function of x), we deduce that, for every $k \in \{0, \ldots, 2^m - 2\}$ different from a power of 2, we have

$\sum_{\substack{1\le i\le 2^n-1\\ i\equiv k\ [\mathrm{mod}\ 2^m-1]}} a_i\omega^i = 0$. This completes the proof, by uniqueness of the univariate representation of a function from $\mathbb{F}_{2^n}$ to itself (here, a function of ω). □

Note that this result extends to any function f from $\mathbb{F}_{2^n}$ to itself.

Bent functions whose restrictions to the vectorspaces $\omega\mathbb{F}_{2^m}$ are all linear have already been investigated in [15] and [18]. Since the exponents congruent with powers of 2 modulo $2^m - 1$ are called Niho exponents, we shall call these functions *Niho bent functions*. We have seen in Sect. 5.4.3.1 five examples of infinite classes of Niho bent functions are known up to affine equivalence or more exactly, four examples since one of the classes is the generalization of one of the others.

8.2.1 A Natural Extension of Class $\mathcal{H}$

Since class $\mathcal{H}$ is the set of bent functions whose restrictions to the $\omega\mathbb{F}_{2^m}$'s are linear, a natural extension to consider is the set of those bent functions whose restrictions to the $\omega\mathbb{F}^\star_{2^m}$'s are affine. Clearly, such functions are the sums of an element of class $\mathcal{H}$ and of a function which is constant on each $\omega\mathbb{F}^\star_{2^m}$ (note that, since bent functions have algebraic degree at most m, we can assume this function has even Hamming weight, and therefore has the form $\sum_{\omega\in S} 1_{\omega\mathbb{F}_{2^m}}$, where $1_{\omega\mathbb{F}_{2^m}}$ is the indicator of $\omega\mathbb{F}_{2^m}$).

Proposition 8.2.2 ([8]). *Let h be an element of $\mathcal{H}$ (that is, a Boolean function whose restriction to every $\omega\mathbb{F}_{2^m}$, $\omega \in \mathbb{F}^\star_{2^m}$, is linear). Let S be any subset of U (the multiplicative subgroup of $\mathbb{F}^\star_{2^n}$ of order 2^m+1) and let $g = \sum_{\omega\in S} 1_{\omega\mathbb{F}_{2^m}}$. Then $g+h$ is bent if and only if g is constant and h is bent, or g is bent and h is linear or S equals a singleton $\{\omega_0\}$ or its complement and h is Niho bent.*

Proof. We may without loss of generality assume that $g(0) = 0$, that is, S has even size (up to replacing g by $g+1$). As shown in [9], denoting then by g_ω the value of g on $\omega\mathbb{F}^\star_{2^m}$, we have:

$$\forall c \in \mathbb{F}_{2^n},\quad \widehat{\chi_{g+h}}(c) = 1 - \sum_{\omega\in U}\chi(g_\omega) + 2^m \sum_{\omega\in I(c)}\chi(g_\omega), \tag{8.4}$$

where $I(c) = \{\omega \in U \mid \forall t \in \omega\mathbb{F}_{2^m},\ h(t) = Tr_1^n(ct)\}$

$$\text{and}\quad \widehat{\chi_h}(c) = 2^m(\#I(c) - 1). \tag{8.5}$$

According to (8.4), $g+h$ can be bent only if $1 - \sum_{\omega\in U}\chi(g_\omega) \equiv 0 \pmod{2^m}$ that is, $\sum_{\omega\in U}\chi(g_\omega) = 1 + \epsilon 2^m$ with $\epsilon \in \{0, \pm 1\}$.

If $\epsilon = 1$, then $g = 0$.

If $\epsilon = 0$, then g is bent (it belongs to the PS_{ap} class) and $g+h$ is bent if and only if, for every c, we have $\sum_{\omega \in I(c)} \chi(g_\omega) \in \{-1, 1\}$. Necessarily $\#I(c)$ must then be odd, and according to (17.28), $\widehat{\chi_h}(c)$ is then non-negative. According to Lemma 12.1.23, h is then linear. Conversely, if g is bent and h is linear then $g + h$ is bent.

If $\epsilon = -1$, then $g_\omega = 0$ for a single ω, that is, $g = 1_{\omega_0 \mathbb{F}_{2^m}} + 1$. We know from [4] that if a bent function f is affine on an m-dimensional affine space E then $f + 1_E$ is bent too. Then taking $E = \omega_0 \mathbb{F}_{2^m}$, we see that $g + h$ is bent if and only if h is Niho bent (indeed, the restrictions of h and $g + h$ to $\omega_0 \mathbb{F}_{2^m}$ are affine). □

Remark 8.2.3. We can see that the corresponding bent functions $g+h$ are not really new: they are equal to "known" bent functions added with affine functions.

8.2.2 Determining the Duals of Some Bent Functions in Univariate Form

In the following, we present the computation of the dual of bent functions in the Class $\mathcal{H}$ carried out in [8].

8.2.2.1 On the Duals of the Known Binomial Bent Functions via Niho Exponents

In the following we are interested in the binomial bent function given in by

$$f(t) = Tr_1^m(at^{2^m+1}) + Tr_1^n(bt^{(2^m-1)\frac{1}{4}+1}), \forall t \in \mathbb{F}_{2^n}.$$

Dobbertin et al. showed that if $b^{2^m+1} = a$ then function f is bent. We have shown in [8] that the converse is true.

Proposition 8.2.4 ([8]). *Let f be defined as*

$$\forall t \in \mathbb{F}_{2^n}, \quad f(t) = Tr_1^m(at^{2^m+1}) + Tr_1^n(bt^{(2^m-1)\frac{1}{4}+1}) \tag{8.6}$$

with m odd, $a \in \mathbb{F}_{2^m}^\star$ and $b \in \mathbb{F}_{2^n}^\star$. Then, f is bent if and only if $b^{2^m+1} = a$.

In the next theorem, we provide the dual of the cubic bent function of Niho type which has been introduced by Dobbertin et al. and left open in [15]

Theorem 8.2.5 ([8]). *Let $n = 2m$ with m odd and f be defined as*

$$\forall t \in \mathbb{F}_{2^n}, \quad f(t) = Tr_1^m(at^{2^m+1}) + Tr_1^n(bt^{(2^m-1)\frac{1}{4}+1})$$

where $a \in \mathbb{F}_{2^m}^\star$ and $b \in \mathbb{F}_{2^n}^\star$ are such that $b^{2^m+1} = a$ and $b^4 \neq a^2$. Let v be such that $Tr_m^n(v) = 1$ and $b^4 = a^2 v^{2^m-1}$. Then the dual of f is such that

$$\tilde{f}(a^{\frac{1}{2}}w) = Tr_1^m\left(\left(v^{\frac{2^m+1}{2}} + 1 + Tr_m^n(v^{2^m}w)\right)\left(\frac{Tr_m^n(vw) + v^{\frac{2^m+1}{2}}}{Tr_m^n(v^{-1})}\right)^{\frac{1}{3}}\right).$$

It has algebraic degree $\frac{m+3}{2}$. Hence, for $m > 3$, $\tilde{f}$ is EA-inequivalent to the functions introduced in [15].

Remark 8.2.6. Function $\tilde{f}$ in Theorem 8.2.5 is affinely equivalent to the bivariate function $g(x,y) = xy^{1/3}$. The function $y \in \mathbb{F}_{2^m} \mapsto y^{1/3} \in \mathbb{F}_{2^m}$ is a permutation and $\tilde{f}$ belongs then to the completed Maiorana–McFarland class (but we knew this already since the dual of a function in the completed Maiorana–McFarland class belongs to this same class (see e.g. [5])).

8.2.2.2 On the Duals of the Known Bent Functions with 2^r Niho Exponents

We have seen in Sect. 5.4.3.1 that an extension of the second class of Niho bent from [15] has the form:

$$Tr_1^n\left(at^{2^m+1} + \sum_{i=1}^{2^{r-1}-1} t^{(2^m-1)\frac{i}{2^r}+1}\right)$$

with $r > 1$ satisfying $\gcd(r,m) = 1$ and $a \in \mathbb{F}_{2^n}$ is such that $a + a^{2^m} = 1$.

Recall that the class $\mathcal{H}$ introduced in the previous section is defined as the set of (bent) functions g satisfying

$$g(x,y) = \begin{cases} Tr_1^m\left(xH\left(\frac{y}{x}\right)\right), & \text{if } x \neq 0 \\ Tr_1^m(\mu y), & \text{if } x = 0\,, \end{cases} \tag{8.7}$$

where $\mu \in \mathbb{F}_{2^m}$ and H is a mapping from $\mathbb{F}_{2^m}$ to itself satisfying the following necessary and sufficient conditions

$$G : z \mapsto H(z) + \mu z \text{ is a permutation on } \mathbb{F}_{2^m} \tag{8.8}$$

$$z \mapsto G(z) + \beta z \text{ is 2-to-1 on } \mathbb{F}_{2^m} \text{ for any } \beta \in \mathbb{F}_{2^m}^*. \tag{8.9}$$

As proved in [8], condition (8.9) implies condition (8.8) and, thus, is necessary and sufficient for g being bent.

In the following proposition, we show that the bent function given above has the form of (8.7) and calculate the corresponding function G. By showing that G satisfies conditions (8.8) and (8.9), we give an alternative proof of the bentness.

Proposition 8.2.7 ([3, 7]). *Let $r > 1$ be a positive integer with $\gcd(r,m) = 1$, $a \in \mathbb{F}_{2^n}$ with $a + a^{2^m} = 1$ and Boolean function f over $\mathbb{F}_{2^n}$ be defined as*

$$f(t) = Tr_1^n\Big(at^{2^m+1} + \sum_{i=1}^{2^{r-1}-1} t^{(2^m-1)\frac{i}{2^r}+1}\Big).$$

Take any $u \in \mathbb{F}_{2^n} \setminus \mathbb{F}_{2^m}$ and $v \in \mathbb{F}_{2^m}^*$. Then for any $x, y \in \mathbb{F}_{2^m}$,

$$f(ux + vy) = \begin{cases} Tr_1^m(xH(y/x)), & if \quad x \neq 0 \\ Tr_1^m(vy), & if \quad x = 0 \end{cases}$$

and mapping G such that $G(z) = H(z) + vz$ can be expressed by

$$G^{2^r}(z) = (u + u^{2^m})^{2^r-1} vz + \frac{u^{2^m+2^r} + u^{2^{m+r}+1}}{u + u^{2^m}}$$

and satisfies conditions (8.8) and (8.9).

Given a function f, we give below an interesting relationship between the Walsh transforms of f in its univariate and the Walsh transforms of f in its bivariate representation, when choosing a specific basis.

Lemma 8.2.8 ([3, 7]). *Take any $u \in \mathbb{F}_{2^n}$ with $u + u^{2^m} = 1$. Then the pair $(u, 1)$ makes up a basis of $\mathbb{F}_{2^n}$ as a two-dimensional vectorspace over $\mathbb{F}_{2^m}$ and for any $w \in \mathbb{F}_{2^n}$,*

$$\hat{\chi}_f(w) = \hat{\chi}_g\big(Tr_m^n(uw), Tr_m^n(w)\big),$$

where $g(x, y) = f(ux + y)$ for any $x, y \in \mathbb{F}_{2^m}$.

Proof. Any $w, t \in \mathbb{F}_{2^n}$ can be decomposed as $w = u\alpha + \beta$ and $t = ux + y$ with the uniquely defined $\alpha, \beta, x, y \in \mathbb{F}_{2^m}$. Then

$$\begin{aligned} Tr_m^n(wt) &= \alpha x Tr_m^n(u^2) + \alpha y Tr_m^n(u) + \beta x Tr_m^n(u) \\ &= (\alpha + \beta)x + \alpha y = Tr_m^n(uw)x + Tr_m^n(w)y. \end{aligned}$$

Therefore, the Walsh transform of a Boolean function f over $\mathbb{F}_{2^n}$ can be expressed at point w as

$$\begin{aligned} \hat{\chi}_f(w) &= \sum_{t \in \mathbb{F}_{2^n}} (-1)^{f(t)+Tr_1^n(wt)} \\ &= \sum_{x,y \in \mathbb{F}_{2^m}} (-1)^{g(x,y)+Tr_1^m\left(Tr_m^n(uw)x+Tr_m^n(w)y\right)} \\ &= \hat{\chi}_g\big(Tr_m^n(uw), Tr_m^n(w)\big) \end{aligned}$$

as claimed. □

Now, we can compute the univariate representation of the dual function of $f(t)$.

Theorem 8.2.9 ([3, 7]). *Let $n = 2m$, $r > 1$ be a positive integer with $\gcd(r, m) = 1$ and bent Boolean function f over $\mathbb{F}_{2^n}$ be defined as*

$$f(t) = Tr_1^n\Big(at^{2^m+1} + \sum_{i=1}^{2^{r-1}-1} t^{(2^m-1)\frac{i}{2^r}+1}\Big),$$

where $a \in \mathbb{F}_{2^n}$ with $a + a^{2^m} = 1$. Take any $u \in \mathbb{F}_{2^n}$ with $u + u^{2^m} = 1$. Then the dual of $f(t)$ is equal to

$$\tilde{f}(w) = Tr_1^m\Big(\big(u(1 + w + w^{2^m}) + u^{2^{n-r}} + w^{2^m}\big) \times (1 + w + w^{2^m})^{1/(2^r-1)}\Big).$$

Moreover, if $d < m$ is a positive integer defined uniquely by $dr \equiv 1 \pmod m$ then the algebraic degree of $\tilde{f}(w)$ is equal to $d + 1$.

Note that $\tilde{f}(w)$ belongs to the completed class of $\mathcal{M}$ since this is the dual of a bent function $f(t)$ also belonging to this class (see, e.g., [5]). Moreover, $\tilde{f}(w)$ does not belong to class $\mathcal{H}$ since its restriction to any multiplicative coset of $\mathbb{F}_{2^m}$ (except when taking $\mathbb{F}_{2^m}$ itself) is not linear. In other words, $\tilde{f}(w)$ is not a Niho bent function. Also note that the dual of $f(t)$ does not depend on the chosen value of $u \in \mathbb{F}_{2^n}$ as long as $u + u^{2^m} = 1$.

8.3 Functions in Class $\mathcal{H}$ and o-Polynomials

We observe now that Condition (8.3) (which implies Condition (8.2)) is equivalent to the fact that G is an *o-polynomial.*[1]

Definition 8.3.1. Let m be any positive integer. A permutation polynomial G over $\mathbb{F}_{2^m}$ is called an o-polynomial (an oval polynomial) if, for every $\gamma \in \mathbb{F}_{2^m}$, the function

[1]The notion of o-polynomial comes from Finite Projective Geometry. First of all, a projective space of dimension n over a finite field $\mathbb{F}_q$ is a set of any non-zero subspace of ${\mathbb{F}_q}^{n+1}$ with respect to inclusion. This space is denoted by $PG_n(q)$. Let consider the case of projective space of dimension 2 (finite projective plane) over $\mathbb{F}_{2^n}$ i.e. $PG_2(2^n)$. A k-arc in $PG_2(2^n)$ is a set of k points no three collinear (i.e. there exists no line that contains any three points). The maximum cardinality of an arc in $PG_2(2^n)$ is $2^n + 2$. An *oval* of $PG_2(2^n)$ is an arc of cardinality $2^n + 1$. A *hyperoval* of $PG_2(2^n)$ is an arc of maximum cardinality (i.e. a set of $2^n + 2$ points no three collinear). Now certain types of polynomial give rise to hyperovals in $PG_2(2^n)$. More precisely, a polynomial f such that $D(f) = \{(1, t, f(t)), t \in \mathbb{F}_{2^n}\} \cup \{(0, 1, 0), (0, 0, 1)\}$ is a hyperoval is called an *o-polynomial.* A hyperoval of $PG_2(2^n)$ can then be represented by $D(f)$ where f is an o-polynomial. There is thus a close connection between hyperovals and o-polynomials.

$$z \in \mathbb{F}_{2^m} \mapsto \begin{cases} \frac{G(z+\gamma)+G(\gamma)}{z} & \text{if } z \neq 0 \\ 0 & \text{if } z = 0 \end{cases}$$

is a permutation of $\mathbb{F}_{2^m}$.

Note that some authors like Dobbertin in [14] add the condition "$G(0) = 0$, $G(1) = 1$" to the definition of o-polynomials; we do not include it since if it is not satisfied by an o-polynomial G, we can replace G by the o-polynomial $\frac{G(z)+G(0)}{G(1)+G(0)}$, which satisfies it.

Lemma 8.3.2 ([8]). *Any function G from $\mathbb{F}_{2^m}$ to $\mathbb{F}_{2^m}$ satisfies Condition* (8.3) *if and only if it is an o-polynomial.*

Proof. For every $\beta, \gamma \in \mathbb{F}_{2^m}$, the equation $G(z) + \beta z = G(\gamma) + \beta\gamma$ is satisfied by γ. Thus, if Condition (8.3) is satisfied, then for every $\beta \in \mathbb{F}_{2^m}{}^\star$ and every $\gamma \in \mathbb{F}_{2^m}$, there exists exactly one $z \in \mathbb{F}_{2^m}^\star$ such that $G(z+\gamma) + \beta(z+\gamma) = G(\gamma) + \beta\gamma$, that is, $\frac{G(z+\gamma)+G(\gamma)}{z} = \beta$. Then, for every $\gamma \in \mathbb{F}_{2^m}$, the function $z \in \mathbb{F}_{2^m}^\star \mapsto \frac{G(z+\gamma)+G(\gamma)}{z} \in \mathbb{F}_{2^m}^\star$ is bijective, that is, G and the function $z \in \mathbb{F}_{2^m} \mapsto \begin{cases} \frac{G(z+\gamma)+G(\gamma)}{z} & \text{if } z \neq 0 \\ 0 & \text{if } z = 0 \end{cases}$ are permutations. Hence, G is an o-polynomial. Conversely, if G is an o-polynomial, then for every $\gamma \in \mathbb{F}_{2^m}$, we have $\frac{G(z+\gamma)+G(\gamma)}{z} \neq 0$ for every $z \neq 0$ and for every $\beta \neq 0$ there exists exactly one nonzero z such that $G(z+\gamma) + G(\gamma) = \beta z$. Then for every $c \in \mathbb{F}_{2^m}$, either the equation $G(z) + \beta z = c$ has no solution, or it has at least a solution γ and then exactly one second solution $z + \gamma$ ($z \neq 0$). This completes the proof. □

A similar property was observed by Maschietti in [20] (as recalled by Dobbertin in [14]) for power functions. Maschietti was interested in cyclic difference sets while we are interested here in difference sets in elementary Abelian 2-groups (it is interesting to see that o-polynomials play a role in both frameworks). The fact that the result of Lemma 8.3.2 is true for general polynomials will have important consequences below.

Note that, according to the proof of Lemma 8.3.2, the property that for every $\gamma \in \mathbb{F}_{2^m}$, the function $z \in \mathbb{F}_{2^m} \mapsto \begin{cases} \frac{G(z+\gamma)+G(\gamma)}{z} & \text{if } z \neq 0 \\ 0 & \text{if } z = 0 \end{cases}$ is a permutation of $\mathbb{F}_{2^m}$ implies that G is a permutation of $\mathbb{F}_{2^m}$.

It has taken 50 years for geometers to find 9 classes of inequivalent o-polynomials listed below. The simplest example of an o-polynomial is the already seen Frobenius automorphism $G(z) = z^{2^i}$ where i is coprime with n. Other known examples are the following:

1. $G(z) = z^6$ where m is odd [23];
2. $G(z) = z^{3\cdot 2^k+4}$, where $m = 2k - 1$ [16];
3. $G(z) = z^{2^k+2^{2k}}$, where $m = 4k - 1$ [16];
4. $G(z) = z^{2^{2k+1}+2^{3k+1}}$, where $m = 4k + 1$ [16];

5. $G(z) = z^{2^k} + z^{2^k+2} + z^{3\cdot 2^k+4}$, where $m = 2k - 1$ [10];
6. $G(z) = z^{\frac{1}{6}} + z^{\frac{1}{2}} + z^{\frac{5}{6}}$ where m is odd [24]; note that $G(z) = D_5\left(z^{\frac{1}{6}}\right)$, where D_5 is the Dickson polynomial of index 5 [19];
7. $G(z) = \frac{\delta^2(z^4+z)+\delta^2(1+\delta+\delta^2)(z^3+z^2)}{z^4+\delta^2 z^2+1} + z^{1/2}$, where $Tr_1^m(1/\delta) = 1$ and, if $m \equiv 2$ [mod 4], then $\delta \notin \mathbb{F}_4$ [26];
8. $G(z) = \frac{1}{Tr_m^n(v)}\left[Tr_m^n(v^r)(z+1) + Tr_m^n\left[(vz+v^{2^m})^r\right]\left(z + Tr_m^n(v)z^{1/2}+1\right)^{1-r}\right] + z^{1/2}$, where m is even, $r = \pm\frac{2^m-1}{3}$, $v \in \mathbb{F}_{2^{2m}}$, $v^{2^m+1} = 1$ and $v \neq 1$ [11].

8.3.1 O-Equivalence

We shall say that two functions G and G' are *o-equivalent* [8] (the reason why we choose such term will come below) if G' can be obtained from G by a sequence of the transformations $G \mapsto G'$ above. This gives a notion of equivalence of functions in class $\mathcal{H}$ which is not a sub-equivalence of the EA-equivalence of bent functions and is not a super-equivalence either.

Note that the general $\mathbb{F}_{2^m}$-linear equivalence between the corresponding bent functions (when one equals the other composed on the right by an $\mathbb{F}_{2^m}$-linear automorphism over $\mathbb{F}_{2^n}$) is included in this notion of o-equivalence: applying to the function $g(x,y) = Tr_1^m\left(xG\left(\frac{y}{x}\right)\right)$ the transformation $(x,y) \mapsto (ax+by, cx+dy)$, where $a,b,c,d \in \mathbb{F}_{2^m}$ are such that $ad \neq bc$, gives the function $g'(x,y)$ equal, if $x \neq 0$, to $Tr_1^m\left((ax+by)G\left(\frac{cx+dy}{ax+by}\right)\right) = Tr_1^m\left(x(a+bz)G\left(\frac{c+dz}{a+bz}\right)\right)$, where $z = \frac{y}{x}$ (and still assuming the convention $\frac{1}{0} = 0$) and if $x = 0$ to $Tr_1^m\left(byG\left(\frac{d}{b}\right)\right)$. This corresponds to the transformation

$$G'(z) = (a+bz)\,G\left(\frac{c+dz}{a+bz}\right) + bz\,G\left(\frac{d}{b}\right). \tag{8.10}$$

If $b = 0$, this transformation reduces to $G'(z) = a\,G\left(\frac{c+dz}{a}\right)$ (with $a \neq 0$ and $d \neq 0$ since $ad \neq bc$) and can be obtained by applying transformations (3) and (4), and if $b \neq 0$, then it corresponds to applying (3), (4) and (5).

Note that, conversely, we obtain transformation (3) with $\lambda' = 0$ by choosing $(a,b,c,d) = (\lambda,0,0,\lambda)$, and transformations (4) and (5) by choosing $(a,b,c,d) = (1,0,\lambda',\lambda)$ and $(0,1,1,0)$.

The equivalences between o-polynomials can lead to two EA (extended affine)-inequivalent Niho bent functions. So this leads to the question: which equivalences between o-polynomials lead to EA-equivalent Niho bent functions and which ones do not?

Let $f(x,y) = Tr_1^m(xG(\frac{y}{x}))$ (with G defined as above) and $f'(x,y) = Tr_1^m(xG'(\frac{y}{x}))$ where G' is defined by (1), (2), (3) and (4), then the associated bent Boolean functions $f(x,y) = Tr_1^m(xG(\frac{y}{x}))$ (with the convention $\frac{1}{0} = 0$) and $f'(x,y) =$

$Tr_1^m(xG'(\frac{y}{x}))$ are EA-equivalent. However, if $G' = G^{-1}$ then the corresponding functions f and f' are in general EA-inequivalent. Note that in [22], the definition of o-equivalence has been a bit modified by excluding the transformation $G \mapsto G^{-1}$ in the definition of the o-equivalence so that the o-equivalence between the o-polynomials implies the EA-equivalence between the associated bent Boolean functions of the class $\mathcal{H}$. A study of o-equivalence has been done deeper by Budaghyan et al [1]. In this article the authors have in fact discussed a group of transformations of order 24 preserving the equivalence of o-polynomials and they showed that these transformations can lead up to up to four EA-inequivalent Niho bent functions.

To each o-polynomial G above correspond, according to Lemma 8.3.2 and to the observations made in Sect. 8.3.1, two Niho bent functions up to EA-equivalence: the one corresponding to G and the one corresponding to G^{-1}. We shall detail these bent functions below. Conversely, to every Niho bent function corresponds an o-polynomial. The question arises then of determining whether the o-polynomials we can deduce from the already known Niho bent functions are new up to o-equivalence.

For each of the six first o-polynomials G of the list above, we have two potentially new n-variable bent functions: $Tr_1^m\left(xG\left(\frac{y}{x}\right)\right)$ and $Tr_1^m\left(xG^{-1}\left(\frac{y}{x}\right)\right)$. For each of these, we have one potentially new bent function. We indicate now the bent functions we can obtain with the 6 first o-polynomials (we do not do the same for the two last o-polynomials since their situation needs to be clarified and since the expression of these bent functions would be complex—they are probably simpler in univariate form):

1. for m odd and $x, y \in \mathbb{F}_{2^m}$:

   - $f(x,y) = Tr_1^m(x^{-5}y^6)$;
   - $f(x,y) = Tr_1^m(x^{\frac{5}{6}}y^{\frac{1}{6}})$.

   The first function has algebraic degree $m - w_2(5) + w_2(6) = m$. Since in $\mathbb{Z}/(2^m - 1)\mathbb{Z}$ we have $\frac{1}{3} = \frac{2^{m+1}-1}{3} = 1 + 2^2 + 2^4 + \cdots + 2^{m-1}$ and therefore $w_2\left(\frac{1}{6}\right) = w_2\left(\frac{1}{3}\right) = \frac{m+1}{2}$ and $w_2\left(\frac{5}{6}\right) = w_2\left(\frac{5}{3}\right) = w_2\left(1 + \frac{2}{3}\right) = w_2\left(4 + 2^3 + 2^5 + \cdots + 2^{m-2}\right) = \frac{m-1}{2}$, the second function has degree m as well, which does not allow proving these two functions are EA-inequivalent. The question is left open.
2. for $m = 2k - 1$ and $x, y \in \mathbb{F}_{2^m}$:

   - $f(x,y) = Tr_1^m(x^{-3\cdot(2^k+1)}y^{3\cdot 2^k+4})$;
   - $f(x,y) = Tr_1^m(x^{-3\cdot(2^{k-1}-1)}y^{3\cdot 2^{k-1}-2})$ (since the inverse of $3\cdot 2^k+4$ [mod 2^m-1] equals $3\cdot 2^{k-1}-2$; indeed, $(3\cdot 2^k+4)(3\cdot 2^{k-1}-2) = 9-8 = 1$ [mod 2^m-1]).

   The first function has degree $m - w_2(3\cdot(2^k+1)) + w_2(3\cdot 2^k+4) = m-4+3 = m-1$ (if $k > 2$) and the second has degree $k + (k-1) = 2k-1 = m$ (if $k > 2$) since $-3\cdot(2^{k-1}-1) = 2^m - 3\cdot 2^{k-1} + 2 = 2^{k-1}(2^k - 1 - 2) + 2$ [mod $2^m - 1$] and $3\cdot 2^{k-1} - 2 = 2^k + 2\cdot(2^{k-2} - 1)$; hence the two functions are EA-inequivalent.

3. for $m = 4k - 1$ and $x, y \in \mathbb{F}_{2^m}$:
   - $f(x,y) = Tr_1^m(x^{1-2^k-2^{2k}}y^{2^k+2^{2k}})$;
   - $f(x,y) = Tr_1^m(x^{2^{3k-1}-2^{2k}+2^k}y^{1-2^{3k-1}+2^{2k}-2^k})$ (since the inverse of 2^k+2^{2k} [mod 2^m-1] equals $1-2^{3k-1}+2^{2k}-2^k$; indeed, $(2^k+2^{2k})(1-2^{3k-1}+2^{2k}-2^k) = 2^{m+1}-1$).

   The first function has degree $(3k-2)+2 = 3k$ since $2^m-2^k-2^{2k} = 2^k(2^{3k-1}-1-2^k)$ and the second has degree $k+2k = 3k$, which does not allow proving these two functions are EA-inequivalent. This question is left open.
4. for $m = 4k + 1$ and $x, y \in \mathbb{F}_{2^m}$:
   - $f(x,y) = Tr_1^m(x^{1-2^{2k+1}-2^{3k+1}}y^{2^{2k+1}+2^{3k+1}})$;
   - $f(x,y) = Tr_1^m(x^{2^{3k+1}-2^{2k+1}+2^k}y^{1-2^{3k+1}+2^{2k+1}-2^k})$ (since the inverse of $2^{2k+1}+2^{3k+1}$ [mod 2^m-1] equals $2^m-2^{3k+1}+2^{2k+1}-2^k$).

   The first function has degree $(2k-1)+2 = 2k+1$ and the second has degree $(k+1)+(2k+1) = 3k+2$; hence the two functions are EA-inequivalent.
5. for $m = 2k - 1$ and $x, y \in \mathbb{F}_{2^m}$:
   - $f(x,y) = Tr_1^m(x^{1-2^k}y^{2^k} + x^{-(2^k+1)}y^{2^k+2} + x^{-3\cdot(2^k+1)}y^{3\cdot 2^k+4})$;
   - $f(x,y) = Tr_1^m\left(y\left(y^{2^k+1}x^{-(2^k+1)} + y^3x^{-3} + yx^{-1}\right)^{2^{k-1}-1}\right)$, since we have $G^{-1}(z) = z\left(z^{2^k+1}+z^3+z\right)^{2^{k-1}-1}$ (see Lemma 8.3.3 below).

   The first function has degree $\max((k-1)+1, (2k-3)+2, (2k-5)+3) = 2k-1 = m$ (if $k > 2$). The second has also (optimal) algebraic degree m since its expansion contains the term $Tr_1^m\left(y^{1+3\cdot(2^{k-1}-1)}x^{3\cdot(1-2^{k-1})}\right) = Tr_1^m\left(y^{2^k+2^{k-1}-2}x^{2+(2^{2k-1}-2^{k-1})-2^k}\right)$. This does not allow proving these two functions are EA-inequivalent. This question is left open.
6. for m odd and $x, y \in \mathbb{F}_{2^m}$:
   - $f(x,y) = Tr_1^m(x^{\frac{5}{6}}y^{\frac{1}{6}} + x^{\frac{1}{2}}y^{\frac{1}{2}} + x^{\frac{1}{6}}y^{\frac{5}{6}})$;
   - $f(x,y) = Tr_1^m\left(x\left[D_{\frac{1}{5}}\left(\frac{y}{x}\right)\right]^6\right)$ where $D_{\frac{1}{5}}$ is the Dickson polynomial of index $\frac{1}{5}$, the inverse of 5 modulo $2^{2m}-1$ (see [19] or Remark 8.3.4 below); note that $\frac{1}{5} = 2^{2m} - 2^{2m-1} + 2^{2m-3} - 2^{2m-5} + \ldots + 2^7 - 2^5 + 2^3 - 2$ [mod $2^{2m}-1$].

   The first function has degree $\max(m, 2, m) = m$, since we already saw that $w_2\left(\frac{1}{6}\right) = \frac{m+1}{2}$ and $w_2\left(\frac{5}{6}\right) = \frac{m-1}{2}$. The questions of determining an explicit expression of the second function and of the determination of its algebraic degree were left open.

Lemma 8.3.3 ([8]). *Let k be any positive integer and $m = 2k-1$. The inverse of function $z \in \mathbb{F}_{2^m} \mapsto z^{2^k} + 2^{2^k+2} + z^{3\cdot 2^k+4} \in \mathbb{F}_{2^m}$ equals:* $z\left(z^{2^k+1}+z^3+z\right)^{2^{k-1}-1}$.

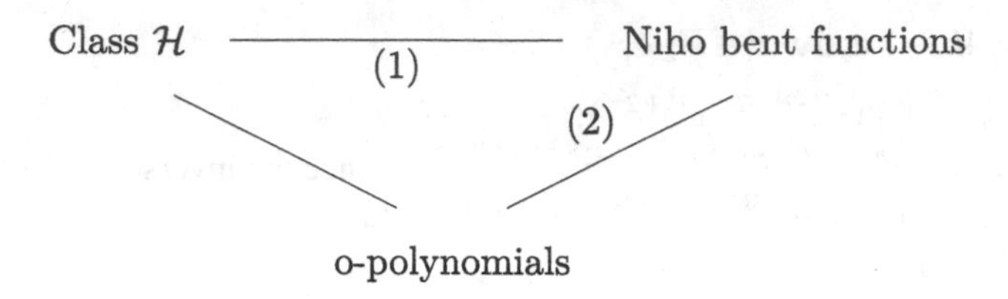

Fig. 8.1 Diagram: correspondences

Remark 8.3.4. Let us recall why the inverse of D_α equals D_β with $\beta\alpha \equiv 1 \pmod{2^n - 1}$ for every α co-prime with $2^n - 1$. Recall that $D_\alpha(D_\beta(y + \frac{1}{y})) = y^{\alpha\beta} + (\frac{1}{y})^{\alpha\beta}$ for every $y \in \mathbb{F}^\star_{2^n}$. Since every element $x \in \mathbb{F}^\star_{2^m}$ can be written as $x = c + \frac{1}{c}$ with $c \in \mathbb{F}_{2^n}$, we have $D_\alpha(D_\beta(x)) = D_\alpha(D_\beta(c + \frac{1}{c})) = c^{\alpha\beta} + (\frac{1}{c})^{\alpha\beta} = c + \frac{1}{c} = x$, proving that $D_\beta = D_\alpha^{-1}$ (note that $D_\alpha(0) = D_\beta(0) = 0$).

Note that it is shown in [10] that, for $m > 2$, all monomials in an o-polynomial have even exponents.

To summarize, the direct connection between the elements of class $\mathcal{H}$ and o-polynomials on one side, and Niho bent functions on the other side, are schematized by the diagram in Fig. 8.1.

1. Correspondence (1) offers a new framework to study Niho bent functions.
2. Connection (2) provides the construction of several potentially new families of bent functions in $\mathcal{H}$ (and thus new bent functions of Niho type) from the classes of o-polynomials in finite projective geometry. Each class of o-polynomials gives rise to several inequivalent classes of bent functions.

An explicit example describing the above correspondences is given in [21].

Finally, the five projectively equivalent o-polynomials of the o-polynomial G on $\mathbb{F}_{2^m}$ that are

$$G^{-1}(z), \rho G(z) + \eta, G(\rho z + \eta), zG(z^{2^m-2}), (G(z^{2^s}))^{2^{m-s}},$$

and some of their composite functions can be divided into four classes

$$\mathcal{O}_{G(z)}, \mathcal{O}_{G^{-1}(z)}, \mathcal{O}_{(zG(z^{2^m-2}))^{-1}}, \mathcal{O}_{(z+zG(z^{2^m-2}+1))^{-1}}.$$

In Table 8.1, we summarize the discussions from [8] and [1] by presenting the known classification of the projectively equivalence o-polynomials of the o-polynomial G on $\mathbb{F}_{2^m}$. The reader notices that except $(G(z^{2^s}))^{2^{m-s}}$, all the other projectively equivalent o-polynomials of the o-polynomial G can be obtained by considering $G^{-1}(z)$, $\rho G(z) + \eta$, $G(\rho z + \eta)$ and $(G(z^{2^s}))^{2^{m-s}}$. The bent functions of the class $\mathcal{H}$ are EA-equivalent whenever the corresponding o-polynomials belong to same class $\mathcal{O}_R$ where $R \in \{G^{-1}(z), \rho G(z) + \eta, G(\rho z + \eta), zG(z^{2^m-2}), (G(z^{2^s}))^{2^{m-s}}\}$. However, the o-polynomials in different classes $\mathcal{O}_R$ may induce EA-inequivalent bent functions from the class $\mathcal{H}$.

Table 8.1 The known classification of the projectively equivalent o-polynomials of the o-polynomial G on $\mathbb{F}_{2^m}$

	Reference
The o-polynomials in $\mathcal{O}_{G(z)}$	
$G(z)$	–
$(G(z^{2^s}))^{2^{m-s}}, s \in \mathbb{N}$	[8]
$\rho G(z) + \eta, \rho \in \mathbb{F}_{2^m}^{\star}, \eta \in \mathbb{F}_{2^m}$	[8]
$G(\rho z + \eta)$	[8]
$zG(z^{2^m-2})$	[1, 8]
$G(z+1)+1$	[1]
$z(G(z^{2^m-2}+1)+1)$	[1]
$z+(z+1)G(z(z+1)^{2^m-2})$	[1]
$(z+1)G((z+1)^{2^m-2})+1$	[1][a]
The o-polynomials in $\mathcal{O}_{G^{-1}(z)}$	
$G^{1}(z)$	[8]
$zG^{-1}(z^{2^m-2})$	[1]
$G^{-1}(z+1)+1$	[1]
$z(G^{-1}(z^{2^m-2}+1)+1)$	[1][a]
$z+(z+1)G^{-1}(z(z+1)^{2^m-2})$	[1]
$(z+1)G^{-1}((z+1)^{2^m-2})+1$	[1][a]
The o-polynomials in $\mathcal{O}_{(zG(z^{2^m-2}))^{-1}}$	
$(zG(z^{2^m-2}))^{-1}$	[1]
$(zG^{-1}(z^{2^m-2}))^{-1}$	[1]
$(z(z^{2^m-2}+(z^{2^m-2}+1)G((z+1)^{2^m-2}))^{-1})^{-1}$	[1][a]
$((z+1)G((z+1)^{2^m-2})+1)^{-1}$	[1]
$(z(z^{2^m-2}+(z^{2^m-2}+1)G^{-1}((z+1)^{2^m-2}))^{-1})^{-1}$	[1][a]
The o-polynomials in $\mathcal{O}_{(z+zG(z^{2^m-2}+1))^{-1}}$	
$(z+zG(z^{2^m-2}+1))^{-1}$	[1]
$(z+zG^{-1}(z^{2^m-2}+1))^{-1}$	[1][a]
$(z+(z+1)G(z(z+1)^{2^m-2}))^{-1}$	[1]
$(z+(z+1)G^{-1}(z(z+1)^{2^m-2}))^{-1}$	[1]
$z(z^{2^m-2}+(z^{2^m-2}+1)G((z+1)^{2^m-2}))^{-1}$	[1][a]
$z(z^{2^m-2}+(z^{2^m-2}+1)G^{-1}((z+1)^{2^m-2}))^{-1}$	[1][a]

[a]Deduced from [1]

8.3.2 Niho Bent Functions and Subiaco/Adelaide Hyperovals

The o-polynomials associated with the Leander–Kholosha bent functions are equivalent to Frobenius automorphisms [7]. The relation between the binomial Niho bent functions above with $d_2 = (2^m-1)\,3+1$ and $6d_2 = (2^m-1)+6$ and the Subiaco and Adelaide classes of hyperovals was found by Helleseth et al in [17]. This allowed to expand the class of bent functions corresponding to Subiaco hyperovals, in the case when $m \equiv 2 \pmod 4$ (see this condition in [15]).

8.3.3 Bent Functions from Other o-Polynomials

Later, in [2, 7, 8, 17], the o-polynomials associated to all known Niho bent functions have been identified and the class of Niho bent functions consisting of 2^r terms has been extended; such an extension is achieved by inserting coefficients of the power terms in the original function; it can then give any Niho bent function. Doing this, a relation to all the existing quadratic o-monomials has been found.

Recently, Budaghyan et al [2] have proved that any univariate Niho bent function is obtained as a sum of functions having the form of Leander–Kholosha bent function with relaxed conditions both on r and coefficients of power terms. In particular, any o-monomial corresponds to a 2^r-term Niho bent function of Leander–Kholosha type with particular coefficients of the power terms. This result allows immediately, knowing the terms of an o-polynomial, to obtain the powers of the additive terms in the polynomial representing the corresponding bent function. Knowing the terms of an o-polynomial, the authors of [2] have showed how to obtain the powers of the additive terms in the polynomial representing the corresponding bent function. They succeeded in finding the explicit form of bent functions obtained from quadratic and cubic o-polynomials. However, it is hard in general, to calculate coefficients explicitly.

8.4 Some Attempts to Generalize the Class $\mathcal{H}$

8.4.1 More $\mathcal{H}$-Like Bent Functions

We have seen in the previous sections that the class H introduced by Dillon in this thesis [12] has been later slightly modified into a class called $\mathcal{H}$ so as to relate it to the so-called Niho bent functions, (up to addition of affine functions) the set of bent functions whose restrictions to the subspaces of the *D*esarguesian spread are linear (see Chap. 14). Moreover, it has been observed that functions in $\mathcal{H}$ are related to o-polynomials. This connection led to derive several classes of bent functions in bivariate trace form.

Very recently, Carlet [6] has also characterized the bentness of those Boolean functions which are linear on the elements of a given spread of $\mathbb{F}_2^m \times \mathbb{F}_2^m$, and pointed out by a formula their general expression. He has given an explicit formula in the case of André's spreads. The bentness of the resulting functions leads to a notion extending that of o-polynomial. He also obtained similar characterizations and formulae for the $\mathcal{H}$-like functions derived from the spreads presented by Wu in [27] (see Chap. 14).

8.4.2 Class $\mathcal{H}$ in Characteristic p

In the continuation of [8], Ceşmelioğlu, Meidl and Pott [25] have studied in 2015 bent functions in characteristic p (see Chap. 13) whose restrictions to spreads are affine. More precisely, Ceşmelioğlu et al. have shown in 2015 that bent functions f from $\mathbb{F}_p^m \times \mathbb{F}_p^m$ to $\mathbb{F}_p$ which are constant or affine on the elements of a given spread of $\mathbb{F}_p^m \times \mathbb{F}_p^m$, either arise from partial spread bent functions, or are Boolean and a generalization of class $\mathcal{H}$ to this spread. Note that bent Boolean functions which are affine on the elements of the Desarguesian spread have been studied in [21]. For spreads of a presemifield S, Ceşmelioğlu et al. showed that a bent function of the second class corresponds to an o-polynomial of a presemifield in the Knuth orbit of S. Unfortunately, the only examples found of related bent functions belong to the completed Maiorana–McFarland class.

References

1. L. Budaghyan, C. Carlet, T. Helleseth, and A. Kholosha. On o-equivalence of Niho bent functions. In *Proceedings of International Workshop on the Arithmetic of Finite Fields (WAIFI), pages 155–168*, 2014.
2. L. Budaghyan, A. Kholosha, C. Carlet, and T. Helleseth. Niho bent functions from quadratic o-monomials. In *Proceedings of ISIT 2014, pages 1827–1831*, 2014.
3. L. Budaghyan, C. Carlet, T. Helleseth, A. Kholosha and S. Mesnager, Further results on Niho bent functions. *IEEE Transactions on Information Theory-IT, Vol 58, No. 11*, pages 6979–6985, 2012.
4. C. Carlet. Two new classes of bent functions. In *Proceedings of EUROCRYPT'93, Lecture Notes in Computer Science 765*, pages 77–101, 1994.
5. C. Carlet. Boolean functions for cryptography and error correcting codes. In Yves Crama and Peter L. Hammer, editors, *Boolean Models and Methods in Mathematics, Computer Science, and Engineering*, pages 257–397. Cambridge University Press, June 2010.
6. C. Carlet. More PS and H-like bent functions. In *Cryptology ePrint Archive, Report 2015/168*, 2015.
7. C. Carlet, T. Helleseth, A. Kholosha, and S. Mesnager. On the duals of bent functions with 2^r Niho exponents. In *IEEE International Symposium on Information Theory, ISIT 2011*, pages 703–707, 2011.
8. C. Carlet and S. Mesnager. On Dillon's class H of bent functions, Niho bent functions and o-polynomials. In *Journal of Combinatorial Theory, Series A, Vol 118, no. 8*, pages 2392–2410, 2011.
9. C. Carlet and S. Mesnager. On Semi-bent Boolean Functions. In *IEEE Transactions on Information Theory-IT, Vol 58 No 5*, pages 3287–3292, 2012.
10. W. E. Cherowitzo. Hyperovals in Desarguesian planes of even order. In *Ann. Discrete Mathematics, Vol. 37*, pages 87–94, 1988.
11. W.E. Cherowitzo, C.M. O'Keefe, and T. Penttila. A unified construction of finite geometries associated with q-clans in characteristic two. In *Advances in Geometry, 3*, pages 1–21, 2003.
12. J. Dillon. Elementary Hadamard difference sets. In *PhD dissertation, University of Maryland.*
13. J. F. Dillon. *Elementary Hadamard Difference Sets*. PhD thesis, University of Maryland, 1974.
14. H. Dobbertin. Uniformly representable permutation polynomials. In *Proceedings of Sequences and their Applications, SETA 01, Discrete Mathematics and Theoretical Computer Science, Springer*, pages 1–22, 2002.

15. H. Dobbertin, G. Leander, A. Canteaut, C. Carlet, P. Felke, and P. Gaborit. Construction of bent functions via Niho Power Functions. In *Journal of Combinatorial theory, Series A 113*, pages 779–798, 2006.
16. D. Glynn. Two new sequences of ovals in finite Desarguesian planes of even order. In *Lecture Notes in Mathematics 1036*.
17. T. Helleseth, A. Kholosha, and S. Mesnager. Niho Bent Functions and Subiaco/Adelaide Hyperovals. In *Proceedings of the 10-th International Conference on Finite Fields and Their Applications (Fq'10), Contemporary Math., AMS. Vol 579*, pages 91–101, 2012.
18. G. Leander and A. Kholosha. Bent functions with 2^r Niho exponents. In *IEEE Trans. Inform. Theory 52 (12)*, pages 5529–5532, 2006.
19. R. Lidl, G. L. Mullen, and G. Turnwald. Dickson Polynomials. In *ser.Pitman Monographs in Pure and Applied Mathematics. Reading, MA: Addison-Wesley, vol. 65*, 1993.
20. A. Maschietti. Difference sets and hyperovals. In *Designs, Codes and Cryptography 14*, pages 89–98, 1998.
21. S. Mesnager. Bent functions from spreads. In *Journal of the American Mathematical Society (AMS), Contemporary Mathematics (Proceedings the 11th International conference on Finite Fields and their Applications Fq11), Volume 632, page 295–316*, 2015.
22. S. Mesnager. Bent vectorial functions and linear codes from o-polynomials. In *Journal Designs, Codes and Cryptography, 77(1), pages 99–116*, 2015.
23. S. E. Payne. Ovali e curve σ nei piani di Galois di caratteristica due. In *Atti dell' Accad. Naz. Lincei Rend. 32 8*, pages 785–790, 1962.
24. S. E. Payne. A new infinite family of generalized quadrangles. In *Congr. Numer. 49*, pages 115–128, 1985.
25. A. Çeşmelioğlu, W. Meidl, and A. Pott. Bent functions, spreads, and o-polynomials. In *SIAM Journal on Discrete Mathematics, 29(2), pages 854–867*, 2015.
26. C.M. O'Keefe W.E. Cherowitzo and T. Penttila. W. Cherowitzo, T. Penttila, I. Pinneri and G. F. Royle. Flocks and ovals. In *Geometriae Dedicata, 60, no. 1*, pages 17–37, 1996.
27. B. Wu. PS bent functions constructed from finite pre-quasifield spreads. In *ArXiv e-prints*, 2013.

Chapter 9
Subclasses of Bent Functions: Hyper-Bent Functions

In this chapter we are interested in an important subclass of bent functions: the so-called *hyperbent functions*.

9.1 Definitions and Properties

In [25], A. Youssef and G. Gong study the Boolean functions f on the field $\mathbb{F}_{2^n}$ (n even) such that $f(x^k)$ is bent for every k co-prime with $2^n - 1$. These functions are called *hyper-bent functions*. Obviously, hyper-bent functions are in particular bent. Therefore they exist only when n is even and, their Hamming weight is even. Consequently, their polynomial form is

$$\forall x \in \mathbb{F}_{2^n}, \quad f(x) = \sum_{j \in \Gamma_n} \mathrm{Tr}_1^{o(j)}(a_j x^j) \tag{9.1}$$

where Γ_n, $o(j)$ are defined as above and $a_j \in \mathbb{F}_{2^{o(j)}}$.

The condition of hyper-bentness seems difficult to satisfy. However, A. Youssef and G. Gong show in [25] that hyper-bent functions exist. Their result is equivalent to the following (the definition of elements of the class $\mathcal{PS}^{\#}_{ap}$ is defined in included in Proposition 9.4.1)

Proposition 9.1.1 ([3]). *All functions of class $\mathcal{PS}^{\#}_{ap}$ are hyper-bent.*

S. Mesnager, *Bent Functions*, DOI 10.1007/978-3-319-32595-8_9

9.2 Hyper-Bent Boolean Functions in Symmetric Cryptography

Hyper-bent functions are both of theoretical and practical interest. In fact, they were initially proposed by Golomb and Gong [12] as a component of S-boxes to ensure the security of symmetric cryptosystems. These functions are currently used in the Data Encryption Standard (DES). The idea behind hyper-bent functions is to maximize the minimum distance to all Boolean functions coming from bijective monomials on $\mathbb{F}_{2^n}$ (that is, bijective functions whose expression is the absolute trace of a single power function), not just the affine monomial functions (that is, functions of the form $\mathrm{Tr}_1^n(ax)+\epsilon; a \in \mathbb{F}_{2^n}, \epsilon \in \mathbb{F}_2$). The first definition of hyper-bent functions was based on a property of the *extended Walsh–Hadamard transform* of Boolean functions (introduced by Golomb and Gong [12]).

Definition 9.2.1.

$$\forall \omega \in \mathbb{F}_{2^n}, \quad \widehat{\chi_f}(\omega, k) = \sum_{x \in \mathbb{F}_{2^n}} (-1)^{f(x)+\mathrm{Tr}_1^n(\omega \cdot x^k)}, \text{ with } \gcd(k, 2^n - 1) = 1.$$

We have the following characterization of hyper-bent functions in terms of the extended Walsh transform:

Proposition 9.2.2. *f is hyper-bent on $\mathbb{F}_{2^n}$ if and only if its extended Hadamard transform takes only the values $\pm 2^{\frac{n}{2}}$.*

Carlet and Gaborit [3] have showed that every hyper-bent function $f : F_{2^n} \to F_2$, can be represented as: $f(x) = \sum_{i=1}^{r} Tr(a_i x^{t_i}) \oplus \epsilon$, where $a_i \in F_{2^n}, \epsilon \in F_2$ and $w_2(t_i) = n/2$. Consequently, all hyper-bent functions defined on $\mathbb{F}_{2^n}$ have algebraic degree exactly $n/2$.

9.3 Hyper-Bent Boolean Functions in Coding Theory

9.3.1 Background on Binary Cyclic Codes

In all this subsection, we refer to [10] and [7]. Let N be a positive integer relatively prime to 2. Let t be the order of 2 modulo N that is, the smallest positive integer a such that $2^a \equiv 1 \pmod{N}$. Let α be a primitive N-th root of unity in $\mathbb{F}_{2^t}$. Let $\mathcal{C}$ be a binary cyclic code of length N with generator polynomial $g(X)$ in the ring $R_N := \mathbb{F}_2[X]/(X^N - 1)$, consisting of the residue classes of $\mathbb{F}_2[X]$ modulo $X^N - 1$. The polynomial $g(X)$ is the unique monic polynomial of minimum degree in $\mathcal{C}$ and $g(X) = \prod_s \prod_{i \in C_s} (X - \alpha^i)$, where s runs through some subset of the 2-cyclotomic cosets C_s modulo N. Let $T = \bigcup_s C_s$ be the union of these 2-cyclotomic cosets. The roots of the unity $Z = \{\alpha^i \mid i \in T\}$ are called *the zeroes* of the code $\mathcal{C}$ and $\{\alpha^i \mid i \notin T\}$ are the non-zeroes of $\mathcal{C}$. The set T is called *the defining set* of $\mathcal{C}$. Every

vector $f = (f_0, f_1, \cdots, f_{N-1})$, identified with the polynomial $f(X) = f_0 \oplus +f_1X \oplus + \cdots \oplus f_{N-1}X^{N-1}$ belongs to $\mathcal{C}$ if and only if $f(\alpha^i) = 0$ for each $i \in T$. The defining set T of $\mathcal{C}$, and hence either the set of zeroes or the set of non-zeroes, completely determines $g(X)$. The dimension of $\mathcal{C}$ is $N - deg(g(X)) = N - \#T$. Now, if we consider binary cyclic codes in the primitive case, more precisely, we assume that $N = 2^n - 1$ for n a positive integer, the order of 2 modulo N equals n. If α is a primitive element of $\mathbb{F}_{2^n}$. then, the vector $f = (f_0, f_1, \cdots, f_{N-1})$ can be identified with the restriction of a Boolean function f to the set $\mathbb{F}_{2^n}^\star$, defined by $f(\alpha^i) = f_i$, for every integer $i \in \{0, \cdots, 2^n - 2\}$.
Given a cyclic code $\mathcal{C}$ of length N and dimension k, we can define the extended cyclic code $\hat{\mathcal{C}}$ of $\mathcal{C}$ as the set of vectors $(f_0 \oplus \cdots \oplus f_{N-1}, f_0, \cdots, f_{N-1})$. The obtained code $\hat{\mathcal{C}}$ is a linear code of length $N + 1$ and dimension k. The vector $(f_0 \oplus \cdots \oplus f_{N-1}, f_0, \cdots, f_{N-1})$ can be identified with a Boolean function f on $\mathbb{F}_{2^n}$ whose algebraic degree is smaller than n.

9.3.2 Extended Cyclic Codes and Hyper-Bent Functions

There exists a relationship between cyclic codes and hyper-bent functions (see [3]). Recall that by definition, a Boolean function on $\mathbb{F}_{2^n}$ is hyper-bent if the function $x \mapsto f(x^i)$ is bent, for every integer i co-prime with $2^n - 1$. This implies that every hyper-bent function belongs to the intersection of all the images of the Reed–Muller codes of order $\frac{n}{2}$ by the mappings $f \mapsto f(x^i)$, where i is co-prime with $2^n - 1$. Consequently, all the hyper-bent functions on $\mathbb{F}_{2^n}$ belong to the extended cyclic code H_n whose zeroes are all the elements of the form α^{ij}, where i is co-prime with $2^n - 1$ and $1 \le j \le 2^n - 2$ with $1 \le w_2(j) \le \frac{n}{2} - 1$. The non-zeroes of H_n are the α^j's such that j is zero or j satisfies $w_2(ij) = \frac{n}{2}$ for any i co-prime with $2^n - 1$. Carlet and Gaborit deduce that all hyper-bent functions on $\mathbb{F}_{2^n}$ have algebraic degree $\frac{n}{2}$. Hence hyper-bent functions belong to $\mathcal{RM}(\frac{n}{2}, n) \setminus \mathcal{RM}(\frac{n}{2} - 1, n)$ (while bent functions belong to the Reed–Muller codes $\mathcal{RM}(\frac{n}{2}, n)$ of order $\frac{n}{2}$).
It has been proved in [3], that functions of the $\mathcal{PS}_{ap}$ (these functions are in fact hyper-bent—see e.g. [3]) are some codewords of weight $2^{n-1} - 2^{\frac{n}{2}-1}$ of a subcode of H_n. The authors deduce that for some n, depending on the factorization of $2^n - 1$, the only hyper-bent functions on n variables are the elements of the class $\mathcal{PS}_{ap}^\#$ (see Proposition 9.4.1). Now, let A_n be the extended cyclic code whose non-zeroes are the power of α whose exponents are all the multiples of $2^{\frac{n}{2}} - 1$. Let B_n be the cyclic code with non-zeroes α^i for i a symmetric element of the ring of integer modulo $2^n - 1$ (i is said to be symmetric if i and $-i$ belong to the same 2-cyclotomic coset modulo $2^n - 1$). We denote by S_n the set of vectors of length 2^n and of weights $2^{n-1} \pm 2^{\frac{n}{2}-1}$. Then we have the following inclusions ($A \subset B$ means that A is a subcode of B):

$$\mathcal{PS}_{ap}^\# = A_n \cap S_n \subset A_n \subset B_n \subset H_n.$$

9.4 A Characterization of Hyper-Bentness

Recall that Dillon has exhibited a subclass of $\mathcal{PS}^-$, denoted by $\mathcal{PS}_{ap}$, whose elements are defined in an explicit form (see Sect. 5.1, Chap. 4). Furthermore, it is well-known (see e.g. [3]) that all the functions of $\mathcal{PS}_{ap}$ are hyper-bent.

Youssef and Gong [25] showed that hyper-bent functions actually exist. The following proposition, due to Carlet and Gaborit [3], is an easy translation of this result, which was originally given in terms of sequences, stated using only the terminology of Boolean functions.

Proposition 9.4.1 ($\mathcal{PS}^{\#}_{ap}$ Class [25, Theorem 1], [3, Proposition 3]). *Let α be a primitive element of $\mathbb{F}_{2^n}$. Let f be a Boolean function defined on $\mathbb{F}_{2^n}$ such that $f(\alpha^{2^m+1}x) = f(x)$ for every $x \in \mathbb{F}_{2^n}$ and $f(0) = 0$. Then f is a hyper-bent function if and only if the weight of the vector $(f(1), f(\alpha), f(\alpha^2), \cdots, f(\alpha^{2^m}))$ equals 2^{m-1}. In this case f is said to belong to the $\mathcal{PS}^{\#}_{ap}$ class.*

Charpin and Gong [4] have derived a slightly different version of the preceding proposition.

Proposition 9.4.2 ([4, Theorem 2]). *Let α be a primitive element of $\mathbb{F}_{2^n}$. Let f be a Boolean function defined on $\mathbb{F}_{2^n}$ such that $f(\alpha^{2^m+1}x) = f(x)$ for every $x \in \mathbb{F}_{2^n}$ and $f(0) = 0$. Denote by U the cyclic subgroup of $\mathbb{F}^*_{2^n}$ of order $2^m + 1$. Let $\zeta = \alpha^{2^m-1}$ be a generator of U. Then f is a hyper-bent function if and only if the cardinality of the set $\{i \mid f(\zeta^i) = 1, 0 \leq i \leq 2^m\}$ equals 2^{m-1}.*

Remark 9.4.3. It is important to point out that bent functions f defined on $\mathbb{F}_{2^n}$ such that $f(\alpha^{2^m+1}x) = f(x)$ for every $x \in \mathbb{F}_{2^n}$ and $f(0) = 0$ are always hyper-bent. A proof of this claim can be found in [4, Proof of Theorem 2] or it can be directly observed that the support *supp(f)* of such a Boolean function f can be decomposed as $\mathrm{supp}(f) = \bigcup_{i \in S} \alpha^i \mathbb{F}^*_{2^m}$, where $S = \{i \mid f(\alpha^i) = 1\}$, that is, f is bent if and only if $\#S = 2^{m-1}$, proving that such bent functions are actually hyper-bent functions according to Proposition 9.4.1.

Finally, Carlet and Gaborit have proved the following more precise statement about the functions considered in Proposition 9.4.1.

Proposition 9.4.4 ([3, Proposition 4]). *Hyper-bent functions as in Proposition 9.4.1 such that $f(1) = 0$ are the elements of the $\mathcal{PS}_{ap}$ class. Those such that $f(1) = 1$ are elements of $\mathcal{PS}^{\#}_{ap}$ and they are the functions of the form $f(x) = g(\delta x)$ for some $g \in \mathcal{PS}_{ap}$ and $\delta \in \mathbb{F}_{2^n} \setminus \{1\}$ such that $g(\delta) = 1$.*

This result can be stated in a simple way: all functions in the so-called class $\mathcal{PS}^{\#}_{ap}$, whose elements are those functions $f(\delta x)$, where $x \in \mathbb{F}_{2^n}$, $\delta \in \mathbb{F}^*_{2^n}$ and $f \in \mathcal{PS}_{ap}$ (in univariate form), are hyper-bent (see [3], where $\mathcal{PS}^{\#}_{ap}$ is introduced and an alternative proof of the fact it is contained in $\mathcal{HB}_n$ is given; see another simple proof given by the author in [19]).

It is shown in [3] that the elements in $\mathcal{PS}^{\#}_{ap}$ are the functions of Hamming weight $2^{n-1} \pm 2^{m-1}$ which can be written as $\sum_{i=1}^{r} Tr_1^n(a_i x^{j_i})$ for $a_i \in F_{2^n}$ and j_i a multiple of $2^m - 1$. Hence, $\mathcal{PS}^{\#}_{ap}$ coincides with the set of bent functions whose polynomial form is the sum of multiple trace terms constructed via Dillon-like exponents $r(2^m - 1)$.

The study of the links between the set of hyper-bent functions and Dillon's class $\mathcal{PS}$, initiated in [3], has been continued in [19], where is denoted by $\mathcal{D}_n$ the set of bent Boolean functions f over $\mathbb{F}_{2^n}$ with $f(x) = \sum_{i \in \Gamma_{n,m}} Tr_1^{o(i)}(a_i x^i)$ where $\Gamma_{n,m}$ is the set of cyclotomic cosets $[i]$ such that $i \equiv 0 \pmod{2^m - 1}$ and such that $f(0) = 0$. It has been highlighted in [19] that $\mathcal{PS}^{\#}_{ap} \cap (\mathcal{PS}^{-}) = \mathcal{D}_n$. This is a direct consequence of the result from [3] recalled above, but [19] clarifies the arguments.

Moreover, it has been shown in [3] by computer experiments, for $n = 4$, that there exist hyper-bent functions which are not in $\mathcal{PS}^{\#}_{ap}$. As a consequence, the set of hyper-bent functions contains strictly $\mathcal{PS}^{\#}_{ap}$, and $\mathcal{PS}_{ap} \subset \mathcal{PS}^{\#}_{ap} \subset \mathcal{HB}_n$. Note that the duals of hyper-bent functions in $\mathcal{PS}^{\#}_{ap}$ are also in $\mathcal{PS}^{\#}_{ap}$ and so are hyper-bent.

As shown in the next subsections, the only known constructions of hyper-bent functions are those in the set $\mathcal{PS}^{\#}_{ap}$ (in practice, in $\mathcal{D}_n$) and except for a sporadic example known for $n = 4$, the set of all known hyper-bent functions is $\mathcal{PS}^{\#}_{ap}$.

9.5 Primary Constructions and Characterization of Hyperbent Functions in Polynomial Forms

9.5.1 Monomial Hyper-Bent Functions via Dillon Exponents

Among all the known monomial bent, only Dillon's function is also hyper-bent. Recall that the monomial Dillon function is the function whose expression is defined with *Dillon exponent* (that is the exponent given in the second row of Table 5.1) as:

$$\forall x \in \mathbb{F}_{2^n}, \quad f_a^{(r)}(x) = Tr_1^n(ax^{r(2^m-1)}), \quad a \in \mathbb{F}_{2^n}^{\star}$$

where $m = \frac{n}{2}$ and r is an integer such that $\gcd(r, 2^m + 1) = 1$. The characterization of the bentness of the monomial functions $f_a^{(r)}$ has been studied by Dillon [8] in the case $r = 1$ and, next by Leander [14] (who refined the result of Dillon using a different point of view) and by Charpin and Gong [4] (who extended the family of Dillon, implying in particular that the original functions were actually hyper-bent) for any integer r co-prime with $2^m + 1$. Thanks to these works, the bent and hyper-bent functions $f_a^{(r)}$ have been completely identified. Furthermore, it has been proved that, up to affine equivalence, we can restrict the study of the bentness of $f_a^{(r)}$ to the case where $a \in \mathbb{F}_{2^m}^{\star}$ (see e.g. [14]). The following theorem summarizes the results related to the bentness of the function $f_a^{(r)}$.

Theorem 9.5.1 ([4, 8]). *Let $n = 2m$, $a \in \mathbb{F}^{\star}_{2^m}$ and $f_a^{(r)}$ be the boolean function defined on $\mathbb{F}_{2^n}$ as follows*

$$\forall x \in \mathbb{F}_{2^n}, \quad f_a^{(r)}(x) = \mathrm{Tr}_1^n(ax^{r(2^m-1)}), \quad \gcd(r, 2^m+1) = 1.$$

1. *$f_a^{(r)}$ is bent if and only if $f_a^{(1)}$ is bent.*
2. *$f_a^{(1)}$ is bent if and only if $K_m(a) = 0$.*
3. *If $f_a^{(r)}$ is bent then its dual function is $f_a^{(r)}$ itself.*
4. *$f_a^{(r)}$ is hyper-bent if and only if $f_a^{(r)}$ is bent.*
5. *The bent functions $f_a^{(r)}$ are in the Partial Spread class PS_{ap}.*

Remark 9.5.2. Note that an alternative direct proof of Dillon's result (point 2. of Theorem 9.5.1) has been proposed recently by Leander in [14] (see also [4]). Leander's proof gives also more information on the spectrum of monomial functions $f_{a,0}$. A small mistake in his proof was rectified in [4]). Note also that the existence of some a in $\mathbb{F}_{2^m}$ which are zeros of Kloosterman sum on $\mathbb{F}_{2^m}$ had been conjectured by Dillon. It has been proved by Lachaud and Wolfmann in [13] that the values of such Kloosterman sums are all the numbers divisible by 4 in the range $[-2^{(m+2)/2}+1, 2^{(m+2)/2}+1]$ (which implies in particular that the family defined by Dillon is never empty).

9.5.2 *Binomial Hyper-Bent Functions via Dillon (Like) Exponents*

In the following, we are interested in the problem which consists in finding two exponents s_1 and s_2 with $o(s_2) < n$ and the corresponding coefficients $a \in \mathbb{F}^{\star}_{2^{o(s_1)}}$ and $b \in \mathbb{F}^{\star}_{2^{o(s_2)}}$ defining bent or hyper-bent functions defined on $\mathbb{F}_{2^n}$ whose expression is of the form

$$\mathrm{Tr}_1^{o(s_1)}(ax^{s_1}) + \mathrm{Tr}_1^{o(s_2)}(bx^{s_2}). \tag{9.2}$$

9.5.2.1 A First Family of Binomial Hyper-Bent Functions $\mathfrak{F}_n$

By computer experiments, for small values of n ($n \leq 16$; because of the complexity of the problem), we have found that the set of all functions of type (9.2) with the exponents $s_1 = 3(2^m - 1)$ and $s_2 = \frac{2^n-1}{3}$ contains bent functions when m is odd. Note that $o(s_1) = n$ and $o(s_2) = 2$. The polynomial form of a function of type (9.2), denoted by $f_{a,b}$, is then of the form:

$$f_{a,b}(x) = \mathrm{Tr}_1^n\left(ax^{2^m-1}\right) + \mathrm{Tr}_1^2\left(bx^{\frac{2^n-1}{3}}\right) \tag{9.3}$$

where, $a \in \mathbb{F}^{\star}_{2^{o(s_1)}} = \mathbb{F}^{\star}_{2^n}$ and $b \in \mathbb{F}^{\star}_{2^{o(s_2)}} = \mathbb{F}^{\star}_4$.

Note that when $b = 0$, the corresponding function $f_{a,0}$ is a monomial bent function if and only if a is a zero of Kloosterman sums on $\mathbb{F}_{2^n}$. Denote by $\mathfrak{F}_n$ the set of the Boolean functions $f_{a,b}$ defined on $\mathbb{F}_{2^n}$ whose polynomial form is given by the above expression (9.3). This infinite class is not contained in the class studied by Charpin and Gong [4] that we have mentioned above. In the following, we present the study of bentness of elements of $\mathfrak{F}_n$. To this end, we investigate a precise characterization of such functions of $\mathfrak{F}_n$ which are hyper-bent, by giving explicit conditions on the coefficients a and b. We first show that $\mathfrak{F}_n$ is a subclass of the well-known Partial Spread class for which the bentness of its functions can be characterized by means of the Hamming weight of their restrictions to a certain set. Next, we investigate the conditions on the choice of a and b for obtaining an explicit family of bent functions. Thanks to the recent works of Charpin, Helleseth and Zinoviev on the Kloosterman sums and cubic sums, we establish an explicit characterization of the bentness of functions belonging to $\mathfrak{F}_n$ in terms of the Kloosterman sums of the coefficient a when m is odd.

Study of the Bentness of the Binomial Family $\mathfrak{F}_n$

- First of all, notice that all the functions are of algebraic degree m, which is maximum (recall that the algebraic degree of any bent Boolean function on $\mathbb{F}_{2^n}$ is at most m).

Proposition 9.5.3 ([15]). *The algebraic degree of any function $f_{a,b}$ of $\mathfrak{F}_n$ is equal to m.*

Proof. The two exponents $2^m - 1$ and $\frac{2^n-1}{3}$ are of 2-weight m since $2^m - 1 = 1 + 2 + 2^2 + \cdots + 2^{m-1}$ and $\frac{2^n-1}{3} = 1 + 4 + \cdots + 4^{m-1}$. Therefore, the two Boolean functions $x \mapsto \mathrm{Tr}_1^n(ax^{2^m-1})$ and $x \mapsto \mathrm{Tr}_1^2(bx^{\frac{2^n-1}{3}})$ are of algebraic degree equal to m. Since $\mathrm{Tr}_1^n(ax^{2^m-1})$ and $\mathrm{Tr}_1^2(bx^{\frac{2^n-1}{3}})$ are two separate parts in the trace representation of $f_{a,b}$, the algebraic degree of $f_{a,b}$ is equal to m. □

- Let us note now that all the Boolean functions of the family $\mathfrak{F}_n$ have the following property

$$\forall a \in \mathbb{F}_{2^n}^\star, \quad \forall b \in \mathbb{F}_4, \quad \forall c \in \mathbb{F}_{2^m}^\star, \quad \forall x \in \mathbb{F}_{2^n}, \quad f_{a,b}(c^3x) = f_{a,b}(x). \tag{9.4}$$

That implies in particular that a Boolean function $f_{a,b}$ of $\mathfrak{F}_n$ is constant on each coset of $C = \{x^3 \mid x \in \mathbb{F}_{2^m}^\star\}$. Denote by H a set of representatives for the equivalence relation $\sim$ defined on $\mathbb{F}_{2^n}^\star$ by $x \sim y$ if and only if $y = xv$ for some $v \in C$. Then, we have

$$supp(f_{a,b}) = \bigcup_{x \in S_{a,b}} xC \text{ where } S_{a,b} := \{x \in H \mid f_{a,b}(x) = 1\} \tag{9.5}$$

When m is odd, every element of $\mathbb{F}_{2^m}^\star$ is a cube and thus we have $C = \mathbb{F}_{2^m}^\star$ (indeed, the map $x \in \mathbb{F}_{2^m}^\star \mapsto x^3$ is a permutation for m odd). On the other hand, recall that every element x of $\mathbb{F}_{2^n}^\star$ has a unique decomposition as: $x = yu$, with $y \in \mathbb{F}_{2^m}^\star$ and $u \in U := \{u \in \mathbb{F}_{2^n} \mid u^{2^m+1} = 1\}$. Therefore, one can take $H = U$ in this case leading to

$$supp(f_{a,b}) = \bigcup_{u \in S_{a,b}} u\mathbb{F}_{2^m}^\star, \text{ with } S_{a,b} = \{u \in U \mid f_{a,b}(u) = 1\}. \tag{9.6}$$

This implies in particular that bent functions belonging to $\mathfrak{F}_n$ are in the Partial Spread class $\mathcal{PS}$ introduced by Dillon [8]. Therefore, for m odd, the question of deciding whether an element $f_{a,b}$ of $\mathfrak{F}_n$ is bent or not can be reduced to compute the Hamming weight of its restriction to U, that is, we have

Proposition 9.5.4 ([15]). *For m odd, the Boolean function $f_{a,b}$ of $\mathfrak{F}_n$ is bent if and only if* $\mathrm{wt}(f_{a,b}|_U) = 2^{m-1}$.

Based on this result, we shall characterize the elements a of $\mathbb{F}_{2^n}$ and $b \in \mathbb{F}_4$ for which $f_{a,b}$ is bent in terms of Kloosterman sum.

- Restriction to the case where $a \in \mathbb{F}_{2^m}^\star$: we show that we can restrict ourselves to study the bentness of $f_{a,b}$ with $a \in \mathbb{F}_{2^m}^\star$ without loss of generality. Let $a \in \mathbb{F}_{2^n}^\star$, $b \in \mathbb{F}_4$, $a' \in \mathbb{F}_{2^m}^\star$ and $b' \in \mathbb{F}_4$. Note that, if $a = a'\lambda^{2^m-1}$ and $b = b'\lambda^{\frac{2^n-1}{3}}$ for some $\lambda \in \mathbb{F}_{2^n}^\star$, the functions $f_{a',b'}$ and $f_{a,b}$ are linearly equivalent. Indeed, one has, for every x in $\mathbb{F}_{2^n}^\star$, $f_{a,b}(x) = f_{a',b'}(\lambda x)$. It follows we can always replace $a \in \mathbb{F}_{2^n}^\star$ by the unique element $a' \in \mathbb{F}_{2^m}^\star$ defined by $a = a'u$ where $u \in U = \{\lambda^{2^m-1} \mid \lambda \in \mathbb{F}_{2^n}^\star\}$. In other words, we have

Proposition 9.5.5 ([15, 17]). *Let $f_{a,b}$ be a Boolean function whose expression is of the form (9.3). Then,*

$$\{(a,b) \mid a \in \mathbb{F}_{2^n}^\star,\ b \in \mathbb{F}_4,\ f_{a,b} \textit{ is bent}\}$$
$$= \{(a'\lambda^{2^m-1}, b'\lambda^{\frac{2^n-1}{3}}) \mid a' \in \mathbb{F}_{2^m}^\star, b' \in \mathbb{F}_4, \lambda \in \mathbb{F}_{2^n}^\star, f_{a',b'} \textit{ is bent}\}.$$

Thanks to the previous proposition, one can restrict oneself to the case $a \in \mathbb{F}_{2^m}^\star$ without loss of generality. Proposition 9.5.4 says that for m odd, it suffices to compute the Hamming weight of the restriction to U of $f_{a,b}$, $(a,b) \in \mathbb{F}_{2^m}^\star \times \mathbb{F}_4^\star$, to decide whether $f_{a,b}$ is bent or not. Our aim in this section is to give a necessary and sufficient condition for the bentness of $f_{a,b}$ in terms of the Kloosterman sum $K_m(a)$. We first reword Proposition 9.5.4. To this end, we introduce the following sum

$$\forall (a,b) \in \mathbb{F}_{2^m}^\star \times \mathbb{F}_4^\star, \qquad \Lambda(a,b) := \sum_{u \in U} \chi(f_{a,b}(u)). \tag{9.7}$$

Then, by noting that $\sum_{u\in U}\chi(f_{a,b}(u)) = \#U - 2\,\mathrm{wt}(f_{a,b}|_U) = 2^m + 1 - 2\,\mathrm{wt}(f_{a,b}|_U)$, we have, for every $a \in \mathbb{F}_{2^m}^\star$ and $b \in \mathbb{F}_4^\star$,

$$f_{a,b} \text{ is bent if and only if } \Lambda(a,b) = 1. \tag{9.8}$$

- The key result is that the sum $\Lambda(a,b)$ can be expressed by means of Kloosterman sums and the cubic sums on $\mathbb{F}_{2^m}$ thanks to Proposition 2.4.3

Proposition 9.5.6 ([15, 17]). *Let β a primitive element of $\mathbb{F}_4$. Let $a \in \mathbb{F}_{2^m}^\star$. Then we have*

$$\Lambda(a,\beta) = \Lambda(a,\beta^2) = \frac{K_m(a) - 2C_m(a,a) - 1}{3},$$
$$\Lambda(a,1) = \frac{K_m(a) + 4C_m(a,a) - 1}{3}.$$

Now, thanks to (9.8), Proposition 9.5.6 , Proposition 2.3.5 and Corollary 2.4.4, we identify the values of a for which the Boolean functions $f_{a,1}, f_{a,\beta}$ or f_{a,β^2} is bent.

Theorem 9.5.7 ([15, 17]). *Let $n = 2m$ be an even integer. Suppose that m is odd, $m > 3$. Let $a \in \mathbb{F}_{2^m}^\star$. Let β be a primitive element of $\mathbb{F}_4$. Let $f_{a,1}, f_{a,\beta}$ and f_{a,β^2} be the Boolean functions on $\mathbb{F}_{2^n}$ whose expression is of the form (9.3). If $K_m(a) = 4$ (in this case $\mathrm{Tr}_1^m(a^{1/3}) = 0$),then $f_{a,1}$, $f_{a,\beta}$ and f_{a,β^2} are bent while, if $K_m(a) \neq 4$, then $f_{a,1}, f_{a,\beta}$ and f_{a,β^2} are not bent.*

Remark 9.5.8. For $m = 3$, we have made an exhaustive search of all $a \in \mathbb{F}_{2^m}$ and $b \in \mathbb{F}_4^\star$ such that $f_{a,b}$ is bent. We have found that the bent Boolean functions of $\mathfrak{F}_6$ are $f_{1,\beta}, f_{1,\beta^2}$ and every Boolean function $f_{au,b}$ with $b \in \mathbb{F}_4^\star$, $u \in U = \{x \in \mathbb{F}_{2^6}^\star \mid x^9 = 1\}$ and $a \in \mathbb{F}_{2^3}^\star$ such that $K_3(a) = 4$.

- Now, recall that if a Boolean function f defined on $\mathbb{F}_{2^n}$ is bent then its dual function $\tilde{f}$ is the Boolean function defined on $\mathbb{F}_{2^n}$ by: $\widehat{\chi_f}(x) = 2^{\frac{n}{2}}\chi(\tilde{f}(x))$. Moreover, it is well-known that if f is bent then, its dual $\tilde{f}$ is also bent and that its own dual is f itself.

Proposition 9.5.9 ([17]). *Let $n = 2m$ be an even integer. Suppose that m is odd. Let $f_{a,b}$ ($a \in \mathbb{F}_{2^m}^\star$ and $b \in \mathbb{F}_4^\star$) be a bent Boolean functions on $\mathbb{F}_{2^n}$ whose expression is of the form (9.3). Then, the dual function of $f_{a,b}$ is equal to $f_{a^{2^m},b^2}$, that is, we have*

$$\forall \omega \in \mathbb{F}_{2^n}, \quad \widehat{\chi_{f_{a,b}}}(\omega) = 2^m\chi(f_{a^{2^m},b^2}(\omega)).$$

Proof. Let w be an element of $\mathbb{F}_{2^n}$. Since every element x of $\mathbb{F}_{2^n}^\star$ has a unique decomposition as : $x = yu$, with $y \in \mathbb{F}_{2^m}^\star$ and $u \in U$ we have

$$\widehat{\chi_{f_{a,b}}}(w) := \sum_{x\in\mathbb{F}_{2^n}} \chi(f_{a,b}(x) + \mathrm{Tr}_1^n(wx)) = 1 + \sum_{u\in U}\sum_{y\in\mathbb{F}_{2^m}^\star} \chi(f_{a,b}(yu)) + \mathrm{Tr}_1^n(wyu)).$$

One has, for m odd, $f_{a,b}(yu) = f_{a,b}(u)$ for every $u \in U$ and $y \in \mathbb{F}_{2^m}^\star$. Thus,

$$\begin{aligned}\widehat{\chi_{f_{a,b}}}(w) &= 1 - \sum_{u\in U}\chi(f_{a,b}(u)) + \sum_{u\in U}\chi(f_{a,b}(u))\sum_{y\in\mathbb{F}_{2^m}}\chi(\mathrm{Tr}_1^n(wyu))\\ &= 1 - \sum_{u\in U}\chi(f_{a,b}(u)) + \sum_{u\in U}\chi(f_{a,b}(u))\sum_{y\in\mathbb{F}_{2^m}}\chi(\mathrm{Tr}_1^m(y\mathrm{Tr}_m^n(wu)))\\ &= 1 - \sum_{u\in U}\chi(f_{a,b}(u)) + 2^m\sum_{\substack{u\in U\\ \mathrm{Tr}_m^n(wu)=0}}\chi(f_{a,b}(u)).\end{aligned}$$

Note first that $\widehat{\chi_{f_{a,b}}}(0) = 1 - \sum_{u\in U}\chi(f_{a,b}(u)) + 2^m\sum_{u\in U}\chi(f_{a,b}(u))$.
Now, if w is an element of $\mathbb{F}_{2^n}^\star$, then, we have $\mathrm{Tr}_m^n(wu) = 0$ if and only if $uw + u^{2^m}w^{2^m} = 0$, that is, $u^{2^m-1} = w^{1-2^m}$. Then, using the fact that $f_{a,b}(u) = f_{a,b}(w^{-1})$, we obtain

$$\widehat{\chi_{f_{a,b}}}(w) = 1 - \sum_{u\in U}\chi(f_{a,b}(u)) + 2^m\chi(f_{a,b}(w^{-1})).$$

Moreover, one has $f_{a,b}(w^{-1}) = \mathrm{Tr}_1^n(a\omega^{1-2^m}) + \mathrm{Tr}_1^2(bw^{\frac{1-2^n}{3}}) = \mathrm{Tr}_1^n(a^{2^m}\omega^{2^m-1}) + \mathrm{Tr}_1^2(b(w^{\frac{2^n-1}{3}})^2) = \mathrm{Tr}_1^n(a^{2^m}\omega^{2^m-1}) + \mathrm{Tr}_1^2(b^2w^{\frac{2^n-1}{3}}) = f_{a^{2^m},b^2}(\omega)$. Hence,

$$\widehat{\chi_{f_{a,b}}}(w) = 1 - \sum_{u\in U}\chi(f_{a,b}(u)) + 2^m\chi(f_{a^{2^m},b^2}(\omega)). \tag{9.9}$$

Now, recall that according to (9.8), $f_{a,b}$ is bent if and only if $\sum_{u\in U}\chi(f_{a,b}(u)) = 1$. Thus, the result follows. □

Remark 9.5.10. Note that one can get criterion (9.8) of bentness in terms of $\Lambda(a,b)$ from formula (9.9), that is, without using Dillon's results.

The following theorem summarizes the results presented above related to the bentness of the functions of the family $\mathfrak{F}_n$.

Theorem 9.5.11 ([18]). *Let $n = 2m$ with m odd ($m > 3$). Let $a \in \mathbb{F}_{2^m}^\star$ and $b \in \mathbb{F}_4^\star$. Let $f_{a,b}$ be the function defined on $\mathbb{F}_{2^n}$ by*

$$f_{a,b}(x) = \mathrm{Tr}_1^n\left(ax^{2^m-1}\right) + \mathrm{Tr}_1^2\left(bx^{\frac{2^n-1}{3}}\right).$$

1. *The algebraic degree of $f_{a,b}$ equals m (hence any bent function $f_{a,b}$ has a maximal algebraic degree).*
2. *$f_{a,b}$ is hyper-bent if and only if $f_{a,b}$ is bent.*
3. *$f_{a,b}$ is bent if and only if $K_m(a) = 4$.*

4. *The bent functions* $f_{a,b}$ *are in the class* $\mathcal{PS}^-$ *and in the Partial Spread class* PS_{ap} *if* $b = 1$.
5. *If* $f_{a,b}$ *is bent then its dual function equals* $f_{a^{2^m},b^2}$.

Remark 9.5.12. Note that according to [6], the condition $K_m(a) = 4$ (m odd) implies that $a = \frac{c}{(1+c)^4}$ for some c in $\mathbb{F}_{2^m}^\star$.

Example 9.5.13. Let $n = 10$. Let us describe the set of bent functions $f_{a,b}$ defined on $\mathbb{F}_{2^{10}}$ of the form $\mathrm{Tr}_1^{10}(ax^{31}) + \mathrm{Tr}_1^2(bx^{341})$ where $a \in \mathbb{F}_{2^{10}}^\star$ and $b \in \mathbb{F}_4^\star$. Let α be a primitive element of $\mathbb{F}_{32} = \mathbb{F}_2(\alpha)$ with $\alpha^5 + \alpha^2 + 1 = 0$. According to table 4 in [5], $E_0 := \{a \in \mathbb{F}_{2^5}^\star, \mathrm{Tr}_1^5(a^{1/3}) = 0\} = \{\alpha^3, \alpha^{21}, \alpha^{14}\}$, $\{a \in \mathbb{F}_{2^5}^\star, K_5(a) = 4\} = \{\alpha^3, \alpha^{21}\}$ and $E_1 := \{a \in \mathbb{F}_{2^5}^\star, \mathrm{Tr}_1^5(a^{1/3}) = 1\} = \{1, \alpha^2, \alpha^9, \alpha^{15}\}$ (recall that, $\mathrm{Tr}_1^5(a^{1/3}) = 1$ implies that $K_5(a) \neq 4$). Then according to Theorem 9.5.11, the functions $f_{\alpha^3,1}, f_{\alpha^3,\beta}, f_{\alpha^3,\beta^2}, f_{\alpha^{21},1}, f_{\alpha^{21},\beta}$ and f_{α^{21},β^2} are bent while $f_{\alpha^{14},1}, f_{\alpha^{14},\beta}, f^{(1)}_{\alpha^{14},\beta^2}, f_{a,1}, f_{a,\beta}$ and f_{a,β^2} are not bent if $a \in \{1, \alpha^2, \alpha^9, \alpha^{15}\}$. Now, the set $\{(a,b) \mid a \in \mathbb{F}_{2^n}^\star,\ b \in \mathbb{F}_4,\ f_{a,b}$ is bent$\}$ is equal to the set $\{(a'\lambda^{2^m-1}, b'\lambda^{\frac{2^n-1}{3}}) \mid a' \in \mathbb{F}_{2^m}^\star, b' \in \mathbb{F}_4, \lambda \in \mathbb{F}_{2^n}^\star, f_{a',b'}$ is bent$\}$. Therefore, we conclude that there exist 198 bent Boolean functions defined over $\mathbb{F}_{2^{10}}$ of the form $\mathrm{Tr}_1^{10}(ax^{31}) + \mathrm{Tr}_1^2(bx^{341})$ (with $b \neq 0$). Such functions are $f_{\alpha^3u,1}, f_{\alpha^3u,\beta}, f_{\alpha^3u,\beta^2}, f_{\alpha^{21}u,1}, f_{\alpha^{21}u,\beta}$ and $f_{\alpha^{21}u,\beta^2}$ where u is an element of the group of 33-rd roots of unity of $\mathbb{F}_{2^{10}}$ and β denotes a primitive element of $\mathbb{F}_4$.

Example 9.5.14. Let $n = 14$. According to Theorem 9.5.11 and to table 4 in [5], we find that there exist 1161 bent Boolean functions $f_{a,b}$ (with $b \neq 0$) defined over the field $\mathbb{F}_{16384}$ of the form $\mathrm{Tr}_1^{14}(cvx^{127}) + \mathrm{Tr}_1^2(bx^{5461})$ where $c \in \{\alpha^{14}, \alpha^{15}, \alpha^{62}\}$, α is a primitive element of $\mathbb{F}_{128}$ satisfying $\alpha^7 + \alpha^3 + 1 = 0$, v runs through the set of 129-st roots of unity of $\mathbb{F}_{2^{14}}$ and $b \in \{1, \beta, \beta^2\}$ were β is a primitive element of $\mathbb{F}_4$.

9.5.2.2 A First Family of Binomial Hyper-Bent Functions $\mathfrak{F}_n$: A Generalization

In the following we show that the characterization of bentness that we obtained in Sect. 9.5.2.1 is also valid for functions of more general form $f_{a,b}^{(r)}$:

$$f_{a,b}^{(r)}(x) = \mathrm{Tr}_1^n(ax^{r(2^m-1)}) + \mathrm{Tr}_1^2(bx^{\frac{2^n-1}{3}}) \tag{9.10}$$

where r is co-prime with $2^m + 1$ and with m odd. In fact, Theorem 9.5.11 can be generalized to any r co-prime with $2^m + 1$. and one can show that the bent Boolean functions $f_{a,b}^{(r)}$ are also hyper-bent and belong to the Partial Spread class PS_{ap} under some condition on the coefficients a and b. As in the case $r = 1$, one can show that, up to affine equivalence, we can restrict the study of the bentness of $f_{a,b}^{(r)}$ to the case where $a \in \mathbb{F}_{2^m}^\star$.

Theorem 9.5.15 ([18]). *Let $n = 2m$ with m odd ($m > 3$). Let $a \in \mathbb{F}_{2^m}^\star$ and $b \in \mathbb{F}_4^\star$. Let $f_{a,b}^{(r)}$ be the function defined on $\mathbb{F}_{2^n}$ by (9.10) $f_{a,b}^{(r)}(x) = \mathrm{Tr}_1^n(ax^{r(2^m-1)}) + \mathrm{Tr}_1^2(bx^{\frac{2^n-1}{3}})$*

1. *$f_{a,b}^{(r)}$ is bent if and only if $K_m(a) = 4$.*
2. *$f_{a,b}^{(r)}$ is hyper-bent if and only if $f_{a,b}^{(r)}$ is bent.*
3. *Bent functions $f_{a,b}^{(r)}$ are in the class $\mathcal{PS}^-$. Moreover, they are elements of the Partial Spread class $\mathcal{PS}_{ap}$ (resp. $\mathcal{PS}_{ap}^\#$) if $b = 1$ (resp. if $b \neq 1$).*
4. *If $f_{a,b}^{(r)}$ is bent then its dual function equals $f_{a^{2^m},b^2}^{(r)}$.*

Proof. Most of the arguments are similar to those used to prove Theorem 9.5.11. Note that $f_{a,b}^{(r)}$ have the following property:

$$\forall c \in \mathbb{F}_{2^m}^\star, \quad \forall x \in \mathbb{F}_{2^n}, \quad f_{a,b}^{(r)}(c^3x) = f_{a,b}^{(r)}(x).$$

Indeed (note that $c^{2^m-1} = 1$ since $c \in \mathbb{F}_{2^m}^\star$),

$$\begin{aligned} f_{a,b}^{(r)}(c^3x) &= \mathrm{Tr}_1^n\left(a(c^3x)^{r(2^m-1)}\right) + \mathrm{Tr}_1^2\left(b(c^3x)^{\frac{2^n-1}{3}}\right) \\ &= \mathrm{Tr}_1^n\left(a(c^{2^m-1})^{3r}x^{r(2^m-1)}\right) + \mathrm{Tr}_1^2\left(b(c^{2^m-1})^{2^m+1}x^{\frac{2^n-1}{3}}\right) = f_{a,b}^{(r)}(x) \end{aligned}$$

That implies in particular that the function $f_{a,b}^{(r)}$ is constant on each coset of $C = \{x^3 \mid x \in \mathbb{F}_{2^m}^\star\}$. Denote by H a set of representatives for the equivalence relation $\sim$ defined on $\mathbb{F}_{2^n}^\star$ by $x \sim y$ if and only if $y = xv$ for some $v \in C$. Then, we have $supp(f_{a,b}^{(r)}) = \bigcup_{x \in S_{a,b}} xC$ where, $S_{a,b} := \{x \in H \mid f_{a,b}^{(r)}(x) = 1\}$. When m is odd, every element of $\mathbb{F}_{2^m}^\star$ is a cube and thus we have $C = \mathbb{F}_{2^m}^\star$ (indeed, the map $x \in \mathbb{F}_{2^m}^\star \mapsto x^3$ is a permutation for m odd). On the other hand, recall that every element x of $\mathbb{F}_{2^n}^\star$ has a unique decomposition as: $x = yu$, with $y \in \mathbb{F}_{2^m}^\star$ and $u \in U$. Therefore, one can take $H = U$ in this case leading to $supp(f_{a,b}^{(r)}) = \bigcup_{u \in S_{a,b}} u\mathbb{F}_{2^m}^\star$, with $S_{a,b} = \{u \in U \mid f_{a,b}^{(r)}(u) = 1\}$. This implies that $f_{a,b}^{(r)}$ is bent if and only if, $\mathrm{wt}(f_{a,b}^{(r)}|_U) = 2^{m-1}$ and that bent functions $f_{a,b}^{(r)}$ are in the well known Partial Spread class $\mathcal{PS}^-$ (which proves the first part of assertion 3)) and that $f_{a,b}^{(r)}$ is bent if and only if $\sum_{u \in U} \chi(f_{a,b}^{(r)}(u)) = 1$ (since $\sum_{u \in U} \chi(f_{a,b}^{(r)}(u)) = \#U - 2\,\mathrm{wt}(f_{a,b}^{(r)}|_U) = 2^m + 1 - 2\#S_{a,b}$). The assertion 1) is then a direct consequence of Proposition 2.4.3 and Corollary 2.4.4. Now, if α is a primitive element of $\mathbb{F}_{2^n}$ then, $f_{a,b}^{(r)}(\alpha^{2^m+1}x) = f_{a,b}^{(r)}(x)$ for every $x \in \mathbb{F}_{2^n}$ (since 3 divides $2^m + 1$ when m is odd) and 0 is not in the support of $f_{a,b}^{(r)}$. The conditions of bentness given by Proposition 9.4.1 are then satisfied. Therefore, $f_{a,b}^{(r)}$ is hyper-bent if and only if $\sum_{u \in U} \chi(f_{a,b}^{(r)}(u)) = 1$ which proves the assertion 2). The second part of the assertion 3) is a direct application of Proposition 9.4.4. Finally, to compute the dual of a bent function $f_{a,b}^{(r)}$ it suffices to compute the Walsh transform $\widehat{\chi_{f_{a,b}^{(r)}}}(w)$

of $f_{a,b}^{(r)}(w)$ for every $w \in \mathbb{F}_{2^n}$. The calculation of $\widehat{\chi_{f_{a,b}^{(r)}}}(w)$ is analogous to the one of $\widehat{\chi_{f_{a,b}^{(1)}}}(w)$. We include the proof for completeness. Let w be an element of $\mathbb{F}_{2^n}$. Since every element x of $\mathbb{F}_{2^n}^\star$ has a unique decomposition as : $x = yu$, with $y \in \mathbb{F}_{2^m}^\star$ and $u \in U$ we have

$\widehat{\chi_{f_{u,b}^{(r)}}}(w) := \sum_{x\in\mathbb{F}_{2^n}} \chi(f_{a,b}^{(r)}(x) + Tr_1^n(wx)) = 1 + \sum_{u\in U}\sum_{y\in\mathbb{F}_{2^m}^\star} \chi(f_{a,b}^{(r)}(yu)) + Tr_1^n(wyu))$.

One has, for m odd, $f_{a,b}^{(r)}(yu) = f_{a,b}^{(r)}(u)$ for every $u \in U$ and $y \in \mathbb{F}_{2^m}^\star$. Thus,

$$\begin{aligned}\widehat{\chi_{f_{a,b}^{(r)}}}(w) &= 1 - \sum_{u\in U} \chi(f_{a,b}^{(r)}(u)) + \sum_{u\in U} \chi(f_{a,b}^{(r)}(u)) \sum_{y\in\mathbb{F}_{2^m}} \chi(Tr_1^n(wyu))\\ &= 1 - \sum_{u\in U} \chi(f_{a,b}^{(r)}(u)) + \sum_{u\in U} \chi(f_{a,b}^{(r)}(u)) \sum_{y\in\mathbb{F}_{2^m}} \chi(Tr_1^m(yTr_m^n(wu)))\\ &= 1 - \sum_{u\in U} \chi(f_{a,b}^{(r)}(u)) + 2^m \sum_{\substack{u\in U\\ Tr_m^n(wu)=0}} \chi(f_{a,b}^{(r)}(u)).\end{aligned}$$

Note first that $\widehat{\chi_{f_{a,b}^{(r)}}}(0) = 1 - \sum_{u\in U} \chi(f_{a,b}^{(r)}(u)) + 2^m \sum_{u\in U} \chi(f_{a,b}^{(r)}(u))$.
Now, if w is an element of $\mathbb{F}_{2^n}^\star$, then, we have $Tr_m^n(wu) = 0$ if and only if $uw + u^{2^m}w^{2^m} = 0$, that is, $u^{2^m-1} = w^{1-2^m}$. Then, using the fact that $f_{a,b}^{(r)}(u) = f_{a,b}^{(r)}(w^{-1})$, we obtain

$$\widehat{\chi_{f_{a,b}^{(r)}}}(w) = 1 - \sum_{u\in U} \chi(f_{a,b}^{(r)}(u)) + 2^m \chi(f_{a,b}^{(r)}(w^{-1})).$$

Moreover, one has

$$\begin{aligned}f_{a,b}^{(r)}(w^{-1}) &= Tr_1^n(a\omega^{r(1-2^m)}) + Tr_1^2(bw^{\frac{1-2^n}{3}})\\ &= Tr_1^n(a^{2^m}\omega^{r(2^m-1)}) + Tr_1^2(b(w^{\frac{2^n-1}{3}})^2)\\ &= Tr_1^n(a^{2^m}\omega^{r(2^m-1)}) + Tr_1^2(b^2w^{\frac{2^n-1}{3}})\\ &= f_{a^{2^m},b^2}^{(r)}(\omega).\end{aligned}$$

Hence, $\widehat{\chi_{f_{a,b}^{(r)}}}(w) = 1 - \sum_{u\in U} \chi(f_{a,b}^{(r)}(u)) + 2^m \chi(f_{a^{2^m},b^2}^{(r)}(\omega))$. We have seen that $f_{a,b}^{(r)}$ is bent if and only if $\sum_{u\in U} \chi(f_{a,b}^{(r)}(u)) = 1$. Thus, the assertion 4) follows. □

9.5.2.3 A First Family of Binomial Hyper-Bent Functions $\mathfrak{F}_n$: A Special Case

In this subsection, we focus on the functions $f_{a,b}$ of the family $\mathfrak{F}_n$ or more generally on functions of the form $f_{a,b}^{(r)}(x) = Tr_1^n(ax^{r(2^m-1)}) + Tr_1^2(bx^{\frac{2^n-1}{3}})$ in the case where m is even. (in this case, 3 divides $2^m - 1$). By exhaustive search, we find that the family $\mathfrak{F}_n$ contains bent functions for larger values of $m \geq 2$. We shall discuss the computational point of view in Sect. 9.5.2.3. From now on we therefore assume that $m \geq 2$. In the following we are interested in treating the even case from a theoretical point of view.
We have seen in Sect. 9.5.2.1 that all the Boolean functions of the family $\mathfrak{F}_n$ of the form : have the following property

$$\forall a \in \mathbb{F}_{2^n}^\star, \quad \forall b \in \mathbb{F}_4, \quad \forall c \in \mathbb{F}_{2^m}^\star, \quad \forall x \in \mathbb{F}_{2^n}, \quad f_{a,b}(c^3x) = f_{a,b}(x). \tag{9.11}$$

That implies in particular that a Boolean function $f_{a,b}$ of $\mathfrak{F}_n$ is constant on each coset of $C = \{x^3 \mid x \in \mathbb{F}_{2^m}^\star\}$. Denote by H a set of representatives for the equivalence relation $\sim$ defined on $\mathbb{F}_{2^n}^\star$ by $x \sim y$ if and only if $y = xv$ for some $v \in C$. Then, we have

$$supp(f_{a,b}) = \bigcup_{x \in S_{a,b}} xC \text{ where } S_{a,b} := \{x \in H \mid f_{a,b}(x) = 1\} \tag{9.12}$$

We have seen that when m is odd, $C = \mathbb{F}_{2^m}^\star$ and one can take $H = U$. Let us now consider the case where m is even. Then, one has $C \neq \mathbb{F}_{2^m}^\star$. Now, unlike in the odd case, a Boolean function $f_{a,b}$ is not constant on any coset $u\mathbb{F}_{2^m}^\star$, $u \in U$, of $\mathbb{F}_{2^m}^\star$. Indeed, for every $y \in \mathbb{F}_{2^m}^\star$, we have

$$f_{a,b}(uy) = Tr_1^n(au^{2^m-1}) + Tr_1^2(by^{\frac{2^n-1}{3}}) \tag{9.13}$$

because $\frac{2^n-1}{3}$ is a multiple of $2^m + 1$ for m even. The algebraic degree of the restriction of $f_{a,b}$ to $u\mathbb{F}_{2^m}^\star$ is hence equal to the 2-weight of $\frac{2^n-1}{3}$, that is, equal to m. The situation is more complicated that in the odd case since the support of $f_{a,b}$ is not of the form (9.6), that is, the study of the bentness of $f_{a,b}$ cannot be done as in the Sect. 9.5.2.1. In particular, it is difficult to answer the question of knowing if a function $f_{a,b}$ is or not in the Partial Spread class. But nevertheless, in this case, we succeed in establishing in Theorem 9.5.16 a necessary condition expressed in terms of Kloosterman sum that an element a has to satisfy so that the function $f_{a,b}$ is bent.

Theorem 9.5.16 ([17]). *Let $f_{a,b} \in \mathfrak{F}_n$, with $a \in \mathbb{F}_{2^m}^\star$, $m \geq 2$ even and $b \in \mathbb{F}_4^\star$. Then, a function $f_{a,b}$ is bent only if $K_m(a) = 4$.*

In fact the result given in Theorem 9.5.16 can be extended for functions $f_{a,b}^{(r)}$ of the form (9.10) (for any integer r co-prime with $2^m + 1$) as follows.

Theorem 9.5.17 ([18]). *Let $n = 2m$ with m even ($m \geq 2$). Let $a \in \mathbb{F}_{2^m}^\star$ and $b \in \mathbb{F}_4^\star$. Let $f_{a,b}^{(r)}$ be the function defined on $\mathbb{F}_{2^n}$ by $f_{a,b}^{(r)}(x) = Tr_1^n(ax^{r(2^m-1)}) + Tr_1^2(bx^{\frac{2^n-1}{3}})$. If $f_{a,b}^{(r)}$ is bent then, $K_m(a) = 4$.*

Proof. Recall that $f_{a,b}^{(r)}$ is bent if and only if $\widehat{\chi_{f_{a,b}^{(r)}}}(w) = \pm 2^m$ for every $w \in \mathbb{F}_{2^n}$. In particular, if $f_{a,b}^{(r)}$ is bent then, we should have $\widehat{\chi_{f_{a,b}^{(r)}}}(0) = \pm 2^m$. Recall that every non-zero element x of $\mathbb{F}_{2^n}$ has a unique decomposition as: $x = yu$ with $y \in \mathbb{F}_{2^m}^\star$ and $u \in U$. Then, the Walsh transform of $f_{a,b}^{(r)}$ at 0 is given by (we use the fact that $y^{2^m-1} = 1$ and $u^{\frac{2^n-1}{3}} = 1$, since 3 divides $2^m - 1$ when m is even):

$$\widehat{\chi_{f_{a,b}^{(r)}}}(0) = \sum_{x \in \mathbb{F}_{2^n}} \chi(f_{a,b}^{(r)}(x)) = 1 + \sum_{x \in \mathbb{F}_{2^n}^\star} \chi(f_{a,b}^{(r)}(x))$$

$$= 1 + \sum_{u \in U} \chi(Tr_1^n(au^{r(2^m-1)})) \sum_{y \in \mathbb{F}_{2^m}^\star} \chi(Tr_1^2(by^{\frac{2^n-1}{3}})).$$

Split $\mathbb{F}_{2^m}^\star$ as $\mathbb{F}_{2^m}^\star = C' \cup \beta C' \cup \beta^2 C'$ where C' the set of the cubic elements of $\mathbb{F}_{2^m}^\star$ and β is an element of $\mathbb{F}_{2^m} \setminus C'$. We thus get

$$\widehat{\chi_{f_{a,b}^{(r)}}}(0) = 1 + \sum_{u \in U} \chi(Tr_1^n(au^{r(2^m-1)})) \sum_{i=0}^{2} \sum_{z \in C'} \chi(Tr_1^2(b(z\beta^i)^{\frac{2^n-1}{3}})).$$

Since z is a cube of an element of $\mathbb{F}_{2^m}^\star$ we have,

$$\sum_{i=0}^{2} \sum_{z \in C'} \chi(Tr_1^2(b(z\beta^i)^{\frac{2^n-1}{3}})) = \sum_{i=0}^{2} \sum_{z \in C'} \chi(Tr_1^2(b\beta^{i\frac{2^n-1}{3}}))$$

$$= \sum_{z \in C'} \sum_{\tau \in \mathbb{F}_4^\star} \chi(Tr_1^2(\tau)) = \sum_{z \in C'} \Big(\sum_{\tau \in \mathbb{F}_4} \chi(Tr_1^2(\tau)) - 1 \Big)$$

$$= -\#C' = -\frac{2^m - 1}{3}.$$

On the other hand, the map $u \mapsto u^r$ is a permutation of U (since $\gcd(r, 2^m + 1) = 1$) and so is the map $u \mapsto u^{2^m-1}$. Hence, using the well known result, $\sum_{u \in U} \chi(Tr_1^n(au)) = 1 - K_m(a)$, we obtain, $\sum_{u \in U} \chi(Tr_1^n(au^{r(2^m-1)})) = \sum_{u \in U} \chi(Tr_1^n(au)) = 1 - K_m(a)$. We thus deduce

$$\widehat{\chi_{f_{a,b}^{(r)}}}(0) = 1 + \frac{2^m - 1}{3}(K_m(a) - 1).$$

Now, if $f_{a,b}^{(r)}$ is bent, then one has necessarily $1 + \frac{2^m-1}{3}(K_m(a) - 1) = \pm 2^m$, that is, $K_m(a) = 4$ or $(2^m - 1)(K_m(a) - 1) = -3(2^m + 1)$. The second equality being impossible since $2^m - 1$ and $2^m + 1$ are co-prime, this proves the result. □

Note that according to [6], the condition $K_m(a) = 4$ (m even) implies that $a = c^3$ for some c such that $Tr_2^m(c) \neq 0$. The previous theorem enables one to exhibit an infinite family of functions of type (9.10) which are not bent. Moreover, one can prove that the study of the bentness of $f_{a,b}^{(r)}$ can be reduced to the case where $b = 1$.

Proposition 9.5.18 ([18]). *Let $n = 2m$ with m even ($m \geq 2$). Let $a \in \mathbb{F}_{2^m}^\star$ and $b \in \mathbb{F}_4^\star$. Let $f_{a,b}^{(r)}$ be the function defined on $\mathbb{F}_{2^n}$ by (9.10). Then, $f_{a,b}^{(r)}$ is bent if and only if, $f_{a,1}^{(r)}$ is bent.*

Proof. Since m is even, $\mathbb{F}_4^\star \subset \mathbb{F}_{2^m}^\star$. In particular, for every $b \in \mathbb{F}_4^\star$, there exists $\alpha \in \mathbb{F}_{2^m}^\star$ such that $\alpha^{\frac{2^n-1}{3}} = b$. For $x \in \mathbb{F}_{2^n}$, we have

$$\begin{aligned} f_{a,b}^{(r)}(x) &:= Tr_1^n(ax^{r(2^m-1)}) + Tr_1^2(bx^{\frac{2^n-1}{3}}) \\ &= Tr_1^n(a(\alpha^{2^m-1})^r x^{r(2^m-1)}) + Tr_1^2(\alpha^{\frac{2^n-1}{3}} x^{\frac{2^n-1}{3}}) \\ &= Tr_1^n(a(\alpha x)^{r(2^m-1)}) + Tr_1^2((\alpha x)^{\frac{2^n-1}{3}}) \\ &= f_{a,1}^{(r)}(\alpha x). \end{aligned}$$

Hence, for every $\omega \in \mathbb{F}_{2^n}^\star$, we have

$$\begin{aligned} \widehat{\chi_{f_{a,b}^{(r)}}}(\omega) &= \sum_{x\in\mathbb{F}_{2^n}} \chi(f_{a,b}^{(r)}(x) + Tr_1^n(\omega x)) \\ &= \sum_{x\in\mathbb{F}_{2^n}} \chi(f_{a,1}^{(r)}(\alpha x) + Tr_1^n(\omega x)) \\ &= \widehat{\chi_{f_{a,1}^{(r)}}}(\omega\alpha^{-1}) \end{aligned}$$

□

The exact value of the Walsh transform $\widehat{\chi_{f_{a,1}^{(r)}}}(\omega)$ of $f_{a,1}^{(r)}$ seems difficult to compute. Nevertheless, we give in the following an expression of $\widehat{\chi_{f_{a,1}^{(r)}}}(\omega)$ for every element ω of $\mathbb{F}_{2^n}^\star$.

Proposition 9.5.19 ([18]). *Let $n = 2m$ with m even. Let $a \in \mathbb{F}_{2^m}^\star$ such that $K_m(a) = 4$. Then for every $\omega \in \mathbb{F}_{2^n}^\star$, we have $\widehat{\chi_{f_{a,1}^{(r)}}}(\omega) = \frac{2}{3}\sum_{u\in U}\sum_{y\in\mathbb{F}_{2^m}} \chi(Tr_1^n(au^{r(2^m-1)}) + Tr_1^n(\omega uy^3)) - 2^m\chi(Tr_1^n(a\omega^{r(1-2^m)}))$.*

Proof. Every non-zero element x of $\mathbb{F}_{2^n}$ has a unique decomposition as: $x = yu$ with $y \in \mathbb{F}_{2^m}^\star$ and $u \in U$. Since $y^{2^m-1} = 1$ and $u^{\frac{2^n-1}{3}} = 1$ (because 3 divides $2^m - 1$ when m is even), we have for every $\omega \in \mathbb{F}_{2^n}$

$$\widehat{\chi_{f_{a,1}^{(r)}}}(\omega) = \sum_{x\in\mathbb{F}_{2^n}} \chi(\mathrm{Tr}_1^n(ax^{r(2^m-1)}) + \mathrm{Tr}_1^2(x^{\frac{2^n-1}{3}}) + \mathrm{Tr}_1^n(wx))$$

$$= 1 + \sum_{u\in U}\sum_{y\in\mathbb{F}_{2^m}^\star} \chi(\mathrm{Tr}_1^n(au^{r(2^m-1)}) + \mathrm{Tr}_1^2(y^{\frac{2^n-1}{3}}) + \mathrm{Tr}_1^n(wuy)).$$

Now, $\mathrm{Tr}_1^2(y^{\frac{2^n-1}{3}}) = 0$ if and only if $y \in C' := \{y^3,\ y \in \mathbb{F}_{2^m}^\star\}$. Then, $\widehat{\chi_{f_{a,1}^{(r)}}}(\omega)$

$$= 1 + \sum_{u\in U}\sum_{y\in C'} \chi(\mathrm{Tr}_1^n(au^{r(2^m-1)}) + \mathrm{Tr}_1^n(wuy))$$

$$- \sum_{u\in U}\sum_{y\notin C'} \chi(\mathrm{Tr}_1^n(au^{r(2^m-1)}) + \mathrm{Tr}_1^n(wuy))$$

$$= 1 + 2\sum_{u\in U}\sum_{y\in C'} \chi(\mathrm{Tr}_1^n(au^{r(2^m-1)}) + \mathrm{Tr}_1^n(wuy))$$

$$- \sum_{u\in U}\sum_{y\in\mathbb{F}_{2^m}^\star} \chi(\mathrm{Tr}_1^n(au^{r(2^m-1)}) + \mathrm{Tr}_1^n(wuy))$$

$$= 1 + 2\sum_{u\in U}\sum_{y\in C'} \chi(\mathrm{Tr}_1^n(au^{r(2^m-1)}) + \mathrm{Tr}_1^n(wuy))$$

$$+ \sum_{u\in U} \chi(\mathrm{Tr}_1^n(au^{r(2^m-1)}))$$

$$- \sum_{u\in U}\sum_{y\in\mathbb{F}_{2^m}} \chi(\mathrm{Tr}_1^n(au^{r(2^m-1)}) + \mathrm{Tr}_1^n(wuy)).$$

Firstly, the maps $x \mapsto x^{2^m-1}$ and $x \mapsto x^r$ being permutations of U (since $\gcd(r, 2^m+1) = 1$) hence, $\sum_{u\in U} \chi(\mathrm{Tr}_1^n(au^{r(2^m-1)})) = \sum_{u\in U} \chi(\mathrm{Tr}_1^n(au)) = 1 - K_m(a)$. Secondly,

$$\sum_{u\in U}\sum_{y\in\mathbb{F}_{2^m}} \chi(\mathrm{Tr}_1^n(au^{r(2^m-1)}) + \mathrm{Tr}_1^n(wuy))$$

$$= \sum_{u\in U} \chi(\mathrm{Tr}_1^n(au^{r(2^m-1)}) \sum_{y\in\mathbb{F}_{2^m}} \chi(\mathrm{Tr}_1^n(wuy))).$$

Using the transitivity rule of trace function, we obtain

$$\sum_{u\in U}\sum_{y\in\mathbb{F}_{2^m}} \chi(\mathrm{Tr}_1^n(au^{r(2^m-1)}) + \mathrm{Tr}_1^n(wuy))$$

$$= \sum_{u\in U} \chi(\mathrm{Tr}_1^n(au^{r(2^m-1)}) \sum_{y\in\mathbb{F}_{2^m}} \chi(\mathrm{Tr}_1^m(\mathrm{Tr}_m^n(\omega u)y)).$$

But $\sum_{y\in\mathbb{F}_{2^m}} \chi(\mathrm{Tr}_1^m(\mathrm{Tr}_m^n(\omega u)y))$

$$= \begin{cases} 2^m \text{ if } \mathrm{Tr}_m^n(\omega u) = 0, \text{ that is, if } u^{2^m-1} = \omega^{1-2^m} \\ 0 \quad \text{othorwisc} \end{cases}$$

Hence, $\sum_{u\in U} \sum_{y\in\mathbb{F}_{2^m}} \chi(\mathrm{Tr}_1^n(au^{r(2^m-1)}) + \mathrm{Tr}_1^n(wuy)) = 2^m \chi(\mathrm{Tr}_1^n(a\omega^{r(1-2^m)}))$.
Moreover, $\sum_{u\in U} \sum_{y\in C'} \chi(\mathrm{Tr}_1^n(au^{r(2^m-1)}) + \mathrm{Tr}_1^n(wuy))$

$$= \frac{1}{3} \sum_{u\in U} \sum_{y\in\mathbb{F}_{2^m}^\star} \chi(\mathrm{Tr}_1^n(au^{r(2^m-1)}) + \mathrm{Tr}_1^n(\omega u y^3))$$
$$= \frac{1}{3} \sum_{u\in U} \sum_{y\in\mathbb{F}_{2^m}} \chi(\mathrm{Tr}_1^n(au^{r(2^m-1)}) + \mathrm{Tr}_1^n(\omega u y^3)) - \frac{1}{3} \sum_{u\in U} \chi(\mathrm{Tr}_1^n(au^{r(2^m-1)}))$$
$$= \frac{1}{3} \sum_{u\in U} \sum_{y\in\mathbb{F}_{2^m}} \chi(\mathrm{Tr}_1^n(au^{r(2^m-1)}) + \mathrm{Tr}_1^n(\omega u y^3)) + \frac{1}{3} K_m(a) - \frac{1}{3}.$$

Collecting the previous calculations, we obtain $\widehat{\chi_{f_{a,1}^{(r)}}}(\omega)$

$$= \frac{1}{3}(4 - K_m(a)) + \frac{2}{3} \sum_{u\in U} \sum_{y\in\mathbb{F}_{2^m}} \chi(\mathrm{Tr}_1^n(au^{r(2^m-1)}) + \mathrm{Tr}_1^n(\omega u y^3)) - 2^m \chi(\mathrm{Tr}_1^n(a\omega^{r(1-2^m)})).$$

The result follows if $K_m(a)$ equals 4. □

Remark 9.5.20. Let $n = 2m$ with m even. Let $a \in \mathbb{F}_{2^m}^\star$. For every $\omega \in \mathbb{F}_{2^n}^\star$, we have

$$\sum_{b\in\mathbb{F}_4^\star} \widehat{\chi_{f_{a,b}^{(1)}}}(\omega) = 4 - K_m(a) - 2^m \chi(\mathrm{Tr}_1^n(a\omega^{2^m-1})).$$

In particular, $\sum_{b\in\mathbb{F}_4} \widehat{\chi_{f_{a,b}^{(1)}}}(\omega) = 4$. Indeed, every element x of $\mathbb{F}_{2^n}^\star$ has a unique decomposition as: $x = yu$, with $y \in \mathbb{F}_{2^m}^\star$ and $u \in U$. Hence, for every $\omega \in \mathbb{F}_{2^n}^\star$ we have $\sum_{b\in\mathbb{F}_4^\star} \widehat{\chi_{f_{a,b}^{(1)}}}(\omega)$

$$= \sum_{x \in \mathbb{F}_{2^n}} \chi(\mathrm{Tr}_1^n(ax^{2^m-1} + \omega x)) \sum_{b \in \mathbb{F}_4^\star} \chi(\mathrm{Tr}_1^2(bx^{\frac{2^n-1}{3}}))$$

$$= 3 - \sum_{x \in \mathbb{F}_{2^n}^\star} \chi(\mathrm{Tr}_1^n(ax^{2^m-1} + \omega x))$$

$$= 3 - \sum_{u \in U} \chi(\mathrm{Tr}_1^n(au^{2^m-1})) \sum_{y \in \mathbb{F}_{2^m}^\star} \chi(\mathrm{Tr}_1^m(\mathrm{Tr}_m^n(\omega u)y))$$

$$= 4 - K_m(a) -$$

$$\sum_{u \in U} \chi(\mathrm{Tr}_1^n(au^{2^m-1})) \sum_{y \in \mathbb{F}_{2^m}} \chi(\mathrm{Tr}_1^m(\mathrm{Tr}_m^n(\omega u)y)).$$

The previous equality follows from the well known result, that is $\sum_{u \in U} \chi(\mathrm{Tr}_1^n(au)) = 1 - K_m(a)$. Now, $\sum_{y \in \mathbb{F}_{2^m}} \chi(\mathrm{Tr}_1^m(\mathrm{Tr}_m^n(\omega u)y))$

$$= \begin{cases} 2^m \text{ if } \mathrm{Tr}_m^n(\omega u) = 0, \text{ that is, if } u^{2^m-1} = \omega^{1-2^m} \\ 0 \quad \text{otherwise.} \end{cases}$$

Thus, $\sum_{b \in \mathbb{F}_4^\star} \widehat{\chi_{f_{a,b}^{(1)}}}(\omega) = 4 - K_m(a) - 2^m \chi(\mathrm{Tr}_1^n(a\omega^{1-2^m}))$.

Now, according to [14],

$$\forall \omega \in \mathbb{F}_{2^n}, \widehat{\chi_{f_{a,0}^{(1)}}}(\omega) = 2^m \chi(\mathrm{Tr}_1^n(a\omega^{2^m-1})) + K_m(a).$$

Hence, $\sum_{b \in \mathbb{F}_4} \widehat{\chi_{f_{a,b}^{(1)}}}(\omega) = 4$.

- Experimental results for m even:

  The functions that we have introduced in [17] are defined for $a \in \mathbb{F}_{2^m}^*$ and $b \in \mathbb{F}_4^*$ as the Boolean functions $f_{a,b}$ with $n = 2m$ inputs given by

$$f_{a,b}(x) = \mathrm{Tr}_1^n\left(ax^{2^m-1}\right) + \mathrm{Tr}_1^2\left(bx^{\frac{2^n-1}{3}}\right). \tag{9.14}$$

When m is even, we have shown that the situation seems more complicated theoretically than when m is odd. Here, we only have a necessary condition to build bent functions from the value 4 of binary Kloosterman sum. To get a better understanding of the situation we conducted some experimental investigations to check whether the Boolean functions constructed with the formula (9.14) were bent or not for all the a's in $\mathbb{F}_{2^m}$ giving a Kloosterman sum with value 4.

First, we show that it is enough to study the bentness of a subset of these functions to get results about all of them.

Next proposition proves that the study of the bentness of $f_{a,b}$ can be reduced to the case where $b = 1$.

Proposition 9.5.21 ([11]). *Let $n = 2m$ with $m \geq 2$ even. Let $a \in \mathbb{F}_{2^m}^*$ and $b \in \mathbb{F}_4^*$. Let $f_{a,b}$ be the function defined on $\mathbb{F}_{2^n}$ by Eq. (9.14). Then $f_{a,b}$ is bent if and only if $f_{a,1}$ is bent.*

Proof. Since m is even, we have the inclusion of fields $\mathbb{F}_4^* \subset \mathbb{F}_{2^m}^*$. In particular, for every $b \in \mathbb{F}_4^*$, there exists $\alpha \in \mathbb{F}_{2^m}^*$ such that $\alpha^{\frac{2^n-1}{3}} = b$. For $x \in \mathbb{F}_{2^n}$, we have

$$\begin{aligned} f_{a,b}(x) &= \mathrm{Tr}_1^n\left(ax^{2^m-1}\right) + \mathrm{Tr}_1^2\left(bx^{\frac{2^n-1}{3}}\right) \\ &= \mathrm{Tr}_1^n\left(a\alpha^{2^m-1}x^{2^m-1}\right) + \mathrm{Tr}_1^2\left(\alpha^{\frac{2^n-1}{3}}x^{\frac{2^n-1}{3}}\right) \\ &= \mathrm{Tr}_1^n\left(a(\alpha x)^{2^m-1})\right) + \mathrm{Tr}_1^2\left((\alpha x)^{\frac{2^n-1}{3}}\right) \\ &= f_{a,1}(\alpha x). \end{aligned}$$

Hence, for every $\omega \in \mathbb{F}_{2^n}^*$, we have

$$\begin{aligned} \widehat{\chi_{f_{a,b}}}(\omega) &= \sum_{x\in\mathbb{F}_{2^n}} (-1)^{f_{a,b}(x)+\mathrm{Tr}_1^n(\omega x)} \\ &= \sum_{x\in\mathbb{F}_{2^n}} (-1)^{f_{a,1}(\alpha x)+\mathrm{Tr}_1^n(\omega x)} \\ &= \widehat{\chi_{f_{a,1}}}(\omega\alpha^{-1}). \end{aligned}$$

□

Second, we know that $K_m(a) = K_m(a^2)$, so the a's in $\mathbb{F}_{2^m}$ giving binary Kloosterman sums with value 4 come in cyclotomic classes. Fortunately, it is enough to check one a per class. Indeed, $f_{a,b}$ is bent if and only if f_{a^2,b^2} is, as proved in the following proposition.

Proposition 9.5.22 ([11]). *Let $n = 2m$ with $m \geq 2$ even. Let $a \in \mathbb{F}_{2^m}^*$ and $b \in \mathbb{F}_4^*$. Let $f_{a,b}$ be the function defined on $\mathbb{F}_{2^n}$ by Eq. (9.14). Then $f_{a,b}$ is bent if and only if f_{a^2,b^2} is bent.*

Proof.

$$\begin{aligned} \widehat{\chi_{f_{a,b}}}(\omega) &= \sum_{x\in\mathbb{F}_{2^n}} (-1)^{f_{a,b}(x)+\mathrm{Tr}_1^n(\omega x)} \\ &= \sum_{x\in\mathbb{F}_{2^n}} (-1)^{\mathrm{Tr}_1^n\left(ax^{2^m-1}\right)+\mathrm{Tr}_1^2\left(bx^{\frac{2^n-1}{3}}\right)+\mathrm{Tr}_1^n(\omega x)} \\ &= \sum_{x\in\mathbb{F}_{2^n}} (-1)^{\mathrm{Tr}_1^n\left(a^2x^{2^{2^m-1}}\right)+\mathrm{Tr}_1^2\left(b^2x^{2\frac{2^n-1}{3}}\right)+\mathrm{Tr}_1^n(\omega^2x^2)} \end{aligned}$$

$$= \sum_{x \in \mathbb{F}_{2^n}} (-1)^{\mathrm{Tr}_1^n\left(a^2 x^{2^m-1}\right)+\mathrm{Tr}_1^2\left(b^2 x^{\frac{2^n-1}{3}}\right)+\mathrm{Tr}_1^n(\omega^2 x)}$$

$$= \sum_{x \in \mathbb{F}_{2^n}} (-1)^{f_{a^2,b^2}(x)+\mathrm{Tr}_1^n(\omega^2 x)}$$

$$= \widehat{\chi_{f_{a^2,b^2}}}(\omega^2).$$

□

In the specific case $b = 1$ that we are interested in, we get $f_{a,1}$ is bent if and only if $f_{a^2,1}$ is, which proves that checking one element of each cyclotomic class is enough.

Finally, as mentioned in Sect. 11.4.2, finding all the a's in $\mathbb{F}_{2^m}$ giving a specific value is a different problem from finding one such $a \in \mathbb{F}_{2^m}$. One can compute the Walsh–Hadamard transform of the trace of the inverse function using a fast Walsh–Hadamard transform. As long as the basis of $\mathbb{F}_{2^m}$ considered as a vectorspace over $\mathbb{F}_2$ is correctly chosen so that the trace corresponds to the scalar product, the implementation is straightforward.

The algorithm that we implemented is described in Algorithm 9.1.

The implementation was made using Sage [21] and Cython [1], performing direct calls to Givaro [9], NTL [20] and gf2x [2] libraries for efficient manipulation of finite field elements and construction of Boolean functions.

In Table 9.1 we give the results of the computations we conducted along with different pieces of information about them. One should remark that all the Boolean functions which could be tested are bent.

Algorithm 9.1: Testing bentness for m even

Input: An even integer $m \geq 2$

Output: A list of couples made of one representative for each cyclotomic class of elements $a \in \mathbb{F}_{2^m}$ such that $K_m(a) = 4$ together with 1 if the corresponding Boolean functions $f_{a,b}$ are bent, 0 otherwise

1 Build the Boolean function $f : x \in \mathbb{F}_{2^n} \mapsto \mathrm{Tr}_1^n(1/x) \in \mathbb{F}_2$
2 Compute the Walsh–Hadamard transform of f
3 Build a list A made of one $a \in \mathbb{F}_{2^m}$ for each cyclotomic class such that $K_m(a) = 4$
4 Initialize an empty list R
5 **foreach** $a \in A$ **do**
6   Build the Boolean function $f_{a,1}$
7   Compute the Walsh–Hadamard transform of $f_{a,1}$
8   **if** $f_{a,1}$ *is bent* **then**
9   Append $(a, 1)$ to R
10   **else**
11   Append $(a, 0)$ to R
12 **return** R

Table 9.1 Test of bentness for m even

m	Number of cyclotomic classes	Time	All bent?
4	1	<1 s	Yes
6	1	<1 s	Yes
8	2	<1 s	Yes
10	3	4 s	Yes
12	6	130 s	Yes
14	8	3000 s	Yes
16	14	82,000 s	Yes
18	20	–	–
20	76	–	–
22	87	–	–
24	128	–	–
26	210	–	–
28	810	–	–
30	923	–	–
32	2646	–	–

Evidence that our computations were correct is given by the fact that the number of cyclotomic classes we found is so. This can be checked using the formula of Proposition 11.1.4. We are looking for elliptic curves with trace t of the Frobenius endomorphism equal to $t = 1 - K_m(a) = -3$. Hence, the number of cycloctomic classes is $H(\Delta)/m$ where $H(\Delta)$ is the Kronecker class number and $\Delta = 9 - 4 \cdot 2^m$. Moreover, for the values we tested, except $m = 12, 30, 32$, this discriminant is fundamental, so that the order $\mathbb{Z}[\alpha]$ is maximal and $H(\Delta) = h(\Delta)$ the classical class number, a quantity even easier to compute.

Unfortunately, we were not able to check bentness of functions for $m > 16$ due to lack of memory. Constructing the Boolean functions in $n = 2m$ variables is the most time consuming part of the test, but the real bottleneck is the amount of memory needed to compute their Walsh–Hadamard transforms. One must indeed perform these computations using integers of size at least $2m + 1$ bits, so, with our implementation, integers of 64 bits as soon as $m \geq 16$. The amount of memory needed is then $64 \cdot 2^{2m} \cdot 2^{-30} = 2^{2m-24}$ GB. For $m = 16$ this represents already 32 GB of memory; for $m = 18$ it would be 512 GB of memory. Therefore, we give in Table 9.2 the fourteen values of a found for $m = 16$, the highest value that we could test. In this table the finite field $\mathbb{F}_{2^{16}}$ is represented as $\mathbb{F}_2[x]/(x^{16} + x^5 + x^3 + x^2 + 1)$. The corresponding Boolean functions in $n = 32$ variables are all bent as we already pointed out.

Finally, we give some open questions:

Question 9.5.23. *Assume m even. Does a bent function $f_{a,b}^{(r)}$ of the form (9.10) belong to the Partial Spread class PS^- ?*

Question 9.5.24. *Assume m even. Are the bent functions $f_{a,b}^{(r)}$ of the from (9.10) also hyper-bent ?*

Table 9.2 The fourteen cyclotomic classes such that $K_{16}(a) = 4$ as elements of $\mathbb{F}_2[x]/(x^{16}+x^5+x^3+x^2+1)$

$x^{14}+x^{11}+x^8+x^6+x^3+x$
$x^{15}+x^{13}+x^{10}+x^8+x^7+x^6+x^5+x^4+x^3+1$
$x^{14}+x^{13}+x^{12}+x^{10}+x^8+x^2+x$
$x^{14}+x^{12}+x^{11}+x^9+x^6+x$
$x^{15}+x^{11}+x^9+x^7+x^6+x^3+x^2+1$
$x^{13}+x^6+x^4+x^2+x+1$
$x^{12}+x^{11}+x^{10}+x^9+x^5+x^3+x^2+x$
$x^{15}+x^{11}+x^7+x^6+x^5+x^4+x^3+x^2$
$x^{15}+x^{13}+x^9+x^8+x^5+x^4+x^3+x$
$x^{15}+x^{11}+x^{10}+x^3$
$x^{13}+x^{10}+x^9+x^7+x^6+x^5+x^3+x^2+x$
$x^{13}+x^{10}+x^9+x^7+x^6+x^5+x^4+x^3+x^2+x$
$x^{15}+x^{13}+x^{10}+x^9+x^8+x^7+x^5+x$
$x^{15}+x^{11}+x^{10}+x^3+x+1$

If the answer to Question 9.5.23 is "no" and the one to Question 9.5.24 is positive, then we will obtain for the first time a family of hyper-bent functions which are not in the class PS^-. Such functions do not exist in the literature.

9.5.2.4 A Second Family of Binomial Hyper-Bent Functions $\mathfrak{G}_n$

By computer experiments, for small values of n ($n \le 14$; because of the complexity of the problem), we have found that the set of all functions of type (9.2) with the exponents $s_1 = 3(2^m - 1)$ and $s_2 = \frac{2^n-1}{3}$ contains bent functions. Note that $o(s_1) = n$ and $o(s_2) = 2$. The polynomial form of a function of type (9.2), denoted by $g_{a,b}$, is then of the form:

$$g_{a,b}(x) = \mathrm{Tr}_1^n\left(ax^{3(2^m-1)}\right) + \mathrm{Tr}_1^2\left(bx^{\frac{2^n-1}{3}}\right), \tag{9.15}$$

where, $a \in \mathbb{F}^\star_{2^{o(s_1)}} = \mathbb{F}^\star_{2^n}$ and $b \in \mathbb{F}^\star_{2^{o(s_2)}} = \mathbb{F}^\star_4$.
Note that when $b = 0$, the function $g_{a,0}$ is never bent. Moreover, we only treat the case where m is odd since when m is even, s_1 is a Dillon exponent (since $\gcd(3, 2^m+1) = 1$ if m is even), that is, a case studied previously. In addition, by computer experiments, we have found that there exist no bent functions for n less than 16 with m even. Therefore, in the following, we assume in this subsection m odd.

Denote by $\mathfrak{G}_n$ the set of the Boolean functions $g_{a,b}$ defined on $\mathbb{F}_{2^n}$ whose polynomial form is given by the above expression (9.15). This infinite class is not contained in the class of functions $\mathfrak{F}_n$ studied in Sect. 9.5.2.1 (since 3 is a divisor of 2^m+1 (m being odd)) nor in the class studied by Charpin and Gong [4] that we have mentioned above. In the following, we present the study of bentness of elements of $\mathfrak{G}_n$. To this end, we investigate a precise characterization of such functions of $\mathfrak{G}_n$ which are hyper-bent, by giving explicit conditions on the coefficients a and b.

We firstly show that one can restrict oneself to study the bentness for some particular forms of functions belonging to $\mathfrak{G}_n$ (Lemma 9.5.27). Afterwards, we show that $\mathfrak{G}_n$ is a subclass of the well known Partial Spread class for which the bentness of its functions can be characterized by means of the Hamming weight of their restrictions to a certain set (Lemma 9.5.28). We show in Proposition 9.5.29 that bent functions of the class $\mathfrak{G}_n$ are also hyper-bent and more precisely, are (up to a linear transformation) elements of the $\mathcal{PS}_{ap}$ class. We prove in Proposition 9.5.31 and Proposition 9.5.32 that, deciding whether an element of $\mathfrak{G}_n$ is bent or not, depends strongly on the Kloosterman sums and also (in some cases) on the cubic sums involving only the coefficient a. Theorem 9.5.35 recapitulates the results of our study in which we prove that the class $\mathfrak{G}_n$ contains hyper-bent functions when $m \not\equiv 3 \pmod 6$, while there is no hyper-bent functions in this class when $m \equiv 3 \pmod 6$; an important point is that this class does not contain other bent functions except those which are hyper-bent. Finally, we show that a bent function of the class $\mathfrak{G}_n$ is normal and we compute its dual function.

The Study of the Bentness of the Binomial Family $\mathfrak{G}_n$

- First let compute the algebraic degree of functions in $\mathfrak{G}_n$.

Proposition 9.5.25 ([16]). *The elements $g_{a,b}$ of $\mathfrak{G}_n$ are all of algebraic degree m.*

Proof. Note that the 2-weights of $3(2^m - 1)$ and $\frac{2^n-1}{3}$ are both equal to m (since $3(2^m - 1) = 1 + 2^2 + 2^3 + \cdots + 2^{m-1} + 2^{m+1}$ and $\frac{2^n-1}{3} = 1 + 4 + \cdots + 4^{m-1}$). Thus, the two Boolean functions $x \mapsto Tr_1^n(ax^{3(2^m-1)})$ and $x \mapsto Tr_1^2(bx^{\frac{2^n-1}{3}})$ are of algebraic degree equal to m. The trace functions $Tr_1^n\left(ax^{3(2^m-1)}\right)$ and $Tr_1^2(bx^{\frac{2^n-1}{3}})$ are two separate parts in the trace representation of $g_{a,b}$, the algebraic degree of $g_{a,b}$ is then equal to m. □

Recall that the algebraic degree of any bent Boolean function on $\mathbb{F}_{2^n}$ is at most m (in the case that $n = 2$, the bent functions have degree 2). Bent functions of $\mathfrak{G}_n$ are then of maximum algebraic degree.

Remark 9.5.26. An integer d is called a *bent exponent* if there exists $a \in \mathbb{F}_{2^n}^\star$ for which the function $x \mapsto Tr_1^n(ax^d)$ is bent. An integer d is a bent exponent then, either $\gcd(d, 2^m - 1) = 1$ or $\gcd(d, 2^m + 1) = 1$, where $m = n/2$ (see for instance [14]). Consequently, unlike the functions $\mathfrak{F}_n$ presented above [15, 17], the monomial functions of the class $\mathfrak{G}_n$ (case $b = 0$) are never bent since the exponent $d = 3(2^m - 1)$ is not co-prime with $2^m - 1$ nor with $2^m + 1$ (because when m is odd then 3 divides $2^m + 1$).

- Now, recall (see Sect. 4.2, Chap. 4) that if f and f' are two n-variable Boolean functions such that f' is linearly equivalent to f (that is, there exists an $\mathbb{F}_2$-linear automorphism L of $\mathbb{F}_{2^n}$ such that $f' = f \circ L$) then, f is bent if and only if f' is bent.

Let $a \in \mathbb{F}_{2^m}^\star$, $\lambda \in \mathbb{F}_{2^n}^\star$ and $b \in \mathbb{F}_4^\star$. Set $a' = a\lambda^{3(2^m-1)}$ and $b' = b\lambda^{\frac{2^n-1}{3}}$. Then we remark that, for every $x \in \mathbb{F}_{2^n}$, we have:

$$g_{a',b'}(x) = \mathrm{Tr}_1^n(a(\lambda x)^{3(2^m-1)}) + \mathrm{Tr}_1^2(b(\lambda x)^{\frac{2^n-1}{3}}) = g_{a,b}(\lambda x). \tag{9.16}$$

This means that $g_{a',b'}$ is linearly equivalent to $g_{a,b}$. Consequently, we are not obliged to consider all the possible values of $a \in \mathbb{F}_{2^n}$ in our study of the bentness of an element of $\mathfrak{G}_n$. Indeed, recall that every element of x in $\mathbb{F}_{2^n}^\star$ admits a unique polar decomposition $x = uy$ where $y \in \mathbb{F}_{2^m}^\star$ and $u \in U := \{u \in \mathbb{F}_{2^n}^\star \mid u^{2^m+1} = 1\}$. Now, m being odd, one can decompose U as follows

$$U = V \cup \zeta V \cup \zeta^2 V \tag{9.17}$$

where $V = \{u^3 \mid u \in U\}$ and $\zeta = \xi^{2^m-1}$ where ξ denotes a primitive element of the field $\mathbb{F}_{2^n}$. Thus, every element $u \in U$ can be uniquely decomposed as $u = \zeta^i v$ with $i \in \{0, 1, 2\}$ and $v \in V$. Therefore, one deduces straightforwardly from (9.16) the following Lemma.

Lemma 9.5.27 ([16]). *Let $n = 2m$ with m odd. Let $a' \in \mathbb{F}_{2^n}^\star$ and $b' \in \mathbb{F}_4^\star$. Suppose that $a' = a\zeta^i v$ with $a \in \mathbb{F}_{2^m}^\star$ $i \in \{0, 1, 2\}$, ζ be a generator of the cyclic group $U := \{u \in \mathbb{F}_{2^n}^\star \mid u^{2^m+1} = 1\}$ and, $v \in V := \{u^3 \mid u \in U\}$. Then, there exists $b \in \mathbb{F}_4^\star$ such that $g_{a',b'}$ is linearly equivalent to $g_{a\zeta^i,b}$.*

Every element $a' \in \mathbb{F}_{2^n}^\star$ can be (uniquely) decomposed as $a' = a\zeta^i v$ with $a \in \mathbb{F}_{2^m}^\star$, $i \in \{0, 1, 2\}$, ζ be a generator of the cyclic group $U := \{u \in \mathbb{F}_{2^n}^\star \mid u^{2^m+1} = 1\}$ and, $v \in V := \{u^3 \mid u \in U\}$. Therefore, according to the preceding Lemma, *one can restrict oneself to study the bentness of $g_{a\zeta^i,b}$ with $a \in \mathbb{F}_{2^m}^\star$ $b \in \mathbb{F}_4^\star$.*

- Now collect the material that we have obtained to study the (hyper)-bentness of functions in $\mathfrak{G}_n$.

Lemma 9.5.28 ([16]). *Let $a \in \mathbb{F}_{2^n}^\star$ and $b \in \mathbb{F}_4^\star$. Suppose that m is odd. Then, a function $g_{a,b}$ of the family $\mathfrak{G}_n$ is bent if and only if $\Gamma(a, b) := \sum_{u \in U} \chi(g_{a,b}(u)) = 1$. Moreover, bent functions $g_{a,b}$ of the family $\mathfrak{G}_n$ belong to the Partial Spread class $\mathcal{PS}^-$.*

Proof. We have $x = yu$, with $y \in \mathbb{F}_{2^m}^\star$ and $u \in U$. Since 3 divides $2^m + 1$ when m is odd, for every $x \in \mathbb{F}_{2^n}^\star$, we have

$$g_{a,b}(x) = g_{a,b}(uy) = \mathrm{Tr}_1^n\left(au^{3(2^m-1)}\right) + \mathrm{Tr}_1^2\left(bu^{\frac{2^n-1}{3}}\right) = g_{a,b}(u). \tag{9.18}$$

The function $g_{a,b}$ is then constant on the cosets $u\mathbb{F}_{2^m}^\star$, $u \in U$. Therefore, the support of $g_{a,b}$ can be decomposed into the disjoint union sets (with the null vector, these sets are vector subspaces of dimension 2^m) as follows

$$supp(g_{a,b}) = \bigcup_{u \in S_{a,b}} u\mathbb{F}_{2^m}^\star \text{ where } S_{a,b} := \{u \in U \mid g_{a,b}(u) = 1\}. \tag{9.19}$$

This implies that the bentness of $g_{a,b}$ is equivalent to the fact that the Hamming weight of the restriction of $g_{a,b}$ to U is equal to 2^{m-1} and that bent functions $g_{a,b}$ of the class $\mathfrak{G}_n$ are in the class $\mathcal{PS}^-$. To conclude, it suffices to note that $\Gamma(a,b) = \#U - 2\,\mathrm{wt}(g_{a,b}|_U)$ ($\#U = 2^m + 1$). □

We have seen that, when m is odd, bent functions $g_{a,b}$, with $a \in \mathbb{F}_{2^n}^\star$ and $b \in \mathbb{F}_4^\star$ are in the class PS^- class. In the following, we will give a more precise statement of Lemma 9.5.28 , in particular, we will see that when m is odd, then bent functions of $\mathfrak{G}_n$ are in the class of hyper-bent functions.
Thus, according to Proposition 9.4.2, Proposition 9.4.4 and Lemma 9.5.28, one can straightforwardly deduce a more precise statement of Lemma 9.5.28.

Proposition 9.5.29 ([16]). *Let $a \in \mathbb{F}_{2^n}^\star$ and $b \in \mathbb{F}_4^\star$. Let $g_{a,b}$ be a Boolean function belonging to the family $\mathfrak{G}_n$,($n = 2m$, m odd). Then $g_{a,b}$ is hyper-bent if and only if $\Gamma(a,b) := \sum_{u \in U} \chi(g_{a,b}(u)) = 1$. Moreover, $g_{a,b}$ is in the class $\mathcal{PS}_{ap}$ if and only if $\mathrm{Tr}_1^n(a) + \mathrm{Tr}_1^2(b) = 0$.*

Proof. If α is a primitive element of $\mathbb{F}_{2^n}$ then, $g_{a,b}(\alpha^{2^m+1}x) = g_{a,b}(x)$ for every $x \in \mathbb{F}_{2^n}$ (since 3 divides $2^m + 1$ when m is odd) and 0 is not in the support of $g_{a,b}$. The conditions for bentness given by Proposition 9.4.2 are then satisfied thanks to Lemma 9.5.28. The second part of the Proposition is a direct application of Proposition 9.4.4. □

According to Lemma 9.5.27 and Proposition 9.5.29, the question of deciding whether an element $g_{a,b}$ of $\mathfrak{G}_n$ is hyper-bent or not can be reduced to computing the sum $\Gamma(a\zeta^i, \beta^j)$ for $(i,j) \in \{0,1,2\}^2$. For that, we shall use Proposition 2.4.7 in Sect. 2.4 (Chap. 2).

Now, recall that $V := \{u^3 \mid u \in U\}$. Let ζ be a generator of the cyclic group U. We introduce the sums

$$\forall a \in \mathbb{F}_{2^m}^\star,\ \forall i \in \{0,1,2\},\quad S_i(a) = \sum_{v \in V} \chi(\mathrm{Tr}_1^n(a\zeta^i v)). \tag{9.20}$$

The sums $S_i(a)$ can be expressed in terms of Kloosterman sums and cubic sums. These expressions can be obtained from Proposition 9 in [15]. For completeness, we include the proof.

Lemma 9.5.30 ([16]). *For every $a \in \mathbb{F}_{2^m}^\star$, we have:*

$$S_0(a) = \frac{1 - K_m(a) + 2C_m(a,a)}{3}, \quad S_2(a) = S_1(a) = \frac{1 - K_m(a) - C_m(a,a)}{3}.$$

Proof. Note firstly that the mapping $x \mapsto x^3$ being 3-to-1 on U, then, thanks to Lemma 2.4.7, one has

$$\sum_{v \in V} \chi(\mathrm{Tr}_1^n(av)) = \frac{1}{3} \sum_{u \in U} \chi(\mathrm{Tr}_1^n(au^3)) = \frac{1}{3}\big(1 - K_m(a) + 2C_m(a,a)\big).$$

Now, since ζ^{2^m-2} is an element of V (because 3 divides (2^m+1)) and the mapping $v \mapsto \zeta^{2^m-2}v^{2^m}$ is a permutation on V, then, we have:

$$S_1(a) = \sum_{v \in V} \chi(\mathrm{Tr}_1^n(a\zeta v)) = \sum_{v \in V} \chi(\mathrm{Tr}_1^n(a\zeta^{2^m} v^{2^m}))$$
$$= \sum_{v \in V} \chi(\mathrm{Tr}_1^n(a\zeta^2(\zeta^{2^m-2}v^{2^m}))) = S_2(a).$$

Next, using the fact that $\sum_{u \in G} \chi(\mathrm{Tr}_1^n(au)) = 1 - K_m(a)$, where G is a cyclic group of order 2^m+1, we obtain :

$$S_0(a) + S_1(a) + S_2(a) = \sum_{u \in U} \chi(\mathrm{Tr}_1^n(au)) = 1 - K_m(a).$$

Therefore, $S_1(a) = \frac{1-K_m(a)-S_0(a)}{2}$. To conclude, it suffices to note that, the mapping $x \mapsto x^3$ being 3-to-1 from U to itself, one has $\sum_{u \in U} \chi(\mathrm{Tr}_1^n(au^3)) = 3S_0(a)$ and that $\sum_{u \in U} \chi(\mathrm{Tr}_1^n(au^3)) = 1 - K_m(a) + 2C_m(a,a)$ according to Lemma 2.4.7.

□

• At this stage, we have the material to study the (hyper)-bentness of Boolean functions belonging to the family $\mathfrak{G}_n$.

Proposition 9.5.31 ([16]). *Let $n = 2m$ be an even integer with m odd. Let $a \in \mathbb{F}_{2^m}^\star$, β be a primitive element of $\mathbb{F}_4$ and ζ be a generator of the cyclic group U of (2^m+1)-st of unity. Suppose that $\mathrm{Tr}_1^m(a^{1/3}) = 0$. For $(i,j) \in \{0,1,2\}^2$, let $g_{a\zeta^i,\beta^j}$ be a Boolean function defined on $\mathbb{F}_{2^n}$ whose expression is of the form (9.15). Suppose that $m \not\equiv 3$ (mod 6). Then, $g_{a\zeta^i,\beta^j}$ is bent if and only if $K_m(a) = 4$.*

Proof. Recall that, $\Gamma(a,b)$ denotes the sum $\sum_{u \in U} \chi(g_{a,b}(u))$.
For $(i,j) \in \{0,1,2\}^2$, we have (using the fact that the mapping $u \mapsto u^{2^m-1}$ is a permutation of U)

$$\Gamma(a\zeta^i, \beta^j) := \sum_{u \in U} \chi\left(g_{a\zeta^i,\beta^j}(u)\right) = \sum_{u \in U} \chi\left(\mathrm{Tr}_1^n(a\zeta^i u^{3(2^m-1)}) + \mathrm{Tr}_1^2(\beta^j u^{\frac{2^n-1}{3}})\right)$$
$$= \sum_{u \in U} \chi\left(\mathrm{Tr}_1^n(a\zeta^i u^3) + \mathrm{Tr}_1^2(\beta^j u^{\frac{2^m+1}{3}})\right).$$

Now, thanks to (9.17), we have seen that every element $u \in U$ can be uniquely decomposed as $u = \zeta^l v$ with $l \in \{0,1,2\}$ and $v \in V := \{u^3 \mid u \in U\}$. Hence, for $(i,j) \in \{0,1,2\}^2$, we have (in the last equality, we use the fact that v is a cube of an element of U which is a group of order 2^m+1)

$$\Gamma(a\zeta^i, \beta^j) = \sum_{l=0}^{2}\sum_{v\in V}\chi\left(\mathrm{Tr}_1^n(a\zeta^{3l+i}v^3) + \mathrm{Tr}_1^2(\beta^j\zeta^{l\frac{2^m+1}{3}}v^{\frac{2^m+1}{3}})\right)$$

$$= \sum_{l=0}^{2}\sum_{v\in V}\chi\left(\mathrm{Tr}_1^n(a\zeta^{3l+i}v^3) + \mathrm{Tr}_1^2(\beta^j\zeta^{l\frac{2^m+1}{3}})\right).$$

Next, $m \not\equiv 3 \pmod 6$ then, integers 3 and $\frac{2^m+1}{3}$ are co-prime. The mapping $x \mapsto x^3$ is then a permutation of V and thus for $(i,j) \in \{0,1,2\}^2$, we have (in the last equality, we use the fact that the mapping $v \mapsto \zeta^{3l}v$ is a permutation of V)

$$\Gamma(a\zeta^i, \beta^j) = \sum_{l=0}^{2}\sum_{v\in V}\chi(\mathrm{Tr}_1^n(a\zeta^{3l+i}v) + \mathrm{Tr}_1^2(\beta^j\zeta^{l\frac{2^m+1}{3}}))$$

$$= \sum_{l=0}^{2}\sum_{v\in V}\chi(\mathrm{Tr}_1^n(a\zeta^{i}v) + \mathrm{Tr}_1^2(\beta^j\zeta^{l\frac{2^m+1}{3}})).$$

But, for every $j \in \{0,1,2\}$, the set $\{\beta^j, \beta^j\zeta^{\frac{2^m+1}{3}} \beta^j\zeta^{2\frac{2^m+1}{3}}\}$ is equal to $\mathbb{F}_4^\star$ (which contains two elements of absolute trace 1 on $\mathbb{F}_4$ and one element of absolute trace 0 on $\mathbb{F}_4$). We thus conclude that

$$\Gamma(a\zeta^i, \beta^j) = -\sum_{v\in V}\chi(\mathrm{Tr}_1^n(a\zeta^i v)) =: -S_i(a). \tag{9.21}$$

Next, since m is odd, the mapping $x \mapsto x^3$ is permutation on $\mathbb{F}_{2^m}$. Hence, every element $a \in \mathbb{F}_{2^m}$ can be (uniquely) written as $a = c^3$ with $c \in \mathbb{F}_{2^m}$. One has

$$C_m(a,a) := \sum_{x\in\mathbb{F}_{2^m}}\chi(\mathrm{Tr}_1^m(ax^3 + ax)) = \sum_{x\in\mathbb{F}_{2^m}}\chi(\mathrm{Tr}_1^m((cx)^3 + ax))$$

$$= \sum_{x\in\mathbb{F}_{2^m}}\chi(\mathrm{Tr}_1^m((cx)^3 + a^{2/3}(cx))) = \sum_{x\in\mathbb{F}_{2^m}}\chi(\mathrm{Tr}_1^m(x^3 + a^{2/3}x)) = C_m(1, a^{2/3}).$$

Now, since $\mathrm{Tr}_1^m(a^{2/3}) = \mathrm{Tr}_1^m(a^{1/3})$ and $\mathrm{Tr}_1^m(a^{1/3}) = 0$ (by hypothesis), one has $C_m(a,a) = 0$, according to Proposition 2.3.5. Therefore, thanks to Lemma 9.5.30, we obtain

$$\Gamma(a\zeta^i, \beta^j) = \frac{K_m(a) - 1}{3}.$$

We conclude thanks to Lemma 9.5.28. □

Proposition 9.5.32 ([16]). *Let $n = 2m$ be an even integer with m odd. Let $a \in \mathbb{F}_{2^m}^\star$, β be a primitive element of $\mathbb{F}_4$ and, ζ be a generator of the cyclic group U of (2^m+1)-st of unity. Suppose that $\mathrm{Tr}_1^m(a^{1/3}) = 1$. For $(i,j) \in \{0,1,2\}^2$, let $g_{a\zeta^i,\beta^j}$*

be a Boolean function on $\mathbb{F}_{2^n}$ whose expression is of the form (9.15). Assume that $m \not\equiv 3 \pmod 6$. Then

1. *The function g_{a,β^j} is not bent for every $j \in \{0, 1, 2\}$.*
2. *For every $i \in \{1, 2\}$ and $j \in \{0, 1, 2\}$, the function $g_{a\zeta^i,\beta^j}$ is bent if and only if $K_m(a) + C_m(a, a) = 4$.*

Proof. We have seen in the proof of Proposition 9.5.31, that $C_m(a, a) = C_m(1, a^{2/3})$. Then, according to Proposition 2.3.5, one has $C_m(a, a) = \epsilon_a \left(\frac{2}{m}\right) 2^{(m+1)/2}$ with $\epsilon_a = \pm 1$ (since $\mathrm{Tr}_1^m(a^{2/3}) = \mathrm{Tr}_1^m(a^{1/3})$ and $\mathrm{Tr}_1^m(a^{1/3}) = 1$, by hypothesis).

1. Let $j \in \{0, 1, 2\}$. According to (9.21), valid only if $m \not\equiv 3 \pmod 6$, and thanks to Lemma 9.5.30, we have that $\Gamma(a, \beta^j) = \frac{K_m(a)-1-\epsilon_a\left(\frac{2}{m}\right)2^{(m+3)/2}}{3}$. Then, according to Lemma 9.5.28, the Boolean function g_{a,β^j} is therefore bent if and only if $K_m(a) = 4 \pm \left(\frac{2}{m}\right) 2^{(m+3)/2}$, which is impossible for $m > 3$, since the Kloosterman sums $K_m(a)$ take values in the range $[-2^{(m+2)/2} + 1, 2^{(m+2)/2} + 1]$, according to Proposition 2.3.2.
2. According to (9.21) and Lemma 9.5.30, for every $i \in \{1, 2\}$ and $j \in \{0, 1, 2\}$, we have $\Gamma(a\zeta^i, \beta^j) = \frac{K_m(a)+C_m(a,a)-1}{3}$. The Boolean function $g_{a\zeta^i,\beta^j}$ is therefore bent if and only if $K_m(a) + C_m(a, a) = 4$, according to Lemma 9.5.28.

□

Remark 9.5.33. Since the cubic sums $C_m(a, a)$ equal $\epsilon_a \left(\frac{2}{m}\right) 2^{(m+1)/2}$ with $\epsilon_a = \pm 1$ (when $\mathrm{Tr}_1^m(a^{1/3}) = 1$, m odd) and the Jacobi symbol $\left(\frac{2}{m}\right)$ equals $(-1)^{\frac{(m^2-1)}{8}}$ (when m is odd) then, the condition $K_m(a) + C_m(a, a) = 4$ on $a \in \mathbb{F}_{2^m}^\star$ says that the Kloosterman sums $K_m(a)$ take the values $4 \pm 2^{(m+1)/2}$.

Proposition 9.5.34 ([16]). *Let $n = 2m$. Suppose that m is odd such that $m \equiv 3 \pmod 6$. Let $a \in \mathbb{F}_{2^m}^\star$, $b \in \mathbb{F}_4$ and, ζ be a generator of the cyclic group U of $(2^m + 1)$-st of unity. For $i \in \{0, 1, 2\}$, let $g_{a\zeta^i,b}$ be a Boolean function on $\mathbb{F}_{2^n}$ whose expression is of the form (9.15). Then, $g_{a\zeta^i,b}$ is not bent.*

Proof. According to Lemma 9.5.27 and Lemma 9.5.28, it suffices to compute the value $\sum_{u \in U} \chi(g_{a\zeta^i,b}(u))$ to decide whether $g_{a\zeta^i,b}$ is bent or not. Note now that (since 9 divides $2^m + 1$ if $m \equiv 3 \pmod 6$)

$$\sum_{u \in U} \chi(g_{a\zeta^i,b}(u)) = \sum_{u \in U} \chi(\mathrm{Tr}_1^n(a\zeta^i u^{3(2^m-1)}) + \mathrm{Tr}_1^2(bu^{3(2^m-1)\cdot\frac{2^m+1}{9}})).$$

The mapping $x \mapsto x^{3(2^m-1)}$ is 3-to-1 from U to itself. Thus, we get that

$$\sum_{u \in U} \chi(g_{a\zeta^i,b}(u)) = 3 \sum_{v \in V} \chi(\mathrm{Tr}_1^n(a\zeta^i v) + \mathrm{Tr}_1^2(bv^{\frac{2^m+1}{9}}))$$

where $V = \{u^3 \mid u \in U\}$. The sum $\sum_{u \in U} \chi(g_{a\zeta^i,b}(u))$ is therefore a multiple of 3 and cannot be equal to 1 implying that $g_{a\zeta^i,b}$ cannot be bent. □

Collecting the results obtained in Proposition 9.5.31, Proposition 9.5.32 and Proposition 9.5.34 we obtain the following characterization of the bentness for Boolean function of the form (9.15).

Theorem 9.5.35 ([16]). *Let $n = 2m$. Suppose that m is odd. Let $a \in \mathbb{F}_{2^m}^\star$. Let β be a primitive element of $\mathbb{F}_4$. For $(i,j) \in \{0,1,2\}^2$, let $g_{a\xi^i,\beta^j}$ be a Boolean function on $\mathbb{F}_{2^n}$ whose expression is of the form (9.15).*

1. Assume $m \not\equiv 3 \pmod 6$. Then, we have:

- *If $\mathrm{Tr}_1^m(a^{1/3}) = 0$ then, for every $(i,j) \in \{0,1,2\}^2$, a function $g_{a\xi^i,\beta^j}$ is (hyper)-bent if and only if $K_m(a) = 4$.*
- *If $\mathrm{Tr}_1^m(a^{1/3}) = 1$ then:*

  (a) g_{a,β^j} is not bent for every $j \in \{0,1,2\}$.
  (b) For every $i \in \{1,2\}$, $g_{a\xi^i,\beta^j}$ is (hyper)-bent if and only if $K_m(a) + C_m(a,a) = 4$.

2. Assume $m \equiv 3 \pmod 6$. Then, for every $i \in \{0,1,2\}$, $g_{a\xi^i,b}$ is not bent for every $a \in \mathbb{F}_{2^m}^\star$ and $b \in \mathbb{F}_4^\star$.

Example 9.5.36. Let us describe for example the set of bent functions $g_{a,b}$ belonging to the class $\mathfrak{G}_{10}$ (with $b \neq 0$), that is, of the form $\mathrm{Tr}_1^{10}(ax^{93}) + \mathrm{Tr}_1^2(bx^{341})$ where $a \in \mathbb{F}_{2^{10}}^\star$ and $b \in \mathbb{F}_4^\star$.
Let α be a primitive element of $\mathbb{F}_{32} = \mathbb{F}_2(\alpha)$ with $\alpha^5 + \alpha^2 + 1 = 0$. Let ξ be a primitive element of $\mathbb{F}_{2^{10}}$. According to table 4 in [5], the set $\{a \in \mathbb{F}_{2^5}^\star, \mathrm{Tr}_1^5(a^{1/3}) = 0\}$ is equal to $\{\alpha^3, \alpha^{21}, \alpha^{14}\}$ and, the set $\{a \in \mathbb{F}_{2^5}^\star, \mathrm{Tr}_1^5(a^{1/3}) = 1\}$ is equal to $\{1, \alpha^2, \alpha^9, \alpha^{15}\}$. The elements a of $\mathbb{F}_{2^5}^\star$ whose the Kloosterman sums $K_5(a)$ on $\mathbb{F}_{2^5}$ equals 4 (those elements a satisfy necessary $\mathrm{Tr}_1^5(a^{1/3}) = 0$) are α^3 and α^{21} while, those such that $K_5(a) + C_5(a,a) = 4$ are 1 and α^9 (more precisely, we have $K_5(1) = 12$ and $K_5(\alpha^9) = -4$).
According to Theorem 9.5.35 and Lemma 9.5.27 we conclude that there exist 330 hyper-bent Boolean functions defined on the field $\mathbb{F}_{2^{10}}$ belonging to the class $\mathfrak{G}_{10}$ (with $b \neq 0$). Such functions are $g_{\alpha^3 v,b}$, $g_{\alpha^{21} v,b}$, $g_{\xi^{31} v,b}$, $g_{\alpha^3\xi^{31} v,b}$, $g_{\alpha^9\xi^{31} v,b}$, $g_{\alpha^{21}\xi^{31} v,b}$, $g_{\xi^{62} v,b}$, $g_{\alpha^3\xi^{62} v,b}$, $g_{\alpha^9\xi^{62} v,b}$, $g_{\alpha^{21}\xi^{62} v,b}$, with $b \in \mathbb{F}_4^\star$ and v runs through the set $\{u^3 \mid u \in U\}$ where U is the cyclic group of 33-rd roots of unity of $\mathbb{F}_{2^{10}}$.

Example 9.5.37. Let $n = 14$ then, according to table 4 in [5], we find that there exist 1935 hyper-bent Boolean functions $g_{a,b}$ (with $b \neq 0$) defined on the field $\mathbb{F}_{16384}$ belonging to the class $\mathfrak{G}_{14}$. Such functions are of the form

- $\mathrm{Tr}_1^{14}(cvx^{381}) + \mathrm{Tr}_1^2(bx^{5461})$, $c \in \{\alpha^{14}, \alpha^{15}, \alpha^{62}\}$,
- $\mathrm{Tr}_1^{14}(c'\xi^{127i}vx^{381}) + \mathrm{Tr}_1^2(bx^{5461})$, $i \in \{1,2\}$, $c' \in \{1, \alpha^{14}, \alpha^{15}, \alpha^{21}, \alpha^{62}, \alpha^{93}\}$,

where α is a primitive element of $\mathbb{F}_{128}$ satisfying $\alpha^7 + \alpha^3 + 1 = 0$, ξ is a primitive element of $\mathbb{F}_{2^{14}}$, v runs through the set $\{u^3 \mid u \in U\}$ where U is the cyclic group of 129-st roots of unity of $\mathbb{F}_{2^{14}}$ and $b \in \{1, \beta, \beta^2\}$ where β is a primitive element of $\mathbb{F}_4$.

Example 9.5.38. Let $n = 18$ then, according to Theorem 9.5.35, there exist no bent Boolean functions in the class $\mathfrak{G}_{18}$.

Now, the dual functions of elements of $\mathfrak{G}_n$ can be explicitly computed as follows.

Proposition 9.5.39 ([16]). *Let $n = 2m$ with m odd. Let $(a, b) \in \mathbb{F}_{2^n}^\star \times \mathbb{F}_4^\star$. The dual function of a bent function $g_{a,b}$ of $\mathfrak{G}_n$ is equal $g_{a^{2^m},b^2}$, that is, we have*

$$\forall \omega \in \mathbb{F}_{2^n}, \quad \widehat{\chi_{g_{a,b}}}(\omega) = 2^m \chi(g_{a^{2^m},b^2}(\omega)).$$

Proof. The arguments are for the most part the same as those used in [17]. Nevertheless, for the sake of completeness, we present below a little shorter proof. Since the function $g_{a,b}$ is assumed to be bent then, according to Lemma 9.5.28, $\sum_{u\in U} \chi(g_{a,b}(u)) = 1$. Given, $\omega \in \mathbb{F}_{2^n}$, since every element x of $\mathbb{F}_{2^n}^\star$ has a unique decomposition as : $x = yu$, with $y \in \mathbb{F}_{2^m}^\star$ and $u \in U$, one has (in the last equality, we use (9.18))

$$\begin{aligned}
\widehat{\chi}_{g_{a,b}}(w) &:= \sum_{x\in\mathbb{F}_{2^n}} \chi(g_{a,b}(x) + \mathrm{Tr}_1^n(wx)) \\
&= 1 + \sum_{u\in U} \sum_{y\in\mathbb{F}_{2^m}^\star} \chi(g_{a,b}(yu) + \mathrm{Tr}_1^n(wyu)) \\
&= 1 + \sum_{u\in U} \sum_{y\in\mathbb{F}_{2^m}^\star} \chi(g_{a,b}(u) + \mathrm{Tr}_1^n(wyu)).
\end{aligned}$$

Note first that $\widehat{\chi_{g_{a,b}}}(0) = 2^m$. Now, if w is an element of $\mathbb{F}_{2^n}^\star$, we have $\mathrm{Tr}_m^n(wu) = 0$ if and only if $uw + u^{2^m} w^{2^m} = 0$, that is, $u^{2^m-1} = w^{1-2^m}$. Classical results about character sums say that

$$\sum_{y\in\mathbb{F}_{2^m}} \chi(\mathrm{Tr}_1^n(\omega uy)) = \sum_{y\in\mathbb{F}_{2^m}} \chi(\mathrm{Tr}_1^m(\mathrm{Tr}_m^n(\omega u)y)) = 2^m$$

if $\mathrm{Tr}_m^n(\omega u) = 0$, that is, if $u^{2^m-1} = \omega^{1-2^m}$ and, is equal to 0 otherwise. Hence, using properties of trace functions, we have

$$\begin{aligned}
\widehat{\chi}_{g_{a,b}}(w) &= 1 + \sum_{u\in U} \chi(g_{a,b}(u)) \left(\sum_{y\in\mathbb{F}_{2^m}} \chi(\mathrm{Tr}_1^n(wyu)) - 1 \right) \\
&= 1 - \sum_{u\in U} \chi(g_{a,b}(u)) + 2^m \sum_{\substack{u\in U \\ \mathrm{Tr}_m^n(wu)=0}} \chi(g_{a,b}(u)) \\
&= 2^m \chi(\mathrm{Tr}_1^n(aw^{3(1-2^m)}) + \mathrm{Tr}_1^2(bw^{\frac{1-2^n}{3}})) \\
&= 2^m \chi(\mathrm{Tr}_1^n(a^{2^m} w^{3(2^m-1)}) + \mathrm{Tr}_1^2(b^{2^m} w^{\frac{2^n-1}{3}})) \\
&= 2^m \chi(\mathrm{Tr}_1^n(a^{2^m} w^{3(2^m-1)}) + \mathrm{Tr}_1^2(b^2 w^{\frac{2^n-1}{3}})) = 2^m \chi(g_{a^{2^m},b^2}(\omega))
\end{aligned}$$

($b^{2^m-2} = 1$ because m is being odd then, 3 divides (2^m+1) and thus divides (2^m-2)). □

Recall that a bent function defined on $\mathbb{F}_{2^n}$ is said to be *normal* if it is constant on an $\frac{n}{2}$-dimensional flat $b+E$ where E is a subspace of $\mathbb{F}_{2^{\frac{n}{2}}}$.

Proposition 9.5.40 ([16]). *The bent functions $g_{a,b}$ of $\mathfrak{G}_n$ (where $n = 2m$ with m odd) are normal.*

Proof. $g_{a,b}$ is constant on each set $u\mathbb{F}_{2^m}^\star$, $u \in U = \{x \in \mathbb{F}_{2^n}^\star \mid x^{2^m+1} = 1\}$. Choose u such that $g_{a,b}(u) = 0$. Then $g_{a,b}$ is constant of the vectorspace $u\mathbb{F}_{2^m}$ (of dimension m) proving that $g_{a,b}$ is normal. □

Remark 9.5.41. By computer experiments, for small values of n ($n \leq 14$, because of the complexity of the problem) we have found that, the family $\mathfrak{G}_n$ does not contain bent functions when $m = \frac{n}{2}$ is even.

The following theorem summarizes the results presented above related to the bentness of the functions $g_{a,b}$ of the family $\mathfrak{G}_n$.

Theorem 9.5.42 ([16]). *Let $n = 2m$ with m odd. Let $a \in \mathbb{F}_{2^m}^\star$ and $b \in \mathbb{F}_4^\star$. Let β be a primitive element of $\mathbb{F}_4$. Let $g_{a,b}$ be the function defined on $\mathbb{F}_{2^n}$ by (9.15).*

1. *The algebraic degree of $g_{a,b}$ equals m (bent functions $g_{a,b}$ are then of maximal algebraic degree).*
2. *$g_{a,b}$ is hyper-bent if and only if $g_{a,b}$ is bent.*
3. *If $g_{a,b}$ is bent then its dual function equals $g_{a^{2^m},b^2}$.*
4. *The bent functions $g_{a,b}$ are in the class $\mathcal{PS}^-$. Moreover, the bent functions $g_{a,b}$ are elements of the Partial Spread class $\mathcal{PS}_{ap}$ (resp. $\mathcal{PS}_{ap}^\#$) if $b = 1$ (resp. if $b \neq 1$).*

Moreover,

*) *Assume $m \not\equiv 3 \pmod 6$.*

- *If $\mathrm{Tr}_1^m(a^{1/3}) = 0$ then, for every $(i,j) \in \{0,1,2\}^2$, $g_{a\zeta^i,\beta^j}$ is bent if and only if $K_m(a) = 4$.*
- *If $\mathrm{Tr}_1^m(a^{1/3}) = 1$ then*
  - a) *g_{a,β^j} is not bent for every $j \in \{0,1,2\}$.*
  - b) *for every $i \in \{1,2\}$, $j \in \{0,1,2\}$, $g_{a\zeta^i,\beta^j}$ is bent if and only if $K_m(a) + C_m(a,a) = 4$.*

**) *Assume $m \equiv 3 \pmod 6$. Then, for every $i \in \{0,1,2\}$, $g_{a\zeta^i,b}$ is not bent.*

9.5.2.5 A Third Family of Binomial Hyper-Bent Functions

Adopting our approach [16, 17] (developed in the previous sections) Wang et al. studied in late 2011 the hyper-bentess of the following binomial family [23, 24] with an additional trace term on $\mathbb{F}_{16}$:

$$f_{a,b}(x) = \mathrm{Tr}_1^n\left(ax^{r(2^m-1)}\right) + \mathrm{Tr}_1^4\left(bx^{\frac{2^n-1}{5}}\right)$$

where the coefficients a are in $\mathbb{F}_{2^m}$ ($m = \frac{n}{2}$), the coefficient b is in $\mathbb{F}_{16}$ and m must verify $m \equiv 2 \pmod 4$. They characterize the hyper-bentness where $r \equiv 0 \pmod 5$ and in the case where $r \not\equiv 0 \pmod 5$ and $(b+1)(b^4+b+1) = 0$ in terms of Kloosterman sums and using the factorization of $x^5 + x + a^{-1}$. We summarize their result in the following theorem

Theorem 9.5.43 ([23, 24]). *Let $n = 2m$ and $m = 2m_1$ with $m_1 \equiv 2 \pmod 4$. and $m_1 \geq 3$. Let $a \in \mathbb{F}_{2^m}^\star$ and $b \in \mathbb{F}_{16}^\star$. Let $f_{a,b}^{(r)}$ be the function defined on $\mathbb{F}_{2^n}$ by*

$$f_{a,b}(x) = \mathrm{Tr}_1^n\left(ax^{2^m-1}\right) + \mathrm{Tr}_1^4\left(bx^{\frac{2^n-1}{5}}\right).$$

1. *If $b = 1$ then $f_{a,1}$ is hyper-bent iff $p(X) = X^5 + X + a^{-1}$ is irreducible over $\mathbb{F}_{2^m}$ and the quadratic form $q(x) = \mathrm{Tr}_1^m\left(x(ax^4 + ax^2 + a^2x)\right)$ over $\mathbb{F}_{2^m}$ is even and $K_m(a) = \frac{4}{3}(2 - 2^{m_1})$.*
2. *if b is a primitive element of $\mathbb{F}_{16}^\star$ such that $\mathrm{Tr}_1^4(b) = 0$ then $f_{a,b}$ is hyper-bent iff $p(X) = X^5 + X + a^{-1}$ is irreducible over $\mathbb{F}_{2^m}$, the quadratic form $q(x) = \mathrm{Tr}_1^m(x(ax^4 + ax^2 + a^2x))$ over $\mathbb{F}_{2^m}$ is even and $K_m(a) = 2 \cdot 2^{m_1} - 4$.*

Moreover, they give all the hyper-bent functions in the case where $a \in \mathbb{F}_{2^{\frac{m}{2}}}$. The reader can refer to the following references of the authors [23, 24].

9.6 Hyper-Bent Functions from Boolean Functions with the Walsh Spectrum Taking the Same Value Twice

As we have seen in the previous sections, most known hyper-bent functions are functions with Dillon exponents and are often characterized by special values of Kloosterman sums. In [22] Tang and Qi have presented a method for characterizing hyper-bent functions over $\mathbb{F}_{2^n}$ with Dillon exponents of the form $c(2^m - 1)$ (where $m = \frac{n}{2}$). More precisely, the authors have provided a method for characterizing hyper-bent functions over $\mathbb{F}_{2^n}$ by a function over $\mathbb{F}_{2^m}$, whose Walsh spectrum takes the same value twice. To this end, they introduced a function f over $\mathbb{F}_{2^n}$ constructed by means of a function g defined over $\mathbb{F}_{2^m}$. Such a function f is defined as follows:

$$f(x) = g\Big(\frac{1}{\lambda_1 + \lambda_2} \cdot \frac{1}{x^{2^m-1} + x^{1-2^m}}\Big) + \mathrm{Tr}_1^m\Big(\frac{\lambda_i}{\lambda_1 + \lambda_2} \cdot \frac{1}{x^{2^m-1} + x^{1-2^m}}\Big) \qquad (9.22)$$

where $\lambda_i \in \mathbb{F}_{2^m}$ ($i = 1, 2$) with $\lambda_1 \neq \lambda_2$. The hyper-bentness of $f(x)$ is therefore characterized by the same Walsh–Hadamard coefficient of $g(y)$ as the following theorem shows (obtained thanks to a result given in [17] on special 2-to-1 mappings).

Theorem 9.6.1 ([22]). *Let f be defined by (9.22) and set $g(0) = 0$. Then the function f is hyper-bent if and only if $\widehat{\chi_g}(\lambda_1) = \widehat{\chi_g}(\lambda_2)$.*

Theorem 9.6.1 offers a method to find hyper-bent functions of the form (9.22). On the Walsh spectra of $g(y)$ there are many existing results, which can be used to find two different elements λ_1 and λ_2 satisfying $\widehat{\chi_g}(\lambda_1) = \widehat{\chi_g}(\lambda_2)$. From the proper choice of a Boolean function $g(y)$, λ_1 and λ_2 a lot of hyper-bent functions $f(x)$ can be obtained.

Tang and Qi [22] obtained several classes of hyper-bent functions from some common functions g over $\mathbb{F}_{2^m}$:

- $g(y) = \mathrm{Tr}_1^m(ay^{-d})$;
- $g(y) = \mathrm{Tr}_1^m(y)$;
- $g(y) = \mathrm{Tr}_1^m(\frac{1}{y})$;
- $g(y) = \mathrm{Tr}_1^m(y^{2^{t-1}-1})$.

Kloosterman sum identities and Walsh spectra of the above common functions are then used to characterize several classes of hyper-bent functions.

References

1. R. Bradshaw, C. Citro, and D. S. Seljebotn. Cython: the best of both worlds. *CiSE 2011 Special Python Issue*, page 25, 2010.
2. R. P. Brent, P. Gaudry, E. Thomé, and P. Zimmermann. Faster multiplication in GF(2)[x]. In A. J. van der Poorten and A. Stein, editors, *ANTS*, volume 5011 of *Lecture Notes in Computer Science*, pages 153–166. Springer, 2008.
3. C. Carlet and P. Gaborit. Hyperbent functions and cyclic codes. In *Journal of Combinatorial Theory, Series A, vol 113, no. 3*, pages 466–482, 2006.
4. P. Charpin and G. Gong. Hyperbent functions, Kloosterman sums and Dickson polynomials. In *IEEE Trans. Inform. Theory (54) 9*, pages 4230–4238, 2008.
5. P. Charpin, T. Helleseth, and V. Zinoviev. The divisibility modulo 24 of Kloosterman sums of $GF(2^m)$, m odd. *Journal of Combinatorial Theory, Series A*, 114:322–338, 2007.
6. P. Charpin, T. Helleseth, and V. Zinoviev. Divisibility properties of Kloosterman sums over finite fields of characteristic two. In *ISIT 2008, Toronto, Canada, July 6–11*, pages 2608–2612, 2008.
7. G. Cohen, I. Honkala, S. Litsyn, and A. Lobstein. Covering codes. In *North Holland*, 1997.
8. J. Dillon. Elementary Hadamard difference sets. In *PhD dissertation, University of Maryland.*
9. J.-G. Dumas, T. Gautier, P. Giorgi, J.-L. Roch, and G. Villard. *Givaro-3.2.13rc1: C++ library for arithmetic and algebraic computations*, September 2008. http://ljk.imag.fr/CASYS/LOGICIELS/givaro/.
10. F. J. MacWilliams and N. J. Sloane. The theory of error-correcting codes. In *Amsterdam, North Holland*, 1977.
11. J-P Flori and S. Mesnager. Dickson polynomials, hyperelliptic curves and hyper-bent functions. In *7th International conference SETA 2012, LNCS 7280, Springer*, pages 40–52, 2012.
12. G. Gong and S. W. Golomb. Transform domain analysis of DES. *IEEE Transactions on Information Theory*, 45(6):2065–2073, 1999.
13. G. Lachaud and J. Wolfmann. The weights of the orthogonals of the extended quadratic binary Goppa codes. In *IEEE Trans. Inform. Theory 36 (3)*, pages 686–692, 1990.
14. G. Leander. Monomial Bent Functions. In *IEEE Trans. Inform. Theory (52) 2*, pages 738–743, 2006.
15. S. Mesnager. A new class of bent Boolean functions in polynomial forms. In *Proceedings of international Workshop on Coding and Cryptography, WCC 2009*, pages 5–18, 2009.

16. S. Mesnager. A new family of hyper-bent Boolean functions in polynomial form. In *Proceedings of Twelfth International Conference on Cryptography and Coding, Cirencester, United Kingdom. M. G. Parker (Ed.): IMACC 2009, LNCS 5921, Springer, Heidelberg*, pages 402–417, 2009.
17. S. Mesnager. A new class of bent and hyper-bent Boolean functions in polynomial forms. In *journal Design, Codes and Cryptography, 59(1–3)*, pages 265–279, 2011.
18. S. Mesnager. Bent and hyper-bent functions in polynomial form and their link with some exponential sums and Dickson polynomials. *IEEE Transactions on Information Theory*, 57(9):5996–6009, 2011.
19. S. Mesnager. Bent functions from spreads. In *Journal of the American Mathematical Society (AMS), Contemporary Mathematics (Proceedings the 11th International conference on Finite Fields and their Applications Fq11), Volume 632, page 295–316*, 2015.
20. V. Shoup. NTL 5.4.2: A library for doing number theory, March 2008. www.shoup.net/ntl.
21. W. A. Stein et al. *Sage Mathematics Software (Version 4.7)*. The Sage Development Team, 2011. http://www.sagemath.org.
22. C. Tang and Y. Qi. Constructing hyper-bent functions from Boolean functions with the Walsh spectrum taking the same value twice. In *Proceedings SETA "Sequences and Their Applications", Lecture Notes in Computer Science Volume 8865, pages 60–71*, 2014.
23. B. Wang, C. Tang, Y. Qi, and Y. Yang. A generalization of the class of hyper-bent Boolean functions in binomial forms. Cryptology ePrint Archive, Report 2011/698, 2011. http://eprint.iacr.org/.
24. B. Wang, C. Tang, Y. Qi, Y. Yang, and M. Xu. A new class of hyper-bent Boolean functions in binomial forms. *CoRR*, abs/1112.0062, 2011.
25. A. M. Youssef and G. Gong. Hyper-bent functions. In Birgit Pfitzmann, editor, *EUROCRYPT*, volume 2045 of *Lecture Notes in Computer Science*, pages 406–419. Springer, 2001.

Chapter 10
Hyper-Bent Functions: Primary Constructions with Multiple Trace Terms

10.1 Hyper-Bent Functions with Multiple Trace Terms via Dillon (Like) Exponents: The Charpin and Gong Family

Let E' be a set of representatives of the cyclotomic cosets modulo $2^m + 1$ for which each coset has the maximal size n. Let f_{a_r} be the function defined on $\mathbb{F}_{2^n}$ by

$$f_{a_r}(x) = \sum_{r \in R} \mathrm{Tr}_1^n(a_r x^{r(2^m-1)}) \tag{10.1}$$

where $a_r \in \mathbb{F}_{2^n}$ and $R \subseteq E'$. Charpin and Gong [1] have studied the bentness of the class of Boolean functions f_{a_r} defined on $\mathbb{F}_{2^n}$ by (10.1) and denoted by $\mathcal{F}_n$ in the case when all the coefficients a_r are in $\mathbb{F}_{2^m}$.

They introduced a tool by means of Dickson polynomials to describe hyper-bent functions f_{a_r}. In particular, when r is co-prime with $2^m + 1$, the functions f_{a_r} are the sum of several Dillon monomial functions; the link between the Dillon monomial hyper-bent functions and the zeros of some Kloosterman sums has been generalized to a link between hyper-bent functions f_{a_r} of this class and some exponential sums where Dickson polynomials are involved. More precisely, one has the following result.

Theorem 10.1.1 ([1]). *Let f_{a_r} be the function defined on $\mathbb{F}_{2^n}$ by (10.1) where $a_r \in \mathbb{F}_{2^m}$. Let g_{a_r} be the related Boolean function defined on $\mathbb{F}_{2^m}$ by $g_{a_r}(x) = \sum_{r \in R} \mathrm{Tr}_1^m(a_r D_r(x))$, where $D_r(x)$ is the Dickson polynomial of degree r. Then f_{a_r} is hyper-bent if and only if $\sum_{x \in \mathbb{F}_{2^m}} \chi(\mathrm{Tr}_1^m(x^{-1}) + g_{a_r}(x)) = 2^m - 2\,\mathrm{wt}(g_{a_r})$.*

By Theorem 10.1.1, Charpin and Gong have characterized the class of binomial hyperbent functions whose expression is of the form $\mathrm{Tr}_1^n\left(a\left(x^{(2^r-1)(2^m-1)} + x^{(2^r+1)(2^m-1)}\right)\right)$, where $a \in \mathbb{F}_{2^m}^\star$ and r is an integer such that $0 < r < m$ and $\{2^r - 1, 2^r + 1\} \subset E'$ (note that the functions of type (9.10) do not belong to this class). Continuing their interesting approach, Gologlu [8] has identified some trace

S. Mesnager, *Bent Functions*, DOI 10.1007/978-3-319-32595-8_10

representation of some hyper-bent functions and proved that the following functions defined on $\mathbb{F}_{2^n}$, are hyper-bent:

- $x \mapsto \sum_{i=1}^{2^{m-1}-1} \mathrm{Tr}_1^n\left(\beta x^{i(2^m-1)}\right)$, $\beta \in \mathbb{F}_{2^m} \setminus \mathbb{F}_2$.
- $x \mapsto \sum_{i=1}^{2^{m-2}-1} \mathrm{Tr}_1^n\left(\beta x^{i(2^m-1)}\right)$ where, m odd and $\beta^{(2^m-4)^{-1}} \in \{x \in \mathbb{F}_{2^m}^\star \mid \mathrm{Tr}_1^m(x) = 0\}$.

10.2 Hyper-Bent Functions with Multiple Trace Terms via Dillon (Like) Exponents: The Family $\mathfrak{H}_n$

In the sequel, n is an even positive integer, $m = \frac{n}{2}$ is an odd integer and E is a set of representatives of the cyclotomic classes modulo $2^n - 1$ for which each class has full size n. We denote by $\mathfrak{H}_n$ the set of Boolean functions $f_{a_r,b}$ defined on $\mathbb{F}_{2^n}$ whose polynomial forms are:

$$f_{a_r,b}(x) := \sum_{r \in R} \mathrm{Tr}_1^n(a_r x^{r(2^m-1)}) + \mathrm{Tr}_1^2(b x^{\frac{2^n-1}{3}}). \tag{10.2}$$

where $R \subseteq E$, all the coefficients a_r are in $\mathbb{F}_{2^m}$ and $b \in \mathbb{F}_4^\star$.

The size of the cyclotomic coset of 2 modulo $2^n - 1$ containing $\frac{2^n-1}{3}$ is equal to 2 (i.e. $o(\frac{2^n-1}{3}) = 2$) and that, the function $f_{a_r,b}$ does not belong to the class considered by Charpin and Gong [1] in Sect. 10.1.

In the following, we show that hyper-bent functions of $\mathfrak{H}_n$ can be described by means of exponential sums involving Dickson polynomials (Theorem 10.2.10 and Theorem 10.2.8). In particular, when b is a primitive element of $\mathbb{F}_4$, we provide a way to transfer the characterization of hyper-bentness of an element of $\mathfrak{H}_n$ to the evaluation of the Hamming weight of some Boolean functions. As a first illustration, we show that the results presented in the binomial case [15, 16] can be deduced. Finally, in the end of the subsection we provide a possibly new infinite family of hyper-bent functions provided that some sets are not empty (Conjecture 10.2.15 and Conjecture 10.2.17).

Study of the Bentness of the Family with Multiple Trace Terms $\mathfrak{H}_n$

For m odd, $2^m + 1$ is a multiple of 3 and thus all exponents for x in (10.2) are multiples of $2^m - 1$. Therefore, every function $f_{a_r,b}$ in $\mathfrak{H}_n$ satisfies

$$\forall x \in \mathbb{F}_{2^n}, \quad f_{a_r,b}(\alpha^{2^m+1} x) = f_{a_r,b}(x).$$

where α denotes any primitive element of $\mathbb{F}_{2^n}$. Furthermore, since every $f_{a_r,b}$ of $\mathfrak{H}_n$ vanishes at 0, one can apply Proposition 9.4.2 to get the following characterization of hyper-bentness for an element of $\mathfrak{H}_n$.

Proposition 10.2.1. *Let $f_{a_r,b} \in \mathfrak{F}_n$. Set $\Lambda(f_{a_r,b}) := \sum_{u\in U} \chi(f_{a_r,b}(u))$ where $U = \{x \in \mathbb{F}_{2^n} \mid x^{2^m+1} = 1\}$. Then, $f_{a_r,b}$ is hyper-bent if and only if $\Lambda(f_{a_r,b}) = 1$. Moreover, a hyper-bent function $f_{a_r,b}$ is in the Partial Spread class PS_{ap} if and only if $b \in \mathbb{F}_2$.*

Proof. The Boolean function $f_{a_r,b}$ satisfies the assumptions of Proposition 9.4.2. Therefore $f_{a_r,b}$ is hyper-bent if and only if its restriction to U has Hamming weight 2^{m-1}, according to Proposition 9.4.2. Now, one has $\Lambda(f_{a_r,b}) = 2^m + 1 - 2|\{u \in U \mid f_{a_r,b}(u) = 1\}|$. Therefore, the Hamming weight of the restriction of $f_{a_r,b}$ to U equals 2^{m-1} if and only if $\Lambda(f_{a_r,b}) = 1$. The second part of the proposition is a direct application of Proposition 9.4.4. Indeed, note that $f_{a_r,b}(1) = \sum_{r\in R} \mathrm{Tr}_1^n(a_r) + \mathrm{Tr}_1^2(b) = \mathrm{Tr}_1^2(b)$ (since $\mathrm{Tr}_1^n(a_r) = 0$ for every $r \in R$ because $a_r \in \mathbb{F}_{2^m}$) and it is clear that the elements b of $\mathbb{F}_4$ whose trace over $\mathbb{F}_4$ equals 0, are the elements of $\mathbb{F}_2$. □

We are interested in characterizing the hyper-bentness of the Boolean functions of the form (10.2). To this end, we introduce some additional notation while underlining some facts.

Let β be a primitive element of $\mathbb{F}_4$. Suppose that $\beta = \alpha^{\frac{2^n-1}{3}}$ for some primitive element α of $\mathbb{F}_{2^n}$. Set $\xi := \alpha^{2^m-1}$ so that ξ is a generator of the cyclic group $U := \{u \in \mathbb{F}_{2^n} \mid u^{2^m+1} = 1\}$. Note that U can be decomposed as : $U = \bigcup_{i=0}^{2} \xi^i V$ where $V := \{u^3,\ u \in U\}$. Next, let us introduce the sums

$$S_i := \sum_{v\in V} \chi(f_{a_r,0}(\xi^i v)), \quad \forall i \in \{0,1,2\} \tag{10.3}$$

First of all, note that

$$S_0 + S_1 + S_2 = \sum_{u\in U} \chi(f_{a_r,0}(u)). \tag{10.4}$$

Next, one has

Lemma 10.2.2 ([13]). $S_1 = S_2$.

Proof. Since the trace map is invariant under the Frobenius automorphism $x \mapsto x^2$, we get applying m times the Frobenius automorphism : $\forall x \in \mathbb{F}_{2^n}$,

$$f_{a_r,0}(x) = \sum_{r\in R} \mathrm{Tr}_1^n\left(a_r^{2^m} x^{2^m r(2^m-1)}\right) = \sum_{r\in R} \mathrm{Tr}_1^n\left(a_r x^{2^m r(2^m-1)}\right) = f_{a_r,0}(x^{2^m})$$

because all the coefficients a_r are in $\mathbb{F}_{2^m}$. Hence,

$$S_1 = \sum_{v\in V} \chi(f_{a_r,0}(\xi^{2^m} v^{2^m})) = \sum_{v\in V} \chi(f_0(\xi^2(\xi^{2^m-2} v^{2^m}))).$$

Now, since m is odd, 3 divides $2^m + 1$ and then divides $2^m - 2$. Hence, ξ^{2^m-2} is a cube of an element of U and the mapping $v \mapsto \xi^{(2^m-2)}v^{2^m}$ is a permutation of V. Consequently, $S_1 = \sum_{v \in V} \chi(f_0(\xi^2 v)) = S_2$. □

Now, for $b \in \mathbb{F}_4^\star$, we establish expressions for $\Lambda(f_{a_r,b}) := \sum_{u\in U} \chi(f_{a_r,b}(u))$ (where U is the group of $(2^m + 1)$-st roots of unity) involving the sums S_i.

Proposition 10.2.3 ([13]). $\Lambda(f_{a_r,\beta}) = \Lambda(f_{a_r,\beta^2}) = -S_0$ *and* $\Lambda(f_{a_r,1}) = S_0 - 2S_1$.

Proof. Introduce for every element c of $\mathbb{F}_4$ $T(c) := \sum_{b\in\mathbb{F}_4} \Lambda(f_{a_r,b})\chi(\mathrm{Tr}_1^2(bc))$. Recall that one has

$$\Lambda(f_b) = \frac{1}{4}\sum_{c\in\mathbb{F}_4} T(c)\chi(\mathrm{Tr}_1^2(bc)). \tag{10.5}$$

Indeed

$$\begin{aligned}&\sum_{c\in\mathbb{F}_4} T(c)\chi(\mathrm{Tr}_1^2(bc))\\ &= \sum_{c\in\mathbb{F}_4}\sum_{d\in\mathbb{F}_4} \Lambda(f_d)\chi(\mathrm{Tr}_1^2(dc))\chi(\mathrm{Tr}_1^2(bc))\\ &= \sum_{d\in\mathbb{F}_4} \Lambda(f_d)\sum_{c\in\mathbb{F}_4} \chi(\mathrm{Tr}_1^2(c(d+b))).\end{aligned}$$

But $\sum_{c\in\mathbb{F}_4} \chi(\mathrm{Tr}_1^2(c(d+b))) = 4$ if $d = b$ (i.e $b + d = 0$) and 0 otherwise. Then, one gets

$$\sum_{c\in\mathbb{F}_4} T(c)\chi(\mathrm{Tr}_1^2(bc)) = 4\Lambda(f_b).$$

Now, note that $T(c) = \sum_{u\in U} \chi(f_0(u)) \sum_{b\in\mathbb{F}_4} \chi\left(\mathrm{Tr}_1^2\left(b\left(c+u^{\frac{2^n-1}{3}}\right)\right)\right)$. Furthermore, one has

$$\sum_{b\in\mathbb{F}_4} \chi\left(\mathrm{Tr}_1^2\left(b\left(c+u^{\frac{2^n-1}{3}}\right)\right)\right) = 0 \text{ if } u^{\frac{2^n-1}{3}} \neq c \text{ and 4 otherwise.}$$

Since, $u^{\frac{2^n-1}{3}} \neq 0$ for every $u \in U$, $T(0) = 0$. Since β is a primitive element of $\mathbb{F}_4$, let suppose from now that $c = \beta^i$, $i \in \{0, 1, 2\}$. Recall that $\beta = \alpha^{\frac{2^n-1}{3}}$ and $\xi = \alpha^{2^m-1}$ for some primitive element α of $\mathbb{F}_{2^n}$. Then $\beta^i = \xi^{i\frac{2^m+1}{3}}$. Hence, $T(\beta^i) = 4\sum_{u\in U,\, u^{\frac{2^n-1}{3}}=\beta^i=\xi^{i\frac{2^m+1}{3}}} \chi(f_0(u))$. Now,

$$u^{\frac{2^n-1}{3}} = \xi^{i\frac{2^m+1}{3}} \iff \left(u^{-2}\xi^{-i}\right)^{\frac{2^m+1}{3}} = 1 \iff u^{-2} \in \xi^i V.$$

That follows from the fact that the only elements x of U such that $x^{\frac{2^m+1}{3}} = 1$ are the elements of V. Next, noting that the map $x \mapsto x^{2^{m-1}}$ is one-to-one from $\xi^i V$ to $\xi^i V$ (because $\xi^{i(2^{m-1}-1)}$ is a cube since $2^{m-1} - 1 \equiv 0 \pmod 3$ for m odd), one gets that $u^{\frac{2^n-1}{3}} = \xi^{i\frac{2^m+1}{3}} \iff u \in \xi^i V$.

Therefore

$$T(\beta^i) = 4 \sum_{v \in V} \chi(f_0(\xi^i v)) = 4S_i.$$

Finally, by the inversion formula (10.5), one gets $\Lambda(f_{a_r,b}) = \frac{1}{4} \sum_{c \in \mathbb{F}_4} T(c)$ $\chi(\mathrm{Tr}_1^2(bc))$ that is,

$$\begin{aligned} \Lambda(f_{a_r,1}) &= S_0\chi(\mathrm{Tr}_1^2(1)) + S_1\chi(\mathrm{Tr}_1^2(\beta)) + S_2\chi(\mathrm{Tr}_1^2(\beta^2)), \\ \Lambda(f_\beta) &= S_0\chi(\mathrm{Tr}_1^2(\beta)) + S_1\chi(\mathrm{Tr}_1^2(\beta^2)) + S_2\chi(\mathrm{Tr}_1^2(1)), \\ \Lambda(f_{a_r,\beta^2}) &= S_0\chi(\mathrm{Tr}_1^2(\beta^2)) + S_1\chi(\mathrm{Tr}_1^2(1)) + S_2\chi(\mathrm{Tr}_1^2(\beta)). \end{aligned}$$

The result follows from Lemma 10.2.2 and from the fact that $\mathrm{Tr}_1^2(1) = 0$ and $\mathrm{Tr}_1^2(\beta) = \mathrm{Tr}_1^2(\beta^2) = 1$. □

From Proposition 10.2.1, Proposition 10.2.3, Lemma 10.2.2 and (10.4), one straight-forwardly deduces the following statement.

Lemma 10.2.4 ([13]). *Let $n = 2m$ be an even integer with m odd. For $b \in \mathbb{F}_4$, let $f_{a_r,b}$ be a function defined by (10.2). Let β be a primitive element of $\mathbb{F}_4$. Let U be the cyclic group of $(2^m + 1)$-st roots of unity and V be the set of the cube of U. Then,*

1. *$f_{a_r,\beta}$ is hyper-bent if and only if $\sum_{v \in V} \chi(f_0(v)) = -1$.*
2. *$f_{a_r,\beta}$ is hyper-bent if and only if f_{a_r,β^2} is hyper-bent.*
3. *$f_{a_r,1}$ is hyperbent if and only if $2\sum_{v \in V} \chi(f_{a_r,0}(v)) - \sum_{u \in U} \chi(f_{a_r,0}(u)) = 1$.*

Now we shall separate the case where $b = 1$ and the case where b is a primitive element of $\mathbb{F}_4$.

- The case where b is a primitive element of $\mathbb{F}_4$

According to Assertion (b) of Lemma 10.2.4, we can suppose that $b = \beta$ without loss of generality. As in the case where $b = 0$ (Theorem 10.1.1), one can establish a characterization of the hyper-bentness of f_β involving the Dickson polynomials. To this end, we begin with proving the following important technical result.

Lemma 10.2.5 ([13]). *Let f_0 be the function defined on $\mathbb{F}_{2^n}$ by $f_0(x) := \sum_{r \in R} \mathrm{Tr}_1^n(a_r x^{r(2^m-1)})$. Let g be the related function defined on $\mathbb{F}_{2^m}$ by $g(x) = \sum_{r \in R} \mathrm{Tr}_1^m(a_r D_r(x))$, where $D_r(x)$ is the Dickson polynomial of degree r. Let U be the cyclic group of $(2^m + 1)$-st roots of unity. Then, for any positive integer p, we have*

$$\sum_{u \in U} \chi\Big(f_0(u^p)\Big) = 1 + 2 \sum_{c \in \mathbb{F}_{2^m}^\star, \mathrm{Tr}_1^m(c^{-1})=1} \chi\Big(g(D_p(c))\Big).$$

Proof. Using the transitivity rule $\mathrm{Tr}_1^n = \mathrm{Tr}_1^m \circ \mathrm{Tr}_m^n$, the fact that the coefficients a_r are in the subfield $\mathbb{F}_{2^m}$ of $\mathbb{F}_{2^n}$ and the fact that the mapping $u \mapsto u^{2^m-1}$ is a permutation of U, one has

$$\sum_{u\in U}\chi\Big(f_0(u^p)\Big) = \sum_{u\in U}\chi\Big(\sum_{r\in R}\mathrm{Tr}_1^m\big(a_r(u^{(2^m-1)rp} + u^{2^m(2^m-1)rp})\big)\Big)$$

$$= \sum_{u\in U}\chi\Big(\sum_{r\in R}\mathrm{Tr}_1^m\big(a_r(u^{rp} + u^{-rp})\big)\Big) = \sum_{u\in U}\chi\Big(\sum_{r\in R}\mathrm{Tr}_1^m\big(a_r D_{rp}(u+u^{-1})\big)\Big)$$

since $u^p + u^{-p} = D_p(u+u^{-1})$. Recall now that every element $1/c$ where $c \in \mathbb{F}_{2^m}^\star$ with $\mathrm{Tr}_1^m(c) = 1$ can be uniquely represented as $u + u^{2^m} = u + u^{-1}$ with $u \in U$. Thus

$$\sum_{u\in U}\chi\Big(f_0(u^p)\Big) = 1 + \sum_{u\in U\setminus\{1\}}\chi\Big(\sum_{r\in R}\mathrm{Tr}_1^m\big(a_r D_{rp}(u+u^{-1})\big)\Big)$$

$$= 1 + 2\sum_{c\in\mathbb{F}_{2^m}^\star,\mathrm{Tr}_1^m(c)=1}\chi\Big(\sum_{r\in R}\mathrm{Tr}_1^m\big(a_r D_{rp}(1/c)\big)\Big)$$

$$= 1 + 2\sum_{c\in\mathbb{F}_{2^m}^\star,\mathrm{Tr}_1^m(c^{-1})=1}\chi\Big(\sum_{r\in R}\mathrm{Tr}_1^m\big(a_r D_{rp}(c)\big)\Big).$$

In the last equality, we use the fact that the map $c \mapsto 1/c$ is a permutation on $\mathbb{F}_{2^m}$. Now, since $D_{rp} = D_r \circ D_p$, one gets

$$\sum_{u\in U}\chi\Big(f_0(u^p)\Big) = 1 + 2\sum_{c\in\mathbb{F}_{2^m}^\star,\mathrm{Tr}_1^m(c^{-1})=1}\chi\Big(g(D_p(c))\Big).$$

□

From Lemma 10.2.4 and Lemma 10.2.5, one deduces the following statement.

Theorem 10.2.6 ([13]). *Let $n = 2m$ be an even integer with m odd. Let β be a primitive element of $\mathbb{F}_4$. Let $f_{a_r,\beta}$ be the function defined on $\mathbb{F}_{2^n}$ by (10.2). Let g be the related function defined on $\mathbb{F}_{2^m}$ by $g(x) = \sum_{r\in R}\mathrm{Tr}_1^m(a_r D_r(x))$, where $D_r(x)$ is the Dickson polynomial of degree r. Then, the three assertions are equivalent*

1. *$f_{a_r,\beta}$ is hyper-bent.*
2. $$\sum_{x\in\mathbb{F}_{2^m}^\star,\mathrm{Tr}_1^m(x^{-1})=1}\chi\Big(g(D_3(x))\Big) = -2.$$
3. $$\sum_{x\in\mathbb{F}_{2^m}^\star}\chi\Big(\mathrm{Tr}_1^m(x^{-1}) + g(D_3(x))\Big) = 2^m - 2\,\mathrm{wt}(g\circ D_3) + 4.$$

Proof. According to Lemma 10.2.5, we have

$$S_0 = \sum_{v \in V} \chi\Big(f_0(v)\Big) = \frac{1}{3} \sum_{u \in U} \chi\Big(f_0(u^3)\Big) = \frac{1}{3}\left(1 + 2 \sum_{x \in \mathbb{F}_{2^m}^\star, \mathrm{Tr}_1^m(x^{-1})=1} \chi\Big(g(D_3(x))\Big)\right).$$

The equivalence between assertions (a) and (b) in Theorem 10.2.10 follows then from assertion (1) of Lemma 10.2.4.

Now, note that the indicator of the set $\{x \in \mathbb{F}_{2^m}^\star \mid \mathrm{Tr}_1^m(x^{-1}) = 1\}$ can be written as $\frac{1}{2}\left(1 - \chi(\mathrm{Tr}_1^m(x^{-1}))\right)$. Therefore,

$$\begin{aligned}
&\sum_{x \in \mathbb{F}_{2^m}^\star, \mathrm{Tr}_1^m(x^{-1})=1} \chi\Big(g(D_3(x))\Big) \\
&\quad = \frac{1}{2}\left(\sum_{x \in \mathbb{F}_{2^m}^\star} \chi\Big(g(D_3(x))\Big) - \sum_{x \in \mathbb{F}_{2^m}^\star} \chi\Big(\mathrm{Tr}_1^m(x^{-1} + g(D_3(x)))\Big)\right) \\
&\quad = \frac{1}{2}\left(\sum_{x \in \mathbb{F}_{2^m}} \chi\Big(g(D_3(x))\Big) - \sum_{x \in \mathbb{F}_{2^m}} \chi\Big(\mathrm{Tr}_1^m(x^{-1} + g(D_3(x)))\Big)\right).
\end{aligned}$$

Now, $f_{a_r,\beta}$ is hyper-bent if and only if $\sum_{x \in \mathbb{F}_{2^m}^\star, \mathrm{Tr}_1^m(x^{-1})=1} \chi\Big(g(D_3(x))\Big) = -2$. Therefore, using the fact that, for a Boolean function h defined on $\mathbb{F}_{2^n}$, $\sum_{x \in \mathbb{F}_{2^n}} \chi(h(x)) = 2^n - 2\,\mathrm{wt}(h)$, we get that f_β is hyper-bent if and only if

$$\sum_{x \in \mathbb{F}_{2^m}} \chi\Big(\mathrm{Tr}_1^m(x^{-1}) + g(D_3(x))\Big) = 4 + 2^m - 2\,\mathrm{wt}(g \circ D_3).$$

□

One also has

Proposition 10.2.7 ([13]). *Let $n = 2m$ be an even integer with m odd. Let d be a positive integer. Suppose that d and $\frac{2^m+1}{3}$ are co-prime. Let β be a primitive element of $\mathbb{F}_4$. Let $f_{a_r,\beta}$ be the function defined by (10.2) and $h_{a_r,\beta}$ be the function whose expression is*

$$\sum_{r \in R} \mathrm{Tr}_1^n(a_r x^{dr(2^m-1)}) + \mathrm{Tr}_1^2(\beta x^{\frac{2^n-1}{3}})$$

where $a_r \in \mathbb{F}_{2^m}$. Then, $f_{a_r,\beta}$ is hyper-bent if and only if $h_{a_r,\beta}$ is hyper-bent.

Proof. According to assertion (a) of Lemma 10.2.4, $h_{a_r,\beta}$ is hyper-bent if and only if $\sum_{v \in V} \chi(h_0(v)) = -1$. Now, $\sum_{v \in V} \chi(h_0(v)) = \sum_{v \in V} \chi(f_0(v^d)) = \sum_{v \in V} \chi(f_0(v))$

since the mapping $v \mapsto v^d$ is then a permutation of V if $\frac{2^m+1}{3}$ and d are co-prime. The result follows again from assertion (a) of Lemma 10.2.4. □

- The case where $b = 1$:

we are interested in characterizing the hyper bentness of the Boolean function $f_{a_r,1}$ whose polynomial form is $f_{a_r,1}(x) = \sum_{r \in R} \mathrm{Tr}_1^n(a_r x^{r(2^m-1)}) + \mathrm{Tr}_1^2(x^{\frac{2^n-1}{3}})$. In this case one can give a characterization hyperbentness, analogous to the assertion (b) of Theorem 10.2.10.

Theorem 10.2.8 ([13]). *Let $n = 2m$ be an even integer with m odd. Let $f_{a_r,1}$ be the Boolean function defined on $\mathbb{F}_{2^n}$ by*

$$f_{a_r,1}(x) = \sum_{r \in R} \mathrm{Tr}_1^n(a_r x^{r(2^m-1)}) + \mathrm{Tr}_1^2(x^{\frac{2^n-1}{3}}).$$

Let g be the related function defined on $\mathbb{F}_{2^m}$ by $g(x) = \sum_{r \in R} \mathrm{Tr}_1^m(a_r D_r(x))$, where $D_r(x)$ is the Dickson polynomial of degree r.
Then, f_1 is hyper-bent if and only if,

$$2 \sum_{x \in \mathbb{F}_{2^m}^\star, \mathrm{Tr}_1^m(x^{-1})=1} \chi\Big(g(D_3(x))\Big) - 3 \sum_{x \in \mathbb{F}_{2^m}^\star, \mathrm{Tr}_1^m(x^{-1})=1} \chi\Big(g(x)\Big) = 2.$$

Proof. Note that

$$2\sum_{v \in V} \chi(f_0(v)) - \sum_{u \in U} \chi(f_0(u)) = \frac{2}{3}\sum_{u \in U} \chi(f_0(u^3)) - \sum_{u \in U} \chi(f_0(u))$$

$$= -\frac{1}{3} + \frac{4}{3} \sum_{x \in \mathbb{F}_{2^m}^\star, \mathrm{Tr}_1^m(x^{-1})=1} \chi\Big(g(D_3(x))\Big) - 2 \sum_{x \in \mathbb{F}_{2^m}^\star, \mathrm{Tr}_1^m(x^{-1})=1} \chi\Big(g(x)\Big)$$

according to Lemma 17.5.5. One then concludes using Lemma 10.2.4 that states that f_1 is hyper-bent if and only if

$$2\sum_{v \in V} \chi(f_0(v)) - \sum_{u \in U} \chi(f_0(u)) = 1.$$

□

One can also prove the similar result to Proposition 10.2.7.

Proposition 10.2.9 ([13]). *Let $n = 2m$ be an even integer with m odd. Suppose that $m \not\equiv 3 \pmod 6$. Let d be a positive integer such that $\gcd(d, 2^m + 1) = 3$. Let β be a primitive element of $\mathbb{F}_4$. Let $f_{a_r,\beta}$ be the function defined by (10.2) and $h_{a_r,1}$ be the function whose expression is*

$$\sum_{r \in R} \mathrm{Tr}_1^n(a_r x^{dr(2^m-1)}) + \mathrm{Tr}_1^2(x^{\frac{2^n-1}{3}}).$$

If $f_{a_r,\beta}$ is hyper-bent then, $h_{a_r,1}$ is hyper-bent.

Proof. Set $h_0(x) := \sum_{r\in R} \mathrm{Tr}_1^n(a_r x^{dr(2^m-1)})$. One has (since $\gcd(d, 2^m+1) = 3$)

$$\sum_{v\in V} \chi(h_0(v)) = \sum_{v\in V} \chi(f_0(v^d)) = \sum_{v\in V} \chi(f_0(v^3)) = \sum_{v\in V} \chi(f_0(v))$$

since the mapping $v \mapsto v^3$ is a permutation when $m \not\equiv 3 \pmod 6$. On the other hand, note that (since $\gcd(d, 2^m+1) = 3$)

$$\sum_{u\in U} \chi(h_0(u)) = \sum_{u\in U} \chi(f_0(u^d)) = \sum_{u\in U} \chi(f_0(u^3)) = 3\sum_{v\in V} \chi(f_0(v)).$$

Now, $\sum_{v\in V} \chi(f_0(v)) = -1$ according to Lemma 10.2.4, since $f_{a_r,\beta}$ is hyper-bent. Hence, $2\sum_{v\in V} \chi(h_0(v)) - \sum_{u\in U} \chi(h_0(u)) = -2-(-3) = 1$, proving that $h_{a_r,1}$ is hyper-bent (according to Lemma 10.2.4). □

The following theorem summarizes the study of the bentness of functions in $\mathfrak{H}_n$

Theorem 10.2.10 ([13]). *Let $n = 2m$ with m odd. Let $b \in \mathbb{F}_4^\star$ and β be a primitive element of $\mathbb{F}_4$. Let $f_{a_r,b}$ be a function of $\mathfrak{H}_n$ defined on $\mathbb{F}_{2^n}$ by (10.2). Let g_{a_r} be the related function defined on $\mathbb{F}_{2^m}$ by $g_{a_r}(x) = \sum_{r\in R} \mathrm{Tr}_1^m(a_r D_r(x))$, where $D_r(x)$ is the Dickson polynomial of degree r.*

1. *$f_{a_r,b}$ is hyper-bent if and only if $f_{a_r,b}$ is bent.*
2. *The bent functions $f_{a_r,b}$ are in the class $\mathcal{PS}^-$. Moreover, the bent functions $f_{a_r,b}$ are elements of the Partial Spread class $\mathcal{PS}_{ap}$ (resp. $\mathcal{PS}_{ap}^{\#}$) if $b = 1$ (resp. if $b \neq 1$).*
3. *The three following assertions are equivalent:*

   (a) *$f_{a_r,\beta}$ is hyper-bent;*

   (b) $$\sum_{x\in\mathbb{F}_{2^m}^\star, \mathrm{Tr}_1^m(x^{-1})=1} \chi\Big(g_{a_r}(D_3(x))\Big) = -2;$$

   (c) $$\sum_{x\in\mathbb{F}_{2^m}^\star} \chi\Big(\mathrm{Tr}_1^m(x^{-1}) + g_{a_r}(D_3(x))\Big) = 2^m - 2\,\mathrm{wt}(g_{a_r}\circ D_3) + 4.$$

4. *$f_{a_r,1}$ is hyper-bent if and only if,*

$$2\sum_{x\in\mathbb{F}_{2^m}^\star, \mathrm{Tr}_1^m(x^{-1})=1} \chi\Big(g_{a_r}(D_3(x))\Big) - 3\sum_{x\in\mathbb{F}_{2^m}^\star, \mathrm{Tr}_1^m(x^{-1})=1} \chi\Big(g_{a_r}(x)\Big) = 2.$$

Note that the previous theorem is valid when the coefficients a_r are elements of $\mathbb{F}_{2^m}$.

Problem 10.2.11. Give an analogous characterization of functions of type (10.2) which are hyper-bent, in the case where some of the coefficients are in $\mathbb{F}_{2^n}$, but not in $\mathbb{F}_{2^m}$.

Corollary 10.2.12 and Corollary 10.2.13 show that we can recover the results given in [14] and [15] using directly the characterizations given by Theorem 10.2.10.

Corollary 10.2.12 ([17]). *Let $n = 2m$ with m odd ($m > 3$). Take in Theorem 10.2.10, $\#R = 1$ and $r = 1$. For simplicity, denote by a the coefficient a_1. Let f_{a,β^i} be the corresponding function (where β be a primitive element of $\mathbb{F}_4$, $i \in \{0, 1, 2\}$) defined on $\mathbb{F}_{2^n}$ by (10.2). Then, the function $f_{a,1}$ is not bent and, the function $f_{a,\beta}$ (resp. f_{a,β^2}) is hyper-bent whenever $K_m(a) = 4$ while, when $K_m(a) \neq 4$, $f_{a,\beta}$ (resp. f_{a,β^2}) is not hyper-bent.*

Proof. According to Lemma 11 in [13], $f_{a,\beta}$ is hyper-bent if and only if f_{a,β^2} is hyper-bent. Moreover, according to Theorem 10.2.10, the function $f_{a,\beta}$ is hyper-bent if and only if

$$\sum_{x\in\mathbb{F}_{2^m}^\star, \mathrm{Tr}_1^m(1/x)=1} \chi(g_a(D_3(x))) = -2$$

where g_a is the related function (defined in Theorem 10.2.10) which is equal to $\mathrm{Tr}_1^m(ax)$ (since the Dickson polynomial of degree 1 is equals X). The Dickson polynomial of degree 3 equals $X^3 + X$ thus,

$$\begin{aligned}\sum_{x\in\mathbb{F}_{2^m}^\star, \mathrm{Tr}_1^m(1/x)=1} \chi(g(D_3(x))) &= \sum_{x\in\mathbb{F}_{2^m}^\star} \chi(\mathrm{Tr}_1^m(a(x^3+x))) \\ &- \sum_{x\in\mathbb{F}_{2^m}^\star, \mathrm{Tr}_1^m(1/x)=0} \chi(\mathrm{Tr}_1^m(a(x^3+x))) \\ &= C_m(a,a) - 1 - \sum_{x\in\mathbb{F}_{2^m}^\star, \mathrm{Tr}_1^m(1/x)=0} \chi(\mathrm{Tr}_1^m(a(x^3+x))) \\ &= C_m(a,a) - 1 - \sum_{x\in\mathbb{F}_{2^m}^\star, \mathrm{Tr}_1^m(1/x)=0} \chi(\mathrm{Tr}_1^m(ax)).\end{aligned}$$

In the last equality, we use the fact that the mapping $x \mapsto D_3(x) := x^3 + x$ is a permutation on the set of $\mathbb{F}_{2^m}^\star$ such that $\mathrm{Tr}_1^m(1/x) = 0$ (see e.g. [2, Lemma 7]). Now, according to Charpin et al. [2],

$$\sum_{x\in\mathbb{F}_{2^m}^\star, \mathrm{Tr}_1^m(1/x)=0} \chi(\mathrm{Tr}_1^m(ax)) = \frac{K_m(a)}{2} - 1.$$

Hence, we get that

$$\sum_{x\in\mathbb{F}_{2^m}^\star, \mathrm{Tr}_1^m(1/x)=1} \chi(g(D_3(x))) = C_m(a,a) - \frac{K_m(a)}{2}.$$

Therefore, $f_{a,\beta}$ (resp. f_{a,β^2}) is hyper-bent if and only if $K_m(a) - 2C_m(a,a) = 4$. The mapping $x \mapsto x^3$ is a permutation on $\mathbb{F}_{2^m}$ for m odd, every element $a \in \mathbb{F}_{2^m}$

can be (uniquely) written as $a = a'^3$ with $a' \in \mathbb{F}_{2^m}$. One has $C_m(a,a) = \sum_{x\in\mathbb{F}_{2^m}} \chi(\mathrm{Tr}_1^m((a'x)^3 + ax) = C_m(1, a^{2/3})$.
Hence, according to Proposition 2.3.5 (note that $\mathrm{Tr}_1^m(a^{2/3}) = \mathrm{Tr}_1^m(a^{1/3})$), the function $f_{a,\beta}$ (resp. f_{a,β^2}) is hyper-bent if and only if,

$$K_m(a) = \begin{cases} 4 \text{ if } \mathrm{Tr}_1^m(a^{1/3}) = 0 \\ 4 \pm \left(\frac{2}{m}\right) 2^{(m+3)/2} \text{ if } \mathrm{Tr}_1^m(a^{1/3}) = 1 \end{cases}$$

However, using Proposition 2.3.2, the value $4 \pm \left(\frac{2}{m}\right) 2^{(m+3)/2}$ does not belong to $[-2^{(m+2)/2} + 1, 2^{(m+2)/2} + 1]$ for every $m > 3$. This proves that if $\mathrm{Tr}_1^m(a^{1/3}) = 0$, then the function $f_{a,\beta}$ (resp. f_{a,β^2}) is hyper-bent whenever $K_m(a) = 4$ while, when $K_m(a) \neq 4$, f_β (resp. f_{a,β^2}) is not hyper-bent. Otherwise, if $\mathrm{Tr}_1^m(a^{1/3}) = 1$ (which implies that $K_m(a) \neq 4$), then the function $f_{a,\beta}$ (resp. f_{a,β^2}) cannot be hyper-bent when $m > 3$. On the other hand, according to Theorem 10.2.10, $f_{a,1}$ is hyper-bent if and only if,

$$2 \sum_{x\in\mathbb{F}_{2^m}^\star, \mathrm{Tr}_1^m(x^{-1})=1} \chi\Big(g_a(D_3(x))\Big) - 3 \sum_{x\in\mathbb{F}_{2^m}^\star, \mathrm{Tr}_1^m(x^{-1})=1} \chi\Big(g_a(x)\Big) = 2.$$

We have seen that

$$\sum_{x\in\mathbb{F}_{2^m}^\star, \mathrm{Tr}_1^m(1/x)=1} \chi(g_a(D_3(x))) = C_m(a,a) - \frac{K_m(a)}{2}.$$

Furthermore, according to [2],

$$\sum_{x\in\mathbb{F}_{2^m}^\star, \mathrm{Tr}_1^m(x^{-1})=1} \chi\Big(g_a(x)\Big) = \sum_{x\in\mathbb{F}_{2^m}^\star, \mathrm{Tr}_1^m(1/x)=1} \chi(\mathrm{Tr}_1^m(ax)) = -\frac{K_m(a)}{2}.$$

Therefore, $f_{a,1}$ is hyper-bent if and only if,

$$K_m(a) + 4C_m(a,a) = 4.$$

Recalling that $C_m(a,a) = C_m(1, a^{2/3})$ and Proposition 2.3.5, we get that $f_{a,1}$ is hyper-bent if and only if,

$$K_m(a) = \begin{cases} 4 \text{ if } \mathrm{Tr}_1^m(a^{1/3}) = 0 \\ 4 \pm \left(\frac{2}{m}\right) 2^{(m+5)/2} \text{ if } \mathrm{Tr}_1^m(a^{1/3}) = 1 \end{cases}.$$

However, again by Proposition 2.3.2, the value $4 \pm \left(\frac{2}{m}\right) 2^{(m+5)/2}$ does not belong to $[-2^{(m+2)/2} + 1, 2^{(m+2)/2} + 1]$ for every $m > 3$. This proves that if $\mathrm{Tr}_1^m(a^{1/3}) = 0$, then the function $f_{a,1}$ is hyper-bent whenever $K_m(a) = 4$ while, when $K_m(a) \neq 4$,

$f_{a,1}$ is not hyper-bent. Otherwise, if $\mathrm{Tr}_1^m(a^{1/3}) = 1$ (which implies that $K_m(a) \neq 4$), then the function $f_{a,1}$ cannot be hyper-bent when $m > 3$. □

Corollary 10.2.13 ([17]). *Let $n = 2m$ with m odd such that $m \not\equiv 3 \pmod 6$. Take in Theorem 10.2.10, $\#R = 1$ and $r = 3$. For simplicity, denote by a the coefficient a_1. Let $f_{a,\beta}$ be the corresponding function (where β be a primitive element of $\mathbb{F}_4$) defined on $\mathbb{F}_{2^n}$ by (10.2). If $\mathrm{Tr}_1^m(a^{1/3}) = 0$, then the function $f_{a,\beta}$ is hyper-bent whenever $K_m(a) = 4$ and if $\mathrm{Tr}_1^m(a^{1/3}) = 1$, then the function $f_{a,\beta}$ is not hyper-bent.*

Proof. Since $m \not\equiv 3 \pmod 6$, the integers $\frac{2^m+1}{3}$ and 3 are co-prime. Applying Proposition 10.2.9 for $d = 3$ we obtain, $f_{a,\beta}$ is hyper-bent if and only if, the function $x \mapsto \mathrm{Tr}_1^n(ax^{(2^m-1)}) + \mathrm{Tr}_1^2(\beta x^{\frac{2^n-1}{3}})$ is hyper-bent. Now according to Corollary 10.2.12, we deduce that $f_{a,\beta}$ is hyper-bent whenever $K_m(a) = 4$ while, when $K_m(a) \neq 4$, f_β is not hyper-bent. Otherwise, if $\mathrm{Tr}_1^m(a^{1/3}) = 1$ (which implies that $K_m(a) \neq 4$), then the function $f_{a,\beta}$ cannot be hyper-bent when $m > 3$. □

The extended Walsh–Hadamard transform of $f_{a,b}$[1] can be expressed as follows.

Proposition 10.2.14. *The notation is as in Theorem 10.2.10 except that we allow b to be equal to zero. In that specific case, we do not suppose m to be odd. Then*

$$\widehat{\chi_{f_{a,b}}}(0,k) = 1 + \Lambda(f_{a,b})\,(-1 + 2^m)\,,$$

and, for $\omega \in \mathbb{F}_{2^n}^*$ non-zero,

$$\widehat{\chi_{f_{a,b}}}(\omega,k) = 1 - \Lambda(f_{a,b}) + 2^m(-1)^{f_{a,b}(\omega^{(2^m-1)/(2k)})}\,.$$

Proof. It is a well-known fact that every non-zero element $x \in \mathbb{F}_{2^n}^*$ has a unique polar decomposition as a product $x = yu$ where y lies in the subfield $\mathbb{F}_{2^m}$ and $u \in U$.

The extended Walsh–Hadamard transform of $f_{a,b}$ at (ω, k) can consequently be expressed as

$$\begin{aligned}
\widehat{\chi_{f_{a,b}}}(\omega,k) &= \sum_{x\in\mathbb{F}_{2^n}} \chi\left(f_{a,b}(x) + \mathrm{Tr}_1^n\left(\omega x^k\right)\right) \\
&= 1 + \sum_{x\in\mathbb{F}_{2^n}^*} \chi\left(f_{a,b}(x) + \mathrm{Tr}_1^n\left(\omega x^k\right)\right) \\
&= 1 + \sum_{u\in U}\sum_{y\in\mathbb{F}_{2^m}^*} \chi\left(f_{a,b}(yu) + \mathrm{Tr}_1^n\left(\omega y^k u^k\right)\right).
\end{aligned}$$

[1]For simplicity, we shall write from now on $f_{a,b}$ instead of $f_{a_r,b}$.

But

$$\begin{aligned}
f_{a,b}(yu) &= \sum_{r\in R} \mathrm{Tr}_1^n\left(a_r (yu)^{r(2^m-1)}\right) + \mathrm{Tr}_1^2\left(b(yu)^{\frac{2^n-1}{3}}\right) \\
&= \sum_{r\in R} \mathrm{Tr}_1^n\left(a_r y^{r(2^m-1)} u^{r(2^m-1)}\right) + \mathrm{Tr}_1^2\left(b y^{(2^m-1)\frac{2^m+1}{3}} u^{\frac{2^n-1}{3}}\right) \\
&= \sum_{r\in R} \mathrm{Tr}_1^n\left(a_r u^{r(2^m-1)}\right) + \mathrm{Tr}_1^2\left(b u^{\frac{2^n-1}{3}}\right) \\
&= f_{a,b}(u),
\end{aligned}$$

so that

$$\begin{aligned}
\widehat{\chi_{f_{a,b}}}(\omega,k) &= 1 + \sum_{u\in U}\sum_{y\in\mathbb{F}_{2^m}^*} \chi\left(f_{a,b}(u) + \mathrm{Tr}_1^n\left(\omega y^k u^k\right)\right) \\
&= 1 + \sum_{u\in U}(-1)^{f_{a,b}(u)} \sum_{y\in\mathbb{F}_{2^m}^*} \chi\left(\mathrm{Tr}_1^n\left(\omega y^k u^k\right)\right) \\
&= 1 + \sum_{u\in U}(-1)^{f_{a,b}(u)} \left(-1 + \sum_{y\in\mathbb{F}_{2^m}} \chi\left(\mathrm{Tr}_1^n\left(\omega y^k u^k\right)\right)\right).
\end{aligned}$$

If $\omega = 0$, then $\widehat{\chi_f}(\omega,k) = 1 + \Lambda(f_{a,b})(-1+2^m)$ as desired. If $\omega \neq 0$, then one uses the transitivity of the trace: $\mathrm{Tr}_1^n(x) = \mathrm{Tr}_1^m\left(\mathrm{Tr}_m^n(x)\right) = \mathrm{Tr}_1^m\left(x + x^{2^m}\right)$, which yields

$$\begin{aligned}
\mathrm{Tr}_1^n\left(\omega y^k u^k\right) &= \mathrm{Tr}_1^m\left(\mathrm{Tr}_m^n\left(\omega y^k u^k\right)\right) \\
&= \mathrm{Tr}_1^m\left(\omega y^k u^k + \left(\omega y^k u^k\right)^{2^m}\right) \\
&= \mathrm{Tr}_1^m\left(\omega y^k u^k + \omega^{2^m} y^k u^{-k}\right) \\
&= \mathrm{Tr}_1^m\left(y^k\left(\omega u^k + \omega^{2^m} u^{-k}\right)\right).
\end{aligned}$$

As k is co-prime with 2^m-1, the map $y \mapsto y^k$ is a permutation of $\mathbb{F}_{2^m}$ and the sum over $\mathbb{F}_{2^m}$ is non-zero if and only if $u^{2k} = \omega^{2^m-1}$. As k is co-prime with 2^m+1, this only occurs for a value of u and we get the final equality

$$\widehat{\chi_{f_{a,b}}}(\omega,k) = 1 - \Lambda(f_{a,b}) + 2^m(-1)^{f_{a,b}\left(\omega^{(2^m-1)/(2k)}\right)}.$$

□

10.2.1 Some Conjectures: Towards New Hyper-Bent Functions

In the following, we make some conjectures that lead to construct new hyper-bent functions. To this end, we introduce some notation. Let $I := \{x \in \mathbb{F}_{2^m}^\star \mid x = c^3 + c, \mathrm{Tr}_1^m(c^{-1}) = 1\}$ and set, for $a, a' \in \mathbb{F}_{2^m}$,

$$S(a, a') := \sum_{x \in I} (-1)^{\mathrm{Tr}_1^m(a(x+x^3)+a'x^5)}.$$

Conjecture 10.2.15. *For every $a \in \mathbb{F}_{2^m}^\star$, the set $\Gamma_a := \{a' \in \mathbb{F}_{2^m}^\star \mid S(a, a') = -1\}$ is non empty.*

By a computer program, we have checked that Conjecture 10.2.15 holds for all $n = 2m$ up to $n = 26$ and for every $a \in \mathbb{F}_{2^m}^\star$ such that $K_m(a) = 4$. Moreover, we have made an exhaustive search by a computer program for $n \in \{10, 14, 18, 22\}$ of all sets Γ_a for each value a such that $K_m(a) = 4$. Let ζ be a primitive element of $\mathbb{F}_{2^{10}}$ (whose minimal polynomial is $x^{10} + x^7 + 1$) and set $\alpha = \zeta^{33}$ (so that α is a primitive element of $\mathbb{F}_{2^5}$). We list in Table 10.1 all the pairs of indices (i, j) such that $K_5(\alpha^i) = 4$ and $\alpha^j \in \Gamma_{\alpha^i}$. We have also found all pairs (i, j) for $n \in \{14, 18, 22\}$. Due to their number, we do not list them like for $n = 10$ but we only give in Table 10.2 the numbers of pairs that we found (including the case where $K_m(a) = 4$ and $S(a, 0) = -1$).

Table 10.1 Exponents i and j such that (α^i, α^j) satisfy Conjecture 10.2.15 for $n = 10$

i=1	j=0, 1, 2, 3, 5, 7, 8, 9, 11, 12, 13, 14, 17, 20, 22, 24, 26, 27, 29
i=2	j=0, 2, 3, 4, 6, 9, 10, 13, 14, 16, 17, 18, 21, 22, 23, 24, 26, 27, 28
i=4	j=0, 1, 3, 4, 5, 6, 8, 11, 12, 13, 15, 17, 18, 20, 21, 23, 25, 26, 28
i=7	j=0, 3, 4, 5, 7, 8, 10, 11,12, 14, 16, 18, 19, 23, 26, 27, 28, 29, 30
i=8	j=0, 2, 3, 5, 6, 8, 9, 10, 11,12, 15 16, 19, 21, 22, 24, 25, 26, 30,
i=14	j=0, 1, 5, 6, 7, 8, 10, 14,15, 16, 20, 21, 22, 23, 24, 25, 27, 28, 29
i=16	j=0, 1, 4, 6, 7, 10, 11 12, 13, 16, 17, 18, 19, 20, 21, 22, 24, 29, 30
i=19	j=0, 4, 5, 6, 7, 8, 9, 13, 14, 15, 17, 18, 19, 21, 25, 27, 29, 2, 30
i=25	j=0, 1, 2, 3, 4, 7, 9, 15, 18, 19, 20, 22, 23, 24, 25, 26, 28, 29, 30
i=28	j=0, 1, 2, 9, 10, 11, 12, 13, 14, 15 16, 17, 19, 20, 23, 25, 27, 28, 30

Table 10.2 Number of exponents such that (α^i, α^j) satisfy Conjecture 10.2.15 for $n \in \{14, 18, 22\}$

n	14	18	22
Number of pairs	882	3978	13,948

Proposition 10.2.16 ([17]). *Let $n = 2m$ with m odd. Suppose that Conjecture 10.2.15 holds. Let $a \in \mathbb{F}_{2^m}^\star$, $a' \in \Gamma_a$ ($\neq \emptyset$) and β is a primitive element of $\mathbb{F}_4$. Then, the function f defined on $\mathbb{F}_{2^n}$ by*

$$f(x) = \mathrm{Tr}_1^n((a + a')x^{3(2^m-1)}) + \mathrm{Tr}_1^n(a'x^{5(2^m-1)}) + \mathrm{Tr}_1^2(\beta x^{\frac{2^n-1}{3}})$$

is hyper-bent.

Proof. We denote by g be the function defined on $\mathbb{F}_{2^m}$ as

$$g(x) = \mathrm{Tr}_1^m((a + a')D_3(x)) + \mathrm{Tr}_1^m(a'D_5(x)).$$

Recall that $D_3(x) = x+x^3$ and, $D_5(x) = x+x^3+x^5$. So $g(x) = \mathrm{Tr}_1^m\big(a(x+x^3)+a'x^5\big)$. Now, according to Theorem 10.2.10,f is hyper-bent if and only if, $\sum_{x\in\mathbb{F}_{2^m}^\star, \mathrm{Tr}_1^m(x^{-1})=1} \chi(g(D_3(x))) = -2$. Now, according to Charpin et al. [2] (Lemma 6), the mapping $x \mapsto D_3(x)$ is 3-to-1 from $\{x \in \mathbb{F}_{2^m} \setminus \mathbb{F}_2 \mid \mathrm{Tr}_1^m(x^{-1}) = 1\}$ to $I := \{x \in \mathbb{F}_{2^m}^\star \mid x = c^3 + c, \mathrm{Tr}_1^m(c^{-1}) = 1\}$. Thus, the above condition of hyper-bentness can be reworded as

$$1 + 3\sum_{x\in I} \chi(g(x)) = -2, \text{ that is, } \sum_{x\in I} \chi(g(x)) = -1.$$

The result follows. □

More generally, let set, for $(a, a') \in \mathbb{F}_{2^m}^\star$ and $a'' \in \mathbb{F}_{2^m}$,

$$\mathcal{S}'(a, a', a'') := \sum_{x\in I} (-1)^{\mathrm{Tr}_1^m(ax+a'x^3+a''x^5)}$$

Conjecture 10.2.17. *The set $\Gamma' := \{(a, a', a'') \in \mathbb{F}_{2^m}^\star \times \mathbb{F}_{2^m}^\star \times \mathbb{F}_{2^m} \mid \mathcal{S}'(a, a', a'') = -1\}$ is not empty.*

By a computer search, we have found that for $n = 10$, there exist 1524 3-tuples $(a, a', a")$ such that $\mathcal{S}'(a, a', a'') = -1$, for $n = 14$, there exist 58790 such 3-tuples $(a, a', a")$ and, for $n = 18$, there exist 1904870 such 3-tuples $(a, a', a")$.

Proposition 10.2.18 ([17]). *Let $n = 2m$ with m odd. Suppose that Conjecture 10.2.17 holds. Let $(a, a', a'') \in \Gamma'$ ($\neq \emptyset$) and β be a primitive element of $\mathbb{F}_4$. Then, the function f defined on $\mathbb{F}_{2^n}$ by*

$$f(x) = \mathrm{Tr}_1^n((a + a')x^{2^m-1}) + \mathrm{Tr}_1^n((a' + a'')x^{3(2^m-1)}) + \mathrm{Tr}_1^n(a''x^{5(2^m-1)}) + \mathrm{Tr}_1^2(\beta x^{\frac{2^n-1}{3}})$$

is hyper-bent.

Proof. We denote g be function defined on $\mathbb{F}_{2^m}$ by

$$g(x) := \mathrm{Tr}_1^m((a + a')D_1(x)) + \mathrm{Tr}_1^m((a' + a'')D_3(x)) + \mathrm{Tr}_1^m(a''D_5(x)).$$

According to the values of Dickson polynomial,

$$g(x) = \mathrm{Tr}_1^m((a+a')x) + \mathrm{Tr}_1^m((a'+a'')(x+x^3)) + \mathrm{Tr}_1^m(a''(x+x^3+x^5))$$
$$= \mathrm{Tr}_1^m(ax + a'x^3 + a''x^5).$$

The mapping $x \mapsto D_3(x)$ is 3-to-1 from $\{x \in \mathbb{F}_{2^m} \setminus \mathbb{F}_2 \mid \mathrm{Tr}_1^m(x^{-1}) = 1\}$ to $I := \{x \in \mathbb{F}_{2^m}^\star \mid x = c^3 + c, \mathrm{Tr}_1^m(c^{-1}) = 1\}$, therefore, the condition of hyper-bentness given by Theorem 10.2.10 can be reworded as

$$1 + 3\sum_{x\in I} \chi(g(x)) = -2, \text{ that is, } \sum_{x\in I} \chi(g(x)) = -1.$$

Equivalently,

$$S'(a,a',a'') := \sum_{x\in I} (-1)^{\mathrm{Tr}_1^m(ax+a'x^3+a''x^5)} = -1.$$

The result follows. □

10.3 Hyper-Bent Functions with Multiple Trace Terms via Dillon (Like) Exponents: The Wang et al. Family

Adopting our approach presented in the previous section, Wang, Tang, Qi, Yang and Xu [22] have studied in late 2011 the following family with an additional trace term on $\mathbb{F}_{16}$:

$$f_{a,b}(x) = \sum_{r\in R} \mathrm{Tr}_1^n\left(a_r x^{r(2^m-1)}\right) + \mathrm{Tr}_1^4\left(bx^{\frac{2^n-1}{5}}\right) \tag{10.6}$$

where the coefficients a_r lie in $\mathbb{F}_{2^m}$, the coefficient b is in $\mathbb{F}_{16}$ and m must verify $m \equiv 2 \pmod 4$ (the set R is defined as above, that is, a subset of representatives of the cyclotomic cosets modulo $2^m + 1$ for which each coset has the maximal size n). We denote by $\mathfrak{W}_n$ the set of Boolean functions $f_{a,b}$ defined on $\mathbb{F}_{2^n}$ by (10.6).

We have provided a finer study of this family by giving results including useful expressions for their extended Walsh–Hadamard transform, their algebraic degrees and their duals.

The divisibility condition on m essentially entails that $2^m \equiv -1 \pmod 5$. A first consequence of this equality is that all functions in this family have the same algebraic degree, even the ones which are not hyper-bent.

Proposition 10.3.1 ([7]). *Let $f_{a,b}$ be a function of $\mathfrak{W}_n$. The algebraic degree of the function $f_{a,b}$ is equal to m.*

Proof. The exponent $2^m - 1$ has 2-weight m since $2^m - 1 = 1+2+2^2+\cdots+2^{m-1}$. Moreover, $m \equiv 2 \pmod 4$ so that $n = 2m$ can be expressed as $n = 8l + 4$. Then

$$\frac{2^n - 1}{5} = \frac{16^{2l+1} - 1}{5} = 3 \times \frac{16^{2l+1} - 1}{15}$$
$$= 3 \times \sum_{i=0}^{2m} 16^i = \sum_{i=0}^{2l} 2^{4i} + \sum_{i=0}^{2l} 2^{4i+1}.$$

Therefore, the 2-weight of $\frac{2^n-1}{5}$ is $4l + 2 = \frac{n}{2} = m$ as well.

Both Boolean functions $x \mapsto \sum_{r \in R} \mathrm{Tr}_1^n \left(a_r x^{r(2^m-1)}\right)$ and $x \mapsto \mathrm{Tr}_1^4 \left(bx^{\frac{2^n-1}{5}}\right)$ are thus of algebraic degree m. Since they are separate parts in the trace representation of $f_{a,b}$, the algebraic degree of $f_{a,b}$ is equal to m as well. □

The divisibility condition on m also implies that $f_{a,b}(xy) = f_{a,b}(y)$ for any x in the subfield $\mathbb{F}_{2^m}$. The extended Walsh–Hadamard spectrum of $f_{a,b}$ can then be expressed with $\Lambda(f_{a,b})$ in a classical manner [12, Theorem 3], [6], thus extending the result of Wang et al. [22, Proposition 3.1] which gives a characterization of the hyper-bentness of $f_{a,b}$ using $\Lambda(f_{a,b})$ but does not provide an explicit expression for its extended Walsh–Hadamard spectrum.

Proposition 10.3.2 ([7]). *Let $f_{a,b}$ be a function of $\mathfrak{W}_n$. Then*

$$\widehat{\chi_{f_{a,b}}}(0, k) = 1 + \Lambda(f_{a,b})\,(-1 + 2^m),$$

and, for $\omega \in \mathbb{F}_{2^n}^*$ non-zero,

$$\widehat{\chi_{f_{a,b}}}(\omega, k) = 1 - \Lambda(f_{a,b}) + 2^m(-1)^{f_{a,b}(\omega^{(2^m-1)/(2k)})}.$$

In particular, $f_{a,b}$ is hyper-bent if and only if $\Lambda(f_{a,b}) = 1$.

The dual of $f_{a,b}$ can then be explicitly computed when $f_{a,b}$ is hyper-bent.

Proposition 10.3.3 ([7]). *If $f_{a,b}$ is hyper-bent, then its dual is f_{a,b^4}, i.e. we have*

$$\forall \omega \in \mathbb{F}_{2^n}, \quad \widehat{\chi_{f_{a,b}}}(\omega) = 2^m \chi_{f_{a,b^4}}(\omega).$$

Proof. Let $u \in U$ be the unique element such that $u^{1-2^m} = u^2 = \omega^{2^m-1}$, that is $u = \omega^{(2^m-1)/2}$. Then $f_{a,b}(u) = f_{a,b}(\omega^{-1})$.

Moreover, since $m \equiv 2 \pmod 4$ $(m > 2)$, 15 divides $2^m - 4$. Hence, $b^{2^m} = b^4$ and it follows that $f_{a,b}(\omega^{-1}) = f_{a,b^4}(\omega)$. □

Extending our approach [13, 16], Wang et al. [22] derived the following characterization of the hyper-bentness property of such functions in two cases

- $b = 1$ and $b^4 + b + 1 = 0$
- $a_r \in \mathbb{F}_{2^{\frac{m}{2}}}$

They showed that hyper-bentness of these functions for the two cases are related to some character sums involving Dickson polynomials of degree r and 5. The following theorem summarizes their results.

Theorem 10.3.4 ([22]). *Suppose $m := \frac{n}{2} \equiv 2 \pmod 4$. Let $R \subseteq E$ where E is a set of representatives of the cyclotomic classes modulo $2^n - 1$ for which each class has the full size n. For $b \in \mathbb{F}_{16}^\star$ and $a_r \in \mathbb{F}_{2^m}^\star$, we denote by $\tilde{g}_{a_r,b}$ the function defined on $\mathbb{F}_{2^n}$ by $\sum_{r\in R} \mathrm{Tr}_1^n(a_r x^{r(2^m-1)}) + \mathrm{Tr}_1^4(b' x^{\frac{2^n-1}{5}})$, and by h_{a_r} the function defined on $\mathbb{F}_{2^m}$ by $\sum_{r\in R} \mathrm{Tr}_1^m(a_r D_r(x))$, where $D_r(x)$ is the Dickson polynomial of degree r. Then,*

1. *If b a primitive element of $\mathbb{F}_{16}$ such that $\mathrm{Tr}_1^4(b) = 0$ then, $\sum_{u\in U} \chi\left(\tilde{g}_{a_r,b}(u)\right) = 1$ if and only if,*

$$\sum_{x\in\mathbb{F}_{2^m}^\star, \mathrm{Tr}_1^m(x^{-1})=1} \chi\left(h_{a_r}(D_5(x))\right) = 2$$

2. *If $b = 1$ then, $\sum_{u\in U} \chi\left(\tilde{g}_{a_r,1}(u)\right) = 1$ if and only if*

$$2 \sum_{x\in\mathbb{F}_{2^m}^\star, \mathrm{Tr}_1^m(x^{-1})=1} \chi\left(h_{a_r}(D_5(x))\right) - 5 \sum_{x\in\mathbb{F}_{2^m}^\star, \mathrm{Tr}_1^m(x^{-1})=1} \chi\left(h_{a_r}(x)\right) = 4.$$

3. *Assume $a_r \in \mathbb{F}_{2^{\frac{m}{2}}}$. If $b \in \{\beta, \beta^2, \beta^3, \beta^4\}$ where β is a primitive 5-th root of unity in $\mathbb{F}_{16}$ then, $\sum_{u\in U} \chi\left(\tilde{g}_{a_r,b}(u)\right) = 1$ if and only if,*

$$\sum_{x\in\mathbb{F}_{2^m}^\star, \mathrm{Tr}_1^m(x^{-1})=1} \chi\left(h_{a_r}(D_5(x))\right) + 5 \sum_{x\in\mathbb{F}_{2^m}^\star, \mathrm{Tr}_1^m(x^{-1})=1} \chi\left(h_{a_r}(x)\right) = -8.$$

4. *Assume $a_r \in \mathbb{F}_{2^{\frac{m}{2}}}$. If b is a primitive element of $\mathbb{F}_{16}$ such that $\mathrm{Tr}_1^4(b) = 1$ then, $\sum_{u\in U} \chi\left(\tilde{g}_{a_r,b}(u)\right) = 1$ if and only if,*

$$3 \sum_{x\in\mathbb{F}_{2^m}^\star, \mathrm{Tr}_1^m(x^{-1})=1} \chi\left(h_{a_r}(D_5(x))\right) - 5 \sum_{x\in\mathbb{F}_{2^m}^\star, \mathrm{Tr}_1^m(x^{-1})=1} \chi\left(h_{a_r}(x)\right) = -4.$$

5. *Assume $a_r \in \mathbb{F}_{2^{\frac{m}{2}}}$. If $b \in \{\beta + \beta^2, \beta + \beta^3, \beta^2 + \beta^4, \beta^3 + \beta^4, \beta + \beta^4, \beta^2 + \beta^3\}$ where β is a primitive 5-th root of unity in $\mathbb{F}_{16}$ then, $\sum_{u\in U} \chi\left(\tilde{g}_{a_r,b}(u)\right) = 1$ if and only if,*

$$\sum_{x\in\mathbb{F}_{2^m}^\star, \mathrm{Tr}_1^m(x^{-1})=1} \chi\left(h_{a_r}(D_5(x))\right) = 2.$$

We deduce the following expressions for $\Lambda(f_{a,b})$.

Theorem 10.3.5. *Suppose* $m := \frac{n}{2} \equiv 2 \pmod 4$. *Let* $R \subseteq E$ *where* E *is a set of representatives of the cyclotomic classes modulo* $2^n - 1$ *for which each class has the full size* n. *For* $b \in \mathbb{F}_{16}^{\star}$ *and* $a_r \in \mathbb{F}_{2^m}^{\star}$, *we denote by* $\tilde{g}_{a_r,b}$ *the function defined on* $\mathbb{F}_{2^n}$ *by* $\sum_{r\in R} \mathrm{Tr}_1^n(a_r x^{r(2^m-1)}) + \mathrm{Tr}_1^4(b' x^{\frac{2^n-1}{5}})$, *and by* g_a *the function defined on* $\mathbb{F}_{2^m}$ *by* $\sum_{r\in R} \mathrm{Tr}_1^m(a_r D_r(x))$, *where* $D_r(x)$ *is the Dickson polynomial of degree* r. *Then,*

1. *If* $b = 1$, *then* $5\Lambda(f_{a,1}) = 4T_1^5(g_a) - 10T_1(g_a) - 3$.
2. *If* b *is a primitive element of* $\mathbb{F}_{16}$ *such that* $\mathrm{Tr}_1^4(b) = 0$, *then* $5\Lambda(f_{a,b}) = 2T_1^5(g_a) + 1$.
3. *If moreover* $a_r \in \mathbb{F}_{2^{\frac{m}{2}}}$, *then*
   (a) *if* b *is a primitive element of* $\mathbb{F}_{16}$ *such that* $\mathrm{Tr}_1^4(b) = 1$, *then* $5\Lambda(f_{a,b}) = -3T_1^5(g_a) + 5T_1(g_a) + 1$;
   (b) *if* b *is a primitive 5-th root of unity, then* $5\Lambda(f_{a,b}) = -T_1^5(g_a) - 5T_1(g_a) - 3$;
   (c) *if* b *is a primitive 3-rd root of unity, then* $5\Lambda(f_{a,b}) = 2T_1(g_a) + 1$.

Recall that $f_{a,b}$ is hyper-bent if and only if $\Lambda(f_{a,b}) = 1$. Therefore, the above theorem gives a characterization of the hyper-bentness of $f_{a,b}$ using $T_1^5(g_a)$ and $T_1(g_a)$. These exponential sums can then be reformulated in terms of the Hamming weight of g_a and related functions using Lemma 2.3.10.

10.4 Hyper-Bent Functions via Dillon-Like Exponents: The General Study

In this section, we shall use the notation in Sect. 2.3.3 when dealing with partial exponential sums.

10.4.1 Extending the Charpin–Gong Criterion

The family of Boolean functions $\mathcal{F}_n$ consists of the functions f_a given in trace representation by Dillon-like exponents, that is

$$f_a(x) = \sum_{r\in R} \mathrm{Tr}_1^n\left(a_r x^{r(2^m-1)}\right) \tag{10.7}$$

where R is a set of representatives of the cyclotomic classes modulo $2^m + 1$ (hence the elements $r(2^m-1)$ yield a set of representatives of the cyclotomic classes modulo $2^n - 1$ of the form $[i(2^m - 1)]$) and the coefficients a_r live in the field $\mathbb{F}_{2^n}$. Departing from the approach of Charpin and Gong, we do not require that the cyclotomic cosets are of maximal size $n = 2m$.

Lemma 10.4.1 ([18]). *Let* f_a *be a Boolean function in* $\mathcal{F}_n$. *Then* $f_a(\alpha^{2^m+1}x) = f_a(x)$.

Proof. We indeed have

$$f_a(\alpha^{2^m+1}x) = \sum_{r\in R} \mathrm{Tr}_1^n\left(a_r(\alpha^{2^m+1}x)^{r(2^m-1)}\right)$$

$$= \sum_{r\in R} \mathrm{Tr}_1^n\left(a_r\alpha^{r(2^n-1)}x^{r(2^m-1)}\right)$$

$$= f_a(x).$$

□

Proposition 9.4.2 can therefore be directly applied to characterize the hyper-bentness of f_a with the partial exponential sum $\Lambda(a) = \Lambda(f_a)$.

Proposition 10.4.2. *Let f_a be a Boolean function in $\mathcal{F}_n$. The function f_a is hyper-bent if and only if $\Lambda(a) = 1$.*

Proof. According to Proposition 9.4.2, f_a is hyper-bent if and only if its restriction to U has Hamming weight 2^{m-1}. Moreover, we have $\Lambda(a) = \#U - 2\,\mathrm{wt}(f_a|_U) = 2^m + 1 - 2\,\mathrm{wt}(f_a|_U)$. Thus, f_a is hyper-bent if and only if $\Lambda(a) = 1$. □

Remark 10.4.3. A hyper-bent function $f_a \in \mathcal{F}_n$ is in $\mathcal{PS}_{ap}$ if and only if $\sum_{r\in R} \mathrm{Tr}_1^n(a_r) = 1$.

In fact, the complete extended Walsh–Hadamard spectrum of f_a can be expressed with $\Lambda(a)$.

Proposition 10.4.4 ([18]). *Let f_a be a Boolean function in $\mathcal{F}_n$ and k an integer co-prime with $2^n - 1$. For $\omega = 0$,*

$$\widehat{\chi_{f_a}}(0,k) = 1 + \Lambda(a)\left(-1 + 2^m\right),$$

and, for $\omega \in \mathbb{F}_{2^n}^*$ non-zero,

$$\widehat{\chi_{f_a}}(\omega,k) = 1 - \Lambda(a) + 2^m \chi_{f_a}\left(\omega^{(2^m-1)/(2k)}\right).$$

Proof. It is a well-known fact that every non-zero element $x \in \mathbb{F}_{2^n}^*$ has a unique polar decomposition as a product $x = yu$ where y lies in the subfield $\mathbb{F}_{2^m}$ and $u \in U$.

The extended Walsh–Hadamard transform of f_a at (ω,k) can consequently be expressed as

$$\widehat{\chi_{f_a}}(\omega,k) = \sum_{x\in\mathbb{F}_{2^n}} \chi\left(f_a(x) + \mathrm{Tr}_1^n\left(\omega x^k\right)\right)$$

$$= 1 + \sum_{x\in\mathbb{F}_{2^n}^*} \chi\left(f_a(x) + \mathrm{Tr}_1^n\left(\omega x^k\right)\right)$$

$$= 1 + \sum_{u\in U}\sum_{y\in\mathbb{F}_{2^m}^*} \chi\left(f_a(yu) + \mathrm{Tr}_1^n\left(\omega y^k u^k\right)\right).$$

But

$$\begin{aligned} f_a(yu) &= \sum_{r\in R} \mathrm{Tr}_1^n \left(a_r (yu)^{r(2^m-1)}\right) \\ &= \sum_{r\in R} \mathrm{Tr}_1^n \left(a_r y^{r(2^m-1)} u^{r(2^m-1)}\right) \\ &= \sum_{r\in R} \mathrm{Tr}_1^n \left(a_r u^{r(2^m-1)}\right) \\ &= f_a(u), \end{aligned}$$

so that

$$\begin{aligned} \widehat{\chi_{f_a}}(\omega,k) &= 1 + \sum_{u\in U} \sum_{y\in\mathbb{F}_{2^m}^*} \chi\left(f_a(u) + \mathrm{Tr}_1^n\left(\omega y^k u^k\right)\right) \\ &= 1 + \sum_{u\in U} \chi_{f_a}(u) \sum_{y\in\mathbb{F}_{2^m}^*} \chi\left(\mathrm{Tr}_1^n\left(\omega y^k u^k\right)\right) \\ &= 1 + \sum_{u\in U} \chi_{f_a}(u) \left(-1 + \sum_{y\in\mathbb{F}_{2^m}} \chi\left(\mathrm{Tr}_1^n\left(\omega y^k u^k\right)\right)\right) \\ &= 1 - \Lambda(a) + \sum_{u\in U} \chi_{f_a}(u) \sum_{y\in\mathbb{F}_{2^m}} \chi\left(\mathrm{Tr}_1^n\left(\omega y^k u^k\right)\right). \end{aligned}$$

If $\omega = 0$, then $\widehat{\chi_f}(\omega,k) = 1 + \Lambda(a)\left(-1 + 2^m\right)$ as desired.
If $\omega \neq 0$, then the transitivity of the trace yields

$$\begin{aligned} \mathrm{Tr}_1^n\left(\omega y^k u^k\right) &= \mathrm{Tr}_1^m(\mathrm{Tr}_m^n\left(\omega y^k u^k\right)) \\ &= \mathrm{Tr}_1^m\left(\omega y^k u^k + \left(\omega y^k u^k\right)^{2^m}\right) \\ &= \mathrm{Tr}_1^m\left(\omega y^k u^k + \omega^{2^m} y^k u^{-k}\right) \\ &= \mathrm{Tr}_1^m\left(y^k\left(\omega u^k + \omega^{2^m} u^{-k}\right)\right). \end{aligned}$$

As a consequence of this equality and of the fact that k is co-prime with $2^m - 1$, the sum over $\mathbb{F}_{2^m}$ is non-zero if and only if $u^{2k} = \omega^{2^m-1}$. As k is co-prime with $2^m + 1$, this only occurs for a value of u. Therefore

$$\widehat{\chi_{f_a}}(\omega,k) = 1 - \Lambda(a) + 2^m \chi_{f_a}\left(\omega^{(2^m-1)/(2k)}\right).$$

□

In particular, Proposition 10.4.2 is a direct corollary to the above proposition.

Remark 10.4.5. Set

$$\bar{f}_a(x) = \sum_{r \in R} \mathrm{Tr}_1^n\left(a_r x^r\right),$$

and let $\overline{\Lambda}(a) = \Lambda(\bar{f}_a)$. The integers $2^m - 1$ and $2^m + 1$ are co-prime and so the (2^m-1)-power map induces a permutation of U. In particular, one has $\Lambda(a) = \overline{\Lambda}(a)$.

We now specailize to the family $\mathcal{G}_n$ of Boolean functions defined as above, but where the coefficients a_r are chosen in the subfield $\mathbb{F}_{2^m}$. The following remark shows that it is enough to restrict to Dillon-like exponents whose cyclotomic coset sizes do not divide m.

Remark 10.4.6. If $t = o(r(2^m - 1))$, then

$$\mathrm{Tr}_1^n\left(a_r x^{r(2^m-1)}\right) = \mathrm{Tr}_1^t\left(\mathrm{Tr}_t^n\left(a_r\right) x^{r(2^m-1)}\right).$$

Suppose now that $a_r \in \mathbb{F}_{2^m}$, e.g. $f_a \in \mathcal{G}_n$. If t divides m, then $\mathrm{Tr}_t^n(a_r) = \mathrm{Tr}_t^m\left(a_r + a_r^{2^m}\right) = 0$ and

$$\mathrm{Tr}_1^n\left(a_r x^{r(2^m-1)}\right) = 0.$$

Otherwise, if $k = \gcd(t, m)$, then $\mathrm{Tr}_t^n(a_r) \in \mathbb{F}_{2^k}$.

Furthermore, Proposition 10.4.4 can be used to compute the dual of f_a in the case where f_a is hyper-bent.

Proposition 10.4.7 ([18]). *Suppose that $f_a \in \mathcal{G}_n$ is hyper-bent. Then it is its own dual, i.e. we have*

$$\widehat{\chi_{f_a}}(\omega) = 2^m \chi_{f_a}(\omega).$$

Proof. If f_a is hyper-bent, then $\Lambda(a) = 1$ and one has

$$\widehat{\chi_{f_a}}(\omega) = 2^m \chi_{f_a}(u),$$

where $u^{1-2^m} = \omega^{2^m-1}$. In particular, one has $f_a(u) = f_a(\omega^{-1})$. One then concludes that $f_a(\omega^{-1}) = f_a(\omega)$ using the facts that $a_r^{2^m} = a_r$ and that $2^m(1-2^m) \equiv 2^m - 1 \pmod{2^n - 1}$. □

For functions f_a in $\mathcal{G}_n$, Remark 10.4.5 combined with the transitivity of the trace yields a useful expression of $\Lambda(a)$ using the partial exponential sum T_1 whose proof we recall here.

Lemma 10.4.8 ([13, Lemma 12]). *Let f_a be a Boolean function in $\mathcal{G}_n$ and l be any positive integer. Let g_a be the Boolean function defined on $\mathbb{F}_{2^m}$ as $g_a(x) = \sum_{r \in R} \mathrm{Tr}_1^m\left(a_r D_r(x)\right)$. Then $\Lambda(f_a(x^l)) = 1 + 2T_1(g_a \circ D_l)$.*

Proof. Using the facts that the (2^m-1)-power map induces a permutation of U, that $a_r^{2^m} = a_r$ and that $D_r(x+x^{-1}) = x^r + x^{-r}$ for any $x \in \mathbb{F}_{2^n}$, one gets

$$\begin{aligned}
\Lambda(f_a(x^l)) &= \sum_{u\in U} \chi\left(\sum_{r\in R} \mathrm{Tr}_1^n\left(a_r\left(u^{2^m-1}\right)^{lr}\right)\right)\\
&= \sum_{u\in U} \chi\left(\sum_{r\in R} \mathrm{Tr}_1^n\left(a_r u^{lr}\right)\right)\\
&= \sum_{u\in U} \chi\left(\sum_{r\in R} \mathrm{Tr}_1^m\left(\left(a_r u^{lr}\right) + \left(a_r u^{lr}\right)^{2^m}\right)\right)\\
&= \sum_{u\in U} \chi\left(\sum_{r\in R} \mathrm{Tr}_1^m\left(a_r\left(u^{lr} + u^{-lr}\right)\right)\right)\\
&= \sum_{u\in U} \chi\left(\sum_{r\in R} \mathrm{Tr}_1^m\left(a_r D_r(D_l(u+u^{-1}))\right)\right).
\end{aligned}$$

To conclude, recall that the map $x \mapsto x + x^{-1}$ is 2-to-1 from $U \setminus \{1\}$ to $\mathcal{T}_1$ to obtain

$$\begin{aligned}
\Lambda(f_a(x^l)) &= 1 + 2\sum_{t\in\mathcal{T}_1} g_a(D_l(t))\\
&= 1 + 2T_1(g_a \circ D_l).
\end{aligned}$$

□

The following extension of the Charpin–Gong criterion [1, Theorem 7] is then straightforward.

Theorem 10.4.9 ([18]). *Let f_a be a Boolean function in $\mathcal{G}_n$. Let g_a be the Boolean function defined on $\mathbb{F}_{2^m}$ as $g_a(x) = \sum_{r\in R} \mathrm{Tr}_1^m(a_r D_r(x))$. Then f_a is hyper-bent if and only if $T_1(g_a) = 0$. Moreover, if f_a is hyper-bent, then it is in the $\mathcal{PS}_{ap}$ class.*

Proof. This is a direct consequence of Proposition 10.4.2 and Lemma 10.4.8. □

10.4.2 Hyper-Bentness Criterion for Functions in $\mathcal{H}_n$

The above approach yields criteria for hyper-bentness of Boolean functions f_a in the families $\mathcal{F}_n$, respectively $\mathcal{G}_n$, involving only one exponential sum over $U \subset \mathbb{F}_{2^n}$, respectively $\mathcal{T}_1 \subset \mathbb{F}_{2^m}$.

In particular, applying Lemma 2.3.10 to Theorem 10.4.9, one gets a characterization for the hyper-bentness of $f_a \in \mathcal{G}_n$ involving only complete exponential sums over $\mathbb{F}_{2^m}$, or equivalently the Hamming weights of g_a and the related function $x \mapsto \mathrm{Tr}_1^m(1/x) + g_a(x)$ defined over $\mathbb{F}_{2^m}$.

Nonetheless, the restriction that is put on the coefficients a_r in the latter case is not satisfying, namely they should live in the field $\mathbb{F}_{2^n}$ rather than in $\mathbb{F}_{2^m}$. In this subsection, we extend our approach to partially address this issue, that is, allow an additional trace term without any restriction on its coefficient.

We therefore consider a different family of Boolean functions defined as follows. The family of Boolean functions $\mathcal{H}_n$ consists of the functions $f_{a,b}$ defined as

$$f_{a,b}(x) = \sum_{r \in R} \mathrm{Tr}_1^n\left(a_r x^{r(2^m-1)}\right) + \mathrm{Tr}_1^t\left(b x^{s(2^m-1)}\right) \tag{10.8}$$

where R is a set of representatives of the cyclotomic classes modulo $2^m + 1$, the coefficients a_r are in $\mathbb{F}_{2^m}$, s divides $2^m + 1$, i.e. $s(2^m - 1)$ is a Dillon-like exponent, $t = o(s(2^m - 1))$, i.e. t is the size of the cyclotomic coset of s modulo $2^m + 1$, and the coefficient b is in $\mathbb{F}_{2^t}$. Moreover, let $\tau = \frac{2^m+1}{s}$. Remark that $f_{a,0} = f_a$ where $f_a \in \mathcal{G}_n$ is the function defined in the previous subsection. Set

$$\overline{f}_{a,b}(x) = \sum_{r \in R} \mathrm{Tr}_1^n(a_r x^r) + \mathrm{Tr}_1^t(b x^s).$$

Remark 10.4.10. According to Remark 10.4.6, the family $\mathcal{H}_n$ is always strictly larger than the family $\mathcal{G}_n$.

Let $U = \{u \in \mathbb{F}_{2^n}^* \mid u^{2^m+1} = 1\}$ and $V = \{v \in \mathbb{F}_{2^n}^* \mid v^s = 1\}$ its subgroup of order s and $W = \{w \in \mathbb{F}_{2^n}^* \mid w^\tau = 1\}$ its subgroup of order τ. Denote by α a primitive element of $\mathbb{F}_{2^n}$. Then $\zeta = \alpha^{2^m-1}$ is a generator of U, $\rho = \zeta^\tau$ is a generator of V and $\xi = \zeta^s$ is a generator of W.

Remark 10.4.11. Note that $\mathbb{F}_{2^t}^* \supset W$. Indeed, by definition $s(2^m - 1) \equiv 2^t s(2^m - 1) \pmod{2^n - 1}$. Thus, $(2^t - 1)s \equiv 0 \pmod{2^m + 1}$, which implies that $2^t - 1 \equiv 0 \pmod{\tau}$, that is τ divides $2^t - 1$.

Remark 10.4.12. Let us consider the τ-power homomorphism $\phi : x \in \mathbb{F}_{2^n}^* \mapsto x^\tau \in \mathbb{F}_{2^n}^*$. Its kernel is W and so it is τ-to-1.

Furthermore, V and W are subsets of U, so that the restriction of ϕ to U maps U onto V and is again τ-to-1.

A similar statement is clearly true for s, exchanging the sets V and W.

Remark 10.4.13. The set U can be decomposed as

$$U = \bigcup_{i=0}^{\tau-1} \zeta^i V = \bigcup_{i=0}^{s-1} \zeta^i W.$$

Definition 10.4.14. For $i \in \mathbb{Z}$, define $S_i(a)$ and $\overline{S}_i(a)$ to be the partial exponential sums

$$S_i(a) = \sum_{v \in V} \chi\left(f_a(\zeta^i v)\right),$$

$$\overline{S}_i(a) = \sum_{v \in V} \chi\left(\overline{f}_a(\zeta^i v)\right).$$

Moreover, define $\Lambda(a,b) = \Lambda(f_{a,b})$ and $\overline{\Lambda}(a,b) = \Lambda(\overline{f}_{a,b})$.

Remark 10.4.15. The Boolean function $f_{a,b}$ is hyper-bent if and only if $\Lambda(a,b) = 1$. Moreover, Remark 10.4.5 can be extended to $f_{a,b}$ and $\overline{f}_{a,b}$ and yields $\Lambda(a,b) = \overline{\Lambda}(a,b)$. Finally, Proposition 10.4.7 can be extended to show that, if $f_{a,b}$ is hyper-bent, then its dual is $f_{a,b^{2^m}}$.

Remark 10.4.16. Remark that ζ is of order τ so that $S_i(a)$ and $\overline{S}_i(a)$ only depend on the value of i modulo τ.

Remark 10.4.17. One obviously has

$$\sum_{i=0}^{\tau-1} S_i(a) = \Lambda(a,0) = \Lambda(a).$$

In particular, Lemma 10.4.8 yields

$$\sum_{i=0}^{\tau-1} S_i(a) = 1 + 2T_1(g_a).$$

In the particular case where f_a is a monomial function with a Dillon exponent, i.e. $f_a(x) = \mathrm{Tr}_1^n\left(ax^{r(2^m-1)}\right)$ where r is co-prime with 2^m+1, Remark 10.4.17 can be further refined.

Lemma 10.4.18 ([18]). *Suppose that r is co-prime with* $2^m + 1$. *One has*

$$\sum_{i=0}^{\tau-1} S_i(a) = 1 - K_m(a).$$

Proof. The function $u \mapsto u + u^{-1}$ being onto and 2-to-1 from $U \setminus \{1\}$ to $\mathcal{T}_1$, one gets

$$\begin{aligned} K_m(a) &= -2T_1(\mathrm{Tr}_1^m(ax)) \\ &= -\sum_{u \in U,\, u \neq 1} \chi\left(\mathrm{Tr}_1^m\left(a\left(u+u^{-1}\right)\right)\right) \end{aligned}$$

$$= -\sum_{u\in U,\ u\neq 1} \chi\left(\mathrm{Tr}_1^n(au)\right)$$

$$= 1 - \sum_{u\in U} \chi\left(\mathrm{Tr}_1^n(au)\right).$$

Furthermore, the r-power map induces a permutation of U and thus

$$\begin{aligned}\sum_{u\in U} \chi\left(\mathrm{Tr}_1^n(au)\right) &= \sum_{u\in U} \chi\left(\mathrm{Tr}_1^n(au^r)\right)\\ &= \overline{\Lambda}(a)\\ &= \Lambda(a).\end{aligned}$$

□

The two partial exponential sums S_i and $\overline{S}_i$ defined above are closely related.

Lemma 10.4.19 ([18]). *For* $0 \leq i \leq \tau - 1$, *one has*

$$S_i(a) = \overline{S}_{-2i}(a).$$

Proof. First, one has

$$\begin{aligned}S_i(a) &= \sum_{v\in V} \chi\left(f_a(\zeta^i v)\right)\\ &= \sum_{v\in V} \chi\left(\sum_{r\in R} \mathrm{Tr}_1^n\left(a_r (\zeta^i v)^{r(2^m-1)}\right)\right)\\ &= \sum_{v\in V} \chi\left(\sum_{r\in R} \mathrm{Tr}_1^n\left(a_r \left(\zeta^{i(2^m-1)} v^{2^m-1}\right)^r\right)\right).\end{aligned}$$

But $2^m - 1$ is co-prime with s, so that the (2^m-1)-power map induces a permutation of V, as does multiplication by ζ^τ. Moreover, $2^m + 1 \equiv 0 \pmod{\tau}$ implies that $2^m - 1 \equiv -2 \pmod{\tau}$. Hence,

$$S_i(a) = \sum_{v\in V} \chi\left(\sum_{r\in R} \mathrm{Tr}_1^n\left(a_r \left(\zeta^{-2i} v\right)^r\right)\right).$$

□

Remark 10.4.17 can then be extended to express $\Lambda(a, b)$ as a linear combination of the sums S_i.

Proposition 10.4.20 ([18]). *One has*

$$\Lambda(a,b) = \sum_{i=0}^{\tau-1} \chi\left(\mathrm{Tr}_1^t\left(b\xi^i\right)\right) \overline{S}_i(a).$$

Proof. Indeed,

$$\begin{aligned}
\Lambda(a,b) &= \overline{\Lambda}(a,b) \\
&= \sum_{u\in U} \chi\left(\overline{f}_a(u) + \mathrm{Tr}_1^t\left(bu^s\right)\right) \\
&= \sum_{u\in U} \chi\left(\overline{f}_a(u)\right) \chi\left(\mathrm{Tr}_1^t\left(bu^s\right)\right) \\
&= \sum_{i=0}^{\tau-1} \sum_{v\in V} \chi\left(\overline{f}_a(\zeta^i v)\right) \chi\left(\mathrm{Tr}_1^t\left(b\left(\zeta^i v\right)^s\right)\right) \\
&= \sum_{i=0}^{\tau-1} \chi\left(\mathrm{Tr}_1^t\left(b\xi^i\right)\right) \sum_{v\in V} \chi\left(\overline{f}_a(\zeta^i v)\right) \\
&= \sum_{i=0}^{\tau-1} \chi\left(\mathrm{Tr}_1^t\left(b\xi^i\right)\right) \overline{S}_i(a).
\end{aligned}$$

□

We now devise an additional relation between the partial exponential sums S_i and the partial exponential sum T_1. In particular, we express the partial exponential sum S_0 using T_1.

Lemma 10.4.21 ([18]). *Let l be a divisor of τ and let k be the integer $k = \tau/l$. Then*

$$\sum_{i=0}^{k-1} S_{il}(a) = \sum_{i=0}^{k-1} \overline{S}_{il}(a) = \frac{1}{l}\left(1 + 2T_1(g_a \circ D_l)\right).$$

For $l = 1$, it reads

$$\sum_{i=0}^{\tau-1} S_i(a) = \sum_{i=0}^{\tau-1} \overline{S}_i(a) = \left(1 + 2T_1(g_a)\right),$$

which is nothing but Remark 10.4.17. For $l = \tau$, it reads

$$S_0(a) = \overline{S}_0(a) = \frac{1}{\tau}\left(1 + 2T_1(g_a \circ D_\tau)\right).$$

Proof. According to a straightforward extension of Remark 10.4.11, the l-power map is l-to-1 from U onto $\bigcup_{i=0}^{k-1}\zeta^{il}V$. Therefore,

$$\sum_{i=0}^{k-1} S_{il}(a) = \sum_{i=0}^{k-1}\sum_{v\in V}\chi\left(f_a(\zeta^{il}v)\right)$$
$$= \frac{1}{l}\sum_{u\in U}\chi\left(f_a(u^l)\right).$$

One then concludes with Lemma 10.4.8.

The results for $\overline{S}_i$ readily follows from the fact that multiplication by -2 induces a permutation of $\{il\}_{i=0}^{k-1}$ and Lemma 10.4.19. □

Remark 10.4.22. Recall that τ divides 2^m+1, and so does l. Therefore, τ and l are co-prime with 2^m-1. According to Corollary 2.1.6, D_l induces a permutation of $\mathcal{T}_0$, whence the validity of the equality

$$\sum_{i=0}^{k-1} S_{il}(a) = \sum_{i=0}^{k-1}\overline{S}_{il}(a) = \frac{1}{l}\left(1+2\Xi\left(g_a\circ D_l\right)-2T_0(g_a)\right).$$

In the case where $l=\tau$, it reads

$$S_0(a) = \overline{S}_0(a) = \frac{1}{\tau}\left(1+2\Xi\left(g_a\circ D_\tau\right)-2T_0(g_a)\right).$$

To conclude this section, we show how further identities involving the partial exponential sums S_i can be obtained by restricting the field of definition of the coefficients a_r to a strict subfield of $\mathbb{F}_{2^m}$.

Lemma 10.4.23 ([18]). *Let l be a divisor of m and $k = m/l$. Suppose that the coefficients a_r lie in $\mathbb{F}_{2^l}$ and that $2^l \equiv j \pmod{\tau}$, where j is a k-th root of -1 modulo τ. Then*

$$\overline{S}_i(a) = \overline{S}_{ij}(a).$$

Proof. Recall that $2^m \equiv -1 \pmod{\tau}$. Hence, if $2^l \equiv j \pmod{\tau}$, then j is a k-th root of -1 modulo τ.

Since $a_r\in\mathbb{F}_{2^l}$, one has $a_r^{2^l}=a_r$. Recall that $\mathrm{Tr}_1^n\left(x^2\right)=\mathrm{Tr}_1^n(x)$, so that

$$\overline{S}_i(a) = \sum_{v\in V}\chi\left(\overline{f}_a(\zeta^i v)\right)$$
$$= \sum_{v\in V}\chi\left(\sum_{r\in R}\mathrm{Tr}_1^n\left(a_r(\zeta^i v)^r\right)\right)$$

$$= \sum_{v \in V} \chi \left(\sum_{r \in R} \mathrm{Tr}_1^n \left(a_r^{2^l} (\zeta^{2^l i} v^{2^l})^r \right) \right)$$

$$= \sum_{v \in V} \chi \left(\sum_{r \in R} \mathrm{Tr}_1^n \left(a_r (\zeta^{ij} \zeta^{i(2^l - j)} v^{2^l})^r \right) \right).$$

But the (2^l)-power map and multiplication by $\zeta^{i(2^l-j)}$ induce permutations of V and therefore

$$\overline{S}_i(a) = \sum_{v \in V} \chi \left(\sum_{r \in R} \mathrm{Tr}_1^n \left(a_r (\zeta^{ij} v)^r \right) \right)$$

$$= \overline{S}_{ij}(a).$$

□

Remark 10.4.24. In the particular case where $l = m$, note that $2^m \equiv -1 \pmod{\tau}$. Therefore, one has

$$\overline{S}_i(a) = \overline{S}_{-i}(a).$$

One then deduces from Proposition 10.4.20 that

$$\Lambda(a, b) = \chi\left(\mathrm{Tr}_1^t(b)\right) \overline{S}_0(a) + \sum_{i=1}^{\frac{\tau-1}{2}} \left(\chi\left(\mathrm{Tr}_1^t\left(b\xi^i\right)\right) + \chi\left(\mathrm{Tr}_1^t\left(b\xi^{-i}\right)\right) \right) \overline{S}_i(a).$$

Remark 10.4.25. It is a difficult problem to deduce a completely general characterization of hyper-bentness in terms of complete exponential sums from the results of the current section, that is, a characterization valid for any m, s and b. Nevertheless, several powerful applications of these results, valid for infinite families of Boolean functions, will be described in Sect. 10.6.

10.4.3 An Alternate Proof

To provide an alternate proof of Proposition 10.4.20, we introduce different exponential sums.

Proposition 10.4.26 ([18]). *For $c \in \mathbb{F}_{2^t}$, let $\tilde{\Lambda}(a, c)$ be the exponential sum*

$$\tilde{\Lambda}(a, c) = \sum_{b \in \mathbb{F}_{2^t}} \chi\left(\mathrm{Tr}_1^t(bc)\right) \Lambda(a, b).$$

1. *For all* $c \in \mathbb{F}_{2^t}$*, one has*

$$\tilde{\Lambda}(a,c) = 2^t \sum_{u \in U,\ u^s = c} \chi\left(\overline{f}_a(u)\right).$$

2. *If* $c \in \mathbb{F}_{2^t} \setminus W$*, then* $\tilde{\Lambda}(a,c) = 0$*. If* $c \in W$*, that is if* $c = \xi^i$ *for some* i*, then*

$$\tilde{\Lambda}(a,\xi^i) = 2^t \overline{S}_i(a).$$

Proof. 1. Exchanging the summation orders on U and $\mathbb{F}_{2^t}$ yields

$$\begin{aligned}
\tilde{\Lambda}(a,c) &= \sum_{b \in \mathbb{F}_{2^t}} \chi\left(\operatorname{Tr}_1^t(bc)\right) \sum_{u \in U} \chi\left(f_{a,b}(u)\right) \\
&= \sum_{b \in \mathbb{F}_{2^t}} \chi\left(\operatorname{Tr}_1^t(bc)\right) \sum_{u \in U} \chi\left(f_a(u)\right) \chi\left(\operatorname{Tr}_1^t\left(bu^{s(2^m-1)}\right)\right) \\
&= \sum_{u \in U} \chi\left(f_a(u)\right) \sum_{b \in \mathbb{F}_{2^t}} \chi\left(\operatorname{Tr}_1^t\left(b\left(c + u^{s(2^m-1)}\right)\right)\right).
\end{aligned}$$

The sum over $\mathbb{F}_{2^t}$ is non-zero if and only if $c = u^{s(2^m-1)}$ so that

$$\begin{aligned}
\tilde{\Lambda}(a,c) &= 2^t \sum_{u \in U,\ u^{s(2^m-1)} = c} \chi\left(f_a(u)\right) \\
&= 2^t \sum_{u \in U,\ u^s = c} \chi\left(\overline{f}_a(u)\right).
\end{aligned}$$

2. According to Remark 10.4.11, if $c \in \mathbb{F}_{2^t} \setminus W$, then the equation $u^s = c$ has no solutions in U. Therefore, $\tilde{\Lambda}(a,c) = 0$.

Suppose now that $c \in W$ and that $c = \xi^i = \zeta^{is}$ for some i. The kernel of the s-power map is V so that $u^s = \zeta^{is}$ if and only if $u \in \zeta^i V$. Thus, we have

$$\tilde{\Lambda}(a,c) = 2^t \sum_{v \in V} \chi\left(\overline{f}_a(\zeta^i v)\right).$$

□

The partial exponential sum $\Lambda(a,b)$ can now be expressed with $\tilde{\Lambda}(a,c)$.

Lemma 10.4.27 ([18]). *One has*

$$\Lambda(a,b) = \frac{1}{2^t} \sum_{c \in \mathbb{F}_{2^t}} \chi\left(\operatorname{Tr}_1^t(bc)\right) \tilde{\Lambda}(a,c).$$

Proof. Going back to the definition of $\tilde{\Lambda}(a,c)$, one has

$$\begin{aligned}\sum_{c\in\mathbb{F}_{2^t}} \chi\left(\mathrm{Tr}_1^t(bc)\right)\tilde{\Lambda}(a,c) &= \sum_{c\in\mathbb{F}_{2^t}} \chi\left(\mathrm{Tr}_1^t(bc)\right)\sum_{d\in\mathbb{F}_{2^t}} \chi\left(\mathrm{Tr}_1^t(dc)\right)\Lambda(a,d)\\ &= \sum_{d\in\mathbb{F}_{2^t}} \Lambda(a,d)\sum_{c\in\mathbb{F}_{2^t}} \chi\left(\mathrm{Tr}_1^t(bc)\right)\chi\left(\mathrm{Tr}_1^t(dc)\right)\\ &= \sum_{d\in\mathbb{F}_{2^t}} \Lambda(a,d)\sum_{c\in\mathbb{F}_{2^t}} \chi\left(\mathrm{Tr}_1^t((b+d)\,c)\right).\end{aligned}$$

But $\sum_{c\in\mathbb{F}_{2^t}} \chi\left(\mathrm{Tr}_1^t((b+d)\,c)\right) = 0$ if $b \neq d$ and 2^t otherwise. Therefore

$$\sum_{c\in\mathbb{F}_{2^t}} \chi\left(\mathrm{Tr}_1^t(bc)\right)\tilde{\Lambda}(a,c) = 2^t\Lambda(a,b).$$

□

Remark 10.4.28. Proposition 10.4.26 and Lemma 10.4.27 provide an alternate proof of Proposition 10.4.20:

$$\begin{aligned}\Lambda(a,b) &= \frac{1}{2^t}\sum_{c\in\mathbb{F}_{2^t}} \chi\left(\mathrm{Tr}_1^t(bc)\right)\tilde{\Lambda}(a,c)\\ &= \frac{1}{2^t}\left(\sum_{c\in\mathbb{F}_{2^t}\setminus W} \chi\left(\mathrm{Tr}_1^t(bc)\right)\tilde{\Lambda}(a,c) + \sum_{c\in W} \chi\left(\mathrm{Tr}_1^t(bc)\right)\tilde{\Lambda}(a,c)\right)\\ &= \frac{1}{2^t}\sum_{c\in W} \chi\left(\mathrm{Tr}_1^t(bc)\right)\tilde{\Lambda}(a,c)\\ &= \frac{1}{2^t}\sum_{i=0}^{\tau-1} \chi\left(\mathrm{Tr}_1^t(b\xi^i)\right)\tilde{\Lambda}(a,\xi^i)\\ &= \frac{1}{2^t}\sum_{i=0}^{\tau-1} \chi\left(\mathrm{Tr}_1^t(b\xi^i)\right)2^t\overline{S}_i(a)\\ &= \sum_{i=0}^{\tau-1} \chi\left(\mathrm{Tr}_1^t(b\xi^i)\right)\overline{S}_i(a).\end{aligned}$$

10.5 Building Infinite Families of Extension Degrees

In the previous section, we set an extension degree m and studied the corresponding exponents s dividing $2^m + 1$. It is however customary to go the other way around, i.e. set an exponent s, or a given form of exponents, which is valid for an infinite family of extension degrees m and devise characterizations valid for this infinity of extension degrees. In this section, we provide the link between these two approaches.

More precisely, we supposed above that the additional trace term had a Dillon-like exponent, i.e. that is was of the form $s(2^m - 1)$ where s divides $2^m + 1$ and $\tau = \frac{2^m+1}{s}$. Hence, the Dillon-like exponent could be written as $\frac{2^n-1}{\tau} = \frac{2^m+1}{\tau}(2^m-1)$ and the above construction then relied on the fact that τ divided $2^m + 1$, that is that $2^m \equiv -1 \pmod{\tau}$ or equivalently that -1 was in the cyclotomic coset of 1 modulo $2^m + 1$.

The problem we tackle in this section is the following: fix a value for τ and devise the extension degrees m for which τ divides $2^m + 1$. In fact, there are only two possibilities for a given τ: either there is an infinity of such extension degrees, or there is no extension degree at all. Therefore, we focus on the construction and characterization of values of τ for which an infinite number of such extension degrees m exists, starting with prime numbers and then extending our approach to prime powers and finally to odd composite numbers. An even τ can not divide $2^m + 1$ and this last case covers all possibilities for τ.

10.5.1 Prime Case

Let p be an odd prime number and set $\tau = p$. The set of modular integers $\mathbb{Z}/p\mathbb{Z}$ is a field and there exists i such that $2^i \equiv -1 \pmod{p}$ if and only if the multiplicative order of 2 modulo p is even. In this case, $2^m \equiv -1 \pmod{p}$ if and only if $m \equiv o \pmod{2o}$, where $2o$ is the multiplicative order of 2 modulo p. In particular, the family of such extension degrees m is infinite. The size $t = o(s)$ of the cyclotomic coset of $s = (2^m + 1)/p$ modulo $2^m + 1$ is then

$$t = 2o.$$

Furthermore, one has

$$2^m \equiv 2^o \pmod{2^t - 1},$$

so that if $f_{a,b} \in \mathcal{H}_n$ is hyper-bent, then its dual is $f_{a,b^{2^o}}$.

To actually devise such prime numbers, we now focus on the specific case where the multiplicative order of 2 modulo p is maximal, that is where 2 is a primitive root modulo p. In this situation, the above condition becomes

$$2^{\frac{p-1}{2}} \equiv -1 \pmod{p}.$$

This implies that the Legendre symbol $\left(\frac{2}{p}\right)$ of 2 modulo p is -1 and that 2 is a quadratic nonresidue modulo p. It is well-known that the Legendre symbol of 2 modulo an odd prime p is

$$\left(\frac{2}{p}\right) = (-1)^{\frac{p^2-1}{8}} = \begin{cases} 1 & \text{if } p \equiv \pm 1 \pmod 8, \\ -1 & \text{if } p \equiv \pm 3 \pmod 8. \end{cases}$$

Therefore, if 2 is a primitive root modulo p, then one must have $p \equiv \pm 3 \pmod 8$. This gives a practical criterion to discard prime numbers such that 2 is not a primitive element. Further characterizations of primes p such that 2 is a primitive root modulo p can be found in a paper of Park, Park and Kim [19].

For such a prime number p, $2^m \equiv -1 \pmod p$ if and only if $m \equiv \frac{p-1}{2} \pmod{p-1}$. The size $t = o(s)$ of the cyclotomic coset of $s = (2^m + 1)/p$ modulo $2^m + 1$ is then

$$t = p - 1.$$

Finding an infinite number of odd prime numbers for which 2 is a primitive element would thus give an elegant solution to our problem, i.e. finding an infinite family of denominators $\tau = p$ associated with infinite families of extension degrees m. This question is however difficult; it is a special case of Artin's conjecture on primitive roots.

Conjecture 10.5.1 (Artin's Conjecture on Primitive Roots). *Let a be an integer which is neither a perfect square nor -1. Then the number of primes numbers p such that a is a primitive element modulo p is infinite.*

It should be noted that Artin's conjecture has been proved by Hooley [10] under the Generalized Riemann Hypothesis. Heath-Brown [9] has proved unconditionally that there exist at most two exceptional primes for which Artin's conjecture fails; nonetheless, this proof is non-constructive.

From a more computational perspective, the first elements of the sequence of primes such that 2 is a primitive element is sequence `A001122` in OEIS [11] and begins with

$$3,\ 5,\ 11,\ 13,\ 19,\ 29,\ 37,\ 53,\ 59,\ 61,\ 67,\ 83.$$

As mentioned in the beginning of this section, it is not necessary that 2 is a primitive root modulo 2 for 1 and -1 to lie in the same cyclotomic coset modulo p. The list of odd primes p smaller than 100 such that the multiplicative order of 2 modulo p is even and a strict divisor of $p - 1$, together with half the order o of 2, i.e. the smallest integer o such that $2^o \equiv -1 \pmod p$, is

$$(17, 4),\ (41, 10),\ (43, 7),\ (97, 24).$$

Finally, there exist as well primes for which 1 and -1 are not in the same cyclotomic coset modulo p. The list of such primes smaller than 100 is

$$7,\ 23,\ 31,\ 47,\ 71,\ 73,\ 79,\ 89.$$

10.5.2 Prime Power Case

Let p be an odd prime number and $k \geq 2$ a positive integer. Set $\tau = p^k$. The multiplicative group of units modulo p^k is once again cyclic and isomorphic to

$$(\mathbb{Z}/p^k\mathbb{Z})^\times \simeq (\mathbb{Z}/(p-1)\mathbb{Z}) \times (\mathbb{Z}/p\mathbb{Z})^{k-1}.$$

The condition for the prime case is thus still valid; there exists i such that $2^i \equiv -1 \pmod{p^k}$ if and only if the multiplicative order of 2 modulo p^k is even. In this case, $2^m \equiv -1 \pmod{p^k}$ if and only if $m \equiv o \pmod{2o}$, where $2o$ is the multiplicative order of 2 modulo p^k. In particular, the family of such extension degrees is also infinite. The size $t = o(s)$ of the cyclotomic coset of $s = (2^m + 1)/p^k$ modulo $2^m + 1$, is then

$$t = 2o.$$

If $f_{a,b} \in \mathcal{H}_n$ is hyper-bent, then its dual is $f_{a,b^{2^o}}$.

It is a classical result [3, Lemma 1.4.5 and following remarks], that if an integer a is a primitive root modulo p, then a or $a + p$ is a primitive root modulo p^2. Furthermore, if a is a primitive root modulo p^2, then it is modulo p^k for any $k \geq 2$ [3, Lemma 1.4.5 and following remarks]. Conversely, if a is not a primitive root modulo p^i, then it is not a primitive root modulo p^k for any $k \geq i$. The approach of the previous subsection can therefore be extended to any prime power p^k with $k \geq 2$ by just checking that 2 is a primitive root modulo p^2. If it is, then

$$2^{\frac{\phi(p^k)}{2}} \equiv -1 \pmod{\phi(p^k)}$$

for any $k \geq 2$, where ϕ denotes Euler's totient function. In particular, we have $\phi(p^k) = (p-1)p^{k-1}$. In this case, one would choose $m \equiv \frac{\phi(p^k)}{2} \pmod{\phi(p^k)}$. The size $t = o(s)$ of the cyclotomic coset of $s = (2^m + 1)/p^k$ modulo $2^m + 1$, is then

$$t = \phi(p^k).$$

The primes smaller than 100 such that 2 is a primitive root modulo p^2 are

$$3,\ 5,\ 11,\ 13,\ 19,\ 29,\ 37,\ 53,\ 59,\ 61,\ 67,\ 83.$$

From a computational perspective, more can be said. Indeed, if 2 is a primitive root modulo p, but is not modulo p^2, a simple calculation shows that $2^{p-1} \equiv 1 \pmod{p^2}$, that is, p is a *Wieferich prime*. The sequence of such primes is sequence A001220 in the OEIS [11]. Only two of them are currently known: 1093 and 3511; and 2 is not a primitive root for both of these primes. Checking that 2 is a primitive root modulo p is therefore enough to ensure that it is modulo any power of p as long as p is not too large, less than 15 decimal digits according to Dorais and Klyve [4].

The list of odd primes p smaller than 100 such that the multiplicative order of 2 modulo p^2 is even and a strict divisor of $\phi(p^2)$, together with half the order o of 2, i.e. the smallest integer o such that $2^o \equiv -1 \pmod{p^2}$, is

$$(17, 68), \ (41, 410), \ (43, 301), \ (97, 2328).$$

Finally, the list of odd primes p smaller than 100 such that 1 and -1 do not lie in the same cyclotomic coset modulo p^2 is

$$7, \ 23, \ 31, \ 47, \ 71, \ 73, \ 79, \ 89.$$

10.5.3 Composite Case

We now consider the general case of an odd composite number. Suppose that $\tau = p_1^{k_1} \cdots p_r^{k_r}$ is a product of $r \geq 2$ distinct prime powers.

The multiplicative group of units modulo τ is not cyclic anymore and is isomorphic to the product of the cyclic groups corresponding to each prime power:

$$(\mathbb{Z}/\tau\mathbb{Z})^\times \simeq \left(\mathbb{Z}/p_1^{k_1}\mathbb{Z}\right)^\times \times \cdots \times \left(\mathbb{Z}/p_r^{k_r}\mathbb{Z}\right)^\times .$$

The multiplicative order of 2 modulo τ is the least common multiple of its multiplicative orders modulo the prime powers dividing τ. There exists an integer i such that $2^i \equiv -1 \pmod{\tau}$ if and only if there exists such integers for each prime power dividing τ, that is if the multiplicative order of 2 modulo $p_j^{k_j}$ is even for $1 \leq j \leq r$, and if moreover their least common multiple is an odd multiple of each of them, that is if they all have the same 2-adic valuation. In such a situation, $2^m \equiv -1 \pmod{\tau}$ if and only if $m \equiv o \pmod{2o}$, where $2o$ is the multiplicative order of 2 modulo τ. In particular, the family of such extension degrees is still infinite. Recall that the corresponding denominator is τ. The size $t = o(s)$ of the cyclotomic coset of $s = (2^m + 1)/\tau$ modulo $2^m + 1$, is then

$$t = 2o.$$

If $f_{a,b} \in \mathcal{H}_n$ is hyper-bent, then its dual is $f_{a,b^{2^o}}$.

In particular, if 2 is a primitive root modulo each prime power dividing τ, then the multiplicative order of 2 modulo τ is

$$2o = \mathrm{lcm}(\phi(p_1^{k_1}), \ldots, \phi(p_r^{k_r})),$$

and $2^o \equiv -1 \pmod{\tau}$ if and only if $\nu_2(p_1-1) = \cdots = \nu_2(p_r-1)$, where ν_2 denotes the 2-adic valuation. Conditioned by the fact that there exists an infinite number of primes p such that 2 is a primitive root modulo p or modulo p^2 and such that $p-1$ has a given 2-adic valuation, we can construct an infinite number of composite odd numbers addressing our original problem.

The list of suitable odd composite numbers τ smaller than 100, together with half the multiplicative order o of 2 modulo τ, that is the smallest integer such that $2^o \equiv -1 \pmod{\tau}$, is

$$(33, 5),\ (57, 9),\ (65, 6),\ (99, 15).$$

10.6 Applications

In this section, we show how the results of Sect. 10.4 can be applied to several infinite families of Boolean functions in order to obtain characterizations of their hyper-bentness in terms of exponential sums over $\mathcal{T}_1 \subset \mathbb{F}_{2^m}$. Such characterizations can easily be transformed into characterizations involving complete exponential sums over $\mathbb{F}_{2^m}$ using Lemma 2.3.10, or the Hamming weights of g_a and the related function $x \mapsto \mathrm{Tr}_1^m(1/x) + g_a(x)$ defined over $\mathbb{F}_{2^m}$. Much of these applications can be straightforwardly extended to additional cases.

10.6.1 The Case $b = 1$

We first apply results of Sects. 10.4.1 and 10.4.2 to $f_{a,1}$ defined as in Eq. (10.8) in the specific case where $b = 1$.

Since 1 lies in $\mathbb{F}_2$, there exists $\beta \in \mathbb{F}_{2^m} \subset \mathbb{F}_{2^n}$ such that $\mathrm{Tr}_t^n(\beta) = 1$. In particular, $f_{a,1}$ belongs to both families $\mathcal{G}_n$ and $\mathcal{H}_n$. In fact, the discussion in Sect. 10.5, shows that $m \equiv o \pmod{2o}$ and that $t = o(s) = 2o$ where $2o$ is the multiplicative order of 2 modulo $\tau = \frac{2^m+1}{s}$. Hence, $n/t = m/l$ is odd and β can be chosen to be 1. Applying Theorem 10.4.9 shows that $f_{a,1}$ is hyper-bent if and only if

$$\sum_{t \in \mathcal{T}_1} \chi\left(\sum_{r \in R} \mathrm{Tr}_1^m(a_r D_r(t)) + \mathrm{Tr}_1^m(D_s(t))\right) = 0.$$

Applying Lemma 2.3.10, this condition is straightforwardly expressed in terms of complete exponential sums over $\mathbb{F}_{2^m}$, or of the Hamming weights of $g'_a : x \mapsto g_a(x) + \mathrm{Tr}_1^m(D_s(t))$ and the related function $x \mapsto \mathrm{Tr}_1^m(1/x) + g'_a(x)$ defined over $\mathbb{F}_{2^m}$. To summarize, we have the following characterization for the value of $\Lambda(a, 1)$ and so for the hyper-bentness for $f_{a,1}$.

Proposition 10.6.1 ([18]). *Let g'_a be the Boolean function defined on $\mathbb{F}_{2^m}$ as $g'_a(x) = g_a(x) + \mathrm{Tr}_1^m(D_s(x))$. Then*

$$\Lambda(a, 1) = 2T_1(g'_a) + 1.$$

In particular, $f_{a,1}$ is hyper-bent if and only if

$$T_1(g'_a) = 0.$$

We now show how the results of Sect. 10.4.2 can be applied to obtain a different characterization of the hyper-bentness of $f_{a,1}$. According to Proposition 10.4.2, $f_{a,1}$ is hyper-bent if and only if

$$\Lambda(a, 1) = 1.$$

Let ξ be a primitive τ-th root of unity. First, recall that ξ lies in $\mathbb{F}_{2^t}$, that $\mathrm{Tr}_1^t(\xi^2) = \mathrm{Tr}_1^t(\xi)$ and that

$$\sum_{i=0}^{\tau-1} \xi^i = 0.$$

Second, remark that the results of Sect. 10.5 imply that t is even, so that $\mathrm{Tr}_1^t(1) = 0$. Moreover, ξ is a $(2^{t/2} + 1)$-th root of unity so that $\xi + \xi^{-1} \in \mathbb{F}_{2^{t/2}}$ which implies that

$$\mathrm{Tr}_1^t(\xi^i) = \mathrm{Tr}_1^t(\xi^{-i}).$$

Finally, Proposition 10.4.20 reads

$$\Lambda(a, 1) = \overline{S}_0(a) + 2\sum_{i=1}^{\frac{\tau-1}{2}} \chi\left(\mathrm{Tr}_1^t(\xi^i)\right) \overline{S}_i(a).$$

Nonetheless, the trace of ξ^i for $i \neq 0$ depends on the exact value of τ. In the sequel, we deal with some specific cases.

10.6.1.1 Prime Case

For simplicity, we first suppose that $\tau = p$ is a prime and that 2 is a primitive root modulo p. In this case, we have $t = p - 1$ and i is co-prime with p, so that

$$\mathrm{Tr}_1^{p-1}(\xi^i) = \sum_{j=0}^{p-2} \xi^{i2^j} = \sum_{j=1}^{p-1} \xi^{ij} = \sum_{j=1}^{p-1} \xi^j = 1.$$

Therefore

$$\Lambda(a,1) = 2\overline{S}_0(a) - \sum_{i=0}^{p-1} \overline{S}_i(a).$$

Applying Lemma 10.4.21 with $l = 1$ and $l = p$ yields

$$\Lambda(a,1) = \frac{2}{p}(1 + 2T_1(g_a \circ D_p)) - (1 + 2T_1(g_a)).$$

Consequently, we get the following characterization.

Proposition 10.6.2 ([18]). *Suppose that $\tau = p$ is a prime and that 2 is a primitive root modulo p. Then*

$$p\Lambda(a,1) = 4T_1(g_a \circ D_p) - 2pT_1(g_a) - p + 2.$$

In particular, $f_{a,1}$ is hyper-bent if and only if

$$2T_1(g_a \circ D_p) - pT_1(g_a) = p - 1.$$

10.6.1.2 Prime Power Case

We now treat the case where $\tau = p^k$ is a prime power and 2 is a primitive root modulo p^k, including the prime case where $k = 1$. Then $t = \phi(p^k) = (p-1)p^{k-1}$. Remark that in this situation, for every positive integers $i \geq 0$ and $j > 0$ such that $i + j = k$, one has $\left(\xi^{p^i}\right)^{p^j} = \xi^{p^k} = 1$, so that

$$\sum_{l=0}^{p^j-1} \xi^{lp^i} = 0. \tag{10.9}$$

Then

$$\mathrm{Tr}_1^{\phi(p^k)}(\xi^i) = \sum_{j=0}^{\phi(p^k)-1} \xi^{i2^j} = \sum_{1 \leq j \leq p^k-1,\ p \nmid j} \xi^{ij}.$$

If $p^e \mid\mid i$ with $0 \leq e \leq k-1$, then $i = lp^e$ with l co-prime with $p-1$ and

$$
\begin{aligned}
\mathrm{Tr}_1^{\phi(p^k)}\left(\xi^i\right) &= \sum_{1 \leq j \leq p^k-1,\ p \nmid j} \xi^{jlp^e} \\
&= \sum_{1 \leq j \leq p^k-1,\ p \nmid j} \xi^{jp^e} \\
&= \sum_{j=0}^{p^k-1} \xi^{jp^e} + \sum_{j=0}^{p^{k-1}-1} \xi^{jp^{e+1}} \\
&= \sum_{j=0}^{p^k-1} \xi^{jp^e} + \sum_{j=0}^{p^k-1} \xi^{jp^{e+1}} + \sum_{j=p^{k-1}}^{p^k-1} \xi^{jp^{e+1}}.
\end{aligned}
$$

Equation (10.9) shows that the first two sums of the right hand side of the last equality can be split into a multiple of sums equal to zero. If $0 \leq e \leq k-2$, then the third sum is zero as well, so that

$$\mathrm{Tr}_1^{\phi(p^k)}\left(\xi^i\right) = 0.$$

If $e = k-1$, then the third sum reads

$$\sum_{j=p^{k-1}}^{p^k-1} \xi^{jp^k} = \sum_{j=p^{k-1}}^{p^k-1} \xi^j = 1.$$

Therefore

$$\mathrm{Tr}_1^{\phi(p^k)}\left(\xi^i\right) = 1.$$

Summing up the above observations yields

$$
\begin{aligned}
\Lambda(a,1) &= \sum_{i=0}^{p^k-1} \overline{S}_i(a) - 2\sum_{i=1}^{p-1} \overline{S}_{ip^{k-1}}(a) \\
&= 2\overline{S}_0(a) + \sum_{i=0}^{p^k-1} \overline{S}_i(a) - 2\sum_{i=0}^{p-1} \overline{S}_{ip^{k-1}}(a).
\end{aligned}
$$

Applying Lemma 10.4.21 with $l = 1$, $l = p^{k-1}$ and $l = p^k$ then gives

$$\Lambda(a,1) = \frac{2}{p^k}(1 + 2T_1(g_a \circ D_{p^k})) - \frac{2}{p^{k-1}}(1 + 2T_1(g_a \circ D_{p^{k-1}})) + (1 + 2T_1(g_a)).$$

Consequently, we get the following characterization.

Proposition 10.6.3 ([18]). *Suppose that $\tau = p^k$ is a prime power and that 2 is a primitive root modulo p^k. Then*

$$p^k\Lambda(a,1) = 4T_1(g_a \circ D_{p^k}) - 4pT_1(g_a \circ D_{p^{k-1}}) + 2p^kT_1(g_a) + p^k - 2p + 2.$$

In particular, $f_{a,1}$ is hyper-bent if and only if

$$2T_1(g_a \circ D_{p^k}) - 2pT_1(g_a \circ D_{p^{k-1}}) + p^kT_1(g_a) = p - 1.$$

10.6.2 Explicit Values for τ

The previous subsection dealt with a fixed value of $b \in \mathbb{F}_{2^t}^*$ casting as few restrictions as possible on τ. In this subsection we go the other way around and treat the first few possible values of τ for all values of b with as few restrictions as possible on the corresponding infinite family of Boolean functions. Hence, we consider functions $f_{a,b} \in \mathcal{H}_n$ of the form

$$f_{a,b}(x) = \sum_{r \in R} \mathrm{Tr}_1^n\left(a_r x^{r(2^m-1)}\right) + \mathrm{Tr}_1^t\left(bx^{\frac{2^m+1}{\tau}(2^m-1)}\right)$$

for a fixed value of τ as in Eq. (10.8). Recall that such functions are hyper-bent if and only if the associated exponential sum $\Lambda(a,b) = \Lambda(f_{a,b})$ is equal to 1. Thus, the explicit expressions for $\Lambda(a,b)$ that we give in this subsection trivially turn into characterizations for hyper-bentness of $f_{a,b}$.

A large part of the data presented in this subsection has been checked or generated with the mathematical software Sage [20]. In addition to the functionality provided by Sage itself, the computations involved used, for the most, the underlying libraries Givaro [5] for finite field arithmetic, and Pynac [21] for symbolic manipulations. For small values of τ, that is, $\tau = 3$, 5 and 9, we provide all details and corresponding data. For higher values of τ, only the characterizations we obtain are given. The basic algorithm used is an explicit version of the approach taken in the previous subsection. To find characterizations valid for

- an integer τ such that 1 and -1 lie in the same cyclotomic class modulo τ,
- a coefficient $b \in \mathbb{F}_{2^{2o}}^*$ where $2o$ is the multiplicative order of 2 modulo τ,
- a divisor l of o,
- an l-th root r of -1 modulo τ and corresponding extension degrees m,
- and coefficients $a_r \in \mathbb{F}_{2^{\frac{m}{l}}}$,

we proceed as described in Algorithm 10.1.

Algorithm 10.1: Expression for $\Lambda(a, b)$ in terms of the sums $T_1(g_a \circ D_k)$

Input: An integer τ and associated data.
Output: Expression for $\Lambda(a, b)$ in terms of the sums $T_1(g_a \circ D_k)$

1. Compute the traces $\mathrm{Tr}_1^{2o}(b\xi^i)$, where ξ is a primitive element of $\mathbb{F}_{2^{2o}}$
2. Deduce an expression of $\Lambda(a, b)$ in terms of the sums S_i using Proposition 10.4.20
3. Compute the orbits of invertible integers modulo τ under the action of multiplication by r
4. Devise relations between the sums S_i using Lemma 10.4.23
5. Express the sums S_i in terms of the sums $T_1(g_a \circ D_k)$, where k divides τ, using Lemma 10.4.21
6. If possible, deduce an expression of $\Lambda(a, b)$ in terms of the sums $T_1(g_a \circ D_k)$

10.6.2.1 The Case $\tau = 3$

The smallest possible value for τ is $\tau = 3$. This case was originally addressed by the author in 2009 for the binomial case [16] and further in 2010 for the general case [13]. We now show how the characterizations for the general case can be directly deduced from the results of Sect. 10.4.

In this case, we have $t = 2$ and $m \equiv 1 \pmod 2$. Furthermore, if $f_{a,b}$ is hyper-bent, then its dual is f_{a,b^2}.

According to Remark 10.4.24, we have

$$\Lambda(a, b) = \chi\left(\mathrm{Tr}_1^2(b)\right)\overline{S}_0(a) + \left(\chi\left(\mathrm{Tr}_1^2(b\xi)\right) + \chi\left(\mathrm{Tr}_1^2(b\xi^{-1})\right)\right)\overline{S}_1(a).$$

Note that ξ is a primitive 3-rd root of unity and that $\xi + \xi^{-1} = 1$, so that

$$\Lambda(a, b) = \chi\left(\mathrm{Tr}_1^2(b)\right)\overline{S}_0(a) + \chi\left(\mathrm{Tr}_1^2(b\xi)\right)\left(1 + \chi\left(\mathrm{Tr}_1^2(b)\right)\right)\overline{S}_1(a).$$

Moreover, we have $\mathbb{F}_4^* = \langle\xi\rangle$. Thus, if $b = 1$, then $\Lambda(a, 1) = \overline{S}_0(a) - 2\overline{S}_1(a)$, and if $b = \xi$ or $b = \xi^{-1}$, that is if b is a primitive 3-rd root of unity or equivalently a primitive element of $\mathbb{F}_4$, then $\Lambda(a, b) = -\overline{S}_0(a)$. Applying Lemma 10.4.21 with $l = 1$ and $l = 3$ then gives the following theorem and the corresponding characterizations for hyper-bentness.

Theorem 10.6.4 ([13]). *Let $\tau = 3$ and $m \equiv 1 \pmod 2$. Then*

1. *If $b = 1$, then $3\Lambda(a, 1) = 4T_1(g_a \circ D_3) - 6T_1(g_a) - 1$.*
2. *If b is a primitive element of $\mathbb{F}_4$, then $3\Lambda(a, b) = -2T_1(g_a \circ D_3) - 1$.*

10.6.2.2 The Case $\tau = 5$

The next possible value for τ is $\tau = 5$. This case was originally addressed by Wang et al. in late 2011 for the general case [22], but they also gave specific treatments for the binomial case [23, 24]. We now show how their characterizations for the general case can be directly deduced from the results of Sect. 10.4.

In this case, we have $t = 4$ and $m \equiv 2 \pmod 4$. Furthermore, if $f_{a,b}$ is hyper-bent, then its dual is f_{a,b^4}.

According to Remark 10.4.24, we have

$$\begin{aligned}\Lambda(a,b) &= \chi\left(\mathrm{Tr}_1^4(b)\right)\overline{S}_0(a)\\ &\quad+\left(\chi\left(\mathrm{Tr}_1^4(b\xi)\right)+\chi\left(\mathrm{Tr}_1^4(b\xi^{-1})\right)\right)\overline{S}_1(a)\\ &\quad+\left(\chi\left(\mathrm{Tr}_1^4(b\xi^2)\right)+\chi\left(\mathrm{Tr}_1^4(b\xi^{-2})\right)\right)\overline{S}_2(a).\end{aligned}$$

Introduce $\gamma = \xi + \xi^{-1} \in \mathbb{F}_4$. Then

$$\begin{aligned}\Lambda(a,b) &= \chi\left(\mathrm{Tr}_1^4(b)\right)\overline{S}_0(a)\\ &\quad+\chi\left(\mathrm{Tr}_1^4(b\xi)\right)\left(1+\chi\left(\mathrm{Tr}_1^4(b\gamma)\right)\right)\overline{S}_1(a)\\ &\quad+\chi\left(\mathrm{Tr}_1^4(b\xi^2)\right)\left(1+\chi\left(\mathrm{Tr}_1^4(b\gamma^2)\right)\right)\overline{S}_2(a).\end{aligned}$$

Next, recall that ξ is a 5-th root of unity, so that $\sum_{i=0}^{4}\xi^i = 0$. In particular, we have $\gamma + \gamma^2 = 1$ and

$$\mathrm{Tr}_1^4(b\gamma) + \mathrm{Tr}_1^4(b\gamma^2) = \mathrm{Tr}_1^4(b),$$

what can be used to refine the above expression.

Here, we rather explicitly compute all the traces $\mathrm{Tr}_1^4(b\xi^i)$. The finite field $\mathbb{F}_{16}$ is represented as $\mathbb{F}_2[x]/(C_4(x))$ where $C_4(x) = x^4 + x + 1$ is the 4-th *Conway polynomial*.[2] We denote the class of x modulo $C_4(x)$ by β; this is a primitive element of $\mathbb{F}_{16}$. Let $\xi = \beta^3$ be a 5-th root of unity. The traces $\mathrm{Tr}_1^4(\beta^j\xi^i)$ are given in Table 10.3. The expression of $\Lambda(a,\beta^j)$ as a sum of the partial exponential sums $\overline{S}_i$, together with the minimal polynomial m_j of β^j, are given in Table 10.4.

Table 10.3 Traces $\mathrm{Tr}_1^4(\beta^j\xi^i)$ for $\tau = 5$

$j\backslash i$	0	1	2	3	4
0	0	1	1	1	1
1	0	0	1	0	1
2	0	0	0	1	1
3	1	1	1	1	0
4	0	1	0	1	0
5	0	0	1	1	0
6	1	1	1	0	1
7	1	0	1	0	0

$j\backslash i$	0	1	2	3	4
8	0	1	1	0	0
9	1	1	0	1	1
10	0	1	0	0	1
11	1	1	0	0	0
12	1	0	1	1	1
13	1	0	0	1	0
14	1	0	0	0	1

[2]Conway polynomial $C_{p,n}$ for the finite field $\mathbb{F}_{p^n}$ is a particular irreducible polynomial of degree n over $\mathbb{F}_p$ that can be used to define a standard representation of $\mathbb{F}_{p^n}$ as a splitting field of $\mathbb{F}_p$.

Table 10.4 $\Lambda(a, \beta^j)$ for $\tau = 5$

j	$\Lambda(a, \beta^j)$	m_j
0	$\overline{S}_0 - 2\overline{S}_1 - 2\overline{S}_2$	$x + 1$
1	$\overline{S}_0$	$x^4 + x + 1$
2	$\overline{S}_0$	$x^4 + x + 1$
3	$-\overline{S}_0 - 2\overline{S}_2$	$x^4 + x^3 + x^2 + x + 1$
4	$\overline{S}_0$	$x^4 + x + 1$
5	$\overline{S}_0 + 2\overline{S}_1 - 2\overline{S}_2$	$x^2 + x + 1$
6	$-\overline{S}_0 - 2\overline{S}_1$	$x^4 + x^3 + x^2 + x + 1$
7	$-\overline{S}_0 + 2\overline{S}_1$	$x^4 + x^3 + 1$
8	$\overline{S}_0$	$x^4 + x + 1$
9	$-\overline{S}_0 - 2\overline{S}_1$	$x^4 + x^3 + x^2 + x + 1$
10	$\overline{S}_0 - 2\overline{S}_1 + 2\overline{S}_2$	$x^2 + x + 1$
11	$-\overline{S}_0 + 2\overline{S}_2$	$x^4 + x^3 + 1$
12	$-\overline{S}_0 - 2\overline{S}_2$	$x^4 + x^3 + x^2 + x + 1$
13	$-\overline{S}_0 + 2\overline{S}_1$	$x^4 + x^3 + 1$
14	$-\overline{S}_0 + 2\overline{S}_2$	$x^4 + x^3 + 1$

Moreover, if the coefficients a_r lie in $\mathbb{F}_{2^l}$, where $l = m/2$, then $l \equiv 1 \pmod 2$ and $2^l \equiv \pm 2 \pmod 5$. Lemma 10.4.23 tells that either $\overline{S}_1(a) = \overline{S}_2(a)$ or $\overline{S}_1(a) = \overline{S}_3(a)$. But $\overline{S}_2(a) = \overline{S}_3(a)$, so that one always has

$$\overline{S}_1(a) = \overline{S}_2(a).$$

Finally applying Lemma 10.4.21 for $l = 1$ and $l = 5$ gives the following theorem which summarizes the above discussion.

Theorem 10.6.5 ([22]). *Let* $\tau = 5$ *and* $m \equiv 2 \pmod 4$.

1. *If* $b = 1$, *then* $5\Lambda(a, b) = 4T_1(g_a \circ D_5) - 10T_1(g_a) - 3$.
2. *If* b *is a primitive element of* $\mathbb{F}_{16}$ *such that* $\mathrm{Tr}_1^4(b) = 0$, *i.e. with minimal polynomial* $x^4 + x + 1$, *then* $5\Lambda(a, b) = 2T_1(g_a \circ D_5) + 1$.
3. *Suppose moreover that* $a_r \in \mathbb{F}_{2^{\frac{m}{2}}}$.
   (a) *If* b *is a primitive 3-rd root of unity, i.e. with minimal polynomial* $x^2 + x + 1$, *then* $5\Lambda(a, b) = 2T_1(g_a \circ D_5) + 1$.
   (b) *If* b *is a primitive 5-th root of unity, i.e. with minimal polynomial* $x^4 + x^3 + x^2 + x + 1$, *then* $5\Lambda(a, b) = -T_1(g_a \circ D_5) - 5T_1(g_a) - 3$.
   (c) *If* b *is a primitive element of* $\mathbb{F}_{16}$ *such that* $\mathrm{Tr}_1^4(b) = 1$, *i.e. with minimal polynomial* $x^4 + x^3 + 1$, *then* $5\Lambda(a, b) = -3T_1(g_a \circ D_5) + 5T_1(g_a) + 1$.

10.6.2.3 The Case $\tau = 7$

For $\tau = 7$, 1 and -1 do not lie in the same cyclotomic coset modulo 7, hence the next suitable value for τ is $\tau = 9$.

10.6.2.4 The Case $\tau = 9$

In the case $\tau = 9$, we have $t = 6$ and $m \equiv 3 \pmod 6$. Furthermore, if $f_{a,b}$ is hyper-bent, then its dual is f_{a,b^8}.

According to Remark 10.4.24, we have

$$\begin{aligned}\Lambda(a,b) = {} & \chi\left(\mathrm{Tr}_1^6(b)\right)\overline{S}_0(a) \\ & + \left(\chi\left(\mathrm{Tr}_1^6(b\xi)\right) + \chi\left(\mathrm{Tr}_1^6(b\xi^8)\right)\right)\overline{S}_1(a) + \left(\chi\left(\mathrm{Tr}_1^6(b\xi^2)\right) + \chi\left(\mathrm{Tr}_1^6(b\xi^7)\right)\right)\overline{S}_2(a) \\ & + \left(\chi\left(\mathrm{Tr}_1^6(b\xi^3)\right) + \chi\left(\mathrm{Tr}_1^6(b\xi^6)\right)\right)\overline{S}_3(a) + \left(\chi\left(\mathrm{Tr}_1^6(b\xi^4)\right) + \chi\left(\mathrm{Tr}_1^6(b\xi^5)\right)\right)\overline{S}_4(a).\end{aligned}$$

Introduce $\gamma = \xi^8 + \xi \in \mathbb{F}_8$. Note that we have

$$\begin{aligned}\gamma^2 &= \xi^2 + \xi^7, \\ \gamma^3 &= \xi + \xi^3 + \xi^6 + \xi^8, \\ \gamma^4 &= \xi^4 + \xi^5, \\ \gamma^5 &= \xi^3 + \xi^4 + \xi^5 + \xi^6, \\ \gamma^6 &= \xi^2 + \xi^3 + \xi^6 + \xi^7.\end{aligned}$$

Thus, we have

$$\begin{aligned}\Lambda(a,b) = {} & \chi\left(\mathrm{Tr}_1^6(b)\right)\overline{S}_0(a) \\ & + \chi\left(\mathrm{Tr}_1^6(b\xi)\right)\left(1 + \chi\left(\mathrm{Tr}_1^6(b\gamma)\right)\right)\overline{S}_1(a) \\ & + \chi\left(\mathrm{Tr}_1^6(b\xi^2)\right)\left(1 + \chi\left(\mathrm{Tr}_1^6(b\gamma^2)\right)\right)\overline{S}_2(a) \\ & + \chi\left(\mathrm{Tr}_1^6(b\xi^3)\right)\left(1 + \chi\left(\mathrm{Tr}_1^6(b(\gamma^3+\gamma))\right)\right)\overline{S}_3(a) \\ & + \chi\left(\mathrm{Tr}_1^6(b\xi^4)\right)\left(1 + \chi\left(\mathrm{Tr}_1^6(b\gamma^4)\right)\right)\overline{S}_4(a).\end{aligned}$$

Next, recall that $\sum_{i=0}^{8} \xi^i = 0$. Hence, we have $\gamma^2 + \gamma^3 + \gamma^4 = 1$ and

$$\mathrm{Tr}_1^6(b\gamma) + \mathrm{Tr}_1^6(b\gamma^2) + \mathrm{Tr}_1^6(b(\gamma+\gamma^3)) + \mathrm{Tr}_1^6(b\gamma^4) = \mathrm{Tr}_1^6(b),$$

what can be used to refine the above expression for $\Lambda(a,b)$.

Here, we rather explicitly compute all the traces $\mathrm{Tr}_1^6(b\xi^i)$. The finite field $\mathbb{F}_{64}$ is represented as $\mathbb{F}_2[x]/(C_6(x))$ where $C_6(x) = x^6 + x^4 + x^3 + x + 1$ is the 6-th Conway polynomial. We denote the class of x modulo $C_6(x)$ by β; this is a primitive element of $\mathbb{F}_{64}$. Let $\xi = \beta^7$ be a 9-th root of unity. The traces $\mathrm{Tr}_1^6(\beta^j\xi^i)$ are given in Table 10.5. The expression of $\Lambda(a,\beta^j)$ as a sum of the partial exponential sums $\overline{S}_i$, together with the minimal polynomial m_j of β^j, are given in Tables 10.6 and 10.7.

Table 10.5 Traces $\mathrm{Tr}_1^6\left(\beta^j\xi^i\right)$ for $\tau = 9$

$j\backslash i$	0	1	2	3	4	5	6	7	8
0	0	0	0	1	0	0	1	0	0
1	0	0	0	1	1	0	1	1	0
2	0	0	0	1	0	1	1	0	1
3	1	0	0	1	1	1	0	1	1
4	0	1	0	1	0	0	1	1	0
5	0	1	1	1	1	0	1	0	1
6	1	1	0	0	0	1	1	1	1
7	0	0	1	0	0	1	0	0	0
8	0	0	1	1	0	1	1	0	0
9	0	0	1	0	1	1	0	1	0
10	0	0	1	1	1	0	1	1	1
11	1	0	1	0	0	1	1	0	0
12	1	1	1	1	0	1	0	1	0
13	1	0	0	0	1	1	1	1	1
14	0	1	0	0	1	0	0	0	0
15	0	1	1	0	1	1	0	0	0
16	0	1	0	1	1	0	1	0	0
17	0	1	1	1	0	1	1	1	0
18	0	1	0	0	1	1	0	0	1
19	1	1	1	0	1	0	1	0	1
20	0	0	0	1	1	1	1	1	1
21	1	0	0	1	0	0	0	0	0
22	1	1	0	1	1	0	0	0	0
23	1	0	1	1	0	1	0	0	0
24	1	1	1	0	1	1	1	0	0
25	1	0	0	1	1	0	0	1	0
26	1	1	0	1	0	1	0	1	1
27	0	0	1	1	1	1	1	1	0
28	0	0	1	0	0	0	0	0	1
29	1	0	1	1	0	0	0	0	1
30	0	1	1	0	1	0	0	0	1
31	1	1	0	1	1	1	0	0	1
32	0	0	1	1	0	0	1	0	1
33	1	0	1	0	1	0	1	1	1
34	0	1	1	1	1	1	1	0	0
35	0	1	0	0	0	0	0	1	0
36	0	1	1	0	0	0	0	1	1
37	1	1	0	1	0	0	0	1	0
38	1	0	1	1	1	0	0	1	1
39	0	1	1	0	0	1	0	1	0
40	0	1	0	1	0	1	1	1	1
41	1	1	1	1	1	1	0	0	0
42	1	0	0	0	0	0	1	0	0
43	1	1	0	0	0	0	1	1	0
44	1	0	1	0	0	0	1	0	1
45	0	1	1	1	0	0	1	1	1
46	1	1	0	0	1	0	1	0	0
47	1	0	1	0	1	1	1	1	0
48	1	1	1	1	1	0	0	0	1
49	0	0	0	0	0	1	0	0	1
50	1	0	0	0	0	1	1	0	1
51	0	1	0	0	0	1	0	1	1
52	1	1	1	0	0	1	1	1	0
53	1	0	0	1	0	1	0	0	1
54	0	1	0	1	1	1	1	0	1
55	1	1	1	1	0	0	0	1	1
56	0	0	0	0	1	0	0	1	0
57	0	0	0	0	1	1	0	1	1
58	1	0	0	0	1	0	1	1	0
59	1	1	0	0	1	1	1	0	1
60	0	0	1	0	1	0	0	1	1
61	1	0	1	1	1	1	0	1	0
62	1	1	1	0	0	0	1	1	1

Moreover, if the coefficients a_r lie in $\mathbb{F}_{2^l}$, where $l = m/3$, then 2^l is -1, 2 or -4 modulo 9 when l is respectively 0, 1 and 2 modulo 3. In the last two cases, Lemma 10.4.23 tells that

$$\overline{S}_1(a) = \overline{S}_2(a) = \overline{S}_4(a).$$

Table 10.6 $\Lambda(a, \beta^j)$ for $\tau = 9$—Part I

j	$\Lambda(a, \beta^j)$	m_j
0	$\overline{S}_0 + 2\overline{S}_1 + 2\overline{S}_2 - 2\overline{S}_3 + 2\overline{S}_4$	$x + 1$
1	$\overline{S}_0 + 2\overline{S}_1 - 2\overline{S}_3$	$x^6 + x^4 + x^3 + x + 1$
2	$\overline{S}_0 + 2\overline{S}_2 - 2\overline{S}_3$	$x^6 + x^4 + x^3 + x + 1$
3	$-\overline{S}_0 - 2\overline{S}_4$	$x^6 + x^5 + x^4 + x^2 + 1$
4	$\overline{S}_0 - 2\overline{S}_3 + 2\overline{S}_4$	$x^6 + x^4 + x^3 + x + 1$
5	$\overline{S}_0 - 2\overline{S}_1 - 2\overline{S}_3$	$x^6 + x + 1$
6	$-\overline{S}_0 - 2\overline{S}_1$	$x^6 + x^5 + x^4 + x^2 + 1$
7	$\overline{S}_0 + 2\overline{S}_1 + 2\overline{S}_3$	$x^6 + x^3 + 1$
8	$\overline{S}_0 + 2\overline{S}_1 - 2\overline{S}_3$	$x^6 + x^4 + x^3 + x + 1$
9	$\overline{S}_0 + 2\overline{S}_1 - 2\overline{S}_2 + 2\overline{S}_3 - 2\overline{S}_4$	$x^3 + x + 1$
10	$\overline{S}_0 - 2\overline{S}_2 - 2\overline{S}_3$	$x^6 + x + 1$
11	$-\overline{S}_0 + 2\overline{S}_1$	$x^6 + x^5 + x^2 + x + 1$
12	$-\overline{S}_0 - 2\overline{S}_2$	$x^6 + x^5 + x^4 + x^2 + 1$
13	$-\overline{S}_0 - 2\overline{S}_4$	$x^6 + x^5 + x^4 + x + 1$
14	$\overline{S}_0 + 2\overline{S}_2 + 2\overline{S}_3$	$x^6 + x^3 + 1$
15	$\overline{S}_0 + 2\overline{S}_3 - 2\overline{S}_4$	$x^6 + x^4 + x^2 + x + 1$
16	$\overline{S}_0 + 2\overline{S}_2 - 2\overline{S}_3$	$x^6 + x^4 + x^3 + x + 1$
17	$\overline{S}_0 - 2\overline{S}_2 - 2\overline{S}_3$	$x^6 + x + 1$
18	$\overline{S}_0 - 2\overline{S}_1 + 2\overline{S}_2 + 2\overline{S}_3 - 2\overline{S}_4$	$x^3 + x + 1$
19	$-\overline{S}_0 - 2\overline{S}_1$	$x^6 + x^5 + x^4 + x + 1$
20	$\overline{S}_0 - 2\overline{S}_3 - 2\overline{S}_4$	$x^6 + x + 1$
21	$-\overline{S}_0 + 2\overline{S}_1 + 2\overline{S}_2 + 2\overline{S}_4$	$x^2 + x + 1$
22	$-\overline{S}_0 + 2\overline{S}_2$	$x^6 + x^5 + x^2 + x + 1$
23	$-\overline{S}_0 + 2\overline{S}_1$	$x^6 + x^5 + 1$
24	$-\overline{S}_0 - 2\overline{S}_4$	$x^6 + x^5 + x^4 + x^2 + 1$
25	$-\overline{S}_0 + 2\overline{S}_1$	$x^6 + x^5 + x^2 + x + 1$
26	$-\overline{S}_0 - 2\overline{S}_1$	$x^6 + x^5 + x^4 + x + 1$
27	$\overline{S}_0 + 2\overline{S}_1 - 2\overline{S}_2 - 2\overline{S}_3 - 2\overline{S}_4$	$x^3 + x^2 + 1$
28	$\overline{S}_0 + 2\overline{S}_3 + 2\overline{S}_4$	$x^6 + x^3 + 1$
29	$-\overline{S}_0 + 2\overline{S}_4$	$x^6 + x^5 + 1$
30	$\overline{S}_0 - 2\overline{S}_1 + 2\overline{S}_3$	$x^6 + x^4 + x^2 + x + 1$
31	$-\overline{S}_0 - 2\overline{S}_1 + 2\overline{S}_2 - 2\overline{S}_4$	$x^6 + x^5 + x^3 + x^2 + 1$

The corresponding expressions for $\Lambda(a, b)$, obtained after applying Lemma 10.4.21 for $l = 1$, $l = 3$ and $l = 9$, are given in Table 10.8, where m_b is the minimal polynomial of b.

Finally, the following theorem summarizes the above discussion.

Theorem 10.6.6 ([18]). *Let* $\tau = 9$ *and* $m \equiv 3 \pmod 6$.

1. If $b = 1$, *then*

$$9\Lambda(a, b) = 4T_1(g_a \circ D_9) - 12T_1(g_a \circ D_3) + 18T_1(g_a) + 5.$$

Table 10.7 $\Lambda(a, \beta^j)$ for $\tau = 9$—Part II

j	$\Lambda(a, \beta^j)$	m_j
32	$\overline{S}_0 - 2\overline{S}_3 + 2\overline{S}_4$	$x^6 + x^4 + x^3 + x + 1$
33	$-\overline{S}_0 - 2\overline{S}_2$	$x^6 + x^5 + x^4 + x^2 + 1$
34	$\overline{S}_0 - 2\overline{S}_3 - 2\overline{S}_4$	$x^6 + x + 1$
35	$\overline{S}_0 + 2\overline{S}_3 + 2\overline{S}_4$	$x^6 + x^3 + 1$
36	$\overline{S}_0 - 2\overline{S}_1 - 2\overline{S}_2 + 2\overline{S}_3 + 2\overline{S}_4$	$x^3 + x + 1$
37	$-\overline{S}_0 + 2\overline{S}_4$	$x^6 + x^5 + x^2 + x + 1$
38	$-\overline{S}_0 - 2\overline{S}_2$	$x^6 + x^5 + x^4 + x + 1$
39	$\overline{S}_0 - 2\overline{S}_2 + 2\overline{S}_3$	$x^6 + x^4 + x^2 + x + 1$
40	$\overline{S}_0 - 2\overline{S}_1 - 2\overline{S}_3$	$x^6 + x + 1$
41	$-\overline{S}_0 - 2\overline{S}_4$	$x^6 + x^5 + x^4 + x + 1$
42	$-\overline{S}_0 + 2\overline{S}_1 + 2\overline{S}_2 + 2\overline{S}_4$	$x^2 + x + 1$
43	$-\overline{S}_0 + 2\overline{S}_4$	$x^6 + x^5 + 1$
44	$-\overline{S}_0 + 2\overline{S}_4$	$x^6 + x^5 + x^2 + x + 1$
45	$\overline{S}_0 - 2\overline{S}_1 - 2\overline{S}_2 - 2\overline{S}_3 + 2\overline{S}_4$	$x^3 + x^2 + 1$
46	$-\overline{S}_0 + 2\overline{S}_2$	$x^6 + x^5 + 1$
47	$-\overline{S}_0 + 2\overline{S}_1 - 2\overline{S}_2 - 2\overline{S}_4$	$x^6 + x^5 + x^3 + x^2 + 1$
48	$-\overline{S}_0 - 2\overline{S}_1$	$x^6 + x^5 + x^4 + x^2 + 1$
49	$\overline{S}_0 + 2\overline{S}_2 + 2\overline{S}_3$	$x^6 + x^3 + 1$
50	$-\overline{S}_0 + 2\overline{S}_2$	$x^6 + x^5 + x^2 + x + 1$
51	$\overline{S}_0 - 2\overline{S}_1 + 2\overline{S}_3$	$x^6 + x^4 + x^2 + x + 1$
52	$-\overline{S}_0 - 2\overline{S}_2$	$x^6 + x^5 + x^4 + x + 1$
53	$-\overline{S}_0 + 2\overline{S}_2$	$x^6 + x^5 + 1$
54	$\overline{S}_0 - 2\overline{S}_1 + 2\overline{S}_2 - 2\overline{S}_3 - 2\overline{S}_4$	$x^3 + x^2 + 1$
55	$-\overline{S}_0 - 2\overline{S}_1 - 2\overline{S}_2 + 2\overline{S}_4$	$x^6 + x^5 + x^3 + x^2 + 1$
56	$\overline{S}_0 + 2\overline{S}_1 + 2\overline{S}_3$	$x^6 + x^3 + 1$
57	$\overline{S}_0 + 2\overline{S}_3 - 2\overline{S}_4$	$x^6 + x^4 + x^2 + x + 1$
58	$-\overline{S}_0 + 2\overline{S}_1$	$x^6 + x^5 + 1$
59	$-\overline{S}_0 - 2\overline{S}_1 + 2\overline{S}_2 - 2\overline{S}_4$	$x^6 + x^5 + x^3 + x^2 + 1$
60	$\overline{S}_0 - 2\overline{S}_2 + 2\overline{S}_3$	$x^6 + x^4 + x^2 + x + 1$
61	$-\overline{S}_0 + 2\overline{S}_1 - 2\overline{S}_2 - 2\overline{S}_4$	$x^6 + x^5 + x^3 + x^2 + 1$
62	$-\overline{S}_0 - 2\overline{S}_1 - 2\overline{S}_2 + 2\overline{S}_4$	$x^6 + x^5 + x^3 + x^2 + 1$

2. *If b is a primitive 3-rd root of unity, then*

$$9\Lambda(a,b) = -2T_1(g_a \circ D_9) - 6T_1(g_a \circ D_3) + 18T_1(g_a) + 5.$$

3. *Suppose moreover that $a_r \in \mathbb{F}_{2^{\frac{m}{3}}}$ and $\frac{m}{3} \not\equiv 0 \pmod 3$.*

   (a) *If b is a primitive 7-th root of unity with minimal polynomial $x^3 + x + 1$ or a primitive element with minimal polynomial $x^6 + x + 1$, then*

$$9\Lambda(a,b) = 8T_1(g_a \circ D_3) - 6T_1(g_a) + 1.$$

Table 10.8 $\Lambda(a, b)$ for $\tau = 9$—subfield case

m_b	$9\Lambda(a, b)$	$o(b)$
$x + 1$	$4T_1(g_a \circ D_9) - 12T_1(g_a \circ D_3) + 18T_1(g_a) + 5$	1
$x^2 + x + 1$	$-2T_1(g_a \circ D_9) - 6T_1(g_a \circ D_3) + 18T_1(g_a) + 5$	3
$x^3 + x + 1$	$8T_1(g_a \circ D_3) - 6T_1(g_a) + 1$	7
$x^3 + x^2 + 1$	$4T_1(g_a \circ D_9) - 4T_1(g_a \circ D_3) - 6T_1(g_a) - 3$	7
$x^6 + x^3 + 1$	$4T_1(g_a \circ D_3) + 6T_1(g_a) + 5$	9
$x^6 + x^4 + x^2 + x + 1$	$8T_1(g_a \circ D_3) - 6T_1(g_a) + 1$	21
$x^6 + x^5 + x^4 + x^2 + 1$	$-2T_1(g_a \circ D_9) + 2T_1(g_a \circ D_3) - 6T_1(g_a) - 3$	21
$x^6 + x + 1$	$4T_1(g_a \circ D_9) - 4T_1(g_a \circ D_3) - 6T_1(g_a) - 3$	63
$x^6 + x^4 + x^3 + x + 1$	$4T_1(g_a \circ D_9) - 8T_1(g_a \circ D_3) + 6T_1(g_a) + 1$	63
$x^6 + x^5 + 1$	$-2T_1(g_a \circ D_9) - 2T_1(g_a \circ D_3) + 6T_1(g_a) + 1$	63
$x^6 + x^5 + x^2 + x + 1$	$-2T_1(g_a \circ D_9) - 2T_1(g_a \circ D_3) + 6T_1(g_a) + 1$	63
$x^6 + x^5 + x^3 + x^2 + 1$	$-2T_1(g_a \circ D_9) + 2T_1(g_a \circ D_3) - 6T_1(g_a) - 3$	63
$x^6 + x^5 + x^4 + x + 1$	$-2T_1(g_a \circ D_9) + 2T_1(g_a \circ D_3) - 6T_1(g_a) - 3$	63

(b) *If b is a primitive 7-th root of unity with minimal polynomial $x^3 + x^2 + 1$ or a 21-st root of unity with minimal polynomial $x^6 + x^4 + x^2 + x + 1$, then*

$$9\Lambda(a, b) = 4T_1(g_a \circ D_9) - 4T_1(g_a \circ D_3) - 6T_1(g_a) - 3.$$

(c) *If b is a primitive 9-th root of unity with minimal polynomial $x^6 + x^3 + 1$, then*

$$9\Lambda(a, b) = 4T_1(g_a \circ D_3) + 6T_1(g_a) + 5.$$

(d) *If b is a primitive 21-st root of unity with minimal polynomial $x^6 + x^5 + x^4 + x^2 + 1$, or a primitive element with minimal polynomial $x^6 + x^5 + x^3 + x^2 + 1$ or $x^6 + x^5 + x^4 + x + 1$, then*

$$9\Lambda(a, b) = -2T_1(g_a \circ D_9) + 2T_1(g_a \circ D_3) - 6T_1(g_a) - 3.$$

(e) *If b is a primitive element with minimal polynomial $x^6 + x^4 + x^3 + x + 1$, then*

$$9\Lambda(a, b) = 4T_1(g_a \circ D_9) - 8T_1(g_a \circ D_3) + 6T_1(g_a) + 1.$$

(f) *If b is a primitive element with minimal polynomial $x^6 + x^5 + 1$ or $x^6 + x^5 + x^2 + x + 1$, then*

$$9\Lambda(a, b) = -2T_1(g_a \circ D_9) - 2T_1(g_a \circ D_3) + 6T_1(g_a) + 1.$$

10.6.2.5 The Case $\tau = 11$

We now give a few results for $\tau = 11$. In this case, we have $t = 10$ and $m \equiv 5 \pmod{10}$. Furthermore, if $f_{a,b}$ is hyper-bent, then its dual is $f_{a,b^{32}}$. Listing all possible characterizations would not be of high interest, hence we chose to only present results valid when the coefficients a_r are not restricted to a strict subfield of $\mathbb{F}_{2^m}$.

The characterizations valid for $a_r \in \mathbb{F}_{2^m}$, that is, without further restrictions on the field the coefficients a_r lie in, are summarized in the following theorem.

Theorem 10.6.7 ([18]). *Let $\tau = 11$ and $m \equiv 5 \pmod{10}$.*

1. *If $b = 1$, then*

$$11\Lambda(a,b) = 4T_1(g_a \circ D_{11}) - 22T_1(g_a) - 9.$$

2. *If b is a primitive 3-rd root of unity, a primitive 341-st root of unity with minimal polynomial $x^{10}+x^9+x^8+x^3+x^2+x+1$, or a primitive element with minimal polynomial $x^{10}+x^9+x^8+x^4+x^3+x^2+1$ or $x^{10}+x^9+x^8+x^6+x^5+x+1$, then*

$$11\Lambda(a,b) = -2T_1(g_a \circ D_{11}) - 1.$$

10.6.2.6 The Case $\tau = 13$

We now give a few results for $\tau = 13$. In this case, we have $t = 12$ and $m \equiv 6 \pmod{12}$. Furthermore, if $f_{a,b}$ is hyper-bent, then its dual is $f_{a,b^{64}}$. As for $\tau = 11$ we only present results valid when the coefficients a_r are not restricted to a strict subfield of $\mathbb{F}_{2^m}$.

The characterizations valid for $a_r \in \mathbb{F}_{2^m}$, that is without further restrictions on the field the coefficients a_r lie in, are summarized in the following theorem.

Theorem 10.6.8. *Let $\tau = 13$ and $m \equiv 6 \pmod{12}$.*

1. *If $b = 1$, then*

$$13\Lambda(a,b) = 4T_1(g_a \circ D_{13}) - 26T_1(g_a) - 11.$$

2. *If b is a primitive 15-th root of unity with minimal polynomial x^4+x+1, a primitive 819-th root of unity with minimal polynomial $x^{12}+x^9+x^8+x^5+x^4+x+1$, a primitive 1365-th root of unity with minimal polynomial $x^{12}+x^9+x^5+x^2+1$, or a primitive element with minimal polynomial $x^{12}+x^9+x^5+x^4+x^2+x+1$, $x^{12}+x^9+x^8+x^5+1$ or $x^{12}+x^9+x^8+x^6+x^3+x^2+1$, then*

$$13\Lambda(a,b) = 2T_1(g_a \circ D_{13}) + 1.$$

10.6.2.7 The Case $\tau = 17$

In this subsubsection we treat the case $\tau = 17$. In this case, we have $t = 8$ and $m \equiv 4 \pmod 8$. Furthermore, if $f_{a,b}$ is hyper-bent, then its dual is $f_{a,b^{16}}$. Contrary to the cases $\tau = 11$ and $\tau = 13$, 2 is not a primitive root modulo 17, so that t is quite small. Therefore, we provide a complete analysis of this case.

The following theorem summarizes the characterizations valid for $a_r \in \mathbb{F}_{2^m}$, $a_r \in \mathbb{F}_{2^{\frac{m}{2}}}$ and $a_r \in \mathbb{F}_{2^{\frac{m}{4}}}$. In particular, there is none valid when $a_r \in \mathbb{F}_{2^m}$, nor for $b = 1$.

Theorem 10.6.9 ([18]). *Let $\tau = 17$ and $m \equiv 4 \pmod 8$. Suppose moreover that $a_r \in \mathbb{F}_{2^{\frac{m}{2}}}$.*

1. *If b is a primitive element with minimal polynomial $x^8 + x^6 + x^5 + x + 1$ or $x^8 + x^6 + x^5 + x^2 + 1$, then*

$$17\Lambda(a, b) = 2T_1(g_a \circ D_{17}) + 1.$$

2. *Suppose moreover that $a_r \in \mathbb{F}_{2^{\frac{m}{4}}}$.*

   (a) If b is a primitive 15-th root of unity with minimal polynomial $x^4 + x + 1$, a primitive 17-th root of unity with minimal polynomial $x^8 + x^5 + x^4 + x^3 + 1$, or a primitive element with minimal polynomial $x^8 + x^5 + x^3 + x^2 + 1$, then

$$17\Lambda(a, b) = 2T_1(g_a \circ D_{17}) + 1.$$

   (b) If b is a 51-st root of unity with minimal polynomial $x^8 + x^4 + x^3 + x + 1$, then

$$17\Lambda(a, b) = 3T_1(g_a \circ D_{17}) - 17T_1(g_a) - 7.$$

10.6.2.8 The Case $\tau = 33$

To conclude this subsection we treat the case of the composite integer $\tau = 33$, the first suitable value for a composite value of τ. In this case, we have $t = 10$ and $m \equiv 5 \pmod{10}$. Furthermore, if $f_{a,b}$ is hyper-bent, then its dual is $f_{a,b^{32}}$.

The following theorem summarizes the characterizations valid for $a_r \in \mathbb{F}_{2^m}$, and $a_r \in \mathbb{F}_{2^{\frac{m}{5}}}$. In particular, there is none valid when $a_r \in \mathbb{F}_{2^m}$ without further restrictions, nor for $b = 1$.

Theorem 10.6.10 ([18]). *Let $\tau = 33$ and $m \equiv 5 \pmod{10}$. Suppose moreover that $a_r \in \mathbb{F}_{2^{\frac{m}{5}}}$ and $\frac{m}{5} \not\equiv 0 \pmod 5$.*

1. *If b is a primitive 31-st root of unity with minimal polynomial x^5+x^2+1, or a primitive 341-st root of unity with minimal polynomial $x^{10}+x^8+x^4+x^3+x^2+x+1$, then*

$$165\Lambda(a,b) = 8T_1(g_a \circ D_{33}) + 24T_1(g_a \circ D_{11}) - 88T_1(g_a \circ D_3) + 66T_1(g_a) + 5.$$

2. *If b is a primitive 31-st root of unity with minimal polynomial x^5+x^3+1, then*

$$165\Lambda(a,b) = -8T_1(g_a \circ D_{33}) + 48T_1(g_a \circ D_{11}) + 88T_1(g_a \circ D_3) - 198T_1(g_a) - 35.$$

3. *If b is a primitive 31-st root of unity with minimal polynomial $x^5+x^3+x^2+x+1$, a primitive 93-rd root of unity with minimal polynomial $x^{10}+x^8+x^3+x+1$, a primitive 341-st root of unity with minimal polynomial $x^{10}+x^8+x^7+x^5+x^3+x+1$, or a primitive element with minimal polynomial $x^{10}+x^7+1$, $x^{10}+x^7+x^6+x^4+x^2+x+1$ or $x^{10}+x^8+x^7+x^5+1$, then*

$$165\Lambda(a,b) = 24T_1(g_a \circ D_{11}) + 66T_1(g_a) + 45.$$

4. *If b is a primitive 93-rd root of unity with minimal polynomial $x^{10}+x^5+x^4+x^2+1$, or a primitive element with minimal polynomial $x^{10}+x^6+x^5+x^3+x^2+x+1$ or $x^{10}+x^8+x^5+x^4+x^3+x^2+1$, then*

$$165\Lambda(a,b) = -4T_1(g_a \circ D_{33}) + 36T_1(g_a \circ D_{11}) + 44T_1(g_a \circ D_3) - 66T_1(g_a) + 5.$$

5. *If b is a primitive 93-rd root of unity with minimal polynomial $x^{10}+x^8+x^6+x^5+1$, then*

$$165\Lambda(a,b) = 4T_1(g_a \circ D_{33}) + 36T_1(g_a \circ D_{11}) - 44T_1(g_a \circ D_3) - 66T_1(g_a) - 35.$$

6. *If b is a primitive 341-st root of unity with minimal polynomial $x^{10}+x^3+x^2+x+1$ or $x^{10}+x^7+x^4+x^3+1$, then*

$$165\Lambda(a,b) = -8T_1(g_a \circ D_{33}) + 36T_1(g_a \circ D_{11}) + 88T_1(g_a \circ D_3) - 66T_1(g_a) + 25.$$

7. *If b is a primitive 341-st root of unity with minimal polynomial $x^{10}+x^6+x^2+x+1$, or a primitive element with minimal polynomial $x^{10}+x^7+x^3+x+1$, then*

$$165\Lambda(a,b) = 36T_1(g_a \circ D_{11}) - 66T_1(g_a) - 15.$$

8. *If b is a primitive element with minimal polynomial $x^{10}+x^7+x^6+x^5+x^4+x+1$ or $x^{10}+x^7+x^6+x^5+x^4+x^3+x^2+x+1$, then*

$$165\Lambda(a,b) = 4T_1(g_a \circ D_{33}) + 24T_1(g_a \circ D_{11}) - 44T_1(g_a \circ D_3) + 66T_1(g_a) + 25.$$

References

1. P. Charpin and G. Gong. Hyperbent functions, Kloosterman sums and Dickson polynomials. In *IEEE Trans. Inform. Theory (54) 9*, pages 4230–4238, 2008.
2. P. Charpin, T. Helleseth, and V. Zinoviev. Divisibility properties of Kloosterman sums over finite fields of characteristic two. In *ISIT 2008, Toronto, Canada, July 6–11*, pages 2608–2612, 2008.
3. H. Cohen. *A course in computational algebraic number theory*, volume 138 of *Graduate Texts in Mathematics*. Springer-Verlag, Berlin, 1993.
4. F. G. Dorais and D. W. Klyve. A Wieferich prime search up to 6.7×10^{15}. *Journal of Integer Sequences*, 14(9), 2011. Available online at http://www.cs.uwaterloo.ca/journals/JIS/.
5. J.-G. Dumas, T. Gautier, P. Giorgi, J.-L. Roch, and G. Villard. *Givaro-3.2.13rc1: C++ library for arithmetic and algebraic computations*, September 2008. http://ljk.imag.fr/CASYS/LOGICIELS/givaro/.
6. J.-P. Flori and S. Mesnager. An efficient characterization of a family of hyperbent functions with multiple trace terms. Cryptology ePrint Archive, Report 2011/373, 2011. http://eprint.iacr.org/.
7. J.-P. Flori and S. Mesnager. Dickson polynomials, hyperelliptic curves and hyper-bent functions. In *7th International conference SETA 2012, LNCS 7280, Springer*, pages 40–52, 2012.
8. F. Gologlu. Almost Bent and Almost Perfect Nonlinear Functions, Exponential Sums, Geometries and Sequences. In *PhD dissertation, University of Magdeburg*, 2009.
9. D. R. Heath-Brown. Artin's conjecture for primitive roots. *Quart. J. Math. Oxford Ser. (2)*, 37(145):27–38, 1986.
10. C. Hooley. On Artin's conjecture. *J. Reine Angew. Math.*, 225:209–220, 1967.
11. The OEIS Foundation Inc. The On-Line Encyclopedia of Integer Sequences. http://oeis.org.
12. P. Lisoněk. An efficient characterization of a family of hyperbent functions. *IEEE Transactions on Information Theory*, 57(9):6010–6014, 2011.
13. S. Mesnager. Hyper-bent Boolean functions with multiple trace terms. In *Proceedings of International Workshop on the Arithmetic of Finite Fields. M.A. Hasan and T.Helleseth (Eds.): WAIFI 2010, LNCS 6087, pp. 97–113. Springer, Heidelberg (2010).*
14. S. Mesnager. A new class of bent Boolean functions in polynomial forms. In *Proceedings of international Workshop on Coding and Cryptography, WCC 2009*, pages 5–18, 2009.
15. S. Mesnager. A new family of hyper-bent Boolean functions in polynomial form. In *Proceedings of Twelfth International Conference on Cryptography and Coding, Cirencester, United Kingdom. M. G. Parker (Ed.): IMACC 2009, LNCS 5921, Springer, Heidelberg*, pages 402–417, 2009.
16. S. Mesnager. A new class of bent and hyper-bent Boolean functions in polynomial forms. In *journal Design, Codes and Cryptography, 59(1–3)*, pages 265–279, 2011.
17. S. Mesnager. Bent and hyper-bent functions in polynomial form and their link with some exponential sums and Dickson polynomials. *IEEE Transactions on Information Theory*, 57(9):5996–6009, 2011.
18. S. Mesnager and J-P Flori. Hyper-bent functions via Dillon-like exponents. In *IEEE Transactions on Information Theory-IT. Vol. 59 No. 5*, pages 3215–3232, 2013.
19. H. Park, J. Park, and D. Kim. A criterion on primitive roots modulo *p*. *J. KSIAM*, 4(1):29–38, 2000.
20. W. Arthur Stein et al. *Sage Mathematics Software (Version 4.7)*. The Sage Development Team, 2011. http://www.sagemath.org.
21. Pynac.sagemath.net. *Pynac, symbolic computation with Python objects (Version 0.2.2)*, 2011. http://pynac.sagemath.org.

22. B. Wang, C. Tang, Y. Qi, Y. Yang, and M. Xu. A new class of hyper-bent Boolean functions with multiple trace terms. Cryptology ePrint Archive, Report 2011/600, 2011. http://eprint.iacr.org/.
23. B. Wang, C. Tang, Y. Qi, and Y. Yang. A generalization of the class of hyper-bent Boolean functions in binomial forms. Cryptology ePrint Archive, Report 2011/698, 2011. http://eprint.iacr.org/.
24. B. Wang, C. Tang, Y. Qi, Y. Yang, and M. Xu. A new class of hyper-bent Boolean functions in binomial forms. *CoRR*, abs/1112.0062, 2011.

Chapter 11
(Hyper)-Bent Functions, Exponential Sums and (Hyper-)elliptic Curves

11.1 Elliptic Curves and Hyperelliptic Curves

11.1.1 Elliptic Curves Over Finite Fields

In this subsection, we present some classical results about elliptic curves over finite fields, as well as their connections with binary Kloosterman sums.

Let m be a positive integer, $\mathbb{F}_q$ the finite field of characteristic p with $q = p^m$ and $\overline{\mathbb{F}}_q$ its algebraic closure. Let E be an elliptic curve defined over $\mathbb{F}_q$. It can be given by a Weierstrass equation [41, Chapter III] describing its affine part as follows:

$$E : y^2 + a_1xy + a_3y = x^3 + a_2x^2 + a_4x + a_6.$$

Over an algebraically closed field, elliptic curves are classified up to isomorphism by the so-called j-invariant [41, Proposition III.1.4].

There exists an addition on the set of rational points of the curve (i.e. points with coordinates in $\mathbb{F}_q$), giving it a group structure. We denote by O_E the unique point at infinity of E, which is also the neutral point for the addition law, by $[n]$ the multiplication by an integer n on E and by $\mathrm{End}(E) = \mathrm{End}_{\overline{\mathbb{F}}_q}(E)$ the ring of endomorphisms of E over the algebraic closure $\overline{\mathbb{F}}_q$.

The group of rational points of E over an extension $\mathbb{F}_{q^k}$ of $\mathbb{F}_q$ is denoted by $E(\mathbb{F}_{q^k})$; the number of points of this group by $\#E(\mathbb{F}_{q^k})$. When the context is clear, we denote $\#E(\mathbb{F}_q)$ simply by $\#E$. It is a classical result that $\#E = q + 1 - t$ where t is the trace of the *Frobenius automorphism* of E over $\mathbb{F}_q$ [41, Remark V.2.6] and the following theorem has been shown by Hasse.

Theorem 11.1.1 ([41, Theorem V.2.3.1]). *Let t be the trace of the Frobenius automorphism of an elliptic curve over $\mathbb{F}_q$, then*

$$|t| \leq 2\sqrt{q}.$$

S. Mesnager, *Bent Functions*, DOI 10.1007/978-3-319-32595-8_11

For an integer n, we denote by $E[n]$ the n-torsion subgroup of the points of E over $\overline{\mathbb{F}}_q$, i.e.

$$E[n] = \{P \in E(\overline{\mathbb{F}}_q) \mid [n]P = O_E\}.$$

The subgroup of rational points of n-torsion is denoted by $E[n](\mathbb{F}_q) = E[n] \cap E(\mathbb{F}_q)$. The following classical result gives the structure of the groups of torsion points.

Proposition 11.1.2 ([41, Corollary III.6.4]). *Let n be a positive integer.*

- *If $p \nmid n$, then $E[n] \simeq \mathbb{Z}/n\mathbb{Z} \times \mathbb{Z}/n\mathbb{Z}$.*
- *One of the following is true: $E[p^e] \simeq \{0\}$ for all $e \geq 1$ or $E[p^e] \simeq \mathbb{Z}/p^e\mathbb{Z}$ for all $e \geq 1$.*

It can also be shown that a point of E is of n-torsion if and only if its coordinates are roots of a bivariate polynomial called the n-division polynomial of E [3, Section III.4]. In fact one can even choose a univariate polynomial in the x-coordinate that we denote by f_n.

Here we will be interested in *ordinary* elliptic curves which can be defined as follows.

Definition 11.1.3 ([41, Theorem V.3.1]). Let E be an elliptic curve defined over $\mathbb{F}_q$ and t the trace of the Frobenius automorphism of E. E is said to be ordinary if it verifies one of the following equivalent properties:

- $p \nmid t$;
- $E[p] \simeq \mathbb{Z}/p\mathbb{Z}$;
- $\mathrm{End}(E)$ is an order [1] in an imaginary quadratic extension of $\mathbb{Q}$.

If E is not ordinary, it is said to be *supersingular*.

Finally, using classical results of Deuring [12] and Waterhouse [46], the number of ordinary elliptic curves (up to isomorphism) with a given trace t of the Frobenius automorphism (or equivalently a number of points $q + 1 - t$), verifying $|t| \leq 2\sqrt{q}$ and $p \nmid t$, is given in the next proposition. This property indeed implies that $\mathrm{End}(E)$ must be an order $\mathcal{O}$ in $K = \mathbb{Q}[\alpha]$ and contains the order $\mathbb{Z}[\alpha]$ of discriminant Δ where $\alpha = \frac{t+\sqrt{\Delta}}{2}$ and $\Delta = t^2 - 4q$. We denote by $H(\Delta)$ the *Kronecker class number* [9, 40]

$$H(\Delta) = \sum_{\mathbb{Z}[\alpha] \subset \mathcal{O} \subset K} h(\mathcal{O}),$$

[1] An order $\mathcal{O}$ in a number field K is a subring of the ring of integers $\mathcal{O}_K$ which generates the number field over $\mathbb{Q}$. In an imaginary quadratic field, it can be uniquely written as $\mathcal{O} = \mathbb{Z} + f\mathcal{O}_K$ where $f \in \mathbb{N}^*$ is a positive integer and is called the *conductor* of $\mathcal{O}$. Reciprocally, each possible conductor gives an order in an imaginary quadratic field.

where the sum is taken over all the orders $\mathcal{O}$ in K containing $\mathbb{Z}[\alpha]$ and $h(\mathcal{O})$ is the classical class number.

Proposition 11.1.4 ([9, 26, 40]). *Let t be an integer such that $|t| \leq 2\sqrt{q}$ and $p \nmid t$. The number $N(t)$ of elliptic curves over $\mathbb{F}_q$ with $q+1-t$ rational points is given by*

$$N(t) = H(\Delta),$$

where $\Delta = t^2 - 4q$.

It should be noted that $H(\Delta)$ can be computed from the value of the classical class number of (the ring of integers of) K using the following proposition.

Proposition 11.1.5 ([7, 9, 26, 31]). *Let $\mathcal{O}$ be the order of conductor f in K, an imaginary quadratic extension of $\mathbb{Q}$, , $\mathcal{O}_K$ the ring of integers of K and Δ_K the discriminant of (the ring of integers of) K. Then*

$$h(\mathcal{O}) = \frac{fh(\mathcal{O}_K)}{[\mathcal{O}_K^* : \mathcal{O}^*]} \prod_{p|f} \left(1 - \left(\frac{\Delta_K}{p}\right)\frac{1}{p}\right),$$

where $\left(\frac{\cdot}{p}\right)$ is the Kronecker symbol.

Denoting the conductor of $\mathbb{Z}[\alpha]$ by f, $H(\Delta)$ can then be written as

$$H(\Delta) = h(\mathcal{O}_K) \sum_{d|f} \frac{d}{[\mathcal{O}_K^* : \mathcal{O}]} \prod_{p|d} \left(1 - \left(\frac{\Delta_K}{p}\right)\frac{1}{p}\right).$$

We now give specific results to even characteristic. First, E is supersingular if and only if its j-invariant is 0. Second, if E is ordinary, then its Weierstrass equation can be chosen to be of the form

$$E : y^2 + xy = x^3 + bx^2 + a,$$

where $a \in \mathbb{F}_q^*$ and $b \in \mathbb{F}_q$, its j-invariant is then $1/a$; moreover its first division polynomials are given by [3, 28]

$$f_1(x) = 1, \quad f_2(x) = x, \quad f_3(x) = x^4 + x^3 + a, \quad f_4(x) = x^6 + ax^2.$$

The quadratic twist of E is an elliptic curve with the same j-invariant as E, so isomorphic over the algebraic closure $\overline{\mathbb{F}}_q$, but not over $\mathbb{F}_q$ (in fact it becomes so over $\mathbb{F}_{q^2}$). It is unique up to isomorphism and we denote it by $\tilde{E}$. It is given by the Weierstrass equation

$$\tilde{E} : y^2 + xy = x^3 + \tilde{b}x^2 + a,$$

where $\tilde{b}$ is any element of $\mathbb{F}_q$ such that $\mathrm{Tr}_1^m(\tilde{b}) = 1 - \mathrm{Tr}_1^m(b)$ [15]. The trace of its Frobenius automorphism is given by the opposite of the trace of the Frobenius automorphism of E, so that their number of rational points are closely related [3, 15]:

$$\#E + \#\tilde{E} = 2q + 2.$$

11.1.2 Hyperelliptic Curves and Point Counting

In this section we give basic definitions and results for hyperelliptic curves with a special emphasis on point counting on such curves over finite fields of even characteristic. For a general overview of the theory of such curves, with a cryptographic point of view, the reader is referred to the textbooks of Cohen et al. [8] or that of Galbraith [19].

For our purposes, it is enough to consider *imaginary* hyperelliptic curves. Imaginary hyperelliptic curves are *smooth projective curves* whose affine part can be described by an equation of the form

$$H : y^2 + h(x)y = f(x),$$

where $h(x)$ is a polynomial of degree $\leq g$, the *genus* of the curve, and $f(x)$ is a monic polynomial of degree $2g + 1$. They have exactly one point at infinity. Curves for which $h(x) = x^k$, where $0 \leq k \leq g$, are called *Artin–Schreier curves*. The case $g = 1$ corresponds to elliptic curves.

The number of points on a hyperelliptic curve H over the finite field $\mathbb{F}_{2^m}$ is understood as its numbers of points with coordinates in the finite field $\mathbb{F}_{2^m}$, which are also called $\mathbb{F}_{2^m}$*-rational points*. It is denoted by $\#H(\mathbb{F}_{2^m})$. The reference to the finite field is usually omitted when the context makes it clear.

A very important result is that there exist algorithms to compute this number of points in polynomial time and space in m. Such a result has been given by Denef and Vercauteren who extended a previous result of Kedlaya [27] in odd characteristic.

Theorem 11.1.6 ([43, Theorem 4.4.1], [11]). *Let H be an imaginary hyperelliptic curve of genus g defined over $\mathbb{F}_{2^m}$. There exists an algorithm to compute the number of points on H in*

$$O(g^{5+\epsilon}m^{3+\epsilon})$$

bit operations and $O(g^4m^3)$ memory, where $\epsilon \in \mathfrak{R}_+^*$ is any strictly positive real number.

A slightly stronger result is true for Artin–Schreier curves.

Theorem 11.1.7 ([43, Theorem 4.3.1], [10]). *Let H be an Artin–Schreier curve of genus g defined over $\mathbb{F}_{2^m}$. There exists an algorithm to compute the number of points on H in*

$$O(g^{5+\epsilon}m^{3+\epsilon})$$

bit operations and $O(g^3m^3)$ memory, where $\epsilon \in \Re_+^*$ is any strictly positive real number.

Better complexities were recently obtained through the use of complex methods involving deformation theory. For example, Hubrechts obtained the following result.

Theorem 11.1.8 ([25, Theorem 2]). *Let H be an hyperelliptic curve of genus g defined over $\mathbb{F}_{2^m}$. There exists an algorithm to compute the number of points on H in*

$$O(g^{7.376}m^2 + g^{3.376}m^{2.667})$$

bit operations and $O(g^5m^2 + g^3m^{2.5})$ memory.

In fact such algorithms are even more interesting when one wants to compute the number of points on several curves within the same family.

As far as point counting on (hyper)elliptic curves is concerned, the classical Hasse-Weil bounds state that the number of points N over $GF(p^n)$ is of order p^n. For large values of p^n, computing this number seems to be a very hard task. More precisely, if we measure the complexity of the algorithm for determining this number in terms of $\log p$, and n, it is exponential in $\log p$ and in n. Modern point-counting algorithms to compute N use the fact that N can be related to a polynomial known as the characteristic polynomial of the p^n-power Frobenius. This approach allows us to design algorithms with lower complexities in g, $\log p$ and n (the better complexities are polynomial in time and space).

Most of these algorithms have been implemented in the kernel of the closed source computational algebra system MAGMA and are hidden to the community. At present, only a very limited selection of algorithms has efficient and freely-available implementations: PARI/GP includes an implementation of Schoof–Elkies–Atkin, available within the SAGE system together with an implementation of Harvey's variation of Kedlaya's p-adic algorithm.

Finally, let us mention the existence of a quasi-quadratic algorithm described by Lercier and Lubicz [33].

Theorem 11.1.9. *Let H be a hyperelliptic curve of genus g defined over $\mathbb{F}_{2^m}$. There exists an algorithm to compute the cardinality of H in*

$$O(2^{4g+o(1)}g^3m^{2+o(1)})$$

bit operations and $O(2^{3g+o(1)}m^2)$ memory.

Nevertheless, it should be remarked that the time and space complexities of this last algorithm are exponential in the genus of the curve and so it is of practical interest for curves of relatively genera only.

11.2 Exponential Sums and Algebraic Varieties

11.2.1 Kloosterman Sums and Elliptic Curves

The idea to connect Kloosterman sums and elliptic curves goes back to the works of Lachaud and Wolfmann [30], and Katz and Livné [26]. We recall a simple proof of their main result in a simpler and less general formulation here. Indeed, its generalizations which will be covered in the next subsection can be proved in a very similar manner.

Theorem 11.2.1 ([26, 30]). *Let $m \geq 3$ be any positive integer, $a \in \mathbb{F}_{2^m}^*$ and E_a the projective elliptic curve defined over $\mathbb{F}_{2^m}$ whose affine part is given by the equation*

$$E_a : y^2 + xy = x^3 + a.$$

Then

$$\#E_a = 2^m + K_m(a).$$

Proof. Indeed

$$K_m(a) = 1 + \sum_{x \in \mathbb{F}_{2^m}^*} \chi\left(\mathrm{Tr}_1^m(x^{-1} + ax)\right),$$

and

$$\begin{aligned}\sum_{x \in \mathbb{F}_{2^m}^*} \chi\left(\mathrm{Tr}_1^m(x^{-1} + ax)\right) &= \sum_{x \in \mathbb{F}_{2^m}^*} \left(1 - 2\mathrm{Tr}_1^m(x^{-1} + ax)\right)\\ &= 2^m - 1 - 2\#\left\{x \in \mathbb{F}_{2^m}^* \mid \mathrm{Tr}_1^m(x^{-1} + ax) = 1\right\}\\ &= -2^m + 1 + 2\#\left\{x \in \mathbb{F}_{2^m}^* \mid \mathrm{Tr}_1^m(x^{-1} + ax) = 0\right\}.\end{aligned}$$

The additive version of Hilbert's Theorem 90 characterizes elements of trace zero as those which can be written as $t + t^2$. Using the additive version of Hilbert's Theorem 90, we get

$$\sum_{x \in \mathbb{F}_{2^m}^*} \chi\left(\mathrm{Tr}_1^m(x^{-1} + ax)\right) = -2^m + 1 + 2\#\left\{x \in \mathbb{F}_{2^m}^* \mid \exists t \in \mathbb{F}_{2^m},\ t^2 + t = x^{-1} + ax\right\},$$

and applying the substitution $t = t/x$ we get

$$\sum_{x\in\mathbb{F}_{2^m}^*} \chi\left(\mathrm{Tr}_1^m(x^{-1}+ax)\right) = -2^m + 1 + 2\#\left\{x \in \mathbb{F}_{2^m}^* \mid \exists t \in \mathbb{F}_{2^m},\ (t/x)^2 + (t/x) = x^{-1} + ax\right\}$$

$$= -2^m + 1 + 2\#\left\{x \in \mathbb{F}_{2^m}^* \mid \exists t \in \mathbb{F}_{2^m},\ t^2 + xt = x + ax^3\right\}.$$

We recognize the number of points of E_a minus the only point with x-coordinate $x = 0$ and the only point at infinity.

$$\sum_{x\in\mathbb{F}_{2^m}^*} \chi\left(\mathrm{Tr}_1^m(x^{-1}+ax)\right) = -2^m + 1 + \#E_a - 2$$

$$= -2^m - 1 + \#E_a.$$

□

Hence, the necessary and sufficient condition for hyper-bentness of the monomial functions with the Dillon exponent can be reformulated as follows.

Proposition 11.2.2 (Reformulation of the Dillon Criterion). *The notation is as in Theorem 11.2.1. Moreover, let r be an integer such that* $\gcd(r, 2^m + 1) = 1$ *and f_a be the Boolean function with n inputs defined as $f_a(x) = \mathrm{Tr}_1^n\left(ax^{r(2^m-1)}\right)$. Then f_a is hyper-bent if and only if*

$$\#E_a = 2^m.$$

11.2.2 Exponential Sums and Hyperelliptic Curves

In the following two propositions we link exponential sums with cardinalities of hyperelliptic curve, which will be of interest later on.

Proposition 11.2.3 ([17]). *Let $f : \mathbb{F}_{2^m} \to \mathbb{F}_{2^m}$ be a function such that $f(0) = 0$, $g = \mathrm{Tr}_1^m(f)$, and G_f be the (affine) curve defined over $\mathbb{F}_{2^m}$ by*

$$G_f : y^2 + y = f(x).$$

Then

$$\sum_{x\in\mathbb{F}_{2^m}^*} \chi\left(g(x)\right) = -2^m - 1 + \#G_f.$$

Proof. The first step of the proof is to express $\chi(g(x))$ as $1 - 2g(x)$ where $g(x)$ is now understood to be integer-valued:

$$\sum_{x\in\mathbb{F}_{2^m}^*} \chi\left(g(x)\right) = \sum_{x\in\mathbb{F}_{2^m}^*} (1 - 2g(x)).$$

The sum can then be split according to the value of $g(x)$ yielding the equality

$$\sum_{x \in \mathbb{F}_{2^m}^*} \chi(g(x)) = 2^m - 1 - 2\#\{x \in \mathbb{F}_{2^m}^* \mid g(x) = 1\}.$$

We supposed that $g(0) = 0$, so we can include zero in the summation set in the right hand side of the previous equality and deduce

$$\begin{aligned}\sum_{x \in \mathbb{F}_{2^m}^*} \chi(g(x)) &= 2^m - 1 - 2\#\{x \in \mathbb{F}_{2^m} \mid g(x) = 1\} \\ &= 2^m - 1 - 2(2^m - \#\{x \in \mathbb{F}_{2^m} \mid g(x) = 0\}) \\ &= -2^m - 1 + 2\#\{x \in \mathbb{F}_{2^m} \mid g(x) = 0\}.\end{aligned}$$

Using Hilbert's Theorem 90, we get the equivalent formulation

$$\sum_{x \in \mathbb{F}_{2^m}^*} \chi(g(x)) = -2^m - 1 + 2\#\left\{x \in \mathbb{F}_{2^m} \mid \exists t \in \mathbb{F}_{2^m},\ t^2 + t = f(x)\right\}.$$

The last term of the right hand side of the above equality is nothing but the number of $\mathbb{F}_{2^m}$-rational (affine) points of G_f, whence

$$\sum_{x \in \mathbb{F}_{2^m}^*} \chi(g(x)) = -2^m - 1 + \#G_f,$$

which concludes the proof of the proposition. □

Proposition 11.2.4 ([17]). *Let $f : \mathbb{F}_{2^m} \to \mathbb{F}_{2^m}$ be a function, $g = \mathrm{Tr}_1^m(f)$, and H_f be the (affine) curve defined over $\mathbb{F}_{2^m}$ by*

$$H_f : y^2 + xy = x + x^2 f(x),$$

Then

$$\sum_{x \in \mathbb{F}_{2^m}^*} \chi\left(\mathrm{Tr}_1^m(1/x) + g(x)\right) = -2^m + \#H_f.$$

Proof. The proof is quite similar as that of Proposition 11.2.3. It begins with the same sequence of equalities:

$$\begin{aligned}\sum_{x \in \mathbb{F}_{2^m}^*} \chi\left(\mathrm{Tr}_1^m(1/x) + g(x)\right) &= \sum_{x \in \mathbb{F}_{2^m}^*} (1 - 2(\mathrm{Tr}_1^m(1/x) + g(x))) \\ &= 2^m - 1 - 2\#\left\{x \in \mathbb{F}_{2^m}^* \mid \mathrm{Tr}_1^m(1/x) + g(x) = 1\right\}\end{aligned}$$

$$= -2^m + 1 + 2\#\left\{x \in \mathbb{F}_{2^m}^* \mid \mathrm{Tr}_1^m(1/x) + g(x) = 0\right\}$$
$$= -2^m + 1 + 2\#\left\{x \in \mathbb{F}_{2^m}^* \mid \exists t \in \mathbb{F}_{2^m},\ t^2 + t = 1/x + f(x)\right\}.$$

The additional step is then to substitute t by t/x before clearing denominators, which is legal since x is non-zero, before finishing the proof using the same arguments.

$$\sum_{x\in\mathbb{F}_{2^m}^*} \chi\left(\mathrm{Tr}_1^m(1/x) + g(x)\right) = -2^m + 1 + 2\#\left\{x \in \mathbb{F}_{2^m}^* \mid \exists t \in \mathbb{F}_{2^m},\ (t/x)^2 + (t/x) = 1/x + f(x)\right\}$$
$$= -2^m + 1 + 2\#\left\{x \in \mathbb{F}_{2^m}^* \mid \exists t \in \mathbb{F}_{2^m},\ t^2 + xt = x + x^2 f(x)\right\}$$
$$= -2^m + 1 + \#H_f - \#\left\{P \in H_f \mid x = 0\right\}$$
$$= -2^m + \#H_f.$$

□

11.3 Efficient Characterizations of Hyper-Bentness: Reformulation in Terms of Cardinalities of Curves

11.3.1 Efficient Characterizations of Hyper-Bentness: The Charpin and Gong Criterion

Thanks to Proposition 11.2.3 and Proposition 11.2.4 we can now easily deduce the reformulation of the Charpin–Gong criterion given by Lisoněk.

Theorem 11.3.1 (Reformulation of the Charpin–Gong Criterion [17, 35]). *Let E' be a set of representatives of the cyclotomic cosets modulo $2^m + 1$ for which each coset has the maximal size n. Let f_{a_r} be the function of $\mathcal{F}_n$ defined on $\mathbb{F}_{2^n}$ by $f_a(x) = \sum_{r\in R} \mathrm{Tr}_1^n(a_r x^{r(2^m-1)})$ where $a_r \in \mathbb{F}_{2^n}$ and $R \subseteq E'$. Moreover, let H_a and G_a be the (affine) curves defined over $\mathbb{F}_{2^m}$ by*

$$G_a : y^2 + y = \sum_{r\in R} a_r D_r(x),$$
$$H_a : y^2 + xy = x + x^2 \sum_{r\in R} a_r D_r(x).$$

Then f_a is hyper-bent if and only if

$$\#H_a - \#G_a = -1.$$

Proof. According to Proposition 11.2.4, the left hand side of the Charpin–Gong criterion satisfies

$$\sum_{x\in\mathbb{F}_{2^m}^*} \chi\left(\mathrm{Tr}_1^m(x^{-1}) + g_a(x)\right) = -2^m + \#H_a\ ;$$

and, according to Proposition 11.2.3, the right hand side of the Charpin–Gong criterion satisfies

$$\sum_{x\in\mathbb{F}_{2^m}^*} \chi\left(g_a(x)\right) = -2^m - 1 + \#G_a.$$

□

As a consequence of this theorem Lisoněk obtained a polynomial time and space test for hyper-bentness of Boolean functions in the Charpin–Gong family. Let r_{max} the maximal index in R, which can be supposed to be odd, and will be for two reasons:

1. it ensures that the curves H_a and G_a are imaginary hyperelliptic curves;
2. as will be discussed below, r_{max} should be as small as possible for efficiency reasons, so the natural choice for the indices in a cyclotomic coset will be the coset leaders which are odd integers.

In fact, G_a and H_a are even Artin–Schreier curves. Theorems 11.1.7 and 11.1.8 state that there exist efficient algorithms to compute the cardinality of such curves as long as r_{max} is supposed to be relatively small. The polynomial defining H_a (respectively G_a) is indeed of degree $r_{max} + 2$ (respectively r_{max}), so the curve is of genus $(r_{max}+1)/2$ (respectively $(r_{max}-1)/2$). The complexity for testing the hyper-bentness of a Boolean function in this family is then dominated by the computation of the cardinality of a curve of genus $(r_{max} + 1)/2$. Then, applying Theorem 11.1.8 gives the following time and space complexities in m and r_{max}.

Theorem 11.3.2. *Let f_a be a function in the family $\mathcal{F}_n$ defined as above. Let moreover r_{max} be the maximal index in R. Then the hyper-bentness of f_a can be checked in*

$$O(r_{max}^{7.376} m^2 + r_{max}^{3.376} m^{2.667})$$

bit operations and $O(r_{max}^5 m^2 + r_{max}^3 m^{2.5})$ memory.

Therefore, if R is supposed to be fixed, then so are r_{max} and the genera of the curves G_a and H_a, and the complexities of Theorem 11.3.21 are indeed polynomial in m as stated by Lisoněk [35, Theorem 5]. Asymptotically, this is much better than a straightforward application of Theorem 10.1.1 where the exponential sums on $\mathbb{F}_{2^m}$ are naively computed one term at a time. Indeed, for each of the 2^{m-1} terms of the partial exponential sums over $\mathcal{T}_1$, one has to compute the function $g_a(x) = \sum_{r\in R} \mathrm{Tr}_1^m(a_r D_r(x))$. The time complexity of this computation is dominated by the cost of a constant number of multiplications in $\mathbb{F}_{2^m}$. Therefore, the total time complexity is $O(2^m m^{1+\epsilon})$ and the space complexity is $O(m)$, where $\epsilon \in \Re_+^*$ is any strictly positive real number. Testing hyper-bentness through a naive computation of $\Lambda(f_a)$ yields similar complexity, although the arithmetic takes place in $\mathbb{F}_{2^n}$ rather than $\mathbb{F}_{2^m}$.

It should be remarked that if no restriction is cast upon R, then the maximal index r_{max} will obviously depend on m and will in fact grow, at least, as $2^m/m$. It is indeed sufficient to note that this is true when m is prime. Then, each non-trivial cyclotomic coset has indeed size dividing $n = 2m$. It has size 2 if and only if $3r \equiv 0 \pmod{2^m + 1}$ for $0 \leq r \leq 2^m$, i.e. $3r = 2^m + 1$ or $3r = 2(2^m + 1)$. Hence, there are exactly one such class when $3 \mid 2^m + 1$, that is when m is odd, and no such class otherwise. The size of the other cosets is then $2m$, so that the largest coset leader, which is odd, is at least $(2^m - 2)/m$.

Consequently, the time and space complexities of Theorem 11.3.21 will become exponential, whereas the time complexities of the naive approaches will become $O(2^m m^{2+\epsilon})$ (now dominated by the computation of an exponentiation with an arbitrary large exponent), where $\epsilon \in \Re_+^*$ is any strictly positive real number, and their space complexities will not change.

Nonetheless, fixing a set R, i.e. only looking for Boolean functions with a given polynomial form within a large family, is customary in cryptographic applications. Moreover, experimental data provided by Lisoněk [35, Table 1] and in Sect. 11.3.4.3 show that such reformulations also have a practical impact, so that the above approach seems meaningful.

11.3.2 Efficient Characterizations of Hyper-Bentness: Our Criterion

We now show that a similar reformulation can be applied to the different versions of our criterion for Boolean functions with multiple trace terms.

Theorem 11.3.3 (Reformulation of the Hyper-Bentess Criterion for Functions in $\mathfrak{H}_n$ [17]). *Let $f_{a,b}$ be a function of $\mathfrak{H}_n$ defined by $f_{a,b}(x) := \sum_{r \in R} \mathrm{Tr}_1^n(a_r x^{r(2^m-1)}) + \mathrm{Tr}_1^2(b x^{\frac{2^n-1}{3}})$. Moreover, let H_a and G_a be the (affine) curves defined over $\mathbb{F}_{2^m}$ by*

$$G_a : y^2 + y = \sum_{r \in R} a_r D_r(x),$$

$$H_a : y^2 + xy = x + x^2 \sum_{r \in R} a_r D_r(x) \ ;$$

and let H_a^3 and G_a^3 be the (affine) curves defined over $\mathbb{F}_{2^m}$ by

$$G_a^3 : y^2 + y = \sum_{r \in R} a_r D_r(D_3(x)),$$

$$H_a^3 : y^2 + xy = x + x^2 \sum_{r \in R} a_r D_r(D_3(x)).$$

If b is a primitive element of $\mathbb{F}_4$, then $f_{a,b}$ is hyper-bent if and only if

$$\#H_a^3 - \#G_a^3 = 3.$$

If $b = 1$, then $f_{a,1}$ is hyper-bent if and only if

$$\left(\#G_a^3 - \#H_a^3\right) - \frac{3}{2}\left(\#G_a - \#H_a\right) = \frac{3}{2}.$$

Proof. If b is a primitive element of $\mathbb{F}_4$, according to Proposition 11.2.4 the left-hand side of Condition (c)-(iii) of Theorem 10.2.10 satisfies

$$\sum_{x \in \mathbb{F}_{2^m}^*} \chi\left(\mathrm{Tr}_1^m(x^{-1}) + g_a(D_3(x))\right) = -2^m + \#H_a^3,$$

and according to Proposition 11.2.3 the right-hand side of Condition (c)-(iii) of Theorem 10.2.10 satisfies

$$2^m - 2\,\mathrm{w_H}(g_a \circ D_3) + 3 = -2^m + 3 + \#G_a^3,$$

so that the criterion is equivalent to

$$\#H_a^3 - \#G_a^3 = 3.$$

We could also have used Condition (c)-(ii) of Theorem 10.2.10 and that its left-hand side satisfies

$$\begin{aligned}
\sum_{x \in \mathbb{F}_{2^m}^*, \mathrm{Tr}_1^m(x^{-1})=1} \chi\left(g_a \circ D_3(x)\right) &= \frac{1}{2}\left(\sum_{x \in \mathbb{F}_{2^m}^*} \chi\left(g_a \circ D_3(x)\right) - \sum_{x \in \mathbb{F}_{2^m}^*} \chi\left(\mathrm{Tr}_1^m(x^{-1}) + g_a \circ D_3(x)\right)\right) \\
&= \frac{1}{2}\left(\left(-2^m - 1 + \#G_a^3\right) - \left(-2^m + \#H_a^3\right)\right) \\
&= \frac{1}{2}\left(\#G_a^3 - \#H_a^3 - 1\right);
\end{aligned}$$

and deduce the same reformulation.

If $b = 1$, using the previous calculations, the first term in Condition (d) of Theorem 10.2.10 satisfies

$$2 \sum_{x \in \mathbb{F}_{2^m}^*, \mathrm{Tr}_1^m(x^{-1})=1} \chi\left(g_a \circ D_3(x)\right) = \#G_a^3 - \#H_a^3 - 1;$$

and the second term satisfies

$$3 \sum_{x \in \mathbb{F}_{2^m}^*, \mathrm{Tr}_1^m(x^{-1})=1} \chi\left(g_a(x)\right) = \frac{3}{2}\left(\#G_a - \#H_a - 1\right);$$

whence the reformulation. □

Here all the curves are Artin–Schreier curves. So, for a fixed subset of indices R, we also get a test in polynomial time and space in m. However, the complexity of the point counting algorithms also depends on the genera of the curves, and so on the degrees of the polynomials defining them. Denoting by r_{max} the maximal index as above, the genus of H_a^3 (respectively G_a^3) is $(3r_{max} + 1)/2$ (respectively $(3r_{max} - 1)/2$), so approximately three times that of H_a (respectively G_a). Therefore, the associated test will be much slower than for Boolean functions of the family of Charpin and Gong for a given subset R: we have to compute the cardinalities of two curves of genera $(3r_{max} + 1)/2$ and $(3r_{max} - 1)/2$ if b is primitive, or four curves of genera $(3r_{max} + 1)/2$, $(3r_{max} - 1)/2$, $(r_{max} + 1)/2$ and $(r_{max} - 1)/2$ if $b = 1$, instead of two curves of genera $(r_{max} + 1)/2$ and $(r_{max} - 1)/2$. Hence, we propose another reformulation of our criterion involving slightly less computations.

Theorem 11.3.4 (Second Reformulation of the Hyper-Bentness Criterion for Functions in $\mathfrak{H}_n$ [17]). *Let $f_{a,b}$ be a function of the family $\mathfrak{H}_n$ defined as above. If b is a primitive element of $\mathbb{F}_4$, then $f_{a,b}$ is hyper-bent if and only if*

$$\#G_a^3 - \frac{1}{2}\left(\#G_a + \#H_a\right) = -\frac{3}{2}.$$

If $b = 1$, then $f_{a,1}$ is hyper-bent if and only if

$$2\#G_a^3 - \frac{5}{2}\#G_a + \frac{1}{2}\#H_a = \frac{3}{2}.$$

Proof. We use the fact that m is odd, so that the function $x \mapsto D_3(x) = x^3 + x$ is a permutation of the set $\left\{x \in \mathbb{F}_{2^m}^* \mid \mathrm{Tr}_1^m(x^{-1}) = 0\right\}$ (see the papers of Berlekamp, Rumsey and Solomon [14, Theorem 2] and Charpin, Helleseth and Zinoviev [5] for the case of D_3, or more generally the article of Dillon and Dobbertin [13]), and similar arguments as previously.

If b is a primitive element of $\mathbb{F}_4$, then the left hand side in Condition (c)-(ii) of Theorem 10.2.10 satisfies

$$\begin{aligned}\sum_{x \in \mathbb{F}_{2^m}^*, \mathrm{Tr}_1^m(x^{-1})=1} \chi\left(g_a \circ D_3(x)\right) &= \sum_{x \in \mathbb{F}_{2^m}^*} \chi\left(g_a \circ D_3(x)\right) - \sum_{x \in \mathbb{F}_{2^m}^*, \mathrm{Tr}_1^m(x^{-1})=0} \chi\left(g_a \circ D_3(x)\right)\\ &= \sum_{x \in \mathbb{F}_{2^m}^*} \chi\left(g_a \circ D_3(x)\right) - \sum_{x \in \mathbb{F}_{2^m}^*, \mathrm{Tr}_1^m(x^{-1})=0} \chi\left(g_a(x)\right)\end{aligned}$$

$$= \sum_{x\in\mathbb{F}_{2^m}^*} \chi\left(g_a \circ D_3(x)\right)$$

$$-\frac{1}{2}\left(\sum_{x\in\mathbb{F}_{2^m}^*} \chi\left(g_a(x)\right) + \sum_{x\in\mathbb{F}_{2^m}^*} \chi\left(\mathrm{Tr}_1^m(x^{-1}) + g_a(x)\right)\right)$$

$$= (-2^m - 1 + \#G_a^3) - \frac{1}{2}\Big((-2^m - 1 + \#G_a) + (-2^m + \#H_a)\Big)$$

$$= -\frac{1}{2} + \#G_a^3 - \frac{1}{2}\left(\#G_a + \#H_a\right).$$

If $b = 1$, then the first term in Condition (d) of Theorem 10.2.10 satisfies

$$2 \sum_{x\in\mathbb{F}_{2^m}^*, \mathrm{Tr}_1^m(x^{-1})=1} \chi\left(g_a \circ D_3(x)\right) = -1 + 2\#G_a^3 - \left(\#G_a + \#H_a\right).$$

□

Here we discarded the computation of the cardinality of the curve of genus $(3r_{max} + 1)/2$ and we have to compute the cardinalities of three curves of genera $(3r_{max} - 1)/2$, $(r_{max} + 1)/2$ and $(r_{max} - 1)/2$.

We have shown how our criterion can be reformulated in terms of cardinalities of hyperelliptic curves; we now study the practical impact of such reformulations.

To begin with, even though the overall complexity is not changed between the two reformulations we presented, the practical difference is non-negligible. To illustrate this fact, we performed several simulations with Magma v2.17-13 [4]. The computations were performed on an Intel Core2 Quad CPU Q6600 cadenced at 2.40 GHz. The set R of indices used was $R = \{1, 3\}$ and one hundred of couples of coefficients (a_1, a_3) were randomly generated in $\mathbb{F}_{2^m}^*$. The meantimes (in seconds) needed to compute the number of points on the curves G_a, H_a, G_a^3 and H_a^3 for odd integers m between 21 and 59 are presented in Table 11.1. These data show that using the second reformulation is roughly twice as fast as using the first one. It also confirms that testing a function in our family using such a reformulation is much slower than testing a function in the Charpin–Gong family.

Table 11.2 shows how the second reformulation compares with a straightforward application of more classical characterizations involving exponential sums where the given sums are computed one term at a time. The column Λ indicates the meantimes (in seconds) needed to check the hyper-bentness of a function $f_{a,b}$ in our family by computing naively the exponential sum $\Lambda(f_{a,1})$, the column $\mathcal{T}_i$ by computing naively the exponential sums on $\mathcal{T}_i$ of Theorem 10.2.10, and the column $\#H$ by using the second reformulation of the previous section, for $b = 1$ and ten random pairs (a_1, a_3) of coefficients in $\mathbb{F}_{2^m}$ for m from 1 to 29, and only one couple (a_1, a_3) for m from 31 to 59. Two remarks should be made about the data

Table 11.1 Meantimes needed to compute the number of points on G_a, H_a, G_a^3 and H_a^3

m	$\#G_a$	$\#H_a$	$\#G_a^3$	$\#H_a^3$	m	$\#G_a$	$\#H_a$	$\#G_a^3$	$\#H_a^3$
21	0.017	0.488	6.857	13.894	41	0.018	1.868	40.877	108.704
23	0.016	0.576	8.736	16.021	43	0.018	2.575	47.010	128.340
25	0.017	0.653	10.587	20.287	45	0.019	4.986	62.107	176.841
27	0.016	0.912	13.684	25.704	47	0.019	5.663	84.905	210.458
29	0.017	0.869	14.843	27.667	49	0.019	6.532	94.532	234.329
31	0.016	1.026	17.766	34.532	51	0.019	7.982	125.468	242.358
33	0.017	1.166	31.258	59.000	53	0.019	7.676	133.737	249.522
35	0.018	1.317	26.809	57.998	55	0.019	8.437	116.552	275.870
37	0.018	1.562	33.321	79.949	57	0.020	9.504	127.507	305.787
39	0.019	1.893	46.768	99.544	59	0.020	9.881	162.632	360.508

Table 11.2 Meantimes needed to test the hyper-bentness of $f_{a,1}$

m	Λ	T_i	$\#H$	m	Λ	T_i	$\#H$
1	0.000	0.000	0.000	31	23,213.840	29,521.440	18.460
3	0.000	0.000	0.000	33	109,889.470	119,733.320	29.030
5	0.000	0.000	0.000	35	445,344.020	490,439.190	25.750
7	0.001	0.001	0.000	37	–	–	33.631
9	0.003	0.003	0.002	39	–	–	46.898
11	0.019	0.011	0.004	41	–	–	40.585
13	0.073	0.042	0.018	43	–	–	46.713
15	0.301	0.166	0.076	45	–	–	63.693
17	1.165	0.658	0.300	47	–	–	86.434
19	4.571	2.693	1.277	49	–	–	95.525
21	20.863	24.376	6.893	51	–	–	127.055
23	76.744	99.918	8.769	53	–	–	133.471
25	330.874	410.432	10.642	55	–	–	116.726
27	1371.403	1716.147	13.914	57	–	–	127.596
29	5472.347	6794.873	14.799	59	–	–	161.185

exposed in Table 11.2. First, it should be noted that Magma actually uses a *naive* point counting based on exponential sums for m up to 20 where it switches to the Denef–Vercauteren algorithm mentioned in Theorem 11.1.6. Nonetheless, the fact that a naive point counting algorithm has an exponential time complexity and the experimental data provided in Table 11.2 show that using such an algorithm for m greater than 20 would not be beneficial. Second, it is clear that the reformulations in terms of hyperelliptic curves are of practical interest, for relatively small values of m, and for values of m of cryptographic interest.

As a final piece of experimental evidence, the second reformulation made it possible to find hyper-bent functions of cryptographic size in our family, even though the tests are much slower than the corresponding ones for functions in the Charpin–Gong family. A random search on pairs (a_1, a_3) as above indeed showed that the Boolean functions associated with the following coefficients[2] are hyper-bent (the finite field $\mathbb{F}_{2^m}$ is represented as $\mathbb{F}_2[x]$ quotiented by the ideal generated by the m-th binary Conway polynomial):

- for $b = 0$, the pair

$$\begin{aligned} a_1 &= x^{34} + x^{31} + x^{29} + x^{27} + x^{26} + x^{24} + x^{23} + x^{21} + x^{20} + x^{18} + x^{17} + x^{16} \\ &\quad + x^{15} + x^{12} + x^{11} + x^{10} + x^9 + x^8 + x^7 + x^6 + x^4 + x^3 + x^2 + x + 1, \\ a_3 &= x^{32} + x^{29} + x^{27} + x^{25} + x^{24} + x^{23} + x^{21} + x^{20} + x^{18} + x^{16} + x^{12} + x^8 \\ &\quad + x^4 + x, \end{aligned}$$

  in $\mathbb{F}_{2^{35}}$ represented as $\mathbb{F}_2[x]/\left(x^{35} + x^{11} + x^{10} + x^7 + x^5 + x^2 + 1\right)$;
- for $b = 1$, the pair

$$\begin{aligned} a_1 &= x^{27} + x^{26} + x^{25} + x^{24} + x^{22} + x^{21} + x^{20} + x^{19} + x^{18} + x^{17} + x^{16} + x^{15} \\ &\quad + x^{14} + x^{13} + x^{11} + x^7 + x^5 + x^4 + x^2 + 1, \\ a_3 &= x^{30} + x^{29} + x^{27} + x^{26} + x^{22} + x^{20} + x^{17} + x^{16} + x^{15} + x^{12} + x^{10} + x^4 \\ &\quad + x^3 + x^2, \end{aligned}$$

  in $\mathbb{F}_{2^{33}}$ represented as $\mathbb{F}_2[x]/\left(x^{33} + x^{13} + x^{12} + x^{11} + x^{10} + x^8 + x^6 + x^3 + 1\right)$;
- for $b = \beta$ a primitive element of $\mathbb{F}_4$, the pair

$$\begin{aligned} a_1 &= x^{32} + x^{31} + x^{29} + x^{27} + x^{25} + x^{24} + x^{23} + x^{22} + x^{21} + x^{18} + x^{17} + x^{15} \\ &\quad + x^{11} + x^{10} + x^9 + x^3 + x^2 + x, \\ a_2 &= x^{32} + x^{29} + x^{28} + x^{27} + x^{26} + x^{24} + x^{22} + x^{18} + x^{17} + x^{13} + x^{10} + x^8 \\ &\quad + x^7 + x^6 + x^5 + x^4, \end{aligned}$$

  in $\mathbb{F}_{2^{33}}$ represented as $\mathbb{F}_2[x]/\left(x^{33} + x^{13} + x^{12} + x^{11} + x^{10} + x^8 + x^6 + x^3 + 1\right)$.

[2]Recall that the coefficient a_1 and a_3 are defined over $\mathbb{F}_{2^m}$, but that the corresponding Boolean functions have $n = 2m$ inputs.

11.3.3 Efficient Characterizations of Hyper-Bentness: The Wang et al. Criterion

Finally, we extend the previous works to reformulate the characterizations given by Wang et al. in terms of the number of points on hyperelliptic curves and present some numerical results leading to an interesting problem.

Applying Corollary 11.3.9 to Theorem 10.6.5 leads to the following reformulation.

Theorem 11.3.5 ([16]). *The notation is as in Theorem 10.6.5, Proposition 11.2.3 and Proposition 11.2.4*

1. *If* $b = 1$, *then* $5\Lambda(f_{a,1}) = 2(\#G_a^5 - \#H_a^5) - 5(\#G_a - \#H_a)$.
2. *If* b *is a primitive element of* $\mathbb{F}_{16}$ *such that* $\mathrm{Tr}_1^4(b) = 0$, *then* $5\Lambda(f_{a,b}) = \#G_a^5 - \#H_a^5$.
3. *If moreover* $a_r \in \mathbb{F}_{2^{\frac{m}{2}}}$, *then*
   (a) *if* b *is a primitive element of* $\mathbb{F}_{16}$ *such that* $\mathrm{Tr}_1^4(b) = 1$, *then* $10\Lambda(f_{a,b}) = -3(\#G_a^5 - \#H_a^5) + 5(\#G_a - \#H_a)$;
   (b) *if* b *is a primitive 5-th root of unity, then* $10\Lambda(f_{a,b}) = -(\#G_a^5 - \#H_a^5) - 5(\#G_a - \#H_a)$;
   (c) *if* b *is a primitive 3-rd root of unity, then* $5\Lambda(f_{a,b}) = \#G_a^5 - \#H_a^5$.

Applying Corollary 11.3.10 then yields a more practical reformulation for explicit generation of hyper-bent functions.

Theorem 11.3.6 ([16]). *The notation is as in Theorem 11.3.5.*

1. *If* $b = 1$, *then* $5\Lambda(f_{a,1}) = 4\#G_a^5 - 7\#G_a + 3\#H_a$.
2. *If* b *is a primitive element of* $\mathbb{F}_{16}$ *such that* $\mathrm{Tr}_1^4(b) = 0$, *then* $5\Lambda(f_{a,b}) = 2\#G_a^5 - \#G_a - \#H_a$.
3. *If moreover* $a_r \in \mathbb{F}_{2^{\frac{m}{2}}}$, *then*
   (a) *if* b *is a primitive element of* $\mathbb{F}_{16}$ *such that* $\mathrm{Tr}_1^4(b) = 1$, *then* $5\Lambda(f_{a,b}) = -3\#G_a^5 + 4\#G_a - \#H_a$;
   (b) *if* b *is a primitive 5-th root of unity, then* $5\Lambda(f_{a,b}) = -\#G_a^5 - 2\#G_a + 3\#H_a$;
   (c) *if* b *is a primitive 3-rd root of unity, then* $5\Lambda(f_{a,b}) = 2\#G_a^5 - \#G_a - \#H_a$.

The zeta function of a (smooth projective) curve C defined over $\mathbb{F}_q$ is

$$Z(C/\mathbb{F}_q; t) = \exp\left(\sum_{i=1}^{\infty} \frac{\#C(\mathbb{F}_{q^i})}{i} t^i\right).$$

Weil proved that, for a curve of genus g, the zeta function $Z(C/\mathbb{F}_q; t)$ can be written as a rational function

$$Z(C/\mathbb{F}_q; t) = \frac{t^{2g}\chi(1/t)}{(1-t)(1-qt)},$$

where $\chi(t)$ is the characteristic polynomial of the Frobenius endomorphism of the Jacobian of C and that

$$\chi(t) = a_g t^g + \sum_{i=0}^{g-1} a_i (t^{2g-i} + q^{g-i} t^i).$$

In particular, the knowledge of $\chi(t)$ and its factorization over the complex numbers entails that of $\#C(\mathbb{F}_{q^i})$ for all $i \geq 1$. In particular, one has

$$\#C(\mathbb{F}_q) = q + 1 + a_1.$$

Furthermore, the curves we defined are in fact *Artin–Schreier* curves, which are a special kind of imaginary hyperelliptic curves in even characteristic, and Denef and Vercauteren [10, 43] have shown that it is possible to efficiently compute their zeta functions.

Theorem 11.3.7 ([43, Theorem 4.3.1]). *Let C be an Artin–Schreier curve of genus g defined over $\mathbb{F}_{2^m}$. There exists an algorithm to compute the zeta function of C in*

$$O(g^3 m^3 (g^2 + \log^2 m \log\log m) \log gm \log\log gm)$$

bit operations and $O(g^3 m^3)$ memory.

We can therefore compute the number of points of such curves in polynomial time and space in the size of the base field. It should be remarked that the time and space complexities of the above algorithm are also polynomial in the genus of the curve.

If we fix a set $R \subset E$ of indices and suppose that the maximum index $r_{max} \in R$ is odd, then the genera of the curves H_a^5, G_a^5, H_a and G_a are respectively $\frac{5r_{max}+1}{2}$, $\frac{5r_{max}-1}{2}$, $\frac{r_{max}+1}{2}$ and $\frac{r_{max}-1}{2}$. Therefore, even though the overall time and space complexities in m of the point counting algorithm will not change, discarding the computation of the zeta function of the curve H_a^5 by using the reformulation of Theorem 11.3.6, rather than that of Theorem 11.3.5, will have a practical impact.

To illustrate this fact, we performed several simulations with Magma v2.18-2 [4]. The computations were achieved on an Intel Core2 Quad CPU Q6600 cadenced at 2.40 GHz. The set R of indices used was $R = \{1, 3\}$ and ten couples of coefficients (a_1, a_3) were randomly generated in $\mathbb{F}_{2^m}^*$. The meantimes needed to compute the number of points on the curves G_a, H_a, G_a^5 and H_a^3 for integers $m \equiv 2 \pmod 4$ between 6 and 50 are presented in Table 11.3. It should be noted that Magma [4] actually uses a naive point counting algorithm for $m \leq 20$ and switches to the Vercauteren–Kedlaya algorithm for higher values. Nonetheless, the time needed for the naive method growing exponentially, it quickly becomes far less efficient than the Vercauteren–Kedlaya one, even for curves of high genera such as G_a^5 and H_a^5.

We now provide numerical evidence that the characterizations using hyperelliptic curves are more efficient than those involving exponential sums not only

Table 11.3 Meantimes needed to compute the number of points on G_a, H_a, G_a^5 and H_a^5

m	$\#G_a$	$\#H_a$	$\#G_a^5$	$\#H_a^5$	m	$\#G_a$	$\#H_a$	$\#G_a^5$	$\#H_a^5$
6	0.000	0.001	0.000	0.000	30	0.024	1.165	132.982	197.473
10	0.001	0.001	0.000	0.000	34	0.035	1.376	338.97	570.014
14	0.010	0.012	0.020	0.019	38	0.080	1.520	394.670	627.62
18	0.244	0.217	0.309	0.318	42	0.050	2.390	491.030	958.810
22	0.019	0.634	52.533	81.334	46	0.037	5.069	742.901	1111.722
26	0.021	0.850	82.884	143.275	50	0.042	7.814	1022.621	1428.279

Table 11.4 Meantimes needed to test the hyper-bentness of $f_{a,1}$

m	Λ	T_1	$\#G$	m	Λ	T_1	$\#G$
6	0.000	0.000	0.001	22	38.709	56.547	53.490
10	0.012	0.005	0.003	26	660.433	941.750	83.137
14	0.150	0.092	0.041	30	11,271.549	16,141.993	131.745
18	2.462	1.449	0.666	34	212,549.620	277,847.460	328.580

asymptotically, but also for practical values of m. Table 11.4 gives the meantimes needed to test the hyper-bentness of ten randomly chosen functions $f_{a,b}$ with $R = \{1, 3\}$ and $b = 1$ using Magma [4] implementations of Proposition 10.3.2 (denoted by Λ), Theorem 10.6.5 (denoted by T_1) and Theorem 11.3.6 (denoted by $\#G$) on the same hardware as above (for $m = 34$, only one couple was tested). Finally, a random search on such functions using the latter test showed that the following couple (a_1, a_3):

$$\begin{aligned} a_1 &= x^{29} + x^{28} + x^{27} + x^{26} + x^{25} + x^{24} + x^{23} + x^{21} + x^{18} + \\ &\quad x^{16} + x^{15} + x^{14} + x^{12} + x^6 + x^5 + x^3 + x^2 + x, \\ a_3 &= x^{29} + x^{28} + x^{25} + x^{24} + x^{23} + x^{20} + x^{16} + x^{15} + x^{14} + \\ &\quad x^{13} + x^{12} + x^{11} + x^{10} + x^9 + x^8 + x^6 + x^4 + x^3, \end{aligned}$$

where $\mathbb{F}_{2^{30}}$ is represented as $\mathbb{F}_2[x]/(C_{30})$ with C_{30} the 30-th Conway polynomial, gives rise to a hyper-bent function $f_{a,1}$ in $n = 60$ inputs. Finding such a couple would have been quite difficult with a naive approach using exponential sums.

To conclude this section, we investigate the case where $R = \{1, 3\}$ and $a_1 = a_3 = a$ and b is a primitive element of $\mathbb{F}_4$ of trace zero. In this case, the functions of the Wang et al. family are of the form

$$f_{a,b} = \mathrm{Tr}_1^n \left(a \left(x^{3(2^m-1)} + x^{(2^m-1)}\right)\right) + \mathrm{Tr}_1^4 \left(bx^{\frac{2^n-1}{5}}\right),$$

and the associated condition for hyper-bentness is

$$T_1^5(g_a) = 2,$$

or equivalently

$$2\#G_a^5 - \#G_a - \#H_a = 5.$$

For small values of m, numerical investigation pointed out that the associated value ν_a defined as

$$\nu_a = \frac{T_1^5(g_a) - 2}{10} + (-1)^{\frac{m-2}{4}} = \frac{2\#G_a^5 - \#G_a - \#H_a - 5}{20} + (-1)^{\frac{m-2}{4}}$$

takes even integer values with absolute value bounded by a given constant. For $m \in \{6, 10, 14, 18\}$, the constants were respectively 2, 12, 80 and 314. In particular, it is never equal to $(-1)^{\frac{m-2}{4}}$ and the associated family of Boolean functions contains no hyper-bent functions. Proving the above fact is therefore both of practical and theoretical interest.

11.3.4 Algorithmic Generation of Hyper-Bent Functions in the Family $\mathcal{H}_n$ and Hyperelliptic Curves

Recall the fundamental connection between Boolean functions, exponential sums and hyperelliptic curves.

Proposition 11.3.8 ([17, Propositions 3.3 and 3.4]). *Let $\tilde{g} : \mathbb{F}_{2^m} \to \mathbb{F}_{2^m}$ be a function such that $\tilde{g}(0) = 0$ and g be the corresponding Boolean function $g = \mathrm{Tr}_1^m(\tilde{g})$. Let G_g be the (affine) curve defined over $\mathbb{F}_{2^m}$ by*

$$G_g : y^2 + y = \tilde{g}(x),$$

and H_g be the (affine) curve defined over $\mathbb{F}_{2^m}$ by

$$H_g : y^2 + xy = x + x^2\tilde{g}(x).$$

Then

$$\Xi(g) = \#G_g - 2^m,$$
$$\Xi(\mathrm{Tr}_1^m(1/x) + g(x)) = \#H_g - 2^m + 1.$$

We superscript the curves G_g and H_g by r to mean that the corresponding functions $\tilde{g}$ and g are composed with D_r, i.e. $G_g^r = G_{g \circ D_r}$ and $H_g^r = H_{g \circ D_r}$.

Proposition 11.3.8 gives the following reformulation of Lemma 2.3.10 in terms of curves.

Corollary 11.3.9. *The notation is as in Proposition 11.3.8. Then*

$$T_i(g) = \frac{1}{2}\left((\#G_g - 2^m) + \chi(i)(\#H_g - 2^m + 1)\right).$$

When applied to Corollary 2.1.6, we get the following interesting result about curves.

Corollary 11.3.10. *The notation is as in Proposition 11.3.8. Let moreover* $1 \leq r \leq 2^n - 1$ *be an integer such that* $k = \gcd(r, 2^m - 1) = 1$*. Then*

$$\#G_g^r + \#H_g^r = \#G_g + \#H_g.$$

In this subsection, we are interested in the algorithmic generation of hyper-bent functions in the family $\mathcal{H}_n$. Recall that a function $f_{a,b} \in \mathcal{H}_n$ is of the form

$$f_{a,b}(x) = \sum_{r \in R} \mathrm{Tr}_1^n\left(a_r x^{r(2^m-1)}\right) + \mathrm{Tr}_1^t\left(b x^{\frac{2^m+1}{\tau}(2^m-1)}\right),$$

and that it is hyper-bent if and only if the corresponding value $\Lambda(a, b) = \Lambda(f_{a,b}) = 1$.

We show how the results of Sect. 10.6 can be reformulated in terms of hyperelliptic curves. This was done for the Charpin–Gong family [35] (i.e. for the family $\mathcal{F}_n$), in the cases $\tau = 3$ [17] (i.e. for the family $\mathfrak{H}_n$) and $\tau = 5$ [16] (i.e. for the family of Wang et al.), leading in these three cases to both theoretical and practical improvements. In particular, hyper-bent functions were devised which could not have been generated by naive computation of exponential sums.

Here, we generalize these approaches, apply them to the families described in Sect. 10.6 and provide a complexity analysis of the corresponding tests for hyper-bentness. Furthermore, we study how the different available tests behave as τ grows and which one is the fastest for *explicit* generation of hyper-bent functions, that is for moderate values of m where the tests actually permit to generate hyper-bent functions through a random search on the coefficients a_r.

Before taking this quite algorithmic and practical point of view, let us mention that reformulating the previous characterizations in terms of number of points on hyperelliptic curves is also of high theoretical interest. The theory of algebraic curves is rich and can be applied to the study of hyper-bent function through such reformulations. For example, Lachaud and Wolfmann [29, Theorem 3.4] used the theory of elliptic curves to prove that Kloosterman sums take every value divisible by 4 within a given interval and in particular the value zero. A consequence of this result is the existence of hyper-bent monomial functions with the Dillon exponent for every extension degree m, a question which was left as an open problem by Dillon.

11.3.4.1 Characterizations in Terms of Hyperelliptic Curves

1. The family $\mathcal{G}_n$

   To begin with, it should be remarked that Lisoněk criterion for the Charpin–Gong family $\mathcal{F}_n$ [35, Theorem 2] readily extends to the family $\mathcal{G}_n$. Applying Corollary 11.3.9 to Theorem 10.4.9 indeed yields a similar reformulation.

Proposition 11.3.11. *The notation is as in Theorem 10.4.9. Then*

$$\Lambda(a) = \#G_{g_a} - \#H_{g_a}.$$

2. The case $b = 1$

   In the case $b = 1$, we have different characterizations for the hyper-bentness of $f_{a,1}$. Indeed, $f_{a,1}$ lies not only in $\mathcal{H}_n$, but also in $\mathcal{G}_n \subsetneq \mathcal{H}_n$.

   In the formalism of Sect. 10.6.1, applying Corollary 11.3.9 to Proposition 10.6.1 yields the following reformulation which is nothing but a variation of Proposition 11.3.11.

Proposition 11.3.12. *Let g'_a be the Boolean function defined on $\mathbb{F}_{2^m}$ as $g'_a(x) = g_a(x) + \mathrm{Tr}_1^m(D_s(x))$. Then*

$$\Lambda(a,1) = \#G_{g'_a} - \#H_{g'_a}.$$

Recall now that the additional trace term of $f_{a,1}$ involves the Dillon-like exponent $s(2^m - 1)$ and that the extension degree m verifies $m \equiv o \pmod{2o}$ where $2o$ is the multiplicative order of 2 modulo $\tau = \frac{2^m+1}{s}$. In particular, τ divides $2^m + 1$ and is co-prime with $2^m - 1$, so that not only Corollary 11.3.9, but also Corollary 11.3.10, can be applied to Proposition 10.6.3. Doing so, we obtain two different reformulations.

Proposition 11.3.13 ([36]). *Suppose that $\tau = p^k$ is a prime power and that 2 is a primitive root modulo p^k. Then*

$$\begin{aligned} p^k\Lambda(a,1) &= 2\left(\#G^{p^k}_{g_a} - \#H^{p^k}_{g_a}\right) - 2p\left(\#G^{p^{k-1}}_{g_a} - \#H^{p^{k-1}}_{g_a}\right) + p^k\left(\#G_{g_a} - \#H_{g_a}\right), \\ &= 4\#G^{p^k}_{g_a} - 4p\#G^{p^{k-1}}_{g_a} + (p^k + 2p - 2)\#G_{g_a} - (p^k - 2p + 2)\#H_{g_a}. \end{aligned}$$

3. The case $\tau = 3$

   For the record, we recall how Corollaries 11.3.9 and 11.3.10 apply to Theorem 10.6.4 in the case $\tau = 3$.

Proposition 11.3.14 ([36]). *Let $\tau = 3$ and $m \equiv 1 \pmod 2$. Then*

(a) If $b = 1$, then

$$\begin{aligned} 3\Lambda(a,b) &= 2\left(\#G^3_{g_a} - \#H^3_{g_a}\right) - 3\left(\#G_{g_a} - \#H_{g_a}\right), \\ &= 4\#G^3_{g_a} - 5\#G_{g_a} + \#H_{g_a}. \end{aligned}$$

(b) If b is a primitive element of $\mathbb{F}_4$*, then*

$$3\Lambda(a,b) = -\left(\#G_{g_a}^3 - \#H_{g_a}^3\right),$$
$$= -2\#G_{g_a}^3 + \#G_{g_a} + \#H_{g_a}.$$

4. The case $\tau - 5$

For the record, we recall how Corollaries 11.3.9 and 11.3.10 apply to Theorem 10.6.5 in the case $\tau = 5$.

Proposition 11.3.15 ([36]). *Let* $\tau = 5$ *and* $m \equiv 2 \pmod 4$.

(a) If $b = 1$*, then*

$$5\Lambda(a,b) = 2\left(\#G_{g_a}^5 - \#H_{g_a}^5\right) - 5\left(\#G_{g_a} - \#H_{g_a}\right),$$
$$= 4\#G_{g_a}^5 - 7\#G_{g_a} + 3\#H_{g_a}.$$

(b) If b is a primitive element of $\mathbb{F}_{16}$ *such that* $\mathrm{Tr}_1^4(b) = 0$*, i.e. with minimal polynomial* $x^4 + x + 1$*, then*

$$5\Lambda(a,b) = \#G_{g_a}^5 - \#H_{g_a}^5,$$
$$= 2\#G_{g_a}^5 - \#G_{g_a} - \#H_{g_a}.$$

(c) Suppose moreover that $a_r \in \mathbb{F}_{2^{\frac{m}{2}}}$.

i. If b is a primitive 3-rd root of unity, i.e. with minimal polynomial x^2+x+1*, then*

$$5\Lambda(a,b) = \#G_{g_a}^5 - \#H_{g_a}^5,$$
$$= 2\#G_{g_a}^5 - \#G_{g_a} - \#H_{g_a}.$$

ii. If b is a primitive 5-th root of unity, i.e. with minimal polynomial $x^4 + x^3 + x^2 + x + 1$*, then*

$$10\Lambda(a,b) = -\left(\#G_{g_a}^5 - \#H_{g_a}^5\right) - \left(\#G_{g_a} - \#H_{g_a}\right),$$
$$= -2\#G_{g_a}^5 - 4\#G_{g_a} + 6\#H_{g_a}.$$

iii. If b is a primitive element of $\mathbb{F}_{16}$ *such that* $\mathrm{Tr}_1^4(b) = 1$*, i.e. with minimal polynomial* $x^4 + x^3 + 1$*, then*

$$10\Lambda(a,b) = -3\left(\#G_{g_a}^5 - \#H_{g_a}^5\right) + 5\left(\#G_{g_a} - \#H_{g_a}\right),$$
$$= -6\#G_{g_a}^5 + 8\#G_{g_a} - 2\#H_{g_a}.$$

5. The case $\tau = 9$

Applying Corollaries 11.3.9 and 11.3.10 to Theorem 10.6.6 gives the following reformulations for $\tau = 9$.

Proposition 11.3.16 ([36]). *Let $\tau = 9$ and $m \equiv 3 \pmod 6$.*

(a) If $b = 1$, then

$$\begin{aligned} 9\Lambda(a,b) &= 2\left(\#G^9_{g_a} - \#H^9_{g_a}\right) - 6\left(\#G^3_{g_a} - \#H^3_{g_a}\right) + 9\left(\#G_{g_a} - \#H_{g_a}\right), \\ &= 4\#G^9_{g_a} - 12\#G^3_{g_a} + 13\#G_{g_a} - 5\#H_{g_a}. \end{aligned}$$

(b) If b is a primitive 3-rd root of unity, then

$$\begin{aligned} 9\Lambda(a,b) &= -\left(\#G^9_{g_a} - \#H^9_{g_a}\right) - 3\left(\#G^3_{g_a} - \#H^3_{g_a}\right) + 9\left(\#G_{g_a} - \#H_{g_a}\right), \\ &= -2\#G^9_{g_a} - 6\#G^3_{g_a} + 13\#G_{g_a} - 5\#H_{g_a}. \end{aligned}$$

(c) Suppose moreover that $a_r \in \mathbb{F}_{2^{\frac{m}{3}}}$ and $\frac{m}{3} \not\equiv 0 \pmod 3$.

i. If b is a primitive 7-th root of unity with minimal polynomial $x^3 + x + 1$ or a primitive element with minimal polynomial $x^6 + x + 1$, then

$$\begin{aligned} 9\Lambda(a,b) &= 4\left(\#G^3_{g_a} - \#H^3_{g_a}\right) - 3\left(\#G_{g_a} - \#H_{g_a}\right), \\ &= 8\#G^3_{g_a} - 7\#G_{g_a} - \#H_{g_a}. \end{aligned}$$

ii. If b is a primitive 7-th root of unity with minimal polynomial $x^3 + x^2 + 1$ or a 21-st root of unity with minimal polynomial $x^6 + x^4 + x^2 + x + 1$, then

$$\begin{aligned} 9\Lambda(a,b) &= 2\left(\#G^9_{g_a} - \#H^9_{g_a}\right) - 2\left(\#G^3_{g_a} - \#H^3_{g_a}\right) - 3\left(\#G_{g_a} - \#H_{g_a}\right), \\ &= 4\#G^9_{g_a} - 4\#G^3_{g_a} - 3\#G_{g_a} + 3\#H_{g_a}. \end{aligned}$$

iii. If b is a primitive 9-th root of unity with minimal polynomial $x^6 + x^3 + 1$, then

$$\begin{aligned} 9\Lambda(a,b) &= 2\left(\#G^3_{g_a} - \#H^3_{g_a}\right) + 3\left(\#G_{g_a} - \#H_{g_a}\right), \\ &= 4\#G^3_{g_a} + \#G_{g_a} - 5\#H_{g_a}. \end{aligned}$$

iv. If b is a primitive 21-st root of unity with minimal polynomial $x^6+x^5+x^4+x^2+1$, or a primitive element with minimal polynomial $x^6+x^5+x^3+x^2+1$ or $x^6 + x^5 + x^4 + x + 1$, then

$$\begin{aligned} 9\Lambda(a,b) &= -\left(\#G^9_{g_a} - \#H^9_{g_a}\right) + \left(\#G^3_{g_a} - \#H^3_{g_a}\right) - 3\left(\#G_{g_a} - \#H_{g_a}\right), \\ &= -2\#G^9_{g_a} + 2\#G^3_{g_a} - 3\#G_{g_a} + 3\#H_{g_a}. \end{aligned}$$

v. *If b is a primitive element with minimal polynomial $x^6+x^4+x^3+x+1$, then*

$$9\Lambda(a,b) = 2\left(\#G_{g_a}^9 - \#H_{g_a}^9\right) - 4\left(\#G_{g_a}^3 - \#H_{g_a}^3\right) + 3\left(\#G_{g_a} - \#H_{g_a}\right), \\ = 4\#G_{g_a}^9 - 8\#G_{g_a}^3 + 5\#G_{g_a} - \#H_{g_a}.$$

vi. *If b is a primitive element with minimal polynomial x^6+x^5+1 or $x^6+x^5+x^2+x+1$, then*

$$9\Lambda(a,b) = -\left(\#G_{g_a}^9 - \#H_{g_a}^9\right) - \left(\#G_{g_a}^3 - \#H_{g_a}^3\right) + 3\left(\#G_{g_a} - \#H_{g_a}\right), \\ = -2\#G_{g_a}^9 - 2\#G_{g_a}^3 + 5\#G_{g_a} - \#H_{g_a}.$$

6. The case $\tau = 11$
   Applying Corollaries 11.3.9 and 11.3.10 to Theorem 10.6.7 gives the following reformulations for $\tau = 11$.

Proposition 11.3.17 ([36]). *Let $\tau = 11$ and $m \equiv 5 \pmod{10}$.*

(a) If $b = 1$, then

$$11\Lambda(a,b) = 2\left(\#G_{g_a}^{11} - \#H_{g_a}^{11}\right) - 11\left(\#G_{g_a} - \#H_{g_a}\right), \\ = 4\#G_{g_a}^{11} - 13\#G_{g_a} + 9\#H_{g_a}.$$

(b) If b is a primitive 3-rd root of unity, a 341-st root of unity with minimal polynomial $x^{10}+x^9+x^8+x^3+x^2+x+1$, or a primitive element with minimal polynomial $x^{10}+x^9+x^8+x^4+x^3+x^2+1$ or $x^{10}+x^9+x^8+x^6+x^5+x+1$, then

$$11\Lambda(a,b) = -\left(\#G_{g_a}^{11} - \#H_{g_a}^{11}\right), \\ = -2\#G_{g_a}^{11} + \#G_{g_a} + \#H_{g_a}.$$

7. The case $\tau = 13$
   Applying Corollaries 11.3.9 and 11.3.10 to Theorem 10.6.8 gives the following reformulations for $\tau = 13$.

Proposition 11.3.18 ([36]). *Let $\tau = 13$ and $m \equiv 6 \pmod{12}$.*

(a) If $b = 1$, then

$$13\Lambda(a,b) = 2\left(\#G_{g_a}^{13} - \#H_{g_a}^{13}\right) - 13\left(\#G_{g_a} - \#H_{g_a}\right), \\ = 4\#G_{g_a}^{13} - 15\#G_{g_a} + 11\#H_{g_a}.$$

(b) *If b is a primitive* 15*-th root of unity with minimal polynomial* $x^4 + x + 1$, *a primitive* 819*-th root of unity with minimal polynomial* $x^{12} + x^9 + x^8 + x^5 + x^4 + x + 1$, *a primitive* 1365*-th root of unity with minimal polynomial* $x^{12} + x^9 + x^5 + x^2 + 1$, *or a primitive element with minimal polynomial* $x^{12} + x^9 + x^5 + x^4 + x^2 + x + 1$, $x^{12} + x^9 + x^8 + x^5 + 1$ *or* $x^{12} + x^9 + x^8 + x^6 + x^3 + x^2 + 1$, *then*

$$\begin{aligned}13\Lambda(a,b) &= \#G_{g_a}^{13} - \#H_{g_a}^{13},\\ &= 2\#G_{g_a}^{13} - \#G_{g_a} - \#H_{g_a}.\end{aligned}$$

8. The case $\tau = 17$

Applying Corollaries 11.3.9 and 11.3.10 to Theorem 10.6.9 gives the following reformulations for $\tau = 17$.

Proposition 11.3.19 ([36]). *Let* $\tau = 17$ *and* $m \equiv 4 \pmod 8$. *Suppose moreover that* $a_r \in \mathbb{F}_{2^{\frac{m}{2}}}$.

(a) *If b is a primitive element with minimal polynomial* $x^8 + x^6 + x^5 + x + 1$ *or* $x^8 + x^6 + x^5 + x^2 + 1$, *then*

$$\begin{aligned}17\Lambda(a,b) &= \#G_{g_a}^{17} - \#H_{g_a}^{17},\\ &= 2\#G_{g_a}^{17} - \#G_{g_a} - \#H_{g_a}.\end{aligned}$$

(b) *Suppose moreover that* $a_r \in \mathbb{F}_{2^{\frac{m}{4}}}$.

i. *If b is a primitive* 15*-th root of unity with minimal polynomial* x^4+x+1, *a primitive* 17*-th root of unity with minimal polynomial* $x^8+x^5+x^4+x^3+1$, *or a primitive element with minimal polynomial* $x^8 + x^5 + x^3 + x^2 + 1$, *then*

$$\begin{aligned}17\Lambda(a,b) &= \#G_{g_a}^{17} - \#H_{g_a}^{17},\\ &= 2\#G_{g_a}^{17} - \#G_{g_a} - \#H_{g_a}.\end{aligned}$$

ii. *If b is a* 51*-st root of unity with minimal polynomial* $x^8 + x^4 + x^3 + x + 1$, *then*

$$\begin{aligned}34\Lambda(a,b) &= 3\left(\#G_{g_a}^{17} - \#H_{g_a}^{17}\right) - 17\left(\#G_{g_a} - \#H_{g_a}\right),\\ &= 6\#G_{g_a}^{17} - 20\#G_{g_a} + 14\#H_{g_a}.\end{aligned}$$

9. The case $\tau = 33$

Applying Corollaries 11.3.9 and 11.3.10 to Theorem 10.6.10 gives the following reformulations for $\tau = 33$.

Proposition 11.3.20 ([36]). *Let* $\tau = 33$ *and* $m \equiv 5 \pmod{10}$. *Suppose moreover that* $a_r \in \mathbb{F}_{2^{\frac{m}{5}}}$ *and* $\frac{m}{5} \not\equiv 0 \pmod 5$.

(a) *If* b *is a primitive* 31*-st root of unity with minimal polynomial* $x^5 + x^2 + 1$, *or a primitive* 341*-st root of unity with minimal polynomial* $x^{10} + x^8 + x^4 + x^3 + x^2 + x + 1$, *then*

$$\begin{aligned}165\Lambda(a,b) &= 4\left(\#G_{g_a}^{33} - \#H_{g_a}^{33}\right) + 12\left(\#G_{g_a}^{11} - \#H_{g_a}^{11}\right)\\ &\quad - 44\left(\#G_{g_a}^{3} - \#H_{g_a}^{3}\right) + 33\left(\#G_{g_a} - \#H_{g_a}\right),\\ &= 8\#G_{g_a}^{33} + 24\#G_{g_a}^{11} - 88\#G_{g_a}^{3} + 61\#G_{g_a} - 5\#H_{g_a}.\end{aligned}$$

(b) *If* b *is a primitive* 31*-st root of unity with minimal polynomial* $x^5 + x^3 + 1$, *then*

$$\begin{aligned}165\Lambda(a,b) &= -4\left(\#G_{g_a}^{33} - \#H_{g_a}^{33}\right) + 24\left(\#G_{g_a}^{11} - \#H_{g_a}^{11}\right)\\ &\quad + 44\left(\#G_{g_a}^{3} - \#H_{g_a}^{3}\right) - 99\left(\#G_{g_a} - \#H_{g_a}\right),\\ &= -8\#G_{g_a}^{33} + 48\#G_{g_a}^{11} + 88\#G_{g_a}^{3} - 163\#G_{g_a} + 35\#H_{g_a}.\end{aligned}$$

(c) *If* b *is a primitive* 31*-st root of unity with minimal polynomial* $x^5 + x^3 + x^2 + x + 1$, *a primitive* 93*-rd root of unity with minimal polynomial* $x^{10} + x^8 + x^3 + x + 1$, *a primitive* 341*-st root of unity with minimal polynomial* $x^{10}+x^8+x^7+x^5+x^3+x+1$, *or a primitive element with minimal polynomial* $x^{10} + x^7 + 1$, $x^{10} + x^7 + x^6 + x^4 + x^2 + x + 1$ *or* $x^{10} + x^8 + x^7 + x^5 + 1$, *then*

$$\begin{aligned}165\Lambda(a,b) &= 12\left(\#G_{g_a}^{11} - \#H_{g_a}^{11}\right) + 33\left(\#G_{g_a} - \#H_{g_a}\right),\\ &= 24\#G_{g_a}^{11} + 21\#G_{g_a} - 45\#H_{g_a}.\end{aligned}$$

(d) *If* b *is a primitive* 93*-rd root of unity with minimal polynomial* $x^{10} + x^5 + x^4 + x^2 + 1$, *or a primitive element with minimal polynomial* $x^{10} + x^6 + x^5 + x^3 + x^2 + x + 1$ *or* $x^{10} + x^8 + x^5 + x^4 + x^3 + x^2 + 1$, *then*

$$\begin{aligned}165\Lambda(a,b) &= -2\left(\#G_{g_a}^{33} - \#H_{g_a}^{33}\right) + 18\left(\#G_{g_a}^{11} - \#H_{g_a}^{11}\right)\\ &\quad + 22\left(\#G_{g_a}^{3} - \#H_{g_a}^{3}\right) - 33\left(\#G_{g_a} - \#H_{g_a}\right),\\ &= -4\#G_{g_a}^{33} + 36\#G_{g_a}^{11} + 44\#G_{g_a}^{3} - 71\#G_{g_a} - 5\#H_{g_a}.\end{aligned}$$

(e) *If* b *is a primitive* 93*-rd root of unity with minimal polynomial* $x^{10} + x^8 + x^6 + x^5 + 1$, *then*

$$165\Lambda(a,b) = 2\left(\#G_{g_a}^{33} - \#H_{g_a}^{33}\right) + 18\left(\#G_{g_a}^{11} - \#H_{g_a}^{11}\right)$$

$$- 22\left(\#G_{g_a}^3 - \#H_{g_a}^3\right) - 33\left(\#G_{g_a} - \#H_{g_a}\right),$$
$$= 4\#G_{g_a}^{33} + 36\#G_{g_a}^{11} - 44\#G_{g_a}^3 - 31\#G_{g_a} + 35\#H_{g_a}.$$

(f) *If b is a primitive 341-st root of unity with minimal polynomial $x^{10} + x^3 + x^2 + x + 1$ or $x^{10} + x^7 + x^4 + x^3 + 1$, then*

$$\begin{aligned}165\Lambda(a,b) &= -4\left(\#G_{g_a}^{33} - \#H_{g_a}^{33}\right) + 18\left(\#G_{g_a}^{11} - \#H_{g_a}^{11}\right)\\ &\quad + 44\left(\#G_{g_a}^3 - \#H_{g_a}^3\right) - 33\left(\#G_{g_a} - \#H_{g_a}\right),\\ &= -8\#G_{g_a}^{33} + 36\#G_{g_a}^{11} + 88\#G_{g_a}^3 - 91\#G_{g_a} - 25\#H_{g_a}.\end{aligned}$$

(g) *If b is a primitive 341-st root of unity with minimal polynomial $x^{10}+x^6+x^2+x+1$, or a primitive element with minimal polynomial $x^{10}+x^7+x^3+x+1$, then*

$$\begin{aligned}165\Lambda(a,b) &= 18\left(\#G_{g_a}^{11} - \#H_{g_a}^{11}\right) - 33\left(\#G_{g_a} - \#H_{g_a}\right),\\ &= 36\#G_{g_a}^{11} - 51\#G_{g_a} + 15\#H_{g_a}.\end{aligned}$$

(h) *If b is a primitive element with minimal polynomial $x^{10}+x^7+x^6+x^5+x^4+x+1$ or $x^{10}+x^7+x^6+x^5+x^4+x^3+x^2+x+1$, then*

$$\begin{aligned}165\Lambda(a,b) &= 2\left(\#G_{g_a}^{33} - \#H_{g_a}^{33}\right) + 12\left(\#G_{g_a}^{11} - \#H_{g_a}^{11}\right)\\ &\quad - 22\left(\#G_{g_a}^3 - \#H_{g_a}^3\right) + 33\left(\#G_{g_a} - \#H_{g_a}\right),\\ &= 4\#G_{g_a}^{33} + 24\#G_{g_a}^{11} - 44\#G_{g_a}^3 + 41\#G_{g_a} - 25\#H_{g_a}.\end{aligned}$$

11.3.4.2 Asymptotic Complexities

Asymptotically, to test the hyper-bentness of any function in the family $\mathcal{H}_n$ through a naive computation of $\Lambda(f_{a,b})$, that is a partial exponential sum over $U \subset \mathbb{F}_{2^n}$, one has to compute $\#U = 2^m + 1$ summands. For each summand, the computation is dominated by the cost of an exponentiation with an arbitrary large exponent because r_{max}, the maximal index in R, grows exponentially with m. Therefore, the total computation has a time complexity of $O(2^m m^{2+\epsilon})$ and a space complexity of $O(m)$. Using the characterizations given in Sect. 10.6 yields similar complexities, the only notable difference being that the finite field arithmetic occurs in $\mathbb{F}_{2^m}$ rather than $\mathbb{F}_{2^n}$.

As far as the characterizations of the previous subsection are concerned, the situation is not better if one wants to test any function in the family $\mathcal{H}_n$. Indeed, the time and space complexities of the point counting algorithms described in

Theorems 11.1.7 and 11.1.8 are polynomial in the genus of the curve, and so are in the maximal index $r_{max} \in R$.

More precisely, we can suppose that r_{max} is odd, so that it is as small as possible and the curves involved in the characterizations are Artin–Schreier curves. Then, for any odd integer $l \geq 1$, the curves $G^l_{g_a}$ and $H^l_{g_a}$ are of genera respectively $\frac{lr_{max}-1}{2}$ and $\frac{lr_{max}+1}{2}$. Therefore, for a fixed τ and a family $\mathcal{T}_n$ of Boolean functions among the ones studied in the previous subsection (e.g. the family $\mathcal{T}_n = \{f_{a,b} \in \mathcal{H}_n \mid \tau = 9$ and b is a 3-rd root of unity$\}$), one gets the following theorem.

Theorem 11.3.21 ([36]). *For a fixed τ, the hyper-bentness of $f_{a,b} \in \mathcal{T}_n$ defined over $\mathbb{F}_{2^n}$, where $\mathcal{T}_n$ is a family defined as above, can be checked in*

$$O(r_{max}^{7.376}m^2 + r_{max}^{3.376}m^{2.667})$$

bit operations and $O(r_{max}^5 m^2 + r_{max}^3 m^{2.5})$ memory.

As the maximal index r_{max} grows exponentially with m, these complexities are still exponential in the extension degree m. Nonetheless, it is customary in cryptography to restrict to functions of a given form, that is, to fix the set R. In this case, r_{max} does not grow with m. On the one hand, the time complexities of the two naive approaches fall to $O(2^m m^{1+\epsilon})$, the computation of one term being dominated by the cost of a multiplication, and so are still exponential. On the other hand, the time and space complexities of the tests involving hyperelliptic curves become polynomial in m, except for Proposition 11.3.13.

Theorem 11.3.22 ([36]). *For a fixed τ and a fixed R, the hyper-bentness of $f_{a,b} \in \mathcal{T}_n$ defined over $\mathbb{F}_{2^n}$ can be checked in polynomial time and space.*

To conclude, it should be remarked that, except for a few exceptions, all the characterizations devised above include a curve of genus $\frac{\tau r_{max}-1}{2}$. Therefore, the time and space complexities of the tests involving hyperelliptic curves will be polynomial in τ. For example, Proposition 11.3.13, where $b = 1$ and $\tau = p^k$ is a prime power for which 2 is primitive root, yields infinite families of τ for which the corresponding characterization involves a curve of genus exactly $\frac{\tau r_{max}-1}{2}$ as follows: choose p such that 2 is a primitive root modulo p^2, e.g. $p = 3$, and consider the family $\{p^k\}_{k=2}^{+\infty}$.

11.3.4.3 Experimental Results

Although it has been demonstrated that characterizations involving hyperelliptic curves not only yield asymptotically faster algorithms, but also more practical ones for moderate values of m especially interesting in cryptography ([35] for the family $\mathcal{F}_n$, [17] for $\tau = 3$ and [16] for $\tau = 5$), it is not clear that this remains true as τ grows.

- Naive computations:

  Let us first compare the two characterizations of hyper-bentness involving exponential sums that we described in the previous sections, namely the characterization of Proposition 10.4.2 involving $\Lambda(a, b)$ and the characterizations of Sect. 10.6 involving $T_1(g_u \circ D_t)$. Although the arithmetic takes place in $\mathbb{F}_{2^n}$ for the former one, whereas it takes place in $\mathbb{F}_{2^m}$ for the latter ones, this is quite negligible for the small values of m which can be attained in practice. Moreover, the first criterion involves the computation of only one exponential sum over the set U of size $\#U = 2^m + 1$, whereas the second ones involve the computations of several exponentials over the set $\mathcal{T}_1$ of size $\#\mathcal{T}_1 = 2^{m-1}$. For example, the characterization of Proposition 10.6.3, involves three different exponential sums, and the characterizations of Theorem 10.6.10 up to four different ones. Plus, the computation of g_a and its compositions with Dickson polynomials can be more complex than that of $f_{a,b}$.

To confirm these facts and compare the time needed by a direct computation of $\Lambda(a, b)$ as suggested by Proposition 10.4.2 and by a computation based on Proposition 10.6.3 and exponential sums over $\mathcal{T}_1$, we performed different experiments with version 2.18-5 of the Magma software [4] running on an Intel(R) Xeon(R) X5650 CPU cadenced at 2.67GHz for $R = \{1\}$, a_1 randomly chosen in $\mathbb{F}_{2^m}$, $b = 1$, and different values of τ and m. Given a value of τ, some restrictions lie on the extension degree m for the expression of $f_{a,b}$, i.e. m should divide $2^m + 1$. Therefore, for unsuitable values of m, the time needed for a direct computation of $\Lambda(a, b)$ were extrapolated. Nevertheless, the expression of $\Lambda(a, b)$ in Proposition 10.6.3 involving exponential sums on $\mathcal{T}_1$ can always be computed, even though $\Lambda(a, b)$ itself is not well defined in this case. Thus, for Proposition 10.6.3 the computations were performed for every extension degree. The results of these experiments are summarized in Figs. 11.1 and 11.2 where the time needed to compute $\Lambda(a, b)$ directly or through Proposition 10.6.3 for m between 20 and 30 and $\tau = 3$ and 5 are depicted. Similar results can be observed for higher values of τ, e.g. 9, 11 and so on; that is, computing directly $\Lambda(a, b)$ is faster than computing it through the expression given in Proposition 10.6.3.

Finally, it should be noted that the characterization of Proposition 10.4.2, not only covers the family $\mathcal{H}_n$, but the complete family $\mathcal{F}_n$, that is all Boolean functions with Dillon-like exponents without any restriction on the coefficients $a_r \in \mathbb{F}_{2^m}$, nor on the sizes of the cyclotomic cosets of the exponents r.

- Using hyperelliptic curves

Now, the time needed by the two above methods grows exponentially with m, whereas the time needed by methods involving hyperelliptic curves will only grow polynomially for a fixed set R. Nonetheless, the constants involved are much larger for the latter methods than for the former. Therefore, from a practical point of view, such methods are useless if they become more efficient for too large values of m.

Using the same setup as above, we compared the time needed to compute $\Lambda(a, b)$ directly and through the expression given in Proposition 11.3.13. Note that

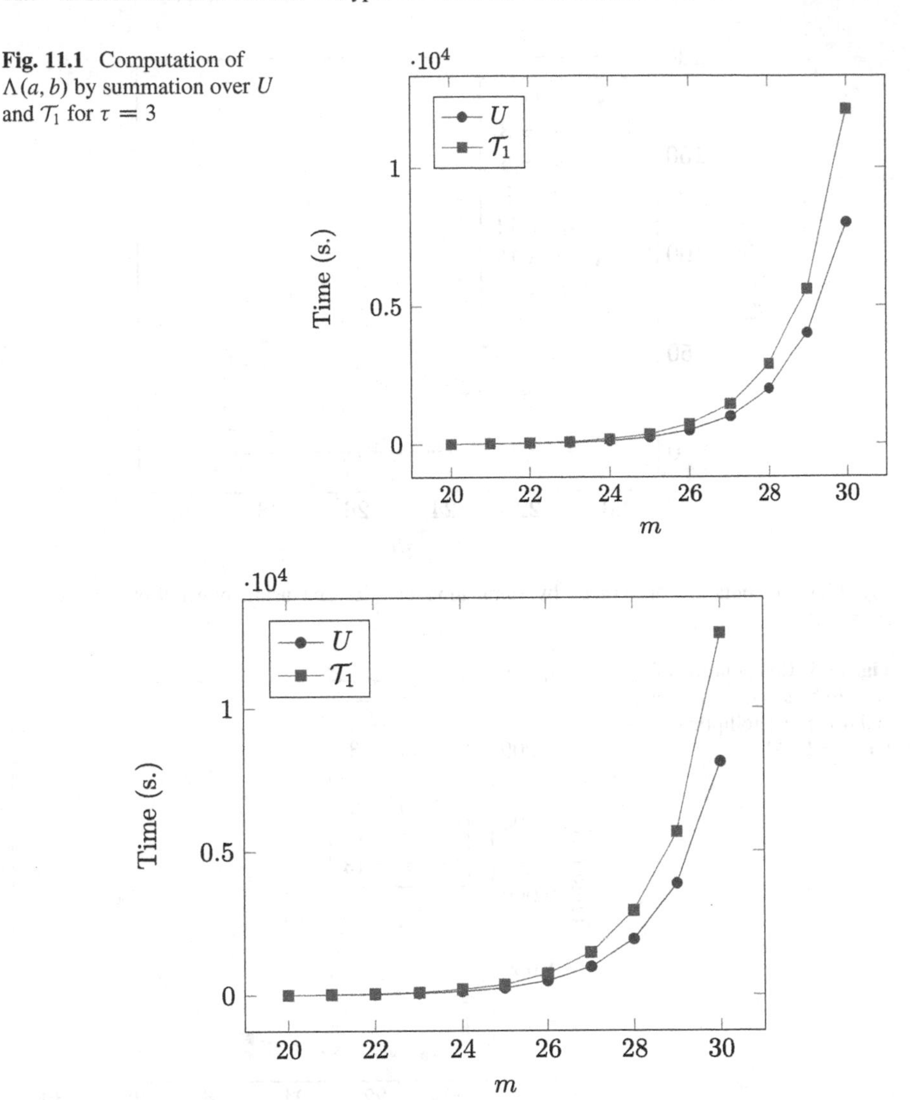

Fig. 11.1 Computation of $\Lambda(a, b)$ by summation over U and $\mathcal{T}_1$ for $\tau = 3$

Fig. 11.2 Computation of $\Lambda(a, b)$ by summation over U and $\mathcal{T}_1$ for $\tau = 5$

the Magma software uses a naive method to compute the number of points on a hyperelliptic curves of genus $g \geq 2$ for $m < 20$ and the Denef–Vercauteren algorithm from $m = 20$ onward. It also implements specialized point counting algorithms for elliptic curves and hyperelliptic curves of genus 2. As above, the definition of $f_{a,b}$ only makes sense for some extension degrees m, whereas the expression given by Proposition 11.3.13 always does. Therefore, timings using hyperelliptic curves where generated for every extension degrees between 20 and 30, even though the correspondence with $\Lambda(a, b)$ is not always valid. Moreover, the aforementioned computations for different values of τ showed that the time needed

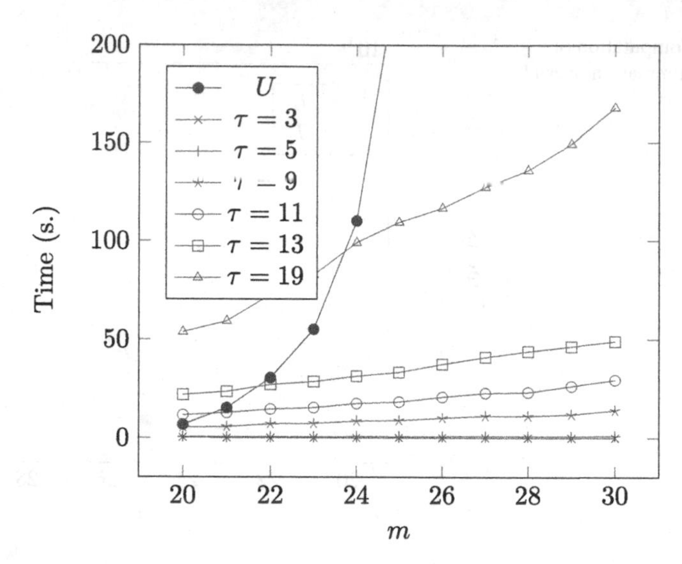

Fig. 11.3 Computation of $\Lambda(a, b)$ by summation over U and using hyperelliptic curves for $R = \{1\}$

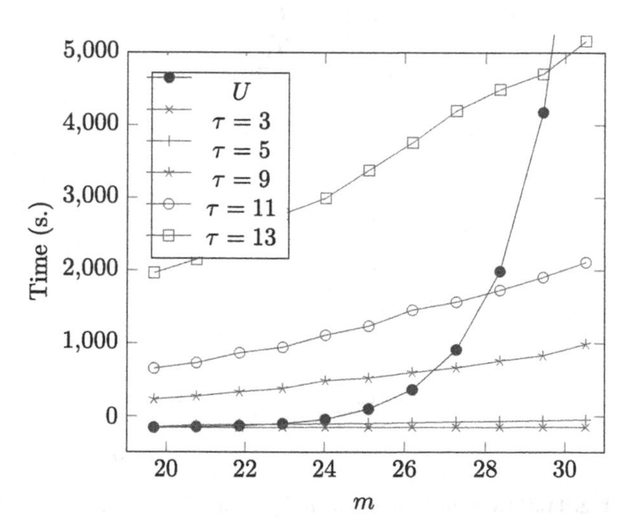

Fig. 11.4 Computation of $\Lambda(a, b)$ by summation over U and using hyperelliptic curves for $R = \{1, 3\}$

for a direct computation of $\Lambda(a, b)$ depends on τ in a negligible way. Hence, we only include timings of such a computation for a given value of τ chosen to be $\tau = 3$. Figure 11.3 gives timings for $R = 1$, that is for binomial functions, for different values of τ, whereas Fig. 11.4 gives similar timings for $R = 1, 3$, that is for trinomial functions.

For $\tau = 3$ and 5, as was already shown in previous works [16, 17, 35], reformulations in terms of hyperelliptic curves yield non-negligible improvements, even for moderate values of m, when $R = \{1\}$ and $R = \{1, 3\}$ are considered. Indeed, such reformulations allow to generate hyper-bent functions which could not have been generated using more naive methods. For $R = \{1\}$, that is for the simplest binomial functions, Fig. 11.3 suggests that this remains true for a few additional values of τ. This fact is confirmed by examples of hyper-bent functions given in Sect. 11.3.4.3.

On the contrary, Fig. 11.4 shows that for trinomial functions with $R = \{1, 3\}$ and τ greater than 5, the crossover happens at extension degrees m for which testing hyper-bentness with a naive method, and using hyperelliptic curves, takes several hundreds of seconds. In particular, it seems hopeless to generate additional hyper-bent trinomials using tests based on hyperelliptic curves.

Nonetheless, the current Magma implementation of point counting over finite fields of even characteristic for hyperelliptic curves is limited to the Denef–Vercauteren algorithm. Algorithms based on deformation theory are limited to odd characteristic. An efficient implementation of such algorithms in even characteristic should provide practical improvements, especially together with the possibility to count points on m different curves at a time with no runtime overhead by implementing multipoint evaluation as described by Hubrechts [25]. Furthermore, all the algorithms mentioned previously were designed not only to compute the number of rational points of a curve over its base field, but its complete zeta function, which is loosely equivalent to the knowledge of the number of points of the curve over the first g extensions of the base field if the curve has genus g. Specializing these algorithms by lowering the needed precision during the computations to only compute the number of points of the curve over the base field, that is, the trace term of the zeta function, should provide both better asymptotic complexities and practical improvements to the runtime.

- Examples of hyper-bent functions

To conclude this subsection, we provide some tuples of coefficients $(a_r)_{r \in R}$ corresponding to hyper-bent functions for different values of τ and b. They were generated through a random search, using the different characterizations proposed in this note. To describe the coefficients a_r, the finite field $\mathbb{F}_{2^m}$ is always represented as $\mathbb{F}_2[x]/(C_m(x))$ where $C_m(x)$ is the m-th Conway polynomial. Furthermore, recall that, although the coefficients a_r live in $\mathbb{F}_{2^m}$, the corresponding Boolean function $f_{a,b}$ is defined over $\mathbb{F}_{2^n}$ where $n = 2m$.

For example, when $b = 1$, we found that the function $f_{a,1}$ is hyper-bent if:

1. $\tau = 3$, $m = 33$, $a_1 = x^{32} + x^{28} + x^{27} + x^{25} + x^{24} + x^{20} + x^{19} + x^{14} + x^{13} + x^9 + x^5 + x^4 + x^2 + 1$;
2. $\tau = 5$, $m = 34$, $a_1 = x^{33} + x^{30} + x^{29} + x^{28} + x^{27} + x^{26} + x^{23} + x^{22} + x^{20} + x^{18} + x^{17} + x^{16} + x^{15} + x^{14} + x^{11} + x^9 + x^2 + 1$;
3. $\tau = 5$, $m = 34$, $a_1 = x^{33} + x^{32} + x^{28} + x^{27} + x^{26} + x^{25} + x^{24} + x^{22} + x^{20} + x^{19} + x^{18} + x^{17} + x^{11} + x^9 + x^6 + x^3 + x^2$, $a_3 = x^{33} + x^{31} + x^{29} + x^{28} + x^{25} + x^{23} + x^{22} + x^{18} + x^{17} + x^{16} + x^{14} + x^8 + x^5 + x^4 + x^2$;
4. $\tau = 9$, $m = 21$, $a_1 = x^{20} + x^{17} + x^{15} + x^{14} + x^{10} + x^9 + x^6 + x^4 + x^2 + x$;
5. $\tau = 9$, $m = 21$, $a_1 = x^{18} + x^{17} + x^{12} + x^{11} + x^5 + x^3 + x + 1$, $a_3 = x^{19} + x^{18} + x^{14} + x^8 + x^7 + x^4 + 1$;

6. $\tau = 9, m = 27, a_1 = x^{26} + x^{25} + x^{23} + x^{22} + x^{20} + x^{16} + x^9 + x^6$;
7. $\tau = 11, m = 15, a_1 = x^{15338}$;
8. $\tau = 11, m = 15, a_1 = x^{1066}, a_3 = x^{19316}$;
9. $\tau = 11, m = 25, a_1 = x^{24} + x^{22} + x^{17} + x^{13} + x^{11} + x^7 + x^5$;
10. $\tau = 13, m = 18, a_1 = x^{253630}$;
11. $\tau = 13, m = 18, a_1 = x^{247490}, a_3 = x^{216257}$;
12. $\tau = 13, m = 30, a_1 = x^{29} + x^{28} + x^{24} + x^{22} + x^{18} + x^{17} + x^{15} + x^6 + x^5 + x^3$.

For $b \neq 1$, here follow some examples of hyper-bent functions $f_{a,b}$:

1. $\tau = 3, m = 29, a_1 = x^{27} + x^{26} + x^{25} + x^{24} + x^{23} + x^{22} + x^{21} + x^{20} + x^{19} + x^{17} + x^{16} + x^{10} + x^8 + x^7 + x^6 + x^5 + x^4 + x^3$, b a primitive element of $\mathbb{F}_4$;
2. $\tau = 3, m = 33, a_1 = x^{29} + x^{26} + x^{24} + x^{23} + x^{20} + x^{18} + x^{17} + x^{16} + x^{15} + x^9 + x^8 + x^7 + x^6$, b a primitive element of $\mathbb{F}_4$;
3. $\tau = 5, m = 30, a_1 = x^{29} + x^{27} + x^{26} + x^{25} + x^{22} + x^{21} + x^{19} + x^{18} + x^{17} + x^{16} + x^{15} + x^{14} + x^{11} + x^{10} + x^9 + x^6 + x^5 + x^3 + x + 1$, b a primitive element of $\mathbb{F}_{16}$ with trace 0;
4. $\tau = 5, m = 34, a_1 = x^{33} + x^{29} + x^{25} + x^{23} + x^{22} + x^{21} + x^{20} + x^{16} + x^{15} + x^{14} + x^{11} + x^8 + x^6 + x^5 + x^4 + x^3$, b a primitive element of $\mathbb{F}_{16}$ with trace 0;
5. $\tau = 9, m = 21, a_1 = x^{20} + x^{19} + x^{17} + x^{15} + x^{14} + x^{13} + x^9 + x^8 + x^7 + x^4 + x^2 + 1$, b a primitive 3-rd root of unity;
6. $\tau = 9, m = 27, a_1 = x^{25} + x^{23} + x^{22} + x^{20} + x^{19} + x^{18} + x^{17} + x^{15} + x^{10} + x^9 + x^7 + x^6 + x^3$, b a primitive 3-rd root of unity;
7. $\tau = 11, m = 25, a_1 = x^{24} + x^{22} + x^{21} + x^{17} + x^{16} + x^{14} + x^{13} + x^{11} + x^{10} + x^6 + x^3 + x^2$, b a primitive 3-rd root of unity;
8. $\tau = 13, m = 18, a_1 = x^{166827}$, b a primitive 15-th root of unity with minimal polynomial $x^4 + x + 1$;
9. $\tau = 13, m = 30, a_1 = x^{26} + x^{23} + x^{22} + x^{21} + x^{20} + x^{17} + x^{16} + x^{15} + x^{14} + x^9 + x^6 + x$, b a primitive 15-th root of unity with minimal polynomial $x^4 + x + 1$.

Not only the above examples show the usefulness of our approach for explicit generation of hyper-bent functions, but also that the families of Boolean functions we consider actually contain hyper-bent functions.

11.4 Values of Binary Kloosterman Sums: Some Methods

11.4.1 Divisibility of Binary Kloosterman Sums

11.4.1.1 Classical Results

Because of their cryptographic interest, divisibility properties of Kloosterman sums have been studied in several recent papers. A nice overview of such results can be found in the Ph.D. thesis of Moloney [37]. Here we cite a few of them which we will explicitly use in search algorithms for binary Kloosterman sums with specific values, especially the values 0 and 4.

Recall that Proposition 2.3.2 states in particular that binary Kloosterman sums are always divisible by 4. Afterwards, several papers studied divisibility properties of binary Kloosterman sums by multiples of 4 and other integers.

The following result, first proved by Helleseth and Zinoviev [24], classifies the values of $K_m(a)$ modulo 8 according to the value of the absolute trace of a.

Proposition 11.4.1 ([24]). *Let $m \geq 3$ be any positive integer and $a \in \mathbb{F}_{2^m}$. Then $K_m(a) \equiv 0 \pmod 8$ if and only if $\mathrm{Tr}_1^m(a) = 0$.*

In the same article, they gave the following sufficient conditions to get certain values of $K_m(a)$ modulo 3.

Proposition 11.4.2 ([24]). *Let $m \geq 3$ be any positive integer and $a \in \mathbb{F}_{2^m}^*$. Suppose that there exists $t \in \mathbb{F}_{2^m}^*$ such that $a = t^4 + t^3$.*

- *If m is odd, then $K_m(a) \equiv 1 \pmod 3$.*
- *If m is even, then $K_m(a) \equiv 0 \pmod 3$ if $\mathrm{Tr}_1^m(t) = 0$ and $K_m(a) \equiv -1 \pmod 3$ if $\mathrm{Tr}_1^m(t) = 1$.*

Furthermore, Charpin, Helleseth and Zinoviev [6] gave additional results about values of $K_m(a)$ modulo 3.

Proposition 11.4.3 ([6]). *Let $m \geq 3$ be any positive integer and $a \in \mathbb{F}_{2^m}^*$. Then we have:*

- *If m is odd, then $K_m(a) \equiv 1 \pmod 3$ if and only if $\mathrm{Tr}_1^m(a^{1/3}) = 0$. This is equivalent to $a = \frac{b}{(1+b)^4}$ for some $b \in \mathbb{F}_{2^m}^*$.*
- *If m is even, then $K_m(a) \equiv 1 \pmod 3$ if and only if $a = b^3$ for some b such that $\mathrm{Tr}_2^m(b) \neq 0$.*

Further divisibility results exist and could be used to refine the tests proposed in this chapter. For example, results up to 64 can be found in a paper of Göloğlu, McGuire and Moloney [22], and results up to 256 in an even more recent paper of Göloğlu, Lisoněk, McGuire and Moloney [21].

Most of these results about divisibility were first proved studying the link between exponential sums and coset weight distribution [6, 24]. However some of them can be proved in a completely different manner as we show in the next subsection.

11.4.1.2 Using Torsion of Elliptic Curves

Theorem 11.2.1 giving the value of $K_m(a)$ as the cardinality of an elliptic curve can indeed be used to deduce divisibility properties of Kloosterman sums from the rich theory of elliptic curves. The *quadratic twist* of the ordinary elliptic curve E_a that we denote by $\tilde{E}_a$ is given by the Weierstrass equation

$$\tilde{E}_a : y^2 + xy = x^3 + bx^2 + a,$$

where $b \in \mathbb{F}_{2^m}$ has absolute trace 1; it has cardinality:

$$\#\tilde{E}_a = 2^m + 2 - K_m(a).$$

First of all, we recall a proof of the divisibility by 4 as it can be found for example in the preprint of Ahmadi and Granger [1]. For $m \geq 3$, $K_m(a) \equiv \#E_a \pmod 4$, so $K_m(a) \equiv 0 \pmod 4$ if and only if $\#E_a \equiv 0 \pmod 4$. This is equivalent to E_a having a non-trivial rational point of 4-torsion. This can also be formulated as both the equation of E_a and its 4-division polynomial $f_4(x) = x^6 + ax^2$ having a rational solution. It is easily seen that $P = (a^{1/4}, a^{1/2})$ is always a non-trivial solution to this problem.

Lisoněk [34] used similar techniques to give a different proof of Proposition 11.4.1. Indeed, for $m \geq 3$, $K_m(a)$ is divisible by 8 if and only if E_a has a non-trivial rational point of 8-torsion. This is easily shown to be equivalent to $\mathrm{Tr}_1^m(a^{1/4}) = \mathrm{Tr}_1^m(a) = 0$.

Finally, it is possible to prove directly that the condition given in Proposition 11.4.2 is not only sufficient, but also necessary, using torsion of elliptic curves.[3]

We use this property in Sect. 11.4.2.3.

Proposition 11.4.4 ([36]). *Let $a \in \mathbb{F}_{2^m}^*$.*

- *If m is odd, then $K_m(a) \equiv 1 \pmod 3$ if and only if there exists $t \in \mathbb{F}_{2^m}$ such that $a = t^4 + t^3$.*
- *If m is even, then:*
  - *$K_m(a) \equiv 0 \pmod 3$ if and only if there exists $t \in \mathbb{F}_{2^m}$ such that $a = t^4 + t^3$ and $\mathrm{Tr}_1^m(t) = 0$;*
  - *$K_m(a) \equiv -1 \pmod 3$ if and only if there exists $t \in \mathbb{F}_{2^m}$ such that $a = t^4 + t^3$ and $\mathrm{Tr}_1^m(t) = 1$.*

Proof. According to Proposition 11.4.2 we only have to show that, if a verifies the given congruence, it can be written as $a = t^4 + t^3$.

- We begin with the case m odd, so that $2^m \equiv -1 \pmod 3$. Then $K_m(a) \equiv 1 \pmod 3$ if and only if $\#E_a \equiv 0 \pmod 3$, i.e. if E_a has a non-trivial rational point of 3-torsion. It implies that the 3-division polynomial of E_a given by $f_3(x) = x^4 + x^3 + a$ has a rational solution, so that there exists $t \in \mathbb{F}_{2^m}$ such that $a = t^4 + t^3$.
- Suppose now that m is even, so that $2^m \equiv 1 \pmod 3$.
  - If $K_m(a) \equiv -1 \pmod 3$, then $\#E_a \equiv 0 \pmod 3$, and as in the previous case we can find $t \in \mathbb{F}_{2^m}$ such that $a = t^4 + t^3$.

[3] We were recently made aware that such a result was proved in a different way also involving elliptic curves by Garashuck and Lisoněk [20] in the case where m is odd.

– If $K_m(a) \equiv 0 \pmod 3$, then $\#E_a \equiv 1 \pmod 3$, but $\#\tilde{E}_a \equiv 0 \pmod 3$. The 3-division polynomial of $\tilde{E}_a$ is also given by $f_3(x) = x^4 + x^3 + a$, so that there exists $t \in \mathbb{F}_{2^m}$ such that $a = t^4 + t^3$.

□

11.4.2 Finding Specific Values of Binary Kloosterman Sums

11.4.2.1 Generic Strategy

In this subsection we present the most generic method to find specific values of binary Kloosterman sums. To this end, one picks random elements of $\mathbb{F}_{2^m}$ and computes the corresponding values until a correct one is found. Before performing any complicated computations, divisibility conditions as those stated in the previous section can be used to restrict the pool of elements to those satisfying certain conditions (but without missing any element giving the value searched for) or to filter out elements which will give inadequate values.

Then, the most naive method to check the value of a binary Kloosterman sum is to compute it as a sum. However, one test would need $O(2^m m \log^2 m \log\log m)$ bit operations and this is obviously highly inefficient. Theorem 11.2.1 tells that this costly computation can be replaced by the computation of the cardinality of an elliptic curve over a finite field of even characteristic. Using p-adic methods à la Satoh [39], also known as canonical lift methods, this can be done quite efficiently in $O(m^2 \log^2 m \log\log m)$ bit operations and $O(m^2)$ memory [23, 32, 44, 45]. Working with elliptic curves also has the advantage that one can check that the current curve is a good candidate before computing its cardinality as follows: one picks a random point on the curve and multiplies it by the targeted order; if it does not give the point at infinity, the curve does not have the targeted cardinality.

Finally, it should be noted that, if ones looks for all the elements giving a specific value, a different strategy can be adopted as noted in the paper of Ahmadi and Granger [1]. Recall that a binary Kloosterman sum can be seen as the Walsh–Hadamard transform of the Boolean function $\mathrm{Tr}_1^m(1/x)$. Therefore, we can construct the Boolean function corresponding to the function $\mathrm{Tr}_1^m(1/x)$ and then use a fast Walsh–Hadamard transform to compute the values of all binary Kloosterman sums. Building the Boolean function costs one multiplication per element, so $O(2^m m \log m \log\log m)$ bit operations and $O(2^m)$ memory. The complexity of the fast Walsh–Hadamard transform is $O(2^m m^2)$ bit operations and $O(2^m m)$ memory [2].

11.4.2.2 Zeros of Binary Kloosterman Sums

When looking for zeros of binary Kloosterman sums, which is of high cryptographic interest as Chap. 10[4] emphasizes, one benefits from even more properties of elliptic curves over finite fields. Indeed, when $K_m(a) = 0$, we get that $\#E_a = 2^m$. Hence all rational points of E_a are of order some power of 2.

In fact, we know even more. As E_a is defined over a field of even characteristic, its complete 2^e-torsion (where e is any strictly positive integer) is of rank 1, whereas the complete l^e-torsion, for a prime l different from 2, is of rank 2, as stated in Proposition 11.1.2. Therefore the rational Sylow 2-subgroup is cyclic, isomorphic to $\mathbb{Z}/2^e\mathbb{Z}$ for some positive integer e. In the case where $K_m(a) = 0$, we even get that the whole group of rational points is isomorphic to $\mathbb{Z}/2^m\mathbb{Z}$. Furthermore, basic group theory tells that E_a will then have 2^{m-1} points of order 2^m.

Finally, it should be noted that, if $2^m \mid \#E_a$, then $\#E_a$ must be equal to 2^m. This is a simple consequence of Hasse theorem (Theorem 11.1.1) giving bounds on the number of rational points of an elliptic curve over a finite field.

These facts have first been used by Lisoněk [34] to develop a probabilistic method to test whether a given a gives a binary Kloosterman zero or not: one takes a random point on E_a and tests whether its order is 2^m or not. This test involves at most m duplications on the curve, hence is quite efficient. Moreover, as soon as $\#E_a = 2^m$, half of its points are generators, so that testing one point on a correct curve gives a probability of success of $1/2$. This led Lisoněk to find zeros of binary Kloosterman sums for m up to 64 in a matter of days.

Afterwards, Ahmadi and Granger [1] proposed a deterministic algorithm to test whether an element $a \in \mathbb{F}_{2^m}$ gives a binary Kloosterman zero or not. From the above discussion, it is indeed enough to compute the size of the Sylow 2-subgroup of E_a to answer that question. This can be efficiently implemented by point halving, starting from a non-trivial point of 4-torsion (remember that such a point always exists on E_a). The complexity of each iteration of their algorithm is dominated by two multiplications in $\mathbb{F}_{2^m}$. So testing a curve with a Sylow 2-subgroup of size 2^e is of complexity $O(e \cdot m \log m \log\log m)$. Furthermore, they showed that the average size of the Sylow 2-subgroup of the curves of the form E_a is 2^3 when m goes to infinity, so that their algorithm has an asymptotic average bit complexity of $O(m \log m \log\log m)$.

[4]We will see also in Chap. 17 that the value 0 of Kloosterman sums gives rise to semi-bent functions in even dimension (see for instance Table 17.2).

11.4.2.3 Implementation for the Value 4

We have seen[5] in Chap. 9 a necessary and sufficient condition to build bent functions from the value 4 of binary Kloosterman sums when m is odd and a necessary only condition when m is even. Unfortunately, the situation is more complicated than in the case of binary Kloosterman zeros.

We are indeed looking for an element $a \in \mathbb{F}_{2^m}$ such that $K_m(a) = 4$. The cardinality of E_a should then be $\#E_a = 2^m + K_m(a) = 4(2^{m-2} + 1)$ which does not ensure to have a completely fixed group structure as was the case when $\#E_a = 2^m$. Moreover, in general, the number $2^{m-2} + 1$ does not verify many divisibility properties leading to an efficient test for the value 4. The cardinality of the twist $\tilde{E}_a$ is given by $\#\tilde{E}_a = 2^m + 2 - K_m(a) = 2(2^{m-1} - 1)$ which does not provide more useful information.

What we can however deduce from these equalities is that, if $K_m(a) = 4$, then:

- $K_m(a) \equiv 4 \pmod 8$, so that $\mathrm{Tr}_1^m(a) = 1$;
- $K_m(a) \equiv 1 \pmod 3$, so that:
  - if m is odd, then a can be written as $t^4 + t^3$;
  - if m is even, then a can be written as t^3 with $\mathrm{Tr}_2^m(t) \neq 0$.

We can use both these conditions to filter out a to be tested as described in Algorithm 11.1 (for m odd).

We implemented this algorithm in Sage [42]. It was necessary to implement a relatively efficient version of point counting in even characteristic, none of them being available. The first implemented algorithm was an extension to even characteristic of Satoh's original algorithm by Fouquet, Gaudry and Harley [18].

Algorithm 11.1: The value 4 of binary Kloosterman sums for m odd

---

Input: A positive odd integer $m \geq 3$
Output: An element $a \in \mathbb{F}_{2^m}$ such that $K_m(a) = 4$

1 $a \leftarrow_R \mathbb{F}_{2^m}$
2 $a \leftarrow a^3(a+1)$
3 **if** $\mathrm{Tr}_1^m(a) = 0$ **then**
4   Go to step 11.1
5 $P \leftarrow_R E_a$
6 **if** $[2^m + 4]P \neq 0$ **then**
7   Go to step 11.1
8 **if** $\#E_a \neq 2^m + 4$ **then**
9   Go to step 11.1
10 **return** a

---

[5] We will see also in Chap. 17 that the value 4 of Kloosterman sums gives rise to semi-bent functions in even dimension (see for instance Table 17.1).

The complexity of this algorithm is $O(m^{3+\epsilon})$ bit operations (or $O(m^5)$ with naive multiplication) and $O(m^3)$ memory, but it is quite simple and there was already an existing implementation in GP/Pari by Yeoh [47] to use as a starting point. The computations in $\mathbb{Z}_{2^m}$, the unique unramified extension of degree m of the 2-adic integers $\mathbb{Z}_2$, were done through the direct library interface to Pari [38] provided in Sage. We also implemented Harley's algorithm [23] as described in Vercauteren's thesis [45] using similar implementation details.

As a result of our experiments, we found that the following value of a for $m = 55$ gives a value 4 of binary Kloosterman sum. The finite field $\mathbb{F}_{2^{55}}$ is represented as $\mathbb{F}_2[x]/(x^{55} + x^{11} + x^{10} + x^9 + x^7 + x^4 + 1)$; a is then given as

$$\begin{aligned} a = {} & x^{53} + x^{52} + x^{51} + x^{50} + x^{47} + x^{43} + x^{41} + x^{38} + x^{37} + x^{35} \\ & + x^{33} + x^{32} + x^{30} + x^{29} + x^{28} + x^{27} + x^{26} + x^{25} + x^{24} \\ & + x^{22} + x^{20} + x^{19} + x^{17} + x^{16} + x^{15} + x^{13} + x^{12} + x^5 . \end{aligned}$$

To conclude, we have seen that Kloosterman sums $K(a)$ are a special type of exponential sums. Their variations are playing an increasing role in cryptography and coding theory. An important problem is to find, given c, all the elements a such that $K(a) = c$, or more simply whether such an a exists and if so to find conditions on such an a. Kloosterman zeroes, corresponding to $c = 0$, are linked to the existence of bent functions in the binary case and the ternary case. We have seen that the value 4 of Kloosterman sums has also been recently linked to the existence of bent functions.

It is well-known that binary and ternary Kloosterman sums are integers while, in characteristic greater than 3, there are Kloosterman sums that are not integral. Lachaud and Wolfmann have identified the range of possible values of binary Kloosterman sums and proved that they take any value congruent to 0 modulo 4 exploiting their connection with elliptic curves. Hence, binary Kloosterman sums are precisely the integers $4\mathbb{Z}$ within the Weil bound, and this gives a positive answer to the existence of Kloosterman zeroes. For ternary Kloosterman sums, a similar congruence has been obtained by Katz and Livné. Several authors have studied other congruences of Kloosterman sums modulo some integers. This was done for binary Kloosterman sums for moduli 3, 8, and 24, for example. For ternary Kloosterman sums, Lisonek proves a divisibility result modulo 27.

It could be interesting to find new congruence results using trace-like functions as well as in studying the distribution of binary and ternary Kloosterman sums. In addition to exploiting connections with coding theory and number of points on algebraic varieties, methods to be investigated include the use of the Gross–Koblitz formula which connects Gauß sums with the p-adic gamma functions

11.5 Bent Functions and Exponential Sums

The nonlinearity and the Walsh spectrum of Boolean functions can be expressed by means of exponential sums when Boolean functions are viewed as functions over Galois fields. One can derive information on the nonlinearity of a Boolean function by estimating the exponential sums involved in its expression. Likewise, divisibility properties of exponential sums can be translated into terms of divisibility properties of the Walsh spectrum of Boolean functions.

Such an approach can yield amazingly simple characterizations of bentness. Despite the large amount of work on bent functions, the benefit of linking the study of bentness with the evaluation of exponential sums has been recognized only recently and has been fruitfully exploited for a subclass of bent functions: the class of hyperbent functions. Thus, a promising and quite unexplored approach to find new (hyper)bent functions is to study the geometric related objects or using methods from analytic number theory.

References

1. O. Ahmadi and R. Granger. An efficient deterministic test for Kloosterman sum zeros. *CoRR*, abs/1104.3882, 2011.
2. J. Arndt. *Matters Computational: Ideas, Algorithms, Source Code*. Springer, 2010.
3. I. Blake, G. Seroussi, and N. P. Smart. *Elliptic curves in cryptography*, volume 265 of *London Mathematical Society Lecture Note Series*. Cambridge University Press, Cambridge, 2000. Reprint of the 1999 original.
4. W. Bosma, J. Cannon, and C. Playoust. The Magma algebra system. I. The user language. *J. Symbolic Comput.*, 24(3–4):235–265, 1997. Computational algebra and number theory (London, 1993).
5. P. Charpin, T. Helleseth, and V. Zinoviev. Divisibility properties of Kloosterman sums over finite fields of characteristic two. In *Information Theory, 2008. ISIT 2008. IEEE International Symposium on*, pages 2608–2612, July 2008.
6. P. Charpin, T.Helleseth, and V. Zinoviev. Divisibility properties of classical binary Kloosterman sums. *Discrete Mathematics*, 309(12):3975–3984, 2009.
7. H. Cohen. *A course in computational algebraic number theory*, volume 138 of *Graduate Texts in Mathematics*. Springer-Verlag, Berlin, 1993.
8. H. Cohen, G. Frey, R. Avanzi, C. Doche, T. Lange, K. Nguyen, and F. Vercauteren, editors. *Handbook of elliptic and hyperelliptic curve cryptography*. Discrete Mathematics and its Applications (Boca Raton). Chapman & Hall/CRC, Boca Raton, FL, 2006.
9. D.A. Cox. *Primes of the form* x^2+ny^2. A Wiley-Interscience Publication. John Wiley & Sons Inc., New York, 1989. Fermat, class field theory and complex multiplication.
10. J. Denef and F. Vercauteren. An extension of Kedlaya's algorithm to Artin-Schreier curves in characteristic 2. In Claus Fieker and David R. Kohel, editors, *ANTS*, volume 2369 of *Lecture Notes in Computer Science*, pages 308–323. Springer, 2002.
11. J. Denef and F. Vercauteren. An extension of Kedlaya's algorithm to hyperelliptic curves in characteristic 2. *J. Cryptology*, 19(1):1–25, 2006.
12. M. Deuring. Die Typen der Multiplikatorenringe elliptischer Funktionenkörper. *Abh. Math. Sem. Hansischen Univ.*, 14:197–272, 1941.

13. J. Dillon and H. Dobbertin. New cyclic difference sets with Singer parameters. *Finite Fields and Their Applications*, 10(3):342–389, 2004.
14. E.Berlekamp, V. Rumsey, and H. Solomon. On the solution of algebraic equations over finite fields. *Information and Control*, 10:553–564, 1967.
15. A. Enge. *Elliptic Curves and Their Applications to Cryptography: An Introduction*. Springer, 1st edition, August 1999.
16. J-P Flori and S. Mesnager. Dickson polynomials, hyperelliptic curves and hyper-bent functions. In *7th International conference SETA 2012, LNCS 7280, Springer*, pages 40–52, 2012.
17. J-P. Flori and S. Mesnager. An efficient characterization of a family of hyper-bent functions with multiple trace terms. In *Journal of Mathematical Cryptology. Vol 7 (1)*, pages 43–68, 2013.
18. M. Fouquet, Pierrick P. Gaudry, and R. Harley. An extension of Satoh's algorithm and its implementation. *J. Ramanujan Math. Soc.*, 15(4):281–318, 2000.
19. S.D. Galbraith. *Mathematics of Public Key Cryptography*. Cambridge University Press, 2011. http://www.math.auckland.ac.nz/~sgal018/crypto-book/crypto-book.html.
20. K. Garaschuk and P. Lisoněk. On binary Kloosterman sums divisible by 3. *Des. Codes Cryptography*, 49(1–3):347–357, 2008.
21. F Göloğlu, P Lisoněk, G McGuire, and R Moloney. Binary Kloosterman sums modulo 256 and coefficients of the characteristic polynomial. *Information Theory, IEEE Transactions on*, PP(99):1, 2012.
22. F Göloğlu, G McGuire, and R Moloney. Binary Kloosterman sums using Stickelberger's theorem and the Gross-Koblitz formula. *Acta Arith.*, 148(3):269–279, 2011.
23. R. Harley. Asymptotically optimal *p*-adic point-counting. Email to NMBRTHRY list, December 2002. http://listserv.nodak.edu/cgi-bin/wa.exe?A2=ind0212&L=nmbrthry&T=0&P=1343.
24. T. Helleseth and V. Zinoviev. On Z_4 linear Goethals codes and Kloosterman sums. *Des. Codes Cryptography*, 17(1–3):269–288, 1999.
25. H. Hubrechts. Point counting in families of hyperelliptic curves in characteristic 2. *LMS J. Comput. Math.*, 10:207–234, 2007.
26. N. Katz and R. Livné. Sommes de Kloosterman et courbes elliptiques universelles en caractéristiques 2 et 3. *C. R. Acad. Sci. Paris. I Math.*, 309(11):723–726, 1989.
27. K.S. Kedlaya. Counting points on hyperelliptic curves using Monsky-Washnitzer cohomology. *J. Ramanujan Math. Soc.*, 16(4):323–338, 2001.
28. N. Koblitz. Constructing elliptic curve cryptosystems in characteristic 2. In A. J. Menezes and S. A. Vanstone, editors, *CRYPTO*, volume 537 of *Lecture Notes in Computer Science*, pages 156–167. Springer, 1990.
29. G. Lachaud and J. Wolfmann. The weights of the orthogonals of the extended quadratic binary Goppa codes. In *IEEE Trans. Inform. Theory 36 (3)*, pages 686–692, 1990.
30. G. Lachaud and J. Wolfmann. Sommes de Kloosterman, courbes elliptiques et codes cycliques en caractéristique 2. *C. R. Acad. Sci. Paris Sér. I Math.*, 305(20):881–883, 1987.
31. S. Lang. *Elliptic functions*, volume 112 of *Graduate Texts in Mathematics*. Springer-Verlag, New York, second edition, 1987. With an appendix by J. Tate.
32. R Lercier, D Lubicz, and F Vercauteren. Point counting on elliptic and hyperelliptic curves. In *Handbook of elliptic and hyperelliptic curve cryptography*, Discrete Math. Appl. (Boca Raton), pages 407–453. Chapman & Hall/CRC, Boca Raton, FL, 2006.
33. R. Lercier and D. Lubicz. A quasi quadratic time algorithm for hyperelliptic curve point counting. *Ramanujan J.*, 12(3):399–423, 2006.
34. P. Lisoněk. On the connection between Kloosterman sums and elliptic curves. In Solomon Wolf Golomb, Matthew Geoffrey Parker, Alexander Pott, and Arne Winterhof, editors, *SETA*, volume 5203 of *Lecture Notes in Computer Science*, pages 182–187. Springer, 2008.
35. P. Lisoněk. An efficient characterization of a family of hyperbent functions. *IEEE Transactions on Information Theory*, 57(9):6010–6014, 2011.

36. S. Mesnager and J-P Flori. Hyper-bent functions via Dillon-like exponents. In *IEEE Transactions on Information Theory-IT. Vol. 59 No. 5*, pages 3215–3232, 2013.
37. R. Moloney. *Divisibility Properties of Kloosterman Sums and Division Polynomials for Edward Curves*. PhD thesis, University College Dublin, may 2011.
38. The PARI Group, Bordeaux. *PARI/GP, version 2.4.3*, October 2010. available from http://pari.math.u-bordeaux.fr/.
39. T. Satoh. The canonical lift of an ordinary elliptic curve over a finite field and its point counting. *J. Ramanujan Math. Soc.*, 15(4):247–270, 2000.
40. R. Schoof. Nonsingular plane cubic curves over finite fields*J. Comb. Theory, Ser. A*, 46(2):183–211, 1987.
41. J.H. Silverman.*The arithmetic of elliptic curves*, volume 106 of *Graduate Texts in Mathematics*. Springer, Dordrecht, second edition, 2009.
42. W.A. Stein et al. *Sage Mathematics Software (Version 4.7)*. The Sage Development Team, 2011. http://www.sagemath.org.
43. F. Vercauteren. *Computing zeta functions of curves over finite fields*. PhD thesis, Katholieke Universiteit Leuven, 2003.
44. F. Vercauteren. Advances in point counting. In *Advances in elliptic curve cryptography*, volume 317 of *London Math. Soc. Lecture Note Ser.*, pages 103–132. Cambridge Univ. Press, Cambridge, 2005.
45. F. Vercauteren. *Computing zeta functions of curves over finite fields*. PhD thesis, Katholieke Universiteit Leuven, 2007.
46. W.C. Waterhouse. Abelian varieties over finite fields. *Ann. Sci. École Norm. Sup. (4)*, 2:521–560, 1969.
47. K-E. Yeoh. GP/Pari implementation of point counting in characteristic 2. http://pages.cs.wisc.edu/~yeoh/nt/satoh-fgh.gp.

Chapter 12
Bent Vectorial Functions

12.1 Bent Vectorial Functions

12.1.1 Vectorial Functions

Let n and r be two positive integers ($n \geq 1$, $r \geq 1$). An (n, r)-function F (that is, a function from $\mathbb{F}_2^n$ to $\mathbb{F}_2^r$) being given, the component functions of F are the Boolean functions $l \circ F$, where l ranges over the set of all the nonzero linear forms over $\mathbb{F}_2^r$. Equivalently, they are the linear combinations of a non-null number of their coordinate functions, that is, the functions of the form $v \cdot F$, $v \in \mathbb{F}_2^r \setminus \{0\}$, where "$\cdot$" denotes the usual inner product in $\mathbb{F}_2^r$ (or any other inner product). The vector spaces $\mathbb{F}_2^n$ and $\mathbb{F}_2^r$ can be identified, if necessary, with the Galois fields $\mathbb{F}_{2^n}$ and $\mathbb{F}_{2^r}$ of orders 2^n and 2^r respectively. Hence, (n, r)-functions can be viewed as functions from $\mathbb{F}_2^n$ to $\mathbb{F}_2^r$ or as functions from $\mathbb{F}_{2^n}$ to $\mathbb{F}_{2^r}$. In the latter case, the component functions are the functions $Tr_1^r(vF(x))$. We recall some basic facts about vectorial functions. Any (n, r)-function F admits a unique representation as a multivariate polynomial over $\mathbb{F}_2^r$, called its *algebraic normal form* (ANF), of the form:

$$F(x_1, \cdots, x_n) = \sum_{u \in \mathbb{F}_2^n} c(u) \left(\prod_{i=1}^{n} x_i^{u_i} \right), \quad c(u) \in \mathbb{F}_2^r.$$

The *algebraic degree* $\deg(F)$ of any (n, r)-function F is by definition the global degree of its ANF (i.e. equals the maximum degree of those monomials whose coefficients are nonzero in its algebraic normal form). It also equals the maximum algebraic degree of the coordinate functions of F or of its component functions. Affine functions (resp. quadratic functions) are functions whose algebraic degree is a most 1 (resp. equals 2). Vectorial cryptographic functions must have high algebraic degree to withstand several kinds of attacks (mainly the higher order differential attack in the case of block ciphers and the Berlekamp–Massey attack in the case of stream ciphers).

S. Mesnager, *Bent Functions*, DOI 10.1007/978-3-319-32595-8_12

If we identify $\mathbb{F}_2^n$ with the finite field $\mathbb{F}_{2^n}$, then, any (n,n)-function F is uniquely expressed as a univariate polynomial over $\mathbb{F}_{2^n}$, of degree at most $2^n - 1$:

$$F(x) = \sum_{j=0}^{2^n-1} c_j x^j , \quad c_j \in \mathbb{F}_{2^n} .$$

The algebraic degree of F is equal to $\max_{j/\, c_j \neq 0} w_2(j)$ where $w_2(j)$ is the *2-weight* of j, that is, the number of nonzero coefficients j_s in the binary expansion $\sum_{s=o}^{n-1} j_s 2^s$ of j.

For every integer r dividing n, an (n,r)-function F can be viewed as a function from $\mathbb{F}_{2^n}$ into itself therefore, admits a unique univariate polynomial representation, which can be put in the form $Tr_r^n(\sum_{j=0}^{2^n-1} c_j x^j)$, where Tr_r^n is the trace function from $\mathbb{F}_{2^n}$ to $\mathbb{F}_{2^r}$ (but without uniqueness if we do not add restrictions on the polynomial inside the brackets).

12.1.2 Vectorial Functions That Are Bent

The notion of balancedness and bentness plays an important role for vectorial Boolean functions in cryptography.

Definition 12.1.1. An (n,r)-function F is called balanced if it takes every value of $\mathbb{F}_2^r$ the same number 2^{n-r} of times. Equivalently, F is balanced if for every $b \in \mathbb{F}_2^r$, the Boolean function ϕ_b defined on $\mathbb{F}_2^n$ by $\phi_b(x) = 1$ if $F(x) = b$ and $\phi_b(x) = 0$ otherwise, has Hamming weight 2^{n-r}.

The balanced vectorial functions can be characterized by the balancedness of their component (Boolean) functions as follows.

Proposition 12.1.2. *An (n,r)-function F is balanced if and only if its component functions are balanced, that is, if for every nonzero $v \in \mathbb{F}_2^r$, the Boolean function $v \cdot F$ on $\mathbb{F}_2^n$ is balanced (i.e has Hamming weight 2^{n-1}).*

The notion of Walsh transform is defined for vectorial functions as well. More precisely, given an (n,r)-function F, the Walsh transform of F is the function which maps any ordered pair $(a,v) \in \mathbb{F}_2^n \times \mathbb{F}_2^r$ to the value at a of the Walsh transform of the component (Boolean) function $v \cdot F$ $(v \neq 0)$, that is: $\widehat{\chi_{v\cdot F}}(a) = \sum_{x\in\mathbb{F}_2^n} (-1)^{v\cdot F(x)+x\cdot a}$, where the same symbol "$\cdot$" is used to denote inner products in $\mathbb{F}_2^r$ and in $\mathbb{F}_2^n$.

Generalized to (n,r)-functions, the nonlinearity is defined as the minimum nonlinearity of all their component functions $v \cdot F, v \in \mathbb{F}_2^r \setminus \{0\}$ and we have:

$$nl(F) = 2^{n-1} - \frac{1}{2} \max_{v\in\mathbb{F}_2^{r*};\, u\in\mathbb{F}_2^n} \left| \sum_{x\in\mathbb{F}_2^n} (-1)^{v\cdot F(x)+u\cdot x} \right| .$$

The nonlinearity represents a measure for the resistance of S-Boxes against linear cryptanalysis [17]. In the case of stream ciphers, another notion of nonlinearity must also be considered: the minimum nonlinearity of all the Boolean functions of the form $\varphi \circ F$ where φ is any non-constant r-variable Boolean function (indeed a fast correlation attack can be performed on the cipher using any such $\varphi \circ F$ as Boolean filtering or combining function), but we shall not be interested in this notion here. The upper bound $2^{n-1} - 2^{n/2-1}$ on the nonlinearity of any n-variable Boolean function is obviously valid for (n, r)-functions.

Definition 12.1.3. Let n be an even integer and r be an integer. An (n, r)-function F is called bent if the upper bound $2^{n-1} - 2^{n/2-1}$ on its nonlinearity $nl(F)$ is achieved with equality.

Bent (n, r)-functions exist only if n is even. But according to Nyberg [23], this condition is not sufficient for the existence of bent (n, r) functions. More precisely, bent (n, r)-functions exist if and only if n is even and $r \leq \frac{n}{2}$. Obviously, the bentness of vectorial functions can be characterized by the bentness of their component (Boolean) functions: an (n, r)-function F is bent if and only if all of the component functions of F are bent, that is, if $\widehat{\chi_{v \cdot F}}(a) = \pm 2^{\frac{n}{2}}$ for all $a \in \mathbb{F}_2^n$ and for all $v \in \mathbb{F}_2^r \setminus \{0\}$. This is equivalent to the fact that, for every $v \in \mathbb{F}_2^r \setminus \{0\}$ and every $a \in \mathbb{F}_2^n \setminus \{0\}$, the function $v \cdot (F(x) + F(x+a))$ is balanced, which is itself (according to Characterization 1) equivalent to saying that the function $F(x)+F(x+a)$ is balanced. Hence, bent functions contribute to an optimal resistance to the differential attack as well. The notion of bent vectorial function is EA-invariant (recall that this means invariant under composition on the left and on the right by affine automorphisms and under addition of affine functions). It is well-known that the algebraic degree of any bent (n, r)-function is at most $\frac{n}{2}$.

From now on, we assume the hypotheses "n is even and $r \leq \frac{n}{2}$" are satisfied on the ordered pair (n, r) when we consider bent (n, r)-functions.

12.1.3 Primary Constructions of Bent Vectorial Functions

Recall that constructions "from scratch" are called primary. On the contrary, secondary constructions will use already constructed functions to build new ones. There exist two general classes of bent Boolean functions, the Maiorana–McFarland class and the PS_{ap} class, which straightforwardly generalize to vectorial functions (this was first observed by Nyberg [23]).

1. Maiorana–McFarland's constructions of vectorial functions:

   Firstly, let us recall the Maiorana–McFarland's constructions of Boolean functions (or see Sect. 5.1.1 in Chap. 5). An n-variable Boolean bent function f belongs to the Maiorana–McFarland class if, up to EA-equivalence and writing its input in the form (x, y), with x, y in $\mathbb{F}_2^{n/2}$, the corresponding output equals $f(x, y) = x \cdot \pi(y) + g(y)$ (where "$\cdot$" is an inner product in $\mathbb{F}_2^{n/2}$), where π is a permutation of $\mathbb{F}_2^{n/2}$ and g is a Boolean function on $\mathbb{F}_2^{n/2}$. The bijectivity of π is a necessary and sufficient condition for the bentness of a Boolean function of the form $x \cdot \pi(y) + g(y)$. It it well known that all the quadratic bent Boolean functions belong to the Maiorana–McFarland class of Boolean (bent) functions. In the following, we shall see that three versions (of different levels of generality) can be given for the extension of this construction to vectorial functions.

- **The strict Maiorana–McFarland class:** We endow $\mathbb{F}_2^{n/2}$ with the structure of the field $\mathbb{F}_{2^{n/2}}$. We identify $\mathbb{F}_2^n$ with $\mathbb{F}_{2^{n/2}} \times \mathbb{F}_{2^{n/2}}$. Any function of the form $F(x,y) = L(x\pi(y)) + G(y)$, where the product $x\pi(y)$ is calculated in $\mathbb{F}_{2^{n/2}}$, where L is any linear or affine function from $\mathbb{F}_{2^{n/2}}$ onto $\mathbb{F}_2^r$, π is any permutation of $\mathbb{F}_{2^{n/2}}$ and G is any $(n/2, r)$-function, is an (n, r) bent function. We call *strict Maiorana–McFarland's class* the set of functions which are EA-equivalent to these functions.

An example is given in [29], whose i-th coordinate is defined as $f_i(x,y) = Tr_1^{\frac{n}{2}}(x\pi_i(y)) + g_i(y)$, $x, y \in \mathbb{F}_{2^{n/2}}$, where g_i is any Boolean function on $\mathbb{F}_{2^{n/2}}$ and where $\pi_i(y) = 0$ if $y = 0$ and $\pi_i(y) = \alpha^{dec(y)+i-1}$ otherwise, with α a primitive element of $\mathbb{F}_{2^{n/2}}$ and $dec(y) = 2^{n/2-1}y_1 + 2^{n/2-2}y_2 + \cdots + y_{n/2}$. This function belongs to the strict Maiorana–McFarland class of bent functions because the function $y \mapsto \begin{cases} 0 \text{ if } y = 0 \\ \alpha^{dec(y)} \text{ otherwise} \end{cases}$ is a permutation from $\mathbb{F}_2^{n/2}$ to $\mathbb{F}_{2^{n/2}}$, and the function $L : x \in \mathbb{F}_{2^{n/2}} \mapsto (Tr_1^{\frac{n}{2}}(x), Tr_1^{\frac{n}{2}}(\alpha x), \cdots, Tr_1^{\frac{n}{2}}(\alpha^{n/2-1}x)) \in \mathbb{F}_2^{n/2}$ is an isomorphism.

- **The extended Maiorana–McFarland class:** Let F be any function of the form

$$F : (x,y) \in \mathbb{F}_2^{n/2} \times \mathbb{F}_2^{n/2} \mapsto \psi(x,y) + G(y) \in \mathbb{F}_2^m,$$

where G is any function from $\mathbb{F}_2^{n/2}$ to $\mathbb{F}_2^m$ and $\psi : \mathbb{F}_2^{n/2} \times \mathbb{F}_2^{n/2} \mapsto \mathbb{F}_2^m$ is such that, for every $y \in \mathbb{F}_2^{n/2}$, the function $x \mapsto \psi(x,y)$ is linear and, for every nonzero $x \in \mathbb{F}_2^{n/2}$, the function $y \mapsto \psi(x,y)$ is balanced. Then F is bent. Indeed, for every nonzero $v \in \mathbb{F}_2^m$ and every $y \in \mathbb{F}_2^{n/2}$, there exists a unique vector v_y such that $v \cdot \psi(x,y) = x \cdot v_y$. The nonzero vector v being fixed, the function $y \mapsto v_y$ is bijective if and only if, for every $x \neq 0$, the function $y \mapsto x \cdot v_y$ is balanced (indeed, a vectorial function is balanced if and only if all its component functions are balanced), that is, the function $y \mapsto v \cdot \psi(x,y)$ is balanced. This property is achieved for every nonzero $v \in \mathbb{F}_2^m$ if and only if, for every nonzero $x \in \mathbb{F}_2^{n/2}$, the function $y \mapsto \psi(x,y)$ is balanced. Then, for any $u, u' \in \mathbb{F}_2^{n/2}$, the value at (u,u') of the Walsh transform $\widehat{\chi_{v\cdot F}}$ of the component function $v \cdot F$ of F is equal to

$$\sum_{(x,y)\in\mathbb{F}_2^{n/2}\times\mathbb{F}_2^{n/2}} (-1)^{v\cdot\psi(x,y)+v\cdot G(y)+u\cdot x+u'\cdot y} = 2^{n/2} \sum_{y\in\mathbb{F}_2^{n/2}/\, v_y=u} (-1)^{v\cdot G(y)+u'\cdot y}$$
$$= \pm 2^{n/2}.$$

We call *extended Maiorana–McFarland's class* the set of functions which are EA-equivalent to these functions. It includes the strict class.

An example of function ψ is $\psi(x,y) = \varphi(x,\pi(y))$, where π is a permutation of $\mathbb{F}_2^{n/2}$ and φ is any $\mathbb{F}_2$-bilinear (non-necessarily symmetric) function such that, for every nonzero $x \in \mathbb{F}_2^{n/2}$ and every nonzero $y \in \mathbb{F}_2^{n/2}$, we have $\varphi(x,y) \neq 0$. Indeed, this condition is necessary and sufficient for the linear function $y \mapsto \varphi(x,y)$ to be balanced over $\mathbb{F}_{2^{n/2}}$ for every nonzero x.
An example of such φ over the field $\mathbb{F}_{2^{n/2}}$ and with $m = n/2$ is obviously $\varphi(x,y) = xy$ but other examples exist. For instance, $\varphi(x,y) = x^4y + wxy^4$, where $w \in \mathbb{F}_{2^{n/2}}$ works, if w is not a cube in $\mathbb{F}_{2^{n/2}}$ (which obliges us to take n divisible by 4) since $x^4y + wxy^4 = 0$ with $x,y \neq 0$ implies $w = (x/y)^3$.
Characterizing all functions $\psi(x,y) = \sum_{i=0}^{n/2-1} \psi_i(y)\, x^{2^i}$ such that the function $y \mapsto \psi(x,y)$ is balanced for all nonzero $x \in \mathbb{F}_2^{n/2}$ is an open problem, as far as we know.

Remark 12.1.4. The (n,r)-functions above are given as defined over $\mathbb{F}_{2^{n/2}} \times \mathbb{F}_{2^{n/2}}$ (it is also the case for other functions, in particular those in the PS_{ap} class, see Sect. 2). And, we know that for r a divisor of n, any (n,r)-function can be viewed as a function from $\mathbb{F}_{2^n}$ to itself and, therefore, can be uniquely expressed as a univariate polynomial over $\mathbb{F}_{2^n}$. So, as mentioned in [3], in the case when $r = n/2$ the univariate representation of such functions can be easily obtained:

- let w be any element in $\mathbb{F}_{2^n} \setminus \mathbb{F}_{2^{n/2}}$, we can write $X = x + wy \in \mathbb{F}_{2^n} = \mathbb{F}_{2^{n/2}} + w\mathbb{F}_{2^{n/2}}$ and we have then $y = \frac{X+X^{2^{n/2}}}{w+w^{2^{n/2}}}$ and $x = \frac{w^{2^{n/2}}X+wX^{2^{n/2}}}{w+w^{2^{n/2}}}$;
- in particular if $n/2$ is odd, we can choose for w a primitive element of $\mathbb{F}_4$ and we have then: $x = w^2X + wX^{2^{n/2}}$ and $y = X + X^{2^{n/2}}$.

For instance, the univariate representation of the simplest Maiorana–McFarland function, that is the function $(x,y) \mapsto xy$, is $(w^2X + wX^{2^{n/2}})(X + X^{2^{n/2}})$, that is, up to addition of linear terms: $X^{1+2^{n/2}}$ if $n/2$ is odd and equals this functions multiplied by a nonzero term if $n/2$ is even.

- **The general Maiorana–McFarland class** is the set of (n,r)-functions such that, for every $v \in \mathbb{F}_2^{r*}$, the component function $v \cdot F$ belongs, up to affine equivalence, to the Maiorana–McFarland class of Boolean bent functions. It straightforwardly includes the extended class. The general class contains all bent quadratic functions, since we know that, up to affine equivalence and addition of a constant, every quadratic n-variable bent Boolean function equals $x_1x_2 + \cdots + x_{n-1}x_n$ [16].

Modifications of the Maiorana–McFarland bent functions have been proposed in [25].

2. Partial Spread constructions of vectorial functions:

We endow $\mathbb{F}_2^{n/2}$ with the structure of the field $\mathbb{F}_{2^{n/2}}$. We identify $\mathbb{F}_2^n$ with $\mathbb{F}_{2^{n/2}} \times \mathbb{F}_{2^{n/2}}$. Recall that Boolean functions of the class PS_{ap} introduced by Dillon [13,

14] are bent. They are defined in an explicit form $f(x,y) = g(\frac{x}{y})$ with $(x,y) \in \mathbb{F}_{2^{n/2}} \times \mathbb{F}_{2^{n/2}}$ and $\frac{x}{y} = 0$ if $y = 0$, where g is a Balanced Boolean function on $\mathbb{F}_{2^{n/2}}$. The balancedness of g is in fact a necessary and sufficient condition for f being bent. Moreover, the dual function $\tilde{f}$ of f is the Boolean function defined, for every $(a,b) \in (\mathbb{F}_{2^n})^2$, by $\tilde{f}(a,b) = g(\frac{b}{a})$, which belongs also to the class of PS_{ap}.

- **The PS_{ap} class of vectorial functions:** any function F over $\mathbb{F}_{2^{n/2}} \times \mathbb{F}_{2^{n/2}}$ defined by $F(x,y) := G(xy^{2^n-2}) = G\left(\frac{x}{y}\right)$ (with $\frac{x}{y} = 0$ if $y = 0$), $x, y \in \mathbb{F}_{2^{n/2}}$, where G is a balanced $(n/2, r)$-function, is a bent (n, r)-function (since for every $v \neq 0$, the component function $v \cdot F$ belongs to the class PS_{ap} of Dillon's functions).
- **A Partial Spread construction:** let us recall a construction of Boolean bent functions proposed by Carlet:

Theorem 12.1.5 ([5]). *Let n and m be two even integers. Let f be a Boolean function on $\mathbb{F}_2^n \times \mathbb{F}_2^m$ such that, for any element y of $\mathbb{F}_2^m$ the Boolean function $f_y : x \in \mathbb{F}_2^n \mapsto f(x,y)$ is bent. Then, f is bent on $\mathbb{F}_2^{n+m}$ if and only if, for any element s of $\mathbb{F}_2^n$, the Boolean function ϕ_s: $y \in \mathbb{F}_2^m \mapsto \widetilde{f_y}(s)$ is bent.*

A way of constructing bent Boolean functions is, after identifying $\mathbb{F}_2^n$ and $\mathbb{F}_2^m$ with the Galois fields $\mathbb{F}_{2^n}$ and $\mathbb{F}_{2^m}$ respectively, to use the following Proposition given by Carlet [3] and which is a consequence of Theorem 12.1.5. For completeness, we include the proof.

Proposition 12.1.6 ([3]). *Let n and m be two positive integers. Let k be a Boolean function on $\mathbb{F}_{2^n} \times \mathbb{F}_{2^m}$. Define a Boolean function h on $\mathbb{F}_{2^n} \times \mathbb{F}_{2^n} \times \mathbb{F}_{2^m} \times \mathbb{F}_{2^m}$ by setting, for every (x,y,z,t) in $\mathbb{F}_{2^n} \times \mathbb{F}_{2^n} \times \mathbb{F}_{2^m} \times \mathbb{F}_{2^m}$, $h(x,y,z,t) = k(\frac{x}{y}, \frac{z}{t})$.*

Assume the following conditions (1) and (2) are satisfied.

1. *For every $x \in \mathbb{F}_{2^n}$, the Boolean function $z \in \mathbb{F}_{2^m} \mapsto k(x,z)$ is balanced;*
2. *For every $z \in \mathbb{F}_{2^m}$, the Boolean function $x \in \mathbb{F}_{2^n} \mapsto k(x,z)$ is balanced.*

Then, the Boolean function h is bent.

Proof. Thanks to hypotheses (1) and (2), for every $(z,t) \in (\mathbb{F}_{2^m})^2$, the Boolean function $g_{z,t} : (x,y) \mapsto k(\frac{x}{y}, \frac{z}{t})$ belongs then to PS_{ap} class and thus is bent. Moreover, its dual $\widetilde{g_{z,t}}$ is defined by $\widetilde{g_{z,t}}(x,y) = k(\frac{y}{x}, \frac{z}{t})$. Therefore, the Boolean function $(z,t) \in (\mathbb{F}_{2^m})^2 \mapsto \widetilde{g_{z,t}}(x,y)$ belongs also to the PS_{ap} class for every (x,y) in $(\mathbb{F}_{2^m})^2$. We then conclude thanks to Theorem 12.1.5. □

The construction given by Proposition 12.1.6 can be straightforwardly extended to vectorial Boolean functions as follows.

Proposition 12.1.7 ([11]). *Let n, m , r be three positive integers such that $r \leq n$ and $r \leq m$. Let K be a function from $\mathbb{F}_{2^n} \times \mathbb{F}_{2^m}$ to $\mathbb{F}_{2^r}$ or to $\mathbb{F}_2^r$ such that*

1. *For every $x \in \mathbb{F}_{2^n}$, the function $y \in \mathbb{F}_{2^m} \mapsto K(x,y)$ is balanced,*
2. *For every $y \in \mathbb{F}_{2^m}$, the function $x \in \mathbb{F}_{2^n} \mapsto K(x,y)$ is balanced.*

Define the function H from $\mathbb{F}_{2^n} \times \mathbb{F}_{2^n} \times \mathbb{F}_{2^m} \times \mathbb{F}_{2^m}$ to $\mathbb{F}_{2^r}$ by setting $H(x, y, z, t) = K(\frac{x}{y}, \frac{z}{t})$. Then H is a bent $(2n + 2m, r)$-function.

Proof. Apply Proposition 12.1.6 to each component function k_λ (where $\lambda \in \mathbb{F}_{2^r}$, $\lambda \neq 0$ or $\lambda \in \mathbb{F}_2^r \setminus \{0\}$) of the function K, that is the functions of the form $k_\lambda(x, y, z, t) = \mathrm{Tr}_1^r(\lambda K(\frac{x}{y}, \frac{z}{t}))$ or $\lambda \cdot K(\frac{x}{y}, \frac{z}{t})$ which is balanced (since a vectorial function is balanced if and only if all its component functions are balanced). □

Example 12.1.8. Let ϕ and ϕ' be two balanced functions from $\mathbb{F}_{2^m}$ to $\mathbb{F}_{2^n}$. Let F be a balanced function from $\mathbb{F}_{2^n}$ to $\mathbb{F}_{2^r}$. Then, the function $K(x, y) := F(\phi(x) + \phi'(y))$ satisfies the conditions (1) and (2) of Proposition 12.1.7. Note that, in general, the corresponding bent function is not the direct sum (see definition below) of functions in x and y.

3. Other primary constructions of bent vectorial functions:

The existence of a bent (n, r)-function is equivalent to the existence of an r-dimensional vectorspace of n-variable Boolean functions whose nonzero elements (the component functions of the vectorial function) are all bent. Let us give some examples of such construction:

- An example derived from the property of some codes: recall that, for given n and $r \leq n$, the binary Reed–Muller code $\mathcal{RM}(r, n)$ of order r and length 2^n consists of all n-variable Boolean functions of algebraic degree at most r and that the Kerdock code $\mathcal{K}(n)$ [16] of same length consists of the binary Reed–Muller code $\mathcal{RM}(1, n)$ of order 1 and length 2^n together with $2^{n-1} - 1$ cosets of $\mathcal{RM}(1, n)$ in the binary Reed–Muller code $\mathcal{RM}(2, n)$ of order 2 and length 2^n. The Boolean functions associated with these cosets are quadratic bent functions, with the property that the sum of any two of them is again a bent function. Consequently, any $(n, 2)$-function whose coordinate functions belong to two distinct cosets, among these $2^{n-1} - 1$ cosets, is a bent vectorial function.
- Given a function F from $\mathbb{F}_{2^n}$ to itself, a nonzero element $a \in \mathbb{F}_{2^n}$ and an integer r dividing n, the (n, r)-function $x \in \mathbb{F}_{2^n} \mapsto \mathrm{Tr}_r^n(aF(x))$ is bent if and only if, for any nonzero $v \in \mathbb{F}_{2^r}$, the Boolean function $x \in \mathbb{F}_{2^n} \mapsto \mathrm{Tr}_1^n(avF(x))$ is bent. Examples of constructions of such bent (n, r)-functions are given in [1] (with r strictly smaller than $n/2$). The authors obtain their results from some specific (n, n)-functions F (and under some assumptions on the nonzero element $a \in \mathbb{F}_{2^n}$ given in [1]):
  - $F(x) = x^{2^i+1} + (x^{2^i} + x + 1)\mathrm{Tr}_1^n(x^{2^i+1})$, for $n \geq 6$ an even integer;
  - $F(x) = \left(x + \mathrm{Tr}_3^n(x^{2(2^i+1)} + x^{4(2^i+1)}) + \mathrm{Tr}_1^n(x)\mathrm{Tr}_3^n(x^{2^i+1} + x^{2^{2i}(2^i+1)})\right)^{2^i+1}$, for n divisible by 6;

  where i is a positive integer not divisible by $n/2$ and such that $n/\gcd(i, n)$ is even. The derived bent (n, r)-functions are CCZ-inequivalent to the quadratic (n, r)-functions $x \in \mathbb{F}_{2^n} \mapsto \mathrm{Tr}_r^n(vx^{2^i+1})$, $v \in \mathbb{F}_{2^r}$, $v \neq 0$.
- An example of bent $(n, n/2)$-function has been found by the first author in common with Leander. Such a function is defined precisely, from $\mathbb{F}_{2^n}$ to $\mathbb{F}_2^{n/2}$,

for n divisible by 2 but not by 4. The output of the function is of the form $(\mathrm{Tr}_1^n(\beta_1 wX^d), \cdots, \mathrm{Tr}_1^n(\beta_{n/2} wX^d)) \in \mathbb{F}_2^{n/2}$, $(X \in \mathbb{F}_{2^n})$; where d is a so-called *Gold exponent*: $d = 2^i + 1$ such that $\gcd(n, i) = 1$, where w is some element of $\mathbb{F}_{2^n} \setminus \mathbb{F}_{2^{n/2}}$ and where $(\beta_1, \cdots, \beta_{n/2})$ is a basis of $\mathbb{F}_{2^{n/2}}$ over $\mathbb{F}_2$. More details concerning this construction can be found in [8].

- **Constructions of $(2m, m)$-bent functions from o-polynomials:**
  First, recall that an m-spread of $\mathbb{F}_{2^{2m}}$ is a set of m-dimensional subspaces which are pairwise only intersecting in the zero vector of $\mathbb{F}_{2^{2m}}$ and whose union equals $\mathbb{F}_{2^{2m}}$. A classical example of m-spread viewed in $\mathbb{F}_{2^{2m}} \approx \mathbb{F}_{2^m} \times \mathbb{F}_{2^m}$ is $\{E_a, E_\infty\}$ where $E_a := \{(x, ax) \mid x \in \mathbb{F}_{2^m}\}$ and $E_\infty := \{(0, y) \mid y \in \mathbb{F}_{2^m}\}$.

In [12], a class of bent functions called class $\mathcal{H}$ has been introduced. The elements of $\mathcal{H}$ are the bent functions defined on $\mathbb{F}_{2^m} \times \mathbb{F}_{2^m}$ and whose restrictions to the m-spread $\{E_a, E_\infty\}$ are linear. More precisely, functions of the class $\mathcal{H}$ are defined in bivariate form as follows.

Definition 12.1.9 ([12]). We call $\mathcal{H}$ the class of functions f defined on $\mathbb{F}_{2^m} \times \mathbb{F}_{2^m}$ of the form

$$f(x, y) = \begin{cases} \mathrm{Tr}_1^m\left(x\psi\left(\frac{y}{x}\right)\right) \text{ if } x \neq 0 \\ \mathrm{Tr}_1^m(\mu y) \text{ if } x = 0 \end{cases} \tag{12.1}$$

where $\psi : \mathbb{F}_{2^m} \to \mathbb{F}_{2^m}$ and $\mu \in \mathbb{F}_{2^m}$ satisfy the following condition:

$$\forall \beta \in \mathbb{F}_{2^m}^\star, \text{the function } z \mapsto G(z) + \beta z \text{ is 2-to-1 on } \mathbb{F}_{2^m} \tag{12.2}$$

where G is defined as $G(z) := \psi(z) + \mu z$.

Condition (12.2) is in fact necessary and sufficient for the bentness property of functions in the form (12.1). We have showed in [12] that f satisfies Condition (12.2) if and only if G is an o-polynomial on $\mathbb{F}_{2^m}$. Now, note that functions f defined by (12.1) can be written as

$$f(x, y) = \mathrm{Tr}_1^m(\mu y + xG(yx^{2^m-2})). \tag{12.3}$$

Indeed (since $1 + x^{2^m-1}$ equals 1 if $x = 0$ and 0 otherwise and $x^{2^m-2} = \frac{1}{x}$), we have:

$$\begin{aligned} f(x, y) &= \mathrm{Tr}_1^m(\mu y(1 + x^{2^m-1}) + x\psi(yx^{2^m-2})) = \mathrm{Tr}_1^m(\mu y + \mu y x^{2^m-1} + x\psi(yx^{2^m-2})) \\ &= \mathrm{Tr}_1^m(\mu y + x(\mu(yx^{2^m-2}) + \psi(yx^{2^m-2}))) = \mathrm{Tr}_1^m(\mu y + xG(yx^{2^m-2})).s \end{aligned}$$

The following theorem shows that the o-polynomials lead to the construction of optimal bent vectorial functions; that is, bent functions from $\mathbb{F}_{2^{2m}}$ (identified to $\mathbb{F}_{2^m} \times \mathbb{F}_{2^m}$) to $\mathbb{F}_{2^m}$.

Theorem 12.1.10. *Let G be an o-polynomial on $\mathbb{F}_{2^m}$. Let F be a function from $\mathbb{F}_{2^m} \times \mathbb{F}_{2^m}$ to $\mathbb{F}_{2^m}$ such that for $(x, y) \in \mathbb{F}_{2^m} \times \mathbb{F}_{2^m}$,*

$$F(x, y) = xG(yx^{2^m-2}),$$

then the vectorial function F is bent.

Proof. The components of F are the Boolean functions $\{v \cdot F\}_{v \in \mathbb{F}_{2^m}^\star}$ defined on $\mathbb{F}_{2^m}$ where "$\cdot$" is a scalar product in $\mathbb{F}_{2^m}$. For $v \in \mathbb{F}_{2^m}^\star$, the Boolean function $v \cdot F$ is defined by $v \cdot F(x,y) = Tr_1^m(vF(x,y)) = Tr_1^m(vxG(yx^{2^m-2})) = Tr_1^m(x(vG)(yx^{2^m-2}))$. Since G is an o-polynomial on $\mathbb{F}_{2^m}$ then vG ($v \neq 0$) is an o-polynomial too. Indeed, it is obvious that if G is a permutation on $\mathbb{F}_{2^m}$ then, for $v \in \mathbb{F}_{2^m}^\star$, the mapping $z \mapsto vG(z)$ is a permutation on $\mathbb{F}_{2^m}$. Now, assume that for every $(\beta', \alpha') \in \mathbb{F}_{2^m}^\star \times \mathbb{F}_{2^m}$, $\#\{z \in \mathbb{F}_{2^m} \mid G(z) + \beta z = \alpha\} \in \{0,2\}$. Let $(\beta, \alpha) \in \mathbb{F}_{2^m}^\star \times \mathbb{F}_{2^m}$. For $v \in \mathbb{F}_{2^m}^\star$, we have $vG(z) + \beta z = \alpha$ if and only if $G(z) + \beta v^{-1} z = \alpha v^{-1}$. Hence (take $\alpha' = \alpha v^{-1}$ and $\beta' = \beta v^{-1}$ in the above assumption; $v^{-1} \neq 0$), $\#\{z \in \mathbb{F}_{2^m} \mid vG(z) + \beta z = \alpha\} \in \{0,2\}$. Hence all the component functions of F are elements of the class $\mathcal{H}$ (by taking $\mu = 0$ in (12.3)) and are therefore bent. Thus F is an $(2m, m)$-bent function. □

Remark 12.1.11. Recall that bent (n, r)-functions exist only for $r \leq \frac{n}{2}$ (n even) and equality can be attained. Note that Theorem 12.1.10 gives rise to bent $(n, \frac{n}{2})$-functions which are rare and sought from a cryptographic point of view.

For vectorial Boolean functions, the most useful concepts of equivalence are the extended affine EA-equivalence and the CCZ-equivalence [9] (note that the EA-equivalence is a particular case of CCZ-equivalence. More precisely:

Definition 12.1.12. Two (n, r)-functions F and F' are called *extended affine equivalent* (EA-equivalent) if there exist an affine automorphism L from $\mathbb{F}_{2^n}$ to $\mathbb{F}_{2^n}$, a linear automorphism L' from $\mathbb{F}_{2^r}$ to $\mathbb{F}_{2^r}$ and an affine function L'' from $\mathbb{F}_{2^n}$ to $\mathbb{F}_{2^r}$ such that $F' = L' \circ F \circ L + L''$.

The nonlinearity is invariant under extended affine equivalence. Moreover, in [12], we have introduced the notion of *o-equivalence*.

Definition 12.1.13 ([12]). Two functions G and G' are said to be *o-equivalent* if one can be obtained from the other by a sequence of the following list of transformations :

1. $G \mapsto G'$ where $G' : z \in \mathbb{F}_{2^m} \mapsto G'(z) := G(\lambda z + \mu)$ with $\lambda \in \mathbb{F}_{2^m}^\star$ and $\mu \in \mathbb{F}_{2^m}$;
2. $G \mapsto G'$ where $G' : z \in \mathbb{F}_{2^m} \mapsto G'(z) := \lambda G(z) + \mu$ with $\lambda \in \mathbb{F}_{2^m}^\star$ and $\mu \in \mathbb{F}_{2^m}$;
3. $G \mapsto G'$ where $G' : z \in \mathbb{F}_{2^m} \mapsto G'(z) := zG(z^{2^m-2})$ (with $G(0) = 0$);
4. $G \mapsto G'$ where $G' : z \in \mathbb{F}_{2^m} \mapsto G'(z) := G(z^{2^j})^{2^{m-j}}$ where $j \in \mathbb{N}$;
5. $G \mapsto G'$ where $G' : z \in \mathbb{F}_{2^m} \mapsto G'(z) := G^{-1}(z)$.

Remark 12.1.14. First note that transformations (1) to (4) translated in terms of the associated bent Boolean functions $f(x,y) = Tr_1^m(xG(\frac{y}{x}))$ (with the convention $\frac{1}{0} = 0$) result in particular cases of EA-equivalence. Moreover, the reason why we have chosen the term "o-equivalent" is explained in [12]. In fact this gives a notion of equivalence of functions in class $\mathcal{H}$ which is not a sub-equivalence of the EA-equivalence of bent functions and is not a super-equivalence either. Indeed, the transformations (1) to (4) translated in terms of the associated bent

Boolean functions $f(x,y) = \mathrm{Tr}_1^m(\mu y + xG(yx^{2^m-2}))$ result in particular cases of EA-equivalence. More precisely, if $f(x,y) = \mathrm{Tr}_1^m(\mu y + xG(yx^{2^m-2}))$ and $f'(x,y) = \mathrm{Tr}_1^m(\mu y + xG'(yx^{2^m-2}))$ are two Boolean functions of the class $\mathcal{H}$ and if G' is obtained from G by a sequence of transformations (1) to (4), then the Boolean functions f and f' are EA-equivalent. But if $G' = G^{-1}$ then the question of knowing whether the corresponding functions f and f' are EA-equivalent is not closed since we have proved the EA-inequivalence for only two o-polynomials of the current list given in Chap. 8 while the question remains open for the others. We shall modify from now on the notion of *o-equivalence*: two functions G and G' are said to be o-equivalent if one can be obtained from the other by a sequence involving only the transformations (1)–(4). By excluding the transformation $G \mapsto G^{-1}$ in the definition of the o-equivalence, one can say now that the o-equivalence between the o-polynomials implies the EA-equivalence between the associated bent Boolean functions of the class $\mathcal{H}$.

In the following, we study the relationship between the o-equivalence for the o-polynomials and the EA-equivalence for the corresponding vectorial functions obtained by Theorem 12.1.10. In fact, it is easy to prove the following statement.

Proposition 12.1.15. *Let G and G' be two o-polynomials on $\mathbb{F}_{2^m}$. Let F and F' be two $(2m, m)$-functions defined by $F(x,y) = xG(yx^{2^m-2})$ and $F'(x,y) = xG'(yx^{2^m-2})$. If G and G' are o-equivalent then the functions F and F' are EA-equivalent.*

Proof. According to Definition 12.1.13 and Remark 12.1.14, we have to check that if G and G' fulfill one of the four transformations $G \mapsto G'$ from (1) to (4), then the corresponding vectorial functions are EA-equivalent. Clearly, we have:

- For transformation (1): $F'(x,y) = xG(\lambda(yx^{2^m-2}) + \mu) = xG((\lambda y + \mu x)x^{2^m-2})$. Hence $F'(x,y) = F(x, \mu x + \lambda y)$.
- For transformation (2): $F'(x,y) = x(\lambda G(yx^{2^m-2}) + \mu) = (\lambda x)G((\lambda y)(\lambda x)^{2^m-2}) + \mu x$. Hence $F'(x,y) = F(\lambda x, \lambda y) + \mu x$.
- For transformation (3): $F'(x,y) = xyx^{2^m-2}G((yx^{2^m-2})^{2^m-2}) = yG(xy^{2m-2}) = F(y,x)$.
- For transformation (4): $F'(x,y) = xG((yx^{2^m-2})^{2^j})^{2^{m-j}} = (x^{2^j}G(y^{2^j}(x^{2^j})^{2^m-2}))^{2^{m-j}}$. Hence $F'(x,y) = (F(x^{2^j}, y^{2^j}))^{2^{m-j}}$.

□

Theorem 12.1.10 provides at least new classes of infinite (optimal) bent vectorial functions thanks to the hard work of the geometers over approximately 40 years. We summarize in Table 12.1 the current list of primary constructions of infinite classes of bent vectorial functions. Note that a few other examples of primary constructions of bent vectorial functions (but not infinite classes) are given in [11].

Remark 12.1.16. A bent Boolean function is not necessarily a component of a vectorial bent function. However, it is important to notice the strong property of the bent functions coming from o-polynomials since all the elements of the class $\mathcal{H}$ are the components of the bent vectorial functions presented in this section.

Table 12.1 The main general primary constructions of bent vectorial functions

General expression	Class of functions	Conditions	References
Bent $(2m, r)$-function: $F(x, y) = L(x\,\pi(y)) + G(y)$; $(x, y) \in \mathbb{F}_{2^m} \times \mathbb{F}_{2^m}$	Strict Maiorana–McFarland	π permutation of $\mathbb{F}_{2^m}$; G is any (m, r)-function.	[3]
Bent $(2m, r)$-function: $F(x, y) = \psi(x, y) + G(y)$; $(x, y) \in \mathbb{F}_{2^m} \times \mathbb{F}_{2^m}$	Extended Maiorana–McFarland	G is any (m, r)-function; $\forall y \in \mathbb{F}_{2^m}, x \mapsto \psi(x, y)$ is linear; $\forall x \in \mathbb{F}_{2^m} \setminus \{0\}$, $y \mapsto \psi(x, y)$ is balanced (that is, if its output is uniformly distributed).	[3]
Bent $(2m, r)$-function: $\forall v \in \mathbb{F}_{2^r}^{\star}$, $v \cdot F$ belongs, up to affine equivalence, to the Maiorana–McFarland class of Boolean bent functions.	General Maiorana–McFarland		[3, 24, 25]
Bent $(2m, r)$-function: $F(x, y) = G\left(\frac{x}{y}\right)$; $(x, y) \in \mathbb{F}_{2^m} \times \mathbb{F}_{2^m}$.	PS_{ap} of vectorial functions	G is a balanced (m, r)-function	[3, 13]
Bent $(2n + 2m, r)$-function: $F(x, y) = K(\frac{x}{y}, \frac{z}{t})$; $(x, y) \in (\mathbb{F}_{2^n} \times \mathbb{F}_{2^n})$; $(z, t) \in \mathbb{F}_{2^m} \times \mathbb{F}_{2^m}$; $r \leq n, r \leq m$.	A Partial Spread construction	K is a $(n + m, r)$-function st. $\forall x \in \mathbb{F}_{2^n}$, $y \in \mathbb{F}_{2^m} \mapsto K(x, y)$ is balanced; $\forall y \in \mathbb{F}_{2^m}$, $x \in \mathbb{F}_{2^n} \mapsto K(x, y)$ is balanced.	[11]
Bent $(2m, m)$-function: $F(x, y) = xG(yx^{2^m-2})$; $(x, y) \in \mathbb{F}_{2^m} \times \mathbb{F}_{2^m}$.	Class $\mathcal{H}$ of vectorial functions	G is an o-polynomial on $\mathbb{F}_{2^m}$	[19]

12.1.4 On the (Non-)existence of Vectorial Monomial Bent Functions of the Form $Tr_r^n(ax^d)$ Where r/n

The bent (n, r)-functions of the form $Tr_r^n(ax^d)$ (where d is a positive integer and $a \in \mathbb{F}_{2^n}^{\star}$ are referred as vectorial monomial bent functions, which are monomial bent functions if $r = 1$, i.e., the Boolean bent functions of the form $\mathrm{Tr}_1^n(ax^d)$ on $\mathbb{F}_{2^n}$ and exist only if $gcd(d, 2^n - 1) \neq 1$. Recently, many works have been done on the characterization and the existence of vectorial monomial bent functions. More precisely, In [28], it was shown by Pasalic and Zhang that $Tr_r^n(ax^d)$ is a vectorial monomial bent function if $\mathrm{Tr}_1^n(ax^d)$ is a monomial bent function and $x \mapsto x^d$ is a permutation on $\mathbb{F}_{2^r}$, that is, $gcd(d, 2^r - 1) = 1$ and some classes of vectorial monomial bent functions with the Kasami exponent, the Leander exponent and the Canteaut–Charpin–Kyureghyan exponent were investigated. However, the condition $gcd(d, 2^r - 1) = 1$ was sufficient for the bentness of the vectorial function $Tr_r^n(ax^d)$ and the authors left open the questions whether that condition is necessary or not for the bentness of the vectorial function $Tr_r^n(ax^d)$ and how to relax the condition $gcd(d, 2^r - 1) = 1$ so that $Tr_r^n(ax^d)$ is bent. Next, in [30], Tang, Qi and Xu have showed that condition is not necessary for the bentness by providing a counter example. In [15], Dong et al. have shown that the particular vectorial monomial function $Tr_{n/2}^n(ax^d)$ is bent function if $gcd(d, 2^n - 1)$ divides $(2^{n/2} + 1)$ and the $\mathrm{Tr}_1^n(ax^d)$ is a monomial bent function with the Gold exponent or the Kassami exponent.

In [22], Muratovic-Ribis, Pasalic and Bajric have proved that there does not exist a vectorial monomial bent function of the form $Tr_r^n(ax^d)$ with the Dillon exponent for $a \in \mathbb{F}_{2^{r\star}}$, $d = s(2^k - 1)$ and $gcd(s, 2^k + 1) = 1$. Very recently, several general constructions of vectorial monomial bent functions have been given by Xu and Wu [31] who gave answers to one open problem proposed by Pasalic and Zhang [28]. In particular, given a monomial bent function $\mathrm{Tr}_1^n(ax^d)$ the authors have presented several conditions which are much closer to the sufficient and necessary conditions for $Tr_r^n(ax^d)$ to be bent than the condition that $gcd(d, 2^r - 1) = 1$. Subsequently, the vectorial monomial bent functions corresponding to the five known classes of bent exponents have been characterized. For vectorial monomial bent functions, the main results are the existence and constructions of the vectorial monomial bent functions corresponding to the five known classes of bent exponents.

In Table 12.2, the reader can find a summary (provided in [31]) on the nonexistence of the vectorial monomial bent functions of the form $Tr_r^n(ax^d)$ where r/n. Note that in Table 12.2, the value r in this column, means that there does not exist vectorial monomial bent functions of the form $Tr_r^n(ax^d)$.

Table 12.2 The nonexistence of the vectorial monomial bent functions of the form $Tr_r^n(ax^d)$ where r/n.

Case	Exponent d	Condition	Value of r
Dillon	$s(2^{n/2}-1)$	$gcd(s, 2^{n/2}+1)=1$	$n/2$
Kassami	$2^{2s}-2^s+1$	$s \in \mathbb{N}, gcd(3,n)=1,$ $gcd(s,n)=1$	r even
Leander	$(2^s+1)^2$	$s \in \mathbb{N}^\star, n=4s$	r even, $r \not| s$
Canteaut–Charpin–Kyureghyan	$2^{2s}+2^s+1$	$s \in \mathbb{N}^\star$, $n=6s$	$\frac{n}{2}$

12.1.5 Secondary Constructions of Bent Vectorial Functions

1. A Maiorana–McFarland-like construction:

In [6] is given the following secondary construction of bent Boolean functions: let r and s be two positive integers with the same parity and such that $r \leq s$, and let $n = r + s$; let ϕ be a function from $\mathbb{F}_2^s$ to $\mathbb{F}_2^r$ and g a Boolean function on $\mathbb{F}_2^s$; let us assume that ϕ is balanced and, for every $a \in \mathbb{F}_2^r$, the set $\phi^{-1}(a)$ is an $(s-r)$-dimensional affine subspace of $\mathbb{F}_2^s$; let us assume additionally if $r < s$ that the restriction of g to $\phi^{-1}(a)$ (viewed as a Boolean function on $\mathbb{F}_2^{n-2r}$ via an affine isomorphism between $\phi^{-1}(a)$ and this vectorspace) is bent; then the function $f_{\phi,g}(x,y) = x \cdot \phi(y) + g(y)$, $x \in \mathbb{F}_2^r$, $y \in \mathbb{F}_2^s$, where "$\cdot$" is an inner product in $\mathbb{F}_2^r$, is bent on $\mathbb{F}_2^n$. This generalizes directly to vectorial functions:

Proposition 12.1.17 ([11]). *Let r and s be two positive integers with the same parity and such that $r \leq \frac{s}{3}$. Let ψ be any (balanced) function from $\mathbb{F}_2^s$ to $\mathbb{F}_{2^r}$ such that, for every $a \in \mathbb{F}_{2^r}$, the set $\psi^{-1}(a)$ is an $(s-r)$-dimensional affine subspace of $\mathbb{F}_2^s$. Let H be any (s,r)-function whose restriction to $\psi^{-1}(a)$ (viewed as an $(s-r,r)$-function via an affine isomorphism between $\psi^{-1}(a)$ and $\mathbb{F}_2^{s-r}$) is bent for every $a \in \mathbb{F}_{2^r}$. Then, the function $F_{\psi,H}(x,y) = x\psi(y) + H(y)$, $x \in \mathbb{F}_{2^r}, y \in \mathbb{F}_2^s$, is a bent function from $\mathbb{F}_2^{r+s}$ to $\mathbb{F}_{2^r}$.*

Proof. Taking $x \cdot y = \mathrm{Tr}_1^r(xy)$ for inner product in $\mathbb{F}_{2^r}$, for every $v \in \mathbb{F}_{2^r}^*$, the function $\mathrm{Tr}_1^r(v\,F_{\psi,H}(x,y))$ is bent, according to the result of [6] recalled above, with $\phi(y) = v\,\psi(y)$ and $g(y) = \mathrm{Tr}_1^r(v\,H(y))$. The condition $r \leq \frac{s}{3}$, more restrictive than $r \leq s$, is meant so that $r \leq \frac{s-r}{2}$, which is necessary for allowing the restrictions of H to be bent. The condition on ψ being easily satisfied (this does not make ψ necessarily affine). Note that it is a simple matter to choose H. □

2. The direct sum of bent functions:

   It is well know that the direct sum $(x,y) \mapsto g(x) + h(y)$ of two bent Boolean functions f, g gives a bent Boolean function. This simple secondary construction can be directly adapted to vectorial functions. Indeed any bent (n,r)-function G and bent (m,r)-function H give a bent $(n+m,r)$-function F by setting, for $(x,y) \in \mathbb{F}_2^n \times \mathbb{F}_2^m$, $F(x,y) := G(x) + H(y)$.

3. An "indirect sum" of bent function construction:

The direct sum of bent Boolean functions is a particular case of a much more general construction introduced in [7], which involves four bent Boolean functions, and which has been recently called the indirect sum. The *indirect sum* does not seem generalizable into a secondary construction of bent vectorial functions involving four bent vectorial functions. But we show however below that it can be adapted to vectorial functions into a rather general construction. Let us first recall what is the indirect sum of bent Boolean functions. It is a particular case of the construction given by Theorem 12.1.5, which has the interest of automatically generating bent functions from bent functions, without that any extra condition be necessary (contrary to Theorem 12.1.5):

Proposition 12.1.18 ([7]). *Let n and m be two even integers. Let f_1 and f_2, be two Boolean functions defined on $\mathbb{F}_2^n$, f_1' and f_2' be two Boolean functions defined on $\mathbb{F}_2^m$. Define the Boolean function h on $\mathbb{F}_2^n \times \mathbb{F}_2^m$ by setting, for every $(x, y) \in \mathbb{F}_2^n \times \mathbb{F}_2^m$:*

$$h(x, y) = f_1(x) + f_1'(y) + (f_1(x) + f_2(x))(f_1'(y) + f_2'(y)).$$

If f_1, f_2, f_1' and f_2' are bent then, h is bent. Moreover, its dual $\tilde{h}$ is obtained from $\widetilde{f_1}$, $\widetilde{f_2}, \widetilde{f_1'}$ and $\widetilde{f_2'}$ by the same formula as h is obtained from f_1, f_2, f_1' and f_2':

$$\tilde{h}(x, y) = \tilde{f}_1(x) + \tilde{f}_1'(y) + (\tilde{f}_1(x) + \tilde{f}_2(x))(\tilde{f}_1'(y) + \tilde{f}_2'(y)).$$

This construction can be extended to vectorial Boolean functions as follows.

Proposition 12.1.19 ([11]). *Let n, m and r be three positive integers such that n and m are even. Let F_1 and F_2 be two (n, r)-functions and $G = (g_1, \ldots, g_{r+1})$ be an $(m, r + 1)$-function. Define the function H from $\mathbb{F}_2^n \times \mathbb{F}_2^m$ to $\mathbb{F}_2^r$ by setting, for every (x, y) in $\mathbb{F}_2^n \times \mathbb{F}_2^m$:*

$$H(x, y) = F_1(x) + G_1(y) + g_1(y)(F_1(x) + F_2(x))$$

where G_1 is the (m, r)-function $(g_2, \ldots, g_{r+1})$.

Assume that

1. *F_1 and F_2 are bent (this requires $r \leq \frac{n}{2}$);*
2. *For every λ in $\mathbb{F}_2^{r+1}$ different from $(1, 0, \ldots, 0)$, the component function $\lambda \cdot G$ is bent.*

Then H is a bent $(n + m, r)$-function.

Proof. Let $\delta \in \mathbb{F}_2^r \setminus \{0\}$. The component function $\delta \cdot H$ of H, that we denote by h_δ, has the form : $h_\delta(x, y) = \delta \cdot F_1(x) + \delta \cdot G_1(y) + g_1(y)(\delta \cdot F_1(x) + \delta \cdot F_2(x))$. This component function falls then in the scope of Proposition 12.1.18 if we take $f_1 = \delta \cdot F_1, f_2 = \delta \cdot F_2, f_1' = \delta \cdot G_1$ and $f_2' = g_1 + \delta \cdot G_1$. The bentness of h_δ is then a straightforward application of Proposition 12.1.18 since the assumptions (1) and (2) imply that f_1, f_2, f_1' and f_2' are bent. □

Remark 12.1.20. The condition on G can be weakened. Indeed, let $G = (g_1, \ldots, g_{r+1})$ be an $(m, r+1)$-function whose component functions $\lambda \cdot G$ are bent for every non zero $\lambda \neq \mu$ for some $\mu \in \mathbb{F}_2^{r+1} \setminus \{0\}$; let L be a linear automorphism of $\mathbb{F}_2^{r+1}$; set $G' = L \circ G$ then, for every $\lambda \in \mathbb{F}_2^{r+1} \setminus \{0\}$, we have: $\lambda \cdot G' = L^\star(\lambda) \cdot G$ for every $\lambda \in \mathbb{F}_2^{r+1} \setminus \{0\}$, where $L^\star$ denotes the adjoint operator of L. Therefore, one can choose L so that $L^\star(\mu) = (1, 0, \ldots, 0)$ and apply Proposition 12.1.19 to G'.

Remark 12.1.21. Obviously, condition (2) of Proposition 12.1.19 is satisfied by any bent $(m, r+1)$-function. The bentness of G is a strictly stronger hypothesis than hypothesis (2) in Proposition 12.1.19 (we shall see below an example of a non-bent function satisfying (2)) but it allows then to build many more bent functions H, since any function $G' := (g'_1, \ldots, g'_{r+1})$ affinely equivalent to G can be taken in Proposition 12.1.19 instead of G. The function g'_1 can in particular be taken equal to any of the $2^{r+1} - 1$ component functions of G. These component functions are all distinct since G is bent. If $g_1 = g'_1$, the functions H corresponding to G and to G' may be affinely equivalent, but if $g_1 \neq g'_1$, they are not, in general. Hence, when applied to a bent function G, Proposition 12.1.19 can lead to $2^{r+1} - 1$ affinely inequivalent functions H, given F_1, F_2 and G.

Remark 12.1.22. If g_1 is not constant, the algebraic degree of H is the maximum value between $\deg(g_1) + \deg(F_1 + F_2)$ and $\deg(G_1)$. In particular, H has algebraic degree $\frac{n+m}{2}$ (the optimum degree for a bent $(n+m, r)$-function) if and only if $\deg(F_1 + F_2) = \frac{n}{2}$ and $\deg(g_1) = \frac{m}{2}$ (which is optimal, since g_1 is the difference between two bent m-variable Boolean functions).

We shall now investigate some non-bent functions G whose component functions are all bent except one and which will lead to corollaries of Proposition 12.1.19.

Example 12.1.23. Let G' be any bent (m, r)-function. Let ℓ be an affine Boolean function on $\mathbb{F}_2^m$. Let G be the $(m, r+1)$-function defined as $G(y) = (\ell(y), G'(y))$. Then, all the component functions of G except its first coordinate function are bent.

Using the particular choice stated in Example 12.1.23, one deduces the following corollary.

Corollary 12.1.24 ([11]). *Let n, m and r be three positive integers such that n and m are even, $r \leq \frac{n}{2}$ and $r \leq \frac{m}{2}$. Let F_1 and F_2 be two bent functions from $\mathbb{F}_2^n$ to $\mathbb{F}_2^r$, G_1 a bent function from $\mathbb{F}_2^m$ to $\mathbb{F}_2^r$ and ℓ be an affine Boolean function on $\mathbb{F}_2^m$. Define the function H from $\mathbb{F}_2^n \times \mathbb{F}_2^m$ to $\mathbb{F}_2^r$ by setting, for every $(x, y) \in \mathbb{F}_2^n \times \mathbb{F}_2^m$:*

$$H(x, y) = F_1(x) + G_1(y) + \ell(y)(F_1(x) + F_2(x))$$

Then H is a bent $(n+m, r)$-function.

This construction does not allow obtaining bent functions of maximal degree $\frac{n+m}{2}$ unless m equals 2. Let us give now another example of function G which does not have this drawback.

Example 12.1.25. Set $m = 2r$ and let us identify $\mathbb{F}_2^r$ with $\mathbb{F}_{2^r}$ and $\mathbb{F}_{2^m}$ with $\mathbb{F}_{2^r} \times \mathbb{F}_{2^r}$. Let us choose for G a function from $\mathbb{F}_{2^r} \times \mathbb{F}_{2^r}$ to $\mathbb{F}_2 \times \mathbb{F}_{2^r}$ of the form $G(y,z) = (\ell(y) + g(z), y\pi(z) + \Gamma(z))$ where π is a permutation of $\mathbb{F}_{2^r}$, Γ is any function from $\mathbb{F}_{2^r}$ to $\mathbb{F}_{2^r}$, ℓ is affine Boolean on $\mathbb{F}_{2^r}$ and g is any Boolean function on $\mathbb{F}_{2^r}$. Let $\lambda = (\eta, \mu) \in \mathbb{F}_2 \times \mathbb{F}_{2^r}$. For every $\mu \neq 0$ and $\eta \in \mathbb{F}_2$, the component function $\lambda \cdot G$ is by definition of the form $\lambda \cdot G(y,z) = \mathrm{Tr}_1^r(\mu y\pi(z) + \mu\Gamma(z)) + \eta\ell(y) + \eta g(z)$, for every (y,z) in $(\mathbb{F}_{2^r})^2$. Hence it belongs to the Maiorana–McFarland class of Boolean bent functions.

Using the particular choice stated in Example 12.1.25, one deduces the following corollary (where we identify $\mathbb{F}_2^s$ with the Galois Field $\mathbb{F}_{2^s}$).

Corollary 12.1.26 ([11]). *Let n and r be two positive integers such that n is even and $r \leq \frac{n}{2}$. Let F_1 and F_2 be two bent functions from $\mathbb{F}_{2^n}$ to $\mathbb{F}_{2^r}$, g a Boolean function over $\mathbb{F}_{2^r}$, Γ a function from $\mathbb{F}_{2^r}$ to itself, π a permutation of $\mathbb{F}_{2^r}$ and ℓ an affine Boolean function over $\mathbb{F}_{2^r}$. Define the function H from $\mathbb{F}_{2^n} \times \mathbb{F}_{2^r} \times \mathbb{F}_{2^r}$ to $\mathbb{F}_{2^r}$ by setting, for every $(x,y,z) \in \mathbb{F}_{2^n} \times \mathbb{F}_{2^r} \times \mathbb{F}_{2^r}$:*

$$H(x,y,z) = F_1(x) + y\pi(z) + \Gamma(z) + (\ell(y) + g(z))(F_1(x) + F_2(x))$$

Then H is a bent $(n+2r, r)$-function.

4. Generalization of the indirect sum construction:

The indirect sum of Boolean functions was a consequence of Theorem 12.1.5. We shall see now that Theorem 12.1.5 leads to a construction of bent vectorial functions which is more general than that of Proposition 12.1.19. In the latter proposition, function H was equal to $F_1(x) + c_1$ (where $c_1 \in \mathbb{F}_2$) for some values of y and $F_2(x) + c_2$ (where $c_2 \in \mathbb{F}_2$) for the other values of y. This generalizes as follows.

Proposition 12.1.27 ([11]). *Let n and m be two even integers and r, k two positive integers. Let $F_1, \cdots, F_k$ be (n,r)-functions and G an (m,r)-function. Let $\varphi_1, \ldots, \varphi_k$ be Boolean functions on $\mathbb{F}_2^m$ whose supports constitute a partition of $\mathbb{F}_2^m$. Let us define the vectorial Boolean function H from $\mathbb{F}_2^n \times \mathbb{F}_2^m$ to $\mathbb{F}_2^r$ by setting, for every (x,y) in $\mathbb{F}_2^n \times \mathbb{F}_2^m$:*

$$H(x,y) = \sum_{i=1}^{k} \varphi_i(y) F_i(x) + G(y).$$

Let us assume that the following conditions (1) and (2) are satisfied:

1. *$F_1, \cdots, F_k$ and G are bent (this requires $r \leq \frac{n}{2}$ and $r \leq \frac{m}{2}$);*
2. $\forall \lambda \in \mathbb{F}_2^r \setminus \{0\}, \forall \epsilon = (\epsilon_1, \cdots \epsilon_k) \in \mathbb{F}_2^k$,
   the Boolean function $y \in \mathbb{F}_2^m \mapsto \sum_{j=1}^{k} \epsilon_j \varphi_j(y) + \lambda \cdot G(y)$ is bent.

Then H is a bent $(n+m, r)$-function.

Proof. Let $\lambda \in \mathbb{F}_2^r \setminus \{0\}$. The component function $\lambda \cdot H$ of H equals : $\sum_{i=1}^k \varphi_i(y)\, \lambda \cdot F_i(x) + \lambda \cdot G(y)$. The function $x \mapsto \lambda \cdot H(x, y)$ is bent for every y in $\mathbb{F}_2^m$ and its dual equals $\sum_{i=1}^k \varphi_i(y)\widetilde{\lambda \cdot F_i}(x) + \lambda \cdot G(y)$. Hence, applying condition (2) with $\epsilon_i = \widetilde{\lambda \cdot F_i}(x)$ proves that H is bent, according to Theorem 12.1.5. □

Remark 12.1.28. Condition (2) of Proposition 12.1.27 implies in particular that for every $j \in \{1, \cdots, k\}$, the Boolean function defined over $\mathbb{F}_2^m$ by $h_j := \varphi_j + \lambda \cdot G$ is bent. Then $wt(h_j)$ has the form $2^{m-1} + \gamma_j 2^{\frac{m}{2}-1}$ with $\gamma_j = \pm 1$. The function $\lambda \cdot G$ is also bent then, $wt(\lambda \cdot G) = 2^{m-1} + \eta 2^{\frac{m}{2}-1}$ with $\eta = \pm 1$. On the other hand, we have $wt(h_j) = wt(\varphi_j) + wt(\lambda \cdot G) - 2wt(\varphi_j(\lambda \cdot G))$. Therefore, thanks to assumptions on φ_j, we have $\sum_{j=1}^k wt(h_j) = 2^m + (k-2)wt(\lambda \cdot G)$. Finally, $\sum_{j=1}^k (2^{m-1} + \gamma_j 2^{\frac{m}{2}-1}) = 2^m + (k-2)(2^{m-1} + \eta 2^{\frac{m}{2}-1})$ and then, $\sum_{j=1}^k \gamma_j = \eta(k-2)$.

The following statement is an example of application of Proposition 12.1.27:

Corollary 12.1.29 ([11]). *Let k, n and r be three positive integers such that n is even and $r \leq \frac{n}{2}$. Let $F_1, \cdots, F_k$ be bent (n, r)-functions, π a permutation of $\mathbb{F}_2^r$ and Γ any (r, r)-function. Let $\varphi_1, \ldots, \varphi_k$ be any Boolean functions on $\mathbb{F}_2^r$ whose supports constitute a partition of $\mathbb{F}_2^r$. Define the function H from $\mathbb{F}_2^n \times \mathbb{F}_2^r \times \mathbb{F}_2^r$ to $\mathbb{F}_2^r$ by setting, for every $(x, y, z) \in \mathbb{F}_2^n \times \mathbb{F}_2^r \times \mathbb{F}_2^r$:*

$$H(x, y, z) = \sum_{i=1}^k \varphi_i(z)F_i(x) + y \cdot \pi(z) + \Gamma(z)$$

Then H is a bent $(n + 2r, r)$-function.

Remark 12.1.30. Corollary 12.1.29 can also be viewed as a generalization to vectorial functions of a secondary construction given in [10] for Boolean functions under the name of "extension of Maiorana–McFarland type" and which can be stated as follows: let π be a permutation on $\mathbb{F}_2^{n/2}$, g be a Boolean function on $\mathbb{F}_2^{n/2}$ and $f_{\pi,g}$ be a related Maiorana–McFarland's bent function that is, $f_{\pi,g}(x, y) = x\pi(y) + g(y)$, $(x, y) \in \mathbb{F}_2^{n/2} \times \mathbb{F}_2^{n/2}$. Let $(h_y)_{y \in \mathbb{F}_2^{n/2}}$ be a collection of bent functions from $\mathbb{F}_2^m$ (for some even integer m) to $\mathbb{F}_2^{n/2}$. Then, the function $(x, y, z) \mapsto h_y(z) + f_{\pi,g}(x, y)$ defined from $\mathbb{F}_2^{n/2} \times \mathbb{F}_2^{n/2} \times \mathbb{F}_2^m$ to $\mathbb{F}_2^{n/2}$, is a bent $(n + m, n/2)$-function.

5. Further generalization of the indirect sum construction:

   Still more generally, Theorem 12.1.5 leads to a construction of vectorial functions as follows:

   Proposition 12.1.31 ([11]). *Let r and s be two positive even integers and m a positive integer such that $m \leq r/2$. Let H be a function from $\mathbb{F}_2^n = \mathbb{F}_2^r \times \mathbb{F}_2^s$ to $\mathbb{F}_2^m$. Assume that, for every $y \in \mathbb{F}_2^s$, the function $H_y : x \in \mathbb{F}_2^r \mapsto H(x, y)$ is a bent (r, m)-function. For every nonzero $v \in \mathbb{F}_2^m$ and every $a \in \mathbb{F}_2^r$ and $y \in \mathbb{F}_2^s$, let us denote by $f_{a,v}(y)$ the value at a of the dual of the Boolean function $v \cdot H_y$, that is,* *the binary value such that* $\sum_{x\in\mathbb{F}_2^r}(-1)^{v\cdot H(x,y)+a\cdot x} = 2^{r/2}(-1)^{f_{a,v}(y)}$. *Then H is bent if and only if, for every nonzero $v \in \mathbb{F}_2^m$ and every $a \in \mathbb{F}_2^r$, the Boolean function $f_{a,v}$ is bent.*

Proof. For every nonzero $v \in \mathbb{F}_2^m$ and every $a \in \mathbb{F}_2^r$ and $b \in \mathbb{F}_2^s$ we have:

$$\sum_{\substack{x\in\mathbb{F}_2^r\\ y\in\mathbb{F}_2^s}}(-1)^{v\cdot H(x,y)+a\cdot x+b\cdot y} = 2^{r/2}\sum_{y\in\mathbb{F}_2^s}(-1)^{f_{a,v}(y)+b\cdot y}.$$

□

Proposition 12.1.31 is very general and not very effective but an effective example can be obtained by choosing every H_y in the Maiorana–McFarland class: $H_y(x,x') = x\,\pi_y(x') + G_y(x')$, $x, x' \in \mathbb{F}_{2^{r/2}}$, where π_y is bijective for every $y \in \mathbb{F}_2^s$. We have then $f_{(a,a'),v}(y) = \mathrm{Tr}_1^{\frac{r}{2}}\left(a'\,\pi_y^{-1}\left(\frac{a}{v}\right) + v\,G_y\left(\pi_y^{-1}\left(\frac{a}{v}\right)\right)\right)$. Then H is bent if and only if, for every $v \in \mathbb{F}_{2^{r/2}}^*$ and every $a, a' \in \mathbb{F}_{2^{r/2}}$, the function $y \mapsto \mathrm{Tr}_1^{\frac{r}{2}}\left(a'\,\pi_y^{-1}(a) + v\,G_y(\pi_y^{-1}(a))\right)$ is bent on $\mathbb{F}_2^s$. A simple possibility for achieving this is for $s = r/2$ to choose π_y^{-1} such that, for every a, the function $y \mapsto \pi_y^{-1}(a)$ is an affine automorphism of $\mathbb{F}_{2^{r/2}}$ (e.g. $\pi_y^{-1}(a) = \pi_y(a) = a + y$) and to choose G_y such that, for every a, the function $y \mapsto G_y(a)$ is bent.

Remark 12.1.32. The secondary constructions given here do not allow constructing a bent function whose number of output bits is strictly larger than the numbers of output bits of the functions used to build it. In particular, they do not allow constructing bent $(n, n/2)$-functions. We leave as an open problem such constructions. Note that, if F is a bent (n, r)-function, then an affine subspace E of dimension strictly larger than $n/2$ of $\mathbb{F}_2^n$ cannot have an image by F included in an affine hyperplane of $\mathbb{F}_2^r$ since we know that an n-variable bent Boolean function cannot be constant on an affine subspace E of dimension strictly more than $n/2$ (see [4]) and if the image of E by F is included in an affine hyperplane of $\mathbb{F}_2^r$, then there exists $v \neq 0$ in $\mathbb{F}_2^r$ such that $v \cdot F$ is constant on E. This means that, in a secondary construction of a bent $(n, n/2)$-function F from two bent functions in smaller numbers of input variables and smaller numbers of output bits, at least one of the bent functions used to build F is inequivalent to any restriction of F (contrary to the constructions of Propositions 12.1.17, 12.1.27 and 12.1.31).

12.2 Vectorial Bent Functions and Kerdock Codes

A collection of $2^{2m-1} - 1$ quadratic Boolean bent functions in $n = 2m$ variables such that the sum of any two of them is again bent, gives rise to a Kerdock codes of length 2^{2m}. The Boolean function $f(x, y) = xy$ (which is a vectorial bent function from $\mathbb{F}_{p^m} \times \mathbb{F}_{p^m}$ to $\mathbb{F}_{p^m}$ and for which all component functions are quadratic) only gives rise to $2^m - 1$ such bent functions.

12.3 Hyper-Bent Vectorial Functions

Hyper-bent functions constitute a subset of bent functions and are harder to approximate than bent functions, making them particularly attractive for cryptographic applications. In the multiple-valued world, up to now, characterization and generation of hyper-bent functions represent an interesting challenging mathematical problem. After a hiatus of several years, research activities on bent vectorial functions have been actively resumed by several researchers since 2011. The importance of the constructions of bent vectorial functions in univariate representation has been highlighted.

In 2013, Moraga et al. [20] showed that multiple-valued hyper-bent functions constitute a reduced subset of the multiple-valued bent functions which have been characterized in a simple manner. They have also introduced the concept of strict hyper-bent functions, and studied some of the properties of these functions.

Identify and specify trace bent functions has been an important research topic lately. What is common to all these approaches is the underlying idea of specifying m bent Boolean functions in a particular way, so that their linear combinations remain bent. The problem of the existence and then the constructions of vectorial bent functions in a monomial form $Tr_m^n(\lambda x^d)$ (for m dividing n) or more generally in a multimonomial form $Tr_m^n(\sum_{i=1}^r \lambda_i x^{d_i})$ have been considered by several authors but treated only in some special cases namely, when the exponents d_i are *Dillon-like exponents*, that is, d_i of the form $r_i(2^m - 1)$. In [15], Dong et al. have studied vectorial bent functions of the form $Tr_m^n(\lambda x^d)$ (where $n = 2m$) based on monomial bent Boolean functions and characterized the bentness property of $d \in \{2^i + 1, 2^{2i} - 2^i + 1\}$. Next, the authors proposed a method to construct vectorial bent functions based on $\mathcal{PS}^-$ and $\mathcal{PS}^+$ bent functions. In [22], Muratović-Ribić, Pasalic and Bajrić have derived three necessary and sufficient conditions for a function of the form $F(x) = Tr_m^n(\sum_{i=1}^r \lambda_i x^{r_i(2^m-1)})$ (where $n = 2m$) to be bent. In particular, the first characterization is a direct consequence of a result of the author [18] which states that $\sum_{u \in U}(-1)^{\mathrm{Tr}_1^m(\lambda F(u))} = 1$ for every $\lambda \in \mathbb{F}_{2^m}^\star$, where U is the he cyclic group of the $(2m+1)$st primitive roots of unity in $\mathbb{F}_{2^n}$ $(n = 2m)$. The second characterization provides an interesting link between the bentness and the evaluation of F on the cyclic group U. The third characterization is stated in terms of the evaluation of certain elementary symmetric polynomials, which can be transformed into some explicit conditions regarding the choice of some coefficients. In [21], the authors have characterized a class of vectorial (hyper)bent functions of the form $F(x) = Tr_m^n(\sum_{i=0}^{2^m} a_i x^{i(2^m-1)})$ (where $n = 2m$) in terms of finding an explicit expression for the coefficients a_i so that F is vectorial hyperbent. These coefficients only depend on the choice of the interpolating polynomial used in the Lagrange interpolation of the elements of U and some prespecified outputs. They also showed that these interpolation polynomials can be chosen in exactly $(2^k + 1)!2^{k-1}$ ways and this is the exact number of vectorial hyperbent functions of the above form. Furthermore, the authors provided a simple optimization method is proposed for selecting the interpolation polynomials that give rise to trace polynomials with a few nonzero coefficients. Since then, a number of authors have focused on the (non)existence and constructions of vectorial bent functions of the form $Tr_m^n(P)$ where $P(x) \in \mathbb{F}_{2^n}[x]$ based on Boolean bent functions of the form $Tr_1^n(P)$. For instance, it has been proved in [26, 27] the nonexistence of some vectorial bent functions with binomial trace representation in the $\mathcal{PS}$. More precisely, it has been proved that for $n = 2m$, $n \equiv 0 \pmod 4$ there are no vectorial bent functions of the form $F(x) = Tr_m^n(x^{2^m-1} + \lambda x^{r(2^m-1)})$ where $1 \leq r \leq 2^m$ and $\lambda \in \mathbb{F}_{2^n}$.

12.4 Bent Boolean Functions Associated to AB Functions

There exist also bent functions associated with some vectorial (n,n)-functions called *almost bent* (AB, for short) . Almost bent functions are those (n,n)-vectorial functions (that is, functions from $\mathbb{F}_{2^n}$ to $\mathbb{F}_{2^n}$) having maximal nonlinearity $2^{n-1} - 2^{\frac{n-1}{2}}$ (n odd)

We recall the notion of the nonlinearity of vectorial functions.

Definition 12.4.1. The nonlinearity of a vectorial function $F : \mathbb{F}_2^n \to \mathbb{F}_2^n$ (resp. $\mathbb{F}_{2^n} \to \mathbb{F}_{2^n}$) equals:

$$nl(F) = \min\{nl(b \cdot F);\ b \in \mathbb{F}_2^n \setminus \{0\}\} \text{ (resp. } \min\{nl(\mathrm{Tr}_1^n(bF));\ b \in \mathbb{F}_{2^n}^*\}).$$

F is almost bent (AB) if $nl(F) = 2^{n-1} - 2^{\frac{n-1}{2}}$, which is the best possible value.

Given such function F, the indicator γ_F of the set $\{(a,b) \in (\mathbb{F}_2^n \setminus \{0\}) \times \mathbb{F}_2^n;\ \exists x \in \mathbb{F}_2^n,\ F(x) + F(x+a) = b\}$ is a bent function. More precisely, we have the following result.

For more details, we send the reader for instance to the Book's chapter [3].

Theorem 12.4.2 ([9]). *Let F be a function from $\mathbb{F}_2^m$ to itself. Then F is AB if and only if $\gamma_F : (\mathbb{F}_{2^m})^2 \to \mathbb{F}_2$ defined by*

$$\gamma_F(a,b) = 1 \Leftrightarrow \begin{cases} a \neq 0 \\ \exists x \mid F(x) + F(x+a) = b \end{cases}$$

is bent. The dual function satisfies

$$\widetilde{\gamma_F}(a,b) = 1 \Leftrightarrow \begin{cases} b \neq 0 \\ \widehat{\chi_{b \cdot F}}(a) \neq 0 \end{cases}.$$

The known AB power functions $F(x) = x^d$, $x \in \mathbb{F}_{2^m}$ are given in Table 12.3 below.

Table 12.3 Known AB power functions x^d on $\mathbb{F}_{2^m}$

Functions	Exponents d	Conditions
Gold	2^i+1	$\gcd(i,m)=1, 1\leq i<m/2$
Kasami	$2^{2i}-2^i+1$	$\gcd(i,m)=1, 2\leq i<m/2$
Welch	2^k+3	$m=2k+1$
Niho	$2^k+2^{\frac{k}{2}}-1$, k even	$m=2k+1$
	$2^k+2^{\frac{3k+1}{2}}-1$, k odd	

The associated bent functions γ_F to known AB functions have been investigated in [2]. We give them below:

- *Gold* : $\gamma_F(a,b)=\mathrm{Tr}_1^m(\frac{b}{a^{2^i+1}})$ with $\frac{1}{0}=0$;
- *Kasami, Welch, Niho* : $F(x+1)+F(x)=q(x^{2^s}+x)$ (q permutation determined by Dobbertin, $\gcd(s,m)=1$);
  $F(x+1)+F(x)=b$ has solutions if and only if $\mathrm{Tr}_1^m(q^{-1}(b))=0$.

$$\text{Then: } \gamma_F(a,b)=\begin{cases}\mathrm{Tr}_1^m(q^{-1}(b/a^d))+1 & \text{if } a\neq 0,\\ 0 & \text{otherwise.}\end{cases}$$

1. Kasami: $s=i$, $q(x)=\frac{x^{2^i+1}}{\sum_{j=1}^{i'}x^{2^{ji}}+\alpha\mathrm{Tr}_1^m(x)}+1$, where

$$i'\equiv 1/i \mod m,\ \alpha=\begin{cases}0 \text{ if } i' \text{ is odd}\\ 1 \text{ otherwise}\end{cases}$$

2. Welch: $s=k$, $q(x)=x^{2^{k+1}+1}+x^3+x+1$.
3. Niho: $s=k/2$ if k is even and $s=(3k+1)/2$ if k is odd, $q(x)=\begin{cases}\frac{1}{g(x^{2^s-1})+1}+1 & \text{if } x\notin\mathbb{F}_2\\ 1 & \text{otherwise}\end{cases}$ where

$$g(x)=x^{2^{2s+1}+2^{s+1}+1}+x^{2^{2s+1}+2^{s+1}-1}+x^{2^{2s+1}+1}+x^{2^{2s+1}-1}+x$$

The functions γ_F associated to Kasami, Welch and Niho functions with $m=7,9$, are neither in the completed MM class, nor in the completed $\mathcal{PS}_{ap}$ class.

The other known infinite classes of AB functions are quadratic; their associated γ_F belong to the completed Maiorana McFarland class.

References

1. L. Budaghyan and C. Carlet. On CCZ-equivalence and its use in secondary constructions of bent functions. In *Proceedings of the International Workshop on Coding and Cryptography WCC 2009*.
2. L. Budaghyan, C. Carlet, and T. Helleseth. On bent functions associated to AB functions. In *Proceedings of IEEE Information Theory Workshop, ITW'11, Paraty, Brazil*, 2011.
3. C. Carlet. Vectorial Boolean Functions for Cryptography. In *Chapter of the monography Boolean Methods and Models, Y. Crama and P. Hammer eds, Cambridge University Press. Preliminary version available at http://www-rocq.inria.fr/codes/Claude.Carlet/pubs.html.*
4. C. Carlet. Two new classes of bent functions. In *Proceedings of EUROCRYPT'93, Lecture Notes in Computer Science 765*, pages 77–101, 1994.
5. C. Carlet. A construction of bent functions. In *Finite Fields and Applications, London Mathematical Society, Lecture Series 233, Cambridge University Press, pages 47–58*, 1996.
6. C. Carlet. On the confusion and diffusion properties of Maiorana–McFarland's and extended Maiorana–McFarland's functions. In *Complexity Issues in Coding and Cryptography, Special Issue dedicated to Prof. Harald Niederreiter on the occasion of his 60th birthday, Journal of Complexity 20*, pages 182–204, 2004.
7. C. Carlet. On the secondary constructions of resilient and bent functions. In *Proceedings of the Workshop on Coding, Cryptography and Combinatorics 2003, published by Birkhäuser Verlag*, pages 3–28, 2004.
8. C. Carlet. Relating three nonlinearity characteristics of vectorial functions and using bent functions to build APN and differentially 4-uniform functions. In *Des. Codes Cryptography, 59 (1–3)*, pages 89–109, 2011.
9. C. Carlet, P. Charpin, , and V. Zinoviev. Codes, bent functions and permutations suitable for DES-like cryptosystems. In *Designs, Codes and Cryptography, 15(2), pp. 125–156*, 1998.
10. C. Carlet, H. Dobbertin, and G. Leander. Normal extensions of bent functions. In *IEEE Trans. Inf. Theory, vol. 50, no. 11*, pages 2880–2885, 2004.
11. C. Carlet and S. Mesnager. On the construction of bent vectorial functions. In *Journal of Information and Coding Theory: Algebraic and Combinatorial Coding Theory Vol 1, no. 2*, pages 133–148, 2010.
12. C. Carlet and S. Mesnager. On Dillon's class H of bent functions, niho bent functions and o-polynomials. In *Journal of Combinatorial Theory, Series A, Vol 118, no. 8*, pages 2392–2410, 2011.
13. J. Dillon. Elementary Hadamard difference sets. In *PhD dissertation, University of Maryland*.
14. J. Dillon. A survey of bent functions. In *NSA Technical Journal Special Issue*, pages 191–215, 1972.
15. D. Dong, X. Zhang, L. Qu, and S. Fu. A note on vectorial bent functions. In *Information Processing Letters, 113 (22), pages 866–870*, 2013.
16. F. J. MacWilliams and N. J. Sloane. The theory of error-correcting codes. In *Amsterdam, North Holland*, 1977.
17. M. Matsui. Linear cryptanalysis method for DES cipher. In *Proceedings of EUROCRYPT'93, Lecture Notes in Computer Science 765*, pages 386–397, 1994.
18. S. Mesnager. Bent and hyper-bent functions in polynomial form and their link with some exponential sums and Dickson polynomials. *IEEE Transactions on Information Theory*, 57(9):5996–6009, 2011.
19. S. Mesnager. Bent vectorial functions and linear codes from o-polynomials. In *Journal Designs, Codes and Cryptography, 77(1), pages 99–116*, 2015.
20. C. Moraga, M. Stanković, R. Stanković, and S. Stojković. Hyper-bent Multiple-Valued Functions. In *Proceedings Computer Aided Systems Theory-EUROCAST 2013 Lecture Notes in Computer Science Vol. 8112, pages 250–257*, 2013.

21. C. Moraga, M. Stanković, R. Stanković, and S. Stojković. Vectorial hyperbent trace functions from the PS_{ap} Class -their exact number and specification. In *Vectorial hyperbent trace functions from the PS_{ap} Class -their exact number and specification. IEEE Transactions on Information Theory 60 (7), pages 4408–4413*, 2014.
22. A. Muratović-Ribić, E. Pasalic, and S. Bajrić. Vectorial bent functions from multiple terms trace Functions. In *IEEE Transactions on Information Theory, 60 (2), pages 1337–1347*, 2014.
23. K. Nyberg. Perfect non-linear S-boxes. In *Proceedings of EUROCRYPT'91, Lecture Notes in Computer Science 547*, pages 378–386, 1992.
24. K. Nyberg. On the construction of highly nonlinear permutations. In *Proceedings of EUROCRYPT'92, Lecture Notes in Computer Science 658*, pages 92–98, 1993.
25. K. Nyberg. New bent mappings suitable for fast implementation. In *of Fast Software Encryption 1993, Lecture Notes in Computer Science 809*, pages 179–184, 1994.
26. E. Pasalic. A note on nonexistence of vectorial bent functions with binomial trace representation in the $\mathcal{PS}$- class. In *Information Processing Letters 115 (2), pages 139–140.*, 2015.
27. E. Pasalic. Corrigendum to A note on nonexistence of vectorial bent functions with binomial trace representation in the PS-class. In *Information Processing Letters 115 (4), page 520*, 2015.
28. E. Pasalic and W.-G. Zhang. On multiple output bent functions. In *Information Processing Letters, 112 (21), pages 811–815*, 2012.
29. T. Satoh, T. Iwata and K. Kurosawa. On cryptographically secure vectorial Boolean functions. In *Proceedings of Asiacrypt 1999, Lecture Notes in Computer Science 1716*, pages 20–28, 1999.
30. C. Tang, Y. Qi, and M. Xu. Multiple output bent functions characterized by families of bent functions. In *Journal of Cryptologic Research, 1(4), pages 321–326*, 2014.
31. Y. Xu and C. Wu. On the primary constructions of vectorial Boolean bent functions. In *Cryptology ePrint Archive: Report 2015/077*, 2015.

Chapter 13
Bent Functions in Arbitrary Characteristic

In 1985, Kumar, Scholtz and Welch in [31] generalized the notion of Boolean bent functions to the case of functions over an arbitrary finite field. In this chapter we consider more generally bent functions in arbitrary characteristic p (where p is a prime integer).

13.1 On p-Ary Bent Functions: Generalities and Constructions

13.1.1 Walsh Transform of a p-Ary Function

Let f be a function from $\mathbb{F}_{p^n}$ to $\mathbb{F}_p$ and $\zeta_p = e^{\frac{2\pi i}{p}}$ is the complex primitive p^{th} root of unity and elements of $\mathbb{F}_p$ are considered as integers modulo p.

We denote by χ_f the *sign* function from $\mathbb{F}_{p^n}$ to $\mathbb{C}$ of f defined as $\chi_f(x) = \zeta_p^{f(x)}$ for all $x \in \mathbb{F}_{p^n}$. The Fourier transform $\hat{\chi}_f$ of the function χ_f is defined as

$$\begin{aligned} \hat{\chi}_f : & \mathbb{F}_{p^n} \to \mathbb{C} \\ w \longmapsto & \hat{\chi}_f(b) = \sum_{x \in \mathbb{F}_{p^n}} \chi_f(x) \zeta_p^{-b \cdot x}, \end{aligned}$$

called *Walsh transform* of f at $b \in \mathbb{F}_{p^n}$ where "$\cdot$" is any scalar product in $\mathbb{F}_{p^n}$. As the notion of a Walsh transform refers to a scalar product, it is convenient to choose the isomorphism such that the canonical scalar product "$\cdot$" in $\mathbb{F}_{p^n}$ coincides with the canonical scalar product in $\mathbb{F}_{p^n}$, which is the trace of the product $b \cdot x := \mathrm{Tr}_p^{p^n}(bx) = tr_n(bx)$.

S. Mesnager, *Bent Functions*, DOI 10.1007/978-3-319-32595-8_13

Thus, the Walsh transform (that we denote by $S_f(b)$) of f at $b \in \mathbb{F}_{p^n}$ is defined by:

$$S_f(b) = \sum_{x \in \mathbb{F}_{p^n}} \zeta_p^{f(x) - tr_n(bx)}$$

and f can be recovered by the inverse transform

$$\zeta_p^{f(x)} = \frac{1}{p^n} \sum_{b \in \mathbb{F}_{p^n}} S_f(b) \zeta_p^{tr_n(bx)}$$

where $\zeta_p = e^{\frac{2\pi i}{p}}$ is the complex primitive p^{th} root of unity.

For every function f from $\mathbb{F}_{p^n}$ to $\mathbb{F}_p$, we have

$$\sum_{w \in \mathbb{F}_{p^n}} S_f(w) = p^n \chi_p(f(0)). \tag{13.1}$$

Set $|z|^2 = z\bar{z}$ where $\bar{z}$ stands for the conjugate of z. Then

$$\sum_{w \in \mathbb{F}_{p^n}} |S_f(w)|^2 = p^{2n}. \tag{13.2}$$

In the book, we shall refer to (13.2) as the *Parseval identity*.

13.1.2 Non-binary Bent Functions

Kumar, Scholtz and Welch have introduced in [31] bent functions over an arbitrary finite field. We start with some definitions.

Definition 13.1.1. A function f is said to *p-ary bent* if all its Walsh coefficients satisfy $|S_f(b)|^2 = p^n$. A bent function f is said to be *regular bent* if for every $b \in \mathbb{F}_{p^n}$, $p^{-\frac{n}{2}} S_f(b) = \zeta_p^{f^\star(b)}$ for some mapping $f^\star : \mathbb{F}_{p^n} \to \mathbb{F}_p$. Such a function $f^\star$ is called the *dual function* .The bent function f is said to be a *weakly regular bent* function if there exists a complex number u with $|u| = 1$ such that $up^{-\frac{n}{2}} S_f(b) = \zeta_p^{f^\star(b)}$ for all $b \in \mathbb{F}_{p^n}$.

Walsh transform coefficients of a p-ary bent function f with odd p satisfy

$$p^{-\frac{n}{2}} S_f(b) = \begin{cases} \pm \zeta_p^{f^\star(b)}, & \text{if } n \text{ is even or } n \text{ odd and } p \equiv 1 \pmod 4, \\ \pm i \zeta_p^{f^\star(b)}, & \text{if } n \text{ is odd and } p \equiv 3 \pmod 4, \end{cases} \tag{13.3}$$

where i is a complex primitive 4-th root of unity. Therefore, regular bent functions can only be found for even n and for odd n with $p \equiv 1 \pmod 4$. Moreover, for a weakly regular bent function, the constant u (defined above) can only be equal to ± 1 or $\pm i$. Weakly regular bent functions always come in pairs, since the dual is bent as well. This does in general not hold for non-weakly regular bent functions, see [9, 10].

13.2 Constructions of Bent Functions in Arbitrary Characteristic

13.2.1 *Known Primary Constructions of Bent Boolean Functions: A Summary*

In this section, we collect all the known primary constructions of bent functions in characteristic 2. We shall list them by separating the univariate case from the bivariate case.

13.2.1.1 Known Infinite Classes of Bent Functions in Univariate Trace Form

Below the current list of the known Infinite classes of bent functions in univariate representation which are small for each n.

Note that finding bent functions in univariate trace form is in general difficult and presents theoretical interest, since it gives more insight on bent functions.

- $f(x) = \mathrm{Tr}_1^n\left(ax^{2^j+1}\right)$, where $a \in \mathbb{F}_{2^n} \setminus \{x^{2^j+1}; x \in \mathbb{F}_{2^n}\}$, $\frac{n}{gcd(j,n)}$ even (the bentness of such function is directly deduced from the characterization recalled above of bent quadratic functions). This class has been generalized in [13, 29, 30, 36, 48] to functions of the form $\mathrm{Tr}_1^n(\sum_{i=1}^{m-1} a_i x^{2^i+1}) + c_m \mathrm{Tr}_1^m(a_m x^{2^m+1})$, $a_i \in \mathbb{F}_2$. Being quadratic, these functions belong to the completed $\mathcal{M}$ class.
- $f(x) = \mathrm{Tr}_1^n\left(ax^{2^{2j}-2^j+1}\right)$, where $a \in \mathbb{F}_{2^n} \setminus \{x^3; x \in \mathbb{F}_{2^n}\}$, $gcd(j,n) = 1$ (Dillon and Dobbertin [16]).
- $f(x) = \mathrm{Tr}_1^n\left(ax^{(2^{n/4}+1)^2}\right)$, where $n \equiv 4$ [mod 8], $a = a'b^{(2^{n/4}+1)^2}$, $a' \in w\mathbb{F}_{2^{n/4}}$, $w \in \mathbb{F}_4 \setminus \mathbb{F}_2$, $b \in \mathbb{F}_{2^n}$ (Leander [32], see also [12]); the functions in this class belong to the completed $\mathcal{M}$ class.
- $f(x) = \mathrm{Tr}_1^n\left(ax^{2^{n/3}+2^{n/6}+1}\right)$, where $6 \mid n$, $a = a'b^{2^{n/3}+2^{n/6}+1}$, $a' \in \mathbb{F}_{2^m}$, $Tr_{m/3}^m(a') = 0$, $b \in \mathbb{F}_{2^n}$ (Canteaut-Charpin-Kyureghyan, [6]); the functions in this class belong to the completed $\mathcal{M}$ class.
- $f(x) = \mathrm{Tr}_1^n\left(a[x^{2^i+1} + (x^{2^i} + x + 1)\mathrm{Tr}_1^n(x^{2^i+1})]\right)$, where $n \geq 6$, m does not divide i, $\frac{n}{\mathrm{gcd}(i,n)}$ even, $a \in \mathbb{F}_{2^n} \setminus \mathbb{F}_{2^i}$, $\{a, a+1\} \cap \{x^{2^i+1}; x \in \mathbb{F}_{2^n}\} = \emptyset$ (Budaghyan-Carlet, [1]). These functions belong to the completed $\mathcal{M}$ class when $a \in \mathbb{F}_{2^m}$.
- $$f(x) = \mathrm{Tr}_1^n\big(a\big[\big(x + Tr_3^n(x^{2(2^i+1)} + x^{4(2^i+1)}\big) + \mathrm{Tr}_1^n(x)Tr_3^n\big(x^{2^i+1} + x^{2^{2i}(2^i+1)}\big)\big)^{2^i+1}\big]\big),$$
  where $6 \mid n$, m does not divide i, $\frac{n}{\mathrm{gcd}(i,n)}$ even, $b + d + d^2 \notin \{x^{2^i+1}; x \in \mathbb{F}_{2^n}\}$ for every $d \in \mathbb{F}_{2^3}$ (Budaghyan-Carlet, [1]). These functions are EA-inequivalent to functions in $\mathcal{M}$.

- Three classes of *Niho bent functions* (Dobbertin–Leander–Canteaut–Carlet–Felke–Gaborit [18]) whose restrictions to the cosets $u\mathbb{F}_{2^m}$ are linear. These functions can be written:

$$f(x) = \mathrm{Tr}_1^m(ax^{2^m+1}) + \mathrm{Tr}_1^n(bx^d);\ b = 0 \text{ or } a = b^{2^m+1} \in \mathbb{F}_{2^m}^\star.$$

  The values of d are such that:

  1. $d = (2^m - 1)\,3 + 1$ (the original condition $\exists u \in \mathbb{F}_{2^n}$ $s.t.\, b = u^5$ if $m \equiv 2$ [mod 4] has been shown not useful by Helleseth–Kholosha–Mesnager [28]),
  2. $d = (2^m - 1)\frac{1}{4} + 1$ (m odd),
  3. $d = (2^m - 1)\frac{1}{6} + 1$ (m even).

  Classes 1 and 3 are not EA-equivalent to $\mathcal{M}$ functions (Budaghyan–Carlet–Helleseth–Kholosha–Mesnager [3]). Class 2 is in completed $\mathcal{M}$ class. In classes 1 (for $m \not\equiv 2$ [mod 4]), 2 and 3, we can up to EA-equivalence fix $b = 1$.
- Extension of the second class of Niho type (Leander–Kholosha, [33]):

$$\mathrm{Tr}_1^m(x^{2^m+1}) + \mathrm{Tr}_1^n\Big(\sum_{i=1}^{2^{r-1}-1} t^{s_i}\Big),$$

  where

  - $r > 1$ and $gcd(r, m) = 1$,
  - $s_i = (2^m - 1)\left(\frac{i}{2^r}[\mod 2^m + 1]\right) + 1, i \in \{1, \cdots, 2^{r-1} - 1\}$.

- *Dillon's and generalized Dillon's functions* [11, 15, 34]: let $gcd(r, 2^m + 1) = 1$ and $a \in \mathbb{F}_{2^n}^*$, then

$$f(x) = \mathrm{Tr}_1^n(ax^{r(2^m-1)})$$

  is bent if and only if

$$K_m(a^{2^m+1}) = 0,$$

  where $K_m(u) = \sum_{x\in\mathbb{F}_{2^m}} \mathrm{Tr}_1^m\left(ux + \frac{1}{x}\right)$, $u \in \mathbb{F}_{2^m}$, is the Kloosterman sum over$\mathbb{F}_{2^m}$.
  This class has been generalized to functions:

$$\mathrm{Tr}_1^n\left(\sum_{r\in R} a_r x^{r(2^m-1)}\right),$$

$$\mathrm{Tr}_1^n\left(ax^{r(2^m-1)} + bx^{\left(\frac{q+1}{e}-l\right)(2^m-1)} + cx^{\left(\frac{q+1}{e}+l\right)(2^m-1)}\right) + \mathrm{Tr}_1^\ell\left(\epsilon x^{\frac{2^n-1}{e}}\right),$$

$$\mathrm{Tr}_1^n\left(\sum_{i\in D} ax^{(ri+s)(2^m-1)}\right) + \mathrm{Tr}_1^\ell\left(\epsilon x^{\frac{2^n-1}{e}}\right),$$

  where $\ell | n$ and $e | 2^\ell - 1$.

- Two explicit classes are given in [21]: $\sum_{i=1}^{2^{m-1}-1} Tr_1^n\left(\beta x^{i(2^m-1)}\right)$, where $\beta \in \mathbb{F}_{2^m} \setminus \mathbb{F}_2$ and $\sum_{i=1}^{2^{m-2}-1} Tr_1^n\left(\beta x^{i(2^m-1)}\right)$, where m is odd, $\beta \in \mathbb{F}_{2^m}^*$, $\mathrm{Tr}_1^m\left(\beta^{(2^m-4)^{-1}}\right) = 0$.
- Classes of (hyper-)bent functions, known as the so-called *Mesnager's functions*:
  1. let $n = 2m$ m odd > 3; $a \in \mathbb{F}_{2^n}^*, b \in \mathbb{F}_{2^2}^*$, then $f(x) = \mathrm{Tr}_1^n(ax^{r(2^m-1)}) + \mathrm{Tr}_1^2(bx^{\frac{2^n-1}{3}})$ is (hyper-)bent if and only if $K(a^{2^m+1}) = 4$ [39, 41]. This class can be extended to the case m even, but no necessary and sufficient condition is known in this case (see [42]).
  2. let $n = 2m$, m odd with $m \not\equiv 3 \pmod 6$, $a \in \mathbb{F}_{2^m}^\star$, β a primitive element of $\mathbb{F}_4$, ζ a generator of the cyclic group U of (2^m+1)-th of unity, then $g_{a\zeta^i,\beta^j} = \mathrm{Tr}_1^n\left(a\zeta^i x^{3(2^m-1)}\right) + \mathrm{Tr}_1^2\left(\beta^j x^{\frac{2^n-1}{3}}\right)$ is (hyper-)bent if $\mathrm{Tr}_1^m(a^{1/3}) = 0$ and $K_m(a) = 4$ or $\mathrm{Tr}_1^m(a^{1/3}) = 1$ and $K_m(a) + C_m(a,a) = 4$ [40].
  3. Generalizations exist by the author [38] and the author with Flori [46, 47] to functions using trace functions in other subfields, including the class of functions $f(x) = \mathrm{Tr}_1^n(ax^{r(2^m-1)}) + \mathrm{Tr}_1^4(bx^{\frac{2^n-1}{5}})$.
- Bent functions have been also obtained by Dillon and McGuire as the restrictions of functions on $\mathbb{F}_{2^{n+1}}$, with $n+1$ odd, to a hyperplane of this field [17].

Note that many of these classes belong to completed Maiorana McFarland class, when viewed in multivariate representation, and their bentness may then be easily explained.

13.2.1.2 Known Infinite Classes of Bent Functions in Bivariate Trace Form

- Maiorana McFarland class can be viewed in bivariate form. Its elements are the functions $f(x,y) = \mathrm{Tr}_1^m(x\pi(y)) + g(y)$; $x, y \in \mathbb{F}_{2^m}$, where π is a permutation on $\mathbb{F}_{2^m}$.
- *Dillon's* $\mathcal{PS}_{ap}$[1] *class*: $f(x,y) = g(xy^{2^m-2}) = g\left(\frac{x}{y}\right)$, where g is any balanced Boolean function on $\mathbb{F}_{2^m}$. The dual equals $g\left(\frac{y}{x}\right)$.

  This class is much larger than the classes above but much smaller than the $\mathcal{M}$ class. It contains, up to EA-equivalence, the generalized Dillon's functions and the so-called Mesnager functions.
- An isolated class [7]: $f(x,y) = \mathrm{Tr}_1^m(x^{2^i+1} + y^{2^i+1} + xy)$, $x, y \in \mathbb{F}_{2^m}$, where n is coprime with 3 and i is coprime with m.
- Bent functions in a bivariate representation related to Dillon's H class obtained from the known o-polynomials.

[1] PS_{ap} class is included in the more general Dillon's PS class, see [15], which has itself been generalized to the GPS class, see [7], which covers all bent functions up to EA-equivalence [22].

1. m odd:
   - $f(x,y) = \mathrm{Tr}_1^m(x^{-5}y^6)$;
   - $f(x,y) = \mathrm{Tr}_1^m(x^{\frac{5}{6}}y^{\frac{1}{6}})$;
2. $m = 2k-1$:
   - $f(x,y) = \mathrm{Tr}_1^m(x^{-3\cdot(2^k+1)}y^{3\cdot 2^k+4})$;
   - $f(x,y) = \mathrm{Tr}_1^m(x^{2^m-3\cdot 2^{k-1}+2}y^{3\cdot 2^{k-1}-2})$;
3. $m = 4k-1$:
   - $f(x,y) = \mathrm{Tr}_1^m(x^{2^m-2^k-2^{2k}}y^{2^k+2^{2k}})$;
   - $f(x,y) = \mathrm{Tr}_1^m(x^{2^{3k-1}-2^{2k}+2^k}y^{2^m-2^{3k-1}+2^{2k}-2^k})$;
4. $m = 4k+1$:
   - $f(x,y) = \mathrm{Tr}_1^m(x^{2^m-2^{2k+1}-2^{3k+1}}y^{2^{2k+1}+2^{3k+1}})$;
   - $f(x,y) = \mathrm{Tr}_1^m(x^{2^{3k+1}-2^{2k+1}+2^k}y^{2^m-2^{3k+1}+2^{2k+1}-2^k})$;
5. $m = 2k-1$:
   - $f(x,y) = \mathrm{Tr}_1^m(x^{1-2^k}y^{2^k} + x^{-(2^k+1)}y^{2^k+2} + x^{-3\cdot(2^k+1)}y^{3\cdot 2^k+4})$;
   - $f(x,y) = \mathrm{Tr}_1^m\left(\frac{x\left(\frac{y}{x}+1+\frac{y^{2^k}}{x^{2^k}}\right)\frac{y^{2^{k-1}}}{x^{2^{k-1}}}}{\frac{y^{2^k}}{x^{2^k}}+\frac{y^2}{x^2}+1}\right)$;
6. m odd :
   - $f(x,y) = \mathrm{Tr}_1^m(x^{\frac{5}{6}}y^{\frac{1}{6}} + x^{\frac{3}{6}}y^{\frac{3}{6}} + x^{\frac{1}{6}}y^{\frac{5}{6}}) = D_5((y/x)^6)$;
   - $f(x,y) = \mathrm{Tr}_1^m\left(x\left[D_{\frac{1}{5}}\left(\frac{y}{x}\right)\right]^6\right)$; $D_{\frac{1}{5}}$ Dickson polynomial.
7. Four functions related to the two extra o-polynomials:
   - $\frac{\delta^2(z^4+z)+\delta^2(1+\delta+\delta^2)(z^3+z^2)}{z^4+\delta^2z^2+1} + z^{1/2}$, where $\mathrm{Tr}_1^m(1/\delta) = 1$ and, if $m \equiv 2$ [mod 4], then $\delta \notin \mathbb{F}_4$;
   - $\frac{\mathrm{Tr}_m^n(v^r)(z+1)+\mathrm{Tr}_m^n\left[(vz+v^{2^m})^r\right]\left(z+\mathrm{Tr}_m^n(v)z^{1/2}+1\right)^{1-r}}{\mathrm{Tr}_m^n(v)} + z^{1/2}$, where m is even, $r = \pm\frac{2^m-1}{3}$, $v \in \mathbb{F}_{2^{2m}}$, $v^{2^m+1} = 1$ and $v \neq 1$.

The Class 1 of Niho bent functions corresponds to the so-called *Subiaco hyperovals*, related to the first of the two classes of o-polynomials recalled just above [28], while Classes 2 and 3 correspond to *Adelaide hyperovals* [27].

- Bent functions associated to AB functions: Functions AB give rise to bent functions. The known AB power functions $F(x) = x^d$, $x \in \mathbb{F}_{2^m}$ are given in Table 18.7. The associated bent functions γ_F are studied in [2]. We give them below:
  - *Gold* : $\gamma_F(a,b) = \mathrm{Tr}_1^m(\frac{b}{a^{2^i+1}})$ with $\frac{1}{0} = 0$;

– *Kasami, Welch, Niho* : $F(x+1)+F(x) = q(x^{2^s}+x)$ (q permutation determined by Dobbertin, $\gcd(s,m)=1$);
  $F(x+1)+F(x)=b$ has solutions if and only if $\mathrm{Tr}_1^m(q^{-1}(b)) = 0$.

$$\text{Then: } \gamma_F(a,b) = \begin{cases} \mathrm{Tr}_1^m(q^{-1}(b/a^d)) + 1 & \text{if } a \neq 0, \\ 0 & \text{otherwise.} \end{cases}$$

1. Kasami: $s = i$, $q(x) = \dfrac{x^{2^i+1}}{\sum_{j=1}^{i'} x^{2^{ji}} + \alpha \mathrm{Tr}_1^m(x)} + 1$, where

$$i' \equiv 1/i \mod m,\ \alpha = \begin{cases} 0 \text{ if } i' \text{ is odd} \\ 1 \text{ otherwise.} \end{cases}$$

2. Welch: $s = k$, $q(x) = x^{2^{k+1}+1} + x^3 + x + 1$.

3. Niho: $s = k/2$ if k is even and $s = (3k+1)/2$ if k is odd, $q(x) = \begin{cases} \frac{1}{g(x^{2^s-1})+1} + 1 & \text{if } x \notin \mathbb{F}_2 \\ 1 & \text{otherwise} \end{cases}$ where

$$g(x) = x^{2^{2s+1}+2^{s+1}+1} + x^{2^{2s+1}+2^{s+1}-1} + x^{2^{2s+1}+1} + x^{2^{2s+1}-1} + x.$$

The functions γ_F associated to Kasami, Welch and Niho functions with $m = 7, 9$, are neither in the completed $\mathcal{M}$ class, nor in the completed $\mathcal{PS}_{ap}$ class.

The other known infinite classes of AB functions are quadratic; their associated γ_F belong to the completed Maiorana McFarland class.

- Several new infinite families of bent functions from new permutations and their duals (Mesnager, [43]).

1. Bent functions obtained by selecting Niho bent functions :

– $f(x) = \mathrm{Tr}_1^m(\lambda x^{2^m+1}) + \mathrm{Tr}_1^n(ax)\mathrm{Tr}_1^n(bx)$; $x \in \mathbb{F}_{2^n}$, $n = 2m$, $\lambda \in \mathbb{F}_{2^m}^\star$ and $(a,b) \in \mathbb{F}_{2^n}^\star \times \mathbb{F}_{2^n}^\star$ such that $a \neq b$ and $\mathrm{Tr}_1^n(\lambda^{-1}b^{2^m}a) = 0$.
  $\tilde{f}(x) = \mathrm{Tr}_1^m(\lambda^{-1}x^{2^m+1}) + \left(\mathrm{Tr}_1^m(\lambda^{-1}a^{2^m+1}) + \mathrm{Tr}_1^n(\lambda^{-1}a^{2^m}x)\right) \times \left(\mathrm{Tr}_1^m(\lambda^{-1}b^{2^m+1}) + \mathrm{Tr}_1^n(\lambda^{-1}b^{2^m}x)\right) + 1$.

– $g(x) = \mathrm{Tr}_1^m(x^{2^m+1}) + \mathrm{Tr}_1^n\left(\sum_{i=1}^{2^{r-1}-1} x^{(2^m-1)\frac{i}{2^r}+1}\right) + \mathrm{Tr}_1^n(\lambda x)\mathrm{Tr}_1^n(\mu x)$; $x \in \mathbb{F}_{2^n}$, $n = 2m$, $(\lambda,\mu) \in \mathbb{F}_{2^m}^\star \times \mathbb{F}_{2^m}^\star$ ($\lambda \neq \mu$).
  $\tilde{g}(x) = \mathrm{Tr}_1^m\left((u(1+x+x^{2^m}) + u^{2^{n-r}} + x^{2^m})(1+x+x^{2^m})^{\frac{1}{2^r-1}}\right) \times \mathrm{Tr}_1^m\left((\lambda+\mu)(1+x+x^{2^m})^{\frac{1}{2^r-1}}\right) + \mathrm{Tr}_1^m\left((u(1+x+x^{2^m}) + u^{2^{n-r}} + x^{2^m} + \lambda)(1+x+x^{2^m})^{\frac{1}{2^r-1}}\right)$

$\times \mathrm{Tr}_1^m\Big(\big(u(1+x+x^{2^m})+u^{2^{n-r}}+x^{2^m}+\mu\big)(1+x+x^{2^m})^{\frac{1}{2^r-1}}\Big)$; where $u \in \mathbb{F}_{2^n}$ satisfying $u + u^{2^m} = 1$.

2. Bent functions obtained by selecting bent Boolean functions of Maiorana–McFarland's class :

   – $f(x,y) = \mathrm{Tr}_1^m(a_1y^dx)\mathrm{Tr}_1^m(a_2y^dx) + \mathrm{Tr}_1^m(a_1y^dx)\mathrm{Tr}_1^m(a_3y^dx) + \mathrm{Tr}_1^m(a_2y^dx)\mathrm{Tr}_1^m(a_3y^dx)$; where $(x,y) \in \mathbb{F}_{2^m} \times \mathbb{F}_{2^m}$, d is a positive integer which is not a power of 2 and $gcd(d, 2^m-1) = 1$, a_i's are pairwise distinct such that $b := a_1 + a_2 + a_3 \neq 0$ and $a_1^{-e} + a_2^{-e} + a_3^{-e} = b^{-e}$ where $e = d^{-1}$ (mod $2^m - 1$).
   $\tilde{f}(x,y) = \mathrm{Tr}_1^m(a_1^{-e}x^ey)\mathrm{Tr}_1^m(a_2^{-e}x^ey) + \mathrm{Tr}_1^m(a_1^{-e}x^ey)\mathrm{Tr}_1^m(a_3^{-e}x^ey) + \mathrm{Tr}_1^m(a_2^{-e}x^ey)\mathrm{Tr}_1^m(a_3^{-e}x^ey)$.
   – $g(x,y) = \mathrm{Tr}_1^m(a^{-11}x^{11}y)\mathrm{Tr}_1^m(a^{-11}c^{-11}x^{11}y)$
   $\mathrm{Tr}_1^m(a^{-11}x^{11}y)\mathrm{Tr}_1^m(c^{11}a^{-11}x^{11}y) + \mathrm{Tr}_1^m(a^{-11}c^{-11}x^{11}y)\mathrm{Tr}_1^m(c^{11}a^{-11}x^{11}y)$; where $(x,y) \in \mathbb{F}_{2^m} \times \mathbb{F}_{2^m}$, $a \in \mathbb{F}_{2^n}^{\star}$ with $n = 2m$ is a multiple of 4 but not of 10, $c \in \mathbb{F}_{2^m}$ is such that $c^4 + c + 1 = 0$.
   $\tilde{g}(x,y) = \mathrm{Tr}_1^m(ay^dx)\mathrm{Tr}_1^m(acy^dx) + \mathrm{Tr}_1^m(ay^dx)\mathrm{Tr}_1^m(ac^{-1}y^dx) + \mathrm{Tr}_1^m(acy^dx)\mathrm{Tr}_1^m(ac^{-1}y^dx)$; with $d = 11^{-1}$ (mod $2^n - 1$).
   – $h(x,y) = (\mathrm{Tr}_1^m(a_1y^dx) + g_1(y))(\mathrm{Tr}_1^m(a_2y^dx) + g_2(y)) + (\mathrm{Tr}_1^m(a_1y^dx) + g_1(y))(\mathrm{Tr}_1^m(a_3y^dx) + g_3(y)) + (\mathrm{Tr}_1^m(a_2y^dx) + g_2(y))(\mathrm{Tr}_1^m(a_3y^dx) + g_3(y))$; where $m = 2r$, $gcd(d, 2^m-1) = 1$, a_1, a_2 and a_3 are three pairwise distinct elements of $\mathbb{F}_{2^m}$ such that $b := a_1+a_2+a_3 \neq 0$ and $a_1^{-e}+a_2^{-e}+a_3^{-e} = b^{-e}$ and for $i \in \{1,2,3\}$, $g_i \in \mathcal{D}_m := \{g : \mathbb{F}_{2^m} \to \mathbb{F}_2 \mid g(ax) = g(x), \forall (a,x) \in \mathbb{F}_{2^r} \times \mathbb{F}_{2^m}\}$.
   $\tilde{h}(x,y) = (\mathrm{Tr}_1^m(a_1^{-e}x^ey) + g_1(x^e))(\mathrm{Tr}_1^m(a_2^{-e}x^ey) + g_2(x^e)) + (\mathrm{Tr}_1^m(a_1^{-e}x^ey) + g_1(x^e))(\mathrm{Tr}_1^m(a_3^{-e}x^ey) + g_3(x^e)) + (\mathrm{Tr}_1^m(a_2^{-e}x^ey) + g_2(x^e))(\mathrm{Tr}_1^m(a_3^{-e}x^ey) + g_3(x^e))$ where $e = d^{-1}$ (mod $2^m - 1$).

3. Self-dual bent functions obtained by selecting functions from Maiorana–McFarland completed class[2]:

   – $g(x) = \mathrm{Tr}_1^{4k}(a_1x^{2^k+1})\mathrm{Tr}_1^{4k}(a_2x^{2^k+1}) + \mathrm{Tr}_1^{4k}(a_1x^{2^k+1})\mathrm{Tr}_1^{4k}(a_3x^{2^k+1}) + \mathrm{Tr}_1^{4k}(a_2x^{2^k+1})\mathrm{Tr}_1^{4k}(a_3x^{2^k+1})$; where $x \in \mathbb{F}_{2^{4k}}$, $k \geq 2$, a_1, a_2, a_3 be three pairwise distinct nonzero solutions in $\mathbb{F}_{2^{4k}}$ of the equation $\lambda^{2^{3k}} + \lambda = 1$ such that $a_1 + a_2 + a_3 \neq 0$.

4. Bent functions obtained by selecting functions from PS_{ap} :

   – $f(x,y) = \mathrm{Tr}_1^m(a_1y^{2^m-2}x)\mathrm{Tr}_1^m(a_2y^{2^m-2}x) + \mathrm{Tr}_1^m(a_1y^{2^m-2}x)\mathrm{Tr}_1^m(a_3y^{2^m-2}x) + \mathrm{Tr}_1^m(a_2y^{2^m-2}x)\mathrm{Tr}_1^m(a_3y^{2^m-2}x)$; where $(x,y) \in \mathbb{F}_{2^m} \times \mathbb{F}_{2^m}$, the a_i's are pairwise distinct in $\mathbb{F}_{2^m}$ such that $a_1 + a_2 + a_3 \neq 0$.
   $\tilde{f}(x,y) = f(y,x)$.

[2]The Maiorana–McFarland completed class is the smallest possible complete class containing the class of Maiorana–McFarland which is globally invariant under the action of the general affine group and under the addition of affine functions.

5. Bent functions obtained by combining Niho bent functions and self-dual bent functions:
   - $f(x) = \mathrm{Tr}_1^{2k}(x^{2^{2k}+1}) + \mathrm{Tr}_1^{4k}(ax)\mathrm{Tr}_1^{2k}(x^{2^{2k}+1}) + \mathrm{Tr}_1^{4k}(ax)\mathrm{Tr}_1^{4k}(\lambda_2(x+\beta)^{2^k+1}) + \mathrm{Tr}_1^{4k}(ax)$; where $x \in \mathbb{F}_{2^{4k}}$ $(k \geq 2)$, $\lambda_2 \in \mathbb{F}_{2^{4k}}$ such that $\lambda_2 + \lambda_2^{2^{3k}} = 1$, $a \in \mathbb{F}_{2^{4k}}^\star$ is a solution of $a^{2^{2k}} + \lambda_2^{2^{-k}} a^{2^{-k}} + \lambda_2 a^{2^k} = 0$ and $\beta \in \mathbb{F}_{2^{4k}}$ such that $\mathrm{Tr}_1^{4k}(\beta a) = \mathrm{Tr}_1^{2k}(a^{2^{2k}+1}) + \mathrm{Tr}_1^{4k}(\lambda_2 a^{2^k+1})$.
     $\tilde{f}(x) = \mathrm{Tr}_1^{2k}(x^{2^{2k}+1}) + \left(\mathrm{Tr}_1^{2k}(x^{2^{2k}+1}) + \mathrm{Tr}_1^{4k}(\lambda_2 x^{2^k+1}) + \mathrm{Tr}_1^{4k}(\beta x)\right) \times \left(\mathrm{Tr}_1^{4k}(a^{2^k}x) + \mathrm{Tr}_1^{2k}(a^{2^{2k}+1})\right)$.

- Several new infinite families of bent functions from new permutations and their duals (Mesnager, [37]).
  Permutations satisfying $(\mathcal{A}_m)$ were introduced in [37].

Definition 13.2.1. Let m be a positive integer. Three permutations ϕ_1, ϕ_2 and ϕ_3 of $\mathbb{F}_{2^m}$ are said to satisfy $(\mathcal{A}_m)$ if the two following conditions hold

1. Their sum $\zeta = \phi_1 + \phi_2 + \phi_3$ is a permutation of $\mathbb{F}_{2^m}$.
2. $\zeta^{-1} = \phi_1^{-1} + \phi_2^{-1} + \phi_3^{-1}$.

1. Let m be a positive integer. Let L be a linear permutation on $\mathbb{F}_{2^m}$. Let f be a Boolean function over $\mathbb{F}_{2^m}$ such that $\mathcal{L}_f^0 := \{\alpha \in \mathbb{F}_{2^m} \mid D_\alpha f = 0\}$ is of dimension at least two over $\mathbb{F}_2$. Let $(\alpha_1, \alpha_2, \alpha_3)$ be any 3-tuple of pairwise distinct elements of $\mathcal{L}_f^0$ such that $\alpha_1 + \alpha_2 + \alpha_3 \neq 0$. Then the Boolean function g defined in bivariate representation on $\mathbb{F}_{2^m} \times \mathbb{F}_{2^m}$ by $g(x, y) = \mathrm{Tr}_1^m(xL(y)) + f(y)\Big(\mathrm{Tr}_1^m(L(\alpha_1)x)\mathrm{Tr}_1^m(L(\alpha_2)x)$
   $+ \mathrm{Tr}_1^m(L(\alpha_1)x)\mathrm{Tr}_1^m(L(\alpha_3)x) + \mathrm{Tr}_1^m(L(\alpha_2)x)\mathrm{Tr}_1^m(L(\alpha_3)x)\Big)$ is bent and its dual function $\tilde{g}$ is given by $\tilde{g}(x, y) = \mathrm{Tr}_1^m(L^{-1}(x)y)$
   $+ f(L^{-1}(x))\Big(\mathrm{Tr}_1^m(\alpha_1 y)\mathrm{Tr}_1^m(\alpha_2 y) + \mathrm{Tr}_1^m(\alpha_1 y)\mathrm{Tr}_1^m(\alpha_3 y) + \mathrm{Tr}_1^m(\alpha_2 y)\mathrm{Tr}_1^m(\alpha_3 y)\Big)$.
2. Let $m = 2k$. Let $a \in \mathbb{F}_{2^k}$ and $b \in \mathbb{F}_{2^m}$ such that $b^{2^k+1} \neq a^2$. Set $\alpha = b^{2^k+1} + a^2$ and $\rho = a + b^{2^k}$. Let g_1, g_2 and g_3 be three Boolean functions over $\mathbb{F}_{2^k}$. Then the Boolean function h defined in bivariate representation on $\mathbb{F}_{2^m} \times \mathbb{F}_{2^m}$ by

$$\begin{aligned} h(x, y) = {} & \mathrm{Tr}_1^m(axy + bxy^{2^k}) + \mathrm{Tr}_1^m(xg_1(\mathrm{Tr}_k^m(\rho y)))\mathrm{Tr}_1^m(xg_2(\mathrm{Tr}_k^m(\rho y))) \\ & + \mathrm{Tr}_1^m(xg_1(\mathrm{Tr}_k^m(\rho y)))\mathrm{Tr}_1^m(xg_3(\mathrm{Tr}_k^m(\rho y))) \\ & + \mathrm{Tr}_1^m(xg_2(\mathrm{Tr}_k^m(\rho y)))\mathrm{Tr}_1^m(xg_3(\mathrm{Tr}_k^m(\rho y))) \end{aligned}$$

is bent and its dual function $\tilde{h}$ is given by

$$\begin{aligned} \tilde{h}(x, y) = {} & \mathrm{Tr}_1^m\left(\alpha^{-1}(axy + bx^{2^k}y)\right) \\ & + \mathrm{Tr}_1^m\left(\alpha^{-1}(a+b)yg_1\left(\mathrm{Tr}_k^m(x)\right)\right)\mathrm{Tr}_1^m\left(\alpha^{-1}(a+b)yg_2\left(\mathrm{Tr}_k^m(x)\right)\right) \end{aligned}$$

$$+\mathrm{Tr}_1^m\Big(\alpha^{-1}(a+b)yg_1\left(\mathrm{Tr}_k^m(x)\right)\Big)\mathrm{Tr}_1^m\Big(\alpha^{-1}(a+b)yg_3\left(\mathrm{Tr}_k^m(x)\right)\Big)$$
$$+\mathrm{Tr}_1^m\Big(\alpha^{-1}(a+b)yg_2\left(\mathrm{Tr}_k^m(x)\right)\Big)\mathrm{Tr}_1^m\Big(\alpha^{-1}(a+b)yg_3\left(\mathrm{Tr}_k^m(x)\right)\Big).$$

3. Let n be a multiple of m where m is a positive integer and $n \neq m$. Let ϕ_1, ϕ_2 and ϕ_3 be three permutations over $\mathbb{F}_{2^m}$ satisfying $(\mathcal{A}_m)$. Let (a_1, a_2, a_3) be a 3-tuple of $\mathbb{F}_{2^m}^\star$ such that $a_1 + a_2 + a_3 \neq 0$. Set

$$g(x,y) = \mathrm{Tr}_1^n(x\phi_1(y))\mathrm{Tr}_1^n(x\phi_2(y)) + \mathrm{Tr}_1^n(x\phi_1(y))\mathrm{Tr}_1^n(x\phi_3(y)) + \mathrm{Tr}_1^n(x\phi_2(y))\mathrm{Tr}_1^n(x\phi_3(y))$$

if $(x,y) \in \mathbb{F}_{2^n} \times \mathbb{F}_{2^m}$ and

$$g(x,y) = \mathrm{Tr}_1^n(a_1xy^{2^n-2})\mathrm{Tr}_1^n(a_2xy^{2^n-2}) + \mathrm{Tr}_1^n(a_1xy^{2^n-2})\mathrm{Tr}_1^n(a_3xy^{2^n-2}) + \mathrm{Tr}_1^n(a_2xy^{2^n-2})\mathrm{Tr}_1^n(a_3xy^{2^n-2})$$

if $(x,y) \in \mathbb{F}_{2^n} \times \mathbb{F}_{2^n} \setminus \mathbb{F}_{2^m}$. Then g is bent and its dual function $\tilde{g}$ is defined by

$$\tilde{g}(x,y) = \mathrm{Tr}_1^n(\phi_1^{-1}(x)y)\mathrm{Tr}_1^n(\phi_2^{-1}(x)y) + \mathrm{Tr}_1^n(\phi_1^{-1}(x)y)\mathrm{Tr}_1^n(\phi_3^{-1}(x)y) + \mathrm{Tr}_1^n(\phi_2^{-1}(x)y)\mathrm{Tr}_1^n(\phi_3^{-1}(x)y)$$

if $(x,y) \in \mathbb{F}_{2^m} \times \mathbb{F}_{2^n}$ and

$$\tilde{g}(x,y) = \mathrm{Tr}_1^n(a_1x^{2^n-2}y)\mathrm{Tr}_1^n(a_2x^{2^n-2}y) + \mathrm{Tr}_1^n(a_1x^{2^n-2}y)\mathrm{Tr}_1^n(a_3x^{2^n-2}y) + \mathrm{Tr}_1^n(a_2x^{2^n-2}y)\mathrm{Tr}_1^n(a_3x^{2^n-2}y)$$

if $(x,y) \in \mathbb{F}_{2^n} \setminus \mathbb{F}_{2^m} \times \mathbb{F}_{2^n}$.

4. Let n be a multiple of m where m is a positive integer and n/m. Let ϕ_1, ϕ_2 and ϕ_3 be three permutations over $\mathbb{F}_{2^m}$ satisfying $(\mathcal{A}_m)$. Let $a \in \mathbb{F}_{2^m}^\star$ and $c \in \mathbb{F}_{2^n}$ such that $c^4 + c + 1 = 0$. Let d be the inverse of 11 modulo $2^n - 1$. Set

$$g(x,y) = \mathrm{Tr}_1^n(x\phi_1(y))\mathrm{Tr}_1^n(x\phi_2(y)) + \mathrm{Tr}_1^n(x\phi_1(y))\mathrm{Tr}_1^n(x\phi_3(y)) + \mathrm{Tr}_1^n(x\phi_2(y))\mathrm{Tr}_1^n(x\phi_3(y))$$

if $(x,y) \in \mathbb{F}_{2^n} \times \mathbb{F}_{2^m}$ and

$$g(x,y) = \mathrm{Tr}_1^n(axy^d)\mathrm{Tr}_1^n(acxy^d) + \mathrm{Tr}_1^n(axy^d)\mathrm{Tr}_1^n(ac^{-1}xy^d) + \mathrm{Tr}_1^n(acxy^d)\mathrm{Tr}_1^n(ac^{-1}xy^d)$$

if $(x,y) \in \mathbb{F}_{2^n} \times \mathbb{F}_{2^n} \setminus \mathbb{F}_{2^m}$. Then g is bent and its dual function $\tilde{g}$ is defined by

$$\tilde{g}(x,y) = \mathrm{Tr}_1^n(\phi_1^{-1}(x)y)\mathrm{Tr}_1^n(\phi_2^{-1}(x)y) + \mathrm{Tr}_1^n(\phi_1^{-1}(x)y)\mathrm{Tr}_1^n(\phi_3^{-1}(x)y) + \mathrm{Tr}_1^n(\phi_2^{-1}(x)y)\mathrm{Tr}_1^n(\phi_3^{-1}(x)y)$$

if $(x,y) \in \mathbb{F}_{2^m} \times \mathbb{F}_{2^n}$ and

$$\tilde{g}(x,y) = \mathrm{Tr}_1^n(a^{-11}x^{11}y)\mathrm{Tr}_1^n(a^{-11}c^{-11}x^{11}y) + \mathrm{Tr}_1^n(a^{-11}x^{11}y)\mathrm{Tr}_1^n(a^{-11}c^{11}x^{11}y) + \mathrm{Tr}_1^n(a^{-11}c^{-11}x^{11}y)\mathrm{Tr}_1^n(a^{-11}c^{11}x^{11}y)$$

if $(x,y) \in \mathbb{F}_{2^n} \setminus \mathbb{F}_{2^m} \times \mathbb{F}_{2^n}$.

5. Let n be a multiple of m where m is a positive integer and n/m. Let ϕ_1, ϕ_2 and ϕ_3 be three permutations over $\mathbb{F}_{2^m}$ satisfying $(\mathcal{A}_m)$. Let $\alpha \in \mathbb{F}_{2^m}^\star$. Let d be a positive integer such that d and $2^n - 1$ are coprime. Denote by e the inverse of d modulo $2^n - 1$. Set

$$g(x,y) = \mathrm{Tr}_1^n(x\phi_1(y))\mathrm{Tr}_1^n(x\phi_2(y)) + \mathrm{Tr}_1^n(x\phi_1(y))\mathrm{Tr}_1^n(x\phi_3(y)) + \mathrm{Tr}_1^n(x\phi_2(y))\mathrm{Tr}_1^n(x\phi_3(y))$$

if $(x,y) \in \mathbb{F}_{2^n} \times \mathbb{F}_{2^m}$ and $g(x,y) = \mathrm{Tr}_1^n(\alpha x y^d)$ if $(x,y) \in \mathbb{F}_{2^n} \times \mathbb{F}_{2^n} \setminus \mathbb{F}_{2^m}$. Then g is bent and its dual function $\tilde{g}$ is defined by

$$\tilde{g}(x,y) = \mathrm{Tr}_1^n(\phi_1^{-1}(x)y)\mathrm{Tr}_1^n(\phi_2^{-1}(x)y) + \mathrm{Tr}_1^n(\phi_1^{-1}(x)y)\mathrm{Tr}_1^n(\phi_3^{-1}(x)y) + \mathrm{Tr}_1^n(\phi_2^{-1}(x)y)\mathrm{Tr}_1^n(\phi_3^{-1}(x)y)$$

if $(x,y) \in \mathbb{F}_{2^m} \times \mathbb{F}_{2^n}$ and $\tilde{g}(x,y) = \mathrm{Tr}_1^n(\alpha^{-e}x^e y)$ if $(x,y) \in \mathbb{F}_{2^n} \setminus \mathbb{F}_{2^m} \times \mathbb{F}_{2^n}$.

6. Let $n = 2m$ where m is a positive integer. Let ϕ_1, ϕ_2 and ϕ_3 be three permutations over $\mathbb{F}_{2^m}$ satisfying $(\mathcal{A}_m)$. Let d be a positive integer such that $d+1$ and $2^n - 1$ are coprime. Let $\lambda \in \mathbb{F}_{2^m}^\star$. Set

$$g(x,y) = \mathrm{Tr}_1^n(x\phi_1(y))\mathrm{Tr}_1^n(x\phi_2(y)) + \mathrm{Tr}_1^n(x\phi_1(y))\mathrm{Tr}_1^n(x\phi_3(y)) + \mathrm{Tr}_1^n(x\phi_2(y))\mathrm{Tr}_1^n(x\phi_3(y))$$

if $(x,y) \in \mathbb{F}_{2^n} \times \mathbb{F}_{2^m}$ and $g(x,y) = \mathrm{Tr}_1^n\left(\lambda xy\left(\mathrm{Tr}_m^n(y)\right)^d\right)$ if $(x,y) \in \mathbb{F}_{2^n} \times \mathbb{F}_{2^n} \setminus \mathbb{F}_{2^m}$. Then g is bent and its dual function $\tilde{g}$ is defined by

$$\tilde{g}(x,y) = \mathrm{Tr}_1^n(\phi_1^{-1}(x)y)\mathrm{Tr}_1^n(\phi_2^{-1}(x)y) + \mathrm{Tr}_1^n(\phi_1^{-1}(x)y)\mathrm{Tr}_1^n(\phi_3^{-1}(x)y) + \mathrm{Tr}_1^n(\phi_2^{-1}(x)y)\mathrm{Tr}_1^n(\phi_3^{-1}(x)y)$$

if $(x,y) \in \mathbb{F}_{2^m} \times \mathbb{F}_{2^n}$ and $\tilde{g}(x,y) = \mathrm{Tr}_1^n\left(\lambda^{-\frac{1}{d+1}}x\left(\mathrm{Tr}_m^n(x)\right)^{-\frac{d}{d+1}}y\right)$ if $(x,y) \in \mathbb{F}_{2^n} \setminus \mathbb{F}_{2^m} \times \mathbb{F}_{2^n}$.

7. Bent functions from involutions (see [44], [45]).

13.2.2 Constructions of Bent Functions in Odd Characteristic

We discuss in this subsection primary constructions of p-ary bent functions in univariate form.

Let F be a mapping from $\mathbb{F}_{p^n}$ to itself. F is said to be *planar* if for any nonzero $b \in \mathbb{F}_{p^n}$ the mapping $F(x+b) - F(x)$ is bijective on $\mathbb{F}_{p^n}$. We know only one example of a nonquadratic planar function called *Coulter–Matthews function* [14] which is defined over $\mathbb{F}_{3^n}$ by $F(x) = x^{\frac{3^k+1}{2}}$, with $gcd(k,n) = 1$ and k odd. Except this class, all the other known *planar functions* are quadratic (see [5]) which means that they can be represented by the so-called *Dembowski–Ostrom polynomials* (see [14]). In [8], it has been proven that every planar function gives a family of p-ary bent functions. More precisely, a mapping F from $\mathbb{F}_{p^n}$ to itself is planar if and only if, for every nonzero $a \in \mathbb{F}_{p^n}$ the function $tr_n(aF)$ is bent. The bent functions coming from the Coulter–Matthews planar functions and the (quadratic) p-ary bent functions $tr_n(aF)$ obtained from Dembowski–Ostrom polynomials are weakly regular bent (see [19, 49]).

In 2006, Helleseth and Kholosha [24] have exhibited a p-ary family of bent functions defined as follows: let f be the p-ary function from $\mathbb{F}_{p^n}$ to $\mathbb{F}_p$ defined as $f(x) = tr_n(ax^{r(p^m-1)})$ with $n = 2m$, $r > 0$ an arbitrary integer with $gcd(r, p^m+1) = 1$ for odd prime p such that $p^m > 3$, $a \in \mathbb{F}_{p^n} \setminus \{0\}$. Then, it is proven that f is bent if and only if, $K_n^{(p)}(a^{p^m+1}) = -1$ where $K_n^{(p)}$ denotes the Kloosterman sum defined over $\mathbb{F}_{p^n}$ by $K_n^{(p)}(a) = \sum_{c \in \mathbb{F}_{p^n} \setminus \{0\}} \zeta_p^{tr_n(c+bc^{-1})}$. Moreover, if the condition holds then f is a regular bent function. If we take $p = 2$ and without loss of generality assume $a \in \mathbb{F}_{2^m}$, then we have exactly the same result as above for any m in the binary case giving the so-called Dillon class of bent functions [11, 15, 32]. For more details about the monomial f, we send the reader to [24].

A ternary weakly regular bent function has been isolated and studied by several authors [20, 23, 24]. Such a function f is defined from $\mathbb{F}_{3^n}$ to $\mathbb{F}_3$ (where $n = 2m$ with m odd) by $f(x) = tr_n(ax^{\frac{3^n-1}{4}+3^m+1})$. The corresponding Walsh transform coefficients have been given.

In 2010, Helleseth and Kholosha [25] discovered another class of bent functions consisting of two terms (so called binomial functions). This class is the only infinite class of nonquadratic p-ary functions, in a univariate representation over fields of arbitrary odd characteristic, that has been proven to be bent. A function of their class is defined as follows: let f be the p-ary function from $\mathbb{F}_{p^n}$ to $\mathbb{F}_p$ defined as $f(x) = tr_n(x^{p^{3k}+p^{2k}-p^k+1} + x^2)$ for $n = 4k$. Helleseth and Kholosha have proved that f is weakly regular bent and found the exact value of its Walsh transform coefficients. Their proof is based on a few new results in the area of exponential sums and polynomials over finite fields that may also be interesting as independent problems.

In Table 13.1 we present all known univariate polynomials representing infinite classes of p-ary bent functions. We shall denote ζ a primitive element of $\mathbb{F}_{3^n}$ and the letters “R” and “WR” refer to regular and weakly regular bent functions respectively.

Table 13.1 Nonquadratic *p*-ary bent functions

n	d or $F(x)$	a	deg	Comments
	$\frac{3^k+1}{2}$, $gcd(k,n)=1$, k odd	$a \neq 0$	$k+1$	tern, R, WR
$2m$	$r(3^m-1)$, $gcd(r, 3^m+1)=1$	$K_n^{(p)}(a^{3m+1})=0$	n	tren, R
$2m$	$\frac{3^n-1}{4}+3^m+1$, m odd	$\zeta^{\frac{3^m+1}{4}}$	n	tern, WR
$4k$	$x^{p^{3k}+p^{2k}-p^k+1}+x^2$		$(p-1)k+2$	WR

The first three families in the table are monomials of the form $tr_n(ax^d)$ while the last one is a binomial bent function in the form $tr_n(F(x))$. When the value of n is not specified in the table it means that n is arbitrary. Obviously, all the exponents d and coefficients a can be replaced with their cyclotomic equivalents. Moreover, the numerous examples of binomial ternary bent functions consisting of two Dillon type exponents are skipped. Note that it was proved by Budaghyan et al. [4] that all bent functions in Table 13.1 possibly except for Dillon type, do not belong to the completed Maiorana–McFarland class. We also mention that for a long time many researchers believed that all p-ary bent functions are weakly regular. However, some counter examples were found. In particular, ternary function f from $\mathbb{F}_{3^6}$ to $\mathbb{F}_3$ given by $f(x) = tr_6(\zeta^7 x^{98})$ where ζ is a primitive element of $\mathbb{F}_{3^6}$ is bent and not weakly regular bent (see [26]). Investigating an infinite class of non-weakly regular bent functions in a univariate representation is an interesting open problem.

In 2013, several new classes of binary and p-ary regular bent functions (including binomials, trinomials, and functions with multiple trace terms) have been given by Li, Helleseth, Tang, and Kholosha [34]. The bentness of all their functions has been determined by some exponential sums over finite fields, most of which have close relations with the well-known Kloosterman sums. In 2014, Li, Tang and Helleseth [35] have proposed new quadratic bent functions in polynomial form. Their constructions give rise to new Boolean bent, generalized Boolean bent and p-ary bent functions. Based on $\mathbb{Z}_4$ valued quadratic forms, a simple method providing several new constructions of generalized Boolean bent functions is given. From these generalized Boolean bent functions, Li et al. have also presented a method to transform them into Boolean bent and semi-bent functions. Moreover, many new p-ary bent functions can also be obtained by applying similar methods.

The author [39, 41] has exhibited in 2009 the first infinite family of binomial hyperbent functions obtained by adding functions defined on a subfield of $\mathbb{F}_{2^n}$ to the generalized Dillon's functions. Such functions are of the shape

$$f_{a,b}^{(r)}(x) = \mathrm{Tr}_1^n\left(ax^{r(2^m-1)}\right) + \mathrm{Tr}_1^2\left(bx^{\frac{2^n-1}{3}}\right),$$

where $a \in \mathbb{F}_{2^n}^\star$, $b \in \mathbb{F}_4^\star$ and $n = 2m$, m odd, $K_n(a^{2^m+1}) = 4$. In 2011, Zheng, Zeng, and Hu [50], have extended the bent functions $f_{a,b}^{(1)}$ in characteristic p. The authors have more precisely studied the bentness of the p-ary binomial function of the form

$$f(x) = tr_n(ax^{p^m-1}) + tr_2(bx^{\frac{p^n-1}{4}}),$$

where $n = 2m$, m odd and p prime such that $p \equiv 3 \pmod 4$. Necessary and sufficient conditions of f being bent by means of an exponential sum and two sequences related to a and b, respectively are given. In the ternary case, the authors have characterized the bentness of f $(p = 3)$ in terms of the Hamming weight of a sequence.

References

1. L. Budaghyan and C. Carlet. On CCZ-equivalence and its use in secondary constructions of bent functions. In *Post-proceedings of WCC 2009, Des. Codes Cryptography 59(1–3)*, pages 69–87, 2011.
2. L. Budaghyan, C. Carlet, and T. Helleseth. On bent functions associated to AB functions. In *Proceedings of IEEE Information Theory Workshop, ITW'11, Paraty, Brazil*, 2011.
3. L. Budaghyan, C. Carlet, T. Helleseth, A. Kholosha, and S. Mesnager. Further results on Niho bent functions. In *IEEE Transactions on Information Theory-IT, Vol 58, no.11*, pages 6979–6985, 2012.
4. L. Budaghyan, C. Carlet, T. Helleseth, and A. Kholosha. Generalized bent functions and their relation to Maiorana-McFarland class. In *Proceedings of the 2012 IEEE International Symposium on Information Theory (ISIT), pages 1217–1220*, 2012.
5. L. Budaghyan and T. Helleseth. New commutative semifields by new PN multinomials. In *Cryptography and Communications, 3(1), pages 1–16*, 2011.
6. A. Canteaut, P. Charpin, and G. Kyureghyan. A New Class of Monomial Bent Functions. In *Finite Fields and Their Applications, Vol 14, no. 1*, pages 221–241, 2008.
7. C. Carlet. Boolean functions for cryptography and error correcting codes. In Yves Crama and Peter L. Hammer, editors, *Boolean Models and Methods in Mathematics, Computer Science, and Engineering*, pages 257–397. Cambridge University Press, June 2010.
8. C. Carlet and S. Dubuc. On generalized bent and q-ary perfect nonlinear functions. In *In Proceedings of the Fifth International Conference "Finite Fields and Applications" pages 81–94*, 2001.
9. A. Çeşmelioğlu, W. Meidl, A. Pott. On the dual of (non)-weakly regular bent functions and self-dual bent functions. Adv. Math. Commun., 7, pp. 425–440, 2013.
10. A. Çeşmelioğlu, W. Meidl, A. Pott. There are infinitely many bent functions for which the dual is not bent. Preprint 2015.
11. P. Charpin and G. Gong. Hyperbent functions, Kloosterman sums and Dickson polynomials. In *IEEE Trans. Inform. Theory (54) 9*, pages 4230–4238, 2008.
12. P. Charpin and G. Kyureghyan. Cubic monomial bent functions: A subclass of $\mathcal{M}$. In *SIAM, J. Discr. Math., Vol.22, no.2*, pages 650–665, 2008.
13. P. Charpin, E. Pasalic, and C. Tavernier. On bent and semi-bent quadratic Boolean functions. In *IEEE Transactions on Information Theory, vol. 51, no. 12*, pages 4286–4298, 2005.
14. R. S. Coulter and R. W. Matthews. Planar functions and planes of Lenz-Barlotti class II. In *Des. Codes Cryptogr., 10 (2), pages 167–184*, 1997.
15. J. Dillon. Elementary Hadamard difference sets. In *PhD dissertation, University of Maryland.*
16. J. F. Dillon and H. Dobbertin. New cyclic difference sets with Singer parameters. In *Finite Fields and Their Applications Volume 10, Issue 3*, pages 342–389, 2004.
17. J. F. Dillon and G. McGuire. Near bent functions on a hyperplane. In *Finite Fields and Their Applications Vol. 14, Issue 3*, pages 715–720, 2008.
18. H. Dobbertin, G. Leander, A. Canteaut, C. Carlet, P. Felke, and P. Gaborit. Construction of bent functions via Niho Power Functions. In *Journal of Combinatorial therory, Serie A 113*, pages 779–798, 2006.

19. K. Feng and J. Luo. Value distribution of exponential sums from perfect nonlinear functions and their applications. In *IEEE Trans. Inf. Theory, vol. 53, no. 9*, pages 3035–3041, 2007.
20. G.Gong, T. Helleseth, H. Hu, and A. Kholosha. On the dual of certain ternary weakly regular bent functions. In *IEEE Transactions on Information Theory, 58(4), pages 2237–2243*, 2012.
21. F. Gologlu. Almost Bent and Almost Perfect Nonlinear Functions, Exponential Sums, Geometries and Sequences. In *PhD dissertation, University of Magdeburg*, 2009.
22. P. Guillot. Completed GPS covers all bent functions. In *Journal of Combinatorial Theory, Series A 93, pages 242–260*, 2001.
23. T. Helleseth, H. D. L. Hollmann, A. Kholosha, Z. Wang, and Q. Xiang. Proofs of two conjectures on ternary weakly regular bent functions. In *IEEE Transactions on Information Theory, 55 (11), pages 5272–5283*, 2009.
24. T. Helleseth and A. Kholosha. Monomial and quadratic bent functions over the finite fields of odd characteristic. In *IEEE Transactions on Information Theory, 52(5), pages 2018–2032*, 2006.
25. T. Helleseth and A. Kholosha. New binomial bent functions over the finite fields of odd characteristic. In *IEEE Transactions on Information Theory, 56(9), pages 4646–4652*, 2010.
26. T. Helleseth and A. Kholosha. On generalized bent functions. In *Information Theory and Applications Workshop (ITA), pages 1–6*, 2010.
27. T. Helleseth and A. Kholosha. Private communication. 2013.
28. T. Helleseth, A. Kholosha, and S. Mesnager. Niho Bent Functions and Subiaco/Adelaide Hyperovals. In *Proceedings of the 10-th International Conference on Finite Fields and Their Applications (Fq'10), Contemporary Math., AMS. Vol 579*, pages 91–101, 2012.
29. H. Hu and D. Feng. On quadratic bent functions in polynomial forms. In *IEEE Trans. Inform. Theory 53 (7)*, pages 2610–2615, 2007.
30. K. Khoo, G. Gong, and D. R. Stinson. A new characterization of semibent and bent functions on finite fields. In *Des. Codes. Cryptogr. vol. 38, no. 2*, pages 279–295, 2006.
31. P. V. Kumar, R. A. Scholtz, and L.R. Welch. Generalized bent functions. In *J. Combin. Theory. Ser. A. Vol. 40, No 1, pages 90–107*, 1985.
32. G. Leander. Monomial Bent Functions. In *IEEE Trans. Inform. Theory (52) 2*, pages 738–743, 2006.
33. G. Leander and A. Kholosha. Bent functions with 2^r Niho exponents. In *IEEE Trans. Inform. Theory 52 (12)*, pages 5529–5532, 2006.
34. N. Li, T. Helleseth, X. Tang, and A. Kholosha. Several New Classes of Bent Functions From Dillon Exponents. In *IEEE Transactions on Information Theory 59 (3), pages 1818–1831*, 2013.
35. N. Li, X. Tang, and T. Helleseth. New constructions of quadratic bent functions in polynomial form. In *IEEE Transactions on Information Theory 60 (9), pages 5760–5767*, 2014.
36. W. Ma, M. Lee, and F. Zhang. A new class of bent functions. In *EICE Trans. Fundamentals, vol. E88-A, no. 7, pages 2039–2040*, 2005.
37. S. Mesnager. Further constructions of infinite families of bent functions from new permutations and their duals. In *Journal Cryptography and Communications (CCDS), Springer. To appear.*
38. S. Mesnager. Hyper-bent Boolean functions with multiple trace terms. In *Proceedings of International Workshop on the Arithmetic of Finite Fields. M.A. Hasan and T.Helleseth (Eds.): WAIFI 2010, LNCS 6087, pp. 97–113. Springer, Heidelberg (2010).*
39. S. Mesnager. A new class of bent Boolean functions in polynomial forms. In *Proceedings of international Workshop on Coding and Cryptography, WCC 2009*, pages 5–18, 2009.
40. S. Mesnager. A new family of hyper-bent Boolean functions in polynomial form. In *Proceedings of Twelfth International Conference on Cryptography and Coding, Cirencester, United Kingdom. M. G. Parker (Ed.): IMACC 2009, LNCS 5921, Springer, Heidelberg*, pages 402–417, 2009.
41. S. Mesnager. A new class of bent and hyper-bent Boolean functions in polynomial forms. In *journal Design, Codes and Cryptography, 59(1–3)*, pages 265–279, 2011.
42. S. Mesnager. Bent and hyper-bent functions in polynomial form and their link with some exponential sums and Dickson polynomials. *IEEE Transactions on Information Theory*, 57(9):5996–6009, 2011.

43. S. Mesnager. Several new infinite families of bent functions and their duals. In *IEEE Transactions on Information Theory-IT, Vol. 60, No. 7, Pages 4397–4407*, 2014.
44. S. Mesnager. On constructions of bent functions from involutions. *Proceedings of 2016 IEEE International Symposium on Information Theory*, ISIT 2016, 2016.
45. Mesnager, S., Cohen.G., and Madore. D.: On existence (based on an arithmetical problem) and constructions of bent functions. Proceedings of the fifteenth International Conference on Cryptography and Coding, Oxford, United Kingdom, IMACC 2015, LNCS, Springer, Heidelberg, pages 3–19, 2015.
46. S. Mesnager and J-P Flori. On hyper-bent functions via Dillon-like exponents. In *IEEE International Symposium on Information Theory, IMT, Cambridge, MA, USA, July 1–6*, pages 1758–1762, 2012.
47. S. Mesnager and J-P Flori. Hyper-bent functions via Dillon-like exponents. In *IEEE Transactions on Information Theory-IT. Vol. 59 No. 5*, pages 3215–3232, 2013.
48. N. Y. Yu and G. Gong. Construction of quadratic bent functions in polynomial forms. In *IEEE Trans. Inform. Theory, Vol 7, No. 52, pages 3291–3299*, 2006.
49. J. Yuan, C. Carlet, and C. Ding. The weight distribution of a class of linear codes from perfect nonlinear functions. In *IEEE Transactions on Information Theory 52 (2), pages 712–717*, 2006.
50. D. Zheng, X. Zeng, and L. Hu. A family of p-ary binomial bent functions. In *EICE Trans. Fundamentals, Vol. 94 A, No. 9, pp. 1868–1872*, 2011.

Chapter 14
Bent Functions and (Partial-)spreads

14.1 Spread and Partial Spreads

14.1.1 Introduction to Spread and Partial Spreads

Partial spreads and spreads play an important role in some constructions of bent functions.

Definition 14.1.1. For a group G of order M^2, a partial spread is a family $S = \{H_1, H_2, \cdots, H_N\}$ of subgroups of order M which satisfy $H_i \cap H_j = \{0\}$ for all $i \neq j$.

Definition 14.1.2. With the previous notation, if $N = M + 1$ (which implies $\cup_{i=1}^{M+1} H_i = G$) then S is called a spread.

- The subgroups of a spread are called the *spread elements* of the spread.

Definition 14.1.3. Let $n = 2m$ be an even integer. A partial spread of $\mathbb{F}_{p^n}$ is a set of pairwise supplementary m-dimensional subspaces of $\mathbb{F}_{p^n}$. A partial spread is a spread if the union of its elements equals $\mathbb{F}_{p^n}$

Hence a collection $\{E_1, \cdots, E_s\}$ of $\mathbb{F}_{p^n}$ is a partial spread of $\mathbb{F}_{p^n}$ ($n = 2m$) if

1. $E_i \cap E_j = \{0\}$ for $i \neq j$;
2. $dim_{\mathbb{F}_p} E_i = m$, $\forall i \in \{1, \cdots, s\}$;

It is a spread if, in addition to the two previous properties, it holds

3. $\bigcup_{i=1}^{s} E_i = \mathbb{F}_{p^n}$.

S. Mesnager, *Bent Functions*, DOI 10.1007/978-3-319-32595-8_14

14.1.2 Some Classical Examples of Spreads

We start with the most classical spread which is the so-called the *Desarguesian spread* but other full or partial spreads exist (see [11, 17]).

1. The Desarguesian spread:

   In characteristic $p \geq 2$, one consider the additive group $(\mathbb{F}_{p^n}, +)$ of the finite field $\mathbb{F}_{p^n}$ with $n = 2m$. Then two representations of the Desarguesian spread are

   - in $\mathbb{F}_{p^n}$ (univariate form): $\{u\mathbb{F}_{p^n}, u \in U\}$ where $U := \{u \in \mathbb{F}_{p^n} \mid u^{p^m+1} = 1\}$
   - in $\mathbb{F}_{p^n} \approx \mathbb{F}_{p^m} \times \mathbb{F}_{p^m}$ (bivariate form): $\{E_a, a \in \mathbb{F}_{p^m}\} \cup \{E_\infty\}$ where $E_a := \{(x, ax)\,;\, x \in \mathbb{F}_{p^m}\}$ and $E_\infty := \{(0, y)\,;\, y \in \mathbb{F}_{p^m}\} = \{0\} \times \mathbb{F}_{p^m}$.

2. André's spreads:

   One example which generalizes the Desarguesian spread has been introduced by J. André in the fifties and independently by Bruck later. André's spreads are defined as follows:
   Let k be a divisor of m. Let $N_k{}^m$ be the norm map : $N_k{}^m$: $\mathbb{F}_{2^m} \to \mathbb{F}_{2^k}$, $x \mapsto N_k{}^m(x) := x^{(\frac{2^m-1}{2^k-1})}$.
   Let Φ be any function $\Phi : \mathbb{F}_{2^k} \to \mathbb{Z}/(m/k)\mathbb{Z}$, $\phi = \Phi \circ N_k{}^m$. Then, the families ($\mathbb{F}_2$-vector spaces):

   $$\{\{(x, x^{2^{k\phi(z)}} z), x \in \mathbb{F}_{2^m}\}, z \in \mathbb{F}_{2^m}\}$$

   and

   $$\{(0, y), y \in \mathbb{F}_{2^m}\}$$

   form together a spread of $\mathbb{F}_{2^m} \times \mathbb{F}_{2^m}$.

   Indeed, these subspaces have trivial pairwise intersection: suppose that $x^{2^{k\phi(y)}} y = x^{2^{k\phi(z)}} z$ for some nonzero elements x, y,z of $\mathbb{F}_{2^m}$, then we have $N_k{}^m(x^{2^{k\phi(y)}} y) = N_k{}^m(x^{2^{k\phi(z)}} z)$, that is, $N_k{}^m(x^{2^{k\phi(y)}})N_k{}^m(y) = N_k{}^m(x^{2^{k\phi(z)}})N_k{}^m(z)$; equivalently, $N_k{}^m(x)N_k{}^m(y) = N_k{}^m(x)N_k{}^m(z)$ (since $x \mapsto x^{2^{k\phi(z)}}$ is in the Galois group of $\mathbb{F}_{2^m}{}^2$ over $\mathbb{F}_{2^k}$), implying that $N_k{}^m(y) = N_k{}^m(z)$ and $\phi(y) = \phi(z)$, which together with $x^{2^{k\phi(y)}} y = x^{2^{k\phi(z)}} z$ implies then $y = z$.

   Note that ϕ can be any function from $\mathbb{F}_{2^m}$ to $\mathbb{Z}/(m/k)\mathbb{Z}$ which is constant on any coset of the subgroup U of $\mathbb{F}_{2^m}^\star$ of order $\frac{2^m-1}{2^k-1}$.

3. Other spreads from quasifields and semifields:
   Other spreads can be constructed from *quasifields* and *semifields*. We start by defining those algebraic objects.

   Let $q = p^n$. Let $\mathbb{F}_q$ be the Galois field with q elements. Denote $+$ the addition and endow $\mathbb{F}_q$ with a *magma* that we denote $\star$.

Definition 14.1.4. $(\mathbb{F}_q^\star, \star)$ is a quasigroup if for every a and b in $\mathbb{F}_q^\star$, there exist unique elements x and y in $\mathbb{F}_q^\star$ such that

- $a \star x = b$,
- $y \star a = b$.

Definition 14.1.5. $(\mathbb{F}_q^\star, \star)$ is a loop if it is a quasigroup and there exist an identity element e such that $e \star a = a \star e = a$ for every $a \in \mathbb{F}_q$.

Definition 14.1.6. $(\mathbb{F}_q, +, \star)$ is a left quasifield (*resp.* left prequasifield) if

(a) $(\mathbb{F}_q, +)$ is an abelian group
(b) $(\mathbb{F}_q^\star, \star)$ is a loop (*resp.* quasigroup)
(c) For every a, b and c in $\mathbb{F}_q$, $a \star (b + c) = a \star b + a \star c$ (left distributivity)
(d) $0 \star x = 0$ for every $x \in \mathbb{F}_q$.

In a right quasifield, we have the right distributivity instead of the left distributivity.

Definition 14.1.7. $(\mathbb{F}_q, +, \star)$ is a semifield (*resp.* presemifield) if it is both a left and right quasifield (*resp.* prequasifield).

Definition 14.1.8. Two presemifileds $(\mathbb{F}_q, +, \star)$ and $(\mathbb{F}_q, +, \circ)$ are called *isotopic* if $M(x) \star N(y) = L(x \circ y)$, $\forall x, y \in \mathbb{F}_q$, for some linearized polynomials M, N and L.

Every presemifield is isotopic to a semifield.

We define a $\mathbb{F}_2$-bilinear form $B : \mathbb{F}_{2^m} \times \mathbb{F}_{2^m} \to \mathbb{F}_2$ by $B(x, y) = \mathrm{Tr}_1^m(xy)$, and an alternating form on $(\mathbb{F}_{2^m} \times \mathbb{F}_{2^m}) \times (\mathbb{F}_{2^m} \times \mathbb{F}_{2^m})$ by

$$\langle (x, y), (x', y') \rangle = B(x, y') - B(y, x').$$

Let $S = (\mathbb{F}_{2^m}, +, \star)$ be a presemifield with respect to operation $\star$. *Dual presemifield* $S^d = (\mathbb{F}_{2^m} +, \star)$ is defined by operation

$$x \star y = y \star x.$$

The classification of finite fields has been done Moore in 1893 but the commutative finite semifields are not classified.

Note that there exists a correspondence between presemifields and the translation planes established by Lenz–Barlotti (see https://www.math.uni-kiel.de/geometrie/klein/math/geometry/barlotti.html). An excellent reference on (partial-) spreads and related topics (such as Finite Geometries [12]) can be found in is the book [18].

Below, we present some classical prequasifields.

(a) Dempwolff–Muller prequasifield : assume k and m are odd integers with $gcd(k, m) = 1$. Let $e = 2^{m-1} - 2^{k-1} - 1$, $L(x) = \sum_{i=0}^{k-1} x^{2i}$ and define a multiplication in $\mathbb{F}_{2^m}$ as

$$x \odot y = x^e L(xy).$$

Endowed by the operation defined above, $(\mathbb{F}_{2^m}, +, \odot)$ becomes a prequasifield, called the Dempwolff–Muller prequasifield [13].

(b) The Knuth prequasifield: Assume m is odd. For any $\beta \in \mathbb{F}_{2^m}^\star$, define a multiplication in $\mathbb{F}_{2^m}$ by

$$x \odot y = xy + x^2 \mathrm{Tr}_1^m(\beta y) + y^2 \mathrm{Tr}_1^m(\beta x).$$

Endowed by the operation defined above, $(\mathbb{F}_{2^m}, +, \odot)$ becomes a prequasifield, called the Knuth presemifiled [24].

(c) The Kantor presemifiled: Assume m is odd. Define a multiplication in $\mathbb{F}_{2^m}$ by

$$x \odot y = x^2 y + \mathrm{Tr}_1^m(xy) + x\mathrm{Tr}_1^m(y).$$

Endowed by the operation defined above, $(\mathbb{F}_{2^m}, +, \odot)$ becomes a presemifiled, called the Kantor presemifiled [19].

One can construct spreads via the quasifields and semifields. Kantor (see for instance [20] and the references therein) has shown how a spread can be derived from any prequasifield. Let A_0 be $m \times m$ zero matrix and A_s ($s \in \mathbb{F}_p^m \setminus \{0\}$), be invertible $m \times m$-matrices over $\mathbb{F}_p$ such that $A_r - A_s$is invertible whenever $r \neq s$. Such matrices induce a binary operation $\star$ on $\mathbb{F}_p^m$ defined by: $x \star s = xA_s$. Thus, $(\mathbb{F}_p^m, +, \star)$ is a right prequasifield (but the left distributive law in general does not hold).
The subsets

$$\{\{(x, x \star z), x \in \mathbb{F}_p^m\}, z \in \mathbb{F}_p^m\}$$

and

$$\{(0, y), y \in \mathbb{F}_p^m\}$$

form together a spread of $\mathbb{F}_p^n$ where $n = 2m$.

In the following, we present a known general construction of spread. Let consider a spread of $\mathbb{F}_{2^n}$ (where $n = 2m$) whose 2^m+1 elements are the subspaces

$$\{(x, L_z(x)), x \in \mathbb{F}_{2^m}\}, z \in \mathbb{F}_{2^m}$$

and

$$\{(0, y), y \in \mathbb{F}_{2^m}\}$$

where $x \mapsto L_z(x)$ is linear

The property of being a spread corresponds to the fact that for all $x \in \mathbb{F}_{2^m}^\star$ the mapping $z \mapsto L_z(x)$ is a permutation on $\mathbb{F}_{2^m}$.

Many examples of spreads based of semifields and quasifields are constructed in this manner :
$(\mathbb{F}_{2^m}, +, \star)$, where $\star$ is defined by $x \star z = L_z(x)$.

Let k be a divisor of m. Let ${N_k}^m$ be the norm map : ${N_k}^m$: $\mathbb{F}_{2^m} \to \mathbb{F}_{2^k}$, $x \mapsto {N_k}^m(x) := x^{(\frac{2^m-1}{2^k-1})}$.
Let Φ be any function $\Phi : \mathbb{F}_{2^k} \to \mathbb{Z}/(m/k)\mathbb{Z}$, $\phi = \Phi \circ {N_k}^m$.

For André's spreads, take $(\mathbb{F}_{2^m}, +, \star)$, with $x \star z = L_z(x) := x^{2^{k\phi(x)}}$

To a presemifield $S = (\mathbb{F}_{2^m}, +, \star)$, one can associate a spread, that is, a collection of subspaces $\{(0, y) \mid y \in \mathbb{F}_{2^m}\}$ and $\{(x, x \star z) \mid m \in \mathbb{F}_{2^m}\}$, $z \in F$. Transpose presemifield $S^t = (\mathbb{F}_{2^m}, +, \circ)$ of the presemifield S is defined as a presemifield whose associated spread is orthogonal (dual) to the spread of S with respect to the alternating form $\langle \cdot, \cdot \rangle$, that is,

$$\langle (x, x \star z), (y, y \circ z) \rangle = 0$$

for any $x, y, z \in \mathbb{F}_{2^m}$. It is equivalent to

$$B(x, y \circ z) = B(x \star z, y).$$

Definition 14.1.9. A presemifield is called symplectic, if its associated spread is symplectic (that is, every subspace from spread is isotropic with respect to the alternating form $\langle \cdot, \cdot \rangle$). This means

$$0 = \langle (x, x \circ z), (y, y \circ z) \rangle = B(x, y \circ z) - B(x \circ z, y)$$

for any $x, y, z \in \mathbb{F}_{2^m}$. Equivalently,

$$B(x, y \circ z) = B(x \circ z, y) \tag{14.1}$$

for any $x, y, z \in \mathbb{F}_{2^m}$.

Using operations S^d and S^t one can get at most 6 isotopy classes of presemifields, which is called the Knuth [25, 26] orbit $\mathcal{K}(S)$ of the presemifield S:

$$\mathcal{K}(S) = \{[S], [S^d], [S^t], [S^{dt}], [S^{td}], [S^{dtd}] = [S^{tdt}]\}.$$

Definition 14.1.10. A presemifield $S = (\mathbb{F}_{2^m}, +, \star)$ is called commutative, if the operation $\star$ of multiplication is commutative. A presemifield S is commutative if and only if $S = S^d$, and a presemifield S is symplectic if and only if $S = S^t$.

Therefore, Knuth orbit of a commutative (symplectic) presemifield contains at most three classes. If presemifield S is commutative then S^{td} is symplectic, and if S is symplectic then S^{dt} is commutative. If presemifield S is commutative then its transpose S^t is dual to symplectic presemifield S^{td}. For further details or complement, we send the reader to [1, 20, 22, 23].

Definition 14.1.11. If $L : \mathbb{F}_{2^m} \rightarrow \mathbb{F}_{2^m}$ is a $\mathbb{F}_2$-linear map, its adjoint operator L^{ad} with respect to the form B is defined as a unique linear operator satisfying the following condition:

$$B(L(x), y) = B(x, L^{ad}(y)), \quad \text{for all } x, y \in \mathbb{F}_{2^m}.$$

Equality (14.1) means that all right multiplication mappings $R_z(x) = x \circ z$ of a symplectic presemifield are self-adjoint with respect to B.

Starting from a symplectic presemifield $(\mathbb{F}_{2^m}, +, \circ)$, one can construct a commutative presemifield in the following way [20, 23]. Consider the linear map $L_z : \mathbb{F}_{2^m} \rightarrow \mathbb{F}_{2^m}$, $L_z(x) = z \circ x$. Let L_z^{ad} be the adjoint operator of L_z with respect to the form B:

$$B(z \circ x, y) = B(L_z(x), y) = B(x, L_z^{ad}(y)).$$

We introduce new operation $\star$ by

$$z \star y = L_z^{ad}(y).$$

Therefore,

$$B(z \circ x, y) = B(x, z \star y).$$

Then $(\mathbb{F}_{2^m} +, \star)$ is a commutative presemifield. Similarly, starting from commutative presemifield $(\mathbb{F}_{2^m}, +, \star)$ and putting $L_z(x) = z \star x$, one can get a symplectic presemifield $(\mathbb{F}_{2^m} +, \circ)$:

$$B(z \star x, y) = B(L_z(x), y) = B(x, L_z^{ad}(y)) = B(x, z \circ y).$$

14.2 Bent Functions from the Desarguesian Spread

In this section, we consider sets of bent Boolean functions defined with respect to the Desarguesian spread. We have firstly considered the set of Boolean functions which are constant on the Desarguesian spread and identified its logical relations with Dillon's partial spread classes, the class of [6] and the set of hyper-bent functions. Next, we present several facts on bent Boolean functions, which are linear or affine on the Desarguesian spread, giving more insight on them.

14.2.1 Bent Functions Whose Restrictions to Multiplicative Cosets $u\mathbb{F}_{2^m}^\star$ ($u \in U$) Are Constant

In this subsection, we study the set of bent functions whose restrictions to each multiplicative cosets $u\mathbb{F}_{2^m}^\star$ is constant. Our aim is to clarify the relationships between the different subsets of bent functions introduced in the literature from 1974 to 2006; more precisely, the connections between the partial spread classes introduced by Dillion (1974), the class $\mathcal{PS}_{ap}^{\#}$ defined by Carlet and Gaborit [6] and the set of hyperbent functions introduced by Youssef and Gong [25] (Chap. 9). The reader notices that most of the statements can be directly obtained from the characterization of the class $\mathcal{PS}_{ap}^{\#}$ and its connection with the set of hyperbent functions given by Carlet and Gaborit in [6]. It is shown in [6] that the elements in $\mathcal{PS}_{ap}^{\#}$ are the functions of Hamming weight $2^{n-1} \pm 2^{m-1}$ which can be written as $\sum_{i=1}^{r} \mathrm{Tr}_1^n(a_i x^{j_i})$ for $a_i \in F_{2^n}$ and j_i a multiple of $2^m - 1$. Hence, $\mathcal{PS}_{ap}^{\#}$ coincides with the set of bent functions whose polynomial form is the sum of multiple trace terms constructed via Dillon-like exponents $r(2^m - 1)$.

Notation 14.2.1. *We denote by* Ω_n *the set of Boolean functions* f *defined on* $\mathbb{F}_{2^n}$ *by* $f(x) = \sum_{i \in \Gamma_{n,m}} \mathrm{Tr}_1^{o(i)}(a_i x^i)$ *where* $\Gamma_{n,m}$ *is the set of cyclotomic cosets* $[i]$ *such that* $i \equiv 0 \pmod{2^m - 1}$.

The study of the links between the set of hyper-bent functions and Dillon's class $\mathcal{PS}$, initiated in [6], has been continued in [31]. We shall denoted by $\mathcal{D}_n$ (see below) the set of bent Boolean functions f in Ω_n which vanish at 0.

Let us begin with the following statement.

Proposition 14.2.2 ([31]). *Let* $n = 2m$. *Let* f *a Boolean function defined on* $\mathbb{F}_{2^n}$ *such that* $f(0) = 0$. *The three following assertions are equivalent:*

1. $f \in \Omega_n$;
2. $\forall u \in U$, *the restriction of* f *to* $u\mathbb{F}_{2^m}^\star$ *is constant (that is,* $f(uy) = f(u), \forall y \in \mathbb{F}_{2^m}^\star$*);*
3. $\forall \omega \in \mathbb{F}_{2^n}$ *the restriction of* f *to* $\omega\mathbb{F}_{2^m}^\star$ *is constant (that is,* $f(\omega y) = f(\omega), \forall y \in \mathbb{F}_{2^m}^\star$*).*

Proof. Let us prove that (1) implies (2). For that, assume that $f(x) = \sum_{i \in \Gamma_{n,m}} \mathrm{Tr}_1^{o(i)}(a_i x^i)$ with $i \equiv 0 \pmod{2^m - 1}$. Then, $\forall u \in U, \forall y \in \mathbb{F}_{2^m}^\star$, $f(uy) = \sum_{i \in \Gamma_{n,m}} \mathrm{Tr}_1^{o(i)}(a_i u^i y^i) = \sum_{i \in \Gamma_{n,m}} \mathrm{Tr}_1^{o(i)}(a_i u^i) = f(u)$ because $y^i = 1$ since $y \in \mathbb{F}_{2^m}^\star$ and the exponents i are divisible by $2^m - 1$.
Let us prove that (2) implies (3). Fix $\omega \in \mathbb{F}_{2^n}^\star$, using the polar decomposition $\omega = uz$ with $u \in U$ and $z \in \mathbb{F}_{2^m}^\star$, we have $\forall y \in \mathbb{F}_{2^m}^\star, f(\omega y) = f(uzy) = f(u) = f(uz) = f(\omega)$ (using (2) in two equalities).
Let us prove that (3) implies (1). Assume that $f(x) = \sum_{i=1}^{2^n-2} \mathrm{Tr}_1^{o(i)}(a_i x^i) + \mathrm{Tr}_1^{o(2^n-1)}(a_{2^n-1} x^{2^n-1})$ (since $f(0) = 0$). Then (since $y \in \mathbb{F}_{2^m}^\star \subset \mathbb{F}_{2^n}^\star$ and $y^i = 1$ for $i \equiv 0 \pmod{2^m - 1}$), we have $\forall \omega \in \mathbb{F}_{2^n}, \forall y \in \mathbb{F}_{2^m}^\star$,

$$
\begin{aligned}
f(\omega y) &= \sum_{i=1}^{2^n-2} \mathrm{Tr}_1^{o(i)}(a_i\omega^i y^i) + \mathrm{Tr}_1^{o(2^n-1)}(a_{2^n-1}\omega^{2^n-1}) \\
&= \sum_{i\in[1,2^n-2] | i\equiv 0 \pmod{2^m-1}} \mathrm{Tr}_1^{o(i)}(a_i\omega^i) \\
&+ \sum_{i\in[1,2^n-2] | i\not\equiv 0 \pmod{2^m-1}} \mathrm{Tr}_1^{o(i)}(a_i\omega^i y^i) + \mathrm{Tr}_1^{o(2^n-1)}(a_{2^n-1}\omega^{2^n-1}).
\end{aligned}
$$

Now, note that for $y = 1$, we have

$$
\begin{aligned}
f(\omega) = &\sum_{i\in[1,2^n-2] | i\equiv 0 \pmod{2^m-1}} \mathrm{Tr}_1^{o(i)}(a_i\omega^i) + \mathrm{Tr}_1^{o(2^n-1)}(a_{2^n-1}\omega^{2^n-1}) \\
&+ \sum_{i\in[1,2^n-2] | i\not\equiv 0 \pmod{2^m-1}} \mathrm{Tr}_1^{o(i)}(a_i\omega^i).
\end{aligned}
$$

But using the hypothesis we have $\forall y \in \mathbb{F}_{2^m}^\star, \forall \omega \in \mathbb{F}_{2^n}, f(\omega y) + f(\omega) = 0$. Therefore, $\forall \omega \in \mathbb{F}_{2^n}$, $\forall y \in \mathbb{F}_{2^m}^\star$, $\sum_{i\in[1,2n-2] | i\not\equiv 0 \pmod{2^m-1}} \mathrm{Tr}_1^{o(i)}(a_i(y^i + 1)\omega^i) = 0$. Now, using the uniqueness of the polar decomposition, we obtain $\forall i \in [1, 2^n - 2]$, $i \not\equiv 0 \pmod{2^m - 1}, a_i(y^i + 1) = 0$. In particular, if y is a primitive element β of $\mathbb{F}_{2^m}$ then $y^i = \beta^i \neq 1$. Hence, $\forall i \in [1, 2^n - 2], i \not\equiv 0 \pmod{2^m - 1}, a_i = 0$, which proves (1). □

Now, we define the set $\mathcal{D}_n$ as follows.

Notation 14.2.3. *We denote by $\mathcal{D}_n$ the set of bent functions f in Ω_n such that $f(0) = 0$.*

Note that the set $\mathcal{D}_n$ is exactly the set of functions belonging to $\mathcal{PS}_{ap}^{\#}$ which vanish at 0. We will see later that $\mathcal{D}_n$ plays an important role in establishing the connections between the partial spread classes introduced in the literature. The reader notices that $\mathcal{D}_n$ is the set of bent functions whose polynomial form is the sum of multiple trace terms constructed via Dillon-like exponents (that is, exponents of the form $r(2^m - 1)$).
To prove Proposition 14.2.5, we need Lemma 14.2.4.

Lemma 14.2.4 ([31]). *Let g be a Boolean function on $\mathbb{F}_{2^n}$ with $n = 2m$. Then g is bent if and only if $\forall \omega \in \mathbb{F}_{2^n}$, $\widehat{\chi_g}(\omega) \equiv 2^m \pmod{2^{m+1}}$.*

Proof. The necessary condition is clear (since if g is bent, then by definition, $\widehat{\chi_g}(\omega) = \pm 2^m, \forall \omega \in \mathbb{F}_{2^n}$ and $\pm 2^m \equiv 2^m \pmod{2^{m+1}}$). Conversely, assume $\widehat{\chi_g}(\omega) \equiv 2^m \pmod{2^{m+1}}, \forall \omega \in \mathbb{F}_{2^n}$. Then, $\forall \omega \in \mathbb{F}_{2^n}$, $\widehat{\chi_g}(\omega) = 2^m(1 + 2\kappa(\omega))$, where $\kappa(\omega) \in \mathbb{Z}$. By Parseval relation, we have $\sum_{\omega\in\mathbb{F}_{2^n}}(2^m(1 + 2\kappa(\omega)))^2 = 2^{2n}$, that is $\sum_{\omega\in\mathbb{F}_{2^n}}(1 + 2\kappa(\omega))^2 = 2^n$. Therefore, $(1 + 2\kappa(\omega)) = \pm 1, \forall \omega \in \mathbb{F}_{2^n}$. This proves that $\widehat{\chi_g}(\omega) = \pm 2^m$ for all $\omega \in \mathbb{F}_{2^n}$, that is, g is bent. □

Proposition 14.2.5 ([31]). *Let f be a Boolean function in Ω_n such that $f(0) = 0$. Then, the three following assertions are equivalent:*

1. *f is bent;*
2. *$\sum_{u\in U}(-1)^{f(u)} = 1$;*
3. *$\#\{u \in U \mid f(u) = 1\} = 2^{m-1}$.*

Proof. Let us compute the Walsh transform of $f \in \mathcal{D}_n$. Let $\omega \in \mathbb{F}_{2^n}$. We have

$$\begin{aligned}
\widehat{\chi_f}(\omega) &= \sum_{x\in\mathbb{F}_{2^n}} (-1)^{\sum_{i\in\Gamma_{n,m}} \mathrm{Tr}_1^{o(i)}(a_i x^i)+\mathrm{Tr}_1^n(\omega x)} \\
&= 1 + \sum_{x\in\mathbb{F}_{2^n}^\star} (-1)^{\sum_{i\in\Gamma_{n,m}} \mathrm{Tr}_1^{o(i)}(a_i x^i)+\mathrm{Tr}_1^n(\omega x)} \\
&= 1 + \sum_{u\in U}\sum_{y\in\mathbb{F}_{2^m}^\star} (-1)^{\sum_{i\in\Gamma_{n,m}} \mathrm{Tr}_1^{o(i)}(a_i y^i u^i)+\mathrm{Tr}_1^n(\omega y u)} \\
&= 1 + \sum_{u\in U}\sum_{y\in\mathbb{F}_{2^m}^\star} (-1)^{\sum_{i\in\Gamma_{n,m}} \mathrm{Tr}_1^{o(i)}(a_i u^i)+\mathrm{Tr}_1^n(\omega y u)} \\
&= 1 + \sum_{u\in U}(-1)^{\sum_{i\in\Gamma_{n,m}} \mathrm{Tr}_1^{o(i)}(a_i u^i)} \sum_{y\in\mathbb{F}_{2^m}^\star} (-1)^{\mathrm{Tr}_1^n(\omega y u)} \\
&= 1 + \sum_{u\in U}(-1)^{\sum_{i\in\Gamma_{n,m}} \mathrm{Tr}_1^{o(i)}(a_i u^i)} \sum_{y\in\mathbb{F}_{2^m}} (-1)^{\mathrm{Tr}_1^n(\omega y u)} \\
&\quad - \sum_{u\in U}(-1)^{\sum_{i\in\Gamma_{n,m}} \mathrm{Tr}_1^{o(i)}(a_i u^i)}.
\end{aligned}$$

- If $\omega = 0$ then, $\widehat{\chi_f}(0) = 1 + (2^m - 1)\sum_{u\in U}(-1)^{\sum_{i\in\Gamma_{n,m}} \mathrm{Tr}_1^{o(i)}(a_i u^i)}$.
- If $\omega \neq 0$ then $\sum_{y\in\mathbb{F}_{2^m}} \chi(\mathrm{Tr}_1^n(\omega y u)) = \sum_{y\in\mathbb{F}_{2^m}} \chi(\mathrm{Tr}_1^m(\mathrm{Tr}_m^n(\omega u)y))$

$$= \begin{cases} 2^m \text{ if } \mathrm{Tr}_m^n(\omega u) = 0, \text{ that is, if } u^{2^m-1} = \omega^{1-2^m} \\ 0 \quad \text{otherwise.} \end{cases}$$

Since $x \mapsto x^{2^m-1}$ is a permutation of U then,

$$\sum_{y\in\mathbb{F}_{2^m}} \chi(\mathrm{Tr}_1^n(\omega y u)) = \begin{cases} 2^m \text{ if } u = \omega^{-1} \\ 0 \quad \text{otherwise.} \end{cases}$$

Therefore,

$$\widehat{\chi_f}(\omega) = 1 - \sum_{u\in U}(-1)^{\sum_{i\in\Gamma_{n,m}} \mathrm{Tr}_1^{o(i)}(a_i u^i)} + 2^m(-1)^{f(\omega^{-1})}.$$

Now, assume that $\sum_{u\in U}(-1)^{f(u)} = 1$. Then, for every $\omega \in \mathbb{F}_{2^n}^\star$, $\widehat{\chi_f}(\omega) = (-1)^{f(\omega^{-1})}2^m \in \{\pm 2^m\}$ and $\widehat{\chi_f}(0) = 2^m$. The function f is thus bent. Conversely, if f is bent then, thanks to Lemma 14.2.4, we have $\sum_{u\in U}(-1)^{f(u)} = 1$. The equivalence (1) $\iff$ (2) follows. The equivalence (2) $\iff$ (3) comes simply from the equality $\sum_{u\in U}(-1)^{f(u)} = (2^m+1) - 2\#\{u \in U \mid \sum_{u\in U}(-1)^{f(u)} = 1\}$. The proof [1] of the proposition follows. □

The next corollary is an easy consequence of Proposition 14.2.2 and Proposition 14.2.5.

Corollary 14.2.6. *Let f be a function in $\mathcal{D}_n$. Then f belongs to the class $\mathcal{PS}^-$.*

Proof. Let $f \in \mathcal{D}_n$. According to Proposition 14.2.2, the restriction of f to $u\mathbb{F}_{2^m}^\star$ is constant for $u \in U$. Hence, $f = \sum_{u|f(u)=1} \mathbf{1}_{u\mathbb{F}_{2^m}^\star} = \sum_{u|f(u)=1} \mathbf{1}_{u\mathbb{F}_{2^m}}$ (since $f(0) = \#\{u \mid f(u) = 1\} \pmod 2 = 2^{m-1} \pmod 2 = 0$, according to Proposition 14.2.5). Consequently, f can be written as the sum of 2^{m-1} indicators of vector spaces $u_i\mathbb{F}_{2^m}$ of dimension m such that $u_i\mathbb{F}_{2^m} \cap u_j\mathbb{F}_{2^m} = \{0\}$ which means that f is in $\mathcal{PS}^-$. □

Proposition 14.2.7 ([31]). *Let e be a positive integer co-prime with $2^n - 1$. Let $f \in \mathcal{D}_n$ and f' be the Boolean function on $\mathbb{F}_{2^n}$ defined as $f'(x) := f(x^e)$. Then $f' \in \mathcal{D}_n$.*

Proof. Set $f(x) = \sum_{i\in\Gamma_{n,m}} \mathrm{Tr}_1^{o(i)}(a_ix^i)$ with $\forall i, d_i \equiv 0 \pmod{2^m-1}$. By definition, $f'(x) = \sum_{i\in\Gamma_{n,m}} \mathrm{Tr}_1^{o(i)}(a_ix^{ed_i})$. It is clear that $ed_i \equiv 0 \pmod{2^m-1}$. Moreover, for all i we have $o(ed_i) = o(i)$. Indeed, by definition, $o(ed_i)$ is the smallest positive integer such that $ed_i \times 2^{o(ed_i)} \equiv ed_i \pmod{2^n-1}$. Since e is co-prime with 2^n-1, we have $d_i \times 2^{o(ed_i)} \equiv d_i \pmod{2^n-1}$, which shows that $o(ed_i) \geq o(i)$. Conversely, by definition, $o(i)$ is the smallest positive integer such that $d_i \times 2^{o(i)} \equiv ed_i \pmod{2^n - 1}$. Multiplying by e, we obtain $o(i) \geq o(ed_i)$. Now, let us prove that f' is bent. According to Proposition 14.2.5, f' is bent if and only if $\#\{u \in U \mid f(u^e) = 1\} = 2^{m-1}$. Now, the mapping $x \mapsto x^e$ is a permutation of U (since $gcd(e, 2^m+1) = 1$). Hence, $\#\{u \in U \mid f(u^e) = 1\} = \#\{u \in U \mid f(u) = 1\} = 2^m - 1$, according to Proposition 14.2.5 (since f is bent). □

From the definition of hyper-bent functions and the previous proposition, we obtain the following straightforward statement.

Corollary 14.2.8 ([31]). *Let f be a function in $\mathcal{D}_n$. Then, f belongs to $\mathcal{HB}_n$.*

Now, we are interested in the bivariate representation of elements of $\mathcal{D}_n$. We prove the following statement.

Proposition 14.2.9 ([31]). *Functions f in $\mathcal{D}_n$ such that $f(1) = 0$ can be defined by $(y,z) \in \mathbb{F}_{2^m}\times\mathbb{F}_{2^m} \mapsto g(\frac{z}{y})$ with $g(0) = 0$ and g balanced on $\mathbb{F}_{2^m}$ (i.e. $wt(g) = 2^{m-1}$).*

[1] Note that the proof is simpler in this way compared to the proof obtained by showing (1) $\Rightarrow$ (2) $\Rightarrow$ (3) $\Rightarrow$ (1).

Proof. Let (v, w) be a basis of $\mathbb{F}_{2^n}$ as $\mathbb{F}_{2^m}$-vector space. Every $x \in \mathbb{F}_{2^n}$ can be uniquely expressed as $x = vy + wz$ with $(y, z) \in \mathbb{F}_{2^m} \times \mathbb{F}_{2^m}$. Then, for $f \in \mathcal{D}_n$ we have,

$$f(x) = \sum_{i\in\Gamma_{n,m}} \mathrm{Tr}_1^{o(i)}(a_i(vy + wz)^i)$$

$$= \begin{cases} \sum_{i\in\Gamma_{n,m}} \mathrm{Tr}_1^{o(i)}(a_i((v + \frac{wz}{y})^i y^i) & \text{if } y \neq 0 \text{ and } z \neq 0 \\ \sum_{i\in\Gamma_{n,m}} \mathrm{Tr}_1^{o(i)}(a_i w^i) = f(w) & \text{if } y = 0 \text{ and } z \neq 0 \\ \sum_{i\in\Gamma_{n,m}} \mathrm{Tr}_1^{o(i)}(a_i v^i) = f(v) & \text{if } y \neq 0 \text{ and } z = 0 \\ 0 & \text{if } y = z = 0. \end{cases}$$

Now, let us choose $\gamma \in \mathbb{F}_{2^n} \setminus \mathbb{F}_{2^m}$ such that $f(\gamma) = 0$. Such an element γ exists. Indeed, according to Proposition 14.2.6, f belongs to the class $\mathcal{PS}^-$. Hence, the Hamming weight of f is equal to $2^{n-1} - 2^{m-1}$. Thus, there are $2^n - (2^{n-1} - 2^{m-1}) = 2^{n-1} + 2^{m-1} > 2^m$ elements x in $\mathbb{F}_{2^n}$ such that $f(x) = 0$. Therefore, the set $\{x \in \mathbb{F}_{2^n} \mid f(x) = 0\}$ is not included in $\mathbb{F}_{2^m}$, which proves the existence of γ.
Now, choose $v = 1$ and $w = \gamma$. The set $\{1, \gamma\}$ is a basis of the two-dimensional $\mathbb{F}_{2^m}$-vector space $\mathbb{F}_{2^n}$. The bivariate representation of an element f of $\mathcal{D}_n$ such that $f(1) = 0$ equals

$$f'(y, z) = \begin{cases} g(\frac{z}{y}) & \text{if } y \neq 0 \text{ and } z \neq 0 \\ f(\gamma) & \text{if } y = 0 \text{ and } z \neq 0 \\ f(1) = 0 & \text{if } y \neq 0 \text{ and } z = 0 \\ 0 & \text{if } y = z = 0. \end{cases}$$

Since $g(0) = f(v) = 0, f'(y, z) = g(\frac{z}{y})$. Now, we prove that g is balanced. Recall that $wt(f') = 2^{n-1} - 2^{m-1}$ (since $f \in \mathcal{PS}^-$). We remark that $f'(y, z)$ is constant on $E_a := \{(x, ax) \mid x \in \mathbb{F}_{2^m}^\star\}$; $a \in \mathbb{F}_{2^m}^\star$ (since $\forall x \in \mathbb{F}_{2^m}^\star, \forall a \in \mathbb{F}_{2^m}^\star$, $f'(x, ax) = g(\frac{ax}{x}) = g(a)$). Now,

$$\mathbb{F}_{2^m} \times \mathbb{F}_{2^m} = \{0\} \times \mathbb{F}_{2^m}^\star \cup \mathbb{F}_{2^m}^\star \times \{0\} \bigcup_{a\in\mathbb{F}_{2^m}^\star} E_a \cup \{(0, 0)\}.$$

For every

$$(y, z) \in \{0\} \times \mathbb{F}_{2^m}^\star \cup \mathbb{F}_{2^m}^\star \times \{0\} \cup \{(0, 0)\},$$

we have $f'(y, z) = g(0) = 0$. Moreover,

$$\begin{aligned}2^{n-1} - 2^{m-1} &= wt(f') \\ &= \sum_{a \in \mathbb{F}_{2^m}^\star} \#\{(x, ax) \in E_a \mid f'(x, ax) = g(a) = 1\} \\ &= (2^m - 1)\#\{a \in \mathbb{F}_{2^m}^\star \mid g(a) = 1\}.\end{aligned}$$

Hence, $\#\{a \in \mathbb{F}_{2^m}^\star \mid g(a) = 1\} = \frac{2^{n-1}-2^{m-1}}{2^m-1} = 2^{m-1}$. But $g(0) = 0$, thus $wt(g) := \#\{x \in \mathbb{F}_{2^m} \mid g(x) = 1\} = 2^{m-1}$, proving that g is balanced on $\mathbb{F}_{2^m}$. □

Proposition 14.2.10 ([31]). *Let f be a function in $\mathcal{PS}_{ap}$. Then f belongs to $\mathcal{D}_n$.*

Proof. By definition of $\mathcal{PS}_{ap}$, there exists a basis (v, w) of the $\mathbb{F}_{2^m}$- vector space $\mathbb{F}_{2^n}$ and there exists a balanced function g on $\mathbb{F}_{2^m}$ with $g(0) = 0$ such that for every $(x, y) \in \mathbb{F}_{2^m} \times \mathbb{F}_{2^m}$, $f(vx + wy) = g(xy^{2^m-2})$. Let $z \in u\mathbb{F}_{2^m}^\star$, then there exists $\tau \in \mathbb{F}_{2^m}^\star$ such that $z = u\tau = (v\alpha + w\beta)\tau$. Then, $f(z) = f(v(\alpha\tau) + w(\beta\tau)) = g(\alpha\tau(\beta\tau)^{2^m-2}) = g(\alpha\beta^{2^m-2}\tau^{2^m-1}) = g(\alpha\beta^{2^m-2})$.

The restriction of f to $u\mathbb{F}_{2^m}^\star$ is thus constant. The proposition follows using the definition of $\mathcal{D}_n$. □

Proposition 14.2.11 ([31]). *Every function $f \in \mathcal{D}_n$ is of the form $f(x) = f'(\delta x)$ for a function $f' \in \mathcal{PS}_{ap}$ and $\delta \in \mathbb{F}_{2^n}^\star$.*

Proof. By Proposition 14.2.9, the functions $f \in \mathcal{D}_n$ such that $f(1) = 0$ belong to $\mathcal{PS}_{ap}$. Now we are interested in functions $f \in \mathcal{D}_n$ such that $f(1) = 1$. One can prove that there exists necessarily $\delta \in \mathbb{F}_{2^n}^\star$ such that $f(\frac{1}{\delta}) = 0$. Define f' by $f'(x) = f(\frac{x}{\delta})$. According to the expression of elements in $\mathcal{D}_n$ (see Notation 14.2.3 and Notation 14.2.1) we have

$$f'(x) = \sum_{i \in \Gamma_{n,m}} \mathrm{Tr}_1^{o(i)}(a_i(\frac{x}{\delta})^i) = \sum_{i \in \Gamma_{n,m}} \mathrm{Tr}_1^{o(i)}(a'_i x^i)$$

where $a_i' = \frac{a_i}{\delta^i}$. Moreover, since the bentness is affine invariant, we deduce that f is bent if and only if f' is bent. Hence, $f' \in \mathcal{D}_n$ and $f'(1) = f(\frac{1}{\delta}) = 0$. Therefore, f' belong to $\mathcal{PS}_{ap}$. Consequently, $f(x) = f'(\delta x)$ with $f' \in \mathcal{PS}_{ap}$ and $\delta \in \mathbb{F}_{2^n}^\star$. Conversely, let $f' \in \mathcal{PS}_{ap}$ and $\delta \in \mathbb{F}_{2^n}^\star$. Set $f(x) := f'(\delta x)$. Since $f' \in \mathcal{PS}_{ap} \subset \mathcal{D}_n$, then using arguments as above, we deduce that $f \in \mathcal{D}_n$. □

To conclude this part, we deduce that functions in $\mathcal{D}_n$ can be described in terms of bent functions from the class $\mathcal{PS}_{ap}$ and that the functions of $\mathcal{PS}_{ap}^\#$ can be described in terms of elements of $\mathcal{D}_n$ and $1 + \mathcal{D}_n := \{1 + f(x) \mid f \in \mathcal{D}_n\}$. More precisely, the following statement follows directly from Proposition 14.2.11 and Proposition 4 in [6].

Proposition 14.2.12 ([31]). *Using the above notation, we have:*

- $\mathcal{D}_n = \{f(\delta x) \mid f \in \mathcal{PS}_{ap}, \delta \in \mathbb{F}_{2^n}^\star\}$;
- $\mathcal{PS}_{ap}^\# = \mathcal{D}_n \cup (1 + \mathcal{D}_n)$.

In the following, we provide an alternative simple proof of a result of Carlet and Gaborit given in [6].

Proposition 14.2.13 ([31]). *The class $\mathcal{PS}^{\#}_{ap}$ is contained in $\mathcal{HB}_n$.*

Proof. Let $f \in \mathcal{PS}^{\#}_{ap}$. From Proposition 14.2.11 and Proposition 14.2.12, we have:

- if $f \in \mathcal{D}_n$, then according to Corollary 14.2.8, f is hyper-bent.
- if $f \in \mathcal{PS}^{\#}_{ap} \setminus \mathcal{D}_n$ then $f = 1 + f'$ with $f' \in \mathcal{D}_n \subset \mathcal{H}_n$. To complete the proof, we have to prove that the complement of a hyper-bent function is also hyper-bent. Inspecting the Walsh transform of hyper-bent functions shows that this class is closed with respect to taking complements. More precisely, for all $\omega \in \mathbb{F}_{2^n}$ and for all k with $gcd(k, 2^n - 1) = 1$ we have

$$\begin{aligned} f' \text{ is hyper-bent} &\iff \widehat{\chi_{f'}}(\omega, k) = \pm 2^m \\ &\iff \sum_{x\in\mathbb{F}_{2^n}} (-1)^{f'(x)+\mathrm{Tr}_1^n((\omega x^k))} = \pm 2^m \\ &\iff -\sum_{x\in\mathbb{F}_{2^n}} (-1)^{f(x)+\mathrm{Tr}_1^n((\omega x^k))} = \pm 2^m \\ &\iff f \text{ is hyper-bent.} \end{aligned}$$

□

Now, an open question left in the literature is whether there exists or not hyper-bent functions which are not in $\mathcal{PS}^-$. The following proposition gives a positive answer.

Proposition 14.2.14 ([31]). *We have $\mathcal{PS}^{\#}_{ap} \cap (\mathcal{PS}^-)^c \neq \emptyset$ where $(\mathcal{PS}^-)^c$ denotes the complementary set of $\mathcal{PS}^-$ in the set of bent functions.*

Proof. The set $\mathcal{PS}^{\#}_{ap}$ contains hyper-bent functions which are of Hamming weight $2^{n-1} + 2^{m-1}$ (since they are the complements of functions in $\mathcal{D}_n$ which are of Hamming weight $2^{n-1} - 2^{m-1}$). Thus, $\mathcal{PS}^{\#}_{ap}$ contains hyper-bent functions which are not in $\mathcal{PS}^-$ (since all the functions in $\mathcal{PS}^-$ are of Hamming weight $2^{n-1} - 2^{m-1}$). □

Remark 14.2.15. By computer experiments, Carlet and Gaborit [6] have found that for $n = 4$ there exist hyper-bent functions which are not in $\mathcal{PS}^{\#}_{ap}$. Consequently, the set of hyper-bent functions contains strictly $\mathcal{PS}^{\#}_{ap}$.

Now, it is easy to prove the following statement.

Proposition 14.2.16 ([31]). *Using the previous notation, we have $\mathcal{PS}^{\#}_{ap} \cap \mathcal{PS}^- = \mathcal{D}_n$*

Proof. The result comes from the fact that $(1 + \mathcal{D}_n) \cap \mathcal{PS}^- = \emptyset$ and $\mathcal{PS}^{\#}_{ap} = \mathcal{D}_n \cup (1 + \mathcal{D}_n)$, according to Proposition 14.2.12. □

The previous proposition is in fact a direct consequence of the result from [6] recalled above, but the author has clarified the arguments in [31].

Remark 14.2.17. Note that we have $\mathcal{D}_n \subset \mathcal{H}_n \cap \mathcal{PS}^-$.

Collecting the previous results, we have:

(*) $\mathcal{PS}_{ap} \subset \mathcal{D}_n \subset \mathcal{PS}^{\#}_{ap} \subset \mathcal{H}_n$;
(**) $\mathcal{PS}^{\#}_{ap} \cap \mathcal{PS}^- = \mathcal{D}_n$;
(***) $\mathcal{D}_n \subset \mathcal{H}_n \cap \mathcal{PS}^-$.

This gives, schematically, the following figure (Fig. 14.1).

Now, we prove that the duals of hyper-bent functions in $\mathcal{D}_n$ are also hyper-bent. Note that the dual of any hyper-bent function is not hyper-bent in general.

Proposition 14.2.18 ([31]). *Let $f \in \mathcal{D}_n$. Then the dual $\tilde{f}$ of f satisfies $\tilde{f}(\omega) = f(\omega^{\frac{2^m-1}{2}})$ for every $\omega \in \mathbb{F}_{2^n}$ (that is, $\forall \omega \in \mathbb{F}_{2^n}, \widehat{\chi_f}(\omega) = 2^m(-1)^{f(\omega^{\frac{2^m-1}{2}})}$). Moreover, $\tilde{f} \in \mathcal{D}_n$.*

Proof. Let $f \in \mathcal{D}_n$ then (using the polar decomposition and the fact that the restriction to each coset $u\mathbb{F}^{\star}_{2^m}$ of functions in $\mathcal{D}_n$ is constant) for every $\omega \in \mathbb{F}_{2^n}$, we have

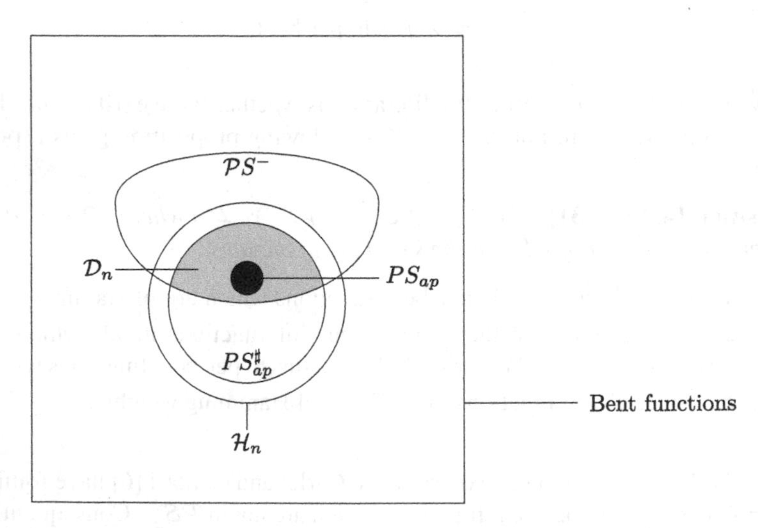

Fig. 14.1 Venn diagram

$$\begin{aligned}
\widehat{\chi_f}(\omega) &= 1 + \sum_{x\in\mathbb{F}_{2^n}^\star} (-1)^{f(x)+Tr_1^n(\omega x)} \\
&= 1 + \sum_{u\in U}\sum_{y\in\mathbb{F}_{2^m}^\star} (-1)^{f(uy)+Tr_1^n(\omega uy)} \\
&= 1 + \sum_{u\in U}\sum_{y\in\mathbb{F}_{2^m}^\star} (-1)^{f(u)+Tr_1^m(Tr_m^n(\omega uy))} \\
&= 1 + \sum_{u\in U}\sum_{y\in\mathbb{F}_{2^m}^\star} (-1)^{f(u)+Tr_1^m(Tr_m^n(\omega u)y)} \\
&= 1 - \sum_{u\in U}(-1)^{f(u)} + \sum_{u\in U}(-1)^{f(u)} \sum_{y\in\mathbb{F}_{2^m}} (-1)^{Tr_1^m(Tr_m^n(\omega u)y)};
\end{aligned}$$

- if $\omega = 0$ then $\widehat{\chi_f}(0) = 1 - \sum_{u\in U}(-1)^{f(u)} + 2^m \sum_{u\in U}(-1)^{f(u)} = 2^m$, according to Proposition 14.2.5;
- if $\omega \neq 0$ then $\widehat{\chi_f}(\omega) = 1 - \sum_{u\in U}(-1)^{f(u)} + 2^m(-1)^{f(\omega^{\frac{2^m-1}{2}})}$ (since $Tr_m^n(\omega u) = 0$ if and only if $u = \omega^{\frac{2^m-1}{2}}$ since the polar decomposition of w has the form $w = w^{(2^m+1)2^{m-1}} w^{(2^m-1)2^{m-1}}$).

Now, recall that $\sum_{u\in U}(-1)^{f(u)} = 1$ since functions in $\mathcal{D}_n$ are bent (we shall use Proposition 14.2.5). Therefore, $\widehat{\chi_f}(0) = 2^m$ and $\widehat{\chi_f}(\omega) = 2^m(-1)^{f(\omega^{\frac{2^m-1}{2}})}$, $\forall\omega \in \mathbb{F}_{2^n}^\star$, which completes the proof. □

Remark 14.2.19. Note that, up to affine equivalence, the only known constructions of hyper-bent functions are those in the set $\mathcal{D}_n$ [9, 10, 27, 29, 30, 33]. Their complements are in $\mathcal{PS}_{ap}^\#$.

14.2.2 Bent Functions Whose Restrictions to the m-Spreads $u\mathbb{F}_{2^m}$ ($u \in U$) Are Linear

This section is dealing with bent functions g defined on $\mathbb{F}_{2^n}$ such that their restrictions to each multiplicative coset $u\mathbb{F}_{2^m}^\star$ is linear for every $u \in U$. More precisely, we are interested in Boolean functions g defined over $\mathbb{F}_{2^n}$ such that there exists a mapping a from U to $\mathbb{F}_{2^m}$ satisfying $g(uy) = Tr_1^m(a(u)y)$, for every $(u, y) \in U \times \mathbb{F}_{2^m}^\star$. Bent functions g defined on $\mathbb{F}_{2^n}$ (identified to $\mathbb{F}_{2^m} \times \mathbb{F}_{2^m}$) whose restrictions to the m-spread $\{E_a, E_\infty\}$ are linear have already been completely identified in [7]. In fact such functions g are in the class $\mathcal{H}$ (introduced in [7]) defined as the class of (bent) functions of the form:

$$g(x,y) = \begin{cases} \mathrm{Tr}_1^m\left(xH\left(\frac{y}{x}\right)\right) \text{ if } x \neq 0 \\ \mathrm{Tr}_1^m(\mu y) \text{ if } x = 0 \end{cases} \tag{14.2}$$

with $H(z) := G(z) + \mu z$, where $\mu \in \mathbb{F}_{2^m}$ and G is a permutation polynomial on $\mathbb{F}_{2^m}$ satisfying (14.10):

$$\forall \beta \in \mathbb{F}_{2^m}^\star, \text{function } z \mapsto G(z) + \beta z \text{ is 2-to-1 on } \mathbb{F}_{2^m}. \tag{14.3}$$

Condition (14.10) is necessary and sufficient for g to be bent. It has been showed in [7] that Condition (14.10) implies that G is a permutation on $\mathbb{F}_{2^m}$ and that such a condition is equivalent to saying that for every $\gamma \in \mathbb{F}_{2^m}$, the function H_γ: $z \in \mathbb{F}_{2^m} \mapsto \begin{cases} \frac{G(z+\gamma)+G(\gamma)}{z} \text{ if } z \neq 0 \\ 0 \text{ if } z = 0 \end{cases}$ is a permutation on $\mathbb{F}_{2^m}$. In other words, G is an *oval polynomial* (or o-polynomial, for short) on $\mathbb{F}_{2^m}$. The notion of o-polynomials comes from finite projective geometry since they are closely related to some hyperovals of the projective plane $PG_2(2^m)$ over $\mathbb{F}_{2^m}$. A *hyperoval* of $PG_2(2^m)$ is a set of $2^m + 2$ points no three of them collinear. A hyperoval of $PG_2(2^m)$ can then be represented by $D(f) = \{(1, t, G(t)), t \in \mathbb{F}_{2^n}\} \cup \{(0, 1, 0), (0, 0, 1)\}$ where G is an o-polynomial on $\mathbb{F}_{2^m}$.

Viewed in their univariate representation, functions g of the class $\mathcal{H}$ are in fact the known so-called Niho bent functions. Recall that a positive integer s (always understood modulo $2^n - 1$) is said to be a *Niho exponent*, and x^s a *Niho power function*, if the restriction of x^s to $\mathbb{F}_{2^m}$ is linear or in other words $s \equiv 2^j \pmod{2^m - 1}$ for some $j < n$. Without loss of generality, we can assume that s is in the normalized form, with $j = 0$, and then we have a unique representation $s = (2^m - 1)d + 1$ with $2 \leq d \leq 2^m$. There exist, up to equivalence, five classes of bent functions via Niho exponents (see for instance [7]).
We recall the direct connection between the elements of the class $\mathcal{H}$ and the o-polynomials together with the correspondence between those elements and the Niho bent functions schematized by the following diagram given in Chap. 8.

1. The correspondence (1) offers a new framework to study the Niho bent functions.
2. The connection (2) provides the construction of several potentially new families of bent functions in $\mathcal{H}$ (and thus new bent functions of type Niho) from the classes of o-polynomials in finite projective geometry.

The diagram (Fig. 14.2) seems simple but, in practice, it needs some non-trivial calculations to obtain explicit expressions !

14.2.2.1 An Explicit Example

Let us explain the above correspondences (Fig. 14.2) through an explicit simple example. Let $a \in \mathbb{F}_{2^m}^\star$. Let $n = 2m$ be an even integer. Define $f : \mathbb{F}_{2^n} \to \mathbb{F}_2$ as : $\forall x \in \mathbb{F}_{2^n}, \quad f(x) = \mathrm{Tr}_1^m(ax^{2^m+1})$. Note that, for every $x \in \mathbb{F}_{2^m}$, we have $x^{2^m} = x$.

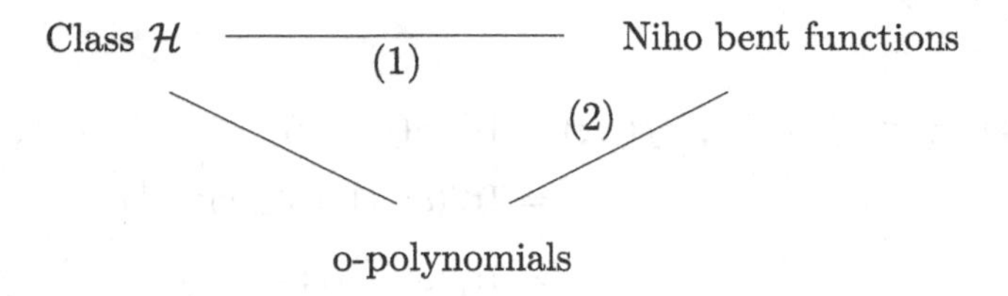

Fig. 14.2 Diagram: correspondences

Thus, $\forall x \in \mathbb{F}_{2^m}, f(x) = Tr_1^m(ax^2) = Tr_1^m(a^{2^{m-1}}x)$, that is, the restriction of f to $\mathbb{F}_{2^m}$ is linear. Now, f can be written (via a Niho exponent) as

$$f(x) = Tr_1^n(\beta x^{(2^m-1)\frac{1}{2}+1})$$

for some $\beta \in \mathbb{F}_{2^n}$ where $\frac{1}{2}$ denotes the inverse of 2 modulo $2^m + 1$. Indeed, every element $a \in \mathbb{F}_{2^m}^\star$ can be written as $a = \alpha + \alpha^{2^m}$ with $\alpha \in \mathbb{F}_{2^n} \setminus \mathbb{F}_{2^m}$. Hence $\forall x \in \mathbb{F}_{2^n}, f(x) = Tr_1^m((\alpha + \alpha^{2^m})x^{2^m+1})$. Remark that $(\alpha + \alpha^{2^m})x^{2^m+1} = Tr_m^n(\alpha x^{2^m+1})$. Hence, by the chain rule $Tr_1^m \circ Tr_m^n = Tr_1 n$, one gets that $\forall x \in \mathbb{F}_{2^n}, f(x) = Tr_1^n(\alpha x^{2^m+1})$. Now,

$$\begin{aligned} 2((2^m-1)(2^{m-1}+1)+1) &= 2(2^{n-1}+2^m-2^{m-1}-1+1) \\ &\equiv 2^n + 2^m \pmod{2^n-1} \\ &\equiv 1 + 2^m \pmod{2^n-1}. \end{aligned}$$

Hence

$$\begin{aligned} \forall x \in \mathbb{F}_{2^n}, \quad f(x) &= Tr_1^n(\alpha x^{2((2^m-1)(2^{m-1}+1)+1)}) \\ &= Tr_1^n(\alpha^{2^{n-1}} x^{(2^m-1)(2^{m-1}+1)+1}). \end{aligned}$$

Note that $2(2^{m-1}+1) = 2^m + 2 \equiv 1 \pmod{2^m+1}$ proving that $2^{m-1}+1$ is the inverse of 2 modulo $2^m + 1$.
Now, let $\zeta \in \mathbb{F}_{2^n}^\star$ be such that $Tr_m^n(\zeta) = 1$, in particular $\zeta \notin \mathbb{F}_{2^m}$. Define a function $\tilde{f}$ from $\mathbb{F}_{2^m} \times \mathbb{F}_{2^m}$ to $\mathbb{F}_2$ by : $\forall (y,z) \in \mathbb{F}_{2^m} \times \mathbb{F}_{2^m}, \tilde{f}(y,z) = f(y + \zeta z)$. Let us prove that $\tilde{f}$ is in the class $\mathcal{H}$.

For the sake of readability, let us denote by τ the quantity $\zeta^{2^{n-1}+2^{m-1}}$.

Firstly, note that $\tilde{f}$ can be written as $Tr_1^m(\mu z)$ if $y = 0$ and $Tr_1^m(y\psi(z/y))$ if $y \neq 0$ for a function $\psi : \mathbb{F}_{2^m} \to \mathbb{F}_{2^m}$. Indeed, for $y = 0$, one has

$$\begin{aligned} \forall z \in \mathbb{F}_{2^m}, \quad \tilde{f}(0,z) &= Tr_1^m(a(\zeta z)^{2^m+1}) = Tr_1^m(a\zeta^{2^m+1}z^2) \\ &= Tr_1^m(a^{2^{m-1}}\tau z). \end{aligned}$$

Next,

$$\forall (y,z) \in \mathbb{F}_{2^m}^{\star} \times \mathbb{F}_{2^m}, \quad \tilde{f}(y,z) = \mathrm{Tr}_1^m(a(y+\zeta z)^{2^m+1})$$
$$= \mathrm{Tr}_1^m(ay^2(1+\zeta z/y)^{2^m+1})$$
$$= \mathrm{Tr}_1^m(a^{2^{m-1}} y(1+\zeta z/y)^{2^{n-1}+2^{m-1}}).$$

Hence, if we set $\mu = a^{2^{m-1}}\tau$ and $\psi(t) = a^{2^{m-1}}(1+\zeta t)^{2^{n-1}+2^{m-1}}$, one has

$$\forall (y,z) \in \mathbb{F}_{2^m} \times \mathbb{F}_{2^m}, \quad \tilde{f}(y,z) = \begin{cases} \mathrm{Tr}_1^m(y\psi(z/y)) & \text{if } y \neq 0 \\ \mathrm{Tr}_1^m(\mu z) & \text{if } y = 0. \end{cases}$$

Secondly, set

$$G(t) := \psi(t) + \mu t = a^{2^{m-1}}(1+\zeta t)^{2^{n-1}+2^{m-1}} + a^{2^{m-1}}\tau t.$$

Note then that

$$G(t) = a^{2^{m-1}}\pi(t)$$

where π is the mapping defined on $\mathbb{F}_{2^m}$ by $\pi(t) = (1+\zeta t)^{2^{m-1}+2^{n-1}} + \zeta^{2^{m-1}+2^{n-1}}t$ for every $t \in \mathbb{F}_{2^m}$. The mapping π is a permutation on $\mathbb{F}_{2^m}$. Indeed, expanding the expression of π, we get

$$\pi(t) = 1 + \zeta^{2^{n-1}}t^{2^{n-1}} + \zeta^{2^{m-1}}t^{2^{m-1}} + \zeta^{2^{m-1}+2^{n-1}}t^{2^{m-1}+2^{n-1}} + \zeta^{2^{m-1}+2^{n-1}}t.$$

Now $t^{2^{n-1}} = t^{2^{m-1}}$ since $2^{n-1} = 2^m \cdot 2^{m-1}$ and $t^{2^{m-1}+2^{n-1}} = t^{2^{m-1}+2^{m-1}} = t^{2^m} = t$. Hence

$$\pi(t) = 1 + \mathrm{Tr}_m^n(\zeta^{2^{m-1}})t^{2^{m-1}} = 1 + (\mathrm{Tr}_m^n\zeta)^{2^{m-1}}t^{2^{m-1}} = 1 + t^{2^{m-1}}.$$

The map $t \in \mathbb{F}_{2^m} \mapsto t^{2^{m-1}}$ is the inverse map of the Frobenius mapping over $\mathbb{F}_{2^m}$. Therefore, π is a permutation on $\mathbb{F}_{2^m}$ as well as G.
Moreover, let $(\alpha, \beta) \in \mathbb{F}_{2^m} \times \mathbb{F}_{2^m}^{\star}$. Then the equation

$$G(t) + \beta t = \alpha \tag{14.4}$$

has either 0 or two solutions in $\mathbb{F}_{2^m}$. The number of solution in $\mathbb{F}_{2^m}$ of the Eq. (14.4) in t is equal to the number of $t \in \mathbb{F}_{2^m}$ such that $\pi(t) + a^{-2^{m-1}}\beta t = a^{-2^{m-1}}\alpha$ which is equal either to 0 or to 2. Indeed,

$$\pi(t) + \beta t = \alpha \iff t^{2^{m-1}} + \beta t = \alpha + 1.$$

Set $\ell(t) := t^{2^{m-1}} + \beta t$. As ℓ is a $\mathbb{F}_2$-linear map from $\mathbb{F}_{2^m}$ to itself (viewed as vector space of $\mathbb{F}_2$ on both sides) with kernel $ker\ell =< \beta^{\frac{1}{2^m-1}} >$ the claim follows. Hence $\#\{t \in \mathbb{F}_{2^m} \mid \ell(t) = \alpha + 1\}$ is equal to 0 if $\alpha + 1 \notin \ell(\mathbb{F}_{2^m})$ and equal to $2^{\dim_{\mathbb{F}_2} \ker(\ell)}$ otherwise. Note then that $\ker(\ell)$ is the set of roots of the polynomial $t^{2^{m-1}} + \beta t$ which is equal to the one-dimensional vector space generated by β^{-2} (since $t^{2^{m-1}} + \beta t = 0 \iff t = 0$ or $t^{2^{m-1}-1} = t^{2^{m-1}-2^m} = t^{-2^{m-1}} = \beta$). Therefore, for every $(\alpha, \beta) \in \mathbb{F}_{2^m} \times \mathbb{F}_{2^m}^\star$, the equation $G(t) + \beta t = \alpha$ has 0 or 2 solutions in $\mathbb{F}_{2^m}$, proving that $\tilde{f}$ belongs to the class $\mathcal{H}$. Now, according to the calculation above, $\pi(t) = 1 + t^{2^{m-1}}$. Thus, the o-polynomial associated to $\tilde{f}$ is $G(t) = a^{2^{m-1}}(1 + t^{2^{m-1}})$. We have shown that $\tilde{f}$ is an element of $\mathcal{H}$ and is thus bent.

Let us compute its dual function. Take ζ and $\tilde{f}$ as before. Note that, $\{1, \zeta\}$ is a $\mathbb{F}_{2^m}$-basis of $\mathbb{F}_{2^n}$ by the choice of ζ. Let us establish a relation between the Walsh transforms of f and $\tilde{f}$, respectively. In fact, we will show below the following relationship:

$$\widehat{\chi_f}(\alpha + \zeta\beta) = \widehat{\chi_{\tilde{f}}}(\beta, \alpha + \beta). \tag{14.5}$$

By definition, for every $w \in \mathbb{F}_{2^n}$, we have

$$\widehat{\chi_f}(w) = \sum_{x \in \mathbb{F}_{2^n}} \chi(f(x) + \mathrm{Tr}_1^n(wx)).$$

Write $w = \alpha + \zeta\beta$. Therefore

$$\widehat{\chi_f}(\alpha + \zeta\beta) = \sum_{y \in \mathbb{F}_{2^m}} \sum_{z \in \mathbb{F}_{2^m}} \chi(f(y + \zeta z) + \mathrm{Tr}_1^n((\alpha + \zeta\beta)(y + \zeta z))).$$

Now

$$\begin{aligned}
\mathrm{Tr}_1^n((\alpha + \zeta b)(y + \zeta z)) &= \mathrm{Tr}_1^n(\alpha y + \zeta\alpha z + \zeta\beta y + \zeta^2\beta z) \\
&= \mathrm{Tr}_1^m(\mathrm{Tr}_m^n(\alpha y + \zeta\alpha z + \zeta\beta y + \zeta^2\beta z)) \\
&= \mathrm{Tr}_1^m(\alpha y \mathrm{Tr}_m^n(1) + \alpha z \mathrm{Tr}_m^n(\zeta) + \beta y \mathrm{Tr}_m^n(\zeta) + \beta z \mathrm{Tr}_m^n(\zeta^2)) \\
&= \mathrm{Tr}_1^m(y\beta + z(\alpha + \beta)).
\end{aligned}$$

Hence

$$\widehat{\chi_f}(\alpha + \zeta\beta) = \sum_{y \in \mathbb{F}_{2^m}} \sum_{z \in \mathbb{F}_{2^m}} \chi(\tilde{f}(y, z) + \mathrm{Tr}_1^m(y\beta + z(\alpha + \beta))) = \widehat{\chi_{\tilde{f}}}(\beta, \alpha + \beta).$$

Now, we want to show that the dual $f^\star$ of the function $\tilde{f}$ is given by $f^\star(w) = \mathrm{Tr}_1^m(a^{-1}w^{2^m+1}) + 1$. We have got $G(t) = a^{2^{m-1}}(1 + t^{2^{m-1}}) = a^{1/2}(1 + t^{1/2})$ and $\mu = a^{1/2}\tau$. Now, according to the characterization of the dual of functions in the class $\mathcal{H}$ given in [7], the dual function $\tilde{f}^\star$ of $\tilde{f}$ is defined as follows:

$$\tilde{f}^\star(\alpha, \beta) = 1 \iff G(z) + (\beta + \mu)z = \alpha \text{ has no solutions in } \mathbb{F}_{2^m} \text{ (in } z).$$

Now, using the criterion, that a polynomial $a'x^2 + b'x + c'$, with $a' \in \mathbb{F}_{2^n}$, $b' \in \mathbb{F}_{2^n}$, $c' \in \mathbb{F}_{2^n}^\star$, has a solution in $\mathbb{F}_{2^m}$ if and only if, $\mathrm{Tr}_1^m(\frac{a'c'}{b'^2}) = 0$, we deduce (the criterion is applied for $a' = a^{-1/2}\beta + \tau$, $b' = 1$ and $c' = a^{-1/2}\alpha + 1$)

$$G(z) + (\beta + \mu)z = \alpha \iff z^{1/2} + (a^{-1/2}\beta + \tau)z = a^{-1/2}\alpha + 1.$$

This latter equation has no solution in $\mathbb{F}_{2^m}$ if and only if

$$\mathrm{Tr}_1^m((a^{-1/2}\beta + \beta^{2^{n-1}+2^{m-1}})(a^{-1/2}\alpha + 1)) = 1$$

yielding

$$\tilde{f}^\star(\alpha, \beta) = \mathrm{Tr}_1^m((a^{-1/2}\beta + \beta^{2^{n-1}+2^{m-1}})(a^{-1/2}\alpha + 1)).$$

Now, if $w = \alpha + \zeta\beta$, then we obtain the expected relation (14.5).

Thus the dual function $f^\star$ of f can be expressed as follows:

$$f^\star(w) = \tilde{f}^\star(\beta, \alpha + \beta) = \mathrm{Tr}_1^m((a^{-1/2}(\beta + \alpha) + \tau)(a^{-1/2}\beta + 1)).$$

Note then that $\mathrm{Tr}_1^m(a^{-1/2}\beta(a^{-1/2}\beta + 1)) = \mathrm{Tr}_1^m((a^{-1/2}\beta)^2 + a^{-1/2}\beta) = 0$. Hence

$$\begin{aligned} f^\star(w) &= \mathrm{Tr}_1^m((a^{-1/2}\alpha + \tau)(a^{-1/2}\beta + 1)) \\ &= \mathrm{Tr}_1^m(a^{-1}\alpha\beta + a^{-1/2}\alpha + a^{-1/2}\tau\beta + \tau). \end{aligned}$$

Now, one can express α and β with respect to w:

$$\beta = \mathrm{Tr}_m^n(w) = w + w^{2^m} \text{ and } \alpha = \mathrm{Tr}_m^n(\zeta^{2^m} w) = \zeta^{2^m} w + \zeta w^{2^m}.$$

Hence,

$$\begin{aligned} &a^{-1}\alpha\beta + a^{-1/2}\alpha + a^{-1/2}\tau\beta \\ &= a^{-1}(\zeta^{2^m} w^2 + \zeta w^{2^m+1} + \zeta^{2^m} w^{2^m+1} + \zeta w^{2^{m+1}}) \\ &\quad + a^{-1/2}(\zeta^{2^m} w + \zeta w^{2^m}) + a^{-1/2}\tau(w + w^{2^m}). \end{aligned}$$

Collect together the linear terms:

$$\begin{aligned} &a^{-1}(\zeta^{2^m} w^2 + \zeta w^{2^{m+1}}) + a^{-1/2}(\zeta^{2^m} w + \zeta w^{2^m}) + a^{-1/2}\tau(w + w^{2^m}) \\ &= \mathrm{Tr}_m^n(a^{-1}\zeta^{2^m} w^2) + \mathrm{Tr}_m^n(a^{-1/2}\zeta^{2^m} w) + \mathrm{Tr}_m^n(a^{-1/2}\tau w). \end{aligned}$$

Thus

$$\begin{aligned}
&\mathrm{Tr}_1^m\left(a^{-1}(\zeta^{2^m}w^2+\zeta w^{2^m+1})+a^{-1/2}(\zeta^{2^m}w+\zeta w^{2^m})+a^{-1/2}\tau(w+w^{2^m})\right)\\
&=\mathrm{Tr}_1^n(a^{-1}\zeta^{2^m}w^2)+\mathrm{Tr}_1^n(a^{-1/2}\zeta^{2^m}w)+\mathrm{Tr}_1^n(a^{-1/2}\tau w)\\
&=\mathrm{Tr}_1^n\left(a^{-1}\zeta^{2^m}(1+\zeta^{2^m}+\zeta^{2^n})w^2\right).
\end{aligned}$$

Now, $1+\zeta^{2^m}+\zeta^{2^n}=1+\zeta^{2^m}+\zeta=1+\mathrm{Tr}_m^n(\zeta)=0$. Therefore, it remains only the constant term and the quadratic term :

$$f^\star(w)=\mathrm{Tr}_1^m(a^{-1}(\zeta^{2^m}+\zeta)w^{2^m+1}+\tau)=\mathrm{Tr}_1^m(a^{-1}w^{2^m+1})+\mathrm{Tr}_1^m(\tau),$$

because $\zeta^{2^m}+\zeta=1$. Now note that $\tau^2=\zeta^{1+2^m}=\zeta+\zeta^2$ (recall that $\tau:=\zeta^{2^{n-1}+2^{m-1}}$). Hence τ^2 belongs to $\mathbb{F}_{2^m}$ and is the image under the Hilbert transform of an element of $\mathbb{F}_{2^n}\setminus\mathbb{F}_{2^m}$ (since $\mathrm{Tr}_m^n(\zeta)=1$, ζ cannot belong to $\mathbb{F}_{2^m}$) and thus its absolute trace over $\mathbb{F}_{2^m}$ is equal to 1. Finally, we get that the univariate representation of the dual $f^\star$ of the function $\tilde{f}$ is given by $f^\star(w)=\mathrm{Tr}_1^m(a^{-1}w^{2^m+1})+1$.

Therefore, with respect to the correspondences (1) and (2) in Fig. 14.2, we have that the Niho bent function $f(x)=\mathrm{Tr}_1^m(ax^{(2^m+1)})$ corresponds to the function $\tilde{f}(y,z)=f(y+\zeta z)$ in $\mathcal{H}$ and to the o-polynomial $G(t)=a^{2^{m-1}}(1+t^{2^{m-1}})$.

14.2.3 Bent Functions Whose Restrictions to the Multiplicative Cosets $u\mathbb{F}_{2^m}^\star$ ($u\in U$) Are Affine

Notation 14.2.20. *We denote by $\mathcal{A}_n$ the set of Boolean functions f on $\mathbb{F}_{2^n}$ such that the restriction of f to $u\mathbb{F}_{2^m}^\star$ is affine for every $u\in U$.*

Functions in $\mathcal{A}_n$ have been firstly investigated in [7]. In this section, we complete the results of [7] by giving a complete classification of functions in $\mathcal{A}_n$. Note that $f\in\mathcal{A}_n$ means that there exists a mapping $a:U\to\mathbb{F}_{2^m}$ and a Boolean function $b:U\to\mathbb{F}_2$ such that $f(uy)=\mathrm{Tr}_1^m(a(u)y)+b(u)$, $\forall u\in U,\forall y\in\mathbb{F}_{2^m}^\star$.

Moreover, the reader can notice that if we would consider the functions over $\mathbb{F}_{2^n}$ such that the restrictions to $u\mathbb{F}_{2^m}$ (instead of $u\mathbb{F}_{2^m}^\star$) is affine for every $u\in U$, then the constant term of such functions f would be the same (equals $f(0)=b(u)$) which is a too restrictive condition.

In the sequel, we denote by $f_{a,b}$ a function in $\mathcal{A}_n$ represented as in Notation 14.2.20.

Therefore, we have the following natural decomposition:

$$f_{a,b}=f_{a,0}+f_{0,b}.$$

Then $f_{a,0}$ is a Boolean function defined on $\mathbb{F}_{2^n}$ such that its restrictions to $u\mathbb{F}_{2^m}^\star$ ($u \in U$) are linear and $f_{0,b}$ is a Boolean function on $\mathbb{F}_{2^n}$ such that its restrictions to $u\mathbb{F}_{2^m}^\star$ ($u \in U$) are constant.

Remark 14.2.21. $f_{a,b} \in \mathcal{A}_n$ if and only if $1+f_{a,b} \in \mathcal{A}_n$. Indeed, it is easy to see that if $f_{a,b} \in \mathcal{A}_n$ then, $1+f_{a,b}(uy) = \mathrm{Tr}_1^m(u(u)y) + b'(u), \forall u \in U, \forall y \in \mathbb{F}_{2^m}^\star$ with $b'(u) := b(u)+1$, which means that $1+f_{a,b} \in \mathcal{A}_n$. The converse is trivial. Moreover, we have $1+f_{a,b} = f_{a,b+1}$.

Notation 14.2.22. *For* $\epsilon \in \{0,1\}$, *set*

$$\mathcal{A}_n^\epsilon := \{f \in \mathcal{A}_n \mid f_{a,b}(0) = \epsilon\}.$$

We have

$$\mathcal{A}_n^\epsilon = \mathcal{A}_n^0 \cup \mathcal{A}_n^1 = \mathcal{A}_n^0 \cup (1+\mathcal{A}_n^0)$$

where $1+\mathcal{A}_n^0$ is the complement of functions in $\mathcal{A}_n^0$. In the following, we are identify the functions in $\mathcal{A}_n$ which are bent. Since bentness is affine invariant, it suffices to study the bent functions in $\mathcal{A}_n^0$. To this end, we first compute the Walsh transform a function in $\mathcal{A}_n^0$.

Proposition 14.2.23 ([31]). *Let* $f_{a,b}$ *be a function in* $\mathcal{A}_n^0$. *Then for all* ω *in* $\mathbb{F}_{2^n}$ *we have* $\widehat{\chi_{f_{a,b}}}(\omega) = 1 - \sum_{u\in U}(-1)^{b(u)} + 2^m \sum_{u\in U \mid a(u)+\mathrm{Tr}_m^n(\omega u)=0}(-1)^{b(u)}$.

Proof. Let $f_{a,b} \in \mathcal{A}_n^0$. For all $\omega \in \mathbb{F}_{2^n}$, we have

$$\begin{aligned}
\widehat{\chi_{f_{a,b}}}(\omega) &= 1 + \sum_{x\in\mathbb{F}_{2^n}^\star}(-1)^{f_{a,b}(x)+\mathrm{Tr}_1^n(\omega x)}\\
&= 1 + \sum_{u\in U}\sum_{y\in\mathbb{F}_{2^m}^\star}(-1)^{\mathrm{Tr}_1^m(a(u)y)+b(u)+\mathrm{Tr}_1^m(\mathrm{Tr}_m^n(\omega u)y)}\\
&= 1 + \sum_{u\in U}\Big(\sum_{y\in\mathbb{F}_{2^m}}(-1)^{\mathrm{Tr}_1^m(a(u)y)+b(u)+\mathrm{Tr}_1^m(\mathrm{Tr}_m^n(\omega u)y)} - (-1)^{b(u)}\Big)\\
&= 1 - \sum_{u\in U}(-1)^{b(u)} + \sum_{u\in U}(-1)^{b(u)}\sum_{y\in\mathbb{F}_{2^m}}(-1)^{\mathrm{Tr}_1^m((a(u)+\mathrm{Tr}_m^n(\omega u))y)}.
\end{aligned}$$

The result follows after noticing that $\sum_{y\in\mathbb{F}_{2^m}}\chi(\mathrm{Tr}_1^m\big((a(u)+\mathrm{Tr}_m^n(\omega u))y\big)$

$$= \begin{cases} 2^m \text{ if } a(u)+\mathrm{Tr}_m^n(\omega u) = 0\\ 0 \quad \text{otherwise.}\end{cases}$$

□

To prove the main result of this subsection, we need the two following statements. Lemma 14.2.24 is a known result (see e.g. [4]).

Lemma 14.2.24. *Let f be a Boolean function on $\mathbb{F}_{2^n}$ such that $\widehat{\chi_f}(\omega) \geq 0$ for all $\omega \in \mathbb{F}_{2^n}$. Then f is linear.*

Proof. We have $\sum_{\omega\in\mathbb{F}_{2^n}} \widehat{\chi_f}(\omega) = (-1)^{f(0)}2^n$. Hence, $\left(\sum_{\omega\in\mathbb{F}_{2^n}} \widehat{\chi_f}(\omega)\right)^2 = 2^{2n}$. But $\left(\sum_{\omega\in\mathbb{F}_{2^n}} \widehat{\chi_f}(\omega)\right)^2 = \sum_{\omega\in\mathbb{F}_{2^n}} (\widehat{\chi_f}(\omega))^2 + \sum_{\omega\neq\omega'} \widehat{\chi_f}(\omega)\widehat{\chi_f}(\omega')$. Now, according to the Parseval identity, we get $\sum_{\omega\neq\omega'} \widehat{\chi_f}(\omega)\widehat{\chi_f}(\omega') = 0$. Since $\widehat{\chi_f}(\omega) \geq 0, \forall\omega \in \mathbb{F}_{2^n}$, $\widehat{\chi_f}(\omega)\widehat{\chi_f}(\omega') = 0, \forall\omega \neq \omega'$. The set $\{\omega \in \mathbb{F}_{2^n} \mid \widehat{\chi_f}(\omega) \neq 0\}$ is then reduced to a singleton $\{\omega_0\}$. Now, according to Fourier inversion formula :

$$(-1)^{f(x)} = \frac{1}{2^n}\sum_{\omega\in\mathbb{F}_{2^n}} \widehat{\chi_f}(\omega)(-1)^{\mathrm{Tr}_1^n(\omega x)}$$

we get

$$(-1)^{f(x)} = \frac{1}{2^n}\widehat{\chi_f}(\omega_0)(-1)^{\mathrm{Tr}_1^n(\omega_0 x)}. \tag{14.6}$$

Hence, we obtain $\widehat{\chi_f}^2(\omega_0) = 2^{2n}$, that is $\widehat{\chi_f}(\omega_0) = 2^n$ (since $\widehat{\chi_f}(\omega_0) \geq 0$). Finally, using Eq. (14.6), we obtain $f(x) = \mathrm{Tr}_1^n(\omega_0 x)$, which proves that f is linear. □

Lemma 14.2.25 is a known result (see. e.g. [4]). We include a slightly different proof.

Lemma 14.2.25. *Let $n = 2m$ and E be a vector space of dimension m. Let f be a function in $\mathbb{F}_{2^n}$ such that f is bent and the restriction of f to E is linear. Then $f + \mathbf{1}_E$ is bent (where $\mathbf{1}_E$ denotes the characteristic function of E).*

Proof. Set $g := f + \mathbf{1}_E$. Let $\mu \in \mathbb{F}_{2^n}$ such that $f(x) = \mathrm{Tr}_1^n(\mu x), \forall x \in E$. Let us compute the Walsh transform of g. For every $\omega \in \mathbb{F}_{2^n}$, we have

$$\begin{aligned}
\widehat{\chi_g}(\omega) &= \sum_{x\in\mathbb{F}_{2^n}} (-1)^{f(x)+\mathbf{1}_E(x)+\mathrm{Tr}_1^n(\omega x)} \\
&= \sum_{x\in\mathbb{F}_{2^n}\setminus E} (-1)^{f(x)+\mathrm{Tr}_1^n(\omega x)} - \sum_{x\in E}(-1)^{f(x)+\mathrm{Tr}_1^n(\omega x)} \\
&= \sum_{x\in\mathbb{F}_{2^n}} (-1)^{f(x)+\mathrm{Tr}_1^n(\omega x)} - 2\sum_{x\in E}(-1)^{\mathrm{Tr}_1^n((\mu+\omega)x)} \\
&= \widehat{\chi_f}(\omega) - 2\#E\ \mathbf{1}_{E^\perp}(\mu + \omega) \\
&= \widehat{\chi_f}(\omega) - 2^{m+1}\ \mathbf{1}_{E^\perp}(\mu + \omega). \\
&\equiv \widehat{\chi_f}(\omega) \pmod{2^{m+1}} \\
&\equiv 2^m \pmod{2^{m+1}} \text{ (since } f \text{ is bent, according to Lemma 14.2.4).}
\end{aligned}$$

□

We give the main result of this section in which we identify the bent functions in $\mathcal{A}_n$, that is, the bent functions f defined on $\mathbb{F}_{2^n}$ such that the restriction of f to $u\mathbb{F}_{2^m}^{\star}$ is affine of every $u \in U$.

Theorem 14.2.26 ([31]). *The bent functions defined on $\mathbb{F}_{2^n}$ ($n = 2m$) such that their restrictions to each multiplicative cosets $u\mathbb{F}_{2^m}^{\star}$ ($u \subset U$) is affine belong to the following classes:*

1. *functions which are the sum of a function from the class $\mathcal{PS}_{ap}^{\#}$ and an affine function.*
2. *Niho bent functions or functions which are the sum of a Niho bent function and the function $1 + \mathbf{1}_{u_0\mathbb{F}_{2^m}}$ or the sum of a Niho bent function and the function $\mathbf{1}_{u_0\mathbb{F}_{2^m}}$ where $u_0 \in U$.*

Proof. According to the discussion above, it suffices to treat the case of bent function in $\mathcal{A}_n^0$. So, let $f_{a,b}$ be a bent function in $\mathcal{A}_n^0$. Since $f_{a,b}$ is bent (equivalently, $\widehat{\chi_{f_{a,b}}}(\omega) = \pm 2^m, \forall \omega \in \mathbb{F}_{2^n}$), we have $\widehat{\chi_{f_{a,b}}}(\omega) \equiv 0 \pmod{2^m}$. But $\widehat{\chi_{f_{a,b}}}(\omega) \equiv 1 - \sum_{u \in U}(-1)^{b(u)} \pmod{2^m}$, by Proposition 14.2.23. Hence,

$$\sum_{u \in U}(-1)^{b(u)} \equiv 1 \pmod{2^m}.$$

Now, since the multiplicative group U is of order $2^m + 1$ then, $\sum_{u \in U}(-1)^{b(u)} \in \{1, 2^m + 1, -2^m + 1\}$. Consequently, we have to consider three cases.

- Case 1: $\sum_{u \in U}(-1)^{b(u)} = 1$.

According to Proposition 14.2.23,

$$\widehat{\chi_{f_{a,b}}}(\omega) = 2^m \sum_{u \in U \mid a(u) + \mathrm{Tr}_m^n(\omega u) = 0} (-1)^{b(u)}, \forall \omega \in \mathbb{F}_{2^n}.$$

Now, recall that $f_{a,b}$ can be decomposed as $f_{a,b} = f_{a,0} + f_{0,b}$, where $f_{a,0}$ (resp. $f_{0,b}$) is such that the restrictions to $u\mathbb{F}_{2^m}^{\star}$ ($u \in U$) are linear (resp. constant). On one hand, we have $f_{0,b}(uy) = f_{0,b}(u) = b(u)$ and we have proved previously (in the proof of Proposition 14.2.5) that $f_{0,b}$ is bent if and only if $\sum_{u \in U}(-1)^{b(u)} = 1$. Therefore, $f_{0,b}$ is a bent function whose restrictions to $u\mathbb{F}_{2^m}^{\star}$ ($u \in U$) are constant. This means that the function $f_{0,b}$ belongs to the class $\mathcal{PS}_{ap}^{\#}$. On the other hand, $\widehat{\chi_{f_{a,b}}}(\omega) = 2^m \sum_{u \in E_\omega}(-1)^{b(u)}, \forall \omega \in \mathbb{F}_{2^n}$ where $E_\omega := \{u \in U \mid a(u) + \mathrm{Tr}_m^n(\omega u) = 0\}$. Hence, $f_{a,b}$ is bent if and only if $\sum_{u \in E_\omega}(-1)^{b(u)} \equiv 1 \pmod 2$. Now, since $(-1)^{b(u)} \equiv 1 \pmod 2$ for all $u \in U$ (because $(-1)^{b(u)} \in \{-1, 1\}, \forall u \in U$), we obtain $\#E_\omega \equiv 1 \pmod 2$, which means that $\#E_\omega$ is odd. Hence, since $f_{a,0}(uy) = \mathrm{Tr}_1^m(a(u)y)$, we have $\widehat{\chi_{f_{a,0}}}(\omega) = 1 - \sum_{u \in U}(-1)^0 + 2^m \sum_{u \in E_\omega}(-1)^0$ (by Proposition 14.2.23), that is, $\widehat{\chi_{f_{a,0}}}(\omega) = 1 - \#U + 2^m \#E_\omega = 2^m(\#E_\omega - 1)$. Now, since $\#E_\omega$ is odd for every $\omega \in \mathbb{F}_{2^n}$, then $\#E_\omega \geq 1, \forall \omega \in \mathbb{F}_{2^n}$, which implies that $\widehat{\chi_{f_{a,0}}} \geq 0, \forall \omega \in \mathbb{F}_{2^n}$. According to Lemma 14.2.24, we conclude that the function $f_{a,0}$ is linear (that is, the function $u \mapsto a(u)$ is constant on U).

- Case 2: $\sum_{u\in U}(-1)^{b(u)} = 2^m + 1$.

Since the set U is of cardinality $2^m + 1$, the function $u \mapsto b(u)$ is necessary identically null. Therefore, the function $f_{a,b} = f_{a,0}$ is a Niho bent function (by definition of $f_{a,0}$ and using the fact that $f_{a,b}$ is bent if and only if $f_{a,0}$ is bent).

- Case 3: $\sum_{u\in U}(-1)^{b(u)} = 1 - 2^m$.

In this case, there exists a unique $u_o \in U$ such that $b(u_0) = 0$ and $b(u) = 1$ for every $u \in U \setminus \{u_0\}$. Indeed, for $\epsilon \in \{0,1\}$ denote by B_ϵ the set $\{u \in U \mid b(u) = \epsilon\}$. We have $\sum_{u\in U}(-1)^{b(u)} = \sum_{u\in B_0}(-1)^{b(u)} + \sum_{u\in B_1}(-1)^{b(u)} = 1-2^m = \#B_0 - \#B_1$. Hence, $\#B_0 = 1$ and $\#B_1 = 2^m$ (since $\#B_0 + \#B_1 = \#U = 2^m + 1$).
Now, $f_{0,b}(uy) = \mathbf{1}_U(u) + \mathbf{1}_{u_0\mathbb{F}_{2^m}}(uy)$, i.e. $f_{0,b} = \mathbf{1}_{u_0\mathbb{F}_{2^m}} + 1$ which implies that $f_{a,b} = f_{a,0} + \mathbf{1}_U + \mathbf{1}_{u_0\mathbb{F}_{2^m}}$. To conclude, we have to prove that $f_{a,0}$ is bent. From the previous equality, we have for every $y \in \mathbb{F}_{2^m}^\star$:

$$\begin{aligned} f_{a,0}(u_0y) &= f_{a,b}(u_0y) + 1 + \mathbf{1}_{u_0\mathbb{F}_{2^m}}(u_0y) \\ &= f_{a,b}(u_0y) \\ &= \mathrm{Tr}_1^m(a(u_0)y) + b(u_0) \\ &= \mathrm{Tr}_1^m(a(u_0)y). \end{aligned}$$

Therefore, the restriction of $f_{a,b}$ to $u_0\mathbb{F}_{2^m}$ is linear (since $f(0) = 0$, by hypothesis). Now, applying Lemma 14.2.25 to the m-dimensional vector space $E := u_0\mathbb{F}_{2^m}$, we deduce that $f_{a,b} + \mathbf{1}_{u_0\mathbb{F}_{2^m}}$ is bent and thus the function $f_{a,0}$ is also bent (since bentness is affine invariant). □

14.3 Bent Functions from Other Spreads

Dillon [14] has also introduced bent functions obtained using, more generally, sets of subgroups of a group. This extension to subgroups has been pushed further in [15, 21]. It has also been shown that the work of Dillon can be extended to odd characteristic (see [28, 32]).

14.3.1 Bent Functions from the Class $\mathcal{PS}$-Like

Recently, finite prequasifield spreads from finite geometry have been revisited by Wu [35] (unpublished) to give explicit forms (more easily usable in applications) of the related functions in $\mathcal{PS}$ and of their duals, thanks to the determination of the compositional inverses of certain parametric permutation polynomials. In particular, Wu has considered the Dempwolff–Muller prequasifield, the Knuth presemifield and the Kantor presemifield to obtain the $\mathcal{PS}$ corresponding bent functions. The constructed functions and their dual functions are in a similar shape as the $\mathcal{PS}_{ap}$ functions, but are more complex.

Let us present some classes. Firstly, for any $a, x \in \mathbb{F}_{2^m}^\star, y \in \mathbb{F}_{2^m}$, define $\odot\frac{y}{x} := a$ if $y = a \odot x$. For $x = 0$, $\odot\frac{y}{x} := 0$. It is not difficult to see that the function f defined on $\mathbb{F}_{2^m} \times \mathbb{F}_{2^m}$ by $f(x, y) = g(\odot\frac{y}{x})$ is a bent function form the class $\mathcal{PS}^-$. Moreover, for the prequasifield $(\mathbb{F}_{2^m}, +, \odot)$, there exists a bivariate polynomial $F(x, y)$ over $\mathbb{F}_{2^m}$ such that $x \odot y = F(x, y)$. One can view $F(x, y)$ as a univariate polynomial $F_y(x)$ by considering the variable y to be a parameter. Then $F_y(x)$ has the following property.

Lemma 14.3.1 ([35]). *$F_0(x) = 0$ and for any $y \in \mathbb{F}_{2^m}^\star$, $F_y(x)$ is a permutation polynomial over $\mathbb{F}_{2^m}$.*

The reader notice that for any $a, x \in \mathbb{F}_{2^m}$, $y \in \mathbb{F}_{2^m}$, if $y = a \odot x = F(a, x) = F_x(a)$, one have $\odot\frac{y}{x} = a = F_x^{-1}(y)$. Notice that this means that if we want to explicitly represent the division operation $\odot\frac{y}{x}$ by a bivariate polynomial over $\mathbb{F}_{2^m}$, we need only to compute the compositional inverse of the parametric permutation polynomial $F_y(x)$ derived from the multiplication operation of the prequasifield and reverse the role of x and y in its formula afterwards. In [35], Wu has related the problem of constructing $\mathcal{PS}$ of bent functions with the problem of computing compositional inverses of special parametric permutation polynomials over finite fields.

Now, we give three classes of $\mathcal{PS}$ bent functions constructed form prequasifield spreads presented in [35]:

1. The $\mathcal{PS}_{\mathcal{D}-\mathcal{M}}$ class related to the Dempwolff–Muller prequasifield:

   Assume k and m are odd integers with $gcd(k, m) = 1$. Let $e = 2^{m-1} - 2^{k-1} - 1$, $L(x) = \sum_{i=0}^{k-1} x^{2i}$ and define a multiplication in $\mathbb{F}_{2^m}$ as

$$x \odot y = x^e L(xy).$$

   We consider the Dempwolff–Muller prequasifield $(\mathbb{F}_{2^m}, +, \odot)$.

$$x \odot y = \frac{1}{xD_d(\frac{y^2}{x^{2^k+1}})}$$

   where $D_d(x)$ is the Dickson polynomial on $\mathbb{F}_{2^m}$ of order d.

   Functions of the subclass the $\mathcal{PS}_{\mathcal{D}-\mathcal{M}}$ class are functions of $\mathcal{PS}$ defined as follows:

$$f(x, y) = g\Big(\frac{1}{xD_d(\frac{y^2}{x^{2^k+1}})}\Big)$$

   where g is a balanced Boolean function over $\mathbb{F}_{2^m}$ such that $g(0) = 0$.

2. The $\mathcal{PS}_{Knu}$ class related to the Knuth prequasifield:
   Assume m is odd. For any $\beta \in \mathbb{F}_{2^m}^\star$, define a multiplication in $\mathbb{F}_{2^m}$ by

$$x \odot y = xy + x^2 \mathrm{Tr}_1^m(\beta y) + y^2 \mathrm{Tr}_1^m(\beta x).$$

   We consider the Knuth presemifiled $(\mathbb{F}_{2^m}, +, \odot)$. Then we have (see [35]),

$$\odot \frac{y}{x} = (1 + \mathrm{Tr}_1^m(\beta x))\frac{y}{x} + x\mathrm{Tr}_1^m(\beta \frac{y}{x}) + x\mathrm{Tr}_1^m(\beta x) C_{\frac{1}{\beta x}}(\frac{y}{x^2})$$

   where $C_a(z)$ is the polynomial on $\mathbb{F}_{2^m}$ defined as follows: $C_a(z) := \sum_{i=0}^{m-1} c_i^{2^i}$ with coefficients given by

$$c_0 = \frac{1}{a}(1, 3, 5, \cdots, m-3),$$

   and for all i such that $1 \leq i \leq m-1$:

$$c_i = \begin{cases} 1 + \frac{1}{a}(1, 3, 5, \cdots, i-2, i+1, i+3, \cdots, m-1) \text{ if i is odd} \\ \frac{1}{a}(0, 2, 4, \cdots, i-2, i+1, i+3, \cdots, m-2) \text{ if i is even} \end{cases}$$

   Functions of the subclass the $\mathcal{PS}_{Knu}$ class are functions of $\mathcal{PS}$ defined as follows:

$$f(x, y) = g\Big((1 + \mathrm{Tr}_1^m(\beta x))\frac{y}{x} + x\mathrm{Tr}_1^m(\beta \frac{y}{x}) + x\mathrm{Tr}_1^m(\beta x) C_{\frac{1}{\beta x}}(\frac{y}{x^2})\Big)$$

   where g is a balanced Boolean function over $\mathbb{F}_{2^m}$ such that $g(0) = 0$.
3. The $\mathcal{PS}_{Kan}$ class related to the Kantor presemifiled: Assume m is odd. Define a multiplication in $\mathbb{F}_{2^m}$ by

$$x \odot y = x^2 y + \mathrm{Tr}_1^m(xy) + x\mathrm{Tr}_1^m(y).$$

   We consider the Kantor presemifiled $(\mathbb{F}_{2^m}, +, \odot)$. Then we have (see [35])

$$\odot \frac{y}{x} = \frac{\mathrm{Tr}_1^m(x)}{x}\Big((xy)^{2^{m-1}} + \sum_{i=0}^{\frac{m-1}{2}} (xy)^{2^{2i}-1} + (\sum_{i=0}^{\frac{m-3}{2}} x^{2^{2i}})\mathrm{Tr}_1^m(xy)\Big)$$

   Functions of the subclass the $\mathcal{PS}_{Kan}$ class are functions of $\mathcal{PS}$ defined as follows:

$$f(x, y) = g\Big(\frac{\mathrm{Tr}_1^m(x)}{x}\Big((xy)^{2^{m-1}} + \sum_{i=0}^{\frac{m-1}{2}} (xy)^{2^{2i}-1} + (\sum_{i=0}^{\frac{m-3}{2}} x^{2^{2i}})\mathrm{Tr}_1^m(xy)\Big)\Big)$$

   where g is a balanced Boolean function over $\mathbb{F}_{2^m}$ such that $g(0) = 0$.

Very recently, Carlet has similarly studied in [5] the $\mathcal{PS}$ functions related to the André spreads and given the trace representation of the $\mathcal{PS}$ corresponding bent functions and of their duals.

14.3.2 Bent Functions Which Are Linear on Elements of Classical Spreads

In the continuation of [7], Ceşmelioğlu, Meidl and Pott [34] have showed in 2015 that bent functions f from $\mathbb{F}_p^m \times \mathbb{F}_p^m$ to $\mathbb{F}_p$ which are constant or affine on the elements of a given spread of $\mathbb{F}_p^m \times \mathbb{F}_p^m$, either arise from partial spread bent functions, or are Boolean and a generalization of Dillon's class H (that is, the class $\mathcal{H}$ for which we can use an arbitrary spread, not just the classical Desarguesian one). Note that bent Boolean functions which are affine on the elements of the Desarguesian spread have been studied in [31] (see above). For spreads of a presemifield S, Ceşmelioğlu et al. showed that a bent function of the second class corresponds to an o-polynomial of a presemifield in the Knuth orbit of S. This is in complete contrast to the case of finite fields where the Knuth orbit collapses to only one isotopy class of semifields and raises the question of the existence of o-polynomials for presemifields other than the finite field (and equivalently the question of the existence of hyperovals for semifield planes). To this end, the authors gave in [34] a canonical example of an o-polynomial for commutative presemifields (which also defines a hyperoval on the semifield plane), and showed that the corresponding bent functions belong to the completed Maiorana–McFarland class. Using Albert's twisted fields and Kantor's family of presemifields they explicitly present examples of such bent functions.

Independently, Carlet [5] has also characterized bent Boolean functions whose restrictions to the spaces of an André spread are linear. This leads to a notion extending that of o-polynomial. He also obtained similar characterizations for the $\mathcal{H}$-like functions derived from the spreads presented by Wu in [35]. Nevertheless, from those characterizations, no explicit construction has been provided.

In the next section, we present an overview on the bent function which are linear on the elements of some spreads (we shall omit details in the case of the Desarguesian spreads since it has been developed above).

14.4 Overview on Constructions of Bent Functions Linear on the Elements of Spreads

14.4.1 Bent Functions Linear on the Elements of the Desarguesian Spreads: The So-called Class $\mathcal{H}$

Recall that in his thesis [14], Dillon introduced the class of bent functions denoted by H. Functions of the class H are defined in bivariate representation as

$$f(x, y) = Tr_1^m(y + x\psi(yx^{2^m-2})), \tag{14.7}$$

where $x, y \in \mathbb{F}_{2^m}$ and ψ is a permutation of $\mathbb{F}_{2^m}$ such that $\psi(x) + x$ does not vanish and for any $\beta \in \mathbb{F}_{2^m}^\star$, the function $\psi(x) + \beta x$ is 2-to-1 (i.e. the the pre-image of any element of $\mathbb{F}_{2^m}$ is either a pair or the empty set). The condition that $\psi(x) + x$ does not vanish is required only for (14.7) to have a particular feature but is not necessary for bentness. Dillon was just able to exhibit bent functions in H that also belong to the completed Maiorana–McFarland class. In [7], Carlet and the second author have been extended the class H into a class denoted by $\mathcal{H}$ defined as follows.

Definition 14.4.1 ([7]). The bent functions f of the class $\mathcal{H}$ are defined as

$$f(x,y) = \begin{cases} Tr_1^m\left(x\psi\left(\frac{y}{x}\right)\right) & \text{if } x \neq 0 \\ Tr_1^m(\mu y) & \text{if } x = 0 \end{cases} \tag{14.8}$$

where $\mu \in \mathbb{F}_{2^m}$ and ψ is a mapping from $\mathbb{F}_{2^m}$ to itself satisfying the following conditions:

$$G := \psi(z) + \mu z \text{ is a permutation on } \mathbb{F}_{2^m} \tag{14.9}$$

$$\text{For every } \beta \in \mathbb{F}_{2^m}^\star, \text{ the function } z \mapsto G(z) + \beta z \text{ is 2-to-1 on } \mathbb{F}_{2^m}. \tag{14.10}$$

Note that Condition (14.8) express the fact that functions f of the class $\mathcal{H}$ are linear on the elements of the Desarguesian spreads, while conditions (14.9) and (14.10) express the bentness property of f. But it has been proven in [7] that Condition (14.10) implies Condition (14.9) and is necessary and sufficient for f being bent. Functions in $\mathcal{H}$ and in the Dillon class H are the same up to the addition of a linear term (namely, the term $Tr_1^m((\mu + 1)y)$). Any mapping G on $\mathbb{F}_{2^m}$ that satisfies Condition (14.10) is called an *oval polynomial* (in brief, "o-polynomial").

14.4.2 Bent Functions Linear on the Elements of Other Spreads: $\mathcal{H}$-Like Functions

In the line of the generalization done by Wu [35] of the well-known class *partial spread* $\mathcal{PS}$[2] of Dillon into class $\mathcal{PS}$-like, the class $\mathcal{H}$ has been also generalized in [5] into class $\mathcal{H}$-like by considering other spreads.

We recall the construction of bent functions from [5]. Let $L_z : \mathbb{F}_{2^m} \to \mathbb{F}_{2^m}$ be a linear function for any $z \in \mathbb{F}_{2^m}$. Consider a spread whose elements are the subspace

[2]Recall that the general partial spreads class $\mathcal{PS}$, introduced by Dillon, equals the union of $\mathcal{PS}^-$ and $\mathcal{PS}^+$. Dillon has applied the construction to the Desarguesian spread and deduced the subclass of $\mathcal{PS}^-$ denoted by $\mathcal{PS}_{ap}$ whose elements are constant on the elements of the Desarguesian spread. Functions f of the class $\mathcal{PS}_{ap}$ are given in bivariate form as $f(x,y) = g(xy^{2^m-2})$ where $x, y \in \mathbb{F}_{2^m}$ and g is any balanced Boolean function on $\mathbb{F}_{2^m}$ which vanishes at 0.

$\{(0, y) \mid y \in \mathbb{F}_{2^m}\}$ and 2^m subspaces $\{(x, L_z(x)) \mid x \in \mathbb{F}_{2^m}\}$. We have seen that these subspaces form a spread if and only if the mapping $z \mapsto L_z(x) = y$ is a permutation of $\mathbb{F}_{2^m}$. Denote by Γ_x the compositional inverse of the permutation L_z (we have $\Gamma_x(y) = z$). A Boolean function on $\mathbb{F}_{2^m} \times \mathbb{F}_{2^m}$ is linear on the elements of the spread if and only if there exists a function $G : \mathbb{F}_{2^m} \to \mathbb{F}_{2^m}$ and an element $\mu \in \mathbb{F}_{2^m}$ such that, for every $y \in \mathbb{F}_{2^m}$,

$$f(0, y) = \mathrm{Tr}_1^m(\mu y), \tag{14.11}$$

and for every $x, z \in \mathbb{F}_{2^m}$,

$$f(x, L_z(x)) = \mathrm{Tr}_1^m(G(z)x). \tag{14.12}$$

Up to EA-equivalence, one can assume that $\mu = 0$. Indeed, one can add the linear function $g(x, y) = \mathrm{Tr}_1^m(\mu y)$ to f; this changes μ into 0 and $G(z)$ into $G(z) + L_z^{ad}(\mu)$, where L_z^{ad} is the adjoint operator of L_z, since for $y = L_z(x)$ one has $\mathrm{Tr}_1^m(\mu y) = B(\mu, y) = B(\mu, L_z(x)) = B(L_z^{ad}(\mu), x) = \mathrm{Tr}_1^m(L_z^{ad}(\mu)x)$.

We take $\mu = 0$ in expression (14.11), and relation (14.12) becomes

$$f(x, y) = \mathrm{Tr}_1^m(G(z)x) = \mathrm{Tr}_1^m(G(\Gamma_x(y))x). \tag{14.13}$$

Theorem 14.4.2 ([5]). *Consider a spread of $\mathbb{F}_{2^m} \times \mathbb{F}_{2^m}$ whose elements are 2^m subspaces of the form $\{(x, L_z(x)) \mid x \in \mathbb{F}_{2^m}\}$, where, for every $z \in \mathbb{F}_{2^m}$, function L_z is linear, and the subspace $\{(0, y) \mid y \in \mathbb{F}_{2^m}\}$. For every $x \in \mathbb{F}_{2^m}^\star \mathbb{F}_{2^m}$, let us denote by Γ_x the compositional inverse of the permutation $L_z : z \mapsto L_z(x) = y$. A Boolean function defined by Eq.* (14.13) *is bent if and only if G is a permutation and, for every $b \neq 0$ the function $G(z) + L_z^{ad}(b)$ is 2-to-1, where L_z^{ad} is the adjoint operator of L_z.*

14.4.2.1 The Case of the André's Spreads

Carlet [5] has applied the construction given by the previous theorem by considering the André's spreads and deduced the related $\mathcal{H}$-like functions. Using the notation of Sect. 14.1.2, in the case of André's spreads, L_z is given by $L_z(x) = x^{2^{k\phi(z)}} z$, $\Gamma_x(y) = \frac{y}{x^{2^{k\phi(y/x)}}}$, and $L_z^{ad}(b) = (bz)^{2^{m-k\phi(z)}}$. Consequently, Relation (14.13) becomes:

$$f(x, y) = \mathrm{Tr}_1^m\Big(G\Big(\frac{y}{x^{2^{k\phi(y/x)}}}\Big)x\Big), \forall x, y \in \mathbb{F}_{2^m}. \tag{14.14}$$

Functions of the form (14.14) are thus linear on the elements of the André's spreads. Such functions f (with the convention $\frac{1}{0} = 0$) are bent if and only if, G satisfy the following Condition (14.15) and Condition (14.16)

$$z \mapsto G(z) \text{ is a permutation polynomial of } \mathbb{F}_{2^m}, \tag{14.15}$$

and

$$z \mapsto G(z) + (bz)^{2^{m-k\phi(z)}} \text{ is 2-to-1 for every } b \in \mathbb{F}_{2^m}^{\star}. \tag{14.16}$$

Any polynomial satisfies Condition (14.15) and Condition (14.16) is called a *ϕ-polynomial.* In particular, when ϕ is null, this notion corresponds to that of o-polynomials. Therefore, the class of bent functions of $\mathcal{H}$ gives rise to o-polynomial and the class of bent functions of $\mathcal{H}$-like gives rise to ϕ-polynomials.

14.4.2.2 The Case of Spreads Based on Prequasifields

Carlet [5] has applied the construction given by the previous theorem by considering some spreads based on prequasifields and deduced the related $\mathcal{H}$-like functions. In the following, we shall use the notation of Sect. 14.1.2.

1. In the case of the spread derived from the Dempwolff–Muller prequasifield, we have, $\Gamma_x(y) = \frac{1}{xD_d(\frac{y^2}{x^{2^k+1}})}$, where D_d is the Dickson polynomial of index the inverse d of $2^k - 1$ modulo $2^n - 1$, and $L_z^{ad}(b) = \sum_{i=0}^{k-1}(bz^e)^{2^{-i}}z$. Consequently, Relation (14.13) becomes:

$$f(x, y) = \mathrm{Tr}_1^m\Big(G\Big(\frac{1}{xD_d(\frac{y^2}{x^{2^k+1}})}\Big)x\Big), \forall x, y \in \mathbb{F}_{2^m}. \tag{14.17}$$

   Functions of the form (14.17) are thus linear on the elements of the spread derived from the Dempwolff–Muller prequasifield. Such functions f are bent if and only if, G satisfy the following Condition (14.18) and Condition (14.19)

$$z \mapsto G(z) \text{ is a permutation polynomial of } \mathbb{F}_{2^m}, \tag{14.18}$$

   and

$$z \mapsto G(z) + \sum_{i=0}^{k-1}(bz^e)^{2^{-i}}z \text{ is 2-to-1 for every } b \in \mathbb{F}_{2^m}^{\star}. \tag{14.19}$$

2. In the case of the spreads derived from the Knuth presemifield, we have $\Gamma_x(y) = (1 + \mathrm{Tr}_1^m(\beta x))\frac{y}{x} + x\mathrm{Tr}_1^m(\beta\frac{y}{x}) + x\mathrm{Tr}_1^m(\beta x)C_{\frac{1}{\beta x}}(\frac{y}{x^2})$, where $\beta \in \mathbb{F}_{2^m}^{\star}$, $C_a(x) = \sum_{i=0}^{m-1} c_i x^{2^i}$, with $c_0 = \frac{1}{a^{2^i}} + \frac{1}{a^{3\cdot 2^i}} + \cdots + \frac{1}{a^{(m-3)\cdot 2^i}}$, $c_i = 1 + \frac{1}{a^{2^i}} + \frac{1}{a^{3\cdot 2^i}} + \cdots + \frac{1}{a^{(i-2)\cdot 2^i}} + \frac{1}{a^{(i+1)\cdot 2^i}} + \cdots + \frac{1}{a^{(m-1)\cdot 2^i}}$ if i odd and $c_i = 1 + \frac{1}{a^{2\cdot 2^i}} + \frac{1}{a^{4\cdot 2^i}} + \cdots + \frac{1}{a^{(i-2)\cdot 2^i}} + \frac{1}{a^{(i+1)\cdot 2^i}} + \cdots + \frac{1}{a^{(m-2)\cdot 2^i}}$ if i even. We have $L_z^{ad}(b) = bz + b^{2^{m-1}}\mathrm{Tr}_1^m(\beta z) + \beta\mathrm{Tr}_1^m(b^{2^{m-1}}z)$. Consequently, Relation (14.13) becomes:

$$f(x, y) = \mathrm{Tr}_1^m\Big(G\Big(\Big(1 + \mathrm{Tr}_1^m(\beta x)\Big)\frac{y}{x} + x\mathrm{Tr}_1^m(\beta\frac{y}{x}) + x\mathrm{Tr}_1^m(\beta x)C_{\frac{1}{\beta x}}(\frac{y}{x^2})\Big)x\Big), \forall x, y \in \mathbb{F}_{2^m}. \tag{14.20}$$

Functions of the form (14.20) are thus linear on the elements of the spread derived from the Knuth presemifield. Such functions f are bent if and only if, G satisfy the following Condition (14.21) and Condition (14.22)

$$z \mapsto G(z) \text{ is a permutation polynomial of } \mathbb{F}_{2^m}, \tag{14.21}$$

and

$$z \mapsto G(z) + bz + b^{2^{m-1}}\mathrm{Tr}_1^m(\beta x) + \beta\mathrm{Tr}_1^m(b^{2^{m-1}}z) \text{ is 2-to-1 for every } b \in \mathbb{F}_{2^m}^\star. \tag{14.22}$$

3. In the case of the spreads derived from the Kantor presemifield, there are two cases (since the Kantor presemifield leads to two spreads (see Sect. 14.3); the corresponding mapping Γ_x in the first case (resp. in the second case) was determined by Wu [35] (and Carlet [5]), respectively. Functions which are linear on the elements of the first spread derived from the Kantor presemifield are of the form

$$f(x,y) = \mathrm{Tr}_1^m\Big(G\Big(\Big((xy)^{2^{m-1}} + \sum_{i=0}^{\frac{m-1}{2}}(xy)^{2^{2i}-1} + \sum_{i=0}^{\frac{m-3}{2}} x^{2^{2i}}\mathrm{Tr}_1^m(xy)\Big)\frac{\mathrm{Tr}_1^m(x)}{x} + x^{2^{m-1}-1}y^{2^{m-1}} + x^{2^{m-1}-1}\mathrm{Tr}_1^m(xy)\Big)x\Big). \tag{14.23}$$

Such functions f are bent if and only if, G satisfy the following Condition (14.24) and Condition (14.25)

$$z \mapsto G(z) \text{ is a permutation polynomial of } \mathbb{F}_{2^m}, \tag{14.24}$$

and

$$z \mapsto G(z) + bz^2 + z\mathrm{Tr}_1^m(b) + \mathrm{Tr}_1^m(bz) \text{ is 2-to-1 for every } b \in \mathbb{F}_{2^m}^\star. \tag{14.25}$$

Functions which are linear on the elements of the second spread derived from the Kantor presemifield are of the form

$$f(x,y) = \mathrm{Tr}_1^m\Big(G\Big(\frac{y}{x^2} + \mathrm{Tr}_1^m\Big(\frac{1}{x}\Big)\Big(\frac{\mathrm{Tr}_1^m(\frac{y}{x^2})}{x^2} + \frac{\mathrm{Tr}_1^m(\frac{y}{x})}{x}\Big) + \Big(\mathrm{Tr}_1^m(\frac{1}{x}) + 1\Big) \Big(\frac{\mathrm{Tr}_1^m(\frac{y}{x^2}) + \mathrm{Tr}_1^m(\frac{y}{x})}{x^2} + \frac{\mathrm{Tr}_1^m(\frac{y}{x^2})}{x}\Big)\Big)x\Big). \tag{14.26}$$

Such functions f are bent if and only if, G satisfy the following Condition (14.27) and Condition (14.28)

$$z \mapsto G(z) \text{ is a permutation polynomial of } \mathbb{F}_{2^m}, \tag{14.27}$$

and

$$z \mapsto G(z) + (bz)^{2^{m-1}} + z\mathrm{Tr}_1^m(b) + b\mathrm{Tr}_1^m(z) \text{ is 2-to-1 for every } b \in \mathbb{F}_{2^m}^\star. \quad (14.28)$$

14.5 Bent Functions Linear on the Elements of Sympletic Presemifields

Applying the class $\mathcal{H}$'s construction to a larger class of spreads gives more numerous $\mathcal{H}$-like bent functions which is interesting theoretically and may be useful for applications and developments (in particular in coding theory). In this section, we investigate further generalizations of the class $\mathcal{H}$ by studying bent functions which are linear on the elements of sympletic presemifields.

14.5.1 Two Explicit Constructions

We have seen in Sect. 14.4.2 that the construction of the class $\mathcal{H}$ has been extended by considering more general spreads. Unfortunately, no explicit construction of such polynomial G in Theorem 14.4.2 was provided.

By considering spreads based on symplectic presemifields, we derive from Theorem 14.4.2, an explicit construction of bent functions and we compute its dual function. To this end, we provide a mapping G satisfying the required conditions. We therefore answer to an open question addressed in [5].

Theorem 14.5.1 ([2]). *Let $(F, +, \circ)$ be a symplectic presemifield, and $(F, +, \star)$ be its corresponding commutative presemifield. Consider a spread of $\mathbb{F}_{2^m} \times \mathbb{F}_{2^m}$ whose elements are subspaces $\{(0, y) \mid y \in \mathbb{F}_{2^m}\}$ and $\{(x, z \circ x) \mid x \in \mathbb{F}_{2^m}\}$, $z \in \mathbb{F}_{2^m}$. For every $x \in \mathbb{F}_{2^m}^\star\mathbb{F}_{2^m}$, denote by Γ_x the inverse of the permutation $z \mapsto z \circ x = y$, and set $G(z) = z \star z$. For every $c \in \mathbb{F}_{2^m}$, let Z_c be the image of the map $z \mapsto z \star (z + c)$, and χ_c be the characteristic function of the set $Z_c \times \{c\}$. Then a Boolean function defined by Eq.* (14.13) *is bent, and its dual function is*

$$\tilde{f} = 1 + \sum_{c \in \mathbb{F}_{2^m}} \chi_c.$$

Proof. We put $L_z(x) = z \circ x$ and use Theorem 14.4.2. We have to show that the equation

$$G(z) + L_z^{ad}(b) = a \quad (14.29)$$

has 0 or 2 solutions in $\mathbb{F}_{2^m}$, for every $b \in \mathbb{F}_{2^m}^\star$, $a \in \mathbb{F}_{2^m}$. Since $L_z^{ad}(b) = z \star b$ and $G(z) = z \star z$, the Eq. (14.29) becomes

$$z \star z + z \star b = a,$$

or

$$z \star (z+b) = a.$$

Denote

$$H_b(z) = G(z) + L_z^{ad}(b) = z \star z + z \star b.$$

We note that $H_b(z)$ is a linear map (over $\mathbb{F}_2$), since operation $*$ is commutative, and $\ker H_b = \{0, b\}$. Therefore, the equation $H_b(z) = a$ has 0 or 2 solutions in F.

It remains to prove that G is a permutation. Suppose that the linear map G is not invertible. Then there exists $a \in \mathbb{F}_{2^m}^\star$ such that $G(a) = 0$. Therefore $a \star a = 0$, a contradiction.

Now, we compute the dual function. The Walsh transform of the function $f(x, y)$ is given by

$$\begin{aligned}
\widehat{\chi_f}(a,b) &= \sum_{x,y \in \mathbb{F}_{2^m}} (-1)^{f(x,y)+\mathrm{Tr}_1^m(ax+by)} \\
&= \sum_{x,y \in \mathbb{F}_{2^m}} (-1)^{\mathrm{Tr}_1^m(G(\Gamma_x(y))x+ax+by)} \\
&= 2^m \delta_0(b) + \sum_{x \in \mathbb{F}_{2^m}^\star \mathbb{F}_{2^m},\, z \in \mathbb{F}_{2^m}} (-1)^{\mathrm{Tr}_1^m(G(z)x+ax+bL_z(x))} \\
&= 2^m(\delta_0(b) - 1) + \sum_{z \in \mathbb{F}_{2^m},\, x \in \mathbb{F}_{2^m}} (-1)^{\mathrm{Tr}_1^m((G(z)+a+L_z^{ad}(b))x)} \\
&= 2^m(\delta_0(b) - 1 + \#\{z \in \mathbb{F}_{2^m}, G(z) + a + L_z^{ad}(b) = 0\}) \\
&= 2^m(\delta_0(b) - 1 + \#\{z \in \mathbb{F}_{2^m}, z \star z + z \star b = a\}) \\
&= 2^m(-1)^{1+\sum_{c \in \mathbb{F}_{2^m}} \chi_c},
\end{aligned}$$

which complete the proof. □

Denote $R_x(z) = z \circ x = y$. Let $\Gamma_x = R_x^{-1}$ be the inverse function, so $z = R_x^{-1}(y)$. Let $G(z) = z \star z$. Then function from (14.13) can be rewritten as

$$f(x,y) = \mathrm{Tr}_1^m(G(z)x) = B(z \star z, x) = B(z, z \circ x) = B(z, y) = B(\Gamma_x(y), y) = \mathrm{Tr}_1^m(R_x^{-1}(y)y). \tag{14.30}$$

If the multiplication $\star$ in a commutative presemifield $(F, +, \star)$ is given by

$$x \star y = xy + \sum_{i<j} a_{ij}(x^{2^i}y^{2^j} + x^{2^j}y^{2^i}),$$

then $G(z) = z \star z = z^2$.

Now we consider spreads of symplectic presemifields and prove a result similar to Theorem 14.5.1.

Theorem 14.5.2 ([2]). *Let* $(\mathbb{F}_{2^m}, +, \circ)$ *be a symplectic presemifield. Consider a spread of* $\mathbb{F}_{2^m} \times \mathbb{F}_{2^m}$ *whose elements are subspaces* $\{(0, y) \mid y \in \mathbb{F}_{2^m}\}$ *and* $\{(x, x \circ z) \mid x \in \mathbb{F}_{2^m}\}$, $z \in \mathbb{F}_{2^m}$. *For every* $x \in \mathbb{F}^{\star}_{2^m}\mathbb{F}_{2^m}$, *denote by* Γ_x *the inverse of the permutation* $z \mapsto x \circ z = y$, *and put* $G(z) = \sqrt{z}$. *For every* $c \in \mathbb{F}_{2^m}$, *let* Z_c *be the image of the map* $z \mapsto \sqrt{z} + c \circ z$, *and* χ_c *be the characteristic function of the set* $Z_c \times \{c\}$. *Then a Boolean function defined by Eq.* (14.13) *is bent, and its dual function is*

$$\tilde{f} = 1 + \sum_{c \in \mathbb{F}_{2^m}} \chi_c.$$

Proof. Set $L_z(x) = x \circ z$ and use again Theorem 14.4.2. We have to show that the equation

$$G(z) + L_z^{ad}(b) = a$$

has 0 or 2 solutions in $\mathbb{F}_{2^m}$, for any nonzero $b \in \mathbb{F}_{2^m}$ and any $a \in \mathbb{F}_{2^m}$. It is equivalent to showing that

$$\dim \operatorname{Im}\left(G(z) + L_z^{ad}(b)\right) = m - 1,$$

since $G(z) + L_z^{ad}(b)$ is a linear function. Note that $L_z^{ad}(b) = L_z(b)$ since right multiplication by z in a symplectic semifield is self-adjoint. Denote $M(z) = L_z^{ad}(b) = b \circ z$. Let $(\mathbb{F}_{2^m}, +, \star)$ be the commutative presemifield corresponding to the symplectic presemifield $(\mathbb{F}_{2^m}, +, \circ)$. Note that the adjoint function $L_z^{ad}(b)$ was calculated as adjoint of a function in b. Now we calculate the adjoint of $M(z) = L_z^{ad}(b)$ as function in z and show that $M^{ad}(z) = b \star z$. Indeed,

$$B(M^{ad}(z), x) = B(z, M(x)) = B(z, L_x^{ad}(b)) = B(z, b \circ x) = B(b \star z, x).$$

We note that $G^{ad}(z) = z^2$ since

$$B(x, G^{ad}(z)) = B(G(x), z) = B(\sqrt{x}, z) = \operatorname{Tr}_1^m(\sqrt{x}z) = \operatorname{Tr}_1^m(xz^2) = B(x, z^2).$$

Therefore,

$$(G(z) + M(z))^{ad} = G^{ad}(z) + M^{ad}(z) = z^2 + b \star z.$$

Then

$$\begin{aligned}\dim \mathrm{Im}\Big(G(z) + L_z^{ad}(b)\Big) &= \dim \mathrm{Im}\Big(G(z) + M(z)\Big)\\ &= \dim \mathrm{Im}\Big((G(z) + M(z))^{ad}\Big)\\ &= \dim \mathrm{Im}(z^2 + z \star b)\\ &= m - 1.\end{aligned}$$

Hence, the Boolean function defined by Eq. (14.13) is bent. Its dual (bent) function can be computed as in proof of Theorem 14.5.1. □

The following statement shows that when the mapping G (involved in the expression of a function f which is linear on the elements of a spread derived from a presemifiled) is linear then corresponding bent function f belong the completed well-know class of Maiorana–McFarland.

Proposition 14.5.3 ([2]). *Let $(F, +, \circ)$ be a presemifield. Consider a spread of $\mathbb{F}_{2^m} \times \mathbb{F}_{2^m}$ whose elements are subspaces $\{(0, y) \mid y \in \mathbb{F}_{2^m}\}$ and $\{(x, z \circ x) \mid x \in \mathbb{F}_{2^m}\}$, $z \in \mathbb{F}_{2^m}$. For every $x \in \mathbb{F}_{2^m}^\star$, denote by Γ_x the inverse of the permutation $z \mapsto z \circ x = y$. Let $G(z)$ be a linear function. If the function $f(x, y)$ defined by Eq. (14.13) is bent then it belongs to the completed Maiorana–McFarland class.*

Proof. It is well-known (see [3, 8, 14]) that a bent function f defined over a vectorspace of dimension n belongs to the completed Maiorana–McFarland class if and only if, there exists an $n/2$-dimensional subspace V for which the second derivative is null on V, that is, $D_aD_bf(x) = f(x+a+b)-f(x+a)-f(x+b)+f(x) = 0$ for all $a, b \in V$ We show that for $V = \{(0, y) \mid y \in \mathbb{F}_{2^m}\}$ the second derivative $D_{(0,\alpha)}D_{(0,\beta)}f(x, y) = 0$ for all $(0, \alpha)$, $(0, \beta) \in V$. It is clear for $x = 0$ since $f(0, y) = 0$ for any $y \in \mathbb{F}_{2^m}$. Let $x \neq 0$. Then by linearity of G and R_x^{-1} (where R_x^{-1} stands for the compositional inverse of R_x defined by $R_x(z) = z \circ x$), we obtain

$$\begin{aligned}D_{(0,\alpha)}D_{(0,\beta)}f(x, y) &= f(x, y + \alpha + \beta) - f(x, y + \alpha) - f(x, y + \beta) + f(x, y)\\ &= B(G(R_x^{-1}(y + \alpha + \beta)), x) - B(G(R_x^{-1}(y + \alpha)), x)\\ &\quad -B(G(R_x^{-1}(y + \beta)), x) + B(G(R_x^{-1}(y)), x)\\ &= 0.\end{aligned}$$

□

14.5.2 On Oval Polynomials for Presemifields

In the following, we introduce the notion of an o-polynomial for a presemifield [2]

Definition 14.5.4. Let $S = (\mathbb{F}_{2^m}, +, \star$ be a presemifield. A mapping $G : \mathbb{F}_{2^m} \to \mathbb{F}_{2^m}$ is said to be an o-polynomial for S if G is a permutation and the function $x \mapsto G(x) + x \star b$ is 2-to-1 function for any nonzero $b \in \mathbb{F}_{2^m}$.

Consider affine semifield plane $\{(x, y) \mid x, y \in \mathbb{F}_{2^m}\}$. If $G(x)$ is an o-polynomial then "curve" $y = G(x)$ intersects with "line" $y = x \star b + a$ in one point if $b = 0$, and 0or 2 points if $b \neq 0$.

In the case of o-polynomial which are linear, one have the following result.

Theorem 14.5.5 ([2]). *Let $S = (\mathbb{F}_{2^m} +, \star)$ be a presemifield, and $S^t = (\mathbb{F}_{2^m} +, \circ)$ be its corresponding transpose presemifield. Let $G : \mathbb{F}_{2^m} \to \mathbb{F}_{2^m}$ be a linear o-polynomial for the presemifield S. Then the adjoint map G^{ad} is an o-polynomial for the presemifield S^t.*

Proof. The mapping G^{ad} is clearly a permutation. Define $L_b(z) = z \star b$. Note that $L_b^{ad}(z) = z \circ b$ since

$$B(L_b^{ad}(z), x) = B(z, L_b(x)) = B(z, x \star b) = B(z \circ b, x).$$

It is given that $G(z) + z \star b = G(z) + L_b(z)$ is a linear 2-to-1 function for any nonzero $b \in \mathbb{F}_{2^m}$. Hence

$$\dim \mathrm{Im}(G(z) + L_b(z)) = m - 1.$$

Now, we have

$$\begin{aligned} \dim \mathrm{Im}\Big(G^{ad}(z) + z \circ b\Big) &= \dim \mathrm{Im}\Big(G^{ad}(z) + L_b^{ad}(z)\Big) \\ &= \dim \mathrm{Im}\Big((G(z) + L_b(z))^{ad}\Big) \\ &= \dim \mathrm{Im}\Big(G(z) + L_b(z)\Big). \end{aligned}$$

Therefore, $\dim \mathrm{Im}\Big(G^{ad}(z) + z \circ b\Big) = m - 1$. Equivalently, $z \mapsto G^{ad}(z) + z \circ b$ is 2-to-1 function for any nonzero $b \in \mathbb{F}_{2^m}$, which completes the proof. □

Example 14.5.6. Recall the construction of Kantor-Williams presemifields [20, 22]. Let $F = \mathbb{F}_{2^m}$ with $m > 1$ odd. Let $F = F_0 \supset F_1 \supset \cdots \supset F_n$ be a chain of subfields. For $i \in \{1, \cdots, n\}$, denote by Tr_i^m the trace function from F to F_i. Let $\zeta_i \in F^*$. The commutative Kantor presemifield is given by operation

$$x \star y = xy + \left(x \sum_{i=1}^{n} Tr_i^m(\zeta_i y) + y \sum_{i=1}^{n} Tr_i^m(\zeta_i x)\right)^2,$$

and the corresponding Kantor-Williams symplectic presemifield is given by operation

$$x \circ y = xy + y^{2^{m-1}} \sum_{i=1}^{n} Tr_i^m(\zeta_i x) + \sum_{i=1}^{n} \zeta_i Tr_i^m(xy^{2^{m-1}}).$$

By setting $G(z) = z \star z = z^2$, the function $f(x, y) = \mathrm{Tr}_1^m(\Gamma_x(y)y)$ is bent. This expression is in implicit form, to make it explicit we have to find the explicit expression for $\Gamma_x(y)$. Let us calculate Γ_x when the chain of subfields is of length 1, reduced to $F \supset F_1$ where $F_1 = \mathbb{F}_{2^k}$. Set $T := Tr_k^m$ and $\zeta := \zeta_1$. Then

$$z \circ x = zx + \sqrt{x}T(\zeta z) + \zeta T(z\sqrt{x}) = y,$$

$$z = \frac{y}{x} + \frac{\sqrt{x}}{x}T(\zeta z) + \frac{\zeta}{x}T(z\sqrt{x}),$$

$$T(\zeta z) = T(\frac{\zeta y}{x}) + T(\frac{\zeta\sqrt{x}}{x})T(\zeta z) + T(\frac{\zeta^2}{x})T(z\sqrt{x}),$$

$$T(z\sqrt{x}) = T(\frac{y\sqrt{x}}{x}) + T(\zeta z) + T(\frac{\zeta\sqrt{x}}{x})T(z\sqrt{x}).$$

Therefore, we obtain the following linear system:

$$\begin{cases} (T(\frac{\zeta\sqrt{x}}{x}) + 1)T(\zeta z) + T(\frac{\zeta^2}{x})T(z\sqrt{x}) = T(\frac{\zeta y}{x}), \\ 1 \cdot T(\zeta z) + (T(\frac{\zeta\sqrt{x}}{x}) + 1)T(z\sqrt{x}) = T(\frac{y\sqrt{x}}{x}). \end{cases}$$

The corresponding determinant equals $(T(\frac{\zeta\sqrt{x}}{x}) + 1)^2 + T(\frac{\zeta^2}{x}) = 1$. So

$$T(\zeta z) = T(\frac{\zeta y}{x})(T(\frac{\zeta\sqrt{x}}{x}) + 1) + T(\frac{y\sqrt{x}}{x})T(\frac{\zeta^2}{x}),$$

$$T(z\sqrt{x}) = (T(\frac{\zeta\sqrt{x}}{x}) + 1)T(\frac{y\sqrt{x}}{x}) + 1 \cdot T(\frac{\zeta y}{x}).$$

Therefore,

$$\Gamma_x(y) = z = \frac{y}{x} + \frac{\sqrt{x}}{x}\left(T(\frac{\zeta y}{x})T(\frac{\zeta\sqrt{x}}{x}) + T(\frac{\zeta y}{x}) + T(\frac{y\sqrt{x}}{x})T(\frac{\zeta^2}{x})\right)$$
$$+\frac{\zeta}{x}\left(T(\frac{\zeta\sqrt{x}}{x})T(\frac{y\sqrt{x}}{x}) + T(\frac{y\sqrt{x}}{x}) + T(\frac{\zeta y}{x})\right).$$

Let us give some remarks.

Remark 14.5.7. For finite fields endowed with the usual additive and multiplicative operations, an o-polynomial gives rise to a hyperoval in finite geometry. It is generally known that the function $G(z) = z \star z$ (from Theorem 14.5.1) gives rise to a hyperoval for commutative semifield planes (see, for example, [16]). The function $G(z) = \sqrt{z}$ (from Theorem 14.5.2) gives rise to hyperovals for semifields which are transpose to commutative semifields (and equivalently they are dual to symplectic semifields [20, 23]). To the best of our knowledge, this fact was never noticed before.

Remark 14.5.8. In the case of finite fields endowed with the usual operations, we know that if $G(z)$ is an o-polynomial, then its compositional inverse $G^{-1}(z)$ is an o-polynomial as well (for instance, see [7]). However, we discover this fact is not true in general in the case of proper semifield, (that is, a finite semifield which is not a field). This means that $G^{-1}(z)$ might not to be o-polynomial neither for the semifield or its transpose. The reader can find examples to this fact in [2].

14.6 Known vs. Unknown Bent Functions

It is written by W. Kantor in a note[3] available online[4] "there are approximately 2^{106} different bent Boolean functions over $\mathbb{F}_2^8$, ignoring the equivalence of such functions. Moreover, of these fewer than 2^{77} arise from constructions in print, which in turn are dominated by two types of constructions introduced: Maiorana–McFarland bent functions and partial spread bent functions. This means that already in a small dimension the known types of constructions are woefully inadequate for describing "most"bent functions. This gap in knowledge increases exponentially as the dimension of the underlying vector space grows. There are many inequivalent spreads from which to choose partial spreads, but the number of known ones does not help at all to deal with the aforementioned gap".

References

1. K. Abdukhalikov. Symplectic spreads, planar functions and mutually unbiased bases. In *J. Algebraic Combin. 41, No. 4*, pages 1055–1077, 2015.
2. K. Abdukhalikov and S. Mesnager. Bent functions linear on elements of some classical spreads and presemifileds spreads. *International Journal Cryptography and Communications (CCDS), Springer. To appear*, 2016.
3. A. Canteaut, M. Daum, H. Dobbertin, and G. Leander. Finding nonnormal bent functions. In *J. Discrete Appl. Math., 154*, pages 202–218, 2006.
4. C. Carlet. Boolean functions for cryptography and error correcting codes. In Yves Crama and Peter L. Hammer, editors, *Boolean Models and Methods in Mathematics, Computer Science, and Engineering*, pages 257–397. Cambridge University Press, June 2010.
5. C. Carlet. More PS and H-like bent functions. In *Cryptology ePrint Archive, Report 2015/168*, 2015.
6. C. Carlet and P. Gaborit. Hyperbent functions and cyclic codes. In *Journal of Combinatorial Theory, Series A, vol 113, no. 3*, pages 466–482, 2006.

[3]In this note, the author can find a brief summary of some of what is known about spreads for use in John Dillon's fundamental partial spread construction of bent functions.

[4]http://darkwing.uoregon.edu/~kantor/PAPERS/Bent+spreadsFinal.pdf.

7. C. Carlet and S. Mesnager. On Dillon's class H of bent functions, niho bent functions and o-polynomials. In *Journal of Combinatorial Theory, Series A, Vol 118, no. 8*, pages 2392–2410, 2011.
8. A. Cesmelioglu, W. Meidl, and A.Pott. Generalized Maiorana–McFarland Class and normality of p-ary bent functions. In *Finite Fields and Their Applications, Vol. 24, pages 105–117*, 2013.
9. P. Charpin and G. Gong. Hyperbent functions, Kloosterman sums and Dickson polynomials. In *IEEE Trans. Inform. Theory (54) 9*, pages 4230–4238, 2008.
10. P. Charpin and G. Gong. Hyperbent functions, Kloosterman sums and Dickson polynomials. In *ISIT 2008, Toronto, Canada, July 6–11*, pages 1758–1762, 2008.
11. P. Dembowski. Finite Geometries. In *Springer, Berlin-Göttingen-Heidelberg*, 1968.
12. P. Dembowski. Finite geometries. In *Springer*, 1968.
13. U. Dempwolff and P. Muller. Permutation polynomials and translation planes of even order. In *Adv. Geom. 13(2)*, pages 293–313, 2013.
14. J. Dillon. Elementary Hadamard difference sets. In *PhD dissertation, University of Maryland*.
15. X. D. Hou. q-ary bent functions constructed from chain rings. In *J. Finite Fields Appl., 4, pages 55–61*, 1998.
16. Jha, V., Wene, G.: An oval partition of the central units of certain semifield planes. Discrete Math. 155, (1–3), pages 127–134, 1996.
17. N. Johnson, V. Jha, and M. Biliotti. Handbook of finite translation planes. In *Pure and Applied Mathematics, vol. 289. London: Chapman & Hall/CRC*, 2007.
18. N. Johnson, V. Jha, and M. Biliotti. Handbook of finite translation planes. In *CRC, Taylor and Francis Group*, 2007.
19. W. Kantor. Spreads, translation planes and Kerdock sets II. In *SIAM J. Algebraic and Discrete Methods*, pages 308–318, 1982.
20. W. M. Kantor. Commutative semifields and symplectic spreads. In *J. Algebra 270*, pages 96–114, 2003.
21. W. M. Kantor. Bent functions generalizing Dillon's partial spread functions. In *arXiv 1211.2600*, 2012.
22. W. M. Kantor and M. E. Williams. Symplectic semifield planes and $\mathbb{Z}_4$-linear codes. In *Trans. Amer. Math. Soc. 356*, pages 895–938, 2004.
23. N. Knarr. Quasifields of symplectic translation planes. In *J. Combin. Theory Ser. A 116, No. 5*, pages 1080–1086, 2009.
24. D. Knuth. A class of projective planes . In *Trans. Amer. Math. Soc., 115*, pages 541–549, 1965.
25. D. E. Knuth. Finite semifields and projective planes. In *J. Algebra 2*, pages 182–217, 1965.
26. M. Lavrauw and O. Polverino. Finite semifields and Galois geometry. In *J. Current Research Topics in Galois Geometry, Nova Science Publishers*, pages 129–157, 2011.
27. N. Li, T. Helleseth, X. Tang, and A. Kholosha. Several New Classes of Bent Functions From Dillon Exponents. In *IEEE Transactions on Information Theory 59 (3), pages 1818–1831*, 2013.
28. P. Lisoněk and H. Y. Lu. Bent functions on partial spreads. In *Designs, Codes and Cryptography, Vol 73, Issue 1, pages 209–216*, 2014.
29. S. Mesnager. A new family of hyper-bent Boolean functions in polynomial form. In *Proceedings of Twelfth International Conference on Cryptography and Coding, Cirencester, United Kingdom. M. G. Parker (Ed.): IMACC 2009, LNCS 5921, Springer, Heidelberg*, pages 402–417, 2009.
30. S. Mesnager. A new class of bent and hyper-bent Boolean functions in polynomial forms. In *journal Design, Codes and Cryptography, 59(1–3)*, pages 265–279, 2011.
31. S. Mesnager. Bent functions from spreads. In *Journal of the American Mathematical Society (AMS), Contemporary Mathematics (Proceedings the 11th International conference on Finite Fields and their Applications Fq11), Volume 632, page 295–316*, 2015.
32. S. Mesnager. On p-ary bent functions from (maximal) partial spreads. In *International conference Finite field and their Applications Fq12, New York, July*, 2015.

33. S. Mesnager and J-P Flori. Hyper-bent functions via Dillon-like exponents. In *IEEE Transactions on Information Theory-IT. Vol. 59 No. 5*, pages 3215–3232, 2013.
34. A. Ceşmelioğlu, W. Meidl, and A. Pott. Bent functions, spreads, and o-polynomials. In *SIAM Journal on Discrete Mathematics, 29(2), pages 854–867*, 2015.
35. B. Wu. PS bent functions constructed from finite pre-quasifield spreads. In *ArXiv e-prints*, 2013.

Chapter 15
Various Cryptographic and Algebraic Generalizations of Bent Functions

15.1 Partially Bent Functions

Since bent functions can never be balanced (which makes them improper for a direct cryptographic use) a research on super-classes of the class of bent functions, whose elements can have high nonlinearities, but can also be balanced (and possibly, with other cryptographic properties such as resiliency) has been investigated. In the following subsections, we shall briefly present the main super-classes of bent functions introduced and studied in the literature.

A first super-class which can be balanced and having high nonlinearities, has been obtained as the set of those functions which achieve the following bound (initially conjecture by Preneel et al. [49] by and proved later by Carlet in [4]) for a given Boolean function f on $\mathbb{F}_2^n$:

$$N_{\Delta_f} \times N_{\widehat{\chi_f}} \geq 2^n \qquad (15.1)$$

where

$$N_{\Delta_f} = \# \left\{ b \in \mathbb{F}_2^n \mid \Delta_f(b) := \sum_{x \in \mathbb{F}_2^n} (-1)^{D_f(b)} \neq 0 \right\},$$

and

$$N_{\widehat{\chi_f}} = \# \left\{ b \in \mathbb{F}_2^n \mid \widehat{\chi_f}(b) \neq 0 \right\}.$$

Moreover, we have $N_{\Delta_f} \times N_{\widehat{\chi_f}} = 2^n$ if and only if, for every $b \in \mathbb{F}_2^n$, the derivative $D_b f$ is either balanced or constant. This property is also equivalent to the fact that there exist two linear subspaces E (of even dimension) and E' of $\mathbb{F}_2^n$, whose direct

S. Mesnager, *Bent Functions*, DOI 10.1007/978-3-319-32595-8_15

sum equals $\mathbb{F}_2^n$, and Boolean functions g, bent on E, and h, affine on E', such that: $\forall x \in E,\ \forall y \in E', f(x+y) = g(x) \oplus h(y)$.

Note that the above Relation (15.1) expresses in fact some trade-off between the number of non-balanced derivatives (that is, of nonzero auto-correlation[1]) of a Boolean function and the number of nonzero values of its Walsh transform.

Functions f satisfying Relation (15.1) with equality $N_{\Delta_f} \times N_{\widehat{\chi_f}} = 2^n$ are called *partially-bent functions*.[2]

Bent functions have balanced derivatives $D_a f$ for all nonzero $a \in \mathbb{F}_2^n$ hence are partially bent. Partially bent functions may be balanced and highly nonlinear. In fact, they have good properties (Walsh spectrum easier to calculate, potential good nonlinearity and good resiliency order) which are favorable for cryptographic applications in stream and block ciphers. They also share with quadratic functions (note that every quadratic function is partially-bent) all of their nice properties (see [4]). In particular, the values of the Walsh transform equal 0 or $\pm 2^{dim(E')+dim(E)/2}$. Nevertheless, when they are not bent, partially-bent functions have (by definition) nonzero linear structures and so do not give full satisfaction. The class of *plateaued* functions (which will be introduced in the next subsection) is a natural extension of that of partially-bent functions. Partially bent functions have been generalized in some directions. Firstly, in [52], Relation (15.1) has been firstly generalized by Quisquater et al. Next, the notions of bent functions and the partially bent functions have been also extended to the ring $\mathbb{Z}_N$, for a positive integer N called *generalized bent* functions and *generalized partially bent* functions over $\mathbb{Z}_N$, respectively. Several results concerning partially bent functions obtained by Carlet have been generalized. Based on the definition of generalized partially bent functions and using the theory of linear transformations, the relationship between generalized partially bent functions over ring $\mathbb{Z}_N$ and generalized bent functions over ring $\mathbb{Z}_N$ has been studied in [63]. In particular, it is proved that when N is a prime integer, some generalized partially bent functions can be decomposed as the sum of a generalized bent function and an affine function. Such a decomposition facilitates the construction of partially bent functions and generalized partially bent functions. Approaches to construct those generalized bent functions have been derived [63]. Furthermore, the notion of partially bent has been extended in characteristic p. A p-ary function $f : \mathbb{F}_p^n \rightarrow \mathbb{F}_p$ is said to be *p-ary partially bent* if, for all $a \in \mathbb{F}_p^n$ the derivative $D_a f(x) := f(x+a) - f(x)$ of f in direction a is balanced or constant, that is, every value in $\mathbb{F}_p$ is taken on p^{n-1} times. Properties of partially bent functions have been deeply studied very recently [11] in particular with appropriate generalizations of relative difference sets and difference sets.

[1]The auto-correlations of a Boolean function f in direction of $b \in \mathbb{F}_2^n$ is defined by $\sum_{x\in\mathbb{F}_2^n}(-1)^{D_f(b)} = \sum_{x\in\mathbb{F}_2^n}(-1)^{f(x+b)+f(x)}$.

[2]The reader should be careful not to be confused with the notion of the so-called *partial bent functions* studied by Guillot in [27].

15.2 Rotation Symmetric (RS) Bent Functions and Idempotent Bent Functions

A Boolean function f is said to be symmetric if its output is invariant under any permutation of its input bits, i.e., $f(x_0, x_1, \cdots, x_{n-1}) = f(x_{\tau(0)}, x_{\tau(1)}, \cdots, x_{\tau(n-1)})$, for any permutation τ of $\{0, 1, \cdots, n-1\}$. The definition implies that f takes the same value for vectors with the same Hamming weight. For example, the majority function (which is a very simple function introduced by Ding et al. [1] defined by $f(x) = 1$ if $wt(x) \geq \lceil n/2 \rceil$ and 0 otherwise) is a symmetric Boolean function.

Rotation symmetric (RS) Boolean functions have been introduced by Filiol and Fontaine in [22] and [23] under the name of idempotent functions and by Pieprzyk and Qu [46] under their final name. They are those Boolean functions which are invariant under cyclic shifts of input coordinates: $f(x_{n-1}, x_0, x_1, \ldots, x_{n-2}) = f(x_0, x_1, \ldots, x_{n-1})$. In other words, the support of an RS function is a cyclic (but not necessarily linear) code.

The original motivation for the study of RS functions is that the rotation symmetry seems to increase the probability of finding interesting functions by random search and gives nice structure. The reader notices that such class of Boolean functions is of interest because of its search space ($\approx 2^{\frac{2^n}{n}}$) smaller compared to the whole space ($\approx 2^{2^n}$), which allows investigating functions for a larger number of variables, and also because of the more compact representation of RS functions. The study of rotation symmetric Boolean functions for good cryptographic properties was initiated in [22], which can be regarded as a generalization of symmetric Boolean functions. Moreover, the class of those functions is extremely rich in terms of cryptographically significant Boolean functions but this fact has been demonstrated by experimental search. Recent research shows that the class of rotation symmetric Boolean functions is potentially rich in functions of cryptographic significance. For instance, Kavut, Maitra and Yucel [31] have found Boolean functions on 9 variables with nonlinearity 241 which solved an almost three-decade old open problem. Hence, RS structure allowed obtaining Boolean functions in odd number of variables beating the best known nonlinearities.

Rotation symmetric Boolean functions are invariant under the action of the cyclic group, which is really another kind of good candidates with optimal algebraic immunity. So far, many rotation symmetric Boolean functions have been obtained from modification of the majority function. For quadratic RS functions we have the following characterization for being rotation symmetric. The result is more or less known. A proof is given by Carlet in [6].

Lemma 15.2.1 ([6]). *Let $f(x) = \sum_{0 \leq i < j \leq n-1} a_{i,j} x_i x_j + \ell(x)$ be any quadratic Boolean function, where $a_{i,j} \in \mathbb{F}_2$ and ℓ is affine. Let M be the associated matrix (see [21]), whose term located at row i and column j equals $a_{i,j}$ if $i < j$, $a_{j,i}$ if $i > j$ and 0 if $i = j$. Then f is RS if and only if M is circulant (i.e. each row of M is a cyclic shift of the previous row) and ℓ is RS.*

15.2.1 Rotation Symmetric (RS) Bent Functions

Rotation symmetric Boolean functions are linked to a notion introduced by Filiol and Fontaine in [22]: a Boolean function f on $\mathbb{F}_{2^n}$ is an idempotent if it satisfies $f(x) = f(x^2), \quad$ for all $x \in \mathbb{F}_{2^n}$. RS functions and idempotents are closely related. The link is that for any Boolean function $f(x)$ over $\mathbb{F}_{2^n}$, and every normal basis $(\alpha, \alpha^2, \dots, \alpha^{2^{n-1}})$ of $\mathbb{F}_{2^n}$, the function

$$(x_0, \dots, x_{n-1}) \mapsto f\left(\sum_{i=0}^{n-1} x_i \alpha^{2^i}\right)$$

is RS if and only if f is an idempotent. But knowing an infinite class of idempotent bent functions is not equivalent to knowing an infinite class of RS bent functions, since there is no expression valid for an infinite number of values of n of the decomposition of $\left(\sum_{i=0}^{n-1} x_i \alpha^{2^i}\right)^j$ over the normal basis $(\alpha, \alpha^2, \dots, \alpha^{2^{n-1}})$, except for j null or equal to a power of 2.

Working on RS bent functions and on univariate bent functions have comparable interests and drawbacks:

- The functions can have simple structure and representation;
- They can be fastly computed and are then better suited for applications;
- Finding new classes of RS or univariate bent functions is difficult and gives more knowledge on finite fields and their applications;
- Most often, these new RS or univariate bent functions are not new as bent functions: they belong to well-known general classes (Maiorana–McFarland, $\mathcal{PS}_{ap}$, etc.). Their discovery does not increase the number of known bent functions;
- However, their particular structure may lead in some cases to new bent functions (e.g. Niho bent functions) or give insight on known bent functions.

RS bent functions and bent idempotents are much rarer than general bent functions. Note that the dual of an RS bent function is an RS bent function.

15.2.2 Infinite Classes of RS Bent Functions

Finding RS bent functions is difficult and has theoretical and practical interest. Some examples of infinite classes of RS bent functions available in the literature. As in the case of bent functions in trace representations, many of the known classes of RS bent functions belong to completed Maiorana–McFarland class, and their bentness may then be easily explained. In the following we present them. Further details can be found in [6] and references within.

1. Quadratic RS bent functions have been characterized Gao et al. [26] by the fact that some related polynomial $P(x)$ over $\mathbb{F}_2$ such that $X^n P(\frac{1}{X})$ is co-prime with $X^n + 1$. More precisely, given m elements $c_1, c_2, \cdots c_m$ of $\mathbb{F}_2$, the function $\sum_{i=1}^{m-1} c_i(\sum_{j=0}^{n-1} x_j x_{i+j}) + c_m(\sum_{j=0}^{m-1} x_j x_{m+j})$ is bent if and only if the polynomial $\sum_{i=1}^{m-1} c_i(X^i + X^{n-i}) + c_m X^m$ is co-prime with $X^n + 1$, equivalently, the linearized polynomial $L(X) = \sum_{i=1}^{m-1} c_i(X^{2^i} + X^{2^{n-i}}) + c_m X^{2^m}$ is a permutation polynomial (a necessary condition is that $c_m = 1$). Deduced examples of bent quadratic RS functions can therefore be obtained from such polynomials. A first example of such polynomial is with $c_i = 0$ for $i \neq m$. A second example with $c_i = 1$ for $i = 1, \ldots, n-1$ gives rise to two infinite classes

$$\sum_{j=0}^{m-1} x_j x_{m+j}$$

and

$$\sum_{i=1}^{m-1}\Big(\sum_{j=0}^{n-1} x_j x_{i+j}\Big) + \Big(\sum_{j=0}^{m-1} x_j x_{m+j}\Big)$$

of quadratic RS bent functions. As explained in [6], more examples can be found. In particular for n not divisible by 3, the following function

$$\sum_{\substack{0\le u,v,w\le m \\ u+v+w=m,\, 2u+v\in\{1,\ldots,m-1\}}} \frac{m!}{u!v!w!}\Big(\sum_{j=0}^{n-1} x_j x_{2u+v+j}\Big) + \Big(\sum_{j=0}^{m-1} x_j x_{m+j}\Big),$$

where the coefficients are taken modulo 2 is RS bent. Moreover, when n is a power of 2, then according to [60, Proposition 3.1], the function $\sum_{i=1}^{m-1} c_i(\sum_{j=0}^{n-1} x_j x_{i+j}) + c_m(\sum_{j=0}^{m-1} x_j x_{m+j})$ is bent if and only if $\sum_{i=0}^{n-1} c_i = 1$, that is, $c_m = 1$.
2. Two infinite classes of cubic RS bent functions belonging to the completed Maiorana–McFarland class are:
   - $\sum_{i=0}^{n-1}(x_i x_{t+i} x_{m+i} + x_i x_{t+i}) + \sum_{i=0}^{m-1} x_i x_{m+i}$, where $m/gcd(m,t)$ is odd [26];
   - $\sum_{i=0}^{n-1} x_i x_{i+r} x_{i+2r} + \sum_{i=0}^{2r-1} x_i x_{i+2r} x_{i+4r} + \sum_{i=0}^{m-1} x_i x_{i+m}$, where $m = 3r$ [25].

Further details on the constructions of quadratic and cubic rotation symmetric bent functions presented above can be found in [26].

15.2.3 Univariate RS Functions (Idempotents)

A Boolean function f on $\mathbb{F}_{2^n}$ is said to an idempotent if $f(x) = f(x^2)$ for all $x \in \mathbb{F}_{2^n}$.

Let f be a Boolean function written in its univariate representation in terms of trace functions: $f(z) = \sum_{j\in\Gamma_n} \mathrm{Tr}_1^{o(j)}(a_j z^j) + a_{2^n-1} z^{2^n-1}$, where:

- $a_{2^n-1} \in \mathbb{F}_2$ equals the Hamming weight of f modulo 2,
- Γ_n is the set of integers obtained by choosing one element in each cyclotomic coset of 2 mod $2^n - 1$,
- $o(j)$ is the size of the cyclotomic coset containing j
- and $a_j \in \mathbb{F}_{2^{o(j)}}$. Then f is an idempotent is and only if every coefficient a_j in every term $\mathrm{Tr}_1^{o(j)}(a_j z^j)$ belongs simply to $\mathbb{F}_2$.

In fact, there is a bijective correspondence between idempotent functions and RS functions, by decomposing x over a normal basis. Indeed , as explained in [8], the reader notices that for any Boolean function $f(z)$ over $\mathbb{F}_{2^n}$, choosing a normal basis $(\alpha, \ldots, \alpha^{2^{n-1}})$ of $\mathbb{F}_{2^n}$ and decomposing z over this basis gives a Boolean function f over $\mathbb{F}_{2^n}$, which is RS if and only if f is an idempotent. Hence, each idempotent over $\mathbb{F}_{2^n}$ can be uniquely obtained from some RS function over $\mathbb{F}_{2^n}$ and vice versa. But since this correspondence needs the choice of a normal basis, it is not equivalent to have an infinite class of RS functions and to have an infinite class of idempotents. In some papers, the authors make an abuse of language by using the term of RS functions for idempotent functions.

Some known bent functions are clearly idempotents under some conditions. Below, we give some examples of bent idempotents:

- The function $f'(z) = Tr_1^m(z^{2^m+1})$ and the function $f'(z) = Tr_1^m(z^{2^m+1}) + \sum_{i=1}^{m-1} Tr_1^n(z^{2^i+1})$ are bent quadratic idempotents.
  More generally, given $c_1, \ldots, c_m$ in $\mathbb{F}_2$, the function equal to $c_m \mathrm{Tr}_1^m(x^{2^m+1}) + \sum_{i=1}^{m-1} c_i \mathrm{Tr}_1^n(x^{2^i+1})$ is bent if and only if $gcd(\sum_{i=1}^{m-1} c_i(X^i + X^{n-i}) + c_m X^m, X^n + 1) = 1$ [41]. Examples of bent idempotents have been deduced. They were described in [6].
- The Niho bent functions [20]: $Tr_1^m(az^{2^m+1}) + Tr_1^n(bz^d)$ are bent idempotents when the coefficients a and b equal 1.
- The second class of know Niho bents extended by Leander and Kholosha [34] gives rise also a bent idempotent.
- The generalized Dillon and (the so-called) Mesnager functions are potentially bent idempotents, under conditions involving Kloosterman sums.
  - For every m such that $K_m(1)$ is null, $f(x) = Tr_1^n(x^{r(2^m-1)})$ is bent when $\gcd(r, 2^m + 1) = 1$.
  - For every m odd such that $K_m(1) = 4, f(x) = Tr_1^n(x^{r(2^m-1)}) + Tr_1^2(x^{\frac{2^n-1}{3}})$ is bent when $\gcd(r, 2^m + 1) = 1$.

But the condition $K_m(1) = 0$ never happens as shown in [39] and it can be checked by computer that the condition $K_m(1) = 4$ never happens as well for $5 \le m \le 20$.

- For $n = 2m = 6r$, $r \ge 1$, $Tr_1^n(z^{1+2^r+2^{2r}}) + Tr_1^{2r}(z^{1+2^{2r}+2^{4r}}) + Tr_1^m(z^{1+2^t}) = Tr_1^r((z+z^{2^{3r}})^{1+2^r+2^{2r}}) + Tr_1^m(z^{1+2^t})$ is a bent idempotent [25].

Further details on the known bent idempotents can be found in [25].

15.2.4 Idempotent Bent Functions and Secondary Constructions of Rotation Symmetric

In [8], Carlet, Gao and Liu have studied more in detail the relationship between rotation symmetric (RS) functions and idempotents, in univariate and bivariate representations, and deduced a construction of bent RS functions from semi-bent RS functions. The authors deduced the first infinite classes found of idempotent and RS bent functions of algebraic degree more than 3. Secondary constructions of rotation symmetric and idempotent bent functions are given in [6]. They provide constructions of idempotent bent functions, for m odd: defining the m-variable idempotent functions $f_1(x) = Tr_1^m(x) + Tr_1^m(x^{2^{(m-1)/2}+1})$ and $f_2(x) = Tr_1^m(x^3)$, function $h(x,y) = f_1(x) + f_1(y) + (f_1 + f_2)(x)\,(f_1 + f_2)(y)$ is a bent idempotent with algebraic degree 4. Similarly, given two RS functions $f_1^*(x) = \sum_{i=0}^{m-1}(x_i + x_i x_{(m-1)/2+i})$ and $f_2^*(x) = \sum_{i=0}^{m-1} x_i x_{1+i}$, where the subscripts are taken modulo m, function $h^*(x_0, y_1, x_2, y_3, \ldots, x_{n-2}, y_{n-1}) = f_1^*(x_0, \ldots, x_{m-1}) + f_1^*(y_0, \ldots, y_{m-1}) + (f_1^* + f_2^*)(x_0, \ldots, x_{m-1})(f_1^* + f_2^*)(y_0, \ldots, y_{m-1})$ is an RS bent function of algebraic degree 4.

In this subsection we recall of main results from [8] and an extension given by Carlet in [6].

15.2.4.1 Bivariate Representation of Idempotents

Assume m odd. Let w be an element of $\mathbb{F}_{2^2} \setminus \mathbb{F}_2$, we have $w^2 = w + 1$, $w^4 = w$, and we can take (w, w^2) for basis of $\mathbb{F}_{2^n}$ over $\mathbb{F}_{2^m}$, since we have $\frac{w^2}{w} = w \notin \mathbb{F}_{2^m}$ for m odd. Any element of $\mathbb{F}_{2^n}$ can then be written in the form $xw + yw^2$, where $x, y \in \mathbb{F}_{2^m}$. Note that, given a normal basis $(\alpha, \ldots, \alpha^{2^{m-1}})$ of $\mathbb{F}_{2^m}$, a natural normal basis of $\mathbb{F}_{2^n}$ over $\mathbb{F}_2$ is:

$$(\alpha w, \alpha^2 w^2, \alpha^4 w, \alpha^8 w^2, \ldots, \alpha^{2^{m-1}} w, \alpha w^2, \alpha^2 w, \ldots, \alpha^{2^{m-1}} w^2). \tag{15.2}$$

Since $(xw + yw^2)^2 = y^2 w + x^2 w^2$, the shift $z \mapsto z^2$ corresponds to the mapping $(x, y) \mapsto (y^2, x^2)$. Given a function $f(x, y)$ in bivariate form, the related Boolean function over $\mathbb{F}_2^n$ obtained by decomposing the input $xw + yw^2$ over the basis (15.2) is then RS if and only if $f(x, y) = f(y^2, x^2)$. Note that the case when m even is left open.

As explained in [6], one considers a more general situation. To this end, let m and k be two co-prime integers and $n = mk$. Let α be a normal element of $\mathbb{F}_{2^m}$ over $\mathbb{F}_2$ and w a normal element of $\mathbb{F}_{2^k}$ over $\mathbb{F}_2$. It is kwown that αw is a normal element of $\mathbb{F}_{2^n}$ over $\mathbb{F}_2$. One then gets the normal bases $(\alpha, \ldots, \alpha^{2^{m-1}})$ of $\mathbb{F}_{2^m}$ over $\mathbb{F}_2$, $(w, \ldots, w^{2^{k-1}})$ of $\mathbb{F}_{2^k}$ over $\mathbb{F}_2$ and

$$(\alpha w, \alpha^2 w^2, \ldots, \alpha^{2^{i \pmod m}} w^{2^{i \pmod k}}, \ldots, \alpha^{2^{n-1} \pmod m} w^{2^{n-1} \pmod k})$$

of $\mathbb{F}_{2^n}$ over $\mathbb{F}_2$. Therefore, any element of $\mathbb{F}_{2^n}$ can be written in the form $\sum_{i=0}^{k-1} x_i w^{2^i}$, where $x_i \in \mathbb{F}_{2^m}$. Since $(\sum_{i=0}^{k-1} x_i w^{2^i})^2 = \sum_{i=0}^{k-1} x_i^2 w^{2^{i+1} \pmod k}$, the univariate shift $z \mapsto z^2$ corresponds to the mapping

$$(x_0, \ldots, x_{k-1}) \mapsto \rho_k(x_0^2, \ldots, x_{k-1}^2),$$

where $\rho_k(x_0, \ldots, x_{k-1}) = (x_{k-1}, x_0, \ldots, x_{k-2})$ is the cyclic shift over $\mathbb{F}_{2^m}^k$.

15.2.4.2 Weak Idempotents and the Related RS and Weak RS Functions

Let $f(x_0, \ldots, x_{k-1})$ be a Boolean function in k-variate form (where $x_i \in \mathbb{F}_{2^m}$). Then the related Boolean function over $\mathbb{F}_2^n$ obtained by decomposing $\sum_{i=0}^{k-1} x_i w^{2^i}$ over (15.2) is then RS if and only if

$$f(x_0, \ldots, x_{k-1}) = f(\rho_k(x_0^2, \ldots, x_{k-1}^2)).$$

We have the following statement.

Proposition 15.2.2 ([6]). *Let m and k be two co-prime integers and $n = mk$. Let α be a normal element of $\mathbb{F}_{2^m}$ over $\mathbb{F}_2$ and w a normal element of $\mathbb{F}_{2^k}$ over $\mathbb{F}_2$. Then the n-variable Boolean idempotents are those polynomials $f(z)$ representing Boolean functions over $\mathbb{F}_{2^n}$ whose associate k-variate expressions, defined as $f(x_0, \ldots, x_{k-1}) = f(\sum_{i=0}^{k-1} x_i w^{2^i})$, satisfy $f(x_0, \ldots, x_{k-1}) = f(\rho_k(x_0^2, \ldots, x_{k-1}^2))$. In particular, if $k = 2$, the n-variable Boolean idempotents are those polynomials $f(z)$ representing Boolean functions over $\mathbb{F}_{2^n}$ whose associate bivariate expressions $f(x, y) = f(wx + w^2 y)$ satisfy $f(x, y) = f(y^2, x^2)$.*

From the previous proposition, a weaker notion of idempotence has been introduced.

Definition 15.2.3. Under the hypotheses of Proposition 15.2.2, any polynomial $f(z)$ whose k-variate expression satisfies $f(x_0, \ldots, x_{k-1}) = f(x_0^2, \ldots, x_{k-1}^2)$ is called *k-weak-idempotent*.

Note that the condition $f(x_0, \ldots, x_{k-1}) = f(x_0^2, \ldots, x_{k-1}^2)$ is equivalent to $f(x_0, \ldots, x_{k-1}) = f(x_0^{2^k}, \ldots, x_{k-1}^{2^k})$ since m and k are co-prime.

It has been shown a relationship between weak-idempotents and idempotents. More precisely, one has the following statement.

Proposition 15.2.4 ([6]). *The set of n-variable idempotent functions is included in that of k-weak-idempotents. An idempotent is a k-weak-idempotent invariant under the shift* ρ_k.

Examples of bent 2-weak idempotents can be obtained by considering all the functions derived from o-polynomials with coefficients equal to 1.

The corresponding definition at the bit level is obtained by decomposing the univariate representation over the normal basis (15.2) and the k-variate representation over the basis $(\alpha, \ldots, \alpha^{2^{m-1}})$:

Definition 15.2.5. Let m and k be two co-prime integers and $n = mk$. A Boolean function

$$f(x_{0,0}, y_{1,1}, \ldots, x_{n-1,n-1})$$

(where each first index is reduced modulo k and each second index is reduced modulo m) over $\mathbb{F}_2^n$ is k-weak-RS if it is invariant under the cyclic shift by k positions.

In the case where $n = 2m$ with m odd, a function $f(x_0, y_1, x_2, y_3, \ldots, x_{n-2}, y_{n-1})$ (where each index is reduced modulo m; we skip the first index) over $\mathbb{F}_2^n$ is 2-weak-RS if it is invariant under the transformation $\begin{cases} x_j \mapsto x_{j+1} \\ y_j \mapsto y_{j+1} \end{cases}$.
Such 2-weak-RS function is RS if and only if, in bivariate form, it is invariant under $(x, y) \mapsto (y, x)$. The following statement provides a necessary and sufficient condition for a Boolean function $f(x_{0,0}, y_{1,1}, \ldots, x_{n-1,n-1})$ to be RS in terms of weak RS.

Proposition 15.2.6 ([6]). *A Boolean function* $f(x_{0,0}, y_{1,1}, \ldots, x_{n-1,n-1})$ *is RS if and only if it is m-weak-RS and k-weak-RS.*

For instance, direct sum $f(x)+g(y)$ where f and g are RS m-variable functions gives rise to 2-weak-RS functions. Note that the function derived from the direct sum is RS if f coincides with g. The direct sum allows then constructing, for $n = 2m$, an n-variable weak idempotent from two m-variable idempotents.

15.2.4.3 A Secondary Construction of RS and Idempotent Functions

The indirect sum allows constructing, for $n = 2m$, an n-variable weak idempotent h from four m-variable idempotents f_1, f_2, g_1, g_2:

$$h(x, y) = f_1(x) + g_1(y) + (f_1 + f_2)(x)\,(g_1 + g_2)(y);\; x, y \in \mathbb{F}_{2^m}.$$

If $f_1 = g_1$ and $f_2 = g_2$ then we obtain the idempotent $h(x,y) = f_1(x) + f_1(y) + (f_1 + f_2)(x)\,(f_1 + f_2)(y)$. This gives also a secondary construction of an RS n-variable function from two RS m-variable functions ($n = 2m$, m odd). This function is bent if the two functions are near-bent. The secondary construction given by the following statement is closely related to a result dealing with a modification of the indirect sum given in Chap. 6 (concerning a modified indirect sum introduced recently in [65]).

Proposition 15.2.7 ([8]). *Let f_1 and f_2 be two m-variable RS near-bent functions, m odd, and let $n = 2m$. If the Walsh supports of f_1 and f_2 are complementary, then $h(x_0, y_1, x_2, y_3, \ldots, x_{n-2}, y_{n-1}) = f_1(x_0, \ldots, x_{m-1}) + f_1(y_0, \ldots, y_{m-1}) + (f_1 + f_2)(x_0, \ldots, x_{m-1})(f_1 + f_2)(y_0, \ldots, y_{m-1})$ is bent RS.*

We have (see [5]) the Walsh transform $W_h(a_0, b_1, a_2, b_3, \ldots, a_{n-2}, b_{n-1})$ of h is equal to

$$\frac{1}{2}W_{f_1}(a)\left[W_{f_1}(b) + W_{f_2}(b)\right] + \frac{1}{2}W_{f_2}(a)\left[W_{f_1}(b) - W_{f_2}(b)\right].$$

A case of application of the construction of Proposition 15.2.7 happens with the bent quadratic function involved in the definition of the Kerdock code. Another example, found in [8], of such a pair (f_1, f_2) gives rise to an infinite class of idempotent bent functions of algebraic degree 4 coming from two m-variable idempotent functions f_1 and f_2 (defined in the theorem below) which are near-bent functions with complementary Walsh supports.

Theorem 15.2.8 ([8]). *Let $n = 2m$, m odd. We define the m-variable idempotent functions $f_1(x) = Tr_1^m(x) + Tr_1^m(x^{2^{(m-1)/2}+1})$ and $f_2(x) = Tr_1^m(x^3)$. Then $h(x,y) = f_1(x) + f_1(y) + (f_1 + f_2)(x)\,(f_1 + f_2)(y)$ is a bent idempotent with algebraic degree 4.*

Similarly, one can define the RS functions $f_1^*(x) = \sum_{i=0}^{m-1}(x_i + x_i x_{(m-1)/2+i})$ and $f_2^*(x) = \sum_{i=0}^{m-1} x_i x_{1+i}$, where the subscripts are taken modulo m. Then function $h^*(x_0, y_1, x_2, y_3, \ldots, x_{n-2}, y_{n-1}) = f_1^*(x_0, \ldots, x_{m-1}) + f_1^*(y_0, \ldots, y_{m-1}) + (f_1^* + f_2^*)(x_0, \ldots, x_{m-1})(f_1^* + f_2^*)(y_0, \ldots, y_{m-1})$ is a RS bent function of algebraic degree 4.

15.2.5 A Transformation on Rotation Symmetric Bent Functions

In [8], is introduced a natural way of transforming a rotation symmetric bent function into an idempotent Boolean function $f'(z) = f(z, z^2, \cdots, z^{2^{n-1}})$ over $\mathbb{F}_{2^n}$ leading to another RS Boolean function. More precisely, let $f(x_0, \cdots, x_{n-1}) = \sum_{u \in \mathbb{F}_2^n} a_u x^u$ be any Boolean RS function over $\mathbb{F}_2^n$, then $f'(z) = f(z, z^2, \cdots, z^{2^{n-1}}) = \sum_{u \in \mathbb{F}_2^n} a_u z^{\sum_{i=0}^{n-1} u_i 2^i}$ is a Boolean idempotent, and any idempotent Boolean function can be obtained this way. This transformation, contrary to the decomposition of an idempotent over a normal basis, gives infinite classes from infinite classes. Also, note that the trace representation of f' is directly deduced from the algebraic normal form of f, but the authors have showed that f and f', which have the same algebraic degree, are in general not affinely equivalent to each other. If f is a quadratic RS function, then f is bent if and only if f' is bent. In this case, f and f' are affinely equivalent, since we know that two quadratic bent functions are affinely equivalent up to the addition of a constant. For non-quadratic functions, all cases can happen; examples of an infinite class of cubic bent RS functions f such that f' is not bent, of an infinite class of cubic bent idempotents f' such that f is not bent, and of infinite classes of bent RS functions f such that f' is bent are known (see them e.g. in [6]).

In particular a class of RS bent functions f, such that f' is not, is the following: let $f(x) = \sum_{i=0}^{n-1}(x_i x_{t+i} x_{m+i} + x_i x_{t+i}) + \sum_{i=0}^{m-1} x_i x_{m+i}$ over $\mathbb{F}_2^n$, where $n = 2m$ and $0 < t < m$ is such that $m/gcd(m,t)$ is odd. Then f is bent and $f'(z) = \mathrm{Tr}_1^n(z^{1+2^t+2^m}) + \mathrm{Tr}_1^n(z^{1+2^t}) + \mathrm{Tr}_1^m(z^{1+2^m})$ is not bent.

A class of RS functions f such that f' is bent and f is not is the following: $f'(z) = Tr_1^m(x^{1+2^m}) + Tr_1^n(x^d)$; $d = (2^m - 1)/4 + 1$ is bent (that is, the second known binomial Niho bent function) but $f(x) = \sum_{i=0}^{n-1} x_i x_{1+i} x_{m+i} + \sum_{i=0}^{m-1} x_i x_{m+i}$ writes in bivariate form $x \cdot \pi(y)$ where π is not a permutation, and is then not bent.

Since all situations can happen concerning the bentness of an RS function f and of the related idempotent f', interesting questions have been highlighted by Carlet et al. :

- Does searching for RS bent functions f provide larger probability of success when we choose them such that f' is bent?
- Same question when searching for bent idempotents f' when f is chosen bent.

The functions f and f' being in general not equivalent, the bent functions found with the help of this transformation are likely to be new under affine equivalence.

15.2.5.1 Relationship Between the Bentness of f and f'

Carlet et al. [8] have studied the relationship between the bentness of f and that of f': it has been checked with infinite classes of RS functions that f can be bent when f' is not and that f' can be bent when f is not; it is shown that if f is quadratic then it is bent if and only if f' is bent and we study classes of bent RS non-quadratic functions f for which f' is bent.

- Quadratic functions.

  The characterizations recalled above for the bentness of quadratic RS functions and bent idempotents, given respectively in [26] and [41], are the same. Then:

  Theorem 15.2.9 ([6]). *If f is a quadratic RS function, then f is bent if and only if f' is bent.*

- An infinite class of cubic bent RS functions f such that f' is not bent.

Let

$$f_t(x) = \sum_{i=0}^{n-1}(x_i x_{t+i} x_{m+i} + x_i x_{t+i}) + \sum_{i=0}^{m-1} x_i x_{m+i}$$

over $\mathbb{F}_2^n$, where $n = 2m$ and $0 < t < m$ is such that $m/gcd(m,t)$ is odd. Then we have recalled that f is bent and it is shown in [8] that

$$f_t'(z) = \mathrm{Tr}_1^n(z^{1+2^t+2^m}) + \mathrm{Tr}_1^n(z^{1+2^t}) + \mathrm{Tr}_1^m(z^{1+2^m})$$

is not bent.

- An infinite class of cubic bent idempotents f' such that f is not bent.

  Let

$$f'(z) = Tr_1^m(x^{1+2^m}) + Tr_1^n(x^d);\ d = (2^m - 1)/4 + 1$$

be the second Niho bent function given in [20], then, as shown in [8]

$$f(x) = \sum_{i=0}^{n-1} x_i x_{1+i} x_{m+i} + \sum_{i=0}^{m-1} x_i x_{m+i}$$

can be written in the Maiorana McFarland class form where π is not a permutation; and f is then not bent.

- Infinite classes of bent RS functions f such that f' is bent.

  A first example is given by Theorem 15.2.8. Let us give another example.

  Let

$$f(x) = \sum_{i=0}^{n-1} x_i x_r x_{i+2r} + \sum_{i=0}^{2r-1} x_i x_{i+2r} x_{i+4r} + \sum_{i=0}^{m-1} x_i x_{i+m},$$

where $n = 2m = 6r$ with $r \geq 1$.

We know that f and $f'(z) = \mathrm{Tr}_1^n(z^{1+2^r+2^{2r}}) + \mathrm{Tr}_1^{2r}(z^{1+2^{2r}+2^{4r}}) + \mathrm{Tr}_1^m(z^{1+2^t}) = \mathrm{Tr}_1^r((z+z^{2^{3r}})^{1+2^r+2^{2r}}) + \mathrm{Tr}_1^{\nu(t)}(z^{1+2^t})$ are bent.

The investigations in [8] suggest that searching RS bent functions f provides larger probability of success when we choose them such that f' is bent, and vice-versa. The transformation $f \mapsto f'$ is however not an equivalence between RS bent functions; the bent functions are likely to be new under affine equivalence. Finally, we inform the reader that in [6], are presented many interesting open problem devoted to RS bent functions.

15.3 Homogeneous Bent Functions

Homogeneous bent functions have been isolated by Qu, Seberri and Pieprzyk [51]. A bent function is called homogeneous if all monomials of its algebraic normal form are of the same degree. Searching for *homogeneous bent* functions having some degree of symmetry was motivated by cryptographic applications. Loosely speaking, the symmetry property ensures that in repeated evaluations of the functions (such as in cryptographic algorithms) partial evaluations can be reused. As a consequence cryptographic algorithms which are designed using such symmetric Boolean functions have a fast implementation. The design of hashing functions is one of the applications of these ideas. Homogeneous bent functions have been studied in only few papers. Firstly, Qu et al. [51] have enumerated all homogeneous bent functions of degree 3 of six variables (it turns out that there are exactly 30 of them) and posed the problem of classifying the homogeneous bent functions with more variables. Further study has been done in some papers namely co-authored by Charnes, Rotteler and Beth in [12] and next in the nice paper [13] proving that there exist homogeneous bent functions of degree three in $2m$ variables for $m > 2$. In [13], the authors have in fact established an interesting connection between invariant theory and the theory of bent functions. Such a connection has allowed to the construction of Boolean functions with a prescribed symmetry group action. Besides the quadratic bent functions the only other previously known homogeneous bent functions until 2002 are the six variable degree three functions constructed in [50] having interesting combinatorial structures. Afterward, Charnes et al. have shown that these bent functions arise as invariants under an action of the symmetric group on four letters and determine the stabilizer which turns out to be a matrix group of order 10752. They have applied the machinery of invariant theory in order to construct for the first time homogeneous bent functions of degree three in 8, 10, and 12 variables. The authors have demonstrated a connection between these functions and 1-designs as well as certain graphs. Such an approach has then provided a great computational advantage over the unstructured search problem and yields Boolean functions which have a concise description in terms of certain designs and graphs. The question of linear equivalence of the constructed bent functions and the study the properties of the associated elementary Abelian difference sets have also been considered in [13]. Since then, essentially results of nonexistence of homogeneous bent functions have been derived. Using difference sets, the authors [13] have proved that there exist no homogeneous bent functions of degree m in $2m$ variables for $m > 3$ (In other words, the upper bound on the degree of homogeneous bent function is no more than $m - 1$). In [43], the authors have obtained a tighter bound on the degree of homogenous bent functions and proved that for any nonnegative integer k, there exists a positive integer N such that for $m \geq N$ there exist no homogeneous bent functions in $2m$ variables having degree $m - k$ or more, where N is the least integer satisfying a condition involving k. Since then some conjectures on the degree of the homogeneous bent functions have been formulated. Partial results towards a conjectured nonexistence of homogeneous rotation symmetric bent functions having degree greater than 2 have been obtained recently [42].

15.4 Generalized Bent Functions: The $\mathbb{Z}_p$-Valued Bent Functions

In 1985, Kumar, Scholtz and Welch [32] have generalized the notion of bent Boolean functions (introduced by Rothaus) to to p ary bent functions from $\mathbb{Z}_p^m$ to $\mathbb{Z}_p$ where p is an integer and $\mathbb{Z}_l$ denotes the ring of integers modulo l, that is $\mathbb{Z}_l := \{0, 1, \cdots, l-1\}$. Let $f : \mathbb{Z}_p^m \to \mathbb{Z}_p$ be a p-ary function in m variables. The Fourier transform of f is then defined to be the function F given by $F(\lambda) := \frac{1}{\sqrt{p^m}} \sum_{y \in \mathbb{Z}_p^m} f(y)\omega^{-\lambda \cdot y}$, $\lambda \in \mathbb{Z}_p^m$ (where ω be a primitive pth root of unity and $\lambda \cdot y$ denotes the inner product of two vectors λ and y). Then f is said to be bent if the Fourier transform coefficients of ω^f all have unit magnitude. Equivalently, f is bent if and only if its Hadamard transform W_f defined by $W_f(\lambda) := \sum_{y \in \mathbb{Z}_p^m} \omega^{f(y) - \lambda \cdot y}$ satisfies $|W_f(\lambda)| = p^{m/2}$ for every $\lambda \in \mathbb{Z}_p^m$. The properties of generalized bent functions have been firstly examined in [32]. Below we present some properties provided by Kumar et al.

1. (Property 1) every affine or linear translate of a bent function is also bent (recall that an affine (resp. linear) function g from $\mathbb{Z}_p^m$ to $\mathbb{Z}_p$ is of the form $g(x) = f(x) + z \cdot x + c$ where f is a mapping from $\mathbb{Z}_p^m$ to $\mathbb{Z}_p$ and $(z, c) \in \mathbb{Z}_p^m \times \mathbb{Z}_p$ with $c \neq 0$ (resp. $c = 0$).
2. (Property 2) a bent function remains bent under a linear or affine transformation in coordinates.
3. (Property 3) if f and g are bent functions over $\mathbb{Z}_p^m$ and $\mathbb{Z}_p^n$, respectively, the function $f + g$ over $\mathbb{Z}_p^{m+n}$ defined by $f + g(x_1, x_2, \cdots, x_{n+m}) = f(x_1, \cdots, x_m) + g(x_{m+1}, \cdots, x_{m+n})$ is a bent function.
4. (Property 4) a function f with values in $\mathbb{Z}_p$ is bent if and only if the matrix H whose (x, y)th entry $w^{f(x-y)}$ where $w := e^{\frac{2\pi i}{p}}$ is a generalized Hadamard matrix (recall that a $n \times n$ matrix H whose entries are integral powers of a complex primitive nth root of unity and which satisfies $HH^\star = nI$ (here "$\star$" means the complex conjugation) is called a generalized Hadamard matrix).
5. (Property 5) if f is a bent function defined on $\mathbb{Z}_p^m$, the Fourier coefficients of γ^f have unit magnitude for every choice of complex primitive pth root γ of unity.
6. (Property 6) let m be odd and $p \equiv 2 \pmod 4$ satisfying either one of the following two conditions:

   - $p = 2$
   - $p \neq 2$ but there exists an integer b satisfying $2^b \equiv -1 \pmod{p/2}$.

   Then bent functions over $\mathbb{Z}_p^m$ do not exist.
7. (Property 7) let f be a bent function over $\mathbb{Z}_p^m$ and let p and m satisfy any one of the following conditions:

   - $p = r^k$, r prime, $p \neq 2$,
   - $p = \prod_{i=1}^{s} r_i^{k_i}$, $s > 1$, r_i prime for all $i \in \{1, 2, \cdots, s\}$, with the primes r_i being such that for each integer i, $i \in \{1, 2, \cdots, s\}$ there exists an integer b_i

for which

$$r_i^{b_i} \equiv -1 \pmod{p/r_i^{k_i}}.$$

Then each Fourier coefficient $F(\lambda)$, $\lambda \in \mathbb{Z}_p^m$ of ω^f is a root of unity.

In fact the generalized versions share many properties such as the connection with Hadamard matrices, in common with their binary counterparts (note that property 1 above can be easily checked from the definition of generalized bent functions and the proofs to Properties 2–4 are identical to those given by Rothaus [56] for the binary case). On the other hand, some other properties of generalized bent functions are specific to the general case in the sense that the corresponding binary results are uninteresting.

Bent functions on $\mathbb{Z}_p^m$ are known to exist except for the case when m is odd and $p \equiv 2 \pmod 4$. When m is odd, $p \equiv 2 \pmod 4$ and in addition, $2^b \equiv -1 \pmod{p/2}$ for some $b \in \mathbb{Z}$, bent functions on $\mathbb{Z}_p^m$ do not exist.

For every value of p and m (other than m odd and $p \equiv 2 \pmod 4$), Kumar et al have provided constructions for bent functions over $\mathbb{Z}_p^m$.

In [33] Langevin has compared binary bent functions and the generalized bent functions from the metric and degree point of view. He provided an upper bound on the covering radius of the affine Reed–Muller codes defined over a finite and commutative ring as well as a bound on the degree of a generalized bent function. In 1998, Hou [29] derived a construction of p-ary bent functions based on a ring structure of $\mathbb{Z}_p^m$ and proposed q-ary bent functions constructed from chain rings (that is, a finite local principal ideal ring).

15.5 Generalized Bent Functions from Coding Point of View

From the viewpoint of cyclic codes over a Galois ring, Schmidt [58] introduced in 2006 another generalization of bent functions in connection to constructions of quaternary constant-amplitude codes for multicode CDMA systems. These generalized bent functions are from $\mathbb{Z}_2^m$ to $\mathbb{Z}_{2^s}$ (where s is a positive integer). A generalized Boolean function $f : \mathbb{Z}_2^m \to \mathbb{Z}_{2^s}$ is said to be bent if its Hadamard transform W_f defined by $W_f(\lambda) := \sum_{y\in\mathbb{Z}_2^m}(-1)^{\lambda\cdot y}\omega^{f(y)}$ (where ω be a primitive 2^s-th root of unity) satisfies $|W_f(\lambda)| = 2^{m/2}$ for every $\lambda \in \mathbb{Z}_2^m$. Note that several nonexistence results of generalized bent functions have been proved in the literature (see [30] and the references therein). Schmidt considered in detail the case when $s = 2$ and studied the relations between generalized bent functions, constant amplitude codes, and the available $\mathbb{Z}_4$-linear codes. Next, Solé and Tokareva [61] have showed how generalizations of the notion of bent function involving the ring $\mathbb{Z}_4$ could produce, by Gray map or by base 2 expansion, bent Boolean functions in the classical sense. The approach of Kumar et al. [32] and that of Schmidt do not seem to be equivalent. Both approaches are inspiring. Schmidt's definition gives better

$\mathbb{Z}_4$ cyclic codes constructions and Kumar et al approach allows a nice analogue of Maiorana McFarland construction. In both cases an analogue of Dillon construction is lacking. In [61] Solé and Tokareva have showed the close connections between Boolean bent and quaternary bent functions. More precisely, direct links between Boolean bent functions, generalized bent functions due to Schmidt and quaternary bent functions [32] have been explored. Note that the authors have also studied in [61] the Gray images of bent functions and notions of generalized nonlinearity for functions that are relevant to generalized linear cryptanalysis. Since 2011, some advances have been made in this topic showing that it is possible to construct many generalized Boolean bent and p-ary bent functions, some of them are based on $\mathbb{Z}_4$-valued quadratic forms.

In the following, we start by introducing basic notions in this topic.

15.5.1 Galois Ring

Let $h(x) \in \mathbb{Z}_4[x]$ of degree m be a primitive irreducible polynomial. The ring $\mathbb{Z}_4(x)/(h(x))$ is denoted by $\mathbb{R} = GR(4, m)$ and called the Galois ring (which is the Galois extension of degree m over $\mathbb{Z}_4$). Let $\mu : \mathbb{Z}_4 \to \mathbb{Z}_2$ be the modulus 2 reduction. Naturally, the mapping μ induces a homomorphism from the Galois ring $\mathbb{R}$ to the finite field $\mathbb{F}_{2^m}$ of order 2^m. In the literature, sometimes, $\mu(x)$ is denoted by $\bar{x}$.

For every element $z \in \mathbb{R}$ it can be uniquely expressed in the form $z = x + 2y$, $x, y \in \mathbb{T}_m$ where $\mathbb{T}_m$ is the Teichmüller set of $\mathbb{R}$ defined by $\mathbb{T}_m := \{z \in \mathbb{R} \mid z^{2^m} = z\}$. If ζ is a root of $h(x)$ satisfying $\zeta^{2^m-1} = 1$, then $\mathbb{R} = \mathbb{Z}_4[\zeta]$ and $\mathbb{T}_m = \{0, 1, \cdots, \zeta^{2^m-2}\}$. Note that $\mathbb{T}_m$ is not closed under the usual addition. For any $x, y \in \mathbb{T}_m$, there exists a unique $z \in \mathbb{T}_m$ such that $z = x + y + 2\sqrt{xy}$. One can define a new operation $\oplus$ as $x \oplus y = x + y + 2\sqrt{xy}$. Clearly, $\mu(\mathbb{T}_m) = \mathbb{F}_{2^m}$ which means that the Teichmüller set of $\mathbb{T}_m$ is isomorphic to the finite field $\mathbb{F}_{2^m}$ (more precisely, $(\mathbb{T}_m, \oplus, \cdot)$ is isomorphic to $\mathbb{F}_{2^m}$).

One can define a trace function $TR_1^m(\cdot)$ from the Galois ring $\mathbb{R}$ to $\mathbb{Z}_4$ by

$$TR_1^m(x + 2y) := \sum_{j=0}^{m-1} (x^{2^j} + 2y^{2^j}), x, y \in \mathbb{T}_m.$$

There exists a relationship between the trace function $TR_1^m(\cdot)$ over $\mathbb{Z}_4$ and the trace function $Tr_1^m(\cdot)$ over $\mathbb{Z}_2$. The connection between both trace functions is given by the following:

1. $\overline{TR_1^m(x)} = \mathrm{Tr}_1^m(\bar{x})$;
2. $2TR_1^m(x) = 2\mathrm{Tr}_1^m(\bar{x})$.

It should be noted that the addition operation in L is not closed. Especially, for any $x, y \in \mathbb{T}_m$, there exists a unique element z in $\mathbb{T}_m$ such that $z = x + y + 2\sqrt{xy}$.

15.5.2 Generalized Bent Function $\mathbb{Z}_4$-Valued

Let $f : \mathbb{T}_m \to \mathbb{Z}_4$ be a generalized Boolean function in m variables and i be a complex primitive 4-th root of unity. Then the Walsh transform of f is given by

$$\hat{f}(\omega) = \sum_{i \in \mathbb{T}_m} i^{f(x)+2TR_1^m(\omega x)}, \omega \in \mathbb{T}_m.$$

Definition 15.5.1. Let $f : \mathbb{T}_m \to \mathbb{Z}_4$ be a generalized Boolean function in m variables. f is said to be bent if $|\hat{f}(\omega)| = 2^{\frac{m}{2}}$ for all $\omega \in \mathbb{T}_m$.

Note that generalized Boolean bent functions exist for both even m and odd m, while Boolean bent functions only exist for even m.

The connections between Boolean bent and generalized Boolean bent functions have been revealed by Solé and Tokareva [61] who have showed the following result.

Proposition 15.5.2 ([61]). *Let $f(x)$ be a generalized Boolean function over $\mathbb{Z}_4$ with its 2-adic expansion $f(x) = a(\bar{x}) + 2b(\bar{x})$ where $x \in \mathbb{T}_m$ and $a(\cdot)$, $b(\cdot)$ be two Boolean functions over $\mathbb{F}_{2^m}$. Let $\Phi_f(y,z) = a(y)z + b(y)$ be a Boolean function in two variables $y \in \mathbb{F}_{2^m}$ and $z \in \mathbb{Z}_2$. Then we have*

- $2|\hat{f}(\omega)|^2 = \hat{b}^2(\bar{\omega}) + \widehat{a+b}^2(\bar{\omega})$, $\forall \omega \in \mathbb{T}_m$;
- *if $g(x)$ is bent, then $\Phi_f(y,z)$ is either bent (m odd) or semi-bent (m even);*
- *assume m even. Then $f(x)$, $x \in \mathbb{T}_m$ if and only if the functions $b(x)$ and $a(x)+b(x)$, $x \in \mathbb{F}_{2^m}$ are bent.*

15.5.3 $\mathbb{Z}_4$-Valued Quadratic Forms

The $\mathbb{Z}_4$-valued quadratic form bent functions have been characterized. Let us recall some background on $\mathbb{Z}_4$-valued quadratic forms. A symmetric form on $\mathbb{T}_m$ is a mapping $B : \mathbb{T}_m \times \mathbb{T}_m \to \mathbb{Z}_2$ having the following properties.

1. $B(x,y) = B(y,x)$;
2. $B(x \oplus y, z) = B(x,z) + B(y,z)$.

Moreover, B is said to be alternating if $B(x,x) = 0$ for all $x \in \mathbb{T}_m$.

Given a symmetric bilinear form B, the rank of B denoted by $rank(B)$ is defined as follows: $rank(B) = m - dim_{\mathbb{Z}_2}(rad(B))$ where $rad(B) := \{x \in \mathbb{T}_m \mid B(x,y) = 0, \forall y \in \mathbb{T}_m\}$

Definition 15.5.3. $\mathbb{Z}_4$-valued quadratic form is a mapping $F : \mathbb{T}_m \to \mathbb{Z}_4$ satisfying the two following conditions

1. $F(0) = 0$;
2. $f(x \oplus y) = F(x) + F(y) + 2B(x,y)$ where $B : \mathbb{T}_m \times \mathbb{T}_m \to \mathbb{Z}_2$ is a symmetric bilinear form.

Table 15.1 Distribution of $\{\hat{F}(\omega), \omega \in \mathbb{T}_m\}$ for F alternating

Value	Frequency
0	$2^m - 2^r$
$\pm 2^{m-\frac{r}{2}}$	$2^{r-1} \pm 2^{\frac{r}{2}-1}$

Table 15.2 Distribution of $\{\hat{F}(\omega), \omega \in \mathbb{T}_m\}$ for F nonalternating, for r odd

Valuc	Frequency
0	$2^m - 2^r$
$\pm(1+\sqrt{-1})2^{m-\frac{r+1}{2}}$	$2^{r-2} \pm 2^{\frac{r-3}{2}}$
$\pm(1-\sqrt{-1})2^{m-\frac{r+1}{2}}$	$2^{r-2} \pm 2^{\frac{r-3}{2}}$

Table 15.3 Distribution of $\{\hat{F}(\omega), \omega \in \mathbb{T}_m\}$ for F nonalternating, for r even

Value	Frequency
0	$2^m - 2^r$
$\pm 2^{m-\frac{r}{2}}$	$2^{r-2} \pm 2^{\frac{r}{2}-1}$
$\pm 2^{m-\frac{r}{2}}\sqrt{-1}$	2^{r-2}

Moreover, F is called alternating if B is alternating. Also, the rank of F is defined as $rank(F) = rank(B)$.

For a $\mathbb{Z}_4$-valued quadratic form $F : \mathbb{T}_m \to \mathbb{Z}_4$, the Walsh transform

$$\hat{F}(\omega) = \sum_{x \in \mathbb{T}_m} (\sqrt{-1})^{F(x)+2TR_1^m(\omega x)}$$

was investigated where ω ranges $\mathbb{T}_m$.

Helleseth and Kumar have shown in [28] that the Walsh distribution completely depends on its rank when F is alternating. In 2009, Schmidt [59] has developed the theory of $\mathbb{Z}_4$-valued quadratic forms and established a similar result for F being nonalternating.

If F is an alternating $\mathbb{Z}_4$-valued quadratic form of rank r, then the distribution of $\{\hat{F}(\omega), \omega \in \mathbb{T}_m\}$ is given by Table 15.1.

If F is a nonalternating $\mathbb{Z}_4$-valued quadratic form of rank r, then the distribution of $\{\hat{F}(\omega), \omega \in \mathbb{T}_m\}$ is given by the following tables according if r is odd or even (see Tables 15.2 and 15.3.

From the above distributions, it not difficult to deduce the following result.

Proposition 15.5.4 ([59]). *Let F be a $\mathbb{Z}_4$-valued quadratic form. Then $F(x)$ is generalized Boolean bent if and only if $F(x)$ is of full rank.*

15.5.4 Generalized Bent Functions and $\mathbb{Z}_4$-Valued Bent Function

In [36], Li, Tang and Helleseth have provided new constructions of quadratic bent functions in polynomial forms. Their constructions give rise to Boolean bent and generalized Boolean bent functions. More precisely, Li et al. have investigated a class of generalized Boolean bent functions of the form

$$Q(X) = TR_1^m(x + 2\sum_{i=1}^{\lfloor \frac{m-1}{2} \rfloor} c_i x^{1+2^{ki}}), c_i \in \mathbb{Z}_2, x \in \mathbb{T}_m. \tag{15.3}$$

Recall that where $TR_1^m(\cdot)$ denotes the trace function from the Galois ring $\mathbb{R}$ to $\mathbb{Z}_4$ and $\mathbb{T}_m$ the Teichmüller set of $\mathbb{R}$.

Using the theory of $\mathbb{Z}_4$-valued quadratic forms, Li, Tang, and Helleseth have provided the following necessary and sufficient condition concerning the bentness of $Q(x)$.

Theorem 15.5.5 ([36]). *The function $Q(x)$ defined by (15.3) is generalized Boolean bent if and only if $gcd(c(x^k), x^m - 1) = 1$ where $c(x) \in \mathbb{F}_2[x]$ defined by $c(x) = 1 + \sum_{i=1}^{\lfloor \frac{m-1}{2} \rfloor}(c_i x^i + c_i x^{m-i})$.*

Li et al. have investigated the bentness property of $Q(x)$ for specific values of c_i for $1 \leq i \leq \lfloor \frac{m-1}{2} \rfloor$. They deduced from Theorem 15.5.5 the following series of corollaries.

Corollary 15.5.6 ([36]). *Let m and k be positive integers. Set $Q(x) := TR_1^m(x + 2x^{1+2^{2k}} + 2x^{1+2^{3k}}), x \in \mathbb{T}_m$. Then $Q(x)$ is bent if and only if $gcd(m, 3k) = gcd(m, k)$.*

Corollary 15.5.7 ([36]). *Let m, k and t be positive integers with even (resp. odd) m and $t < \frac{m-2}{4}$ (resp. $\frac{m-2}{2}$). Set $Q(x) := TR_1^m(x + 2\sum_{i=0}^{t} x^{1+2^{(2i+1)k}}), x \in \mathbb{T}_m$. Then $Q(x)$ is bent if and only if $gcd(m, (2t+1)k = gcd(m, k)$, and $gcd(m, (2t+3)k) = gcd(m, k)$.*

Corollary 15.5.8 ([36]). *Let m,k and t be positive integers with even (resp. odd) m and $t < \frac{m}{4}$ (resp. $\frac{m-1}{2}$). Set $Q(x) := TR_1^m(x+2x^{1+2^k} + 2\sum_{i=1}^{t} x^{1+2^{2ik}}), x \in \mathbb{T}_m$. Then $Q(x)$ is bent if and only if $gcd(m, (2t-1)k) = gcd(m, k)$, and $gcd(m, (2t+3)k = gcd(m, k)$.*

Corollary 15.5.9 ([36]). *Let m,k and t be positive integers with $1 \leq t < \frac{m}{2}$. Set $Q(x) := TR_1^m(x + 2\sum_{i=1}^{t} x^{1+2^{ik}}), x \in \mathbb{T}_m$. Then $Q(x)$ is bent if and only if $gcd(m, (2t+1)k) = gcd(m, k)$.*

Note that for distinct concrete values on c_i, $1 \leq i \leq \lfloor \frac{m-1}{2} \rfloor$, different generalized Boolean bent functions can be obtained if $gcd(c(x^k), x^{m-1}) = 1$ where $c(x)$ is defined by $c(x) = 1 + \sum_{i=1}^{\lfloor \frac{m-1}{2} \rfloor}(c_i x^i + c_i x^{m-i}) \in \mathbb{F}_2[x]$.

Moreover, the authors have deduced from the above result a generalized Boolean bent function given by the following statement which easily comes from Theorem 15.5.5.

Theorem 15.5.10 ([36]). *Let $h(x) \in \mathbb{Z}_2[x]$ with even degree $s < 2m$. If $h(x)$ can be written as $h(x) = x^{s/2}(1 + x^{s/2} + x^{-s/2} + \sum_{i=1}^{r}(x^{n_i} + x^{-n_i}))$ where r is a positive integer and n_i, $1 \leq i \leq r$ are integers satisfying $1 \leq n_1 < n_2 \cdots < n_r < s/2$, then for any positive integer k, the generalized Boolean function $Q(X) = TR_1^m(x + \sum_{i=1}^{r} x^{1+2^{n_i k}} + 2x^{1+2^{s/2k}})$ is bent if and only if $gcd(h(x^k), x^m - 1) = 1$.*

From Theorem 15.5.10, new generalized Boolean bent functions have been provided. In [37].

The authors pointed out the following facts. From [58], Schmidt derived the conditions for $a \in \mathbb{R}$ and $b \in \mathbb{T}_m$ such that $TR_1^m(ax + 2bx^3)$ is generalized Boolean bent. Next, it is proposed as an open problem in [61] to characterize the functions of the form $g_{a,b}(x) = TR_1^m(ax + 2bx^{1+2^k})$ which are bent. This problem has been solved by Li , Tang and Helleseth for odd $m/gcd(m,k)$ in [35]. More precisely, the following result it has been proven.

Theorem 15.5.11 ([35]). *Let m and k be positive integers such that $m/gcd(m,k)$ is odd. Let $a \in \mathbb{R}$ and $b \in \mathbb{T}_m$. Then $g_{a,b}(x) = TR_1^m(ax + 2bx^{1+2^k})$ is generalized Boolean bent if the following conditions hold*

1. $\bar{a} \neq 0$ and $\bar{b} = 0$;
2. $\bar{a}\bar{b} \neq 0$ and $\bar{b}^{2^k} x^{1+2^k} + \bar{c}^{2^{k+1}} x + \bar{b} = 0$ has either zero or two roots in $\mathbb{F}_{2^m}$.

Moreover, the number of (a,b) such that $g_{a,b}(x)$ is generalized Boolean bent has also been given in [35]. However, the case of $m/gcd(m,k)$ being even still remains open

From the obtained generalized Boolean bent functions together with the links between generalized Boolean bent and Boolean bent functions, Li et al. [36] have presented a method to transform them into binary bent and semi-bent functions. Their main result is given by the following statement.

Theorem 15.5.12 ([36]). *Let m and k be positive integers with even (resp. odd) m and let $p(x)$ be defined on $\mathbb{F}_{2^m}$ by $p(x) = \sum_{i=1}^{(m-1)/2} \mathrm{Tr}_1^m(x^{2^i+1})$ if m is odd and $p(x) = \sum_{i=1}^{m/2-1} \mathrm{Tr}_1^m(x^{2^i+1}) + \mathrm{Tr}_1^{m/2}(x^{2^{m/2}+1})$ if m is even. Let $f_Q(x)$ be the binary function defined by $f_Q(x) = p(x) + \sum_{i=1}^{\lfloor \frac{m-1}{2} \rfloor} \mathrm{Tr}_1^m(c_i x^{1+2^{ki}})$ where $c_i \in \mathbb{F}_2$ and $x \in \mathbb{F}_{2^m}$. Then f_Q is bent (resp. semi-bent) if and only if $gcd(c(x^k), x^m - 1) = 1$, where $c(x)$ is defined by $c(x) = 1 + \sum_{i=1}^{\lfloor \frac{m-1}{2} \rfloor}(c_i x^i + c_i x^{m-i}) \in \mathbb{F}_2[x]$.*

By the same techniques in [36] and [37], many new p-ary bent functions $f(x)$ of the form $f(x) = \sum_{i=0}^{\lfloor \frac{m-1}{2} \rfloor} tr_m(c_i x^{1+p^{ki}})$ have been obtained, and a simple method to construct them.

To summarize the recent work (2014) due to Li, Tang and Helleseth [37], the authors have in fact presented a construction of quadratic bent functions in polynomial form which gives rise to new Boolean bent, generalized Boolean bent and p-ary bent functions. Next, based on $\mathbb{Z}_4$-valued quadratic forms, they provided a method leading to several new constructions of generalized Boolean bent functions. From these generalized Boolean bent functions they presented a method to transform them into Boolean bent and semi-bent functions. Many new p-ary bent functions can also be obtained by applying similar methods.

Further, Wu and Lin generalized the work of Li, Tang and Helleseth (2012) [36]. A new construction of quaternary bent functions from quaternary quadratic forms over Galois rings of characteristic 4 is proposed in [38]. Based on their construction, several new classes of quaternary bent functions have been obtained, and as a consequence, several new classes of quadratic binary bent and semi-bent functions in polynomial forms were derived.

15.6 ℤ-Bent Functions

Except the remarkable representation theorems due to Carlet and Guillot (see [9, 10]), not much is known about the structure of bent functions in general. A main obstacle in the study of bent functions (which can alternatively be defined as ± 1-valued functions on $\mathbb{F}_2^n$ with ± 1-valued Fourier transform) is the lack of recurrence laws. To embed bent functions into a recursive context one needs more general structures. To this end Dobbertin [17][3] (see also [18, 19]) has the idea of considering integer-valued functions f on $\mathbb{F}_2^n$ whose normalized Fourier transform $2^{-m}\sum_{x\in\mathbb{F}_2{}^n} f(x)(-1)^{a\cdot x}$, let us denote it by $\hat{f}(a)$, is also integer-valued. This has given rise to the notion of the so-called *ℤ-bent function* (ℤ denotes the set of integers). These ℤ-bent functions introduced by Dobbertin can be separated into different levels depending on the size of the maximal absolute value attained by f and $\hat{f}$. Dobbertin [19] presented a natural nested chain consisting of an increasing sequence of sets W_r, $r \geq 0$ defined as follows: $W_0 = \{\pm 1\}$ and for $r \neq 0$, $W_r = \{w \in \mathbb{Z} \mid -2^{r-1} \leq w \leq 2^{r-1}\}$. We have $W_r \pm W_r = W_{r+1}$ for $r > 0$ and $\cup_{r\geq 0} W_r$. The union of all W_r-bent functions gives the set of all ℤ-bent functions. Dobbertin has used the term ℤ-bent function of level r instead of W_r-bent function. In this recursion, the usual bent functions are precisely the ℤ-bent functions of level zero. The following definition (which includes the level) can be found in [19].

Definition 15.6.1. A function $f : \mathbb{F}_2^n \to W_r$ is said to be a ℤ-bent function of size $\frac{n}{2}$ and level r if $\hat{f}$ is also a mapping into W_r. In this case, since the Fourier transform is self-inverse, $\hat{f}$ is also a ℤ-bent function of size k and level r which is called the dual of f.

It is shown how ℤ-bent functions of lower level can be built up recursively by gluing together ℤ-bent functions of higher level. This recursion comes down at level zero, containing the usual bent functions. In fact, ℤ-bent functions of level r on n variables can be used to construct all ℤ-bent functions of level $r-1$ on $(n+2)$ variables. This is referred to as "gluing" technique. Continuing in this way eventually all ℤ-bent functions of level 0 on $(n+2r)$ variables are obtained (which are the same as classical bent functions on $(n+2r)$ variables). Higher level ℤ-bent functions form building blocks that can be glued together under certain conditions to get larger ℤ-bent functions of lower level [19]. Dobbertin believed in fact that a deeper understanding of the structure of bent functions requires the study of ℤ-bent

[3]The study of bent functions embedded into the recursive framework of ℤ-bent functions has been initiated by Dobbertin in 2005.

functions and their role as building blocks. The reader can find in [19] a study of bent functions in the framework of $\mathbb{Z}$-bent functions as well as and some guidelines for further research.

In 2013 Gangopadhyay et al. [24] have generalized the construction of partial spreads $\mathcal{PS}$ bent functions to partial spreads $\mathbb{Z}$-bent functions of arbitrary level. They also identified the natural analogue of the class $\mathcal{PS}_{ap}$. Their motivation was to find new primary constructions of bent functions following the gluing process. To this end, the authors have showed how these partial spreads $\mathbb{Z}$-bent functions give rise to a new construction of (classical) bent functions. Based on a suitable subclass of those $\mathbb{Z}$-bent functions, they demonstrated a variety of bent functions that can efficiently be constructed. In particular they constructed by this way a bent function on eight variables which is inequivalent to all Maiorana–McFarland as well as $\mathcal{PS}_{ap}$ type bents. It is also shown in [24] that all bent functions on six variables, up to equivalence can be obtained by their construction.

15.7 Negabent Functions

Let $f(\mathbf{x}) : \mathbb{F}_2^n \to \mathbb{F}_2$ be a Boolean function of n variables $\mathbf{x} = (x_0, x_1, \ldots, x_{n-1})$, and consider its mapping to a $2 \times 2 \times \ldots \times 2$ n-dimensional array of the form $\hat{f} = (-1)^{f(\mathbf{x})}$, $\mathbf{x} \in \mathbb{F}_2^n$. For instance, if $n = 2$ and $f(\mathbf{x}) = x_0x_1$ then $\hat{f} = \begin{pmatrix} 1 & 1 \\ 1 & -1 \end{pmatrix}$. The n-dimensional periodic Fourier transform, $\hat{F}$, of $\hat{f}$ is obtained by the action of $H^{\otimes n}$ on $\hat{f}$, where $H = \begin{pmatrix} 1 & 1 \\ 1 & -1 \end{pmatrix}$ is the Hadamard matrix, and the action of $H^{\otimes n}$ is also referred to as the Hadamard (or Walsh–Hadamard) transform. A Boolean function, f, is defined to be bent if the elements of the periodic Fourier transform of $\hat{f}$ all have equal magnitude. For instance, our example function $f = x_0x_1$ is bent because $\hat{F} = H^{\otimes 2}\hat{f} = \begin{pmatrix} 2 & 2 \\ 2 & -2 \end{pmatrix}$, and all spectral entries have magnitude 2. Observe that H is a unitary matrix up to a multiplicative factor of $\frac{1}{\sqrt{2}}$. Whilst it is natural to consider just periodic Fourier properties of $\hat{f}$, we could, more generally, consider its continuous Fourier spectrum. This is conventional in the context of telecommunications but such continuous spectral properties are not typically assessed in the context of cryptography. As a step in the direction of continuous Fourier, it is reasonable to consider the n-dimensional negaperiodic Fourier transform, $\tilde{F}$, of $\hat{f}$. This is obtained by the action of $N^{\otimes n}$ on $\hat{f}$, where $N = \begin{pmatrix} 1 & i \\ 1 & -i \end{pmatrix}$ is the Negahadamard matrix, where $i^2 = -1$, and the action of $N^{\otimes n}$ is also referred to as the Negahadamard transform [15, 44, 54]. A Boolean function, f, is defined to be *negabent* if the elements of the negaperiodic Fourier transform of $\hat{f}$ all have equal magnitude [55, 57]. For instance, $f = x_0x_1$ is not negabent because $\tilde{F} = N^{\otimes 2}\hat{f} = \begin{pmatrix} 2+2i & 0 \\ 0 & 2-2i \end{pmatrix}$. However $f = x_0 + x_1$ is negabent as $\tilde{F} = N^{\otimes 2}\hat{f} = \begin{pmatrix} -2i & 2 \\ 2 & 2i \end{pmatrix}$. Observe that N is a unitary matrix up to a multiplicative factor of $\frac{1}{\sqrt{2}}$. Just as it is interesting to construct Boolean functions which are bent or near-bent, it is also interesting to construct Boolean functions which are negabent or near-negabent. All affine Boolean functions are negabent. A more challenging problem is to construct Boolean functions that are both bent and negabent [16, 45] - these are called bent-negabent functions. For instance, for $n = 4, f = x_0x_1 + x_1x_2 + x_2x_3$ is bent-negabent. There is a convenient mapping between bent and negabent functions, as follows:

Lemma 15.7.1. *Let $f(\mathbf{x}) : \mathbb{F}_2^n \rightarrow \mathbb{F}_2$ be a bent function. Then $f(\mathbf{x}) + \sigma(\mathbf{x})$ is a negabent Boolean function, where $\sigma(\mathbf{x}) = \sum_{i<j} x_i x_j$ (σ is sometimes referred to as the clique function as it comprises all quadratic terms). Conversely, if $f(\mathbf{x})$ is negabent then $f(\mathbf{x}) + \sigma(\mathbf{x})$ is bent for n even, and almost bent for n odd.*

Let $\hat{f}$ describe the coefficients of a multivariate polynomial, $s(\mathbf{z})$, in $\mathbf{z} = (z_0, z_1, \ldots, z_{n-1})$. Then computing the Hadamard transform of $\hat{f}$ is equivalent to embedding $s(\mathbf{z})$ in a modulus $(z_0^2 - 1)(z_1^2 - 1)\ldots(z_{n-1}^2 - 1)$, then computing the 2^n residues of $s(\mathbf{z})$ mod $(z_0 \pm 1)(z_1 \pm 1)\ldots(z_{n-1} \pm 1)$. For instance, for $n = 3$, $f = x_0x_1x_2 + x_0x_2 + x_1$ then $s(\mathbf{z}) = 1 + z_0 - z_1 - z_0z_1 + z_2 - z_0z_2 - z_1z_2 - z_0z_1z_2$, and the $2^3 = 8$ residues mod $(z_0^2 - 1)(z_1^2 - 1)(z_2^2 - 1)$ are $-2, 2, 6, 2, 2, -2, 2, -2$. Similarly computing the Negahadamard transform of $\hat{f}$ is equivalent to embedding $s(\mathbf{z})$ in a modulus $(z_0^2 + 1)(z_1^2 + 1)\ldots(z_{n-1}^2 + 1)$, then computing the 2^n residues of $s(\mathbf{z})$ mod $(z_0 \pm i)(z_1 \pm i)\ldots(z_{n-1} \pm i)$, $i^2 = -1$. For instance, for $s(\mathbf{z}) = 1+z_0-z_1-z_0z_1+z_2-z_0z_2-z_1z_2-z_0z_1z_2$, the 2^3 residues mod $(z_0^2+1)(z_1^2+1)(z_2^2+1)$ are $4i+2, -2, 2, 2, -2, -2, 2, 4i-2$.

The aperiodic autocorrelation of $\hat{f}$ can be computed as the coefficients of $a(\mathbf{z})$, where

$$a(\mathbf{z}) = s(\mathbf{z})s^*(\mathbf{z}^{-1}),$$

where $*$ means complex conjugate, and $\mathbf{z}^{-1} = (z_0^{-1}, z_1^{-1}, \ldots, z_{n-1}^{-1})$.

We can embedded this multiplication in a modulus as follows,

$$\bar{z}a(\mathbf{z}) = \bar{z}s(\mathbf{z})s^*(\mathbf{z}^{-1}) = \bar{z}s(\mathbf{z})s^*(\mathbf{z}^{-1}) \bmod (z_0^4 - 1)(z_1^4 - 1)\ldots(z_{n-1}^4 - 1),$$

where $\bar{z} = \prod_{j=0}^{n-1} z_j$. The point here is that the moduli $(z_j^4 - 1)$, $\forall j$, have high enough degree so that there is no modular reduction after $a(\mathbf{z})$ has been 'shifted' by $\bar{z}$. Now observe that $(z_j^4 - 1) = (z_j^2 - 1)(z_j^2 + 1) = (z_j - 1)(z_j + 1)(z_j - i)(z_j + i)$. It is straightforward to see that computing the residues of $\bar{z}a(\mathbf{z})$, mod $(z_j - 1)(z_j + 1)$, $\forall j$, realises, to within a phase shift (due to $\bar{z}$), the power spectrum of the Hadamard transform of $\hat{f}$ - this maps to periodic autocorrelation. Similarly, computing the residues of $\bar{z}a(\mathbf{z})$, mod $(z_j - i)(z_j + i)$, $\forall j$, realises, to within a phase shift, the power spectrum of the Negahadamard transform of $\hat{f}$ - this maps to negaperiodic autocorrelation. Thus, evaluating the residues of $\bar{z}a(\mathbf{z})$, mod $(z_j - 1)(z_j - i)(z_j + 1)(z_j + i)$, computes the power spectrum of the 2^n transforms $H \otimes H \otimes \ldots \otimes H$, $N \otimes H \otimes \ldots \otimes H$, $H \otimes N \otimes \ldots \otimes H$, $\ldots N \otimes N \otimes \ldots \otimes N$ of $\hat{f}$ - this maps to aperiodic autocorrelation. So aperiodic properties of a Boolean function, f, can be assessed by considering all $\{H,N\}^{\otimes n}$ transforms of $\hat{f}$. In the extreme, if the aperiodic autocorrelation is a δ-function, i.e. if $a(\mathbf{z}) = 2^n$, then all 2^n transforms of $\hat{f}$ are flat. This ideal is impossible but motivates us to find functions, f, such that as large a number as possible of the $\{H,N\}^{\otimes n}$ transforms of $\hat{f}$ are flat, i.e. not just bent (flat wrt $H^{\otimes n}$) and negabent (flat wrt $N^{\otimes n}$). A further generalisation is to consider the aperiodic properties of a Boolean function and all its sub-functions obtained by fixing one or more variables of the function - this maps, spectrally, to the power spectrum wrt $\{I,H,N\}^{\otimes n}$, where $I = \begin{pmatrix} 1 & 0 \\ 0 & 1 \end{pmatrix}$ [14].

Yet another motivation for considering the Negahadamard matrix is due to the role of $\{I,H,N\}$ in stabilising the Pauli group where, for ease of discussion, we assume that H and N are normalized multiplicatively by $\frac{1}{\sqrt{2}}$, *i.e.* are unitary. The Pauli group is generated by the unitary matrices $\{X,Z,Y\}$ and the multiplicative factors ± 1, $\pm i$, where $X = \begin{pmatrix} 0 & 1 \\ 1 & 0 \end{pmatrix}$, $Z = \begin{pmatrix} 1 & 0 \\ 0 & -1 \end{pmatrix}$, and $Y = iXZ$. Note that the action of X_j (i.e. X acts at tensor position j) on $\hat{f}$ is to replace x_j with $x_j + 1$ in the Boolean input to f. Similarly, the action of Z_j replaces f with $f + x_j$, and the action of Y_j performs both actions. Observe that $\{I,H,N\}$ stabilize $\{X,Z,Y\}$ as $HXH^{-1} = Z$, $HZH^{-1} = X$, $HYH^{-1} = -Y$, $NXN^{-1} = -iY$, $NZN^{-1} = X$, and $NYN^{-1} = -Z$. Note also that the Pauli matrices, $\{I,X,Z,Y\}$, form a basis for the set of 2×2 unitary matrices. Finally, up to left diagonal matrices, $\{I,H,N\}$ forms a matrix group, as follows. Let $\lambda = \frac{\omega^5}{\sqrt{2}}\begin{pmatrix} 1 & i \\ 1 & -i \end{pmatrix} = \omega^5 N$. Then $\lambda^2 = \frac{\omega^3}{\sqrt{2}}\begin{pmatrix} 1 & 0 \\ 0 & -i \end{pmatrix}\begin{pmatrix} 1 & 1 \\ 1 & -1 \end{pmatrix} = \omega^3 \begin{pmatrix} 1 & 0 \\ 0 & -i \end{pmatrix} H$, and $\lambda^3 = I$. So $\{I,\lambda,\lambda^2\}$ is a matrix group of order 3 and, up to left diagonals, is equal to $\{I,N,H\}$. Moreover $\{I,\lambda,\lambda^2\}$ is a subgroup of the local Clifford group. Instead of considering transforms with respect to $\{I,H,N\}^{\otimes n}$, one could consider transforms with respect to $\{I,\lambda,\lambda^2\}^{\otimes n}$ - only phase properties of the spectra would alter, but magnitude properties would remain unchanged.

Yet another motivation for considering the Negahadamard matrix is due to $\{I,H,N\}$ being an optimal mutually unbiased basis (MUB), where the 3 bases are the pairwise orthogonal length $\delta = 2$ rows of I, H, and N, respectively [2, 64]. Let p be a row of I, H, or N, and q be another row, such that p and q come from distinct matrices. Then $\{I,H,N\}$ is a MUB because $|< p, q >|^2 = a$, where a is a constant. The MUB is then optimal if it comprises a maximum of $\delta + 1 = 3$ orthogonal bases, in which case $a = \frac{1}{\delta} = \frac{1}{2}$.

There are further motivations from graph theory and quantum information. The operation of local complementation on a simple graph can be interpreted as the action of N on a quadratic Boolean function where each quadratic term is an edge in the graph. Such quadratics are mathematically identical to quantum stabiliser states, in which case the action of unitary N defines a local equivalence between such states [3, 14, 15, 53, 55].

15.8 Bent Functions on a Finite Group

Logachev, Sal'nikov and Yashenko [40] have introduced in 1997 the notion of a bent function on a finite Abelian group which in the case of the elementary Abelian 2-group coincides with the well-known notion of a Boolean bent function. Using methods of the theory of characters and commutative harmonic analysis they obtained some properties of bent functions which generalize the corresponding properties of Boolean bent functions. They have also constructed some classes of bent functions.

In 2002, Solodovnikov [62] has proposed the most general approach to algebraic generalizations of bent functions by considering bent functions from a finite Abelian group into a finite Abelian group. To this end, he introduced the notions of an absolutely non-homomorphic function, a minimal function (farthest from homomorphisms) and a bent function, and proved that the class of bent functions coincides with the class of *absolutely non-homomorphic* functions, a function is uniquely determined by the distances to homomorphisms with shifts, and that in the primary case the bent functions are absolutely minimal.

In [7], Carlet and Ding have considered the perfect nonlinearity from an Abelian group to another Abelian group. Let us give more details. Let $(A, +)$ and $(B, +)$ be two Abelian groups. The (Hamming) distance between two functions f and g from A to B, denoted by $d(A, B)$, is defined to be $d(f, g) := |\{x \in A \mid f(x) - g(x) \neq 0\}|$. One way of measuring the nonlinearity of a function f from $(A, +)$ to $(B, +)$ is to use the minimum distance between f and all affine functions from $(A, +)$ to $(B, +)$ (f is said to be affine if f can be written as $f = g + b$ with b is some constant in B and g is a linear map, (that is, $g(x + y) = g(x) + g(y)$ for all $(x, y) \in A^2$). With this approach the nonlinearity of f is defined to be $\mathcal{N}_f = min_{l \in L} d(f, l)$ where L denotes the set of all affine functions from $(A, +)$ to $(B, +)$. This measure of nonlinearity is related to linear cryptanalysis but it is not useful in general cases. A robust measure of the nonlinearity of functions is related to differential cryptanalysis and uses the derivatives $D_a f(x) := f(x + a) - f(x)$. It may be defined by

$$P_f = max_{a \in A, a \neq 0} max_{b \in B} Pr(D_a f(x) = b)$$

where $Pr(E)$ denotes the probability of the occurrence of event E: The smaller the value of P_f the higher the corresponding nonlinearity of f (if f is linear, then $P_f = 1$). Note that both nonlinearity measures are relative to the two operations of the two Abelian groups. P_f is lower bounded by $\frac{1}{|B|}$. This lower bound can be considered as an upper bound for the nonlinearity of f. Functions having a smallest possible value of P_f are important for applications in coding theory and cryptography. A function f from $(A, +)$ to $(B, +)$ is said to be perfect nonlinear if $P_f = \frac{1}{|B|}$. In the case of Boolean functions (that is, functions from $\mathbb{F}_2^n$ to $\mathbb{F}_2$) *perfect nonlinear* functions are also called bent. It has been proved that a function $f : A \to B$ has perfect nonlinearity if and only if, for every $a \in A \setminus \{0\}$, the derivative $D_a f$ is balanced (a function $g : A \to B$ is said to be balanced if the size of $g^{-1}(b) := \{a \in A \mid g(a) = b\}$ is the same for every $b \in B$ (this size is then $\frac{|A|}{|B|}$)). This is possible only if $|B|$ divides $|A|$.

Perfect nonlinear functions from a finite group A to another one B are those functions $f : A \to B$ such that all nonzero $a \in A$ the derivatives $d_a f : x \mapsto f(ax)f(x)^{-1}$ is balanced. In the case where both A and B are Abelian groups, $f : A \to B$ is perfect nonlinear if and only f is bent,that is, for all nonprincipal characters χ of B the (discrete) Fourier transform of $\chi \circ f$has a constant magnitude equal to $|A|$. More precisely, the equivalence between bentness and perfect nonlinearity has been recently extended by Carlet and Ding [7] and Pott [48] to the general case; $f : A \to B$ (where A and B are two finite Abelian groups) is perfect nonlinear if and only if, for all $a \in A$ and for all nonprincipal characters χ of B, $|\widehat{\chi \circ f}(a)|^2 = |A|$ where for $\phi : A \to \mathbb{C}$, $\hat{\phi}$ denotes its discrete Fourier transform defined by $\hat{\phi}(a) = \sum_{x \in \mathbb{Z}_2^m} \phi(x)(-1)^{a \cdot x}$. Using the theory of linear representations, Poinsot [47] has exhibited in 2012 similar bentness-like characterizations in the cases where A and/or B are (finite) non Abelian groups. This work has extended the concept of bent functions to the framework of non Abelian groups.

References

1. G. Xiao C. Ding and W. Shan. The Stability Theory of Stream Cipher. In *Berlin, Germany: Springer-Verlag, LNCS 561*, 1991.
2. A. Cabello, M. G. Parker, G. Scarpa, and S. Severini. Exclusivity structures and graph representatives of local complementation orbits. In *J. Math. Phys. 54, 073507, DOI: 10.1063/1.4813438, (arXiv.1211.4250)*, 2013.
3. A. Cabello, M. G. Parker, G. Scarpa, and S. Severini. Exclusivity structures and graph representatives of local complementation orbits. In *J. Math. Phys. 54, 073507, DOI: 10.1063/1.4813438, (arXiv.1211.4250)*, 2013.
4. C. Carlet. Partially-bent functions. In *Designs Codes and Cryptography, 3, pages 135–145*, 1993.
5. C. Carlet. Boolean functions for cryptography and error correcting codes. In Yves Crama and Peter L. Hammer, editors, *Boolean Models and Methods in Mathematics, Computer Science, and Engineering*, pages 257–397. Cambridge University Press, June 2010.
6. C. Carlet. Open problems on binary bent functions. In *Proceedings of the conference "Open problems in mathematical and computational sciences", September 18–20, 2013, in Istanbul, Turkey, pages 203–241*, 2014.
7. C. Carlet and C. Ding. Highly Nonlinear Mappings. In *Special Issue: "Complexity Issues in Coding and Cryptography" of the Journal of Complexity, 20(2–3), pages 205–244*, 2004.
8. C. Carlet, G. Gao, and W. Liu. A secondary construction and a transformation on rotation symmetric functions, and their action on bent and semi-bent functions. In *J. Comb. Theory, Ser. A 127, pages 161–175*, 2014.

9. C. Carlet and P. Guillot. A characterization of binary bent functions. In *J. Comb. Theory Ser. A 76, pages 328–335*, 1996.
10. C. Carlet and P. Guillot. An alternate characterization of the bentness of binary functions, with uniqueness. In *Des. Codes Cryptogr. 14, pages 133–140*, 1998.
11. A. Cesmelioglu, W. Meidl, and A. Topuzoglu. Partially bent functions and their properties. In *Applied Algebra and Number Theory, page 22*, 2014.
12. C. Charnes, M. Rotteler, and T. Beth. On homogeneous bent functions. In *Proceedings Applied Algebra, Algebraic Algorithms and Error Correcting Codes (AAECC-14). Lecture Notes in Computer Science, Vol. 2227, pages 249–259*, 2001.
13. C. Charnes, M. Rotteler, and T. Beth. Homogeneous Bent Functions, Invariants, and Designs. In *Designs, Codes and Cryptography, Vol. 26, Issue 1–3, pages 139–154*, 2002.
14. L. E. Danielsen, T. A. Gulliver, and M. G. Parker. Aperiodic Propagation Criteria for Boolean Functions. In *Inform. Comput., 204, 5, pages 741–770*, 2006.
15. L. E. Danielsen and M. G. Parker. Spectral orbits and peak-to-average power ratio of Boolean functions with respect to the $\{I, H, N\}^n$ transform. In *Proceedings of SETA 2004, Lecture Notes in Computer Science, LNCS 3486, pages. 373–388*, 2005.
16. L.E. Danielsen and M. G. Parker. The selfnegadual properties of generalised quadratic Boolean functions. In *Des. Codes Cryptogr., Vol. 66, Issue 1–3, pages 305–316*, 2013.
17. H. Dobbertin. Bent functions embedded into the recursive framework of Z-bent functions. In *preprint, January*, 2005.
18. H. Dobbertin and G. Leander. Cryptographer's Toolkit for Construction of 8-Bit Bent Functions. In *Cryptology ePrint Archive, Report 2005/089, available at* http://eprint.iacr.org/, 2005.
19. H. Dobbertin and G. Leander. Bent functions embedded into the recursive framework of Z-bent functions. In *Des. Codes Cryptography 49(1–3), pages 3–22*, 2008.
20. H. Dobbertin, G. Leander, A. Canteaut, C. Carlet, P. Felke, and P. Gaborit. Construction of bent functions via Niho Power Functions. In *Journal of Combinatorial therory, Serie A 113*, pages 779–798, 2006.
21. F. J. MacWilliams and N. J. Sloane. The theory of error-correcting codes. In *Amsterdam, North Holland*, 1977.
22. E. Filiol and C. Fontaine. Highly nonlinear balanced Boolean functions with a good correlation-immunity. In *Proceedings of EUROCRYPT'98, Lecture Notes in Comput. Sci., vol. 1403, pp. 475–488*, 1998.
23. E. Filiol and C. Fontaine. On some cosets of the first Reed-Muller code with high minimum weight. In *IEEE Trans. Inform. Theory 45 (4) pages 1237–1243*, 1999.
24. S. Gangopadhyay, A. Joshi, G. Leander, and R. K. Sharma. A new construction of bent functions based on Z-bent functions. In *J. Designs, Codes and Cryptography. Vol. 66, I. 1–3, pages 243–256*, 2013.
25. G. Gao, C. Carlet, and W. Liu. Results on Constructions of Rotation Symmetric Bent and Semi-bent Functions. In *Lecture Notes in Computer Science Volume 8865, pages 21–33*, 2014.
26. G. Gao, X. Zhang, W. Liu, and C. Carlet. Constructions of quadratic and cubic rotation symmetric bent functions. In *IEEE Trans. Inform. Theory, No 58, pages 4908–4913*, 2012.
27. P. Guillot. Partial bent functions. In *Proceedings of the World Multi-conference on Systemics, Cybernetics and Informatics, SCI 2000*, 2000.
28. T. Helleseth and P. V. Kumar. Sequences with low correlation. In *Handbook of Coding Theory, Part 3: Applications, V. S. Pless, W. C. Huffman, and R. A. Brualdi, Eds. Amsterdam, The Netherlands: Elsevier, chapter. 21*, pages 1765–1853, 1998.
29. X. D. Hou. q-ary bent functions constructed from chain rings. In *Finite Fields and Applications, Vol. 4, No. 1, pages 55–61*, 1998.
30. Y. Jiang and Y. Deng. New results on nonexistence of generalized bent functions. In *Des. Codes Cryptogr. 75, pages 375–385*, 2015.
31. S. Kavut, S. Maitra, and M. D. Yucel. Search for Boolean functions with excellent profiles in the rotation symmetric class. In *IEEE Trans. Inform. Theory 53 (5), pages 1743–1751*, 2007.

32. P. V. Kumar, R. A. Scholtz, and L.R. Welch. Generalized bent functions. In *J. Combin. Theory. Ser. A. Vol. 40, No 1, pages 90–107*, 1985.
33. P. Langevin. On generalized bent functions. In *Eurocod '92," CISM Courses and Lectures, Vol. 339, pages 147–157,Springer-Verlag*, 1992.
34. G. Leander and A. Kholosha. Bent functions with 2^r Niho exponents. In *IEEE Trans. Inform. Theory 52 (12)*, pages 5529–5532, 2006.
35. N. Li, X. Tang, and T. Helleseth. Several classes of codes and sequences derived from a Z4 valued quadratic form.In *IEEE Transactions on Information Theory vol. 57 (11), pages 7618–7628*, 2011.
36. N. Li, X. Tang, and T. Helleseth. New classes of generalized Boolean bent functions over Z4. In *IEEE International Symposium on Information Theory Proceedings (ISIT 2012)*, 2012.
37. N. Li, X. Tang, and T. Helleseth. New constructions of quadratic bent functions in polynomial form. In *IEEE Transactions on Information Theory 60 (9), pages 5760–5767*, 2014.
38. B. Lin and D. Lin. New constructions of quaternary bent functions. In *arXiv:1309.0199*, 2013.
39. Petr Lisonek and Marko J. Moisio. On zeros of Kloosterman sums. *Des. Codes Cryptography*, 59(1–3):223–230, 2011.
40. O. A. Logachev, A. A. Sal'nikov, and V. V. Yashenko. Bent functions on a finite Abelian group (in Russian). In *Diskr. Mat., Volume 9, Issue 4, pages 3–20*, 1997.
41. W. Ma, M. Lee, and F. Zhang. A new class of bent functions. In *EICE Trans. Fundamentals, vol. E88-A, no. 7, pages 2039–2040*, 2005.
42. Q. Meng, L. Chena, and F-W. Fub. On homogeneous rotation symmetric bent functions. In *Discrete Applied Mathematics, Vol. 158, Issue 10, pages 111–117*, 2010.
43. Q. Meng, H. Zhang, M. Yang, , and J. Cui. On the degree of homogeneous bent functions. In *Discrete Applied Mathematics, Vol. 155 (5), pages 665–669*, 2007.
44. M. G. Parker. The constabent properties of Golay-Davis-Jedwab sequences. In *Int. Symp. Information Theory, Sorrento, p. 302, June 25–30*, 2000.
45. M. G. Parker and A. Pott. On Boolean functions which are bent and negabent. In *Sequences, Subsequences, and Consequences, International Workshop, SSC 2007, Los Angeles, CA, USA, May 31–June 2, Lecture Notes in Computer Science, LNCS 4893*, 2007.
46. J. Pieprzyk and C. Qu. Fast Hashing and rotation symmetric functions. In *J. Univers. Comput. Sci. No. 5, pages 20–31*, 1999.
47. L. Poinsot. Non Abelian bent functions. In *Journal Cryptogr Commun 4(1)*, pages 1–23, 2012.
48. A. Pott. Nonlinear functions in Abelian groups and relative difference sets. In *Discrete Applied Mathematics, vol. 138, issue 1–2*, pages 177–193, 2004.
49. B. Preneel, R. Govaerts, and J. Vandevalle. Boolean functions satisfying higher order propagation criteria. In *Proceedings of EUROCRYPT'91, Lecture Notes in Computer Sciences 547, pages 141–152*, 1991.
50. C. Qu, J. Seberry, and J. Pieprzyk. On the symmetric property of homogeneous Boolean functions. In *Proceedings of the Australian Conference on Information Security and Privacy (ACISP. Lecture Notes in Computer Science, Vol. 1587, Springer, pages 26–35*, 1999.
51. C. Qu, J. Seberry, and J. Pieprzyk. Homogeneous Bent Functions. In *Discrete Appl. Math. 102 (1–2), pages 133–139*, 2000.
52. M. Quisquater, B. Preneel, and J. Vandewalle. A new inequality in discrete Fourier theory. In *IEEE Trans. on Inf. Theory, Vol. 49, pages 2038–2040*, 2003.
53. C. Riera, S. Jacob, and M. G. Parker. From Graph States to Two-Graph States. In *Designs, Codes and Cryptography, vol. 48, (2)*, 2008.
54. C. Riera and M. G. Parker. Generalised Bent Criteria for Boolean Functions (I). In *One and Two-Variable Interlace Polynomials: A Spectral Interpretation", International Workshop, Proceedings of WCC2005, Bergen, Norway, March 2005, Revised Selected Papers, Lecture Notes in Computer Science, LNCS 3969, pages 397–411*, 2005.
55. C. Riera and M. G. Parker. Generalised Bent Criteria for Boolean Functions (I). In *IEEE Trans Inform. Theory, 52, 9, pages 4142–4159*, 2006.
56. O.S. Rothaus. On "bent" functions. In *J. Combin.Theory Ser A 20*, pages 300–305, 1976.

57. K. U. Schmidt, M. G. Parker, and A. Pott. Negabent Functions in the Maiorana–McFarland Class. In *Proceedings Sequences and Their Applications - SETA 2008, Lecture Notes in Computer Science, LNCS 5203*, 2008.
58. K.U. Schmidt. Quaternary constant-amplitude codes for multicode. In *CDMA and IEEE Trans. on Inf. Theory, (ISIT 2007) Proc. 2007. pages 2781–2785*, 2006.
59. K.U. Schmidt. Z4-valued quadratic forms and quaternary sequence families. In *IEEE Trans. Inf. Theory, vol. 55, no. 12, pages 5803–5810*, 2009.
60. R. Singh, B. Sarma, and A. Saikia. Public Key Cryptography Using Permutation P-Polynomials Over Finite Fields. In *IACR Cryptology ePrint Archive 2009: 208*, 2009.
61. P. Solé and N. Tokareva. Connections between quaternary and binary bent functions. In *Cryptology ePrint Archives, available online*: http://www.eprint.iacr.org/2009/544, 2009.
62. V. I. Solodovnikov. Bent functions from a finite Abelian group into a finite Abelian group (in Russian). In *Diskr. Mat., Volume 14, Issue 1, Pages 99–113*, 2002.
63. X. Wang and J. Zhou. Generalized partially bent functions. In *Proceedings Future Generation Communication and Networking (FGCN 2007), Vol. 1, pages 16–21*, 2007.
64. G. Wu and M. G. Parker.A complementary construction using mutually unbiased bases.In *Cryptography and Communications, Vol. 6, Issue 1, pages 3–25. DOI, (arxiv.org/abs/1309.0157)*, 2014.
65. F. Zhang, C. Carlet, Y. Hu, and W. Zhang. New Secondary Constructions of Bent Functions. In *Preprint*, 2013.

Chapter 16
Plateaued Functions: Generalities and Characterizations

16.1 Plateaued Boolean Functions

The so-called plateaued functions in n variables (or r-plateaued functions) have been introduced in 1999 by Zheng and Zhang in [61] for $0 < r < n$. They were firstly studied by these authors in [62, 63] and further by Carlet and Prouff in [16] as good candidates for designing cryptographic functions. The Walsh–Hadamard spectrum is a very important tool do define and design plateaued functions. An n-variable Boolean function is said to be *r-plateaued* if the values of its Walsh transform belong to the set $\{0, \pm 2^{\frac{n+r}{2}}\}$ for some fixed r, $0 \leq r \leq n$. Consequently, plateaued functions have low Hadamard transform, which provides protection against fast correlation attacks [42] and linear cryptanalysis [41]. It has been shown in [61] that plateaued functions are significant in cryptography as they possess various desirable characteristics such as high nonlinearity, resiliency, low additive autocorrelation, high algebraic degree and satisfy propagation criteria. Plateaued functions bring together various nonlinear characteristics. They include three significant classes of Boolean functions: the well-known bent functions, the near-bent functions and the semi-bent functions. More precisely, bent functions are exactly 0-plateaued functions, near-bent (also called semi-bent in odd dimension) are 1-plateaued functions and semi-bent functions are 2-plateaued functions. 0-plateaued functions and 2-plateaued functions on $\mathbb{F}_{2^n}$ exist when n is even, while 1-plateaued functions on $\mathbb{F}_{2^n}$ exist when n is odd.

For $r \in \{0, 1, 2\}$, r-plateaued functions have been actively studied and have attracted much attention due to their cryptographic, algebraic and combinatorial properties.

Bent functions are exactly the 0-plateaued functions. Another special family of plateaued functions defined in even dimension is the set of semi-bent functions. The notion of *semi-bent function* has been introduced in 1994, by Chee, Lee and Kim [24]. Nevertheless, these functions had been previously investigated in [3] under the name of three-valued almost optimal Boolean functions. Very recently,

S. Mesnager, *Bent Functions*, DOI 10.1007/978-3-319-32595-8_16

the development of the theory of semi-bent functions has increased. For very recent results on the treatment of semi-bent functions, we refer to [14, 45–47, 52]. The motivation for their study is firstly related to their use in cryptography (we recall that in the design of cryptographic functions, various characteristics need be considered simultaneously). Indeed, unlike bent functions, semi-bent functions can also be balanced and resilient. They also possess various desirable characteristics such as low autocorrelation, a maximal non-linearity among balanced plateaued functions, satisfy the propagation criteria and have high algebraic degree. Secondly, beside their practical use in cryptography, they are also widely used in code division multiple access (CDMA) communication systems for sequence design (see e.g. [28–30, 32, 34, 35, 54]). In this context, families of maximum-length sequences (maximum-length linear feedback shift-register sequences) having three-valued cross-correlation are used. Such sequences have received a lot of attention since the late sixties and can be generated by a semi-bent function [22]. Up to 2011, the main constructions of semi-bent functions in even dimension are either quadratic functions [57] or derived from power polynomials $\mathrm{Tr}_1^n(x^d)$ for a suitably chosen d (see [22]). Since then, several constructions of semi-bent have been proposed in the literature. The principal engine of this progress is the result of several important observations in connection with the construction of bent functions [13, 44, 51]. We shall describe this more precisely in Sect. 17.5.2.

This chapter is devoted to certain plateaued functions. Special attention is directed to semi-bent functions in the next chapter. We review what is known in this context and investigate constructions.

16.1.1 Near-Bent (1-Plateaued Functions)

Near-bent functions (or 1-plateaued functions) on $\mathbb{F}_{2^n}$ exist only when n is odd. They are defined as follows.

Definition 16.1.1. Let n be an odd integer. A Boolean function on $\mathbb{F}_{2^n}$ is said to be near-bent if its Walsh transform satisfies $\widehat{\chi_f}(a) \in \{0, \pm 2^{\frac{n+1}{2}}\}$ for all $a \in \mathbb{F}_{2^n}$.

Note that a function from $\mathbb{F}_{2^n} \to \mathbb{F}_{2^n}$ is said to be almost bent if it has Walsh Fourier spectrum $\{0, \pm 2^{\frac{n+1}{2}}\}$, that is, the same as a near-bent function. The difference between an almost bent function and a near-bent function is that almost bent functions map $\mathbb{F}_{2^n} \to \mathbb{F}_{2^n}$, whereas near-bent functions map $\mathbb{F}_{2^n} \to \mathbb{F}_2$. In this context, $f : \mathbb{F}_{2^n} \to \mathbb{F}_{2^n}$ is almost bent if and only if each of the Boolean functions $x \mapsto \mathrm{Tr}_1^n(vf(x))$ is near-bent, for all $v \in \mathbb{F}_{2^n}^\star$.

Thanks to Parseval's identity, one can determine the number of occurrences of each value of the Walsh transform of a near-bent function (see Table 16.1).

Again from Parseval's identity, it is straightforward to see that the support of the Walsh transform $\widehat{\chi_f}$ of a near-bent function f on $\mathbb{F}_{2^n}$ is of cardinality 2^{n-1} (that is, $\#supp(\widehat{\chi_f}) = 2^{n-1}$).

Table 16.1 Walsh spectrum of near-bent functions (1-plateaued) f with $f(0) = 0$

Value of $\widehat{\chi_f}(\omega)$, $\omega \in \mathbb{F}_{2^n}$	Number of occurrences
0	2^{n-1}
$2^{\frac{n+1}{2}}$	$2^{n-2} + 2^{\frac{n-3}{2}}$
$-2^{\frac{n+1}{2}}$	$2^{n-1} - 2^{\frac{n-3}{2}}$

In the particular case of quadratic functions, there exists a criterion on near-bentness involving the dimension of the linear kernel (see e.g. [22]). More precisely, it is well-known (see Section 8.5.2 in [8]) that a quadratic Boolean function f over $\mathbb{F}_{2^n}$ has for Walsh support the set of elements $\alpha \in \mathbb{F}_{2^n}$ such that $\mathrm{Tr}_1^n(\alpha x) + f(x)$ is constant on E_f, where $E_f := \{x \in \mathbb{F}_{2^n}, \mid \forall y \in \mathbb{F}_{2^n}, f(x+y)+f(x)+f(y)+f(0) = 0\}$ is the linear kernel of f. It has been proved that f is near-bent over $\mathbb{F}_{2^n}$, if and only if E_f has dimension 1 (i.e. has size 2). Note that from Theorem 2.2.2, it is easy to see that quadratic Boolean function f is near-bent if and only if the rank of f is $n-1$, that is, $k_f = 1$.

Several constructions of quadratic near-bent functions have been obtained in the literature. We give a list of the known families on $\mathbb{F}_{2^n}$, n odd:

- $f(x) = \mathrm{Tr}_1^n(x^{2^i+1})$, $gcd(i,n) = 1$ [28];
- $f(x) = \sum_{i=1}^{\frac{n-1}{2}} \mathrm{Tr}_1^n(x^{1+2^i})$ [1];
- $f(x) = \sum_{i=1}^{\lfloor\frac{n-1}{2}\rfloor} c_i \mathrm{Tr}_1^n(x^{1+2^i})$, $c_i \in \mathbb{F}_2$ [22];
- $f(x) = \mathrm{Tr}_1^n(x^{2^i+1} + x^{2^j+1} + x^{2^t+1})$, $1 \le i < j \le t \le \frac{n-1}{2}$, $i+j = t$, $gcd(n,i) = gcd(n,j) = gcd(n,i+j) = 1$ [22];
- $f(x) = \sum_{i=1}^{\frac{n-1}{2}} c_i \mathrm{Tr}_1^n(x^{1+2^i})$, $c_i \in \mathbb{F}_2$, $gcd(x^n+1, c(x)) = x+1$ where $c(x) = \sum_{i=1}^{\frac{n-1}{2}} c_i(x^i + x^{n-i})$ [35];
- $f(x) = \mathrm{Tr}_1^n(x^{2^i+1}) + \mathrm{Tr}_1^n(x^{2^j+1})$, $gcd(n,i+j) = gcd(n,i-j)$ [35];
- $f(x) = \sum_{i=0}^{r} \mathrm{Tr}_1^n(x^{1+2^{k+id}})$, $gcd(2k+rd, n) = 1$ [35];
- $f(x) = \sum_{i=1}^{\frac{q-1}{2}} \mathrm{Tr}_1^n(x^{1+2^{pi}}) + \mathrm{Tr}_1^n(x^{1+2^q})$, $n = pq$, $3 \not| p$, p odd, q odd, $gcd(p,q) = 1$ [27].

Because bent functions exist in even dimensions, and near-bent functions exist in odd dimensions, one can move up and down between bent and near-bent functions. The four possibilities are discussed in [39], see also some results in [3]. In [40], Leander and McGuire have considered the problem on going up from a near-bent function to a bent function and proposed constructions. In particular, it has been shown that two n-variable functions g and h (n odd) are near-bent with complementary Walsh supports (i.e. $supp(\widehat{\chi_g}) \cap supp(\widehat{\chi_h}) = \emptyset$) if and only if the $(n+1)$-variable function $x \mapsto f(x, x_{n+1}) = g(x) + x_{n+1}h(x)$; $x \in \mathbb{F}_2^n$, $x_{n+1} \in \mathbb{F}_2$ is bent. The restrictions to a $(2n)$-bent function to any hyperplan and to the complement of this hyperplan (view as $(2n-1)$-Booleans functions) are near-bent. The problem of the construction of $(2n)$-bent functions from two $(2n-1)$-near-bent functions has also been considered by Wolfmann with a different point of view in [58]. Some progress on this question has been made very recently in [59] and [60]. In particular, Wolfmann [60] has introduced a way to construct new bent functions starting from a near-bent functions having a specific derivative or from a bent function such that the sum of the two components is a Boolean function of degree 1. Some open problems have been presented by Wolfmann [60] in the continuation of his interesting approach.

In 2005, Charpin et al. [22] have proved that some classes of near-bent functions can been derived via the composition with nonpermutation linear polynomials. In fact, the composition of any linear permutation polynomial P with a quadratic near-bent function gives rise again to a near-bent function $x \mapsto f(P(x))$. However, it is not necessary for P to be a permutation polynomial in order for $f \circ P$ to be near-bent. In fact, one may choose a linear mapping P from $\mathbb{F}_{2^n}$ to $\mathbb{F}_{2^n}$ which is still near-bent. In [22] some nonpermutation linear polynomials that preserve the near-bentness property (when composed with a quadratic near-bent function) have been exhibited. For more details on the treatment of near-bent functions we send the reader to [22].

Finally, very few secondary constructions of near-bent functions (that is, constructions of new near-bent functions from two or several already known ones) have been proposed in the literature. The following statement shows that secondary constructions of near-bent functions can be derived under a condition involving the derivatives functions.

Theorem 16.1.2. *Let n be an odd integer. Let f and g be two near-bent functions over $\mathbb{F}_{2^n}$. Assume that there exists an element a of $\mathbb{F}_{2^n}$ such that $D_a f = D_a g$. Then the function $h = f + D_a f(f + g)$ is a near-bent function on $\mathbb{F}_{2^n}$.*

Proof. Let us compute the Walsh transform of h for every $\omega \in \mathbb{F}_{2^n}$. We have

$$\widehat{\chi_h}(\omega) = \sum_{x \in \mathbb{F}_{2^n}} \chi(h(x) + \mathrm{Tr}_1^n(\omega x)) = \sum_{x \in \mathbb{F}_{2^n}} \chi(f(x) + D_a f(x)(f+g)(x) + \mathrm{Tr}_1^n(\omega x)).$$

Now, one can split the sum depending whether $D_a f$ is equal to 1 or not (recall that $D_a f(x) = f(x) + f(x+a)$):

$$\begin{aligned}
\widehat{\chi_h}(\omega) &= \sum_{x \in \mathbb{F}_{2^n} | D_a f = 0} \chi(f(x) + \mathrm{Tr}_1^n(\omega x)) + \sum_{x \in \mathbb{F}_{2^n} | D_a f = 1} \chi(g(x) + \mathrm{Tr}_1^n(\omega x)) \\
&= \frac{1}{2}\Big(\sum_{x \in \mathbb{F}_{2^n}} \chi(f(x) + \mathrm{Tr}_1^n(\omega x)) + \sum_{x \in \mathbb{F}_{2^n}} \chi(f(x+a) + \mathrm{Tr}_1^n(\omega x))\Big) \\
&+ \frac{1}{2}\Big(\sum_{x \in \mathbb{F}_{2^n}} \chi(g(x) + \mathrm{Tr}_1^n(\omega x)) - \sum_{x \in \mathbb{F}_{2^n}} \chi(g(x+a) + \mathrm{Tr}_1^n(\omega x))\Big).
\end{aligned}$$

Hence,

$$\widehat{\chi_h}(\omega) = \frac{1}{2}\Big(\sum_{x\in\mathbb{F}_{2^n}} \chi(f(x) + \mathrm{Tr}_1^n(\omega x)) + \sum_{x\in\mathbb{F}_{2^n}} \chi(f(x) + \mathrm{Tr}_1^n(\omega(x+a)))\Big)$$

$$+\frac{1}{2}\Big(\sum_{x\in\mathbb{F}_{2^n}} \chi(g(x) + \mathrm{Tr}_1^n(\omega x)) - \sum_{x\in\mathbb{F}_{2^n}} \chi(g(x) + \mathrm{Tr}_1^n(\omega(x+a)))\Big)$$

$$= \frac{1}{2}\Big(\widehat{\chi_f}(\omega)(1 + \chi(\mathrm{Tr}_1^n(\omega a)))\Big) + \frac{1}{2}\Big(\widehat{\chi_g}(\omega)(1 - \chi(\mathrm{Tr}_1^n(\omega a)))\Big).$$

Now, f and g being near-bent, therefore if $\mathrm{Tr}_1^n(\omega a) = 0$, then $\widehat{\chi_h}(\omega) = \widehat{\chi_f}(\omega) \in \{0, \pm 2^{\frac{n+1}{2}}\}$. And if $\mathrm{Tr}_1^n(\omega a) = 1$, then $\widehat{\chi_h}(\omega) = \widehat{\chi_g}(\omega) \in \{0, \pm 2^{\frac{n+1}{2}}\}$, which completes the proof. □

16.1.2 Semi-bent (2-Plateaued Functions)

Semi-bent functions (or 2-plateaued functions) on $\mathbb{F}_{2^n}$ exist only when n is even. So, in this section n denotes an even integer and we set $m = \frac{n}{2}$.

In 1994, the notion of *semi-bent function* has been introduced by Chee, Lee and Kim [24] at Asiacrypt' 94. In fact, these functions had been previously investigated under the name of three-valued almost optimal Boolean functions in [3]. Moreover, they are particular cases of the so-called plateaued functions [61, 62]. Semi-bent functions exist for even or odd number of variables. When n is even, the semi-bent functions are those Boolean functions whose Hadamard transform takes values 0 and $\pm 2^{\frac{n+2}{2}}$. They are balanced (up to the addition of a linear function) and have maximal non-linearity among balanced plateaued functions. The maximum-length sequences, also called m-sequences (maximum-length linear feedback shift -register sequences), have received a lot of attention since the late sixties. In terms of linear-feedback shift register (LFSR) synthesis they are usually generated by certain power polynomials over a finite field and in addition are characterized by a low cross correlation and high nonlinearity. Such a sequence is said to be generated by a semi-bent function [22]. Families of maximum-length sequences having three-valued cross-correlation have a wide range of applications in cryptography and code division multiple access (CDMA) communication systems for sequence design [28–30, 32, 34, 35, 54], etc. However, almost all families of semi-bent functions have been derived from power polynomials $\mathrm{Tr}_1^n(x^d)$ for a suitably chosen d (see [22] and [57] for the construction of quadratic semi-bent functions in even dimension).

Semi-bent functions are defined as follows.

Definition 16.1.3. Let n be an even integer. A Boolean function on $\mathbb{F}_{2^n}$ is said to be semi-bent if its Walsh transform satisfies $\widehat{\chi_f}(a) \in \{0, \pm 2^{\frac{n+2}{2}}\}$ for all $a \in \mathbb{F}_{2^n}$.

Table 16.2 Walsh spectrum of semi-bent functions (2-plateaued) f with $f(0) = 0$

Value of $\widehat{\chi_f}(\omega)$, $\omega \in \mathbb{F}_{2^n}$	Number of occurrences
0	$2^{n-1} + 2^{n-2}$
$2^{\frac{n+2}{2}}$	$2^{n-3} + 2^{\frac{n-4}{2}}$
$-2^{\frac{n+2}{2}}$	$2^{n-3} - 2^{\frac{n-4}{2}}$

Thanks to Parseval's identity, one can determine the number of occurrences of each value of the Walsh transform of a semi-bent function (see Table 16.2).

Using the relationship between the nonlinearity and the Walsh spectrum, it is immediate to see that the nonlinearity of a semi-bent function on $\mathbb{F}_{2^n}$ equals $2^{n-1} - 2^{\frac{n}{2}}$. In addition, the possible values of the Hamming weight of a semi-bent function are 2^{n-1}, $2^{n-1} - 2^m$ and $2^{n-1} + 2^m$.

Many recent progresses have been made on the treatment of semi-bent functions. In the next chapter, we focus on the constructions of such functions. A complete survey on semi-bent functions can be found in [49].

16.2 Special Plateaued Functions via Linear Translators

This section[1] deals with two important subclasses of plateaued functions in even dimension, that is, bent and semi-bent functions derived from linear translators.

We recall definitions of linear translator and linear structure.

Definition 16.2.1. Let $n = rk$, $1 \leq k \leq n$. Let f be a function from $\mathbb{F}_{2^n}$ to $\mathbb{F}_{2^k}$, $\gamma \in \mathbb{F}_{2^n}^*$ and b be a constant of $\mathbb{F}_{2^k}$. Then γ is a *b-linear translator* of f if $f(x) + f(x + u\gamma) = ub$ for all $x \in \mathbb{F}_{2^n}$ and $u \in \mathbb{F}_{2^k}$. If $f(x) + f(x + \gamma) = b$ for all $x \in \mathbb{F}_{2^n}$, then γ is called a *b-linear structure* of f.

The notion of *b- linear translator* is well known in the literature (see for example [37]). The notion of *b-linear structure is usually given for functions $f : \mathbb{F}_{2^n} \to \mathbb{F}_2$, that is $k = 1$ (see for example [18]).*

Remark 16.2.2. Note that being b-linear translator is stronger than being b-linear structure if $k > 1$ and they are the same if $k = 1$. For example, let $f : \mathbb{F}_{2^4} \to \mathbb{F}_{2^2}$ be a function defined as $f(x) = Tr_2^4(x^2 + \gamma x)$ where $\gamma \in \mathbb{F}_{2^4} \backslash \mathbb{F}_{2^2}$. Then, γ is a 0-linear structure of f but it is not a 0-linear translator of f as $f(x + u\gamma) \neq f(x)$ for $u \in \mathbb{F}_{2^2} \backslash \mathbb{F}_2$.

The notions of linear structures, linear translators and derivatives are related.

[1] In the literature, usually, Boolean functions are denoted by small letters (like f) and vector Boolean functions by capital letters (like F). Nevertheless, for the sake of simplicity of notation, we also denote Boolean functions (vector Boolean functions) by capital letters (small letters) when it seems more appropriate in this section.

Definition 16.2.3. Let $F : \mathbb{F}_{2^n} \to \mathbb{F}_{2^m}$. For $a \in \mathbb{F}_{2^n}$, the function D_aF given by $D_aF(x) = F(x) + F(x + a)$, $\forall x \in \mathbb{F}_{2^n}$ is called the derivative of F in the direction of a.

Note that $D_\gamma f(x) = b$ for each $x \in \mathbb{F}_{2^n}$ if and only if γ is a b-linear structure of f. Similarly, $D_{u\gamma} f(x) = ub$ for each $x \in \mathbb{F}_{2^n}$ and each $u \in \mathbb{F}_{2^k}$ if and only if γ is a b-linear translator of f.

The concept of a linear translator exists for a p-ary function (see for instance [37]) but it was introduced in cryptography, mainly for Boolean functions (see for instance [19]). Functions with linear structures are considered as weak for some cryptographic applications. For instance, a recent attack on hash functions proposed in [5] exploits a similar weakness of the involved mappings. All Boolean functions using a linear translator have been characterized by Lai [38]. Further, Charpin and Kyureghyan have characterized the functions in univariate variables from $\mathbb{F}_{p^n}$ to $\mathbb{F}_p$ of the form $Tr_{\mathbb{F}_{p^n}/\mathbb{F}_p}(F(x))$, where $F(x)$ is a function over $\mathbb{F}_{p^n}$ and $Tr_{\mathbb{F}_{p^n}/\mathbb{F}_p}$ denotes the trace function from $\mathbb{F}_{p^n}$ to $\mathbb{F}_p$. The result of Lai in [38] has been formulated recently by Charpin and Sarkar [23].

The next sessions are organized as follows. In the first part of the section, we show how one can construct mainly bent and semi-bent functions in the Maiorana–McFarland class using Boolean functions having linear structures (linear translators) systematically. Although most of these results are rather direct applications of some recent results, using linear structures (linear translators) allows us to have certain flexibilities to control extra properties of these plateaued functions. In the second part of the section, using the results of the first part and exploiting these flexibilities, we modify many secondary constructions. Therefore, we obtain new secondary constructions of bent and semi-bent functions not belonging to the Maiorana–McFarland class. Instead of using bent (semi-bent) functions as ingredients, our secondary constructions use only Boolean (vectorial Boolean) functions with linear structures (linear translators) which are easy to choose. Moreover, all of them are explicit and we also determine the duals of the bent functions in our constructions. We show how these linear structures should be chosen in order to satisfy the corresponding conditions coming from using derivatives and quadratic/cubic functions in our secondary constructions.

For a Boolean map, linear structures or linear translators are not desirable and are generally considered as a defect. In this section, we show that one can recycle such Boolean functions to get Boolean functions with optimal or very high nonlinearity. More precisely, we show that one can obtain primary constructions of bent and semi-bent functions from Boolean maps having linear structures or linear translator in Sects. 16.2.1, 16.2.2 and 16.2.3. All the primary constructions proposed belong to the well-known class of Maiorana–McFarland. However, an important feature of the bent functions presented here is that their dual functions can be explicitly computed. Next, we focus on secondary constructions presented in [15] and in [7] (see also [50]). Note that several primary constructions have been derived in [50] and in [43] from the Carlet's result [7, Theorem 3] which has been completed in [50, Theorem 4]. We shall show how to obtain secondary constructions by reusing bent functions. The secondary constructions presented below are very explicit and they use Boolean functions (vectorial Boolean functions) with certain linear structures (linear translators) as ingredients instead of bent or semi-bent functions. The conditions on such linear structures (linear translators) in the considered secondary constructions are easily satisfied. Finally, we show that one can construct bent functions from bent functions of Sects. 16.2.1 and 16.2.2 by adding a quadratic or cubic function appropriately chosen.

16.2.1 Constructions of Bent and Semi-bent Boolean Functions from the Class of Maiorana–McFarland Using One Linear Structure

A function $H : \mathbb{F}_{2^m} \times \mathbb{F}_{2^m} \to \mathbb{F}_2$ is said to be in the class of Maiorana–McFarland if it can be written in bivariate form as

$$H(x, y) = \mathrm{Tr}_1^m (x\phi(y)) + h(y) \tag{16.1}$$

where ϕ is a map from $\mathbb{F}_{2^m}$ to $\mathbb{F}_{2^m}$ and h is a Boolean function on $\mathbb{F}_{2^m}$. It is well-known that we can choose ϕ so that H is bent or H is semi-bent. Indeed, it is well-known that bent functions of the form (16.1) come from one-to-one maps while 2-to-1 maps lead to semi-bent functions.

Proposition 16.2.4 ([8, 25, 47]). *Let H be defined by (16.1). Then,*

1. *H is bent if and only if ϕ is a permutation and its dual function is* $\tilde{H}(x, y) = \mathrm{Tr}_1^m \left(y\phi^{-1}(x)\right) + h(\phi^{-1}(x))$.
2. *H is semi-bent if ϕ is* 2*-to-*1.

As a first illustration of Proposition 16.2.4, let us consider a class of maps from $\mathbb{F}_{2^m}$ to itself: $\phi : y \mapsto y + \gamma f(y)$ where γ is a linear structure of f. This class has the property that it only contains one-to-one maps or 2-to-1 maps. Therefore, by Proposition 16.2.4, one can obtain the following infinite families of bent and semi-bent functions.

Proposition 16.2.5 ([36]). *Let f and h be two Boolean functions over $\mathbb{F}_{2^m}$. Let H be the Boolean function defined on $\mathbb{F}_{2^m} \times \mathbb{F}_{2^m}$ by*

$$H(x, y) = \mathrm{Tr}_1^m(xy + \gamma x f(y)) + h(y),\ \gamma \in \mathbb{F}_{2^m}.$$

H is bent (resp. semi-bent) if and only if γ is a 0*-linear (resp.* 1*-linear) structure of f. Furthermore, if H is bent, then its dual is*

$$\tilde{H}(x, y) = \mathrm{Tr}_1^m\left(yx + \gamma y f(x)\right) + h(x + \gamma f(x)).$$

Proof. Properties of $\phi : y \mapsto y + \gamma f(y)$ are well-known and firstly developed in [17, 18] (see also [20, 37]). Bijectivity is given by Theorem 2 of [17]. For the 2-to-1 property, see Theorems 3,6 in [18]. The proof is then immediately obtained. Also, note that since ϕ is an involution (see also [20, 21, 37]), we have $\tilde{H}(x,y) = Tr_1^m\big(y\phi(x)\big) + h(\phi(x))$. □

In order to show that the hypotheses of Proposition 16.2.5 hold in certain cases, we give the following direct applications of Theorems 3, 4 in [17].

Example 16.2.6. Let $\gamma \in \mathbb{F}_{2^m}^\star$ and $\beta \in \mathbb{F}_{2^m}$ such that $Tr_1^m(\beta\gamma) = 0$ (resp. $Tr_1^m(\beta\gamma) = 1$). Let $H : \mathbb{F}_{2^m} \to \mathbb{F}_{2^m}$ be an arbitrary mapping and h be any Boolean function on $\mathbb{F}_{2^m}$. Then the function g defined over $\mathbb{F}_{2^m} \times \mathbb{F}_{2^m}$ by

$$g(x,y) = Tr_1^m(xy + \gamma x Tr_1^m(H(y^2 + \gamma y) + \beta y)) + h(y)$$

is bent (resp. semi-bent).

Example 16.2.7. Let $0 \le i \le m-1$, $i \notin \{0, \frac{m}{2}\}$ and $\delta, \gamma \in \mathbb{F}_{2^m}$ such that $\delta^{2^i-1} = \gamma^{1-2^{2i}}$. Let h be any Boolean function on $\mathbb{F}_{2^m}$ and g be the Boolean function defined on $\mathbb{F}_{2^m} \times \mathbb{F}_{2^m}$ by

$$g(x,y) = Tr_1^m(xy + \gamma x Tr_1^m(\delta y^{2^i+1})) + h(y).$$

If $Tr_1^m(\delta\gamma^{2^i+1}) = 0$ (resp. $Tr_1^m(\delta\gamma^{2^i+1}) = 1$) then g is bent (resp. semi-bent).

Observe that if we compose ϕ at left by a linearized permutation polynomial L, any output has the same number of preimages under ϕ than under $L \circ \phi$. Hence, one can slightly generalize Proposition 16.2.5 as follows.

Proposition 16.2.8 ([36]). *Let f and h be two Boolean functions over $\mathbb{F}_{2^m}$ and $\gamma \in \mathbb{F}_{2^m}$. Let L be a linearized permutation polynomial of $\mathbb{F}_{2^m}$. The Boolean function H defined by*

$$H(x,y) = Tr_1^m\,(xL(y) + L(\gamma)xf(y)) + h(y)$$

is bent (resp. semi-bent) if and only if γ is a 0-linear (resp. 1-linear) structure of f. Moreover, if H is bent then its dual function $\tilde{H}$ is given by

$$\tilde{H}(x,y) = Tr_1^m(yL^{-1}(x) + \gamma y f(L^{-1}(x)) + h(L^{-1}(x) + \gamma f(L^{-1}(x))).$$

16.2.2 Constructions of Bent and Semi-bent Boolean Functions from the Class of Maiorana–McFarland Using Two Linear Structures

In this section we consider the functions H of the form (16.1).

$$H(x,y) = \mathrm{Tr}_1^m(x\phi(y)) + h(y) \text{ with } \phi(y) = \pi_1(\pi_2(y) + \gamma f(\pi_2(y)) + \delta g(\pi_2(y))) \tag{16.2}$$

where f, g and h are Boolean functions over $\mathbb{F}_{2^m}$, $\gamma, \delta \in \mathbb{F}_{2^m}^\star$, $\gamma \neq \delta$ and π_1, π_2 are permutations of $\mathbb{F}_{2^m}$ (not necessarily linear).

The class (16.2) contains the functions involved in Proposition 16.2.6 and in Proposition 16.2.8 (which corresponds to the case where $f = g$). In the line of Sect. 16.2.1, we study the cases where γ and δ are linear structures of the Boolean functions involved in ϕ. Then one can exhibit conditions of bentness or semi-bentness as those of Propositions 16.2.6 and 16.2.8 that we present in the following two propositions. We indicate that, despite their similarities with Proposition 16.2.6 and 16.2.8, we obtain bent functions that do not fall in the scope of Proposition 16.2.6 and 16.2.8.

Proposition 16.2.9 ([36]). *Let H be defined by Eq. (16.2). Then H is bent if one of the following conditions holds:*

1. *γ is a 0-linear structure of f, δ is a 0-linear structure of f and g,*
2. *γ is a 0-linear structure of f, δ is a 1-linear structure of f and $\delta + \gamma$ is a 0-linear structure of g,*
3. *δ is a 0-linear structure of g, γ is a 0-linear structure of f and g,*
4. *δ is a 0-linear structure of g, γ is a 1-linear structure of g and $\delta + \gamma$ is a 0-linear structure of f,*
5. *δ is a 1-linear structure of f, γ is a 1-linear structure of f and g,*
6. *γ is a 1-linear structure of g, δ is a 1-linear structure of f and g.*

Moreover, if H is bent then its dual is $\tilde{H}(x,y) = \mathrm{Tr}_1^m\left(y\phi^{-1}(x)\right) + h(\phi^{-1}(x))$ where $\phi^{-1} = {\pi_2}^{-1} \circ \rho^{-1} \circ {\pi_1}^{-1}$ and ρ^{-1} is given explicitly in the Appendix as Proposition 16.2.10. In particular, choosing $\pi_1(x) = L(x)$ as a linearized permutation polynomial and π_2 as the identity, we get that

$$H(x,y) = \mathrm{Tr}_1^m(xL(y) + L(\gamma)xf(y) + L(\delta)xg(y)) + h(y) \tag{16.3}$$

is bent in the conditions above and $\tilde{H}(x,y) = \mathrm{Tr}_1^m\left(y\rho^{-1}(L^{-1}(x))\right) + h(\rho^{-1}(L^{-1}(x))$.

Proof. We give the proof only for case (1) since the other cases are very similar. It suffices to show that $\rho : y \mapsto y + \gamma f(y) + \delta g(y)$ is a permutation. Suppose that $\rho(y) = \rho(z)$, i.e.,

$$y + \gamma f(y) + \delta g(y) = z + \gamma f(z) + \delta g(z). \tag{16.4}$$

Taking f of both sides we obtain $f\big(y+\gamma f(y)+\delta g(y)\big)=f\big(z+\gamma f(z)+\delta g(z)\big)$. Since γ and δ are 0-linear structures of f, we have

$$f(y)=f(z). \tag{16.5}$$

Combining Eqs. (16.4) and (16.5), we get $y+\delta g(y)=z+\delta g(z)$. Taking g of both sides we obtain $g(y+\delta g(y))=g(z+\delta g(z))$. Since δ is a 0-linear structure of g, we conclude

$$g(y)=g(z). \tag{16.6}$$

Combining Eqs. (16.4), (16.5) and (16.6), we reach that $y=z$. For the dual function, ρ^{-1} is written explicitly in following Proposition 16.2.10 and the proof for ρ^{-1} for case (i) is given. □

The following proposition is related to Proposition 16.2.9.

Proposition 16.2.10 ([36]). *Let H be defined by Eq. (16.2), γ and δ be defined as in Proposition 16.2.9. Then the dual of H is $\tilde{H}(x,y)=\mathrm{Tr}_1^m\left(y\phi^{-1}(x)\right)+h(\phi^{-1}(x))$ where $\phi^{-1}={\pi_2}^{-1}\circ\rho^{-1}\circ{\pi_1}^{-1}$ and $\rho^{-1}(x)$ is given as follows.*

1. *If γ is a 0-linear structure of f, δ is a 0-linear structure of f and g, then*

$$\rho^{-1}(x)=x+\gamma f(x)+\delta\left[g(x)(1+f(x))+g(x+\gamma)f(x)\right].$$

2. *If γ is a 0-linear structure of f, δ is a 1-linear structure of f and $\delta+\gamma$ is a 0-linear structure of g, then*

$$\rho^{-1}(x)=x+\gamma\left[g(x)+f(x)(1+g(x)+g(x+\gamma))\right]+\delta\left[g(x)(1+f(x))+g(x+\gamma)f(x)\right].$$

3. *If δ is a 0-linear structure of g, γ is a 0-linear structure of f and g, then*

$$\rho^{-1}(x)=x+\gamma\left[f(x)(1+g(x))+f(x+\delta)g(x)\right]+\delta g(x).$$

4. *If δ is a 0-linear structure of g, γ is a 1-linear structure of g and $\delta+\gamma$ is a 0-linear structure of f, then*

$$\rho^{-1}(x)=x+\gamma\left[f(x)(1+g(x))+f(x+\delta)g(x)\right]+\delta\left[f(x)(1+g(x))+(1+f(x+\delta))g(x)\right].$$

5. *If δ is a 1-linear structure of f or δ is a 0-linear structure of g, then*

$$\rho^{-1}(x)=x+\gamma\left[f(x)(1+g(x+\delta))+\big(1+f(x)\big)g(x)\right]+\delta f(x).$$

6. *If γ is a 1-linear structure of g, δ is a 1-linear structure of f and g, then*

$$\rho^{-1}(x)=x+\gamma g(x)+\delta\left[f(x)(1+g(x))+f(x+\gamma)g(x)\right].$$

Proof. We give only the proof for the case (1). Assume that γ is a 0-linear structure of f, δ is a 0-linear structure of f and g, then we claim that

$$\rho^{-1}(x) = \begin{cases} x & \text{if } f(x) = 0 \text{ and } g(x) = 0 \\ x + \delta & \text{if } f(x) = 0 \text{ and } g(x) = 1 \\ x + \gamma & \text{if } f(x) = 1 \text{ and } g(x + \gamma) = 0 \\ x + \gamma + \delta & \text{if } f(x) = 1 \text{ and } g(x + \gamma) = 1 \end{cases} \tag{16.7}$$

Let $\rho(y) = a$. Then,

$$y + \gamma f(y) + \delta g(y) = a \tag{16.8}$$

Taking f of both sides gives $f(y + \gamma f(y) + \delta g(y)) = f(a)$. Since γ and δ are 0-linear structures of f, we get

$$f(y) = f(a). \tag{16.9}$$

Note that, $\big(f(a), g(a)\big) \in \{(0,0), (0,1), (1,0), (1,1)\}$. These four cases correspond to the cases in Eq. (16.7). We prove only the first case in Eq. (16.7) and the proofs of other cases are similar. Hence, we assume that $(f(a), g(a)) = (0,0)$. Then, by Eq. (16.9), $f(y) = 0$ and by Eq. (16.8), $y + \delta g(y) = a$. Taking g of both sides and using that δ is a 0-linear structure of g, we obtain that $g(y + \delta g(y)) = g(y) = g(a)$. As $g(a) = 0$ by our assumption, we get $g(y) = 0$ and putting $f(y) = g(y) = 0$ in Eq. (16.8) we conclude that $y = a$.

Finally, the Eq. (16.7) can be written in the form

$$\rho^{-1}(x) = x + \gamma f(x) + \delta\left[g(x)(1 + f(x)) + g(x + \gamma) f(x)\right].$$

□

Remark 16.2.11. The converse of Proposition 16.2.9 is not always true. For example, for $f(x) = Tr_1^3(x^3 + \alpha^5 x)$, $g(x) = Tr_1^3(\alpha x^3 + \alpha^5 x)$, $\gamma = \alpha$ and $\delta = \alpha^3$ where α is a primitive element of $\mathbb{F}_{2^3}$, ϕ is a permutation but none of the conditions given in Proposition 16.2.9 is satisfied.

The following result shows in which cases ϕ is 2-to-1 and hence H is semi-bent.

Proposition 16.2.12 ([36]). *Let H be defined by (16.2). Then H is semi-bent if one of the following conditions holds:*

1. *γ, δ are 1-linear structures of f and γ is a 0-linear structure of g,*
2. *δ is a 1-linear structure of f and γ, δ are 0-linear structures of g,*
3. *γ, δ are 0-linear structures of f and δ is a 1-linear structure of g,*
4. *δ is a 0-linear structure of f and γ, δ are 1-linear structures of g,*

5. *γ is a 0-linear structure of f, δ is a 1-linear structure of f and $\gamma+\delta$ is a 1-linear structure of g,*
6. *γ is a 1-linear structure of g, δ is a 0-linear structure of g and $\gamma+\delta$ is a 1-linear structure of f.*

In particular, choosing $\pi_1(x) = L(x)$ as a linearized permutation polynomial and π_2 as the identity, we get that $H(x,y) = \mathrm{Tr}_1^m(xL(y) + L(\gamma)xf(y) + L(\delta)xg(y)) + h(y)$ is semi-bent in the conditions above.

Proof. We give the proof for case (1) only since the other cases are similar. Now, we need to show that $\rho(y) : y \mapsto y + \gamma f(y) + \delta g(y)$ is 2-to-1. Let $\rho(y) = a$ for some $a \in \mathbb{F}_{2^m}$. Then, $y \in \{a, a+\gamma, a+\delta, a+\gamma+\delta\}$. As γ is a 1-linear structure of f and 0-linear structure of g, we have $\rho(a) = \rho(a+\gamma)$ and $\rho(a+\delta) = \rho(a+\gamma+\delta)$. Moreover, $\rho(a+\delta) = a+\delta+\gamma f(a+\delta)+\delta g(a+\delta) = a+\delta+\gamma+\gamma f(a)+\delta g(a+\delta)$ where we use that δ is a 1-linear structure of f. We observe that $\rho(a) = a+\gamma f(a)+\delta g(a) \neq \rho(a+\delta)$. Indeed, otherwise if the equality holds, then $\gamma+\delta+\delta\big(g(a)+g(a+\delta)\big) = 0$. This is a contradiction as $\gamma \neq \delta$ and $\gamma \neq 0$. This implies that $\rho^{-1}(a) = \{a, a+\gamma\}$ or $\rho^{-1}(a) = \{a+\delta, a+\gamma+\delta\}$ which shows that ρ is 2-to-1. □

Remark 16.2.13. The converse of Proposition 16.2.12 is not always true. For example, for $f(x) = \mathrm{Tr}_1^3(\alpha^4x^3 + \alpha^4x)$, $g(x) = \mathrm{Tr}_1^3(\alpha x^3 + \alpha^2 x)$, $\gamma = \alpha$ and $\delta = \alpha^3$ where α is a primitive element of $\mathbb{F}_{2^3}$, ϕ is 2-to-1 but none of the conditions given in Proposition 16.2.12 is satisfied.

16.2.3 Constructions of Bent and k-Plateaued Functions Using Linear Translators

In the preceding sections, we have shown that one can construct bent and semi-bent functions from Boolean functions having linear structures, that is, having constant derivatives. An extension of these constructions is to consider Boolean maps taking its values in a subfield of the ambient field instead of Boolean functions in (16.1). In that case, the natural notion replacing linear structures is the notion of linear translators. We still adopt the approach of the preceding sections and aim to construct bent functions in the class of Maiorana–McFarland. To this end, one can apply results on permutations constructed from Boolean maps having linear translators presented in [37] and obtain the following infinite families of bent and plateaued functions.

Proposition 16.2.14 ([36]). *Let m be a positive integer and k be a divisor of m. Let f be a function from $\mathbb{F}_{2^m}$ to $\mathbb{F}_{2^k}$ and h be a Boolean function on $\mathbb{F}_{2^m}$. Let H be the function defined on $\mathbb{F}_{2^m} \times \mathbb{F}_{2^m}$ by*

$$H(x,y) = \mathrm{Tr}_1^m(xy + \gamma x f(y)) + h(y), \quad \gamma \in \mathbb{F}_{2^m}^{\star}.$$

1. *If γ is a c-linear translator of f where $c \in \mathbb{F}_{2^m}$ and $c \neq 1$, then H is bent and its dual function is given as*

$$\tilde{H}(x,y) = \mathrm{Tr}_1^m\left(y\left(x+\gamma\frac{f(x)}{1+c}\right)\right) + h\left(x+\gamma\frac{f(x)}{1+c}\right).$$

Moreover, $H(x,y) = \mathrm{Tr}_1^m(xL(y)+L(\gamma)xf(y))+h(y)$ where L is an $\mathbb{F}_{2^k}$-linearized permutation polynomial, is also bent under these conditions and its dual is

$$\tilde{H}(x,y) = \mathrm{Tr}_1^m\left(y\left(L^{-1}(x)+\gamma\frac{f(L^{-1}(x))}{1+c}\right)\right) + h\left(L^{-1}(x)+\gamma\frac{f(L^{-1}(x))}{1+c}\right).$$

2. *If γ is a 1-linear translator of f and $h = 0$ then H is k-plateaued with Walsh transform values*

$$\widehat{\chi_H}(a,b) = \begin{cases} \pm 2^{m+k} & \text{if } Tr_k^m(b\gamma) = 0, \\ 0 & \text{otherwise.} \end{cases}$$

Note that Proposition 16.2.14 generalizes partially Proposition 16.2.5 (extending the condition 0-linear structure to c-linear translator with $c \neq 1$). Furthermore, one can derive from Proposition 16.2.9 and Proposition 16.2.12 similar statements if $f : \mathbb{F}_{2^m} \to \mathbb{F}_{2^k}$ instead of being a Boolean function. Indeed, it suffices to change the 0-linear structures (resp. 1-linear structures) with 0-linear translators. (resp. 1-linear translators).

16.2.4 Bent Functions Not Belonging to the Class of Maiorana–McFarland Using Linear Translators

In the following we are now interested in investigating constructions of bent functions that do not necessarily belong to the class of Maiorana- McFarland contrary to the preceding sections. To this end, we are particularly interested in the secondary construction of the form $f(x) = \phi_1(x)\phi_2(x) + \phi_1(x)\phi_3(x) + \phi_2(x)\phi_3(x)$ presented in [7] and next completed in [50]. If ϕ_1, ϕ_2, ϕ_3 and ψ are bent, then f is bent if and only if $\tilde{\psi} + \tilde{\phi}_1 + \tilde{\phi}_2 + \tilde{\phi}_3 = 0$ (where $\psi := \phi_1 + \phi_2 + \phi_3$). In this section, we show that one can reuse Boolean functions of the shape presented in the preceding sections in the construction of [43, 50].

Firstly, one can derive easily bent functions f, whose dual functions are very simple, by choosing functions H_i in the class of Maiorana–McFarland such that the permutation involving in each H_i is built in terms of an involution and a linear translator. More explicitly, each H_i is a Boolean function over $\mathbb{F}_{2^m}$ defined by $H_i(y) = \mathrm{Tr}_1^m\Big(L(y)+L(\gamma_i)h(g(y))\Big)$ where L is a $\mathbb{F}_{2^k}$-linear involution on $\mathbb{F}_{2^m}$ (k being a divisor of m); carefully chosen according to the hypothesis of [21, Corollary 2], g is a function from $\mathbb{F}_{2^m}$ to $\mathbb{F}_{2^k}$, h is a mapping from $\mathbb{F}_{2^k}$ to itself, and γ_1, γ_2 and γ_3 are three pairwise distinct elements of $\mathbb{F}_{2^m}^\star$ which are 0-linear translators of g such that $\gamma_1 + \gamma_2 + \gamma_3 \neq 0$. Bent functions f are therefore obtained from a direct application of [50, Theorem 4] and [21, Corollary 2].

Secondly, we extend a result from [43] by considering two linear structures instead of one. This result uses linear structures as in the first case of Proposition 16.2.9. Similarly, for the other five cases we can construct bent functions and their duals.

Proposition 16.2.15 ([36]). *Let f and g be functions from $\mathbb{F}_{2^m}$ to $\mathbb{F}_2$. For $i \in \{1,2,3\}$ set $\phi_i(y) := y + \gamma_i f(y) + \delta_i g(y)$ where*

1. *δ_1, δ_2, δ_3 are elements of $\mathbb{F}_{2^m}^\star$ which are 0-linear structures of f and g;*
2. *γ_1, γ_2 and γ_3 are elements of $\mathbb{F}_{2^m}^\star$ which are 0-linear structures of f;*
3. *$\gamma_1 + \gamma_2$ and $\gamma_1 + \gamma_3$ are 0-linear structures of g.*

Then the function h defined on $\mathbb{F}_{2^m} \times \mathbb{F}_{2^m}$ by

$$h(x,y) = \mathrm{Tr}_1^m\Big(x\phi_1(y)\Big)\mathrm{Tr}_1^m\Big(x\phi_2(y)\Big) + \mathrm{Tr}_1^m\Big(x\phi_1(y)\Big)\mathrm{Tr}_1^m\Big(x\phi_3(y)\Big) + \mathrm{Tr}_1^m\Big(x\phi_2(y)\Big)\mathrm{Tr}_1^m\Big(x\phi_3(y)\Big)$$

is bent and the dual of h is given by

$$\tilde{h}(x,y) = \mathrm{Tr}_1^m\Big(y\phi_1^{-1}(x)\Big)\mathrm{Tr}_1^m\Big(y\phi_2^{-1}(x)\Big) + \mathrm{Tr}_1^m\Big(y\phi_1^{-1}(x)\Big)\mathrm{Tr}_1^m\Big(y\phi_3^{-1}(x)\Big) + \mathrm{Tr}_1^m\Big(y\phi_2^{-1}(x)\Big)\mathrm{Tr}_1^m\Big(y\phi_3^{-1}(x)\Big),$$

where $\phi_i^{-1}(x) = x + \gamma_i f(x) + \delta_i\left[g(x)\big(1 + f(x)\big) + g(x+\gamma_i)f(x)\right].$

Proof. Let $\psi_i(x,y) = \mathrm{Tr}_1^m\Big(x\phi_i(y)\Big)$. Then by Proposition 16.2.9, ψ_i is bent for $i = 1,2,3$. Let $\gamma = \gamma_1 + \gamma_2 + \gamma_3$ and $\delta = \delta_1 + \delta_2 + \delta_3$. Then, $\psi(x,y) = \mathrm{Tr}_1^m\Big(x(y + \gamma f(y) + \delta g(y))\Big)$ is bent since γ is a 0-linear structure of f and δ is a 0-linear structure of f and g. Now, it remains to show that $\tilde{\psi} = \tilde{\psi}_1 + \tilde{\psi}_2 + \tilde{\psi}_3$. $\tilde{\psi} = \mathrm{Tr}_1^m\Big(x\phi^{-1}(y)\Big)$ and $\phi^{-1}(x)$ is given in Proposition 16.2.10 in the Appendix.

Note that $\tilde{\psi} = \tilde{\psi}_1 + \tilde{\psi}_2 + \tilde{\psi}_3$ if and only if $g(x+\gamma_1)=g(x+\gamma_2)=g(x+\gamma_3)=g(x+\gamma_1+\gamma_2+\gamma_3)$ which means $\gamma_1 + \gamma_2$ and $\gamma_1 + \gamma_3$ are 0-linear structures of g. □

The following five propositions are related to Proposition 16.2.15 in Sect. 16.2.4.

Proposition 16.2.16 ([36]). *Let f and g be functions from $\mathbb{F}_{2^m}$ to $\mathbb{F}_2$. For $i \in \{1,2,3\}$ set $\phi_i(y) := y + \gamma_i f(y) + \delta_i g(y)$ where*

1. *γ_1, γ_2, γ_3 are elements of $\mathbb{F}_{2^m}^\star$ which are 0-linear structures of f;*
2. *δ_1, δ_2 and δ_3 are elements of $\mathbb{F}_{2^m}^\star$ which are 1-linear structures of f;*

3. $\gamma_1 + \delta_1$, $\gamma_2 + \delta_2$, $\gamma_3 + \delta_3$ *are* 0*-linear structures of* g;
4. $\gamma_1 + \gamma_2$ *and* $\gamma_1 + \gamma_3$ *are* 0*-linear structures of* g.

Then the function h *defined on* $\mathbb{F}_{2^m} \times \mathbb{F}_{2^m}$ *by*

$$h(x,y) = \mathrm{Tr}_1^m\Big(x\phi_1(y)\Big)\mathrm{Tr}_1^m\Big(x\phi_2(y)\Big) + \mathrm{Tr}_1^m\Big(x\phi_1(y)\Big)\mathrm{Tr}_1^m\Big(x\phi_3(y)\Big) + \mathrm{Tr}_1^m\Big(x\phi_2(y)\Big)\mathrm{Tr}_1^m\Big(x\phi_3(y)\Big)$$

is bent and the dual of h *is given by*

$$\tilde{h}(x,y) = \mathrm{Tr}_1^m\Big(y\phi_1^{-1}(x)\Big)\mathrm{Tr}_1^m\Big(y\phi_2^{-1}(x)\Big) + \mathrm{Tr}_1^m\Big(y\phi_1^{-1}(x)\Big)\mathrm{Tr}_1^m\Big(y\phi_3^{-1}(x)\Big) + \mathrm{Tr}_1^m\Big(y\phi_2^{-1}(x)\Big)\mathrm{Tr}_1^m\Big(y\phi_3^{-1}(x)\Big)$$

where

$$\phi_i^{-1}(x) = x + \gamma\left[g(x) + f(x)(1 + g(x) + g(x+\gamma))\right] + \delta\left[g(x)(1 + f(x)) + g(x+\gamma)f(x)\right].$$

Proposition 16.2.17 ([36]). *Let* f *and* g *be functions from* $\mathbb{F}_{2^m}$ *to* $\mathbb{F}_2$. *For* $i \in \{1,2,3\}$ *set* $\phi_i(y) := y + \gamma_i f(y) + \delta_i g(y)$ *where*

1. γ_1, γ_2, γ_3 *are elements of* $\mathbb{F}_{2^m}^\star$ *which are* 0*-linear structures of* f *and* g;
2. δ_1, δ_2 *and* δ_3 *are elements of* $\mathbb{F}_{2^m}^\star$ *which are* 0*-linear structures of* g;
3. $\delta_1 + \delta_2$ *and* $\delta_1 + \delta_3$ *are* 0*-linear structures of* f.

Then the function h *defined on* $\mathbb{F}_{2^m} \times \mathbb{F}_{2^m}$ *by*

$$h(x,y) = \mathrm{Tr}_1^m\Big(x\phi_1(y)\Big)\mathrm{Tr}_1^m\Big(x\phi_2(y)\Big) + \mathrm{Tr}_1^m\Big(x\phi_1(y)\Big)\mathrm{Tr}_1^m\Big(x\phi_3(y)\Big) + \mathrm{Tr}_1^m\Big(x\phi_2(y)\Big)\mathrm{Tr}_1^m\Big(x\phi_3(y)\Big)$$

is bent and the dual of h *is given by*

$$\tilde{h}(x,y) = \mathrm{Tr}_1^m\Big(y\phi_1^{-1}(x)\Big)\mathrm{Tr}_1^m\Big(y\phi_2^{-1}(x)\Big) + \mathrm{Tr}_1^m\Big(y\phi_1^{-1}(x)\Big)\mathrm{Tr}_1^m\Big(y\phi_3^{-1}(x)\Big) + \mathrm{Tr}_1^m\Big(y\phi_2^{-1}(x)\Big)\mathrm{Tr}_1^m\Big(y\phi_3^{-1}(x)\Big)$$

where $\phi_i^{-1}(x) = x + \gamma\left[f(x)(1 + g(x)) + f(x+\delta)g(x)\right] + \delta g(x)$.

Proposition 16.2.18 ([36]). *Let* f *and* g *be functions from* $\mathbb{F}_{2^m}$ *to* $\mathbb{F}_2$. *For* $i \in \{1,2,3\}$ *set* $\phi_i(y) := y + \gamma_i f(y) + \delta_i g(y)$ *where*

1. γ_1, γ_2, γ_3 *are elements of* $\mathbb{F}_{2^m}^\star$ *which are* 1*-linear structures of* g;

2. δ_1, δ_2 and δ_3 are elements of $\mathbb{F}_{2^m}^\star$ which are 0-linear structures of g;
3. $\gamma_1+\delta_1$, $\gamma_2+\delta_2$, $\gamma_3+\delta_3$ are 0-linear structures of f;
4. $\delta_1+\delta_2$ and $\delta_1+\delta_3$ are 0-linear structures of f.

Then the function h defined on $\mathbb{F}_{2^m}\times\mathbb{F}_{2^m}$ by

$$h(x,y) = \mathrm{Tr}_1^m\big(x\phi_1(y)\big)\mathrm{Tr}_1^m\big(x\phi_2(y)\big) + \mathrm{Tr}_1^m\big(x\phi_1(y)\big)\mathrm{Tr}_1^m\big(x\phi_3(y)\big) + \mathrm{Tr}_1^m\big(x\phi_2(y)\big)\mathrm{Tr}_1^m\big(x\phi_3(y)\big)$$

is bent and the dual of h is given by

$$\tilde{h}(x,y) = \mathrm{Tr}_1^m\big(y\phi_1^{-1}(x)\big)\mathrm{Tr}_1^m\big(y\phi_2^{-1}(x)\big) + \mathrm{Tr}_1^m\big(y\phi_1^{-1}(x)\big)\mathrm{Tr}_1^m\big(y\phi_3^{-1}(x)\big) + \mathrm{Tr}_1^m\big(y\phi_2^{-1}(x)\big)\mathrm{Tr}_1^m\big(y\phi_3^{-1}(x)\big)$$

where

$$\phi_i^{-1}(x) = x + \gamma\,[f(x)(1+g(x)) + f(x+\delta)g(x)] + \delta\,\big[f(x)(1+g(x)) + \big(1+f(x+\delta)\big)g(x)\big].$$

Proposition 16.2.19 ([36]). *Let f and g be functions from $\mathbb{F}_{2^m}$ to $\mathbb{F}_2$. For $i\in\{1,2,3\}$ set $\phi_i(y) := y+\gamma_i f(y)+\delta_i g(y)$ where*

1. γ_1, γ_2, γ_3 are elements of $\mathbb{F}_{2^m}^\star$ which are 1-linear structures of f and g;
2. δ_1, δ_2 and δ_3 are elements of $\mathbb{F}_{2^m}^\star$ which are 1-linear structures of f;
3. $\delta_1+\delta_2$ and $\delta_1+\delta_3$ are 0-linear structures of g.

Then the function h defined on $\mathbb{F}_{2^m}\times\mathbb{F}_{2^m}$ by

$$h(x,y) = \mathrm{Tr}_1^m\big(x\phi_1(y)\big)\mathrm{Tr}_1^m\big(x\phi_2(y)\big) + \mathrm{Tr}_1^m\big(x\phi_1(y)\big)\mathrm{Tr}_1^m\big(x\phi_3(y)\big) + \mathrm{Tr}_1^m\big(x\phi_2(y)\big)\mathrm{Tr}_1^m\big(x\phi_3(y)\big)$$

is bent and the dual of h is given by

$$\tilde{h}(x,y) = \mathrm{Tr}_1^m\big(y\phi_1^{-1}(x)\big)\mathrm{Tr}_1^m\big(y\phi_2^{-1}(x)\big) + \mathrm{Tr}_1^m\big(y\phi_1^{-1}(x)\big)\mathrm{Tr}_1^m\big(y\phi_3^{-1}(x)\big) + \mathrm{Tr}_1^m\big(y\phi_2^{-1}(x)\big)\mathrm{Tr}_1^m\big(y\phi_3^{-1}(x)\big)$$

where $\phi_i^{-1}(x) = x+\gamma\,\big[f(x)(1+g(x+\delta)) + \big(1+f(x)\big)g(x)\big] + \delta f(x)$.

Proposition 16.2.20 ([36]). *Let f and g be functions from $\mathbb{F}_{2^m}$ to $\mathbb{F}_2$. For $i\in\{1,2,3\}$ set $\phi_i(y) := y+\gamma_i f(y)+\delta_i g(y)$ where*

1. *$\gamma_1, \gamma_2, \gamma_3$ are elements of $\mathbb{F}_{2^m}^\star$ which are 1-linear structures of g;*
2. *δ_1, δ_2 and δ_3 are elements of $\mathbb{F}_{2^m}^\star$ which are 1-linear structures of f and g;*
3. *$\gamma_1 + \gamma_2$ and $\gamma_1 + \gamma_3$ are 0-linear structures of f.*

Then the function h defined on $\mathbb{F}_{2^m} \times \mathbb{F}_{2^m}$ by

$$h(x,y) = \mathrm{Tr}_1^m\Big(x\phi_1(y)\Big)\mathrm{Tr}_1^m\Big(x\phi_2(y)\Big) + \mathrm{Tr}_1^m\Big(x\phi_1(y)\Big)\mathrm{Tr}_1^m\Big(x\phi_3(y)\Big) + \mathrm{Tr}_1^m\Big(x\phi_2(y)\Big)\mathrm{Tr}_1^m\Big(x\phi_3(y)\Big)$$

is bent and the dual of h is given by

$$\tilde{h}(x,y) = \mathrm{Tr}_1^m\Big(y\phi_1^{-1}(x)\Big)\mathrm{Tr}_1^m\Big(y\phi_2^{-1}(x)\Big) + \mathrm{Tr}_1^m\Big(y\phi_1^{-1}(x)\Big)\mathrm{Tr}_1^m\Big(y\phi_3^{-1}(x)\Big) + \mathrm{Tr}_1^m\Big(y\phi_2^{-1}(x)\Big)\mathrm{Tr}_1^m\Big(y\phi_3^{-1}(x)\Big)$$

where $\phi_i^{-1}(x) = x + \gamma g(x) + \delta\,[f(x)(1 + g(x)) + f(x+\gamma)g(x)]$.

16.2.5 A Secondary Construction of Bent and Semi-bent Functions Using Derivatives and Linear Translators

In this subsection, we consider a new kind of secondary construction of bent functions [15] presented below.

Theorem 16.2.21 ([15]). *Let f and g be two bent functions over $\mathbb{F}_{2^n}$. Assume that there exists $a \in \mathbb{F}_{2^n}$ such that $D_af = D_ag$. Then the function $h : \mathbb{F}_{2^n} \to \mathbb{F}_2$ defined by $h(x) = f(x) + D_af(x)\big(f(x)+g(x)\big)$ is bent and its dual is $\tilde{h}(x) = \tilde{f}(x) + \mathrm{Tr}_1^n(ax)(\tilde{f}(x) + \tilde{g}(x))$.*

In the line of Theorem 16.2.21 and of the preceding subsections, we shall derive from Theorem 16.2.21 new secondary constructions of bent and semi-bent functions in Theorem 16.2.23 and Theorem 16.2.24. To this end, we will use the following lemma.

Lemma 16.2.22 ([36]). *Let $b \in \mathbb{F}_{2^m}$ and $\mathcal{W} \subseteq \mathbb{F}_{2^m}$ be an $m-1$ dimensional linear subspace with $b \notin \mathcal{W}$. Let $\mu : \mathbb{F}_{2^m} \to \mathbb{F}_2$ be a Boolean function such that b is a 0-linear structure of μ. Choose arbitrary functions $h_1 : \mathbb{F}_{2^m} \to \mathbb{F}_2$ and $u : \mathcal{W} \to \mathbb{F}_2$ and define the Boolean function $h_2 : \mathbb{F}_{2^m} \to \mathbb{F}_2$ by $h_2(w) = u(w) + h_1(w)$ and $h_2(w+b) = u(w) + h_1(w+b) + \mu(w)$ for $w \in \mathcal{W}$. Then $D_bh_1(y) + D_bh_2(y) = \mu(y)$ for all $y \in \mathbb{F}_{2^m}$.*

Proof. We observe that $h_2(w+b) + h_2(w) = h_1(w+b) + h_1(w) + \mu(w)$ for all $w \in \mathcal{W}$ by definition. Using the fact that b is a 0-linear structure of μ we complete the proof. □

Note that Lemma 16.3.2 gives a construction of a Boolean function $h_2 : \mathbb{F}_{2^m} \to \mathbb{F}_2$ with the property $D_b h_1(y) + D_b(h_2(y)) = \mu(y)$ for all $y \in \mathbb{F}_{2^m}$ for given $b \in \mathbb{F}_{2^m}$, $h_1 : \mathbb{F}_{2^m} \to \mathbb{F}_2$ and μ having b with 0-linear structure. The construction uses $m-1$ free variables in the form of the function $u : \mathcal{W} \to \mathbb{F}_2$.

Using Lemma 16.3.2, Theorem 16.2.21 and results from Sect. 16.2.3, we present below a new secondary construction of bent functions.

Theorem 16.2.23 ([36]). *Let $1 \le k < m$ be integers with $k \mid m$. Let f, g be functions from $\mathbb{F}_{2^m}$ to $\mathbb{F}_{2^k}$. Assume that $\gamma, \delta \in \mathbb{F}_{2^m}^{\star}$ are 0-linear translators of f and g, respectively. Further assume that $b \in \mathbb{F}_{2^m}$ is a 0-linear structure of f and g. Let $a \in \mathbb{F}_{2^m}$ be an arbitrary element. For arbitrary function $h_1 : \mathbb{F}_{2^m} \to \mathbb{F}_2$ construct $h_2 : \mathbb{F}_{2^m} \to \mathbb{F}_2$ satisfying $D_b h_1(y) = D_b h_2(y) + \mathrm{Tr}_1^m\big(a(\gamma f(y+b) + \delta g(y+b))\big)$ for all $y \in \mathbb{F}_{2^m}$ using Lemma 16.3.2. Set $F(x, y) := \mathrm{Tr}_1^m(xy + \gamma x f(y)) + h_1(y)$ and $G(x, y) := \mathrm{Tr}_1^m(xy + \delta x g(y)) + h_2(y)$. The function defined by*

$$H(x, y) = F(x, y) + D_{a,b}F(x, y)\big(F(x, y) + G(x, y)\big)$$

is bent and its dual is

$$\begin{aligned}\tilde{H}(x, y) &= \mathrm{Tr}_1^m\big(yx + \gamma y f(x)\big) + h_1(x + \gamma f(x)) \\ &\quad + \mathrm{Tr}_1^m(ax + by)\left[\mathrm{Tr}_1^m\big(y(\gamma f(x) + \delta g(x))\big) + h_1(x + \gamma f(x)) + h_2(x + \delta g(x))\right].\end{aligned}$$

Proof. F and G are bent by Proposition 16.2.14. Using the fact that b is a 0-linear structure of f and g we get that $D_{a,b}F(x, y) = \mathrm{Tr}_1^m\big(xb + a(y + b + \gamma f(y + b))\big) + D_b h_1(y)$ and $D_{a,b}G(x, y) = \mathrm{Tr}_1^m\big(xb + a(y + b + \delta g(y + b))\big) + D_b h_2(y)$. Hence $D_{a,b}F(x, y) = D_{a,b}G(x, y)$ and the proof follows from Theorem 16.2.21 and Proposition 16.2.14. □

Using [57, Theorem 16] instead of Theorem 16.2.21 we obtain the following secondary construction of semi-bent functions.

Theorem 16.2.24 ([36]). *Under notation and assumptions of Theorem 16.2.23 we construct $h_2 : \mathbb{F}_{2^m} \to \mathbb{F}_2$ satisfying $D_b h_1(y) = D_b h_2(y) + \mathrm{Tr}_1^m\big(a(\gamma f(y+b) + \delta g(y+b))\big) + 1$ (instead of $D_b h_1(y) = D_b h_2(y) + \mathrm{Tr}_1^m\big(a(\gamma f(y+b) + \delta g(y+b))\big)$) for all $y \in \mathbb{F}_{2^m}$. Set F and G in the same way. Then the function defined by $H(x, y) = F(x, y) + G(x, y) + D_{a,b}F(x, y) + D_{a,b}FG(x, y)$ is semi-bent.*

Note that Theorem 16.2.24 gives a secondary construction of semi-bent functions of high degree by choosing the arbitrary function $h_1 : \mathbb{F}_{2^m} \to \mathbb{F}_2$ of large degree. Moreover it gives a different construction than the one given in [49, Section 4.2.5] and hence it is an answer to Problem 4 of [49].

16.2.6 A Secondary Construction of Bent Functions Using Certain Quadratic and Cubic Functions Together with Linear Structures

In this subsection we consider Boolean functions that are the sum of a bent function of Sect. 16.2.1 or 16.2.2 and a quadratic or cubic function. We show that one can choose appropriately the quadratic and cubic function so that those Boolean functions are bent again. Furthermore, the dual functions of those bent functions can be explicitly computed as in the preceding subsections. The main results are Theorems 16.2.26, 16.2.28, 16.2.31 and 16.2.32.

Theorem 16.2.26 is based on [16, Lemma 1]. We note that the bent function of Theorem 16.2.26 is different from the two classes of plateaued functions in Section 6 of [16]. First of all we obtain bent functions while two classes of functions in Section 6 of [16] produce only plateaued functions.

Theorem 16.2.31 is a further generalization of Theorem 16.2.26 using cubic functions instead of quadratic functions.

Lemma 16.2.25 ([16]). *Let $w_1, w_2, u \in \mathbb{F}_{2^m}$ with $\{w_1, w_2\}$ linearly independent over $\mathbb{F}_{2^m}$. We have*

$$\sum_{x\in\mathbb{F}_{2^m}} (-1)^{\mathrm{Tr}_1^m(w_1x)\mathrm{Tr}_1^m(w_2x)+\mathrm{Tr}_1^m(ux)} = \begin{cases} 0 & \text{if } u \notin \langle w_1, w_2\rangle = \{0, w_1, w_2, w_1+w_2\}, \\ 2^{m-1} & \text{if } u \in \{0, w_1, w_2\}, \\ -2^{m-1} & \text{if } u = w_1 + w_2. \end{cases}$$

In Lemma 16.2.25, for any given $\mathbb{F}_2$-linearly independent set, the Boolean function on $\mathbb{F}_{2^m}$ given by $x \mapsto \mathrm{Tr}_1^m(w_1x)\mathrm{Tr}_1^m(w_2x)$ is a quadratic function.

Theorem 16.2.26 ([36]). *Let $w_1, w_2, \gamma \in \mathbb{F}_{2^m}$ with $\{w_1, w_2\}$ linearly independent over $\mathbb{F}_2$. Assume that $f, h : \mathbb{F}_{2^m} \to \mathbb{F}_2$ are Boolean functions such that w_1 and w_2 are 0-linear structures of f and h. Moreover, we assume that γ is a 0-linear structure of f. Then the Boolean function F defined on $\mathbb{F}_{2^m} \times \mathbb{F}_{2^m}$ by*

$$F(x, y) = \mathrm{Tr}_1^m(xw_1)\mathrm{Tr}_1^m(xw_2) + \mathrm{Tr}_1^m\big(xy + \gamma xf(y)\big) + h(y) \tag{16.10}$$

is bent and its dual function is

$$\tilde{F}(x, y) = \mathrm{Tr}_1^m(yw_1)\mathrm{Tr}_1^m(yw_2) + \mathrm{Tr}_1^m\big(yx + \gamma yf(x)\big) + h(x + \gamma f(x)).$$

Moreover, $F(x, y) = \mathrm{Tr}_1^m(xw_1)\mathrm{Tr}_1^m(xw_2) + \mathrm{Tr}_1^m\big(xL(y) + L(\gamma)xf(y)\big) + h(y)$ where L is a linearized permutation polynomial of $\mathbb{F}_{2^m}$ is also bent under the same conditions and its dual function is

$$\tilde{F}(x, y) = \mathrm{Tr}_1^m(yw_1)\mathrm{Tr}_1^m(yw_2) + \mathrm{Tr}_1^m\big(yL^{-1}(x) + \gamma yf(L^{-1}(x))\big) + h(L^{-1}(x) + \gamma f(L^{-1}(x))).$$

Proof. One has for every $(a, b) \in \mathbb{F}_{2^m} \times \mathbb{F}_{2^m}$,

$$\widehat{\chi_F}(a,b) = \sum_{y\in\mathbb{F}_{2^m}} (-1)^{h(y)+Tr_1^m(by)} \sum_{x\in\mathbb{F}_{2^m}} (-1)^{Tr_1^m(xw_1)Tr_1^m(xw_2)+Tr_1^m\left(xy+\gamma xf(y)+ax\right)}$$

Let $\phi(y) = y + \gamma f(y)$ and $\mathcal{S} = \sum_{x\in\mathbb{F}_{2^m}} (-1)^{Tr_1^m(xw_1)Tr_1^m(xw_2)+Tr_1^m\left(x(\phi(y)+a)\right)}$. Then by Lemma 16.2.25, we have

$$\mathcal{S} = \begin{cases} 0 & \text{if } \phi(y) + a \notin \{0, w_1, w_2, w_1 + w_2\}, \\ 2^{m-1} & \text{if } \phi(y) + a \in \{0, w_1, w_2\}, \\ -2^{m-1} & \text{if } \phi(y) + a = w_1 + w_2. \end{cases}$$

Now, $f(a) = f(a+w_1) = f(a+w_2) = f(a+w_1+w_2)$ since w_1 and w_2 are 0-linear structures of f. We have two cases, namely $f(a) = 0$ and $f(a) = 1$. Here, only the proof for the case $f(a) = 0$ is given since the proof for the other case is similar.

Assume $f(a) = 0$. Then $\phi(y) + a \in \{0, w_1, w_2\}$ when $y \in \mathcal{A} = \{a, a + w_1, a + w_2\}$ and $\phi(y) + a = w_1 + w_2$ when $y = a + w_1 + w_2$. Hence,

$$\widehat{\chi_F}(a,b) = 2^{m-1}\left[\sum_{y\in\mathcal{A}} (-1)^{h(y)+Tr_1^m(by)} - (-1)^{h(a+w_1+w_2)+Tr_1^m(b(a+w_1+w_2))}\right].$$

Since w_1 and w_2 are 0-linear structures of h, we obtain

$$\widehat{\chi_F}(a,b) = 2^{m-1}\left[(-1)^{h(a)+Tr_1^m(ba)}\right]\mathcal{S}_1$$

where

$$\mathcal{S}_1 = \left[1 + (-1)^{Tr_1^m(bw_1)} + (-1)^{Tr_1^m(bw_2)} - (-1)^{Tr_1^m(b(w_1+w_2))}\right]. \qquad (16.11)$$

Note that

$$\mathcal{S}_1 = \begin{cases} 2 & \text{if } Tr_1^m(bw_1)Tr_1^m(bw_2) = 0, \\ -2 & \text{if } Tr_1^m(bw_1)Tr_1^m(bw_2) = 1. \end{cases}$$

Combining these we obtain that F is bent and its dual $\tilde{F}$ satisfies that

$$\tilde{F}(x,y) = Tr_1^m(yw_1)Tr_1^m(yw_2) + Tr_1^m\left(yx + y\gamma f(x)\right) + h(x + \gamma f(x)).$$

□

Remark 16.2.27. In Theorem 16.2.26, for given $\mathbb{F}_2$-linearly independent subset $\{w_1, w_2\}$, the Boolean function on $\mathbb{F}_{2^m} \times \mathbb{F}_{2^m}$ given by $(x, y) \mapsto Tr_1^m(xw_1)Tr_1^m(xw_2)$ is a quadratic function, which is used as the first summand in the definition of $F(x, y)$

in equation (16.10). In the proof of Theorem 16.2.26, we apply Lemma 16.2.25 for this quadratic function. Note that if $\gamma \neq 0$ and $1 + deg(f)$, $deg(h)$ and 2 are distinct, then the degree of $F(x, y)$ is $max\{1 + deg(f), deg(h), 2\}$, which may be much larger than 2.

In the following we present a straightforward generalization of Theorem 16.2.26.

Theorem 16.2.28 ([36]). *Let $w_1, w_2, \gamma, \delta \in \mathbb{F}_{2^m}$ with $\{w_1, w_2\}$ linearly independent over $\mathbb{F}_2$. Assume that $f, g, h : \mathbb{F}_{2^m} \to \mathbb{F}_2$ are Boolean functions such that w_1 and w_2 are 0-linear structures of f, g and h. Moreover, we assume that γ is a 0-linear structure of f and δ is a 0-linear structure of f and g. Then the Boolean function F defined on $\mathbb{F}_{2^m} \times \mathbb{F}_{2^m}$ by*

$$F(x, y) = \mathrm{Tr}_1^m(xw_1)\mathrm{Tr}_1^m(xw_2) + \mathrm{Tr}_1^m\big(x(L(y) + L(\gamma)f(y) + L(\delta)g(y))\big) + h(y)$$

is bent and its dual function is

$$\tilde{F}(x, y) = \mathrm{Tr}_1^m(yw_1)\mathrm{Tr}_1^m(yw_2) + \mathrm{Tr}_1^m\big(y\rho^{-1}(x)\big) + h(\rho^{-1}(x)) \text{ where}$$

$$\rho^{-1}(x) = L^{-1}(x) + \gamma f(L^{-1}(x)) + \delta\left[g(L^{-1}(x))(1 + f(L^{-1}(x))) + g(L^{-1}(x) + \gamma)f(L^{-1}(x))\right].$$

We now give the analogue of Lemma 16.2.25 which improves Lemma 1 of [16].

Lemma 16.2.29 ([36]). *Let $w_1, w_2, w_3, u \in \mathbb{F}_{2^m}$ with $\{w_1, w_2, w_3\}$ linearly independent over $\mathbb{F}_{2^m}$. We have*

$$\sum_{x \in \mathbb{F}_{2^m}} (-1)^{\mathrm{Tr}_1^m(w_1x)\mathrm{Tr}_1^m(w_2x)\mathrm{Tr}_1^m(w_3x) + \mathrm{Tr}_1^m(ux)} = \begin{cases} 0 & \text{if } u \notin \langle w_1, w_2, w_3 \rangle, \\ 3.2^{m-2} & \text{if } u = 0, \\ 2^{m-2} & \text{if } u \in \{w_1, w_2, w_3, w_1 + w_2 + w_3\}, \\ -2^{m-2} & \text{if } u \in \{w_1 + w_2, w_1 + w_3, w_2 + w_3\}. \end{cases}$$

Proof. Let $\mathcal{T}$ denotes the sum in the statement of the lemma. Let $\mathcal{T}_1$ and $\mathcal{T}_2$ be the sums as

$$\mathcal{T}_1 = \sum_{x \in \mathbb{F}_{2^m} | \mathrm{Tr}_1^m(w_1x) = 0} (-1)^{\mathrm{Tr}_1^m(ux)}$$

and

$$\mathcal{T}_2 = \sum_{x \in \mathbb{F}_{2^m} | \mathrm{Tr}_1^m(w_1x) = 1} (-1)^{\mathrm{Tr}_1^m(w_2x)\mathrm{Tr}_1^m(w_3x) + \mathrm{Tr}_1^m(ux)}.$$

We have that $\mathcal{T} = \mathcal{T}_1 + \mathcal{T}_2$. It is clear that

$$\mathcal{T}_1 = \begin{cases} 0 & \text{if } u \notin \langle 0, w_1 \rangle = \{0, w_1\}, \\ 2^{m-1} & \text{if } u \in \{0, w_1\}. \end{cases}$$

Using Lemma 16.2.25 we obtain that

$$\mathcal{T}_2 = \begin{cases} 0 & \text{if } u \notin \langle w_1, w_2, w_3 \rangle, \\ 2^{m-2} & \text{if } u \in \{0, w_1, w_2, w_3, w_1 + w_2 + w_3\}, \\ -2^{m-2} & \text{if } u \in \{w_1 + w_2, w_1 + w_3, w_2 + w_3\}. \end{cases}$$

Combining $\mathcal{T}_1$ and $\mathcal{T}_2$ we complete the proof. □

Remark 16.2.30. This remark is analogous to Remark 16.2.27. In Theorem 16.2.31, for given $\mathbb{F}_2$-linearly independent subset $\{w_1, w_2, w_3\}$, the Boolean function on $\mathbb{F}_{2^m} \times \mathbb{F}_{2^m}$ given by

$$(x, y) \mapsto \mathrm{Tr}_1^m(xw_1)\mathrm{Tr}_1^m(xw_2)\mathrm{Tr}_1^m(xw_3)$$

is a cubic function, which is used as the first summand in the definition of $F(x, y)$ in Eq. (16.12). In the proof of Theorem 16.2.31, we apply Lemma 16.2.29 for this cubic function. As in Remark 16.2.27, the degree of $F(x, y)$ is $max\{1 + deg(f), deg(h), 3\}$ under suitable conditions, which may be much larger than 3.

Theorem 16.2.31 ([36]). *Let f and h be two Boolean functions on $\mathbb{F}_{2^m}$. Let $w_1, w_2, w_3 \in \mathbb{F}_{2^m}$ be linearly independent and $\gamma \in \mathbb{F}_{2^m}$. Assume that γ is a 0-linear structure of f, and w_1, w_2, w_3 are 0-linear structures of f and h. Then, the function F defined on $\mathbb{F}_{2^m} \times \mathbb{F}_{2^m}$ by*

$$F(x, y) = \mathrm{Tr}_1^m(xw_1)\mathrm{Tr}_1^m(xw_2)\mathrm{Tr}_1^m(xw_3) + \mathrm{Tr}_1^m\big(x(L(y) + L(\gamma)f(y))\big) + h(y) \quad (16.12)$$

is bent and its dual is

$$\tilde{F}(x, y) = \mathrm{Tr}_1^m(yw_1)\mathrm{Tr}_1^m(yw_2)\mathrm{Tr}_1^m(yw_3) + \mathrm{Tr}_1^m\big(y(L^{-1}(x) + \gamma f(L^{-1}(x)))\big)$$
$$+ h(L^{-1}(x) + \gamma f(L^{-1}(x))).$$

Proof. Let $\phi(y) = y + \gamma f(y)$. For every $(a, b) \in \mathbb{F}_{2^m} \times \mathbb{F}_{2^m}$,

$$\widehat{\chi_F}(a, b) = \sum_{y \in \mathbb{F}_{2^m}} (-1)^{h(y) + \mathrm{Tr}_1^m(by)} \sum_{x \in \mathbb{F}_{2^m}} (-1)^{\mathrm{Tr}_1^m(w_1x)\mathrm{Tr}_1^m(w_2x)\mathrm{Tr}_1^m(w_3x) + \mathrm{Tr}_1^m\left(x(\phi(y)+a)\right)}.$$

For the case $f(a) = 0$,

- $\phi(y) + a = 0$ when $y = a$,
- $\phi(y) + a \in \{w_1, w_2, w_3, w_1 + w_2 + w_3\}$ when $y \in \mathcal{A}_1 = \{a + w_1, a + w_2, a + w_3, a + w_1 + w_2 + w_3\}$
- $\phi(y) + a \in \{w_1 + w_2, w_1 + w_3, w_2 + w_3\}$ when $y \in \mathcal{A}_2 = \{a + w_1 + w_2, a + w_1 + w_3, a + w_2 + w_3\}$.

Then, following the steps in proof of Theorem 16.2.26 and using Lemma 16.2.29, we get

$$\widehat{\chi_F}(a,b) = 3.2^{m-2}(-1)^{\mathrm{Tr}_1^m(ba)+h(a)} + 2^{m-2}\sum_{y\in\mathcal{A}_1}(-1)^{\mathrm{Tr}_1^m(by)+h(y)}$$
$$-2^{m-2}\sum_{y\in\mathcal{A}_2}(-1)^{\mathrm{Tr}_1^m(by)+h(y)}$$
$$= 2^{m-2}\left[(-1)^{\mathrm{Tr}_1^m(ba)+h(a)}\right]\mathcal{S}$$

where

$$\mathcal{S} = [3 + \mathcal{S}_1 + \mathcal{S}_2]\,, \tag{16.13}$$

$\mathcal{S}_1 = (-1)^{\mathrm{Tr}_1^m(bw_1)} + (-1)^{\mathrm{Tr}_1^m(bw_2)} + (-1)^{\mathrm{Tr}_1^m(bw_3)} + (-1)^{\mathrm{Tr}_1^m(b(w_1+w_2+w_3))}$ and $\mathcal{S}_2 = (-1)^{\mathrm{Tr}_1^m(b(w_1+w_2))} + (-1)^{\mathrm{Tr}_1^m(b(w_1+w_3))} + (-1)^{\mathrm{Tr}_1^m(b(w_2+w_3))}$. Let $(-1)^{\mathrm{Tr}_1^m(bw_i)} = c_i$ where $c_i \in \mathbb{F}_2$, for $i = 1,2,3$. Then, $3+\mathcal{S}_1+\mathcal{S}_2 = \pm 4$ and hence $\widehat{\chi_F}(a,b) = \pm 2^m$. The proof for the case $f(a) = 1$ is similar. □

As in Theorem 16.2.28, in the following we get a modification of Theorem 16.2.31 using two linear structures instead of one.

Theorem 16.2.32 ([36]). *Let f, g and h be Boolean functions on $\mathbb{F}_{2^m}$. Let $w_1, w_2, w_3 \in \mathbb{F}_{2^m}$ be linearly independent and $\gamma, \delta \in \mathbb{F}_{2^m}$, $\gamma \neq \delta$. Assume that γ is a 0-linear structure of f, δ is a 0-linear structure of f and g. Moreover, assume that w_1, w_2, w_3 are 0-linear structures of f, g and h. Then, the function F defined on $\mathbb{F}_{2^m} \times \mathbb{F}_{2^m}$ by*

$$F(x,y) = \mathrm{Tr}_1^m(xw_1)\mathrm{Tr}_1^m(xw_2)\mathrm{Tr}_1^m(xw_3) + \mathrm{Tr}_1^m\big(x(L(y) + L(\gamma)f(y) + L(\delta)g(y))\big) + h(y)$$

is bent and its dual is

$$\tilde{F}(x,y) = \mathrm{Tr}_1^m(yw_1)\mathrm{Tr}_1^m(yw_2)\mathrm{Tr}_1^m(yw_3) + \mathrm{Tr}_1^m\big(y\rho^{-1}(x)\big) + h(\rho^{-1}(x)) \text{ where}$$

$$\rho^{-1}(x) = L^{-1}(x) + \gamma f(L^{-1}(x)) + \delta\left[g(L^{-1}(x))\big(1 + f(L^{-1}(x))\big) + g(L^{-1}(x)+\gamma)f(L^{-1}(x))\right].$$

16.3 Various Characterizations of Plateaued Functions in Arbitrary Characteristic

Plateaued functions can be characterized by their second-order derivatives. More precisely, it is proved in [16] that a Boolean function f on $\mathbb{F}_{2^n}$ is plateaued if and only if there exists λ (necessarily the amplitude of f) such that , for every $x \in \mathbb{F}_{2^n}$ we have $\sum_{a,b\in\mathbb{F}_{2^n}}(-1)^{D_aD_bf(x)} = \lambda^2$. As a direct consequence, all the quadratic functions are plateaued. Several properties of plateaued functions have been studied. Furthermore, plateaued Boolean functions have been generalized to p-ary plateaued functions (see e.g. [56]) and vectorial p-ary functions (see e.g. [53]) in arbitrary characteristic.

Few characterizations of plateaued functions are given in [62] for Boolean functions, which are direct consequences of the definition. In any characteristic, the author [48] has characterized bent functions and plateaued functions in terms of moments of their Walsh transforms in any characteristic p by considering the ratio of two consecutive Walsh power moments of even orders. Another characterization valid only for Boolean functions was obtained by Carlet and Prouff in [16]. Such a characterization was a starting point for the recent work of Carlet [9, 10] in which he provided several characterizations of plateaued Boolean and vectorial functions in characteristic 2 by means of the value distributions of derivatives and of power moments of the Walsh transform. This allows to derive several characterizations of the so-called APN functions (for the definition, see e.g. [6, p. 417]). Very recently, new results related to the characterization of plateaued p-ary functions as well as plateaued vectorial functions in arbitrary characteristic have been derived by the second author with Özbudak and Sınak [53]. In the following, we present some recent characterizations of p-ary functions presented in [48] and [53].

We start by recalling briefly some notation and background given in Chap. 13 (Sects. 1.3 and 13.1). Let p be a prime integer, $n \geq 1$ be an integer. We denote as usual $\mathbb{F}_{p^n}$ the finite field of size p^n and $\mathbb{F}_{p^n}^{\star}$ the set of nonzero elements of $\mathbb{F}_{p^n}$. Let $\zeta_p = e^{\frac{2\pi i}{p}}$ be the complex primitive p^{th} root of unity. Let f be a p-ary function from $\mathbb{F}_{p^n}$ to $\mathbb{F}_p$. The Walsh transform of f at $b \in \mathbb{F}_{p^n}$ is given by

$$S_f(b) = \sum_{x \in \mathbb{F}_{p^n}} \zeta^{f(x) - tr_n(bx)}. \tag{16.14}$$

where tr_n is the trace function from $\mathbb{F}_{p^n}$ to $\mathbb{F}_p$.

The function f is said to be bent if $|S_f(w)|^2 = p^n$ for every $w \in \mathbb{F}_{p^n}$. Below we give the definition of the so called *s-plateaued function*.

Definition 16.3.1. A p-ary function f from $\mathbb{F}_{p^n}$ to $\mathbb{F}_p$ is said to be s-plateaued if $|S_f(w)| \in \{0, p^{\frac{n+s}{2}}\}$ for all $w \in \mathbb{F}_{p^n}$ where $0 \leq s \leq n$.

It is obvious that bent functions are 0-plateaued functions (in the case where $s = 0$, $|S_f(w)| \in \{0, p^{\frac{n}{2}}\}$ is equivalent to $|S_f(w)| = p^{\frac{n}{2}}$).

Recall the *Parseval identity*:

$$\sum_{w \in \mathbb{F}_{p^n}} |S_f(w)|^2 = p^{2n}. \tag{16.15}$$

The Parseval identity allows to compute the multiplicity of each value of the Walsh transform (when $p = 2$, a more precise statement has been shown in [4]).

The following lemma is useful to prove some results in the next subsections (see for instance [11] and [48]).

Lemma 16.3.2. *Let $f : \mathbb{F}_{p^n} \to \mathbb{F}_p$ be s-plateaued. Then the absolute value of the Walsh transform S_f takes p^{n-s} times the value $p^{\frac{n+s}{2}}$ and $p^n - p^{n-s}$ times the value 0.*

Proof. If N denotes the number of $w \in \mathbb{F}_{p^n}$ such that $|\widehat{\chi_f}(w)| = p^{\frac{n+s}{2}}$, then $\sum_{w\in\mathbb{F}_{p^n}} |\widehat{\chi_f}(w)|^2 = p^{n+s}N$. Now, according to equation (16.15), one must have that $p^{n+s}N = p^{2n}$, that is, $N = p^{n-s}$. The result follows. □

The following tool was introduced in the literature earlier (see, for example in [11] and [48]). The *directional difference* of f at $a \in \mathbb{F}_{p^n}$ is the map

$$\begin{aligned} D_af :&\mathbb{F}_{p^n} \to \mathbb{F}_p \\ &x \longmapsto D_af(x) = f(x+a) - f(x) \quad \forall x \in \mathbb{F}_{p^n}. \end{aligned}$$

Let F be a vectorial function from $\mathbb{F}_{p^n}$ to $\mathbb{F}_{p^m}$. The directional difference of vectorial function F at $a \in \mathbb{F}_{p^n}$ is the map

$$\begin{aligned} D_aF :&\mathbb{F}_{p^n} \to \mathbb{F}_{p^m} \\ &x \longmapsto D_aF(x) = F(x+a) - F(x) \quad \forall x \in \mathbb{F}_{p^n}. \end{aligned}$$

16.3.1 Characterizations of Plateaued p-Ary Functions

In [48], the author has introduced the following tools to characterize plateaued p-ary functions. Let p be a positive prime integer. For any nonnegative integer k, we set

$$S_k(f) = \sum_{w\in\mathbb{F}_{p^n}} |S_f(w)|^{2k} \text{ and } T_k(f) = \frac{S_{k+1}(f)}{S_k(f)}$$

with the convention regarding $k = 0$ that $S_0(f) = p^n$ (in this case, $T_0(f) = \frac{S_1(f)}{S_0(f)} = p^n$). For $k = 1$, $S_1(f) = p^{2n}$ by (16.15).

Let us make a preliminary but important remark : for every integer A and every positive integer k, it holds

$$\sum_{w\in\mathbb{F}_{p^n}} \left(|S_f(w)|^2 - A\right)^2 |\widehat{\chi_f}(w)|^{2(k-1)} = S_{k+1}(f) - 2AS_k(f) + A^2S_{k-1}(f). \tag{16.16}$$

We are now going to deduce from (16.16) a characterization of plateaued functions in terms of moments of the Walsh transform (we shall specialize our characterization to bent functions in a next section).

Theorem 16.3.3 ([48]). *Let n and k be two positive integers. Let f be a function from $\mathbb{F}_{p^n}$ to $\mathbb{F}_p$. Then, the two following assertions are equivalent.*

1. *f is plateaued, that is, there exists a nonnegative integer s such that f is s-plateaued.*
2. $T_{k+1}(f) = T_k(f)$.

Proof.

1. Suppose that f is s-plateaued for some nonnegative integer s, that is, $|\widehat{\chi_f}(w)| \in \{0, p^{\frac{n+s}{2}}\}$. Then, by Lemma 16.3.2,

$$S_k(f) = \sum_{w \in \mathbb{F}_{p^n}} |\widehat{\chi_f}(w)|^{2k} = p^{n-s} \times p^{k(n+s)} = p^{(k+1)n+(k-1)s}$$

$$S_{k+1}(f) = p^{n-s} \times p^{(k+1)(n+s)} = p^{(k+2)n+ks}$$

$$S_{k+2}(f) = p^{n-s} \times p^{(k+2)(n+s)} = p^{(k+3)n+(k+1)s}.$$

Therefore

$$T_k(f) = \frac{p^{(k+2)n+ks}}{p^{(k+1)n+(k-1)s}} = p^{n+s}$$

and

$$T_{k+1}(f) = \frac{p^{(k+3)n+(k+1)s}}{p^{(k+2)n+ks}} = p^{n+s} = T_k(f).$$

2. Suppose $T_{k+1}(f) = T_k(f)$. According to (16.16)

$$\sum_{w \in \mathbb{F}_{p^n}} \left(|\widehat{\chi_f}(w)|^2 - T_k(f)\right)^2 |\widehat{\chi_f}(w)|^{2k}$$

$$= S_{k+2}(f) - 2T_k(f)S_{k+1}(f) + T_k^2(f)S_k(f)$$

$$= S_{k+1}(f)\,(T_{k+1}(f) - 2T_k(f) + T_k(f)) = 0$$

proving that $|S_f(w)| \in \{0, \sqrt{T_k(f)}\}$ for every $w \in \mathbb{F}_{p^n}$. Thus,

$$\sum_{w \in \mathbb{F}_{p^n}} |S_f(w)|^2 = T_k(f)\#\{w \in \mathbb{F}_{p^n} \mid |S_f(w)| = \sqrt{T_k(f)}\}.$$

Now, the Parseval identity (16.15) states that

$$\sum_{w\in\mathbb{F}_{p^n}} |S_f(w)|^2 = p^{2n}.$$

Therefore $T_k(f)$ divides p^{2n} proving that $T_k(f) = p^\rho$ for some positive integer ρ. Now, one has $\#\{w \in \mathbb{F}_{p^n} \mid |S_f(w)| = \sqrt{T_k(f)}\} = p^{2n-\rho} \leq p^n$ which implies that $\rho \geq n$, that is, $\rho = n + s$ for some nonnegative integer s.

□

Remark 16.3.4. Specializing Theorem 16.3.3 to the case where $k = 1$, we get that f is plateaued if and only if $T_2(f) = T_1(f)$, that is

$$S_3(f)S_1(f) - S_2^2(f) = p^{2n}S_3(f) - S_2^2(f) = 0.$$

Remark 16.3.5. In the proof, we have shown more than the sole equivalence between (1) and (2). Indeed, we have shown that if (2) holds then f is s-plateaued and $|S_f(w)| \in \{0, \sqrt{T_k(f)}\}$.

In Theorem 16.3.3, we have considered the ratio of two consecutive sums $S_k(f)$. In fact, one can get a more general result than Theorem 16.3.3. Indeed, for every positive integer k and every nonnegative integer l, we have

$$\sum_{w\in\mathbb{F}_{p^n}} \left(|S_f(w)|^{2l} - A\right)^2 |\widehat{\chi_f}(w)|^{2(k-1)} \tag{16.17}$$

$$= S_{k+2l-1}(f) - 2AS_{k+l-1}(f) + A^2S_{k-1}(f).$$

Then, one can make the same kind of proof as that of Theorem 16.3.3 but with (16.17) in place of (16.16) (the proof being very similar, we omit it).

Theorem 16.3.6 ([48]). *Let n, k and l be positive integers and $f : \mathbb{F}_{p^n} \to \mathbb{F}_p$. Then, the two following assertions are equivalent*

1. *f is plateaued, that is, there exists a nonnegative integer s such that f is s-plateaued.*
2. $\frac{S_{k+2l}(f)}{S_{k+l}(f)} = \frac{S_{k+l}(f)}{S_k(f)}$.

In the following, the results are originated from [48]. We give the characterizations of s-plateaued functions via the sequence of the even moments of their Walsh transforms. We shall give the characterizations of plateaued functions in terms of the moments of their Walsh transforms.

The following seems to be more practical than [48, Theorem 1] in some applications.

Theorem 16.3.7 ([53]). *Let f be a function from $\mathbb{F}_{p^n}$ to $\mathbb{F}_p$. Let s be an integer with $0 \leq s \leq n$ and $i, j \in \mathbb{Z}^+$.*

Then the followings are equivalent.

1. *f is s-plateaued for $s > 0$.*
2. $S_i(f) \cdot S_j(f) = S_{i+1}(f) \cdot S_{j-1}(f)$ *for all $i \geq 1$ and $j \geq 2$.*

Moreover, f is bent if and only if (2) holds for all $i, j \in \mathbb{Z}^+$.

Proof. Suppose that f is s-plateaued for $s > 0$. By Lemma 16.3.2, it is easily seen that

$$S_i(f) \cdot S_j(f) = S_{i+1}(f) \cdot S_{j-1}(f), \quad \forall i \geq 1, j \geq 2.$$

Conversely, for $i = j$ and $A = \frac{S_i(f)}{S_{i-1}(f)}$ in (16.16), the proof is the same as the proof of [48, Theorem 1]. □

In fact, Theorem 16.3.7 is equivalent to [48, Theorem 1], which can be shown as follows.

Corollary 16.3.8 ([53]). *Let f be a function from $\mathbb{F}_{p^n}$ to $\mathbb{F}_p$. Then the followings are equivalent.*

1. $S_i(f) \cdot S_i(f) = S_{i+1}(f) \cdot S_{i-1}(f)$ *for $i \geq 2$.*
2. $S_i(f) \cdot S_j(f) = S_{i+1}(f) \cdot S_{j-1}(f)$ *for all $i, j \geq 2$.*

Proof. Suppose that (*1*) holds. Without loss of generality, we may assume $i < j$ and fix $i \geq 2$. We proceed by induction on j. For $j = i + 1$ and $j = i + 2$, then (2) trivial holds. Let $j = i + 3$. We have

$$\begin{aligned} S_{i+1}(f) \cdot S_{i+1}(f) &= S_{i+2}(f) \cdot S_i(f), \\ S_{i+2}(f) \cdot S_{i+2}(f) &= S_{i+3}(f) \cdot S_{i+1}(f). \end{aligned}$$

It follows that $S_i(f) \cdot S_{i+3}(f) = S_{i+1}(f) \cdot S_{i+2}(f)$. For $j = i + k$, assume that (2) holds. We then have

$$\begin{aligned} S_i(f) \cdot S_{i+k}(f) &= S_{i+1}(f) \cdot S_{i+k-1}(f) \\ S_{i+k-1}(f) \cdot S_{i+k+1}(f) &= S_{i+k}(f) \cdot S_{i+k}(f) \end{aligned}$$

It follows that $S_i(f) \cdot S_{i+k+1}(f) = S_{i+1}(f) \cdot S_{i+k}(f)$. Therefore, (2) holds for $j = i + k + 1$. The converse is obvious for $j = i$. □

For a function f from $\mathbb{F}_{p^n}$ to $\mathbb{F}_p$, the author showed in [48] that $S_2(f) \geq p^{3n}$ and also

$$f \text{ is bent if and only if } S_2(f) = p^{3n}. \tag{16.18}$$

Thus, we deduce that, for a bent function f, the sequence $S_i(f)$ is a simple geometric sequence.

Corollary 16.3.9 ([53]). *Let f be a function from $\mathbb{F}_{p^n}$ to $\mathbb{F}_p$. If f is a bent function, then for all $i \in \mathbb{N}$*

$$S_i(f) = p^{(i+1)n}. \tag{16.19}$$

Proof. By (16.15) and (16.18), $S_1(f) = p^{2n}$ and $S_2(f) = p^{3n}$. Then by Theorem 16.3.7, $S_i(f) = \frac{S_{i-1}(f)^2}{S_{i-2}(f)} = p^{(i+1)n}$ for all $i \geq 3$, recursively. Thus (16.19) holds for all $i \in \mathbb{N}$. □

We also deduce from (16.18) a characterization of s-plateaued functions via the moments of their Walsh transforms.

Theorem 16.3.10 ([53]). *Let f be a function from $\mathbb{F}_{p^n}$ to $\mathbb{F}_p$ and s be an integer with $1 \leq s \leq n$. Then*

$$f \text{ is } s\text{-plateaued if and only if } S_2(f) = p^{3n+s} \text{ and } S_3(f) = p^{4n+2s}.$$

Proof. Assume that f is s-plateaued. By (16.16) with $A = p^{n+s}$ and $i = 0$,

$$\begin{aligned}\textstyle\sum_{w\in\mathbb{F}_{p^n}} \left(|S_f(w)|^2 - p^{n+s}\right)^2 &= S_2(f) - 2p^{n+s}S_1(f) + p^{2n+2s}\cdot S_0(f)\\ &= (p^n - p^{n-s})(-p^{n+s})^2\end{aligned} \tag{16.20}$$

where the last equality of (16.20) follows from Lemma 16.3.2. Therefore, $S_2(f) = p^{3n+s}$ and by Theorem 16.3.7, $S_3(f) = \frac{S_2(f)^2}{S_1(f)} = p^{4n+2s}$.
Conversely, suppose that $S_2(f) = p^{3n+s}$ and $S_3(f) = p^{4n+2s}$. By (16.16) with $A = p^{n+s}$ and $i = 1$,

$$\begin{aligned}\textstyle\sum_{w\in\mathbb{F}_{p^n}} (|S_f(w)|^2 - p^{n+s})^2|S_f(w)|^2 &= S_3(f) - 2p^{n+s}S_2(f) + p^{2n+2s}S_1(f)\\ &= p^{4n+2s} - 2p^{n+s}\cdot p^{3n+s} + p^{2n+2s}\cdot p^{2n} = 0.\end{aligned}$$

Therefore, $|S_f(w)| \in \{0, p^{\frac{n+s}{2}}\}$, which implies that f is s-plateaued. □

We deduce from Theorem 16.3.10 that for an s-plateaued f the sequence $S_i(f)$ is also simple geometric sequence.

Corollary 16.3.11 ([53]). *Let f be a function from $\mathbb{F}_{p^n}$ to $\mathbb{F}_p$. If f is s-plateaued, then for all $i \in \mathbb{Z}^+$*

$$S_i(f) = p^{(i+1)n+(i-1)s}. \tag{16.21}$$

Proof. By Theorem 16.3.10, $S_2(f) = p^{3n+s}$ and $S_3(f) = p^{4n+2s}$. Then by Theorem 16.3.7,

$$S_i(f) = \frac{S_{i-1}(f)^2}{S_{i-2}(f)} = p^{(i+1)n+(i-1)s}$$

for all $i \geq 4$, recursively. Thus (16.21) holds for all $i \in \mathbb{Z}^+$. □

The characterizations of bent and s-plateaued functions in characteristic 2 in terms of the second-order directional differences were given in [16]. In the present section, we will generalize these characterizations to the characteristic p.

Theorem 16.3.12 ([53]). *Let f be a function from $\mathbb{F}_{p^n}$ to $\mathbb{F}_p$. Then f is s-plateaued if and only if*

$$\sum_{a,b\in\mathbb{F}_{p^n}} \zeta_p^{D_bD_af(x)} = \theta \quad \forall x \in \mathbb{F}_{p^n} \tag{16.22}$$

with $\theta = p^{n+s}$ where $n+s$ is even with $0 \leq s \leq n$. In particular, f is bent if and only if $\theta = p^n$ for $s = 0$.

Proof. For a function f,

$$\sum_{a,b\in\mathbb{F}_{p^n}} \zeta_p^{D_bD_af(x)} = \sum_{a,b\in\mathbb{F}_{p^n}} \zeta_p^{f(x+a+b)-f(x+a)-f(x+b)+f(x)} = \theta \quad \forall x \in \mathbb{F}_{p^n}$$

if and only if

$$\sum_{a,b\in\mathbb{F}_{p^n}} \zeta_p^{f(x+a+b)-f(x+a)-f(x+b)} = \theta \cdot \zeta_p^{-f(x)} \quad \forall x \in \mathbb{F}_{p^n}. \tag{16.23}$$

Let $a_1 = x + a$ and $b_1 = x + b$ for $a_1, b_1 \in \mathbb{F}_{p^n}$. Thus, (16.23) is equivalent to

$$\sum_{a_1,b_1\in\mathbb{F}_{p^n}} \zeta_p^{f(a_1+b_1-x)-f(a_1)-f(b_1)} = \theta \cdot \zeta_p^{-f(x)} \quad \forall x \in \mathbb{F}_{p^n}. \tag{16.24}$$

In (16.24), let the left-hand side be $G_1(x)$ and the right-hand side be $G_2(x)$ for all $x \in \mathbb{F}_{p^n}$.

We recall the following well-known property of the Fourier transform:

$$G(x) = 0 \quad \forall x \in \mathbb{F}_{p^n} \quad \text{if and only if} \quad \hat{G}(w) = \sum_{x\in\mathbb{F}_{p^n}} G(x)\zeta_p^{-tr_n(wx)} = 0 \quad \forall w \in \mathbb{F}_{p^n}$$

where $\hat{G}$ is the Fourier transform of G which is a function from $\mathbb{F}_{p^n}$ to $\mathbb{C}$.

Then, for all $w \in \mathbb{F}_{p^n}$ we can write

$$\hat{G}_1(w) = \sum_{x\in\mathbb{F}_{p^n}} G_1(x)\zeta_p^{-tr_n(wx)} = \sum_{x\in\mathbb{F}_{p^n}} G_2(x)\zeta_p^{-tr_n(wx)} = \hat{G}_2(w).$$

The Fourier transform $\hat{G}_1$ of G_1 can be computed in terms of S_f for all $w \in \mathbb{F}_{p^n}$ as follows:

$$\begin{aligned}\hat{G}_1(w) &= \sum_{x\in\mathbb{F}_{p^n}} G_1(x)\zeta_p^{-tr_n(wx)} = \sum_{x\in\mathbb{F}_{p^n}}\sum_{a_1,b_1\in\mathbb{F}_{p^n}} \zeta_p^{f(a_1+b_1-x)-f(a_1)-f(b_1)}\zeta_p^{-tr_n(wx)}\\ &= \sum_{a_1\in\mathbb{F}_{p^n}} \zeta_p^{-f(a_1)-tr_n(wa_1)} \sum_{b_1\in\mathbb{F}_{p^n}} \zeta_p^{-f(b_1)-tr_n(wb_1)} \sum_{x\in\mathbb{F}_{p^n}} \zeta_p^{f(a_1+b_1-x)-tr_n(-w(a_1+b_1-x))}\\ &= (-S_f)(w)\cdot(-S_f)(w)\cdot S_f(-w).\end{aligned}$$

Similarly, for all $w \in \mathbb{F}_{p^n}$

$$\hat{G}_2(w) = \sum_{x\in\mathbb{F}_{p^n}} G_2(x)\zeta_p^{-tr_n(wx)} = \sum_{x\in\mathbb{F}_{p^n}} \theta\cdot\zeta_p^{-f(x)-tr_n(wx)} = \theta\cdot(-S_f)(w).$$

Recall that for all $w \in \mathbb{F}_{p^n}$

$$(-S_f)(w) = \sum_{x\in\mathbb{F}_{p^n}} \zeta_p^{-f(x)-tr_n(wx)} = \overline{\sum_{x\in\mathbb{F}_{p^n}} \zeta_p^{f(x)+tr_n(wx)}} = \overline{\sum_{x\in\mathbb{F}_{p^n}} \zeta_p^{f(x)-tr_n(-wx)}} = \overline{S_f(-w)}.$$

Then since $\hat{G}_1(w) = \hat{G}_2(w)$ for all $w \in \mathbb{F}_{p^n}$,

$$S_f(-w)\cdot\overline{S_f(-w)}\cdot\overline{S_f(-w)} = \theta\cdot\overline{S_f(-w)}. \tag{16.25}$$

Therefore, (16.22) holds if and only if $|S_f(w)|^2 \in \{0, \theta\}$ for all $w \in \mathbb{F}_{p^n}$ where $\theta = p^{n+s}$ and $n + s \geq 2$ is even. In particular, for $s = 0$, f is bent if and only if $|S_f(w)|^2 = p^n$ where n is even. □

Theorem 16.3.12 can be rewritten as follows.

Corollary 16.3.13 ([53]). *Let f be a function from $\mathbb{F}_{p^n}$ to $\mathbb{F}_p$. Then f is s-plateaued if and only if*

$$\sum_{a,b,x\in\mathbb{F}_{p^n}} \zeta_p^{D_bD_af(x)} = p^{2n+s}$$

where $n + s$ is even with $0 \leq s \leq n$.

We remind a link between the second-order directional differences and the fourth power moments of the Walsh transforms in characteristic p in the following proposition (see in [48, Proposition 1], [33, Theorem 10] and [26].

Proposition 16.3.14. *Let f be a function from $\mathbb{F}_{p^n}$ to $\mathbb{F}_p$. Then*

$$S_2(f) = \sum_{w\in\mathbb{F}_{p^n}} |S_f(w)|^4 = p^n \sum_{a,b,x\in\mathbb{F}_{p^n}} \zeta_p^{D_bD_af(x)}.$$

We deduce from Proposition 16.3.14 and Corollary 16.3.13 a new characterization of s-plateaued functions in terms of the fourth power moments of their Walsh transforms.

Theorem 16.3.15. *Let f be a function from $\mathbb{F}_{p^n}$ to $\mathbb{F}_p$ and s be an integer with $0 \le s \le n$. Then f is s-plateaued if and only if*

$$S_2(f) = \sum_{w \in \mathbb{F}_{p^n}} |S_f(w)|^4 = p^{3n+s}.$$

In particular, f is bent if and only if $S_2(f) = p^{3n}$.

Proof. By Proposition 16.3.14 and Corollary 16.3.13, f is s-plateaued if and only if

$$S_2(f) = p^n \sum_{a,b,x \in \mathbb{F}_{p^n}} \zeta_p^{D_b D_a f(x)} = p^{3n+s}.$$

□

Notice that Theorem 16.3.10 is also a direct corollary of Theorem 16.3.15.

Example 16.3.16. Let p be an odd prime and $n \ge 2$ be an integer. Let f be an arbitrary $\mathbb{F}_p$-quadratic form from $\mathbb{F}_{p^n}$ to $\mathbb{F}_p$ defined as

$$f(x) = tr_n(a_0 x^2 + a_1 x^{p+1} + a_2 x^{p^2+1} + \cdots + a_{\lfloor \frac{n}{2} \rfloor} x^{p^{\lfloor \frac{n}{2} \rfloor}+1}).$$

The radical of f given by

$$W = \{x \in \mathbb{F}_{p^n} : f(x+y) = f(x) + f(y), \forall y \in \mathbb{F}_{p^n}\}$$

is an $\mathbb{F}_p$-linear subspace of $\mathbb{F}_{p^n}$. Let $\dim_{\mathbb{F}_p} W = s$. It follows from [64, the proof of Theorem 4.1] that for all $w \in \mathbb{F}_{p^n}$

$$|S_f(w)|^2 = 0 \quad \text{or} \quad p^{2s} \sum_{y_1,\ldots,y_{n-s} \in \mathbb{F}_p} \sum_{z_1,\ldots,z_{n-s} \in \mathbb{F}_p} \zeta_p^{H(y_1,\ldots,y_{n-s}) - H(z_1,\ldots,z_{n-s})}$$

where $H(x_1,\ldots,x_{n-s}) = \frac{1}{2}(x_1^2 + \cdots + x_{n-s-1}^2 + dx_{n-s}^2)$ and $d \in \mathbb{F}_p^\star$. For each pair y_i, z_i where $i = 1,\ldots,n-s$, it is easy to see that

$$\sum_{y_i,z_i \in \mathbb{F}_p} \zeta_p^{\frac{1}{2}(y_i^2 - z_i^2)} = \sum_{t_{i1},t_{i2} \in \mathbb{F}_p} \zeta_p^{\frac{1}{2}(t_{i1}t_{i2})} = \sum_{t_{i2} \in \mathbb{F}_p} \Big(\sum_{t_{i1} \in \mathbb{F}_p} \zeta_p^{\frac{1}{2}t_{i1}} \Big) = p.$$

Therefore we conclude that $|S_f(w)|^2 \in \{0, p^{n+s}\}$ for all $w \in \mathbb{F}_{p^n}$. Moreover, [65, Proposition 5.8] gives an algorithm to construct a such quadratic form f with radical

W of dimension s with $0 \leq s \leq n-1$. In fact, this algorithm holds for any finite field $\mathbb{F}_q$, not necessary that q is a prime. Hence for each odd prime p, integers $n \geq 2$ and s with $0 \leq s \leq n-1$, there exists a quadratic p-ary s-plateaued function f from $\mathbb{F}_{p^n}$ to $\mathbb{F}_p$. For example, if $p = 3$ and $n = 5$, then

- $f(x) = Tr_{\mathbb{F}_{3^5}/\mathbb{F}_3}(x^2 + x^4 + 2x^{10})$ is the quadratic 0 plateaued function,
- $f(x) = Tr_{\mathbb{F}_{3^5}/\mathbb{F}_3}(x^2 + x^4 + x^{10})$ is the quadratic 1-plateaued function,
- $f(x) = Tr_{\mathbb{F}_{3^5}/\mathbb{F}_3}(\xi x^2 + x^4 + 2x^{10})$ is the quadratic 2-plateaued function,
- $f(x) = Tr_{\mathbb{F}_{3^5}/\mathbb{F}_3}(\xi^2 x^2 + 2x^4 + \xi^{28} x^{10})$ is the quadratic 3-plateaued function and
- $f(x) = Tr_{\mathbb{F}_{3^5}/\mathbb{F}_3}(x^2 + 2x^4 + 2x^{10})$ is the quadratic 4-plateaued function

where ξ is a primitive element of $\mathbb{F}_{3^5}$ with $\xi^5 + 2\xi + 1 = 0$.

16.3.2 Characterization of p-Ary Bent (0-Plateaued) Functions

In this section, we shall specialize our study to bent functions and suppose that p is a positive prime integer. In the whole section, n is a positive integer. In Theorem 16.3.3, we have excluded the possibility to for the integer k to be equal to 0 because it does concern both plateaued functions and bent functions. In fact, if we aim to characterize only bent functions, we are going to show that it follows from comparing $T_1(f) = \frac{S_2(f)}{S_1(f)} = \frac{S_2(f)}{p^{2n}}$ to $T_0(f) = \frac{S_1(f)}{S_0(f)} = p^n$.

Theorem 16.3.17 ([48]). *Let n be a positive integer. Let f be a function from $\mathbb{F}_{p^n}$ to $\mathbb{F}_p$. Then*

$$S_2(f) = \sum_{w \in \mathbb{F}_{p^n}} |S_f(w)|^4 \geq p^{3n}$$

and f is bent if and only if $S_2(f) = p^{3n}$.

Proof. If we apply (16.16) with $A = p^n$ at $k = 1$, we get that

$$\sum_{w \in \mathbb{F}_{p^n}} \left(|S_f(w)|^2 - p^n\right)^2 = S_2(f) - 2p^n S_1(f) + p^{2n} S_0(f).$$

Now, $S_0(f) = p^n$ and $S_1(f) = p^{2n}$ (Parseval identity, Eq. 16.15). Hence

$$\sum_{w \in \mathbb{F}_{p^n}} \left(|S_f(w)|^2 - p^n\right)^2 = S_2(f) - p^{3n}. \tag{16.26}$$

Since $\left(|S_f(w)|^2 - p^n\right)^2 \geq 0$ for every $w \in \mathbb{F}_{2^n}$, it implies that $S_2(f) \geq p^{3n}$. Now, f is bent if and only if $|S_f(w)|^2 = p^n$ for every $w \in \mathbb{F}_{p^n}$. Therefore, f is bent if and only if the left-hand side of Eq. (16.26) vanishes, that is, if and only if $S_2(f) = p^{3n}$. □

In characteristic 2, identities have been established involving the Walsh transform of a Boolean function and its directional derivatives (see [2, 8]). For instance, for every Boolean function f, $S_2(f)$ and the second-order derivatives of f have been linked . We now show that one can link $S_2(f)$ and the directional difference.

Proposition 16.3.18 ([48]). *Let n be a positive integer. Let f be a function from $\mathbb{F}_{p^n}$ to $\mathbb{F}_p$. Then*

$$\sum_{w\in\mathbb{F}_{p^n}} |S_f(w)|^4 = p^n \sum_{(a,b,x)\in\mathbb{F}_{p^n}^3} \chi_p(D_aD_bf(x)). \tag{16.27}$$

Proof. Since $|z|^4 = z^2\bar{z}^2$ where $\bar{z}$ stands for the conjugate of z and $\overline{\zeta_p} = \zeta_p^{-1}$, we have

$$\begin{aligned}\sum_{w\in\mathbb{F}_{p^n}} &|S_f(w)|^4 \\ &= \sum_{w\in\mathbb{F}_{p^n}} \sum_{(x_1,x_2,x_3,x_4)\in\mathbb{F}_{p^n}^4} \chi_p\big(f(x_1)-f(x_2)+f(x_3)-f(x_4) \\ &\qquad\qquad -tr_n(w(x_1-x_2+x_3-x_4))\big).\end{aligned}$$

Now,

$$\sum_{w\in\mathbb{F}_{p^n}} \chi_p\big(-tr_n(w(x_1-x_2+x_3-x_4))\big) = \begin{cases} p^n \text{ if } x_1-x_2+x_3-x_4=0 \\ 0 \text{ otherwise.}\end{cases}$$

Hence,

$$\sum_{w\in\mathbb{F}_{p^n}} |S_f(w)|^4 = p^n \sum_{(x_1,x_2,x_3)\in\mathbb{F}_{p^n}^3} \chi_p\big(f(x_1)-f(x_2)+f(x_3)-f(x_1-x_2+x_3)\big).$$

Now note that

$$D_{x_2-x_1}D_{x_3-x_2}f(x_1) = f(x_1)+f(x_3)-f(x_2)-f(x_1+x_3-x_2).$$

Then, since $(x_1,x_2,x_3) \mapsto (x_1,x_2-x_1,x_3-x_2)$ is a permutation of $\mathbb{F}_{p^n}^3$, we get

$$\sum_{w\in\mathbb{F}_{p^n}} |S_f(w)|^4 = p^n \sum_{(a,b,x)\in\mathbb{F}_{p^n}^3} \chi_p\big(D_aD_bf(x)\big).$$

□

Remark 16.3.19. In odd characteristic p, when f is a quadratic form over $\mathbb{F}_{2^n}$, that is, $f(x) = \phi(x, x)$ for some symmetric bilinear map ϕ from $\mathbb{F}_{p^n} \times \mathbb{F}_{p^n}$ to $\mathbb{F}_{p^n}$, then, $f(x+y) = f(x) + f(y) + 2\phi(x, y)$. Let us now compute the directional differences of f at $(a, b) \in \mathbb{F}_{p^n}$:

$$D_b f(x) = f(x+b) - f(x) = f(b) + 2\phi(b, x)$$
$$D_a D_b f(x) = 2\phi(b, x+a) - 2\phi(b, x) = 2\phi(b, a).$$

According to Proposition 16.3.18, one has

$$\begin{aligned} S_2(f) &= p^n \sum_{(a,b,x)\in\mathbb{F}_{p^n}^3} \chi_p(2\phi(b,a)) \\ &= p^{2n} \sum_{b\in\mathbb{F}_{p^n}} \sum_{a\in\mathbb{F}_{p^n}} \chi_p(2\phi(b,a)). \end{aligned}$$

Now, classical results about character sums over finite abelian groups say that

$$\sum_{a\in\mathbb{F}_{p^n}} \chi_p(2\phi(b,a)) = \begin{cases} p^n & \text{if } \phi(b,\bullet) = 0 \\ 0 & \text{otherwise.} \end{cases}$$

Hence,

$$S_2(f) = p^{3n} \#\mathfrak{rad}(\phi)$$

where $\mathfrak{rad}(\phi)$ stands for the radical of ϕ : $\mathfrak{rad}(\phi) = \{b \in \mathbb{F}_{p^n} \mid \phi(b, \bullet) = 0\}$. One can then conclude thanks to Theorem 16.3.17 that f is bent if and only if $\mathfrak{rad}(\phi) = \{0\}$.

Suppose that p is odd and consider now functions of the form

$$f(x) = \mathrm{Tr}_1^n \left(\sum_{\substack{i,j,k=0 \\ i\neq j, j\neq k, k\neq i}}^{n-1} a_{ijk} x^{p^i+p^j+p^k} + \sum_{\substack{i,j=0 \\ i\neq j}}^{n-1} b_{ij} x^{p^i+p^j} \right). \tag{16.28}$$

We are going to characterize bent functions of that form thanks to Theorem 16.3.17 and Proposition 16.3.18. But before, let us note that we can rewrite the expression of f as follows

$$f(x) = \mathrm{Tr}_1^n \left(\sum_{\substack{i,j,k=0 \\ i\neq j, j\neq k, k\neq i}}^{n-1} a_{ijk} x^{p^i+p^j+p^k} \right) + \mathrm{Tr}_1^n \left(\sum_{\substack{i,j=0 \\ i\neq j}}^{n-1} b_{ij} x^{p^i+p^j} \right)$$

$$= \mathrm{Tr}_1^n\left(\sum_{\substack{i,j,k=0\\ i\neq j, j\neq k, k\neq i}}^{n-1} a_{ijk}^{p^{-i}} x^{1+p^{j-i}+p^{k-i}}\right) + \mathrm{Tr}_1^n\left(\sum_{\substack{i,j=0\\ i\neq j}}^{n-1} b_{ij}x^{p^i+p^j}\right)$$

$$= \mathrm{Tr}_1^n\left(x\sum_{\substack{i,j,k=0\\ i\neq j, j\neq k, k\neq i}}^{n-1} a_{ijk}^{p^{-i}} x^{p^{j-i}+p^{k-i}}\right) + \mathrm{Tr}_1^n\left(\sum_{\substack{i,j=0\\ i\neq j}}^{n-1} b_{ij}x^{p^i+p^j}\right).$$

In the second equality, we have used the fact that Tr_1^n is invariant under the Frobenius map $x \mapsto x^p$. Set

$$\psi(x,y) = \frac{1}{2}\sum_{\substack{i,j,k=0\\ i\neq j, j\neq k, k\neq i}}^{n-1} a_{ijk}^{p^{-i}}(x^{p^{j-i}}y^{p^{k-i}} + x^{p^{k-i}}y^{p^{j-i}})$$

$$\phi(x,y) = \frac{1}{2}\mathrm{Tr}_1^n\left(\sum_{\substack{i,j=0\\ i\neq j}}^{n-1} b_{ij}(x^{p^i}y^{p^j} + x^{p^j}y^{p^i})\right),$$

Therefore, a function f of the form (16.28) can be written

$$f(x) = \mathrm{Tr}_1^n(x\psi(x,x)) + \phi(x,x) \tag{16.29}$$

where $\psi : \mathbb{F}_{p^n} \to \mathbb{F}_{p^n}$ is a symmetric bilinear map and $\phi : \mathbb{F}_{p^n} \to \mathbb{F}_{p^n}$ is a symmetric bilinear form. We can now state our characterization.

Theorem 16.3.20 ([48]). *Suppose that p is odd. Let ϕ be a symmetric bilinear form over $\mathbb{F}_{p^n} \times \mathbb{F}_{p^n}$ and ψ be a symmetric bilinear map from $\mathbb{F}_{p^n} \times \mathbb{F}_{p^n}$ to $\mathbb{F}_{2^n}$. Define $f : \mathbb{F}_{p^n} \to \mathbb{F}_p$ by $f(x) = \mathrm{Tr}_1^n(x\psi(x,x)) + \phi(x,x))$ for $x \in \mathbb{F}_{p^n}$. For $(a,b) \in \mathbb{F}_{p^n}$, set $\ell_{a,b}(x) = \mathrm{Tr}_1^n(\psi(a,b)x + a\psi(b,x) + b\psi(a,x))$. For every $a \in \mathbb{F}_{p^n}$, define the vector space $\mathfrak{K}_a = \{b \in \mathbb{F}_{p^n} \mid \ell_{a,b} = 0\}$. Then f is bent if and only if $\{a \in \mathbb{F}_{p^n}, \phi(a,\bullet)|_{\mathfrak{K}_a} = 0\} = \{0\}$.*

Proof. According to Theorem 16.3.17 and Proposition 16.3.18, f is bent if and only if

$$\sum_{(a,b,x)\in\mathbb{F}_{p^n}^3} \chi_p(D_bD_af(x)) = p^{2n}. \tag{16.30}$$

Now, for $(a,b) \in \mathbb{F}_{p^n}^2$,

$$D_af(x) = \mathrm{Tr}_1^n((x+a)\psi(a+x,a+x) - x\psi(x,x))$$
$$+\phi(x+a,x+a) - \phi(x,x)$$

$$= \mathrm{Tr}_1^n\big(a\psi(x,x) + 2x\psi(a,x) + 2a\psi(a,x) + x\psi(a,a) + a\psi(a,a)\big)$$
$$+2\phi(a,x) + \phi(a,a).$$
$$D_bD_af(x) = \mathrm{Tr}_1^n\big(2a\psi(b,x) + a\psi(b,b) + 2b\psi(a,x) + 2x\psi(a,b) + 2b\psi(a,b)$$
$$+2a\psi(a,b) + b\psi(a,a)\big) + 2\phi(a,b))$$
$$= 2\ell_{a,b}(x) + \mathrm{Tr}_1^n(a\psi(b,b) + b\psi(a,a) + 2(a+b)\psi(a,b)) + 2\phi(a,b).$$

Note that, $\ell_{a,b}$ is a linear map from $\mathbb{F}_{p^n}$ to $\mathbb{F}_{2^n}$. Furthermore, for any $a \in \mathbb{F}_{p^n}$ and $b \in \mathfrak{K}_a$, one has

$$\ell_{a,b}(a) = \mathrm{Tr}_1^n(\psi(a,b)a + a\psi(b,a) + b\psi(a,a)) = 0,$$
$$\ell_{a,b}(b) = \mathrm{Tr}_1^n(\psi(a,b)b + a\psi(b,b) + b\psi(a,b)) = 0$$

which implies, summing those two equations, that

$$\mathrm{Tr}_1^n(a\psi(b,b) + b\psi(a,a) + 2(a+b)\psi(a,b)) = 0.$$

Hence,

$$\sum_{(a,b,x)\in\mathbb{F}_{p^n}^3} \chi_p(D_bD_af(x)) = \sum_{(a,b)\in\mathbb{F}_{p^n}^3} \chi_p(2\phi(a,b)) \sum_{x\in\mathbb{F}_{2^n}} \chi_p(2\ell_{a,b}(x))$$
$$= p^n \sum_{a\in\mathbb{F}_{p^n}} \sum_{b\in\mathfrak{K}_a} \chi_p(2\phi(a,b)).$$

Now, for every $a \in \mathbb{F}_{p^n}$, the map $b \in \mathfrak{K}_a \mapsto \phi(a,b)$ is linear over $\mathfrak{K}_a$. Therefore

$$\sum_{b\in\mathfrak{K}_a} \chi_p(2\phi(a,b)) = \begin{cases} \#\mathfrak{K}_a & \text{if } \phi(a,\bullet)|_{\mathfrak{K}_a} = 0 \\ 0 & \text{otherwise} \end{cases}$$

Hence, according to (16.30), f is bent if and only if

$$\sum_{(a,b,x)\in\mathbb{F}_{p^n}^3} \chi_p(D_aD_bf(x)) = p^n \sum_{a\in\mathbb{F}_{p^n},\, \phi(a,\bullet)|_{\mathfrak{K}_a}=0} \#\mathfrak{K}_a = p^{2n},$$

that is, if and only if,

$$\sum_{a\in\mathbb{F}_{p^n},\, \phi(a,\bullet)|_{\mathfrak{K}_a}=0} \#\mathfrak{K}_a = p^n.$$

Now, if $a = 0$, then $\mathfrak{K}_0 = \mathbb{F}_{p^n}$ because $\ell_{0,b} = 0$ for every $b \in \mathbb{F}_{p^n}$. Therefore, f is bent if and only if

$$\sum_{a\in\mathbb{F}_{p^n}^\star,\, \phi(a,\bullet)|_{\mathfrak{K}_a}=0} \#\mathfrak{K}_a = 0$$

which is equivalent to $\#\mathfrak{K}_a = 0$ for every $a \in \mathbb{F}_{2^n}^\star$ such that $\phi(a, \bullet)|_{\mathfrak{K}_a} = 0$. □

16.4 Characterization of Vectorial Bent Functions in Arbitrary Characteristic

Recall that a map F from $\mathbb{F}_{p^n}$ to $\mathbb{F}_{p^n}$ is said to be planar if and only if the function from $\mathbb{F}_{p^n}$ to $\mathbb{F}_{p^n}$ induced by the polynomial $F(X + a) - F(x) - F(a)$ is bijective for every $a \in \mathbb{F}_{p^n}^\star$. Recall that given $f : \mathbb{F}_{p^n} \to \mathbb{F}_p$, the directional difference of f at $a \in \mathbb{F}_{p^n}$ is the map D_af from $\mathbb{F}_{p^n}$ to $\mathbb{F}_p$ defined by

$$\forall x \in \mathbb{F}_{p^n}, \quad D_af(x) = f(x + a) - f(x).$$

We now turn our attention towards maps from $\mathbb{F}_{p^n}$ to $\mathbb{F}_{p^m}$. Let us extend the notion of bentness to those maps as follows.

Let F be a Boolean map from $\mathbb{F}_{p^n}$ to $\mathbb{F}_{p^m}$. For every $\lambda \in \mathbb{F}_{p^n}^\star$, define $f_\lambda : \mathbb{F}_{p^n} \to \mathbb{F}_p$ as : $f_\lambda(x) = \mathrm{Tr}_1^m(\lambda F(x))$ for every $x \in \mathbb{F}_{p^n}$. Then recall that F is said to be bent if and only if f_λ is bent for every $\lambda \in \mathbb{F}_{p^n}^\star$.

Theorem 16.3.17 implies

Theorem 16.4.1 ([48]). *Let F be a map from $\mathbb{F}_{p^n}$ to $\mathbb{F}_{p^m}$. Then, F is bent if and only if*

$$\sum_{\lambda\in\mathbb{F}_{p^m}^\star} S_2(f_\lambda) = p^{3n}(p^m - 1). \tag{16.31}$$

Proof. According to Theorem 16.3.17, for every $\lambda \in \mathbb{F}_{p^m}^\star$, f_λ is bent if and only if $S_2(f_\lambda) = p^{3n}$ which gives (16.31). Conversely, suppose that (16.31) holds. Theorem 16.3.17 states that $S_2(f_\lambda) \geq p^{3n}$ for every $\lambda \in \mathbb{F}_{p^m}^\star$. Thus, one has necessarily, for every $\lambda \in \mathbb{F}_{p^n}^\star$, $S_2(f_\lambda) = p^{3n}$ implying that f_λ is bent for every $\lambda \in \mathbb{F}_{p^n}$, proving that F is bent. □

We now show that one can compute the left-hand side of (16.31) by counting the zeros of the second-order directional differences.

Proposition 16.4.2 ([48]). *Let F be a Boolean map from $\mathbb{F}_{p^n}$ to $\mathbb{F}_{p^m}$. Then*

$$\sum_{\lambda\in\mathbb{F}_{p^m}^\star} S_2(f_\lambda) = p^{n+m}\mathfrak{N}(F) - p^{4n}$$

where $\mathfrak{N}(F)$ is the number of elements of $\{(a,b,x) \in \mathbb{F}_{p^n}^3 \mid D_aD_bF(x) = 0\}$.

Proof. According to Proposition 16.3.18, we have

$$\sum_{\lambda\in\mathbb{F}_{p^m}^\star} S_2(f_\lambda) = p^n \sum_{\lambda\in\mathbb{F}_{p^m}^\star} \sum_{a,b,x\in\mathbb{F}_{p^n}} \chi_p\big(D_aD_bf_\lambda(x)\big).$$

Next, $D_aD_bf_\lambda = \mathrm{Tr}_1^m(\lambda D_aD_bF)$. Therefore

$$\sum_{\lambda\in\mathbb{F}_{p^m}^\star} S_2(f_\lambda) = p^n \sum_{a,b,x\in\mathbb{F}_{p^n}} \sum_{\lambda\in\mathbb{F}_{p^m}^\star} \chi_p\big(\mathrm{Tr}_1^m(\lambda D_aD_bF(x))\big).$$

That is

$$\sum_{\lambda\in\mathbb{F}_{p^m}^\star} S_2(f_\lambda) = p^n \sum_{a,b,x\in\mathbb{F}_{2^n}} \Big(\sum_{\lambda\in\mathbb{F}_{p^m}} \chi_p\big(\mathrm{Tr}_1^m(\lambda D_aD_bF(x))\big)\Big) - p^{4n}.$$

We finally get the result from

$$\sum_{\lambda\in\mathbb{F}_{p^m}} \chi_p\big(\mathrm{Tr}_1^m(\lambda D_aD_bF(x))\big) = \begin{cases} 0 & \text{if } D_aD_bF(x) \neq 0 \\ p^m & \text{if } D_aD_bF(x) = 0 \end{cases}$$

□

We then deduce from Theorem 16.3.17 a characterization of bentness in terms of zeros of the second-order directional differences.

Theorem 16.4.3 ([48]). *Let F be a map from $\mathbb{F}_{p^n}$ to $\mathbb{F}_{p^m}$. Then F is bent if and only if $\mathfrak{N}(F) = p^{3n-m} + p^{2n} - p^{2n-m}$.*

Proof. F is bent if and only if all the functions f_λ, $\lambda \in \mathbb{F}_{p^n}^\star$, are bent. Therefore, according to Proposition 16.3.17, if F is bent then

$$\sum_{\lambda\in\mathbb{F}_{p^m}^\star} S_2(f_\lambda) = (p^m - 1)p^{3n}.$$

Now, according to Proposition 16.4.2, one has

$$\sum_{\lambda\in\mathbb{F}_{p^m}^\star} S_2(f_\lambda) = p^{n+m}\mathfrak{N}(F) - p^{4n}.$$

We deduce from the two above equalities that

$$\begin{aligned}\mathfrak{N}(F) &= p^{-n-m}(p^{4n} + (p^m - 1)p^{3n}) \\ &= p^{3n-m} + p^{2n} - p^{2n-m}.\end{aligned}$$

Conversely, suppose that $\mathfrak{N}(F) = p^{3n-m} + p^{2n} - p^{2n-m}$. Then

$$\sum_{\lambda \in \mathbb{F}_{p^m}^\star} S_2(f_\lambda) = p^{n+m}\mathfrak{N}(F) - p^{4n} = p^{4n} + p^{3n+m} - p^{3n} - p^{4n} = p^{3n}(p^m - 1).$$

We then conclude by Theorem 16.4.1 that F is bent. □

Note that when $a = 0$ or $b = 0$, D_aD_bF is trivially equal to 0. We state below a slightly different version of Theorem 16.4.3 to exclude those trivial cases to characterize the bentness of F.

Corollary 16.4.4 ([48]). *Let F be a map from $\mathbb{F}_{p^n}$ to $\mathbb{F}_{p^m}$. Then F is bent if and only if $\mathfrak{N}^\star(F) = p^n(p^n - 1)(p^{n-m} - 1)$ where $\mathfrak{N}^\star(F)$ is the number of elements of $\{(a, b, x) \in \mathbb{F}_{p^n}^\star \times \mathbb{F}_{2^n}^\star \times \mathbb{F}_{p^n} \mid D_aD_bF(x) = 0\}$.*

Proof. It follows from Proposition 16.4.2 by noting that $\{(a, b, x) \in \mathbb{F}_{p^n}^3 \mid D_aD_bF(x) = 0\}$ contains the set $\{(a, 0, x), a, x \in \mathbb{F}_{p^n},\} \cup \{(0, a, x), a, x \in \mathbb{F}_{p^n}\}$ whose cardinality equals $p^n(1 + 2(p^n - 1)) = 2p^{2n} - p^n$. Hence, the cardinality of $\mathfrak{N}^\star(F)$ equals $p^{3n-m} + p^{2n} - p^{2n-m} - (2p^{2n} - p^n) = p^{3n-m} - p^{2n-m} + p^n - p^{2n} = p^{2n-m}(p^n - 1) + p^n(1 - p^n) = p^n(p^n - 1)(p^{n-m} - 1)$. □

In the particular case of planar functions, Theorem 16.4.4 rewrites as

Corollary 16.4.5 ([48]). *Let $F : \mathbb{F}_{p^n} \to \mathbb{F}_{p^n}$. Then, F is planar if and only if, D_aD_bF does not vanish on $\mathbb{F}_{p^n}$ for every $(a, b) \in \mathbb{F}_{2^n}^\star \times \mathbb{F}_{p^n}^\star$.*

Proof. F is planar if and only if F is bent [31, Lemma 1.1]. Hence, according to Corollary 16.4.4, F is planar if and only if $\mathfrak{N}^\star(F) = 0$ proving the result. □

Now we give a proof of characterization of vectorial bent functions. Let F be a vectorial function from $\mathbb{F}_{p^n}$ to $\mathbb{F}_{p^m}$. In [55, Theorem 2.3], vectorial bent functions arc characterized by using the directional differences: A function F is bent if and only if D_aF is balanced for all $a \in \mathbb{F}_{p^n}^\star$. Recently, the author has characterized vectorial bent functions in [48, Theorem 6] by using the number of zeros of second-order directional differences: A function F is bent if and only if $\mathfrak{N}(F) = p^{3n-m} + p^{2n} - p^{2n-m}$ where

$$\mathfrak{N}(F) = |\{(a, b, x) \in \mathbb{F}_{p^n}^3 | D_bD_aF(x) = 0\}|.$$

It would be interesting to give a direct proof: D_aF is balanced for all $a \in \mathbb{F}_{p^n}^\star$ if and only if $\mathfrak{N}(F) = p^{3n-m} + p^{2n} - p^{2n-m}$ without using the bentness of F. Before proving it, we start with a well-known result.

Lemma 16.4.6 ([53]). *Let $x_1, x_2, \ldots, x_m$ be positive real numbers such that $x_1 + x_2 + \cdots + x_m = n$. We then have $x_1^2 + x_2^2 + \cdots + x_m^2 \geq \frac{n^2}{m}$ and the equality holds if and only if $x_1 = x_2 = \cdots = x_m$.*

The following lemma is similar to Proposition 1, items (*1*) and (2) in [12], but it is valid in characteristic p.

Lemma 16.4.7. *Let G be a vectorial function from $\mathbb{F}_{p^n}$ to $\mathbb{F}_{p^m}$. Then*

$$|\{(x_1,x_2) \subset \mathbb{F}_{p^n}^2 : G(x_1) = G(x_2)\}| \geq p^{2n-m}, \tag{16.32}$$

and the equality holds if and only if G is balanced.

Proof. Let $A_j = \{x \in \mathbb{F}_{p^n} : G(x) = y_j \in \mathbb{F}_{p^m}\}$ and $z_j = |A_j|$ for $j \in \{1,\ldots,p^m\}$. Then we have

$$|\{(x_1,x_2) \in \mathbb{F}_{p^n}^2 : G(x_1) = G(x_2)\}| = \left|\bigcup_{j=1}^{p^m}\{(x_1,x_2) \in \mathbb{F}_{p^n}^2 : x_1,x_2 \in A_j\}\right| = \sum_{j=1}^{p^m}|A_j|^2 = \sum_{j=1}^{p^m} z_j^2.$$

By Lemma 16.4.6, for $\sum_{j=1}^{p^m} z_j = p^n$ and $z_j \geq 0$, we have $\sum_{j=1}^{p^m} z_j^2 \geq p^{2n-m}$. Thus, (16.32) holds. Notice that G is balanced if and only if $z_1 = z_2 = \cdots = z_{p^m}$. The final assertion also follows from Lemma 16.4.6. □

Proposition 16.4.8 ([53]). *Let F be a vectorial function from $\mathbb{F}_{p^n}$ to $\mathbb{F}_{p^m}$. Then*

$$D_aF \text{ is balanced for all } a \in \mathbb{F}_{p^n}^\star \iff \mathfrak{N}(F) = p^{3n-m} + p^{2n} - p^{2n-m} \tag{16.33}$$

where $\mathfrak{N}(F) = |\{(a,b,x) \in \mathbb{F}_{p^n}^3 | D_bD_aF(x) = 0\}|$.

Proof. The second-order directional difference of F at $a, b \in \mathbb{F}_{p^n}$ is

$$D_bD_aF(x) = F(x+a+b) + F(x) - F(x+b) - F(x+a).$$

Notice that $D_bD_aF(x) = 0$ if and only if

$$D_aF(x) = D_aF(x+b). \tag{16.34}$$

First, for $n = m$, let us prove that D_aF is balanced for all $a \in \mathbb{F}_{p^n}^\star$ if and only if $\mathfrak{N}(F) = 2p^{2n} - p^n$. For $a = 0$, it is easy to see that (16.34) holds for all $b, x \in \mathbb{F}_{p^n}$ since D_aF is zero map. Then $|\{(0,b,x) \in \mathbb{F}_{p^n}^3 | D_bD_aF(x) = 0\}| = p^{2n}$. For $a \neq 0$, by Lemma 16.4.7, the number of pairs $(b,x) \in \mathbb{F}_{p^n}^2$ satisfying (16.34) is equal to p^n if and only if D_aF is balanced. Then $|\{(a,b,x) \in \mathbb{F}_{p^n}^3 | a \neq 0, D_bD_aF(x) = 0\}| = p^{2n} - p^n$. Therefore, $\mathfrak{N}(F) = 2p^{2n} - p^n$ if and only if D_aF is balanced for all $a \in \mathbb{F}_{p^n}^\star$.

Now, let $n \neq m$. For $a = 0$, we have $|\{(0,b,x) \in \mathbb{F}_{p^n}^3 | D_bD_aF(x) = 0\}| = p^{2n}$. For $a \neq 0$, by Lemma 16.4.7, the number of pairs $(b,x) \in \mathbb{F}_{p^n}^2$ satisfying (16.34) is equal to p^{2n-m} if and only if D_aF is balanced. Then $|\{(a,b,x) \in \mathbb{F}_{p^n}^3 | a \neq 0, D_bD_aF(x) = 0\}| = (p^n - 1) \cdot p^{2n-m}$. Thus (16.33) holds. □

In [48, Corollary 1], F is bent if and only if $\mathfrak{N}^\star(F) = (p^n - 1)(p^{2n-m} - p^n)$ where

$$\mathfrak{N}^\star(F) = |\{(a,b,x) \in \mathbb{F}_{p^n}^\star \times \mathbb{F}_{p^n}^\star \times \mathbb{F}_{p^n} | D_bD_aF(x) = 0\}|.$$

By Lemma 16.4.7, it is obvious that D_aF is balanced for all $a \in \mathbb{F}_{p^n}^\star$ if and only if $\mathfrak{N}^\star(F) = (p^n - 1)(p^{2n-m} - p^n)$.

16.5 Characterization of Vectorial s-Plateaued Functions

In the present section, we are interested in a special class of vectorial plateaued functions, which are called *vectorial s-plateaued* functions where $s \in \mathbb{N}$. We will provide their characterizations in terms of the moments of their Walsh transforms and the number of zeros of their second-order directional differences. The vectorial plateaued functions in characteristic 2 are defined in [6, Definition 9]. This can be extended to characteristic p as follows.

Definition 16.5.1. Let F be a vectorial function from $\mathbb{F}_{p^n}$ to $\mathbb{F}_{p^m}$. For every $\lambda \in \mathbb{F}_{p^m}^\star$, the component function f_λ from $\mathbb{F}_{p^n}$ to $\mathbb{F}_p$ is defined as $f_\lambda(x) = tr_m(\lambda F(x))$ for all $x \in \mathbb{F}_{p^n}$. Then F is called *vectorial plateaued* if f_λ is plateaued for all $\lambda \in \mathbb{F}_{p^m}^\star$.

We can introduce the notation of vectorial s-plateaued functions in characteristic p.

Definition 16.5.2. Let F be a vectorial function from $\mathbb{F}_{p^n}$ to $\mathbb{F}_{p^m}$ and s be an integer with $0 \le s \le n$. For every $\lambda \in \mathbb{F}_{p^m}^\star$, the component function f_λ from $\mathbb{F}_{p^n}$ to $\mathbb{F}_p$ is defined as $f_\lambda(x) = tr_m(\lambda F(x))$ for all $x \in \mathbb{F}_{p^n}$. Then F is called *vectorial s-plateaued* if f_λ is s-plateaued with the same amplitude s for all $\lambda \in \mathbb{F}_{p^m}^\star$.

Notice that F is said to be *vectorial s-plateaued* if and only if f_λ is s-plateaued with the same amplitude s for all $\lambda \in \mathbb{F}_{p^m}^\star$.

We extract from Theorem 16.3.10 a characterization of vectorial s-plateaued functions in terms of the sum of $S_2(f_\lambda)$ and $S_3(f_\lambda)$ for all $\lambda \in \mathbb{F}_{p^m}^\star$.

Theorem 16.5.3 ([53]). *Let F be a vectorial function from $\mathbb{F}_{p^n}$ to $\mathbb{F}_{p^m}$. Then F is a vectorial s-plateaued function if and only if*

$$\sum_{\lambda\in\mathbb{F}_{p^m}^\star} S_2(f_\lambda) = p^{3n+s}(p^m-1) \text{ and } \sum_{\lambda\in\mathbb{F}_{p^m}^\star} S_3(f_\lambda) = p^{4n+2s}(p^m-1). \quad (16.35)$$

Proof. Suppose that F is vectorial s-plateaued. By Theorem 16.3.10 for all $\lambda \in \mathbb{F}_{p^m}^\star$, f_λ is s-plateaued if and only if $S_2(f_\lambda) = p^{3n+s}$ and $S_3(f_\lambda) = p^{4n+2s}$. Thus (16.35) holds.

Conversely, suppose that (16.35) holds. By (16.16) with $A = p^{n+s}$ and $i = 1$, for all $\lambda \in \mathbb{F}_{p^m}^\star$

$$D_\lambda = \sum_{w\in\mathbb{F}_{p^n}} (|\hat{\chi}_{f_\lambda}(w)|^2 - p^{n+s})^2|\hat{\chi}_{f_\lambda}(w)|^2 = S_3(f_\lambda) - 2p^{n+s}S_2(f_\lambda) + p^{2(n+s)}S_1(f_\lambda).$$

Then by (16.15) and (16.35),

$$\begin{aligned}\textstyle\sum_{\lambda\in\mathbb{F}_{p^m}^\star} D_\lambda &= p^{4n+2s}(p^m-1) - 2p^{n+s}p^{3n+s}(p^m-1) + p^{2n+2s}p^{2n}(p^m-1)\\ &= (p^m-1)\cdot(p^{4n+2s} - 2p^{4n+2s} + p^{4n+2s}) = 0.\end{aligned}$$

Since $D_\lambda \geq 0$ and $\sum_{\lambda\in\mathbb{F}_{p^m}^\star} D_\lambda = 0$, we have $D_\lambda = 0$ for every $\lambda \in \mathbb{F}_{p^m}^\star$. Then, $|\hat{\chi}_{f_\lambda}(w)| \in \{0, p^{\frac{n+s}{2}}\}$ for every $\lambda \in \mathbb{F}_{p^m}^\star$. Therefore F is a vectorial s-plateaued function. □

For a vectorial function F, the relation between the sum of $S_2(f_\lambda)$ for all $\lambda \in \mathbb{F}_{p^m}^\star$ and $\mathfrak{N}(F)$ was given by Mesnager in [48] as follows.

Proposition 16.5.4 ([48]). *Let F be a vectorial function from $\mathbb{F}_{p^n}$ to $\mathbb{F}_{p^m}$. Then*

$$\sum_{\lambda\in\mathbb{F}_{p^m}^\star} S_2(f_\lambda) = p^{n+m}\mathfrak{N}(F) - p^{4n}$$

where $\mathfrak{N}(F) = |\{(a,b,x) \in \mathbb{F}_{p^n}^3 | D_bD_aF(x) = 0\}|$.

We conclude from Proposition 16.5.4 and Theorem 16.5.3 a characterization of vectorial s-plateaued functions.

Theorem 16.5.5 ([53]). *Let F be a vectorial function from $\mathbb{F}_{p^n}$ to $\mathbb{F}_{p^m}$ and $S_3(f_\lambda) = p^{4n+2s}$ for all $\lambda \in \mathbb{F}_{p^m}^\star$. Then F is vectorial s-plateaued if and only if*

$$\mathfrak{N}(F) = p^{3n-m} + p^{2n+s} - p^{2n+s-m}$$

where $\mathfrak{N}(F) = |\{(a,b,x) \in \mathbb{F}_{p^n}^3 | D_bD_aF(x) = 0\}|$.

Proof. By Proposition 16.5.4 and Theorem 16.5.3, $p^{3n+s}(p^m-1) = p^{n+m}\mathfrak{N}(F) - p^{4n}$. Therefore we obtain

$$\mathfrak{N}(F) = p^{-n-m}(p^{4n} + p^{3n+s}(p^m - 1) = p^{3n-m} + p^{2n+s} - p^{2n+s-m}.$$

Conversely, by Proposition 16.5.4,

$$\sum_{\lambda\in\mathbb{F}_{p^m}^\star} S_2(f_\lambda) = p^{n+m}(p^{3n-m} + p^{2n+s} - p^{2n+s-m}) - p^{4n} = p^{3n+s}(p^m - 1).$$

By assumption, $\sum_{\lambda\in\mathbb{F}_{p^m}^\star} S_3(f_\lambda) = p^{4n+2s}(p^m - 1)$. Therefore, by Theorem 16.5.3, F is a vectorial s-plateaued function. □

Example 16.5.6. Let p be an odd prime, $m \geq 2$ and $r \geq 2$ be integers and $q = p^m$. Let f be an arbitrary $\mathbb{F}_q$-quadratic form from $\mathbb{F}_{q^r}$ to $\mathbb{F}_q$ given by

$$f(x) = Tr_{\mathbb{F}_{q^r}/\mathbb{F}_q}(a_0x^2 + a_1x^{q+1} + a_2x^{q^2+1} + \cdots + a_{\lfloor\frac{r}{2}\rfloor}x^{q^{\lfloor\frac{r}{2}\rfloor}+1}).$$

As in Example 16.3.16, by [64, 65], we have an algorithm to construct f with radical

$$W = \{x \in \mathbb{F}_{q^r} : f(x+y) = f(x) + f(y), \forall y \in \mathbb{F}_{q^r}\} \quad (16.36)$$

of prescribed dimension s over $\mathbb{F}_q$ for each given integer s with $0 \le s \le r-1$. For $\lambda \in \mathbb{F}_{p^m}^*$, the component function g_λ from $\mathbb{F}_{p^n}$ to $\mathbb{F}_p$ given by $g_\lambda := tr_m(\lambda f(x))$ is an $\mathbb{F}_p$-quadratic form with radical

$$W_\lambda = \{x \in \mathbb{F}_{p^n} : g_\lambda(x+y) = g_\lambda(x) + g_\lambda(y), \forall y \in \mathbb{F}_{p^n}\} \tag{16.37}$$

where $n = mr$. For a $\mathbb{F}_q$-quadratic function f on $\mathbb{F}_{q^r}$ and $\lambda \in \mathbb{F}_q^*$, the radical W in (16.36) is the set of the roots of the equation

$$a_0x + a_1x^q + (a_1x)^{q^{-1}} + a_2x^{q^2} + (a_2x)^{q^{-2}} + \cdots + a_{\lfloor \frac{r}{2} \rfloor}x^{q^{\lfloor \frac{r}{2} \rfloor}} + \left(a_{\lfloor \frac{r}{2} \rfloor}x\right)^{q^{-\lfloor \frac{r}{2} \rfloor}} \tag{16.38}$$

in $\mathbb{F}_{q^r}$ and W_λ in (16.37) is the set of the roots of the equation

$$\lambda a_0x + \lambda a_1x^q + (\lambda a_1x)^{q^{-1}} + \lambda a_2x^{q^2} + (\lambda a_2x)^{q^{-2}} + \cdots + \lambda a_{\lfloor \frac{r}{2} \rfloor}x^{q^{\lfloor \frac{r}{2} \rfloor}} + \left(\lambda a_{\lfloor \frac{r}{2} \rfloor}x\right)^{q^{-\lfloor \frac{r}{2} \rfloor}} \tag{16.39}$$

(see for example [65, Lemma 2.1]). As $\lambda \in \mathbb{F}_q^*$, it is easy to observe from (16.38) and (16.39) that $W = W_\lambda$. Therefore, we obtain vectorial s-plateaued function F from $\mathbb{F}_{p^n}$ to $\mathbb{F}_{p^m}$ (notice that $F(x) = f(x)$ for all $x \in \mathbb{F}_{p^n}$). This shows existence of an algorithm to construct vectorial s-plateaued functions F for any integer s with $0 \le s \le r-1$. For example, if $q = 9$ and $n = 3$, then $f(x) = Tr_9^{9^3}(x^2 + x^{10})$ is the 0-plateaued function and $f(x) = Tr_9^{9^3}(x^2 + 2x^{10})$ is the 1-plateaued function.

Example 16.5.7. Let p be an odd prime. Let f_1 and f_2 be the quadratic p-ary s_1-plateaued and s_2-plateaued functions from $\mathbb{F}_{p^4}$ to $\mathbb{F}_p$ with $s_1 \ne s_2$, respectively. For any $\theta \in \mathbb{F}_{p^2} \setminus \mathbb{F}_p$, a function F given as

$$F(x) = f_1(x) + \theta f_2(x)$$

is a vectorial plateaued function from $\mathbb{F}_{p^4}$ to $\mathbb{F}_{p^2}$ but it is not a vectorial s-plateaued function for any integer s with $0 \le s \le r-1$. This shows that the vectorial plateaued functions are strictly more general than the vectorial s-plateaued function for any s.

References

1. S. Boztas and P. V. Kumar. Binary sequences with Gold-like correlation but larger linear span. In *IEEE Transactions on Information Theory 40(2)*, pages 532–537, 1994.
2. A. Canteaut, C. Carlet, P. Charpin, and C. Fontaine. Propagation Characteristics and Correlation-Immunity of Highly Nonlinear Boolean Functions. In *EUROCRYPT 2000, Lecture Notes in Comp. Sci. pages 507–522, Springer-Verlag*, 2000.
3. A. Canteaut, C. Carlet, P. Charpin, and C. Fontaine. On cryptographic properties of the cosets of R(1,m). In *IEEE Transactions on Information Theory, vol. 47*, pages 1494–1513, 2001.
4. A. Canteaut and P. Charpin.Decomposing bent functions.In *IEEE Transactions on Information Theory, Vol 49*, pages 2004–2019, 2003.

5. A. Canteaut and M. Naya-Plasencia. Structural weakness of mappings with a low differential uniformity. In *In Conference on Finite Fields and Applications*, 2009.
6. C. Carlet. Vectorial Boolean Functions for Cryptography. In *Chapter of the monography Boolean Methods and Models, Y. Crama and P. Hammer eds, Cambridge University Press. Preliminary version available at* http://www-rocq.inria.fr/codes/Claude.Carlet/pubs.html.
7. C. Carlet. On bent and highly nonlinear balanced/resilient functions and their algebraic immunities. In *AAECC 16, Las Vegas, February 2006, volume 3857 of Lecture Notes in Computer Science, Springer*, pages 1–28, 2006.
8. C. Carlet. Boolean Functions for Cryptography and Error Correcting Codes. In *Chapter of the monography "Boolean Models and Methods in Mathematics, Computer Science, and Engineering" published by Cambridge University Press, Yves Crama and Peter L. Hammer (eds.)*, pages 257–397, 2010.
9. C. Carlet. Boolean and vectorial plateaued functions, and APN functions. In *Preprint*, 2015.
10. C. Carlet. On the properties of vectorial functions with plateaued components and their consequences on APN functions. In *Proceedings of C2SI 2015, pages 63–73*, 2015.
11. C. Carlet and C. Ding. Highly Nonlinear Mappings. In *Special Issue: "Complexity Issues in Coding and Cryptography" of the Journal of Complexity, 20(2–3), pages 205–244*, 2004.
12. C. Carlet and C. Ding. Nonlinearities of S-boxess. In *Finite Fields and Their Applications 13 (1), pages 121–135*, 2007.
13. C. Carlet and S. Mesnager. On Dillon's class H of bent functions, niho bent functions and o-polynomials. In *Journal of Combinatorial Theory, Series A, Vol 118, no. 8*, pages 2392–2410, 2011.
14. C. Carlet and S. Mesnager. On Semi-bent Boolean Functions. In *IEEE Transactions on Information Theory-IT, Vol 58 No 5*, pages 3287–3292, 2012.
15. C. Carlet and J. L. Yucas. Piecewise constructions of bent and almost optimal Boolean functions. In *Des. Codes Cryptography 37*, pages 449–464, 2005.
16. C.Carlet and E. Prouff. On plateaued functions and their constructions. In *Proceedings of Fast Software Encryption 2003, Lecture notes in computer science 2887, pages 54–73*, 2003.
17. P. Charpin and G. M. Kyureghyan. On a class of permutation polynomials over $\mathbb{F}_{2^n}$. In *SETA 2008, in: Lecture Notes in Comput. Sci. 5203, Springer-Verlag, Berlin*, pages 368–376, 2008.
18. P. Charpin and G. M. Kyureghyan. When does $G(x) + \gamma\, Tr(H(x))$ permute $\mathbb{F}_2$? In *Finite Fields: Theory and Applications 15(5)*, pages 615–632, 2009.
19. P. Charpin and G. M. Kyureghyan. Monomial functions with linear structure and permutation polynomials. In *Finite Fields: Theory and Applications - Fq9 - Contemporary Mathematics, AMS. 518*, pages 99–111, 2010.
20. P. Charpin, G. M. Kyureghyan, and V. Suder. Sparse Permutations with Low Differential Uniformity. In *Finite Fields and Their Applications. 28*, pages 214–243, 2014.
21. P. Charpin, S. Mesnager, and S. Sarkar. On involutions of finite fields. In *Proceedings of 2015 IEEE International Symposium on Information Theory, (ISIT)*, 2015.
22. P. Charpin, E. Pasalic, and C. Tavernier. On bent and semi-bent quadratic Boolean functions. In *IEEE Transactions on Information Theory, vol. 51, no. 12*, pages 4286–4298, 2005.
23. P. Charpin and S. Sarkar. Polynomials with linear structure and Maiorana–McFarland construction. In *IEEE Trans. Inf. Theory. 57(6)*, pages 3796–3804, 2011.
24. S. Chee, S. Lee, and K. Kim. Semi-bent Functions. In *Advances in Cryptology-ASIACRYPT94. Proc. 4th Int. Conf. on the Theory and Applications of Cryptology, Wollongong, Australia. Pieprzyk, J. and Safavi-Naini, R., Eds., Lect. Notes Comp. Sci, vol 917*, pages 107–118, 1994.
25. J. Dillon. Elementary Hadamard difference sets. In *PhD dissertation, University of Maryland.*
26. H. Dobbertin, T. Helleseth, P. V. Kumar, and H. M. Martinsen. Ternary m-sequences with three-valued cross-correlation function: new decimations of Welch and Niho type. In *IEEE Transactions on Information Theory, 47(4), pages1473–1481*, 2001.
27. D. Dong, L. Qu, S. Fu, and C. Li. New constructions of semi-bent functions in polynomial forms. In *Mathematical and computer modeling (57)*, pages 1139–1147, 2013.
28. R. Gold. Maximal recursive sequences with 3-valued recursive crosscorrelation functions. In *IEEE Trans. Inform. Theory 14 (1)*, pages 154–156, 1968.

29. T. Helleseth. Some results about the cross-correlation function between two maximal linear sequences. In *Discr. Math, vol. 16*, pages 209–232, 1976.
30. T. Helleseth. Correlation of m-sequences and related topics. In *Proc. SETA'98, Discrete Mathematics and Theoretical Computer Science, C. Ding, T. Helleseth, and H. Niederreiter, Eds. London, U.K.: Springer*, pages 49–66, 1999.
31. T. Helleseth, H. Hollmann, A. Kholosha, Z. Wang, and Q. Xiang. Proofs of two conjectures on ternary weakly regular bent functions. In *IEEE Transactions on Information Theory, Vol. 55, No. 11, pages 5272–5283*, 2009.
32. T. Helleseth and P. V. Kumar. Sequences with low correlation. In *Handbook of Coding Theory, Part 3: Applications, V. S. Pless, W. C. Huffman, and R. A. Brualdi, Eds. Amsterdam, The Netherlands: Elsevier, chapter. 21*, pages 1765–1853, 1998.
33. T. Helleseth, C. Rong, and D. Sandberg. New families of almost perfect nonlinear power mappings. In *IEEE Transactions on Information Theory 45(2), pages 475–485*, 1999.
34. K. Khoo, G. Gong, and D. R. Stinson. A new family of Gold-like sequences. In *IEEE Trans. Inform. Theory Lausanne, Switzerland*, page 181, 2002.
35. K. Khoo, G. Gong, and D. R. Stinson. A new characterization of semibent and bent functions on finite fields. In *Des. Codes. Cryptogr. vol. 38, no. 2*, pages 279–295, 2006.
36. N. Koçak, S. Mesnager, and F. Özbudak. Bent and Semi-bent Functions via Linear Translators. In *Proceedings of the fifteenth International Conference on Cryptography and Coding, Oxford, United Kingdom, IMACC 2015, LNCS, Springer, Heidelberg*, 2015.
37. G. M. Kyureghyan. Constructing permutations of finite fields via linear translators. In *Journal of Combinatorial Theory Series A. 118 (3)*, pages 1052–1061, 2011.
38. X. Lai. Additive and linear structures of cryptographic functions. In *Fast Software Encryption. LNCS, 1008*, pages 75–85, 1995.
39. G. Leander and G. McGuire. Spectra of functions, subspaces of matrices, and going up versus going down. In *Proceedings of AAECC, Lecture Notes in Comput. Sci., Vol. 4851, Springer*, pages 51–66, 2007.
40. G. Leander and G. McGuire. Construction of bent functions from near-bent functions. In *Journal of Combinatorial Theory, Series A 116, pages 960–970*, 2009.
41. M. Matsui. Linear cryptanalysis method for DES cipher. In *Proceedings of EUROCRYPT'93, Lecture Notes in Computer Science 765*, pages 386–397, 1994.
42. W. Meier and O. Staffelbach. Fast correlation attacks on stream ciphers. In *Advances in Cryptology, EUROCRYPT'88, Lecture Notes in Computer Science 330*, pages 301–314, 1988.
43. S. Mesnager. Further constructions of infinite families of bent functions from new permutations and their duals. In *Journal Cryptography and Communications (CCDS), Springer. To appear.*
44. S. Mesnager. A new class of bent and hyper-bent Boolean functions in polynomial forms. In *journal Design, Codes and Cryptography, 59(1–3)*, pages 265–279, 2011.
45. S. Mesnager. Semi-bent functions from Dillon and Niho exponents, Kloosterman sums and Dickson polynomials. In *IEEE Transactions on Information Theory-IT, Vol 57, No 11*, pages 7443–7458, 2011.
46. S. Mesnager. Semi-bent functions with multiple trace terms and hyperelliptic curves. In *Proceeding of International Conference on Cryptology and Information Security in Latin America (IACR), Latincrypt 2012, LNCS 7533, Springer*, pages 18–36, 2012.
47. S. Mesnager. Semi-bent functions from oval polynomials. In *Proceedings of Fourteenth International Conference on Cryptography and Coding, Oxford, United Kingdom, IMACC 2013, LNCS 8308, Springer, Heidelberg*, pages 1–15, 2013.
48. S. Mesnager. Characterizations of plateaued and bent functions in characteristic p. In *Proceedings of the 8th International Conference on SEquences and Their Applications (SETA 2014) Springer, International Publishing Switzerland 2014, LNCS 8865, pages 72–82*, 2014.
49. S. Mesnager. On semi-bent functions and related plateaued functions over the Galois field F_{2^n}. In *In Open Problems in Mathematics and Computational Science, Springer*, pages 243–273, 2014.
50. S. Mesnager. Several new infinite families of bent functions and their duals. In *IEEE Transactions on Information Theory-IT, Vol. 60, No. 7, Pages 4397–4407*, 2014.

51. S. Mesnager. Bent functions from spreads. In *Journal of the American Mathematical Society (AMS), Contemporary Mathematics (Proceedings the 11th International conference on Finite Fields and their Applications Fq11), Volume 632, page 295–316*, 2015.
52. S. Mesnager and G. Cohen. On the link of some semi-bent functions with Kloosterman sums. In *Proceeding of International Workshop on Coding and Cryptology, Y.M. Chee et al. (Eds.): IWCC 2011, LNCS 6639, Springer*, pages 263–272, 2011.
53. S. Mesnager, F. Özbudak, and A. Sınak. Results on Characterizations of Plateaued Functions in Arbitrary Characteristic. In *Proceedings of BalkanCryptSec 2015, Lecture Notes in Computer Science. To appear.*, 2015.
54. Y. Niho. Multi-valued cross-correlation functions between two maximal linear recursive sequences.In *Ph.D. dissertation, Univ. Sothern Calif., Los Angeles*, 1972.
55. K. Nyberg. On the construction of highly nonlinear permutations. In *Proceedings of EUROCRYPT'92, Lecture Notes in Computer Science 658*, pages 92–98, 1993.
56. A. Çeşmelioğlu and W. Meidl. A construction of bent functions from plateaued functions. In *Designs, Codes and Cryptography, 66 (1–3), pages 231–242*, 2013.
57. G. Sun and C.Wu. Construction of Semi-Bent Boolean Functions in Even Number of Variables. In *Chinese Journal of Electronics, vol 18, No 2*, 2009.
58. J. Wolfmann. Cyclic code aspects of bent functions. In *Finite Fields Theory and Applications, Contemporary Mathematics series of the AMS, Vol. 518, Amer. Math Soc.*, pages 363–384, 2010.
59. J. Wolfmann. Bent and near-bent functions. In *arxiv.org/abs/1308.6373*, 2013.
60. J. Wolfmann. Special bent and near-bent functions. In *Advances in Mathematics of Communication, Vol. 8, No. 1*, pages 21–33, 2014.
61. Y. Zheng and X. M. Zhang. Plateaued functions. In *Advances in Cryptology-ICICS 1999 (Lecture Notes in Computer Science). Berlin, Germany: Springer-Verlag, vol.1726*, pages 284–300, 1999.
62. Y. Zheng and X. M. Zhang. Relationships between bent functions and complementary plateaued functions. In *Lecture Notes in Computer Science, vol 1787*, pages 60–75, 1999.
63. Y. Zheng and X. M. Zhang. On plateaued functions. In *IEEE Trans. Inform. Theory 47(3)*, pages 1215–1223, 2001.
64. E. Çakçak and F. Özbudak. Curves related to Coulter's maximal curves. In *Finite Fields and Their Applications.14.1, pages 209–220*, 2008.
65. E. Çakçak and F. Özbudak. Some Artin–Schreier type function fields over finite fields with prescribed genus and number of rational places. In *Journal of Pure and Applied Algebra 210.1, pages 113–135*, 2007.

Chapter 17
Plateaued Boolean Functions: Constructions of Semi-bent Functions

17.1 Semi-bent Functions: Constructions and Characterizations

In the following, we present a general overview of the main known constructions of *semi-bent functions* and investigate new constructions.

17.1.1 On Constructions of Quadratic Semi-bent Functions

The first papers dealing with constructions of semi-bent functions have been dedicated to quadratic functions. For this particular case of functions, there exists a criterion on the semi-bentness involving the dimension of the linear kernel defined above (see e.g. [10]). More precisely, it has been proved that f is semi-bent over $\mathbb{F}_{2^n}$, if and only if its linear kernel E_f (defined previously) has dimension 2. Note that from Theorem 2.2.2, it is easy to see that quadratic Boolean function is semi-bent if and only if the rank of f is $n-2$, that is, $k_f = 2$.

Several constructions of quadratic semi-bent functions have been obtained in the literature. We give a list of the known quadratic semi-bent functions on $\mathbb{F}_{2^n}$, $n = 2m$:

- $f(x) = \sum_{i=1}^{\lfloor \frac{n-1}{2} \rfloor} c_i \mathrm{Tr}_1^n(x^{1+2^i})$, $c_i \in \mathbb{F}_2$, $gcd(\sum_{i=1}^{\frac{n}{2}-1} c_i(x^i + x^{n-i}), x^n + 1) = x^2 + 1$ [10];
- $f(x) = \mathrm{Tr}_1^n(\alpha x^{2^i+1})$, $\alpha \in \mathbb{F}_{2^n}^\star$, i even, m odd [34];
- $f(x) = \mathrm{Tr}_1^n(\alpha x^{2^i+1})$, m even, i odd, $\alpha \in \{x^3, x \in \mathbb{F}_{2^n}^\star\}$ where $\alpha \in \mathbb{F}_{2^n}^\star$ [34];
- $f(x) = \mathrm{Tr}_1^n(\alpha x^{2^i+1})$, m odd, i odd, $gcd(m, i) = 1$, $\alpha \in \{x^3, x \in \mathbb{F}_{2^n}^\star\}$ where $\alpha \in \mathbb{F}_{2^n}^\star$ [34];
- $f(x) = \mathrm{Tr}_1^n(x^{2^i+1} + x^{2^j+1})$, m odd, $1 \le i < j < m$, $gcd(n, i+j) = gcd(n, j-i) = 1)$, $gcd(n, i+j) = gcd(n, j-i) = 2$ [34];

S. Mesnager, *Bent Functions*, DOI 10.1007/978-3-319-32595-8_17

- $f(x) = \sum_{i=1}^{\frac{m-1}{2}} \mathrm{Tr}_1^n(\beta x^{1+4^i})$, m odd, $\beta \in \mathbb{F}_4^\star$ [15];
- $f(x) = \sum_{i=1}^{\frac{m-1}{2}} c_i \mathrm{Tr}_1^n(\beta x^{1+4^i})$, $c_i \in \mathbb{F}_2$, $\beta \in \mathbb{F}_4^\star$, m odd, $gcd(\sum_{i=1}^{\frac{m-1}{2}} c_i(x^i + x^{m-i}), x^m + 1) = x + 1$ [15];
- $f(x) = \sum_{i=1}^{k} \mathrm{Tr}_1^n(\beta x^{1+4^{di}})$ $\beta \in \mathbb{F}_4^\star$, m odd, $d \geq 1$, $1 \leq k \leq \frac{m-1}{2}$, $gcd(k+1, m) = gcd(k, m) = gcd(d, m) = 1$ [15];
- $f(x) = \mathrm{Tr}_1^n(\beta x^{1+4^i} + \beta x^{1+4^j})$ $\beta \in \mathbb{F}_4^\star$, m odd, $1 \leq i < j \leq \lfloor \frac{n}{4} \rfloor$, $gcd(i+j, m) = gcd(j-i, m) = 1$ [15];
- $f(x) = \mathrm{Tr}_1^n(\beta x^{1+4^i} + x^{1+4^j} + x^{1+4^t})$, $\beta \in \mathbb{F}_4^\star$, m odd, $1 \leq i < j < t \leq \lfloor \frac{n}{4} \rfloor$, $i + j = t$, $gcd(i, m) = gcd(j, m) = gcd(j, t) = 1$ [15];
- $f(x) = \mathrm{Tr}_1^n(\beta x^{1+4^i} + \beta x^{1+4^j} + \beta x^{1+4^t})$, $\beta \in \mathbb{F}_4^\star$, $1 \leq i < j < t \leq \lfloor \frac{n}{4} \rfloor$, $i+j = 2t$, $j - i = 3^h p$, $3 \not| p$, $n = 3^k q$, $3 \not| q$, $gcd(2t, m) = 1$, $h \geq k$ [15];
- $f(x) = \mathrm{Tr}_1^n(\beta x^{1+4^i} + \beta x^{1+4^j} + \beta x^{1+4^t})$, $\beta \in \mathbb{F}_4^\star$, m odd, $1 \leq i, j, t \leq \lfloor \frac{n}{4} \rfloor$, $j-i = 2t$, $t \neq i$, $j + i = 3^u p$, $3 \not| p$, $n = 3^v q$, $3 \not| q$, $gcd(2t, m) = 1$, $u \geq v$ [15];
- $f(x) = \mathrm{Tr}_1^n(\beta x^{1+4^i} + \beta x^{1+4^j} + \beta x^{1+4^t})$, $\beta \in \mathbb{F}_4^\star$, $1 \leq i, j, t \leq \lfloor \frac{n}{4} \rfloor$, $j - i = 2t$, $t \neq i$, $j + i = 3^u p$, $3 \not| p$, $n = 3^v q$, $3 \not| q$, $gcd(2t, m) = 1$, $u \geq v$ [15];
- $f(x) = \mathrm{Tr}_1^n(\beta x^{1+4^i} + \beta x^{1+4^j} + \beta x^{1+4^t} + \beta x^{1+4^s})$, $\beta \in \mathbb{F}_4^\star$, $1 \leq i, j, t, s \leq \lfloor \frac{n}{4} \rfloor$, $i < j$, $t < s$, $i+j = t+s = r$, $t \neq i$, $gcd(r, m) = gcd(m, s-i) = gcd(m, s-j) = 1$ [15].

17.1.2 *On Constructions of Semi-bent Functions from Bent Functions*

In the following subsections we are dealing with the construction of semi-bent functions from bent functions. We shall present several such kinds of constructions. A natural problem arises is fo find new primary constructions of bent functions from semi-bent functions.

17.1.2.1 Primary Constructions in Univariate Representation from Niho and Dillon Bent Functions

In this subsection, we consider several Boolean functions in univariate representation (expressed by means of the trace function) with even number of variables. Our main intention is to study the relationship between the semi-bentness property of functions obtained with Dillon and Niho exponents . Recall that a *Dillon exponent* is of the form $r(2^m - 1)$ where r is co-prime with $2^m + 1$. Moreover, a positive integer d (always understood modulo $2^n - 1$) is said to be a *Niho exponent*, and x^d is a *Niho power function*, if the restriction of x^d to $\mathbb{F}_{2^m}$ is linear or in other words $d \equiv 2^j \pmod{2^m - 1}$ for some $j < n$) and some exponential sums (namely, Kloosterman sums.

17.2 Explicit Constructions of Semi-bent Functions in Even Dimension

In this section, we consider several Boolean functions in univariate representation (expressed by means of the trace function) with even number of variables. Our main intention is to study the relationship between the semi-bentness property of functions obtained with Dillon and Niho exponents . Recall that a *Dillon exponent* is of the form $r(2^m-1)$ where r is co-prime with 2^m+1. Moreover, a positive integer d (always understood modulo 2^n-1) is said to be a *Niho exponent*, and x^d is a *Niho power function*, if the restriction of x^d to $\mathbb{F}_{2^m}$ is linear or in other words $d \equiv 2^j \pmod{2^m-1}$ for some $j < n$) and some exponential sums (namely, Kloosterman sums.

Since that is a correspondance between the bent functions of the class $\mathcal{H}$ which are of the form (14.8) (see the definition of the class $\mathcal{H}$ in Chap. 8) and the set of Niho bent functions, a natural question is: does there exist also semi-bent functions of the same form (14.8). Recall that the Walsh transform of a semi-bent function takes only the values 0 and $\pm 2^{m+1}$. Assume without loss of generality that $\mu = 0$ and that $G(0) = 0$. We remark that g is semi-bent if and only if $N_{\alpha,0} \in \{0,2\}$ and $N_{\alpha,\beta} \in \{1,3\}$ ($\beta \neq 0$). If $N_{\alpha,0} \in \{0,2\}$, $\widehat{\chi_g}(\alpha,0) \in \{0,2^{m+1}\}$ and if $N_{\alpha,\beta} \in \{1,3\}$, $\widehat{\chi_g}(\alpha,\beta) \in \{0,2^{m+1}\}$. This is impossible if $n > 2$ because of the Lemma 12.1.23 (see Chap. 2). Therefore, there exists no semi-bent function of the form (14.8) for $n > 2$.

17.2.1 *Explicit Constructions of Semi-bent Functions in Univariate Representation and Their Links with Kloosterman Sums*

The goal of this subsection is to investigate the link between the semi-bentness property of some infinite classes of Boolean functions in univariate representation and some exponential sums (Kloosterman sums and cubic sums) [27, 30]. We shall use the technical results of Sect. 2.4 in Chap. 1.

We consider infinite families of Boolean functions in univariate representation with even number of variables whose expression is given by (17.17). By computer experiments, for small values of n, we have found that the set of functions of the form (17.17) contains semi-bent functions. We investigate criteria involving Kloosterman sums to determine whether a function of the form (17.17) is semi-bent or not:

$$\mathrm{Tr}_1^n\left(ax^{r(2^m-1)}\right) + \mathrm{Tr}_1^2\left(bx^{\frac{2^n-1}{3}}\right) + \mathrm{Tr}_1^n\left(cx^{(2^m-1)\frac{1}{2}+1}\right) \quad + \mathrm{Tr}_1^n\left(dx^{(2^m-1)s+1}\right). \tag{17.1}$$

where r is a positive integer, $s \in \{0, 1/4, 1/6, 3\}$, $a \in \mathbb{F}_{2^n}^\star$, $b \in \mathbb{F}_4$, $c \in \mathbb{F}_{2^n}$ and $d \in \mathbb{F}_2$. If $b \neq 0$, we consider the functions $g_{a,b,c,d}^{(r,s)}$ of the form (17.17) only when m is odd. Note that $o(r(2^m-1)) = n$, $o(\frac{2^n-1}{3}) = 2$, $o((2^m-1)\frac{1}{2}+1) = m$ and $o((2^m-1)s+1) = n$ for $s \in \{1/4, 1/6, 3\}$ (recall that $o(j)$ denotes the size of the cyclotomic coset of 2 modulo 2^n-1 containing j). Moreover, using the transitivity property of the trace function, we have $Tr_1^n(cx^{(2^m-1)\frac{1}{2}+1}) = Tr_1^m(Tr_m^n(c^2)x^{2^m+1}) = Tr_1^m(c'x^{2^m+1})$ where $c' \in \mathbb{F}_{2^m}^\star$. Hence, the polynomial form of $g_{a,b,c,d}^{(r,s)}$ is:

$$Tr_1^n\left(ax^{r(2^m-1)}\right) + Tr_1^2\left(bx^{\frac{2^n-1}{3}}\right) + Tr_1^m(c'x^{2^m+1}) \quad + Tr_1^n\left(dx^{(2^m-1)s+1}\right).$$

So in the sequel, it suffices to use the previous identity to get the polynomial form of all the presented functions.

Now, we introduce the following decomposition

$$\mathbb{F}_{2^n}^\star = \bigcup_{u \in U} u\mathbb{F}_{2^m}^\star.$$

Let $g_{a,b,c,d}^{(r,s)}$ be any Boolean function of the form (17.17); note that the restriction of $g_{a,b,c,d}^{(r,s)}$ to any coset $u\mathbb{F}_{2^m}^\star$ ($u \in U$), is affine. More precisely,

- Assume $b \neq 0$. Thanks to the transitivity of the trace function, we have:

$$\forall y \in \mathbb{F}_{2^m}^\star, g_{a,b,c,d}^{(r,s)}(uy) = Tr_1^m(\alpha_u y) + \beta_u \tag{17.2}$$

with

$$\begin{aligned}\alpha_u &= Tr_m^n\left(du^{(2^m-1)s+1} + cu^{(2^m-1)\frac{1}{2}+1}\right)\\ &= Tr_m^n\left(du^{(2^m-1)s+1} + c\right),\\ \beta_u &= Tr_1^n\left(au^{r(2^m-1)}\right) + Tr_1^2\left(bu^{\frac{2^n-1}{3}}\right).\end{aligned}$$

- Otherwise (that is, if $b = 0$), thanks to the transitivity of the trace function, we have

$$\forall y \in \mathbb{F}_{2^m}^\star, g_{a,0,c,d}^{(r,s)}(uy) = Tr_1^m(\alpha_u y) + \beta_u \tag{17.3}$$

with

$$\begin{aligned}\alpha_u &= Tr_m^n\left(du^{(2^m-1)s+1} + c\right),\\ \beta_u &= Tr_1^n\left(au^{r(2^m-1)}\right).\end{aligned}$$

Therefore, the Walsh transform of a generic element of the form (17.17) can be computed as follows.

Lemma 17.2.1 ([27]). *Using the same notation as in (17.18) or (17.19), for every* $\omega \in \mathbb{F}_{2^n}$*, the Walsh transform of a generic element of the form (17.17) is*

$$\widehat{\chi_{g_{a,b,c,d}^{(r,s)}}}(\omega) = 1 - \sum_{u\in U} \chi(\beta_u) + 2^m \sum_{u\in U} \delta_0(\alpha_u + \mathrm{Tr}_m^n(\omega u))\chi(\beta_u) \tag{17.4}$$

where δ_0 *is the indicator of the singleton* $\{0\}$*, that is,*

$$\delta_0(z) = \begin{cases} 1 \text{ if } z = 0 \\ 0 \text{ otherwise} \end{cases}$$

Proof. Suppose m odd and $b \neq 0$. Let $\omega \in \mathbb{F}_{2^n}$. The Walsh transform of $g_{a,b,c,d}^{(r,s)}$ is defined as

$$\widehat{\chi_{g_{a,b,c,d}^{(r,s)}}}(\omega) = \sum_{x\in\mathbb{F}_{2^n}} \chi(g_{a,b,c,d}^{(r,s)}(x) + \mathrm{Tr}_1^n(wx)).$$

Any element $x \in \mathbb{F}_{2^n}^\star$ having a unique polar decomposition $x = uy$ with $u \in U$ and $y \in \mathbb{F}_{2^m}^\star$, we have :

$$\begin{aligned}
\widehat{\chi_{g_{a,b,c,d}^{(r,s)}}}(\omega) &= 1 + \sum_{u\in U} \sum_{y\in\mathbb{F}_{2^m}^\star} \chi(g_{a,b,c,d}^{(r,s)}(uy) + \mathrm{Tr}_1^n(wuy)) \\
&= 1 + \sum_{u\in U} \sum_{y\in\mathbb{F}_{2^m}^\star} \chi\left(\mathrm{Tr}_1^m\left((\alpha_u + \mathrm{Tr}_m^n(wu))y\right) + \beta_u\right) \\
&= 1 - \sum_{u\in U} \chi(\beta_u) + 2^m \sum_{u\in U} \delta_0(\alpha_u + \mathrm{Tr}_m^n(\omega u))\chi(\beta_u).
\end{aligned}$$

Likewise, one can establish (17.20) by similar calculations when $b = 0$ (for any positive integer m). □

We are now going to investigate several subfamilies of (17.17). We begin with a preliminary technical statement.

Lemma 17.2.2. *Let* $w \in \mathbb{F}_{2^n}^\star$ *and* $c \in \mathbb{F}_{2^n}^\star \setminus \mathbb{F}_{2^m}$*. The number of* $u \in U$ *such that* $\mathrm{Tr}_m^n(wu + c) = 0$ *equals* 0 *or* 2.

Proof. One has

$$\begin{aligned}
\mathrm{Tr}_m^n(wu + c) = 0 &\iff wu + w^{2^m}u^{2^m} + \mathrm{Tr}_m^n(c) = 0 \\
&\iff u^2 + w^{-1}\mathrm{Tr}_m^n(c)u + w^{2^m-1} = 0.
\end{aligned}$$

Now recall that the quadratic equation $X^2 + \alpha X + \beta = 0$, $\alpha \neq 0$, admits 0 or 2 solutions. □

The following result is shown in [14].

Lemma 17.2.3 ([14]). *For every* $w \in \mathbb{F}_{2^n}$,

- *the equation* $\mathrm{Tr}_m^n(wu + u^{\frac{1}{2}}) = 1$ *admits 0 or 2 solutions in U, if m is odd.*
- *the equation* $\mathrm{Tr}_m^n(wu + u^5) = 1$ *admits 0 or 2 solutions in U.*
- *the equation* $\mathrm{Tr}_m^n(wu^3 + u^2) + 1 = 0$ *admits 0 or 2 solutions in U, if m is even.*

Finally, the next result is shown in [18].

Lemma 17.2.4 ([18]). *Let* $r > 1$ *be a positive integer with* $\gcd(r, m) = 1$. *Then, the equation* $\mathrm{Tr}_m^n(wu) + \sum_{i=1}^{2^{r-1}-1} \mathrm{Tr}_m^n(u^{(2^m-1)\frac{i}{2^r}+1}) = 1$ *has 0 or 2 solutions in U for every* $w \in \mathbb{F}_{2^n}$.

In the following we characterize by means of Kloosterman sums the semi-bentness property of functions of the form (17.17) obtained via a Dillon monomial function (that is, a function of the form $\mathrm{Tr}_1^n(ax^{r(2^m-1)})$ where $gcd(r, 2^m + 1) = 1$) and Niho functions. We are going to restrict ourselves the study of the semi-bentness property of $g_{a,b,c,d}^{(r,s)}$ to the case where the coefficient a is in $\mathbb{F}_{2^m}^\star$.

Theorem 17.2.5 ([27]). *Let r be a positive integer such that* $gcd(r, 2^m + 1) = 1$. *Let* $a \in \mathbb{F}_{2^m}^\star$ *and* $c \in \mathbb{F}_{2^n}^\star \setminus \mathbb{F}_{2^m}$. *Then the function* $g_{a,0,c,0}^{(r,0)}$ *is semi-bent if and only if* $K_m(a) = 0$. *Moreover, suppose that* $\mathrm{Tr}_m^n(c) = 1$ *then, each function* $g_{a,0,c,1}^{(r,\frac{1}{4})}$ *(with m odd),* $g_{a,0,c,1}^{(r,3)}$ *and* $g_{a,0,c,1}^{(r,\frac{1}{6})}$ *(with m even) is semi-bent if and only if* $K_m(a) = 0$.

Proof. • Let us study the semi-bentness property of the function $g_{a,0,c,0}^{(r,0)}$. Using the notation of (17.19), one has

$$\alpha_u = \mathrm{Tr}_m^n(c), \quad \beta_u = \mathrm{Tr}_1^n(au^{r(2^m-1)}).$$

According to Lemma 17.5.2,

$$\begin{aligned}\widehat{\chi_{g_{a,0,c,0}^{(r,0)}}}(\omega) &= 1 - \sum_{u\in U} \chi(\beta_u) + 2^m \sum_{u\in U} \delta_0(\alpha_u + \mathrm{Tr}_m^n(\omega u))\chi(\beta_u)\\ &= 1 - \sum_{u\in U} \chi(\mathrm{Tr}_1^n(au^{r(2^m-1)})) + 2^m \sum_{u\in U} \delta_0(\mathrm{Tr}_m^n(\omega u + c))\chi(\mathrm{Tr}_1^n(au^{r(2^m-1)})).\end{aligned}$$

By Lemma 17.2.2, the equation $\mathrm{Tr}_m^n(wu + c) = 0$ admits 0 or 2 solutions in U for every $w \in \mathbb{F}_{2^n}^\star$. Therefore $\sum_{u\in U} \delta_0(\mathrm{Tr}_m^n(\omega u + c))\chi(\mathrm{Tr}_1^n(au^{r(2^m-1)})) \in \{0, \pm 2\}$ for every $w \in \mathbb{F}_{2^n}^\star$. In the case where $w = 0$, since $\mathrm{Tr}_m^n(c) \neq 0$ (because $c \in \mathbb{F}_{2^n}^\star \setminus \mathbb{F}_{2^m}$), $\delta_0(\mathrm{Tr}_m^n(c)) = 0$ one gets, $\sum_{u\in U} \delta_0(\mathrm{Tr}_m^n(c))\chi(\mathrm{Tr}_1^n(au^{r(2^m-1)})) = 0$. Basically, for every $w \in \mathbb{F}_{2^n}$, $\widehat{\chi_{g_{a,0,c,0}^{(r,0)}}}(w) \equiv 1 - \sum_{u\in U} \chi(\mathrm{Tr}_1^n(au^{r(2^m-1)})) \pmod{2^{m+1}}$.

Recall that the function $g_{a,0,c,0}^{(r,0)}$ is semi-bent if and only if $\widehat{\chi_{g_{a,0,c,0}^{(r,0)}}}(w) \in \{0, \pm 2^{m+1}\}$ for every $w \in \mathbb{F}_{2^n}$. Now, since

$$-2^{m+1} < -2^m \leq 1 - \sum_{u \in U} \chi(\mathrm{Tr}_1^n(au^{r(2^m-1)}))$$

and

$$1 - \sum_{u \in U} \chi(\mathrm{Tr}_1^n(au^{r(2^m-1)})) \leq 2^m + 2 < 2^{m+1}$$

then, $g_{a,0,c,0}^{(r,0)}$ is semi-bent if and only

$$\sum_{u \in U} \chi(\mathrm{Tr}_1^n(au^{r(2^m-1)})) = 1.$$

Now, since $\gcd(2^m - 1, 2^m + 1) = 1$, the mapping $u \mapsto u^{2^m-1}$ is a permutation of U. The latter condition became

$$\sum_{u \in U} \chi(\mathrm{Tr}_1^n(au^r)) = 1.$$

We then conclude thanks to Proposition 2.4.2.

- Let us study the semi-bentness property of the function $g_{a,0,c,1}^{(r,\frac{1}{4})}$. Using the notation of (17.19), one has

$$\alpha_u = \mathrm{Tr}_n^m(c) + \mathrm{Tr}_m^n(u^{(2^m-1)\frac{1}{4}+1}) = 1 + \mathrm{Tr}_m^n(u^{\frac{1}{2}});$$
$$\beta_u = \mathrm{Tr}_1^n(au^{r(2^m-1)}).$$

According to Lemma 17.5.2, the Walsh transform of $g_{a,0,c,1}^{(r,\frac{1}{4})}$ is given by

$$\widehat{\chi_{g_{a,0,c,1}^{(r,\frac{1}{4})}}}(\omega) = 1 - \sum_{u \in U} \chi(\mathrm{Tr}_1^n(au^{r(2^m-1)})) + 2^m \sum_{u \in U} \delta_0(1 + \mathrm{Tr}_m^n(u^{\frac{1}{2}} + \omega u))\chi(\mathrm{Tr}_1^n(au^{r(2^m-1)})).$$

Thanks to Lemma 17.2.3 (since m is odd), the equation $\mathrm{Tr}_m^n(u^{\frac{1}{2}} + \omega u) = 1$ has 0 or 2 solutions in U for every $w \in \mathbb{F}_{2^n}$. Therefore $\sum_{u \in U} \delta_0(1 + \mathrm{Tr}_m^n(u^{\frac{1}{2}} + \omega u))\chi(\mathrm{Tr}_1^n(au^{r(2^m-1)})) \in \{0, \pm 2\}$ for every $w \in \mathbb{F}_{2^n}$. Basically, for every $w \in \mathbb{F}_{2^n}$, $\widehat{\chi_{g_{a,0,c,1}^{(r,\frac{1}{4})}}}(w) \equiv 1 - \sum_{u \in U} \chi(\mathrm{Tr}_1^n(au^{r(2^m-1)})) \pmod{2^{m+1}}$. Same arguments used as previously lead to $g_{a,0,c,1}^{(r,\frac{1}{4})}$ being semi-bent if and only if $\sum_{u \in U} \chi(\mathrm{Tr}_1^n(au^{r(2^m-1)})) = 1$. Using the fact that $u \mapsto u^{2^m-1}$ is a permutation of U, we conclude by Proposition 2.4.2.

- Let us study the semi-bentness property of the function $g_{a,0,c,1}^{(r,3)}$. Using the notation of (17.19), one has

$$\alpha_u = \mathrm{Tr}_n^m(u^{(2^m-1)3+1} + cu^{(2^m-1)\frac{1}{2}+1});$$
$$\beta_u = \mathrm{Tr}_1^n(au^{r(2^m-1)}).$$

Note that $\mathrm{Tr}_n^m(u^{(2^m-1)3+1}) = \mathrm{Tr}_n^m(u^{-5})$ and $\mathrm{Tr}_n^m(cu^{(2^m-1)\frac{1}{2}+1}) = \mathrm{Tr}_n^m(c) = 1$. Hence, $\alpha_u = \mathrm{Tr}_n^m(u^{-5}) + 1$. According to Lemma 17.5.2, the Walsh transform of $g_{a,0,c,1}^{(r,3)}$ is given by

$$\widehat{\chi_{g_{a,0,c,1}^{(r,3)}}}(\omega) = 1 - \sum_{u\in U} \chi(\mathrm{Tr}_1^n(au^{r(2^m-1)})) + 2^m \sum_{u\in U} \delta_0(\mathrm{Tr}_n^m(u^{-5} + \omega u) + 1)\times$$
$$\chi(\mathrm{Tr}_1^n(au^{r(2^m-1)}))$$

Thanks to Lemma 17.2.3 , the $\mathrm{Tr}_n^m(u^{-5} + \omega u) = 1$ admits 0 or 2 solutions in U for every $w \in \mathbb{F}_{2^n}$. Same arguments used as previously lead to $g_{a,0,c,1}^{(r,3)}$ being semi-bent if and only if $\sum_{u\in U} \chi(\mathrm{Tr}_1^n(au^{r(2^m-1)})) = 1$. Using the fact that $u \mapsto u^{2^m-1}$ is a permutation of U, we conclude by Proposition 2.4.2.

- Let us study the semi-bentness property of the function $g_{a,0,c,1}^{(r,\frac{1}{6})}$. Using the notation of (17.19), one has

$$\alpha_u = \mathrm{Tr}_n^m(cu^{(2^m-1)\frac{1}{2}+1}) + \mathrm{Tr}_m^n(u^{(2^m-1)\frac{1}{6}+1})$$
$$= 1 + \mathrm{Tr}_m^n(u^{(2^m-1)\frac{1}{6}+1});$$
$$\beta_u = \mathrm{Tr}_1^n(au^{r(2^m-1)}).$$

Note that $u^{(2^m-1)\frac{1}{6}+1} = u^{\frac{2}{3}}$

According to Lemma 17.5.2, the Walsh transform of $g_{a,0,c,1}^{(r,\frac{1}{6})}$ is given by

$$\widehat{\chi_{g_{a,0,c,1}^{(r,\frac{1}{6})}}}(\omega) = 1 - \sum_{u\in U} \chi(\mathrm{Tr}_1^n(au^{r(2^m-1)})) + 2^m \sum_{u\in U} \delta_0(1 + \mathrm{Tr}_m^n(u^{\frac{2}{3}} + \omega u))\chi(\mathrm{Tr}_1^n(au^{r(2^m-1)})).$$

Thanks to Lemma 17.2.3 (since m is even), the equation $\mathrm{Tr}_m^n(u^2 + \omega u^3) + 1 = 0$ admits 0 or 2 solutions in U for every $w \in \mathbb{F}_{2^n}$. But the equation $\mathrm{Tr}_m^n(u^{\frac{2}{3}} + \omega u) = 1$ has 0 or 2 solutions in U if and only if the equation $\mathrm{Tr}_m^n(u^2 + \omega u^3) + 1 = 0$ admits 0 or 2 solutions in U, for every $w \in \mathbb{F}_{2^n}$. Same arguments used as previously lead to $g_{a,0,c,1}^{(r,\frac{1}{6})}$ being semi-bent if and only if $\sum_{u\in U} \chi(\mathrm{Tr}_1^n(au^{r(2^m-1)})) = 1$. We conclude by Proposition 2.4.2. □

Remark 17.2.6. The function $g_{a,0,c,0}^{(r,0)}$ has algebraic degree m, maximal possibly for a semi-bent. Indeed, the exponent $r(2^m - 1)$ is of 2-weight m while the exponent

$(2^m-1)1/2+1$ is of 2-weight 2. Moreover, $\mathrm{Tr}_1^n\left(ax^{r(2^m-1)}\right)$ and $\mathrm{Tr}_1^n\left(cx^{(2^m-1)\frac{1}{2}+1}\right)$ are two separate parts in the trace representation of $g_{a,0,c,0}^{(r,0)}$. Likewise, the functions $g_{a,0,c,1}^{(r,\frac{1}{4})}$, $g_{a,0,c,1}^{(r,3)}$ and $g_{a,0,c,1}^{(r,\frac{1}{6})}$ have algebraic degree equal to m (the exponents $r(2^m-1)$, $(2^m-1)3+1$ and $(2^m-1)1/6+1$ are of 2-weight m while the exponents $(2^m-1)1/2+1$ and $(2^m-1)1/4+1$ are of respectively algebraic degrees 2 and 3 (as observed in [14]).

Theorem 17.2.7 ([27]). *Let $n=2m$ with $m>3$ odd. Let r be a positive integer such that $gcd(r,2^m+1)=1$. Let $a\in\mathbb{F}_{2^m}^\star$, $b\in\mathbb{F}_4^\star$ and $c\in\mathbb{F}_{2^n}^\star\setminus\mathbb{F}_{2^m}$. Then, the function $g_{a,b,c,0}^{(r,0)}$ is semi-bent if and only if $K_m(a)=4$. Moreover, suppose that $\mathrm{Tr}_m^n(c)=1$ then, each function $g_{a,b,c,1}^{(r,\frac{1}{4})}$ and $g_{a,b,c,1}^{(r,3)}$ is semi-bent if and only if $K_m(a)=4$.*

Proof.
- Let us study the semi-bentness property of the function $g_{a,b,c,0}^{(r,0)}$. Using the notation of (17.18), one has

$$\alpha_u=\mathrm{Tr}_m^n(c),\quad \beta_u=\mathrm{Tr}_1^n(au^{r(2^m-1)})+\mathrm{Tr}_1^2(bu^{\frac{2^n-1}{3}}).$$

Since α_u is the same as the one associated to $g_{a,0,c,0}^{(r,0)}$ in the proof of Theorem 17.2.5 then, using the same arguments as those exposed in the beginning of the proof of Theorem 17.2.5, we get that $g_{a,b,c,0}^{(r,0)}$ is semi-bent if and only if

$$\sum_{u\in U}\chi(\mathrm{Tr}_1^n(au^{r(2^m-1)})+\mathrm{Tr}_1^2(bu^{\frac{2^n-1}{3}}))=1.$$

We finally conclude thanks to Corollary 2.4.4.

- Let us study the semi-bentness property of the function $g_{a,b,c,1}^{(r,\frac{1}{4})}$. Using the notation of (17.18), one has

$$\alpha_u=\mathrm{Tr}_n^m(c)+\mathrm{Tr}_m^n(u^{(2^m-1)\frac{1}{4}+1});$$
$$\beta_u=\mathrm{Tr}_1^n(au^{r(2^m-1)})+\mathrm{Tr}_1^2(bu^{\frac{2^n-1}{3}}).$$

Note that, $\mathrm{Tr}_m^n(u^{(2^m-1)\frac{1}{4}+1})=\mathrm{Tr}_m^n(u^{\frac{1}{2}})$. Then, (since $\mathrm{Tr}_m^n(c)=1$, by hypothesis) $\alpha_u=1+\mathrm{Tr}_m^n(u^{\frac{1}{2}})$.

According to Lemma 17.5.2, the Walsh transform of $g_{a,b,c,1}^{(r,\frac{1}{4})}$ is given by

$$\widehat{\chi_{g_{a,b,c,1}^{(r,\frac{1}{4})}}}(\omega)=1-\sum_{u\in U}\chi(\mathrm{Tr}_1^n(au^{r(2^m-1)})+\mathrm{Tr}_1^2(bu^{\frac{2^n-1}{3}}))$$
$$+2^m\sum_{u\in U}\delta_0(1+\mathrm{Tr}_m^n(u^{\frac{1}{2}}+\omega u))\chi(\mathrm{Tr}_1^n(au^{r(2^m-1)})+\mathrm{Tr}_1^2(bu^{\frac{2^n-1}{3}})).$$

Thanks to Lemma 17.2.3 (since m is odd), the equation $Tr_m^n(u^{\frac{1}{2}} + \omega u) = 1$ has 0 or 2 solutions in U for every $w \in \mathbb{F}_{2^n}$. Same arguments being used in the proof of Theorem 17.2.5 lead to $g_{a,b,c,1}^{(r,\frac{1}{4})}$ semi-bent if and only if $\sum_{u\in U} \chi(Tr_1^n(au^{r(2^m-1)})) + Tr_1^2(bu^{\frac{2^n-1}{3}}) = 1$. Using the fact that $u \mapsto u^{2^m-1}$ is a permutation of U, we conclude thanks to Corollary 2.4.4.

- Let us study the semi-bentness property of the function $g_{a,b,c,1}^{(r,3)}$. Using the notation of (17.18), one has

$$\alpha_u = Tr_n^m(u^{(2^m-1)3+1}) + cu^{(2^m-1)\frac{1}{2}+1});$$
$$\beta_u = Tr_1^n(au^{r(2^m-1)}) + Tr_1^2(bu^{\frac{2^n-1}{3}}).$$

Same arguments used as previously lead to $g_{a,b,c,1}^{(r,3)}$ being semi-bent if and only if

$$\sum_{u\in U} \chi(Tr_1^n(au^{r(2^m-1)})) + Tr_1^2(bu^{\frac{2^n-1}{3}}) = 1.$$

We conclude thanks to Corollary 2.4.4. □

Remark 17.2.8. The function $g_{a,b,c,0}^{(r,0)}$ has algebraic degree m, maximal algebraic degree for a semi-bent. Indeed, the two exponents $r(2^m - 1)$ and $\frac{2^n-1}{3}$ are of 2-weight m (since $\frac{2^n-1}{3} = 1 + 4 + \cdots + 4^{m-1}$). Hence, the two Boolean functions $x \mapsto Tr_1^n\left(ax^{r(2^m-1)}\right)$ and $x \mapsto Tr_1^2(bx^{\frac{2^n-1}{3}})$ are of algebraic degree equal to m, while the function $x \mapsto Tr_1^n\left(cx^{(2^m-1)\frac{1}{2}+1}\right)$ is of algebraic degree equals 2. Moreover, $Tr_1^n\left(ax^{r(2^m-1)}\right)$, $Tr_1^2\left(bx^{\frac{2^n-1}{3}}\right)$ and $Tr_1^n\left(cx^{(2^m-1)\frac{1}{2}+1}\right)$ are three separate parts in the trace representation of $g_{a,b,c,0}^{(r,0)}$. Likewise, the functions $g_{a,0,c,1}^{(r,\frac{1}{4})}$ and $g_{a,b,c,1}^{(r,3)}$ have algebraic degree equal to m.

Example 17.2.9. Let us describe for instance, the set of semi-bent Boolean functions $g_{a,b,c,0}^{(1,0)}$ defined on $\mathbb{F}_{2^{10}}$ of the form $Tr_1^{10}(ax^{31}) + Tr_1^2(bx^{341}) + Tr_1^{10}(cx^{528})$ where $a \in \mathbb{F}_{2^5}^\star$, $b \in \mathbb{F}_4^\star$, $c \in \mathbb{F}_{2^{10}}^\star \setminus \mathbb{F}_{2^5}$. Let β be a primitive element of $\mathbb{F}_4$ and α be a primitive element of $\mathbb{F}_{32} = \mathbb{F}_2(\alpha)$ with $\alpha^5 + \alpha^2 + 1 = 0$. Recall that $K_5(a) \equiv 1 \pmod 3$ if and only if $Tr_1^5(a^{1/3}) = 0$. Now, according to table 4 in [9], $E_0 := \{a \in \mathbb{F}_{2^5}^\star, Tr_1^5(a^{1/3}) = 0\} = \{\alpha^3, \alpha^{21}, \alpha^{14}\}$, $\{a \in \mathbb{F}_{2^5}^\star, K_5(a) = 4\} = \{\alpha^3, \alpha^{21}\}$ and $E_1 := \{a \in \mathbb{F}_{2^5}^\star, Tr_1^5(a^{1/3}) = 1\} = \{1, \alpha^2, \alpha^9, \alpha^{15}\}$. Then, according to Theorem 17.2.7, the functions $g_{\alpha^3,1,c,0}^{(1,0)}$, $g_{\alpha^3,\beta,c,0}^{(1,0)}$, $g_{\alpha^3,\beta^2,c,0}^{(1,0)}$, $g_{\alpha^{21},1,c,0}^{(1,0)}$, $g_{\alpha^{21},\beta,c,0}^{(1,0)}$, $g_{\alpha^{21},\beta^2,c,0}^{(1,0)}$ are semi-bent while $g_{\alpha^{14},1,c,0}^{(1,0)}$, $g_{\alpha^{14},\beta,c,0}^{(1,0)}$, $g_{\alpha^{14},\beta^2,c,0}^{(1,0)}$, $g_{a,1,c,0}^{(1,0)}$, $g_{a,\beta,c,0}^{(1,0)}$,$g_{a,\beta^2,c,0}^{(1,0)}$ are not semi-bent if $a \in \{1, \alpha^2, \alpha^9, \alpha^{15}\}$.

Now, we are interested in studying the semi-bentness property of functions of the form $g^{(3,0)}_{a,b,c,0}$ with m odd, $a \in \mathbb{F}^{\star}_{2^n}$, $b \in \mathbb{F}^{\star}_4$ and $c \in \mathbb{F}^{\star}_{2^n}$ (note that the function $x \mapsto Tr^n_1(ax^{3(2^m-1)})$ is not a Dillon monomial function since 3 is not co-prime with $2^m + 1$ when m is odd). To this end, we show that we can identify all the semi-bent functions in the form $g^{(3,0)}_{a,b,c,0}$ by studying only the semi-bentness of $g^{(3,0)}_{a\zeta^i,b,c,0}$ where $a \in \mathbb{F}^{\star}_{2^m}$, $b \in \mathbb{F}^{\star}_4$, $c \in \mathbb{F}^{\star}_{2^n}$, ζ is a generator of the cyclic group U and $i \in \{0, 1\}$. Let $a \in \mathbb{F}^{\star}_{2^m}$, $\lambda \in \mathbb{F}^{\star}_{2^n}$, $b \in \mathbb{F}^{\star}_4$ and $c \in \mathbb{F}^{\star}_{2^n}$. Set $a' = a\lambda^{3(2^m-1)}$, $b' = b\lambda^{\frac{2^n-1}{3}}$ and $c' = c\lambda^{(2^m-1)\frac{1}{2}+1}$. Then remark that, for every $x \in \mathbb{F}_{2^n}$, we have:

$$\begin{aligned} g^{(3,0)}_{a',b',c',0}(x) =& Tr^n_1(a(\lambda x)^{3(2^m-1)}) + Tr^2_1(b(\lambda x)^{\frac{2^n-1}{3}}) \\ &+ Tr^n_1(c\lambda^{(2^m-1)\frac{1}{2}+1}x^{(2^m-1)\frac{1}{2}+1}) \\ =& \, g^{(3,0)}_{a,b,c,0}(\lambda x). \end{aligned}$$

This means that $g^{(3,0)}_{a',b',c',0}$ is linearly equivalent to $g^{(3,0)}_{a,b,c,0}$. Therefore, we don't have to consider all the possible values of $a \in \mathbb{F}^{\star}_{2^n}$ in our study. Indeed, recall that every element of x in $\mathbb{F}^{\star}_{2^n}$ admits a unique polar decomposition $x = uy$ with $y \in \mathbb{F}^{\star}_{2^m}$ and $u \in U$. Now, m being odd, every element $u \in U$ can be uniquely decomposed as $u = \zeta^i v$ with $i \in \{0, 1, 2\}$ and $v \in V = \{u^3 \mid u \in U\}$. One deduces

Lemma 17.2.10 ([27]). *Let $n = 2m$ with m odd. Let $a' \in \mathbb{F}^{\star}_{2^n}$, $b' \in \mathbb{F}^{\star}_4$, $c' \in \mathbb{F}^{\star}_{2^n}$. Suppose that $a' = a\zeta^i v$ with $a \in \mathbb{F}^{\star}_{2^m}$ $i \in \{0, 1, 2\}$, ζ a generator of the cyclic group U and, $v \in V = \{u^3 \mid u \in U\}$. Then, there exist $b \in \mathbb{F}^{\star}_4$ and $c \in \mathbb{F}^{\star}_{2^n}$ such that $g^{(3,0)}_{a',b',c',0}$ is linearly equivalent to $g^{(3,0)}_{a\zeta^i,b,c,0}$.*

Every element $a' \in \mathbb{F}^{\star}_{2^n}$ can be (uniquely) decomposed as $a' = a\zeta^i v$ with a, ζ and v as in the preceding lemma. The property of semi-bentness being affine invariant, one can restrict oneself to study the semi-bentness of $g^{(3,0)}_{a\zeta^i,b,c,0}$ with $a \in \mathbb{F}^{\star}_{2^m}$, $b \in \mathbb{F}^{\star}_4$, $c \in \mathbb{F}^{\star}_{2^n}$ and $i \in \{0, 1\}$.

Theorem 17.2.11 ([27]). *Let $n = 2m$ with m odd, $a \in \mathbb{F}^{\star}_{2^m}$, $b \in \mathbb{F}^{\star}_4$, $c \in \mathbb{F}^{\star}_{2^n} \setminus \mathbb{F}_{2^m}$ and ζ be a generator of U.*

1. *Assume $m \equiv 3 \pmod 6$. Then the functions $g^{(3,0)}_{a,b,c,0}$ and $g^{(3,0)}_{a\zeta,b,c,0}$ are not semi-bent.*
2. *Assume $m \not\equiv 3 \pmod 6$. Then, $g^{(3,0)}_{a\zeta^i,b,c,0}$ is semi-bent if and only if*
   - *$i = 0$ and $K_m(a) = 4$,*
   - *or, $i = 1$ and $K_m(a) + C_m(a, a) = 4$.*

Proof. Let $i \in \{0, 1\}$. Using the notation of (17.18), one has

$$\beta_u = Tr^n_1(a\zeta^i u^{3(2^m-1)}) + Tr^2_1(bu^{\frac{2^n-1}{3}}), \quad \alpha_u = Tr^n_m(c).$$

Same arguments as in the beginning of the proof of Theorem 17.2.5 lead to $g^{(3,0)}_{a\zeta^i,b,c,0}$ being semi-bent if and only if

$$\sum_{u\in U}\chi\left(\mathrm{Tr}_1^n(a\zeta^i u^{3(2^m-1)}) + \mathrm{Tr}_1^2(bu^{\frac{2^n-1}{3}})\right) = 1,$$

equivalently (since the mapping $u \mapsto u^{2^m-1}$ is a permutation on U)

$$\sum_{u\in U}\chi\left(\mathrm{Tr}_1^n(a\zeta^i u^3) + \mathrm{Tr}_1^2(bu^{\frac{2^m+1}{3}})\right) = 1.$$

Assertions (1) and (2) then follow from Corollary 2.4.6. □

Remark 17.2.12. The function $g^{(3,0)}_{a\zeta^i,b,c,0}$ has algebraic degree m, maximal algebraic degree for a semi-bent. Indeed, the exponents $3(2^m - 1)$ and $(2^n - 1)/3$ are of 2-weight m (since $3(2^m - 1) = 1 + 2^2 + 2^3 + \cdots + 2^{m-1} + 2^{m+1}$) while the exponent $(2^m - 1)\frac{1}{2} + 1$ is of 2-weight 2. Moreover, $\mathrm{Tr}_1^n\left(a\zeta^i x^{3(2^m-1)}\right)$, $\mathrm{Tr}_1^2\left(bx^{\frac{2^n-1}{3}}\right)$ and $\mathrm{Tr}_1^n\left(cx^{(2^m-1)\frac{1}{2}+1}\right)$ are three separate parts in the trace representation of $g^{(3,0)}_{a\zeta^i,b,c,0}$.

In the following we characterize by means of Kloosterman sums the semi-bentness property of functions obtained via a Dillon monomial function and 2^r Niho power functions.

Theorem 17.2.13 ([27]). *Let $n = 2m$, r be a positive integer such that $gcd(r, 2^m + 1) = 1$, $\nu > 1$ be a positive integer with $\gcd(\nu, m) = 1$, $\alpha \in \mathbb{F}_{2^n}$ such that $\mathrm{Tr}_m^n(\alpha) = 1$, $a \in \mathbb{F}^\star_{2^m}$ and $b \in \mathbb{F}^\star_4$.*

Let f be the Boolean function defined over $\mathbb{F}_{2^n}$ whose expression is given by

$$\mathrm{Tr}_1^n\left(ax^{r(2^m-1)}\right) + \mathrm{Tr}_1^n\left(\alpha x^{2^m+1} + \sum_{i=1}^{2^{\nu-1}-1} x^{(2^m-1)\frac{i}{2^\nu}+1}\right).$$

Let g be the Boolean function defined over $\mathbb{F}_{2^n}$ (m odd) whose expression is given by $g(x) = f(x) + \mathrm{Tr}_1^2\left(bx^{\frac{2^n-1}{3}}\right)$. Then the function f (Resp. g) is semi-bent if and only if, $K_m(a) = 0$ (Resp. $K_m(a) = 4$).

Proof. • Let us study the semi-bentness property of the function f. Using the notation of (17.19), one has

$$\alpha_u = \mathrm{Tr}_n^m(\alpha^{\frac{1}{2}} + \sum_{i=1}^{2^{\nu-1}-1} u^{(2^m-1)\frac{i}{2^\nu}+1});$$

$$\beta_u = \mathrm{Tr}_1^n(au^{r(2^m-1)}).$$

According to Lemma 17.5.2, the Walsh transform of f is given by

$$\widehat{\chi_f}(\omega) = 1 - \sum_{u\in U} \chi(\mathrm{Tr}_1^n(au^{r(2^m-1)}))$$
$$+ 2^m \sum_{u\in U} \delta_0(\mathrm{Tr}_n^m(\alpha^{\frac{1}{2}} + \sum_{i=1}^{2^{\nu-1}-1} u^{(2^m-1)\frac{i}{2^\nu}+1}) + \omega u))\times$$
$$\chi(\mathrm{Tr}_1^n(au^{r(2^m-1)})).$$

Thanks to Lemma 17.2.4, the equation $\mathrm{Tr}_m^n(wu) + \sum_{i=1}^{2^{\mu-1}-1} \mathrm{Tr}_m^n(u^{(2^m-1)\frac{i}{2^\mu}+1})$ $= 1$ admits 0 or 2 solutions in U for every $w \in \mathbb{F}_{2^n}$. Since $\mathrm{Tr}_m^n(\alpha) = 1$ (by hypothesis) and $\mathrm{Tr}_m^n((\alpha^{\frac{1}{2}})^2) = \mathrm{Tr}_m^n(\alpha^{\frac{1}{2}})$, the equation $\mathrm{Tr}_n^m(\sum_{i=1}^{2^{\nu-1}-1} u^{(2^m-1)\frac{i}{2^\nu}+1}) + \omega u) = 1$ admits 0 or 2 solutions in U for every $w \in \mathbb{F}_{2^n}$. Same arguments used in the proof of Theorem 17.2.5 lead to f being semi-bent if and only $\sum_{u\in U} \chi(\mathrm{Tr}_1^n(au^{r(2^m-1)})) = 1$. We conclude by Proposition 2.4.2.

- Let study us the semi-bentness property of the function g. Using the notation of (17.18), one has

$$\alpha_u = \mathrm{Tr}_n^m(\alpha^{\frac{1}{2}} + \sum_{i=1}^{2^{\nu-1}-1} u^{(2^m-1)\frac{i}{2^\nu}+1});$$
$$\beta_u = \mathrm{Tr}_1^n(au^{r(2^m-1)}) + \mathrm{Tr}_1^2(bu^{\frac{2^n-1}{3}}).$$

Same arguments used as previously lead to g being semi-bent if and only $\sum_{u\in U} \chi(\mathrm{Tr}_1^n(au^{r(2^m-1)})) + \mathrm{Tr}_1^2(bu^{\frac{2^n-1}{3}}) = 1$. We conclude thanks to Corollary 2.4.4. □

17.2.2 Semi-bent Functions in Polynomial Forms with Multiple Trace Terms and Their Link with Dikson Polynomial

In this subsection, we study the relationship between the semi-bentness property of functions in polynomial forms with multiple trace terms and Dickson polynomials.

In the following, we are interested in semi-bent functions whose expression contains multiple trace terms. Let E be a set of representatives of the cyclotomic classes modulo $2^n - 1$ for which each class has full size n. Let $f_{a_r,b,c}$ be the function defined on $\mathbb{F}_{2^n}$ whose polynomial form is given by

$$\sum_{r\in R} \mathrm{Tr}_1^n(a_r x^{r(2^m-1)}) + \mathrm{Tr}_1^2(bx^{\frac{2^n-1}{3}}) + \mathrm{Tr}_1^m(cx^{2^m+1}) \qquad (17.5)$$

where $R \subseteq E$, $a_r \in \mathbb{F}_{2^m}^\star$, $b \in \mathbb{F}_4^\star$ and $c \in \mathbb{F}_{2^m}^\star$. In the following, we will show that semi-bent functions $f_{a_r,b,c}$ of the form (17.21) can be described by means of exponential sums involving the Dickson polynomials. In particular, one can provide a way to transfer the characterization of semi-bentness of a function of the form (17.21) to the evaluation of the Hamming weight of some Boolean functions.

To prove the result of this subsection, we need the following statements (Proposition 17.5.3) and Corollary 17.5.4).

Proposition 17.2.14 ([27]). *For $b \in \mathbb{F}_4^\star$ and $a_r \in \mathbb{F}_{2^m}^\star$, we denote by $g_{a_r,b}$ the function $\sum_{r\in R} \mathrm{Tr}_1^n(a_r x^{r(2^m-1)}) + \mathrm{Tr}_1^2(bx^{\frac{2^n-1}{3}})$ and by $g_{a_r,0}$ the function $\sum_{r\in R} \mathrm{Tr}_1^n(a_r x^{r(2^m-1)})$. Let V be the set of the elements of the cubes of U and ζ be a generator of U. Then, we have the following relations:*

$$\sum_{u\in U} \chi\Big(g_{a_r,\beta}(u)\Big) = \sum_{u\in U} \chi\Big(g_{a_r,\beta^2}(u)\Big) = -\sum_{v\in V} \chi\Big(g_{a_r,0}(v)\Big) \tag{17.6}$$

and

$$\sum_{u\in U} \chi\Big(g_{a_r,1}(u)\Big) = \sum_{v\in V} \chi\Big(g_{a_r,0}(v)\Big) - 2\sum_{v\in V} \chi\Big(g_{a_r,0}(\zeta v)\Big). \tag{17.7}$$

Proof. Introduce for every element b' of $\mathbb{F}_4$, the sum

$$\Lambda(b') := \sum_{b\in\mathbb{F}_4} \sum_{u\in U} \chi\Big(g_{a_r,b}(u)\Big)\chi\Big(\mathrm{Tr}_1^2(bb')\Big).$$

Note that

$$\Lambda(b') = \sum_{u\in U} \chi\Big(g_{a_r,0}(u)\Big) \sum_{b\in\mathbb{F}_4} \chi\Big(\mathrm{Tr}_1^2\Big(b\big(b' + u^{\frac{2^n-1}{3}}\big)\Big)\Big).$$

Furthermore, one has

$$\sum_{b\in\mathbb{F}_4} \chi\Big(\mathrm{Tr}_1^2\Big(b\big(b' + u^{\frac{2^n-1}{3}}\big)\Big)\Big) = \begin{cases} 0 \text{ if } u^{\frac{2^n-1}{3}} \neq b' \\ 4 \text{ otherwise} \end{cases}$$

Since, $u^{\frac{2^n-1}{3}} \neq 0$ for every $u \in U$, $\Lambda(0) = 0$. Since β is a primitive element of $\mathbb{F}_4$, suppose that $b' = \beta^i$ for $i \in \{0, 1, 2\}$. Then, for a generator ζ of U, we have, $\beta^i = \zeta^{i\frac{2^m+1}{3}}$. Hence,

$$\begin{aligned}\Lambda(\beta^i) &= 4 \sum_{u\in U,\, u^{\frac{2^n-1}{3}}=\zeta^{i\frac{2^m+1}{3}}} \chi\Big(g_{a_r,0}(u)\Big)\\ &= 4 \sum_{u\in U,\, (u^{-2}\zeta^{-i})^{\frac{2^m+1}{3}}=1} \chi\Big(g_{a_r,0}(u)\Big)\\ &= 4 \sum_{u\in U,\, u^{-2}\in\zeta^i V} \chi\Big(g_{a_r,0}(u)\Big).\end{aligned}$$

That follows from the fact that the only elements x of U such that $x^{\frac{2^m+1}{3}} = 1$ are the elements of V. Next, note that the map $x \mapsto x^{2^{m-1}}$ is one-to-one from $\zeta^i V$ to $\zeta^i V$ (since $\zeta^{i(2^{m-1}-1)}$ is a cube because $2^{m-1} - 1 \equiv 0 \pmod 3$ for m odd), one gets that $u^{\frac{2^n-1}{3}} = \zeta^{i\frac{2^m+1}{3}}$ if and only if $u \in \zeta^i V$.

Therefore,

$$\Lambda(\beta^i) = 4\sum_{v\in V} \chi\Big(g_{a_r,0}(\zeta^i v)\Big). \tag{17.8}$$

Now, establish an expression of $\sum_{b'\in\mathbb{F}_4} \Lambda(b')\chi\Big(Tr_1^2(bb')\Big)$ involving $\sum_{u\in U} \chi\Big(g_{a_r,b}(u)\Big)$.

$$\begin{aligned}&\sum_{b'\in\mathbb{F}_4} \Lambda(b')\chi\Big(Tr_1^2(bb')\Big)\\ &= \sum_{b'\in\mathbb{F}_4}\sum_{b''\in\mathbb{F}_4}\sum_{u\in U} \chi\Big(g_{a_r,b''}(u)\Big)\chi\Big(Tr_1^2(b''b')\Big)\chi\Big(Tr_1^2(bb')\Big)\\ &= \sum_{b''\in\mathbb{F}_4}\sum_{u\in U} \chi\Big(g_{a_r,b''}(u)\Big)\sum_{b'\in\mathbb{F}_4} \chi\Big(Tr_1^2(b'(b''+b))\Big).\end{aligned}$$

Since,

$$\sum_{b'\in\mathbb{F}_4} \chi\Big(Tr_1^2(b'(b''+b))\Big) = \begin{cases} 4 \text{ if } b''=b\\ 0 \text{ otherwise}\end{cases}$$

then, one gets

$$\sum_{b'\in\mathbb{F}_4} \Lambda(b')\chi\Big(Tr_1^2(bb')\Big) = 4\sum_{u\in U} \chi\Big(g_{a_r,b}(u)\Big)$$

that is,

$$\sum_{u\in U} \chi\Big(g_{a_r,b}(u)\Big) = \frac{1}{4}\sum_{b'\in\mathbb{F}_4} \Lambda(b')\chi\Big(Tr_1^2(bb')\Big). \tag{17.9}$$

Finally, by formula (17.24), one gets (since $\chi(\mathrm{Tr}_1^2(1)) = 1$ and $\chi(\mathrm{Tr}_1^2(\beta)) = \chi(\mathrm{Tr}_1^2(\beta^2)) = -1$)

$$\sum_{u\in U}\chi\Big(g_{a_r,1}(u)\Big) = \sum_{v\in V}\chi\Big(g_{a_r,0}(v)\Big) - \sum_{v\in V}\chi\Big(g_{a_r,0}(\zeta v)\Big) - \sum_{v\in V}\chi\Big(g_{a_r,0}(\zeta^2 v)\Big).$$

$$\sum_{u\in U}\chi\Big(g_{a_r,\beta}(u)\Big) = -\sum_{v\in V}\chi\Big(g_{a_r,0}(v)\Big) - \sum_{v\in V}\chi\Big(g_{a_r,0}(\zeta v)\Big) + \sum_{v\in V}\chi\Big(g_{a_r,0}(\zeta^2 v)\Big).$$

$$\sum_{u\in U}\chi\Big(g_{a_r,\beta^2}(u)) = -\sum_{v\in V}\chi\Big(g_{a_r,0}(v)) + \sum_{v\in V}\chi\Big(g_{a_r,0}(\zeta v)\Big) - \sum_{v\in V}\chi\Big(g_{a_r,0}(\zeta^2 v)\Big).$$

To conclude, note that one has

$$\sum_{v\in V}\chi\Big(g_{a_r,0}(\zeta v)\Big) = \sum_{v\in V}\chi\Big(g_{a_r,0}(\zeta^2 v)\Big). \tag{17.10}$$

Indeed, since the trace function is invariant under the Frobenius automorphism $x \mapsto x^2$, we get, applying m times, the Frobenius automorphism : $\forall x \in \mathbb{F}_{2^n}$,

$$g_{a_r,0}(x) = \sum_{r\in R}\mathrm{Tr}_1^n\left(a_r^{2^m}x^{2^m r(2^m-1)}\right) = \sum_{r\in R}\mathrm{Tr}_1^n\left(a_r x^{2^m r(2^m-1)}\right) = g_{a_r,0}(x^{2^m})$$

because the a_r's are in $\mathbb{F}_{2^m}^\star$. Hence,

$$\begin{aligned}&\sum_{v\in V}\chi\Big(g_{a_r,0}(\zeta v)\Big)\\ &= \sum_{v\in V}\chi\Big(g_{a_r,0}(\zeta^{2^m}v^{2^m})\Big)\\ &= \sum_{v\in V}\chi\Big(g_{a_r,0}(\zeta^2(\zeta^{2^m-2}v^{2^m}))\Big).\end{aligned}$$

Now, since m is odd, 3 divides 2^m+1 and thus divides 2^m-2. Hence, ζ^{2^m-2} is a cube of U and the mapping $v \mapsto \zeta^{(2^m-2)}v^{2^m}$ is a permutation of V. The relation (17.26) follows. □

Corollary 17.2.15 ([27]). *For $b \in \mathbb{F}_4^\star$ and $a_r \in \mathbb{F}_{2^m}^\star$, we denote by $g_{a_r,b}$ the function defined on $\mathbb{F}_{2^n}$ by $\sum_{r\in R} \mathrm{Tr}_1^n(a_r x^{r(2^m-1)}) + \mathrm{Tr}_1^2(bx^{\frac{2^n-1}{3}})$, and by h_{a_r} the function defined on $\mathbb{F}_{2^m}$ by $h_{a_r}(x) = \sum_{r\in R} \mathrm{Tr}_1^m(a_r D_r(x))$, where $D_r(x)$ is the Dickson polynomial of degree r. Then,*

1. $\sum_{u\in U} \chi\Big(g_{a_r,\beta}(u)\Big) = 1$ *if and only if* $\sum_{u\in U} \chi\Big(g_{a_r,\beta^2}(u)\Big) = 1$ *if and only if,*

$$\sum_{x\in\mathbb{F}_{2^m}} \chi\Big(\mathrm{Tr}_1^m(x^{-1}) + h_{a_r}(D_3(x))\Big)$$
$$= 2^m - 2\,\mathrm{wt}(h_{a_r} \circ D_3) + 4.$$

2. $\sum_{u\in U} \chi\Big(g_{a_r,1}(u)\Big) = 1$ *if and only if*

$$3\sum_{x\in\mathbb{F}_{2^m}} \chi\Big(\mathrm{Tr}_1^m(x^{-1}) + h_{a_r}(x)\Big)$$
$$-2\sum_{x\in\mathbb{F}_{2^m}} \chi\Big(\mathrm{Tr}_1^m(x^{-1}) + h_{a_r}(D_3(x))\Big)$$
$$= 4 + 2^m + 4\,\mathrm{wt}(h_{a_r} \circ D_3) - 6\,\mathrm{wt}(h_{a_r}).$$

To prove the corollary, we need the following useful lemma.

Lemma 17.2.16 ([27]). *Keeping the same notations as in Corollary 17.5.4, for any positive integer p, we have*

$$\sum_{u\in U} \chi\Big(g_{a_r,0}(u^p)\Big) = 1 + \sum_{x\in\mathbb{F}_{2^m}} \chi\Big(h_{a_r}(D_p(x))\Big) - \sum_{x\in\mathbb{F}_{2^m}} \chi\Big(\mathrm{Tr}_1^m(x^{-1}) + h_{a_r}(D_p(x))\Big).$$

Proof. We have

$$\sum_{u\in U} \chi\Big(g_{a_r,0}(u^p)\Big) = 1 + 2\sum_{x\in\mathbb{F}_{2^m}^\star, \mathrm{Tr}_1^m(x^{-1})=1} \chi\Big(h_{a_r}(D_p(x))\Big).$$

Now, note that the indicator of the set $\{x \in \mathbb{F}_{2^m}^\star \mid \mathrm{Tr}_1^m(x^{-1}) = 1\}$ can be written as $\frac{1}{2}\left(1 - \chi(\mathrm{Tr}_1^m(x^{-1}))\right)$. Hence, $\sum_{x\in\mathbb{F}_{2^m}^\star, \mathrm{Tr}_1^m(x^{-1})=1} \chi\Big(h_{a_r}(D_p(x))\Big)$

$$= \frac{1}{2} \sum_{x \in \mathbb{F}_{2^m}^\star} \chi\Big(h_{a_r}(D_p(x))\Big)$$

$$- \frac{1}{2} \sum_{x \in \mathbb{F}_{2^m}^\star} \chi\Big(\mathrm{Tr}_1^m(x^{-1}) + h_{a_r}(D_p(x))\Big)$$

$$= \frac{1}{2} \sum_{x \in \mathbb{F}_{2^m}} \chi\Big(h_{a_r}(D_p(x))\Big)$$

$$- \frac{1}{2} \sum_{x \in \mathbb{F}_{2^m}} \chi\Big(\mathrm{Tr}_1^m(x^{-1}) + h_{a_r}(D_p(x))\Big).$$

Therefore,

$$\sum_{u \in U} \chi\Big(g_{a_r,0}(u^p)\Big) = 1 + \sum_{x \in \mathbb{F}_{2^m}} \chi\Big(h_{a_r}(D_p(x))\Big)$$
$$- \sum_{x \in \mathbb{F}_{2^m}} \chi\Big(\mathrm{Tr}_1^m(x^{-1}) + h_{a_r}(D_p(x))\Big).$$

□

Now, we prove Corollary 17.5.4.

Proof. 1. According to Proposition 17.5.3, $\sum_{u \in U} \chi\Big(g_{a_r,\beta}(u)\Big) = 1$ if and only if,

$$\sum_{u \in U} \chi\Big(g_{a_r,\beta^2}(u)\Big) = 1$$

if and only if,

$$\sum_{v \in V} \chi\Big(g_{a_r,0}(v)\Big) = -1.$$

We have

$$\sum_{v \in V} \chi\Big(g_{a_r,0}(v)\Big) = \frac{1}{3} \sum_{u \in U} \chi\Big(g_{a_r,0}(u^3)\Big).$$

Now, take $p = 3$ in Lemma 17.5.5:

$$\sum_{u \in U} \chi\Big(g_{a_r,0}(u^3)\Big) = 1 + \sum_{x \in \mathbb{F}_{2^m}} \chi\Big(h_{a_r}(D_3(x))\Big) - \sum_{x \in \mathbb{F}_{2^m}} \chi\Big(\mathrm{Tr}_1^m(x^{-1}) + h_{a_r}(D_3(x))\Big).$$

Hence $\sum_{u\in U}\chi\Big(g_{a_r,\beta}(u)\Big) = 1$ if and only if, $\sum_{u\in U}\chi\Big(g_{a_r,\beta^2}(u)\Big) = 1$) if and only if,

$$\sum_{x\in\mathbb{F}_{2^m}}\chi\Big(\mathrm{Tr}_1^m(x^{-1}) + h_{a_r}(D_3(x))\Big) = 4 + \sum_{x\in\mathbb{F}_{2^m}}\chi\Big(h_{a_r}(D_3(x))\Big).$$

Now, using the fact that, for a Boolean function f defined on $\mathbb{F}_{2^n}$, $\sum_{x\in\mathbb{F}_{2^n}}\chi\Big(f(x)\Big) = 2^n - 2\,\mathrm{wt}(f)$, we finally get that $\sum_{u\in U}\chi\Big(g_{a_r,\beta}(u)\Big) = 1$ if and only if, $\sum_{u\in U}\chi\Big(g_{a_r,\beta^2}(u)\Big) = 1$ if and only if,

$$\sum_{x\in\mathbb{F}_{2^m}}\chi\Big(\mathrm{Tr}_1^m(x^{-1}) + h_{a_r}(D_3(x))\Big) = 4 + 2^m - 2\,\mathrm{wt}(h_{a_r}\circ D_3).$$

The assertion 1) follows.

2. By Proposition 17.5.3, $\sum_{u\in U}\chi\Big(g_{a_r,1}(u)\Big) = 1$ if and only if,

$$\sum_{v\in V}\chi\Big(g_{a_r,0}(v)\Big) - 2\sum_{v\in V}\chi\Big(g_{a_r,0}(\zeta v)\Big) = 1.$$

Note that we have

$$\sum_{u\in U}\chi\Big(g_{a_r,0}(u)\Big) = \sum_{v\in V}\chi\Big(g_{a_r,0}(v)\Big) + \sum_{v\in V}\chi\Big(g_{a_r,0}(\zeta v)\Big) + \sum_{v\in V}\chi\Big(g_{a_r,0}(\zeta^2 v)\Big).$$

Using relation (17.26) and the fact that $\sum_{v\in V}\chi(g_{a_r,0}(v)) = \frac{1}{3}\sum_{u\in U}\chi(g_{a_r,0}(u^3))$, one gets $\sum_{u\in U}\chi\Big(g_{a_r,1}(u)\Big) = 1$ if and only if, $\frac{2}{3}\sum_{u\in U}\chi(g_{a_r,0}(u^3)) - \sum_{u\in U}\chi(g_{a_r,0}(u)) = 1$.

Now, apply Lemma 17.5.5 for $p = 3$ and $p = 1$:

$$\sum_{u\in U}\chi\Big(g_{a_r,0}(u^3)\Big) = 1 + \sum_{x\in\mathbb{F}_{2^m}}\chi\Big(h_{a_r}(D_3(x))\Big) - \sum_{x\in\mathbb{F}_{2^m}}\chi\Big(\mathrm{Tr}_1^m(x^{-1}) + h_{a_r}(D_3(x))\Big)$$

and (since $D_1(x) = x$)

$$\sum_{u\in U}\chi\Big(g_{a_r,0}(u)\Big) = 1 + \sum_{x\in\mathbb{F}_{2^m}}\chi\Big(h_{a_r}(x)\Big) - \sum_{x\in\mathbb{F}_{2^m}}\chi\Big(\mathrm{Tr}_1^m(x^{-1}) + h_{a_r}(x)\Big).$$

The condition

$$\frac{2}{3}\sum_{u\in U}\chi(g_{a_r,0}(u^3)) - \sum_{u\in U}\chi(g_{a_r,0}(u)) = 1$$

is then equivalent to

$$2/3 + 2/3 \sum_{x\in\mathbb{F}_{2^m}} \chi\Big(h_{a_r}(D_3(x))\Big) - 2/3 \sum_{x\in\mathbb{F}_{2^m}} \chi\Big(\mathrm{Tr}_1^m(x^{-1}) + h_{a_r}(D_3(x))\Big)$$
$$-1 - \sum_{x\in\mathbb{F}_{2^m}} \chi\Big(h_{a_r}(x)\Big) + \sum_{x\in\mathbb{F}_{2^m}} \chi\Big(\mathrm{Tr}_1^m(x^{-1}) + h_{a_r}(x)\Big) = 1.$$

Now,

$$\sum_{x\in\mathbb{F}_{2^m}} \chi\Big(h_{a_r}(D_3(x))\Big) = 2^m - 2\,\mathrm{wt}(h_{a_r} \circ D_3)$$

and

$$\sum_{x\in\mathbb{F}_{2^m}} \chi\Big(h_{a_r}(x)\Big) = 2^m - 2\,\mathrm{wt}(h_{a_r}).$$

The latter condition is equivalent to

$$2/3 + 2/3(2^m - 2\,\mathrm{wt}(h_{a_r} \circ D_3)) - 2/3 \sum_{x\in\mathbb{F}_{2^m}} \chi\Big(\mathrm{Tr}_1^m(x^{-1}) + h_{a_r}(D_3(x))\Big)$$
$$-1 - (2^m - 2\,\mathrm{wt}(h_{a_r})) + \sum_{x\in\mathbb{F}_{2^m}} \chi\Big(\mathrm{Tr}_1^m(x^{-1}) + h_{a_r}(x)\Big) = 1$$

that is,

$$3 \sum_{x\in\mathbb{F}_{2^m}} \chi\Big(\mathrm{Tr}_1^m(x^{-1}) + h_{a_r}(x)\Big) - 2 \sum_{x\in\mathbb{F}_{2^m}} \chi\Big(\mathrm{Tr}_1^m(x^{-1}) + h_{a_r}(D_3(x))\Big)$$
$$= 4 + 2^m + 4\,\mathrm{wt}(h_{a_r} \circ D_3) - 6\,\mathrm{wt}(h_{a_r}).$$

□

Using the previous results, we prove the following characterization of semi-bentness for functions in the form (17.21).

Theorem 17.2.17 ([27]). *Let $n = 2m$ with m odd. Let $b \in \mathbb{F}_4^\star$, β be a primitive element of $\mathbb{F}_4$ and $c \in \mathbb{F}_{2^m}^\star$. Let $f_{a_r,b,c}$ be the function defined on $\mathbb{F}_{2^n}$ whose expression is of the form (17.21). Let h_{a_r} be the related function defined on $\mathbb{F}_{2^m}$ by $h_{a_r}(x) = \sum_{r\in R} \mathrm{Tr}_1^m(a_r D_r(x))$, where $D_r(x)$ is the Dickson polynomial of degree r. Then*

1. $f_{a_r,\beta,c}$ is semi-bent if and only if, $f_{a_r,\beta^2,c}$ is semi-bent, if and only if,

$$\sum_{x\in\mathbb{F}_{2^m}} \chi\Big(\mathrm{Tr}_1^m(x^{-1}) + h_{a_r}(D_3(x))\Big) = 2^m - 2\,\mathrm{wt}(h_{a_r} \circ D_3) + 4.$$

2. $f_{a_r,1,c}$ *is semi-bent if and only if,*

$$3\sum_{x\in\mathbb{F}_{2^m}}\chi\Big(\mathrm{Tr}_1^m(x^{-1})+h_{a_r}(x)\Big)-2\sum_{x\in\mathbb{F}_{2^m}}\chi\Big(\mathrm{Tr}_1^m(x^{-1})+h_{a_r}(D_3(x))\Big)$$
$$=4+2^m+4\,\mathrm{wt}(h_{a_r}\circ D_3)-6\,\mathrm{wt}(h_{a_r}).$$

Proof. For $b\in\mathbb{F}_4^\star$, $a_r\in\mathbb{F}_{2^m}^\star$, denote by $g_{a_r,b}$ the function $\sum_{r\in R}\mathrm{Tr}_1^n(a_rx^{r(2^m-1)})+\mathrm{Tr}_1^2(bx^{\frac{2^n-1}{3}})$ and by $g_{a_r,0}$ the function $\sum_{r\in R}\mathrm{Tr}_1^n(a_rx^{r(2^m-1)})$. Since m is odd, the function $g_{a_r,b}$ is constant on each (multiplicative) coset $u\mathbb{F}_{2^m}^\star$ ($u\in U$) that is, we have:

$$\forall u\in U,\forall y\in\mathbb{F}_{2^m}^\star, g_{a_r,b}(uy)=g_{a_r,b}(u).$$

Using the polar decomposition, the Walsh transform of $f_{a_r,b,c}$ at every $\omega\in\mathbb{F}_{2^n}$ is given by

$$\widehat{\chi_{f_{a_r,b,c}}}(\omega)=\sum_{x\in\mathbb{F}_{2^n}}\chi\Big(f_{a_r,b,c}(x)+\mathrm{Tr}_1^n(xw)\Big)$$
$$=1+\sum_{x\in\mathbb{F}_{2^n}^\star}\chi\Big(f_{a_r,b,c}(x)+\mathrm{Tr}_1^n(xw)\Big)$$
$$=1+\sum_{u\in U}\chi\Big(g_{a_r,b}(u)\Big)\sum_{y\in\mathbb{F}_{2^m}^\star}\chi\Big(\mathrm{Tr}_1^m(cy^{2^m+1}+\mathrm{Tr}_m^n(wu)y)\Big)$$
$$=1+\sum_{u\in U}\chi\Big(g_{a_r,b}(u)\Big)\sum_{y\in\mathbb{F}_{2^m}^\star}\chi\Big(\mathrm{Tr}_1^m(c^{\frac{1}{2}}y+\mathrm{Tr}_m^n(wu)y)\Big)$$
$$=1-\sum_{u\in U}\chi\Big(g_{a_r,b}(u)\Big)$$
$$+\sum_{u\in U}\chi\Big(g_{a_r,b}(u)\Big)\sum_{y\in\mathbb{F}_{2^m}}\chi\Big(\mathrm{Tr}_1^m((c^{\frac{1}{2}}+\mathrm{Tr}_m^n(wu))y)\Big)$$
$$=1-\sum_{u\in U}\chi\Big(g_{a_r,b}(u)\Big)+2^m\sum_{u\in U\mid c^{\frac{1}{2}}+\mathrm{Tr}_m^n(wu)=0}\chi\Big(g_{a_r,b}(u)\Big).$$

Thanks to Lemma 17.2.2, we obtain

$$\widehat{\chi_{f_{a_r,b,c}}}(\omega)\equiv 1-\sum_{u\in U}\chi\Big(g_{a_r,b}(u)\Big)\pmod{2^{m+1}}.$$

But

$$-2^{m+1} < -2^m \leq 1 - \sum_{u \in U} \chi\Big(g_{a_r,b}(u)\Big) \leq 2^m + 2 < 2^{m+1}$$

therefore, $f_{a_r,b,c}$ is semi-bent if and only if, $\sum_{u\in U} \chi\Big(g_{a_r,b}(u)\Big) = 1$. We conclude thanks to Corollary 17.5.4. □

Proposition 17.2.18 ([27]). *Let $n = 2m$ with m odd. For $r \in R$, $a_r \in \mathbb{F}_{2^m}^{\star}$, β a primitive element of $\mathbb{F}_4$ and $c \in \mathbb{F}_{2^m}^{\star}$, let $f_{a_r,\beta,c}$ a function of the form (17.21).*

1. Let d be a positive integer such that $\gcd(d, \frac{2^m+1}{3}) = 1$. Let $h_{a_r,\beta,c}$ be the function

$$h_{a_r,\beta,c}(x) = \sum_{r\in R} \mathrm{Tr}_1^n(a_r x^{dr(2^m-1)})\mathrm{Tr}_1^2(\beta x^{\frac{2^n-1}{3}}) + \mathrm{Tr}_1^m(cx^{2^m+1}).$$

Then, $h_{a_r,\beta,c}$ is semi-bent if and only if, $f_{a_r,\beta,c}$ is semi-bent.

2. Suppose $m \not\equiv 3 \pmod 6$. Let d be a positive integer such that $\gcd(d, 2^m+1) = 3$. Let $h_{a_r,1,c}$ be the function

$$h_{a_r,1,c}(x) = \sum_{r\in R} \mathrm{Tr}_1^n(a_r x^{dr(2^m-1)}) + \mathrm{Tr}_1^2(x^{\frac{2^n-1}{3}}) + \mathrm{Tr}_1^m(cx^{2^m+1}).$$

If $f_{a_r,\beta,c}$ is semi-bent, then $h_{a_r,1,c}$ is semi-bent.

Proof. For two integers r and d, set

$$g_{a_r,0}(x) = \sum_{r\in R} \mathrm{Tr}_1^n(a_r x^{r(2^m-1)});$$

$$h_{a_r,0}(x) = \sum_{r\in R} \mathrm{Tr}_1^n(a_r x^{dr(2^m-1)}).$$

Proof of 1): according to the proof of Theorem 17.5.6 and relation (17.22), $h_{a_r,\beta,c}$ (resp. $f_{a_r,\beta,c}$) is semi-bent if and only if $\sum_{v\in V} \chi\Big(h_{a_r,0}(v)\Big) = -1$ (resp. $\sum_{v\in V} \chi\Big(g_{a_r,0}(v)\Big) = -1$). Now, the integers $\frac{2^m+1}{3}$ and d are co-prime thus, the mapping $v \mapsto v^d$ is then a permutation of V. Therefore,

$$\sum_{v\in V} \chi\Big(h_{a_r,0}(v)\Big) = \sum_{v\in V} \chi\Big(g_{a_r,0}(v^d)\Big) = \sum_{v\in V} \chi\Big(g_{a_r,0}(v)\Big).$$

The result follows.

Proof of 2): the function $f_{a_r,\beta,c}$ is semi-bent thus, according to the proof of Theorem 17.5.6 and relation (17.22), $\sum_{v\in V} \chi\Big(g_{a_r,0}(v)\Big) = -1$. We have to prove that $h_{a_r,1,c}$ is semi-bent, that is, $\sum_{v\in V}\chi\Big(h_{a_r,0}(v)\Big)-2\sum_{v\in V}\chi\Big(h_{a_r,0}(\zeta v)\Big)=1$, according to the proof of Theorem 17.5.6 and relation (17.23). But

$$\sum_{v\in V}\chi\Big(h_{a_r,0}(v)\Big)+\sum_{v\in V}\chi\Big(h_{a_r,0}(\zeta v)\Big)+\sum_{v\in V}\chi\Big(h_{a_r,0}(\zeta^2 v)\Big)=\sum_{u\in U}\chi\Big(h_{a_r,0}(u)\Big)$$

and according to relation (17.26) we have,

$$\sum_{v\in V}\chi\Big(h_{a_r,0}(\zeta v)\Big)=\sum_{v\in V}\chi\Big(h_{a_r,0}(\zeta^2 v)\Big).$$

Therefore, the condition

$$\sum_{v\in V}\chi\Big(h_{a_r,0}(v)\Big)-2\sum_{v\in V}\chi\Big(h_{a_r,0}(\zeta v)\Big)=1$$

is equivalent to

$$2\sum_{v\in V}\chi\Big(h_{a_r,0}(v)\Big)-\sum_{u\in U}\chi\Big(h_{a_r,0}(u)\Big)=1.$$

Now, since $\gcd(d,2^m+1)=3$ and the mapping $v\mapsto v^3$ is a permutation when $m\not\equiv 3 \pmod 6$, one has

$$\sum_{v\in V}\chi\Big(h_{a_r,0}(v)\Big)=\sum_{v\in V}\chi\Big(g_{a_r,0}(v^d)\Big)=\sum_{v\in V}\chi\Big(g_{a_r,0}(v^3)\Big)=\sum_{v\in V}\chi\Big(g_{a_r,0}(v)\Big).$$

On the other hand, note that (since $\gcd(d,2^m+1)=3$)

$$\sum_{u\in U}\chi\Big(h_{a_r,0}(u)\Big)=\sum_{u\in U}\chi\Big(g_{a_r,0}(u^d)\Big)=\sum_{u\in U}\chi\Big(g_{a_r,0}(u^3)\Big)=3\sum_{v\in V}\chi\Big(g_{a_r,0}(v)\Big).$$

Hence, $2\sum_{v\in V}\chi\Big(h_{a_r,0}(v)\Big)-\sum_{u\in U}\chi\Big(h_{a_r,0}(u)\Big)=-2-(-3)=1$, proving that $h_{a_r,1,c}$ is semi-bent. □

To conclude this section, some functions in polynomial form in even dimension are considered in this section. We contribute to the knowledge of the class of semi-bent Boolean functions by deriving explicit criteria by means of Kloosterman sums and exponential sums involving Dickson polynomial for determining whether a function expressed as a sum of trace functions is semi-bent or not. Kloosterman sums are used as a very convenient tool to study the semi-bentness property of several functions. In particular, we have showed that the values 0 and 4 of Kloosterman sums defined on $\mathbb{F}_{2^m}$ give rise to semi-bent functions on $\mathbb{F}_{2^n}$. Tables 17.1 and 17.2 summarize these results.

Table 17.1 Families of semi-bent functions on $\mathbb{F}_{2^n}$ for $K_m(a) = 4$

Class of functions	Property	Conditions	References
$\mathrm{Tr}_1^n\left(ax^{r(2^m-1)}\right) + \mathrm{Tr}_1^2\left(bx^{\frac{2^n-1}{3}}\right) + \mathrm{Tr}_1^n\left(cx^{(2^m-1)\frac{1}{2}+1}\right)$; m odd	Semi-bent	$K_m(a) = 4$	[27]
$\mathrm{Tr}_1^n\left(ax^{3(2^m-1)}\right) + \mathrm{Tr}_1^n\left(cx^{(2^m-1)\frac{1}{2}+1}\right) + \mathrm{Tr}_1^2\left(bx^{\frac{2^n-1}{3}}\right)$; m odd and $m \not\equiv 3 \pmod 6$	Semi-bent	$K_m(a) = 4$	[27]
$\mathrm{Tr}_1^n\left(ax^{r(2^m-1)}\right) + \mathrm{Tr}_1^2\left(bx^{\frac{2^n-1}{3}}\right)$ $+\mathrm{Tr}_1^n\left(cx^{(2^m-1)\frac{1}{2}+1}\right) + \mathrm{Tr}_1^n\left(x^{(2^m-1)\frac{1}{4}+1}\right)$; m odd	Semi-bent	$K_m(a) = 4$	[27]
$\mathrm{Tr}_1^n\left(ax^{r(2^m-1)}\right) + \mathrm{Tr}_1^2\left(bx^{\frac{2^n-1}{3}}\right)$ $+\mathrm{Tr}_1^n\left(cx^{(2^m-1)\frac{1}{2}+1}\right) + \mathrm{Tr}_1^n\left(x^{3(2^m-1)+1}\right)$; $\mathrm{Tr}_m^n c = 1$, m odd	Semi-bent	$K_m(a) = 4$	[27]
$\mathrm{Tr}_1^n\left(ax^{r(2^m-1)}\right) + \mathrm{Tr}_1^n\left(\alpha x^{2^m+1}\right)$ $+\mathrm{Tr}_1^n\left(\sum_{i=1}^{2^{\nu-1}-1} x^{(2^m-1)\frac{i}{2^\nu}+1}\right) + \mathrm{Tr}_1^2\left(bx^{\frac{2^n-1}{3}}\right)$; $\gcd(\nu, m) = 1$, $\alpha \in \mathbb{F}_{2^n}$, $\mathrm{Tr}_m^n \alpha = 1$, m odd	Semi-bent	$K_m(a) = 4$	[27]

Table 17.2 Families of semi-bent functions on $\mathbb{F}_{2^n}$ for $K_m(a) = 0$

Class of functions	Property	Conditions	References
$\mathrm{Tr}_1^n\left(ax^{r(2^m-1)}\right) + \mathrm{Tr}_1^n\left(cx^{(2^m-1)\frac{1}{2}+1}\right)$	Semi-bent	$K_m(a) = 0$	[27]
$\mathrm{Tr}_1^n\left(ax^{r(2^m-1)}\right) + \mathrm{Tr}_1^n\left(cx^{(2^m-1)\frac{1}{2}+1}\right)$ $+\mathrm{Tr}_1^n\left(x^{(2^m-1)\frac{1}{4}+1}\right)$; $\mathrm{Tr}_m^n c = 1$, m odd	Semi-bent	$K_m(a) = 0$	[27]
$\mathrm{Tr}_1^n\left(ax^{r(2^m-1)}\right) + \mathrm{Tr}_1^n\left(cx^{(2^m-1)\frac{1}{2}+1}\right)$ $+\mathrm{Tr}_1^n\left(x^{(2^m-1)3+1}\right)$; $\mathrm{Tr}_m^n c = 1$	Semi-bent	$K_m(a) = 0$	[27]
$\mathrm{Tr}_1^n\left(ax^{r(2^m-1)}\right) + \mathrm{Tr}_1^n\left(cx^{(2^m-1)\frac{1}{2}+1}\right)$ $+\mathrm{Tr}_1^n\left(x^{(2^m-1)\frac{1}{6}+1}\right)$; $\mathrm{Tr}_m^n c = 1$, m even	Semi-bent	$K_m(a) = 0$	[27]
$\mathrm{Tr}_1^n\left(ax^{r(2^m-1)}\right) + \mathrm{Tr}_1^n\left(\alpha x^{2^m+1}\right)$ $+\mathrm{Tr}_1^n\left(\sum_{i=1}^{2^{\nu-1}-1} x^{(2^m-1)\frac{i}{2^\nu}+1}\right)$; $\gcd(\nu, m) = 1$, $\alpha \in \mathbb{F}_{2^n}$, $\mathrm{Tr}_m^n \alpha = 1$	Semi-bent	$K_m(a) = 0$	[27]

17.3 Semi-bent Functions with Multiple Trace Terms and Hyperelliptic Curves

In chronological order, the results of this section were established before those in Chap. 11. More precisely, in the line of the work of Lisonek [20], we have first established the results of this section in the framework of semi-bent functions before extending them for hyper-bent functions.

In this section we provide efficient characterizations of the semi-bentness property of several families of Boolean functions in univariate representation with multiple trace terms expressed by means of trace functions via Dillon-like exponents and Niho exponents with even number of variables. To this end, we have made precise firstly the connection between the semi-bentness property of such functions and some exponential sums involving Dickson polynomials. Next, we gave a link between the property of semi-bentness and the number of rational points on certain hyperelliptic curves. We exploit the connections between semi-bentness property and binary hyperelliptic curves to produce a polynomial complexity test which is of use in constructing semi-bent functions with multiple trace terms. In the following, we present briefly some results of our study but we do not provide proofs since we have already formulated precisely in this book the connection between exponential sums and cardinalities of hyperelliptic curve in Chap. 11 (see Proposition 11.2.3 and Proposition 11.2.4).

In the following, we consider four infinite classes of functions with multiple trace terms defined on $\mathbb{F}_{2^n}$. We denote by E the set of representatives of the cyclotomic classes modulo 2^n-1 for which each class has full size n. Let $f_{a_r,b,c}$ $f'_{a_r,b}$, $\tilde{f}_{a_r,b',c}$ and $\tilde{f}'_{a_r,b'}$ be the functions defined on $\mathbb{F}_{2^n}$ whose polynomial form is given by (17.11), (17.12), (17.13) and (17.14), respectively.

$$f_{a_r,b,c}(x) := \sum_{r\in R} \mathrm{Tr}_1^n(a_r x^{r(2^m-1)}) + \mathrm{Tr}_1^2(bx^{\frac{2^n-1}{3}}) + \mathrm{Tr}_1^m(cx^{2^m+1}), \tag{17.11}$$

$$f'_{a_r,b}(x) := \sum_{r\in R} \mathrm{Tr}_1^n(a_r x^{r(2^m-1)}) + \mathrm{Tr}_1^2(bx^{\frac{2^n-1}{3}}) + \mathrm{Tr}_1^m(x^{2^m+1}) + \mathrm{Tr}_1^n\left(x^{(2^m-1)s+1}\right), \tag{17.12}$$

$$\tilde{f}_{a_r,b',c}(x) := \sum_{r\in R} \mathrm{Tr}_1^n(a_r x^{r(2^m-1)}) + \mathrm{Tr}_1^4(b'x^{\frac{2^n-1}{5}}) + \mathrm{Tr}_1^m(cx^{2^m+1}), \tag{17.13}$$

$$\tilde{f}'_{a_r,b'}(x) := \sum_{r\in R} \mathrm{Tr}_1^n(a_r x^{r(2^m-1)}) + \mathrm{Tr}_1^4(b'x^{\frac{2^n-1}{5}}) + \mathrm{Tr}_1^m(x^{2^m+1}) + \mathrm{Tr}_1^n\left(x^{(2^m-1)s'+1}\right), \tag{17.14}$$

where $R \subseteq E$, $a_r \in \mathbb{F}_{2^m}^\star$, $b \in \mathbb{F}_4^\star$, $b' \in \mathbb{F}_{16}^\star$, $c \in \mathbb{F}_{2^m}^\star$, $s \in \{1/4, 3\}$ and $s' \in \{1/6, 3\}$(the fractions $1/4$ and $1/6$ are understood modulo 2^m+1).

The following statement provides a characterization of the property of semi-bentness for functions of the form (17.11) and (17.12) in terms of cardinalities of hyperelliptic curves.

Theorem 17.3.1 ([28]). *Let $n = 2m$ with m odd. Let $b \in \mathbb{F}_4^\star$, β be a primitive element of $\mathbb{F}_4$ and $c \in \mathbb{F}_{2^m}^\star$. Let $f_{a_r,b,c}$ (resp. $f'_{a_r,b}$) be the function defined on $\mathbb{F}_{2^n}$ whose expression is of the form (17.11) (resp. form (17.12)). Let h_{a_r} be the related function defined on $\mathbb{F}_{2^m}$ by $h_{a_r}(x) = \sum_{r\in R} \mathrm{Tr}_1^m(a_r D_r(x))$, where $D_r(x)$ is the Dickson polynomial of degree r. Moreover, let $H_{a_r}^{(1)}$, $H_{a_r}^{(2)}$ and $H_{a_r}^{(3)}$ be the (affine) curves defined over $\mathbb{F}_{2^m}$ by*

$$H_{a_r}^{(1)} : y^2 + y = \sum_{r \in R} a_r D_r(x),$$

$$H_{a_r}^{(2)} : y^2 + y = \sum_{r \in R} a_r D_r(x + x^3),$$

$$H_{a_r}^{(3)} : y^2 + xy = x + x^2 \sum_{r \in R} a_r D_r(x).$$

a) If β is a primitive element of $\mathbb{F}_4$, then $f_{a_r,\beta,c}$ (resp. $f'_{a_r,\beta}$) is semi-bent if and only if

$$2\#H_{a_r}^{(2)} - (\#H_{a_r}^{(1)} + \#H_{a_r}^{(3)}) = -3.$$

b) If $b = 1$, then $f_{a_r,1,c}$ (resp. $f'_{a_r,1}$) is semi-bent if and only if

$$4\#H_{a_r}^{(2)} - 5\#H_{a_r}^{(1)} + \#H_{a_r}^{(3)} = 3.$$

The following statement provides a characterization of the property of semi-bentness for functions of the form (17.13) and form (17.14) in terms of cardinalities of hyperelliptic curves.

Theorem 17.3.2 ([28]). *Assume $m := \frac{n}{2} \equiv 2 \pmod 4$. Let $R \subseteq E$ where E is a set of representatives of the cyclotomic classes modulo $2^n - 1$ for which each class has the full size n. Let $b' \in \mathbb{F}_{16}^\star$ and $a_r \in \mathbb{F}_{2^m}^\star$. Let $\tilde{f}_{a_r,b',c}$ (resp. $\tilde{f}'_{a_r,b'}$) be the function defined on $\mathbb{F}_{2^n}$ whose expression is of the form (17.13) (resp. form (17.14)). Let h_{a_r} be the related function defined on $\mathbb{F}_{2^m}$ by $h_{a_r}(x) = \sum_{r \in R} \mathrm{Tr}_1^m(a_r D_r(x))$, where $D_r(x)$ is the Dickson polynomial of degree r. Moreover, let $H_{a_r}^{(1)}$, $H_{a_r}^{(3)}$, $\tilde{H}_{a_r}^{(2)}$ and $\tilde{H}_{a_r}^{(3)}$ be the (affine) curves defined over $\mathbb{F}_{2^m}$ by*

$$H_{a_r}^{(1)} : y^2 + y = \sum_{r \in R} a_r D_r(x),$$

$$H_{a_r}^{(3)} : y^2 + xy = x + x^2 \sum_{r \in R} a_r D_r(x),$$

$$\tilde{H}_{a_r}^{(2)} : y^2 + y = \sum_{r \in R} a_r D_r(x + x^3 + x^5),$$

$$\tilde{H}_{a_r}^{(3)} : y^2 + xy = x + x^2 \sum_{r \in R} a_r D_r(x + x^3 + x^5).$$

1. If Let b' a primitive element of $\mathbb{F}_{16}$ such that $\mathrm{Tr}_1^4(b') = 0$, then $\tilde{f}_{a_r,b',c}$ (resp. $\tilde{f}'_{a_r,b'}$) is semi-bent if and only if,

$$\#\tilde{H}_{a_r}^{(2)} - \#\tilde{H}_{a_r}^{(3)} = 5.$$

2. *If* $b' = 1$*, then* $\tilde{f}_{a_r,b',c}$ *(resp.* $\tilde{f}'_{a_r,b'}$*) is semi-bent if and only if*

$$2(\#\tilde{H}^{(2)}_{a_r} - \#\tilde{H}^{(3)}_{a_r}) - 5(\#H^{(1)}_{a_r} - \#H^{(3)}_{a_r}) = 5.$$

3. *Assume* $a_r \in \mathbb{F}_{2^{\frac{m}{2}}}$*. If* $b' \in \{\beta, \beta^2, \beta^3\beta^4\}$ *where* β *is a primitive 5-th root of unity in* $\mathbb{F}_{16}$*, then* $\tilde{f}_{a_r,b',c}$ *(resp.* $\tilde{f}'_{a_r,b'}$*) is semi-bent if and only if,*

$$\#\tilde{H}^{(2)}_{a_r} - \#\tilde{H}^{(3)}_{a_r} + 5(\#H^{(1)}_{a_r} - \#H^{(3)}_{a_r}) = -10.$$

4. *Assume* $a_r \in \mathbb{F}_{2^{\frac{m}{2}}}$*. If* b' *is a primitive element of* $\mathbb{F}_{16}$ *such that* $\mathrm{Tr}_1^4(b') = 1$*, then* $\tilde{f}_{a_r,b',c}$ *(resp.* $\tilde{f}'_{a_r,b'}$*) if and only if,*

$$3\left(\#\tilde{H}^{(2)}_{a_r} - \#\tilde{H}^{(3)}_{a_r}\right) + 5\left(\#H^{(3)}_{a_r} - \#H^{(1)}_{a_r}\right) = -10.$$

5. *Assume* $a_r \in \mathbb{F}_{2^{\frac{m}{2}}}$*. If* $b' \in \{\beta + \beta^2, \beta + \beta^3, \beta^2 + \beta^4, \beta^3 + \beta^4, \beta + \beta^4, \beta^2 + \beta^3\}$ *where* β *is a primitive 5-th root of unity in* $\mathbb{F}_{16}$*, then* $\tilde{f}_{a_r,b',c}$ *(resp.* $\tilde{f}'_{a_r,b'}$*) is semi-bent if and only if,*

$$\#\tilde{H}^{(2)}_{a_r} - \#\tilde{H}^{(3)}_{a_r} = 5.$$

We will not give more details in this section since we have explained in details in Chapter11 the uses of the hyperelliptic curve formalism to reduce computational complexity.

17.4 General Constructions of Semi-bent Functions

In the following, we generalize the constructions given in Sect. 17.2. More precisely, we prove a key result (Theorem 17.5.11, [7]) which gives rise to the construction of several classes of semi-bent functions in even dimension.

The results presented in Sects. 17.2.1 and 17.2.2 have been extended in [2]. In this article [2], a new construction of infinite classes of binary semi-bent functions in polynomial trace was studied. The extension is achieved by inserting mappings h on $\mathbb{F}_{2^n}$ which can be expressed as $h(0) = 0$ and $h(uy) = h_1(u)h_2(y)$ with u ranging over the circle U of unity of $\mathbb{F}_{2^n}$, $y \in \mathbb{F}^*_{2^m}$ and $uy \in \mathbb{F}^*_{2^n}$, where h_1 is a isomorphism on U and h_2 is an arbitrary mapping on $\mathbb{F}^*_{2^m}$. The characterization of the semi-bentness property of the extended family was given in terms of classical binary exponential sums and binary polynomials.

17.4.1 Characterizations of Semi-bent Functions

In the next theorem, given a spread $(E_i)_{i=1,\ldots,2^m+1}$, we characterize when a function whose restriction to every E_i^* is affine (i.e. a function equal to the sum of a function whose restriction to every E_i is linear and of a function whose restriction to every E_i^* is constant) is semi-bent:

Theorem 17.4.1 ([7]). *Let $m \geq 2$ and $n = 2m$. Let $\{E_i, i = 1, \ldots, 2^m + 1\}$ be a spread in $\mathbb{F}_{2^n}$ and h a Boolean function whose restriction to every E_i is linear (possibly null). Let S be any subset of $\{1, \ldots, 2^m + 1\}$ and $g = \sum_{i \in S} 1_{E_i} \pmod 2$ where 1_{E_i} is the indicator of E_i. Then $g + h$ is semi-bent if and only if g and h are bents.*

We call g a $\mathcal{PS}_{ap}$-like bent function.

Proof. We may without loss of generality assume that $g(0) = 0$, that is, S has even size (otherwise, we replace g by $g + 1$). Let us then compute the Walsh–Hadamard transform of $g + h$. We have for all $c \in \mathbb{F}_{2^n}$:

$$\widehat{\chi_{g+h}}(c) = \sum_{x \in \mathbb{F}_{2^n}} \chi((g+h)(x) + Tr_1^n(cx))$$

$$= 1 + \sum_{i=1}^{2^m+1} \sum_{e \in E_i^\star} \chi(g(e) + h(e) + Tr_1^n(ce))$$

since $\bigcup_{i=1}^{2^m+1} E_i^\star = \mathbb{F}_{2^n}^\star$ and $E_i^\star \cap E_j^\star = \emptyset$. Let us denote by g_i the value of g on $E_i^\star$, by h_i the restriction of h to E_i and by $I(c)$ the set $\{i \in [1, \ldots, 2^m + 1]; \forall e \in E_i, h(e) = Tr_1^n(ce)\}$. We have, for every $c \in \mathbb{F}_{2^n}$:

$$\widehat{\chi_{g+h}}(c) = 1 + \sum_{i=1}^{2^m+1} \chi(g_i) \sum_{e \in E_i^\star} \chi(h_i(e) + Tr_1^n(ce))$$

$$= 1 - \sum_{i=1}^{2^m+1} \chi(g_i)$$

$$+ \sum_{i=1}^{2^m+1} \chi(g_i) \sum_{e \in E_i} \chi(h_i(e) + Tr_1^n(ce)).$$

Since h_i is linear on E_i, one has $\sum_{e \in E_i} \chi(h_i(e) + Tr_1^n(ce)) = 2^m$ if $i \in I(c)$ and 0 otherwise. Therefore:

$$\forall c \in \mathbb{F}_{2^n}, \quad \widehat{\chi_{g+h}}(c) = 1 - \sum_{i=1}^{2^m+1} \chi(g_i) + 2^m \sum_{i \in I(c)} \chi(g_i). \tag{17.15}$$

On the other hand, the Walsh–Hadamard transform of h is (take $g = 0$ in the preceding calculation) :

$$\widehat{\chi_h}(c) = 2^m(\#I(c) - 1). \tag{17.16}$$

If g is bent then we know that $\sum_{i=1}^{2^m+1} \chi(g_i) = 1$. If h is bent then, according to (17.28), $\#I(c) \in \{0, 2\}$. Hence, if g and h are bent then, $\forall c \in \mathbb{F}_{2^n}$, $\widehat{\chi_{g+h}}(c) = 2^m \sum_{i\in I(c)} \chi(g_i) \in \{0, \pm 2^{m+1}\}$, proving that $g + h$ is semi-bent.

Conversely, let us assume that $g+h$ is semi-bent and let us show that, necessarily, g and h are bent. According to (17.27), we have $\sum_{i=1}^{2^m+1} \chi(g_i) \equiv 1 \pmod{2^m}$. In other words, $\sum_{i=1}^{2^m+1} \chi(g_i) = 1 + \epsilon 2^m$ with $\epsilon \in \{0, \pm 1\}$. Suppose that $\epsilon \in \{-1, 1\}$, then, for every c, $I(c)$ is non-empty, since if $I(c) = \emptyset$, $\widehat{\chi_{g+h}}(c) = -\epsilon 2^m \notin \{0, \pm 2^{m+1}\}$; this implies that the Walsh–Hadamard transform of h is non-negative and we have seen in Sect. 2.2 (Chap. 2) that h is then linear, say $h(x) = Tr_1^n(ax)$. We have then, according to (17.28): $\#I(c) = 1$ for $c \neq a$ and $\#I(c) = 2^m + 1$ for $c = a$; thus,

$$\begin{aligned}
\widehat{\chi_{g+h}}(a) &= -\epsilon 2^m + 2^m \sum_{i=1}^{2^m+1} \chi(g_i) \\
&= -\epsilon 2^m + 2^m + \epsilon 2^n \\
&= (1 - \epsilon)2^m + \epsilon 2^n \in \{2^n, 2^{m+1} - 2^n\}
\end{aligned}$$

a contradiction with the fact that $g + h$ is semi-bent. Therefore, we have $\epsilon = 0$, $\sum_{i=1}^{2^m+1} \chi(g_i) = 1$, which implies that g is bent. Let us now prove that h is bent. One has necessarily $\sum_{i\in I(c)} \chi(g_i) \in \{-2, 0, 2\}$. Thus, $I(c)$ is of even size for every c, which implies that $\widehat{\chi_h}(c)$ is congruent to 2^m modulo 2^{m+1}, which according to Lemma 1 in [4] implies that h is bent (that is, $\#I(c) \in \{0, 2\}$ for every c). □

Remark 17.4.2. We can modify the hypothesis of Theorem 17.5.11 by assuming that we have only a partial spread. We need then to add a condition on the E_i's, and we have only a sufficient condition for $g + h$ being semi-bent:

Let g be a bent function in the $\mathcal{PS}$ class, equal to the sum modulo 2 of the indicators of $l := 2^{m-1}$ or $2^{m-1} + 1$ pairwise "disjoint" vectorspaces E_i having dimension m, and h a bent function which is linear on each E_i. Assume additionally that for every $c \in \mathbb{F}_{2^n}$ there exist at most 2 indices i such that $\forall e \in E_i$, $h(e) = Tr_1^n(ce)$. Then $g + h$ is semi-bent. Indeed, we have $\widehat{\chi_{g+h}}(c) = \widehat{\chi_h}(c) - 2 \sum_{x\in\mathbb{F}_{2^n}/g(x)=1} (-1)^{h(x)+Tr_1^n(cx)}$ and therefore, since either $l = 2^{m-1}$ and $g(0) = 0$ or $l = 2^{m-1} + 1$ and $g(0) = 1$:

$$\begin{aligned}\widehat{\chi_{g+h}}(c) &= \widehat{\chi_h}(c) - 2\sum_{i=1}^{l}\sum_{e\in E_i}(-1)^{h(e)+Tr_1^n(ce)} + 2^m \\ &= -2^{m+1}\#\{i = 1,\cdots l; \forall e \in E_i,\ h(e) = Tr_1^n(ce)\} \\ &\quad + \widehat{\chi_h}(c) + 2^m\end{aligned}$$

As shown in [3], we have "$\forall e \in E_i; h(e) = Tr_1^n(ce)$" for some i if and only if $\widehat{\chi_h}(u) = 2^m$ for every $u \in c + E_i^{\perp}$ which implies in particular that $\widehat{\chi_h}(c) = 2^m$. Thus we have:

- either $\{i = 1,\cdots l; \forall e \in E_i,\ h(e) = Tr_1^n(ce)\} = \emptyset$ and $\widehat{\chi_{g+h}}(c) = \widehat{\chi_h}(c) + 2^m \in \{0, 2^{m+1}\}$;
- or $\#\{i = 1,\cdots l; \forall e \in E_i,\ h(e) = Tr_1^n(ce)\} \in \{1, 2\}$ and $\widehat{\chi_{g+h}}(c) \in \{0, -2^{m+1}\}$. Hence, $g + h$ is semi-bent.

17.4.2 Constructions of Semi-bent Functions

17.4.2.1 Constructions in Bivariate Form

We identify $\mathbb{F}_{2^m} \times \mathbb{F}_{2^m}$ with $\mathbb{F}_{2^n}$ by considering an orthonormal basis of the $\mathbb{F}_{2^m}$-vectorspace $\mathbb{F}_{2^n}$. We consider the vectorspaces $E_a = \{(x, ax); x \in \mathbb{F}_{2^m}\}$ where $a \in \mathbb{F}_{2^m}$ and $E_\infty = \{(0, y); y \in \mathbb{F}_{2^m}\} = \{0\} \times \mathbb{F}_{2^m}$. The bivariate version of the spread $\{u\mathbb{F}_{2^m}; u \in U\}$ is the spread $\{E_a; a \in \mathbb{F}_{2^m}\} \cup \{E_\infty\}$. It can be directly checked that the E_a's and E_∞ are vectorspaces of dimension m and that we have $E_a \cap E_b = \{0\}$ for every pair (a, b) such that $a \neq b$ and $E_\infty \cap E_a = \{0\}$ for every $a \in \mathbb{F}_{2^m}$. Note that any function g in the $\mathcal{PS}_{ap}$ class can be viewed as the indicator of 2^{m-1} or $2^{m-1}+1$ of these vectorspaces. Moreover, function h having linear restrictions to the E_a's is necessarily defined as, $x, y \in \mathbb{F}_{2^m}$, $h(x, y) = Tr_1^m\left(xH\left(\frac{y}{x}\right)\right)$ if $x \neq 0$ and $h(0, y) = Tr_1^m(\mu y)$ for some mapping H over $\mathbb{F}_{2^m}$ and some $\mu \in \mathbb{F}_{2^m}$. A linear function has bivariate form $\ell(x, y) = Tr_1^m(cx + c'y)$, where $x, y, c, c' \in \mathbb{F}_{2^m}$ and the set denoted by $I(c)$ in Sect. 17.4.1 has to be denoted by $I(c, c')$ here. Then for every $(c, c') \in \mathbb{F}_{2^m} \times \mathbb{F}_{2^m}$ the set $I(c, c')$ equals $\{a \in \mathbb{F}_{2^m}; \forall x \in \mathbb{F}_{2^m}, Tr_1^m(xH(a)) = Tr_1^m(cx + c'ax)\} = \{a \in \mathbb{F}_{2^m}; H(a) = c + c'a\}$ if $c' \neq \mu$ and $\{a \in \mathbb{F}_{2^m}; H(a) = c + c'a\} \cup \{\infty\}$ if $c' = \mu$. Hence, the sets $I(c, c')$ depend on the pre-image of c by the mapping $H + c'Id$ (where Id denotes the identity map). According to (17.28), the necessary and sufficient condition for h being bent is that, denoting $G(x) = H(x) + \mu x$, then G is a permutation and for every $c' \neq 0$ the function $G(x) + c'x$ is 2-to-1. Such bent functions have been first introduced by Dillon in [13]. He could exhibit in the class of such functions only the example of the function h in Corollary 17.4.3 below. But other examples have been found recently in [6] and lead to Corollary 17.4.5.

Corollary 17.4.3 ([7]). *Let g be a function in the $\mathcal{PS}_{ap}$ class (see definition in 5.1 (Chap. 4)). Let i be any integer co-prime with m and $h(x,y) = \mathrm{Tr}_1^m(xy^{2^i-1})$. Then the function $g+h$ is semi-bent.*

Remark 17.4.4. According to [1, Theorem 6], the permutations y^{2^i-1} are the only permutations π such that $x\pi(x)$ is linear.

Corollary 17.4.5 ([7]). *Let g be a function in the $\mathcal{PS}_{ap}$ class. Let h be one of the following functions [6] :*

- $h(x,y) = \mathrm{Tr}_1^m(x^{-5}y^6)$, *$m$ odd;*
- $h(x,y) = \mathrm{Tr}_1^m(x^{\frac{5}{6}}y^{\frac{1}{6}})$, *$m$ odd:*
- $h(x,y) = \mathrm{Tr}_1^m(x^{-3\cdot(2^k+1)}y^{3\cdot 2^k+4})$, $m = 2k-1$;
- $h(x,y) = \mathrm{Tr}_1^m(x^{-3\cdot(2^{k-1}-1)}y^{3\cdot 2^{k-1}-2})$, $m = 2k-1$;
- $h(x,y) = \mathrm{Tr}_1^m(x^{1-2^k-2^{2k}}y^{2^k+2^{2k}})$, $m = 4k-1$;
- $h(x,y) = \mathrm{Tr}_1^m(x^{2^{3k-1}-2^{2k}+2^k}y^{1-2^{3k-1}+2^{2k}-2^k})$, $m = 4k-1$;
- $h(x,y) = \mathrm{Tr}_1^m(x^{1-2^{2k+1}-2^{3k+1}}y^{2^{2k+1}+2^{3k+1}})$, $m = 4k+1$;
- $h(x,y) = \mathrm{Tr}_1^m(x^{2^{3k+1}-2^{2k+1}+2^k}y^{1-2^{3k+1}+2^{2k+1}-2^k})$, $m = 4k+1$;
- $h(x,y) = \mathrm{Tr}_1^m(x^{1-2^k}y^{2^k} + x^{-(2^k+1)}y^{2^k+2} + x^{-3\cdot(2^k+1)}y^{3\cdot 2^k+4})$, $m = 2k-1$;
- $h(x,y) = \mathrm{Tr}_1^m(y(y^{2^k+1}x^{-(2^k+1)} + y^3x^{-3} + yx^{-1})^{2^{k-1}-1})$, $m = 2k-1$;
- $h(x,y) = \mathrm{Tr}_1^m(x^{\frac{5}{6}}y^{\frac{1}{6}} + x^{\frac{1}{2}}y^{\frac{1}{2}} + x^{\frac{1}{6}}y^{\frac{5}{6}})$, *$m$ odd;*
- $h(x,y) = \mathrm{Tr}_1^m(x[D_{\frac{1}{5}}(\frac{y}{x})]^6)$, *$m$ odd, where $D_{\frac{1}{5}}$ is the Dickson polynomial of index* $\frac{1}{5}$.

Then the function $g+h$ is semi-bent.

Remark 17.4.6. There are more bent functions in bivariate form in [6] whose expression are more complex.

17.4.2.2 Constructions in Univariate Form

We apply now Theorem 17.5.11 to the spread $\{u\mathbb{F}_{2^m}\,;\, u \in U\}$ where U is the multiplicative group $\{u \in \mathbb{F}_{2^n}; u^{2^m+1} = 1\}$. In this framework, the functions have to be considered in their univariate form.

- Nonlinear Boolean functions whose restriction to any vectorspace $u\mathbb{F}_{2^m}$ are linear are sums of Niho power functions, that is (see [14]) of functions of the form:

$$\mathrm{Tr}_1^{o((2^m-1)s+1)}\left(a_s x^{(2^m-1)s+1}\right) \text{ with } 1 \le s \le 2^m$$

  We can determine the value of $o((2^m-1)s+1)$ precisely:

Lemma 17.4.7 ([7]). *We have $o((2^m-1)s+1) = m$ if $s = 2^{m-1}+1$ (i.e. if $(2^m-1)s+1$ and 2^m+1 are conjugate) and $o((2^m-1)s+1) = n$ otherwise.*

Proof. $2^i((2^m-1)s+1) \equiv (2^m-1)s+1 \pmod{2^n-1}$ is equivalent to $(2^i-1)((2^m-1)s+1) \equiv 0 \pmod{2^n-1}$ and implies $(2^i-1)((2^m-1)s+1) \equiv 0 \pmod{2^m-1}$. The integers 2^m-1 and $(2^m-1)s+1$ being co-prime then, 2^m-1 divides 2^i-1 and then, m divides i. Now, we have $2^m((2^m-1)s+1) \equiv (1-2^m)s+2^m \pmod{2^n-1}$ is congruent to $(2^m-1)s+1$ modulo 2^n-1 if and only if $(2^m-1)(2s-1) \equiv 0 \pmod{2^n-1}$, that is, $s \equiv 2^{m-1}+1 \pmod{2^m+1}$. Therefore, $o((2^m-1)s+1) = m$ if $s = 2^{m-1}+1$ and n otherwise. □

- Some $\mathcal{PS}_{ap}$ functions can be obtained in the form

$$\sum_{r\in R} \mathrm{Tr}_1^{o((2^m-1)r)}\left(b_r x^{(2^m-1)r}\right) \text{ where } R \subset \{1,\cdots,2^m\}.$$

Collecting results provided in [14] and [8], we get a direct consequence of Theorem 17.5.11:

Corollary 17.4.8. *Let f be a Boolean function of the form:*

$$f(x) = \mathrm{Tr}_1^m(a_0 x^{2^m+1}) + \sum_{i=1}^{L} \mathrm{Tr}_1^n(a_i x^{(2^m-1)s_i+1}) + \sum_{r\in R} \mathrm{Tr}_1^{o((2^m-1)r)}(b_r x^{(2^m-1)r})$$

where L is some non-negative integer, $2 \le s_i \le 2^m$, $s_i \ne 2^{m-1}+1$, $1 \le r \le 2^m$, $a_0 \in \mathbb{F}_{2^m}$, $a_i \in \mathbb{F}_{2^m}$ *and* $b_r \in \mathbb{F}_{2^{o((2^m-1)r)}}$ *(with at least one coefficient* $a_i \ne 0$ *and one coefficient* $b_r \ne 0$*). Assume that:*

1) the number of roots u in $U := \{x \in \mathbb{F}_{2^n}; x^{2^m+1} = 1\}$ *of the equation* $\mathrm{Tr}_m^n(cu) + \sum_{i=1}^{L} \mathrm{Tr}_m^n(a_i u^{2s_i-1}) + a_0^{\frac{1}{2}} = 0$ *is either* 0 *or* 2 *for every* $c \in \mathbb{F}_{2^n}$,
2) the sum $\sum_{u\in U} \chi(\sum_{r\in R} \mathrm{Tr}_1^{o((2^m-1)r)}(b_r u^r))$ *is equal to* 1*. Then, f is semi-bent.*

Proof. Condition 1 is necessary and sufficient to ensure that the Niho part $h(x) := \mathrm{Tr}_1^m(a_0 x^{2^m+1}) + \sum_{i=1}^{L} \mathrm{Tr}_1^n(a_i x^{(2^m-1)s_i+1})$ is bent [14], while condition 2 ensures that the "Dillon" part $g(x) := \sum_{r\in R} \mathrm{Tr}_1^{o((2^m-1)r}(b_r x^{(2^m-1)r})$ is hyper-bent [8]. □

Remark 17.4.9. Condition 2 in Corollary 17.4.8 can be reworded by means of Kloosterman sums in particular cases [13, 17, 23–25] and [26].

Let us specify some infinite families of semi-bent functions in univariate form. Firstly, we give a list of infinite families containing bent functions defined on $\mathbb{F}_{2^n}$ belonging to the class $\mathcal{PS}_{ap}$; here, $K_m(a) := \sum_{x\in\mathbb{F}_{2^m}} \chi\big(\mathrm{Tr}_1^m(ax+\frac{1}{x})\big)$ denotes the binary Kloosterman sums on $\mathbb{F}_{2^m}$ and $C_m(a,a) := \sum_{x\in\mathbb{F}_{2^m}} \chi\big(\mathrm{Tr}_1^m(ax^3+ax)\big)$ denotes the cubic sums on $\mathbb{F}_{2^m}$:

- $g_1(x) = \mathrm{Tr}_1^n(ax^{r(2^m-1)})$; $\gcd(r, 2^m+1) = 1$, $a \in \mathbb{F}_{2^m}^{\star}$ such that $K_m(a) = 0$ [8].

- $g_2(x) = Tr_1^n(ax^{r(2^m-1)}) + Tr_1^2(bx^{\frac{2^n-1}{3}})$; $\gcd(r, 2^m+1) = 1, m > 3$ odd, $b \in \mathbb{F}_4^\star$, $a \in \mathbb{F}_{2^m}^\star$ such that $K_m(a) = 4$ [25].
- $g_3(x) = Tr_1^n(a\zeta^i x^{3(2^m-1)}) + Tr_1^2(\beta^j x^{\frac{2^n-1}{3}})$; m odd and $m \not\equiv 3 \pmod 6$, β is a primitive element of $\mathbb{F}_4$, ζ is a generator of the cyclic group U of (2^m+1)-th of unity, $(i,j) \in \{0,1,2\}^2$, $a \in \mathbb{F}_{2^m}^\star$ such that $K_m(a) = 4$ and $Tr_1^m(a^{1/3}) = 0$ [24].
- $g_4(x) = Tr_1^n(a\zeta^i x^{3(2^m-1)}) + Tr_1^2(\beta^j x^{\frac{2^n-1}{3}})$; m odd and $m \not\equiv 3 \pmod 6$, β is a primitive element of $\mathbb{F}_4$, ζ is a generator of the cyclic group U of (2^m+1)-th of unity, $i \in \{1,2\}, j \in \{0,1,2\}$, $a \in \mathbb{F}_{2^m}^\star$ such that $K_m(a) + C_m(a,a) = 4$ and $Tr_1^m(a^{1/3}) = 1$ [24].
- $g_5(x) = \sum_{i=1}^{2^{m-1}-1} Tr_1^n\left(\beta x^{i(2^m-1)}\right)$; $\beta \in \mathbb{F}_{2^m} \setminus \mathbb{F}_2$ [16].
- $g_6(x) = \sum_{i=1}^{2^{m-2}-1} Tr_1^n\left(\beta x^{i(2^m-1)}\right)$; m odd and $\beta^{(2^m-4)^{-1}} \in \{x \in \mathbb{F}_{2^m}^\star; Tr_1^m(x) = 0\}$ [16].

Secondly, we give a list of known Niho bent functions

- $h_1(x) = Tr_1^m\left(a_1 x^{2^m+1}\right)$; $a_1 \in \mathbb{F}_{2^m}^\star$
- $h_2(x) = Tr_1^n\left(a_1 x^{(2^m-1)\frac{1}{2}+1} + a_2 x^{(2^m-1)3+1}\right)$;
  $a_1 \in \mathbb{F}_{2^n}^\star$, $a_2^{2^m+1} = a_1 + a_1^{2^m} = \beta^5$ for some $\beta \in \mathbb{F}_{2^n}^\star$ [14]
- $h_3(x) = Tr_1^n\left(a_1 x^{(2^m-1)\frac{1}{2}+1} + a_2 x^{(2^m-1)\frac{1}{4}+1}\right)$;
  $a_1 \in \mathbb{F}_{2^n}^\star$ $a_2^{2^m+1} = a_1 + a_1^{2^m}$, m odd [14]
- $h_4(x) = Tr_1^n\left(a_1 x^{(2^m-1)\frac{1}{2}+1} + a_2 x^{(2^m-1)\frac{1}{6}+1}\right)$; $a_1 \in \mathbb{F}_{2^n}^\star$ $a_2^{2^m+1} = a_1 + a_1^{2^m}$, m even [14]
- $h_5(x) = Tr_1^n\left(\alpha x^{2^m+1} + \sum_{i=1}^{2^{r-1}-1} x^{s_i}\right)$, $r > 1$ such that $gcd(r,m) = 1$, $\alpha \in \mathbb{F}_{2^n}$ such that $\alpha + \alpha^{2^m} = 1$, $s_i = (2^m-1)\frac{i}{2^r} \pmod{2^m+1} + 1$, $i \in \{1, \cdots, 2^{r-1}-1\}$ [18]

We obtain new families in univariate form containing semi-bent functions. Each of them are of algebraic degree m (that is, the maximum degree for a semi-bent function).

Remark 17.4.10. The question arises of determining whether all these semi-bent functions are extendable to $(n+2)$-variable bent functions. We checked that all the known secondary constructions of bent functions which increase the number of variables by 2 fail to generate such bent functions from g and h. On the other hand, it is difficult to show that a given semi-bent function is the restriction of any bent function (the algebraic degrees of the semi-bent function and of the indicator of its Walsh–Hadamard support, for instance, do not help since in both cases they are bounded above by a number which is not smaller for the restriction of a bent function than for a semi-bent function). It is a simple matter to show that a given semi-bent function in n variables, for n even (respectively, for n odd), is the restriction of a semi-bent function (respectively of a bent function) in $n+1$ variables if and only if there exists a semi-bent function f' in n variables whose Walsh–Hadamard support $S_{f'} := \{a \in \mathbb{F}_{2^n} \,/\, \widehat{\chi_{f'}}(a) \neq 0\}$ is disjoint from S_f, the Walsh–Hadamard support of f (note that in the case where n is odd, $\{S_f, S_{f'}\}$ is a partition of $\mathbb{F}_{2^n}$). And a semi-bent function in n variables, for n even, is the restriction of a bent function in $n+2$ variables if and only if it is the restriction of a semi-bent function in $n+1$ variables which is the restriction of a bent function in $n+2$ variables. It is probable that there exist semi-bent functions constructed from Theorem 17.4.1 which are the restrictions of no bent function in $n+2$ variables, but we were not able to prove it.

Finally, we give some open problems (Problem 1 has been proposed by Matthew Geoffrey Parker).

Problem 17.4.11. Show that some semi-bent functions obtained in the previous section are not extendable to $(n+2)$-variable bent functions (or deduce new bent functions from them).

Problem 17.4.12. Determine whether there exist spreads which are not linearly equivalent to the spaces $u\mathbb{F}_{2^m}$ and if they exist, deduce related semi-bent functions.

Problem 17.4.13. Find semi-bent functions obtained by applying the result of Remark 17.5.12.

17.5 Semi-bent Functions (in Even Dimension): Constructions and Characterizations

In the following, we present a general overview of the main known constructions of semi-bent functions and investigate new constructions.

17.5.1 On Constructions of Quadratic Semi-bent Functions

The first papers dealing with constructions of semi-bent functions have been dedicated to quadratic functions. In this particular case of functions, there exists a criterion on the semi-bentness involving the dimension of the linear kernel defined above (see e.g. [10]). More precisely, it has been proved that f is semi-bent over $\mathbb{F}_{2^n}$, if and only if its linear kernel E_f (defined previously) has dimension 2. Note that from Theorem 2.2.2, it is easy to see that quadratic Boolean function is semi-bent if and only if the rank of f is $n-2$, that is, $k_f = 2$.

Several constructions of quadratic semi-bent functions have been obtained in the literature. We give a list of the known quadratic semi-bent functions on $\mathbb{F}_{2^n}$, $n = 2m$:

- $f(x) = \sum_{i=1}^{\lfloor \frac{n-1}{2} \rfloor} c_i \mathrm{Tr}_1^n(x^{1+2^i})$, $c_i \in \mathbb{F}_2$, $gcd(\sum_{i=1}^{\frac{n}{2}-1} c_i(x^i + x^{n-i}), x^n + 1) = x^2 + 1$ [10];
- $f(x) = \mathrm{Tr}_1^n(\alpha x^{2^i+1})$, $\alpha \in \mathbb{F}_{2^n}^\star$, i even, m odd [34];
- $f(x) = \mathrm{Tr}_1^n(\alpha x^{2^i+1})$, m even, i odd, $\alpha \in \{x^3, x \in \mathbb{F}_{2^n}^\star\}$ where $\alpha \in \mathbb{F}_{2^n}^\star$ [34];

- $f(x) = \mathrm{Tr}_1^n(\alpha x^{2^i+1})$, m odd, i odd, $gcd(m,i) = 1$, $\alpha \in \{x^3, x \in \mathbb{F}_{2^n}^\star\}$ where $\alpha \in \mathbb{F}_{2^n}^\star$ [34];
- $f(x) = \mathrm{Tr}_1^n(x^{2^i+1} + x^{2^j+1})$, m odd, $1 \leq i < j < m$, $gcd(n, i+j) = gcd(n, j-i) = 1)$, $gcd(n, i+j) = gcd(n, j-i) = 2$ [34];
- $f(x) = \sum_{i=1}^{\frac{m-1}{2}} \mathrm{Tr}_1^n(\beta x^{1+4^i})$, m odd, $\beta \in \mathbb{F}_4^\star$ [15];
- $f(x) = \sum_{i=1}^{\frac{m-1}{2}} c_i \mathrm{Tr}_1^n(\beta x^{1+4^i})$, $c_i \in \mathbb{F}_2$, $\beta \in \mathbb{F}_4^\star$, m odd, $gcd(\sum_{i=1}^{\frac{m-1}{2}} c_i(x^i + x^{m-i}), x^m+1) = x+1$ [15];
- $f(x) = \sum_{i=1}^{k} \mathrm{Tr}_1^n(\beta x^{1+4^{di}})$ $\beta \in \mathbb{F}_4^\star$, m odd, $d \geq 1$, $1 \leq k \leq \frac{m-1}{2}$, $gcd(k+1, m) = gcd(k, m) = gcd(d, m) = 1$ [15];
- $f(x) = \mathrm{Tr}_1^n(\beta x^{1+4^i} + \beta x^{1+4^j})$ $\beta \in \mathbb{F}_4^\star$, m odd, $1 \leq i < j \leq \lfloor \frac{n}{4} \rfloor$, $gcd(i+j, m) = gcd(j-i, m) = 1$ [15];
- $f(x) = \mathrm{Tr}_1^n(\beta x^{1+4^i} + x^{1+4^j} + x^{1+4^t})$, $\beta \in \mathbb{F}_4^\star$, m odd, $1 \leq i < j < t \leq \lfloor \frac{n}{4} \rfloor$, $i+j = t$, $gcd(i,m) = gcd(j,m) = gcd(j,t) = 1$ [15];
- $f(x) = \mathrm{Tr}_1^n(\beta x^{1+4^i} + \beta x^{1+4^j} + \beta x^{1+4^t})$, $\beta \in \mathbb{F}_4^\star$, $1 \leq i < j < t \leq \lfloor \frac{n}{4} \rfloor$, $i+j = 2t$, $j-i = 3^h p$, $3 \not| p$, $n = 3^k q$, $3 \not| q$, $gcd(2t, m) = 1$, $h \geq k$ [15];
- $f(x) = \mathrm{Tr}_1^n(\beta x^{1+4^i} + \beta x^{1+4^j} + \beta x^{1+4^t})$, $\beta \in \mathbb{F}_4^\star$, m odd, $1 \leq i,j,t \leq \lfloor \frac{n}{4} \rfloor$, $j-i = 2t$, $t \neq i$, $j+i = 3^u p$, $3 \not| p$, $n = 3^v q$, $3 \not| q$, $gcd(2t, m) = 1$, $u \geq v$ [15];
- $f(x) = \mathrm{Tr}_1^n(\beta x^{1+4^i} + \beta x^{1+4^j} + \beta x^{1+4^t})$, $\beta \in \mathbb{F}_4^\star$, $1 \leq i,j,t \leq \lfloor \frac{n}{4} \rfloor$, $j-i = 2t$, $t \neq i$, $j+i = 3^u p$, $3 \not| p$, $n = 3^v q$, $3 \not| q$, $gcd(2t, m) = 1$, $u \geq v$ [15];
- $f(x) = \mathrm{Tr}_1^n(\beta x^{1+4^i} + \beta x^{1+4^j} + \beta x^{1+4^t} + \beta x^{1+4^s})$, $\beta \in \mathbb{F}_4^\star$, $1 \leq i,j,t,s \leq \lfloor \frac{n}{4} \rfloor$, $i < j$, $t < s$, $i+j = t+s = r$, $t \neq i$, $gcd(r,m) = gcd(m, s-i) = gcd(m, s-j) = 1$ [15].

17.5.2 On Constructions of Semi-bent Functions from Bent Functions

In the following subsections we are dealing with the construction of semi-bent functions from bent functions. We shall present several such kinds of constructions. A natural problem arises is:

Problem 17.5.1. Find new primary constructions of bent functions from semi-bent functions.

17.5.2.1 Primary Constructions in Univariate Representation from Niho and Dillon Bent Functions

In this subsection, we consider several Boolean functions in univariate representation (expressed by means of the trace function) with even number of variables. Our main intention is to study the relationship between the semi-bentness property of functions obtained with Dillon and Niho exponents . Recall that a *Dillon exponent* is of the form $r(2^m-1)$ where r is co-prime with 2^m+1. Moreover, a positive integer

d (always understood modulo $2^n - 1$) is said to be a *Niho exponent*, and x^d is a *Niho power function*, if the restriction of x^d to $\mathbb{F}_{2^m}$ is linear or in other words $d \equiv 2^j \pmod{2^m - 1}$ for some $j < n$) and some exponential sums (namely, Kloosterman sums.

• Explicit constructions of semi bent functions in univariate representation and their links with Kloosterman sums:

In this subsection, we consider several Boolean functions in univariate representation (expressed by means of the trace function) with even number of variables. Our main intention is to study the relationship between the semi-bentness property of functions obtained with Dillon and Niho exponents . Recall that a *Dillon exponent* is of the form $r(2^m-1)$ where r is co-prime with 2^m+1. Moreover, a positive integer d (always understood modulo $2^n - 1$) is said to be a *Niho exponent*, and x^d is a *Niho power function*, if the restriction of x^d to $\mathbb{F}_{2^m}$ is linear or in other words $d \equiv 2^j \pmod{2^m - 1}$ for some $j < n$) and some exponential sums (namely, Kloosterman sums).

The goal of this subsection is to investigate the link between the semi-bentness property of some infinite classes of Boolean functions in univariate representation and some exponential sums (Kloosterman sums and cubic sums) [27, 30]. We shall use the technical results of Sect. 2.4 in Chap. 1.

We consider infinite families of Boolean functions in univariate representation with even number of variables whose expression is given by (17.17). By computer experiments, for small values of n, we have found that the set of functions of the form (17.17) contains semi-bent functions. We investigate criteria involving Kloosterman sums to determine whether a function of the form (17.17) is semi-bent or not:

$$\mathrm{Tr}_1^n\left(ax^{r(2^m-1)}\right) + \mathrm{Tr}_1^2\left(bx^{\frac{2^n-1}{3}}\right) + \mathrm{Tr}_1^n\left(cx^{(2^m-1)\frac{1}{2}+1}\right) \quad + \mathrm{Tr}_1^n\left(dx^{(2^m-1)s+1}\right), \tag{17.17}$$

where r is a positive integer, $s \in \{0, 1/4, 1/6, 3\}$, $a \in \mathbb{F}_{2^n}^\star$, $b \in \mathbb{F}_4$, $c \in \mathbb{F}_{2^n}$ and $d \in \mathbb{F}_2$. If $b \neq 0$, we consider the functions $g_{a,b,c,d}^{(r,s)}$ of the form (17.17) only when m is odd. Note that $o(r(2^m - 1)) = n$, $o(\frac{2^n-1}{3}) = 2$, $o((2^m - 1)\frac{1}{2} + 1) = m$ and $o((2^m - 1)s + 1) = n$ for $s \in \{1/4, 1/6, 3\}$ (recall that $o(j)$ denotes the size of the cyclotomic coset of 2 modulo $2^n - 1$ containing j). Moreover, using the transitivity property of the trace function, we have $\mathrm{Tr}_1^n(cx^{(2^m-1)\frac{1}{2}+1}) = \mathrm{Tr}_1^m(\mathrm{Tr}_m^n(c^2)x^{2^m+1}) = \mathrm{Tr}_1^m(c'x^{2^m+1})$ where $c' \in \mathbb{F}_{2^m}^\star$. Hence, the polynomial form of $g_{a,b,c,d}^{(r,s)}$ is:

$$\mathrm{Tr}_1^n\left(ax^{r(2^m-1)}\right) + \mathrm{Tr}_1^2\left(bx^{\frac{2^n-1}{3}}\right) + \mathrm{Tr}_1^m(c'x^{2^m+1}) \quad + \mathrm{Tr}_1^n\left(dx^{(2^m-1)s+1}\right).$$

So in the sequel, it suffices to use the previous identity to get the polynomial form of all the presented functions.

Now, we introduce the following decomposition

$$\mathbb{F}_{2^n}^{\star} = \bigcup_{u \in U} u\mathbb{F}_{2^m}^{\star}.$$

Let $g_{a,b,c,d}^{(r,s)}$ be any Boolean function of the form (17.17); note that the restriction of $g_{a,b,c,d}^{(r,s)}$ to any coset $u\mathbb{F}_{2^m}^{\star}$ $(u \in U)$, is affine. More precisely,

- Assume $b \neq 0$. Thanks to the transitivity of the trace function, we have:

$$\forall y \in \mathbb{F}_{2^m}^{\star}, g_{a,b,c,d}^{(r,s)}(uy) = Tr_1^m(\alpha_u y) + \beta_u \tag{17.18}$$

with

$$\alpha_u = Tr_m^n\left(du^{(2^m-1)s+1} + cu^{(2^m-1)\frac{1}{2}+1}\right)$$
$$= Tr_m^n\left(du^{(2^m-1)s+1} + c\right),$$
$$\beta_u = Tr_1^n\left(au^{r(2^m-1)}\right) + Tr_1^2\left(bu^{\frac{2^n-1}{3}}\right).$$

- Otherwise (that is, if $b = 0$), thanks to the transitivity of the trace function, we have

$$\forall y \in \mathbb{F}_{2^m}^{\star}, g_{a,0,c,d}^{(r,s)}(uy) = Tr_1^m(\alpha_u y) + \beta_u \tag{17.19}$$

with

$$\alpha_u = Tr_m^n\left(du^{(2^m-1)s+1} + c\right),$$
$$\beta_u = Tr_1^n\left(au^{r(2^m-1)}\right).$$

Therefore, the Walsh transform of a generic element of the form (17.17) can be computed as follows.

Lemma 17.5.2 ([27]). *Using the same notation as in (17.18) or (17.19), for every $\omega \in \mathbb{F}_{2^n}$, the Walsh transform of a generic element of the form (17.17) is*

$$\widehat{\chi_{g_{a,b,c,d}^{(r,s)}}}(\omega) = 1 - \sum_{u \in U} \chi(\beta_u) + 2^m \sum_{u \in U} \delta_0(\alpha_u + Tr_m^n(\omega u))\chi(\beta_u) \tag{17.20}$$

where δ_0 is the indicator of the singleton $\{0\}$, that is,

$$\delta_0(z) = \begin{cases} 1 \text{ if } z = 0 \\ 0 \text{ otherwise} \end{cases}$$

Proof. Suppose m odd and $b \neq 0$. Let $\omega \in \mathbb{F}_{2^n}$. The Walsh transform of $g_{a,b,c,d}^{(r,s)}$ is defined as

$$\widehat{\chi_{g_{a,b,c,d}^{(r,s)}}}(\omega) = \sum_{x\in\mathbb{F}_{2^n}} \chi(g_{a,b,c,d}^{(r,s)}(x) + Tr_1^n(wx)).$$

Any element $x \in \mathbb{F}_{2^n}^{\star}$ having a unique polar decomposition $x = uy$ with $u \in U$ and $y \in \mathbb{F}_{2^m}^{\star}$, we have :

$$\begin{aligned}\widehat{\chi_{g_{a,b,c,d}^{(r,s)}}}(\omega) &= 1 + \sum_{u\in U}\sum_{y\in\mathbb{F}_{2^m}^{\star}} \chi(g_{a,b,c,d}^{(r,s)}(uy) + Tr_1^n(wuy)) \\ &=1 + \sum_{u\in U}\sum_{y\in\mathbb{F}_{2^m}^{\star}} \chi\left(Tr_1^m\left((\alpha_u + Tr_m^n(wu))y\right) + \beta_u\right) \\ &=1 - \sum_{u\in U}\chi(\beta_u) + 2^m \sum_{u\in U}\delta_0(\alpha_u + Tr_m^n(\omega u))\chi(\beta_u).\end{aligned}$$

Likewise, one can establish (17.20) by similar calculations when $b = 0$ (for any positive integer m). □

To conclude, in 2011, many concrete constructions of semi-bent functions of maximum algebraic degree have been discovered. Indeed, in [27], the semi-bentness of several infinite families functions in polynomial form constructed via Dillon and Niho exponents has been studied in detail. From this study, explicit criteria in terms of Kloosterman sums for deciding whether a function expressed as a sum of trace functions is semi-bent or not have been derived. Kloosterman sums have been used as a very suitable tool to study the semi-bentness property of several functions in univariate representation. In particular, we have showed in [27] that the values 0 and 4 of Kloosterman sums defined on $\mathbb{F}_{2^m}$ give rise to semi-bent functions on $\mathbb{F}_{2^n}$. Below is the list of the known semi-bent functions constructed via the zero of Kloosterman sums:

- $f(x) = Tr_1^n(ax^{r(2^m-1)}) + Tr_1^n(cx^{(2^m-1)\frac{1}{2}+1})$, $K_m(a) = 0$ [27];
- $f(x) = Tr_1^n(ax^{r(2^m-1)}) + Tr_1^n(cx^{(2^m-1)\frac{1}{2}+1}) + Tr_1^n(x^{(2^m-1)\frac{1}{4}+1})$, $Tr_m^n(c) = 1$, m odd, $K_m(a) = 0$ [27];
- $f(x) = Tr_1^n(ax^{r(2^m-1)}) + Tr_1^n(cx^{(2^m-1)\frac{1}{2}+1}) + Tr_1^n(x^{(2^m-1)3+1})$, $K_m(a) = 0$ $Tr_m^n(c) = 1$ [27];
- $f(x) = Tr_1^n(ax^{r(2^m-1)}) + Tr_1^n(cx^{(2^m-1)\frac{1}{2}+1}) + Tr_1^n(x^{(2^m-1)\frac{1}{6}+1})$; $Tr_m^n(c) = 1$, $K_m(a) = 0$, m even [27];
- $f(x) = Tr_1^n(ax^{r(2^m-1)}) + Tr_1^n(\alpha x^{2^m+1}) + Tr_1^n(\sum_{i=1}^{2^{\nu-1}-1} x^{(2^m-1)\frac{i}{2^\nu}+1})$; $\gcd(\nu, m) = 1$, $\alpha \in \mathbb{F}_{2^n}$, $Tr_m^n(\alpha) = 1$, $K_m(a) = 0$ [27].

Below is the list of the known semi-bent functions constructed via the value four of Kloosterman sums:

- $f(x) = Tr_1^n(ax^{r(2^m-1)}) + Tr_1^2(bx^{\frac{2^n-1}{3}}) + Tr_1^n(cx^{(2^m-1)\frac{1}{2}+1})$; m odd , $K_m(a) = 4$ [27];
- $f(x) = Tr_1^n(ax^{3(2^m-1)}) + Tr_1^n(cx^{(2^m-1)\frac{1}{2}+1}) + Tr_1^2(bx^{\frac{2^n-1}{3}})$; m odd and $m \not\equiv 3 \pmod 6$ $K_m(a) = 4$ [27];

- $f(x) = \mathrm{Tr}_1^n(ax^{r(2^m-1)}) + \mathrm{Tr}_1^2(bx^{\frac{2^n-1}{3}}) + \mathrm{Tr}_1^n(cx^{(2^m-1)\frac{1}{2}+1}) + \mathrm{Tr}_1^n(x^{(2^m-1)\frac{1}{4}+1})$, m odd , $K_m(a) = 4$ [27];
- $f(x) = \mathrm{Tr}_1^n(ax^{r(2^m-1)}) + \mathrm{Tr}_1^2(bx^{\frac{2^n-1}{3}}) + \mathrm{Tr}_1^n(cx^{(2^m-1)\frac{1}{2}+1}) + \mathrm{Tr}_1^n(x^{3(2^m-1)+1})$; $\mathrm{Tr}_m^n(c) = 1$, m odd , $K_m(a) = 4$ [27];
- $f(x) = \mathrm{Tr}_1^n(ax^{r(2^m-1)}) + \mathrm{Tr}_1^n(\alpha x^{2^m+1}) + \mathrm{Tr}_1^n(\sum_{i=1}^{2^{\nu-1}-1} x^{(2^m-1)\frac{i}{2^\nu}+1}) + \mathrm{Tr}_1^2(bx^{\frac{2^n-1}{3}})$; $\gcd(\nu, m) = 1$, $\alpha \in \mathbb{F}_{2^n}$, $\mathrm{Tr}_m^n(\alpha) = 1$, m odd, $K_m(a) = 4$ ([27]).

All the families of semi-bent functions presented above are of maximum algebraic degree m and thus are suitable for use in symmetric cryptosystems.

The previous constructions can be generalized, leading to general constructions of semi-bent functions via Dillon-like exponents and Niho exponents.

A positive integer s (always understood modulo $2^n - 1$) is said to be a *Niho exponent* and x^s a Niho power function, if the restriction of x^s to $\mathbb{F}_{2^m}$ is linear. One can show that the restriction of the power function $x \mapsto x^s$ to $\mathbb{F}_{2^m}$ is linear then $s = 2^j$ for some $j < n$. As we consider $\mathrm{Tr}_1^n(x^d)$, without loss of generality, we can assume that s is in the normalized (unique) representation $s = (2^m - 1)d + 1$ with $1 \leq d \leq 2^m$.

• Semi-bent functions in polynomial forms with multiple trace terms and their link with Dikson polynomial:

we study the relationship between the semi-bentness property of functions in polynomial forms with multiple trace terms and Dickson polynomials.

In the following, we are interested in semi-bent functions whose expression contains multiple trace terms. Let E be a set of representatives of the cyclotomic classes modulo $2^n - 1$ for which each class has full size n. Let $f_{a_r,b,c}$ be the function defined on $\mathbb{F}_{2^n}$ whose polynomial form is given by

$$\sum_{r\in R} \mathrm{Tr}_1^n(a_r x^{r(2^m-1)}) + \mathrm{Tr}_1^2(bx^{\frac{2^n-1}{3}}) + \mathrm{Tr}_1^m(cx^{2^m+1}) \tag{17.21}$$

where $R \subseteq E$, $a_r \in \mathbb{F}_{2^m}^\star$, $b \in \mathbb{F}_4^\star$ and $c \in \mathbb{F}_{2^m}^\star$. In the following, we will show that semi-bent functions $f_{a_r,b,c}$ of the form (17.21) can be described by means of exponential sums involving the Dickson polynomials. In particular, one can provide a way to transfer the characterization of semi-bentness of a function of the form (17.21) to the evaluation of the Hamming weight of some Boolean functions.

To prove the result of this subsection, we need the following statements (Proposition 17.5.3) and Corollary 17.5.4).

Proposition 17.5.3 ([27]). *For $b \in \mathbb{F}_4^\star$ and $a_r \in \mathbb{F}_{2^m}^\star$, we denote by $g_{a_r,b}$ the function $\sum_{r\in R} \mathrm{Tr}_1^n(a_r x^{r(2^m-1)}) + \mathrm{Tr}_1^2(bx^{\frac{2^n-1}{3}})$ and by $g_{a_r,0}$ the function $\sum_{r\in R} \mathrm{Tr}_1^n(a_r x^{r(2^m-1)})$. Let V be the set of the elements of the cubes of U and ζ be a generator of U. Then, we have the following relations:*

$$\sum_{u\in U} \chi\Big(g_{a_r,\beta}(u)\Big) = \sum_{u\in U} \chi\Big(g_{a_r,\beta^2}(u)\Big) = -\sum_{v\in V} \chi\Big(g_{a_r,0}(v)\Big) \tag{17.22}$$

and

$$\sum_{u\in U}\chi\Big(g_{a_r,1}(u)\Big)=\sum_{v\in V}\chi\Big(g_{a_r,0}(v)\Big)-2\sum_{v\in V}\chi\Big(g_{a_r,0}(\zeta v)\Big) \tag{17.23}$$

Proof. Introduce for every element b' of $\mathbb{F}_4$, the sum

$$\Lambda(b'):=\sum_{b\in\mathbb{F}_4}\sum_{u\in U}\chi\Big(g_{a_r,b}(u)\Big)\chi\Big(\mathrm{Tr}_1^2(bb')\Big).$$

Note that

$$\Lambda(b')=\sum_{u\in U}\chi\Big(g_{a_r,0}(u)\Big)\sum_{b\in\mathbb{F}_4}\chi\Big(\mathrm{Tr}_1^2\Big(b\Big(b'+u^{\frac{2^n-1}{3}}\Big)\Big)\Big).$$

Furthermore, one has

$$\sum_{b\in\mathbb{F}_4}\chi\Big(\mathrm{Tr}_1^2\Big(b\Big(b'+u^{\frac{2^n-1}{3}}\Big)\Big)\Big)=\begin{cases}0 \text{ if } u^{\frac{2^n-1}{3}}\neq b'\\ 4 \text{ otherwise}\end{cases}$$

Since, $u^{\frac{2^n-1}{3}}\neq 0$ for every $u\in U$, $\Lambda(0)=0$. Since β is a primitive element of $\mathbb{F}_4$, suppose that $b'=\beta^i$ for $i\in\{0,1,2\}$. Then, for a generator ζ of U, we have, $\beta^i=\zeta^{i\frac{2^m+1}{3}}$. Hence,

$$\begin{aligned}\Lambda(\beta^i)&=4\sum_{u\in U,\,u^{\frac{2^n-1}{3}}=\zeta^{i\frac{2^m+1}{3}}}\chi\Big(g_{a_r,0}(u)\Big)\\&=4\sum_{u\in U,\,\left(u^{-2}\zeta^{-i}\right)^{\frac{2^m+1}{3}}=1}\chi\Big(g_{a_r,0}(u)\Big)\\&=4\sum_{u\in U,\,u^{-2}\in\zeta^iV}\chi\Big(g_{a_r,0}(u)\Big).\end{aligned}$$

That follows from the fact that the only elements x of U such that $x^{\frac{2^m+1}{3}}=1$ are the elements of V. Next, note that the map $x\mapsto x^{2^{m-1}}$ is one-to-one from ζ^iV to ζ^iV (since $\zeta^{i(2^{m-1}-1)}$ is a cube because $2^{m-1}-1\equiv 0 \pmod 3$ for m odd), one gets that $u^{\frac{2^n-1}{3}}=\zeta^{i\frac{2^m+1}{3}}$ if and only if $u\in\zeta^iV$.

Therefore,

$$\Lambda(\beta^i)=4\sum_{v\in V}\chi\Big(g_{a_r,0}(\zeta^iv)\Big). \tag{17.24}$$

Now, establish an expression of $\sum_{b' \in \mathbb{F}_4} \Lambda(b')\chi\Big(\mathrm{Tr}_1^2(bb')\Big)$ involving $\sum_{u\in U} \chi\Big(g_{a_r,b}(u)\Big)$.

$$\begin{aligned}
&\sum_{b' \in \mathbb{F}_4} \Lambda(b')\chi\Big(\mathrm{Tr}_1^2(bb')\Big) \\
&= \sum_{b' \in \mathbb{F}_4} \sum_{b'' \in \mathbb{F}_4} \sum_{u \in U} \chi\Big(g_{a_r,b''}(u)\Big)\chi\Big(\mathrm{Tr}_1^2(b''b')\Big)\chi\Big(\mathrm{Tr}_1^2(bb')\Big) \\
&= \sum_{b'' \in \mathbb{F}_4} \sum_{u \in U} \chi\Big(g_{a_r,b''}(u)\Big) \sum_{b' \in \mathbb{F}_4} \chi\Big(\mathrm{Tr}_1^2(b'(b''+b))\Big).
\end{aligned}$$

Since,

$$\sum_{b' \in \mathbb{F}_4} \chi\Big(\mathrm{Tr}_1^2(b'(b''+b))\Big) = \begin{cases} 4 \text{ if } b'' = b \\ 0 \text{ otherwise} \end{cases}$$

then, one gets

$$\sum_{b' \in \mathbb{F}_4} \Lambda(b')\chi\Big(\mathrm{Tr}_1^2(bb')\Big) = 4\sum_{u \in U} \chi\Big(g_{a_r,b}(u)\Big)$$

that is,

$$\sum_{u \in U} \chi\Big(g_{a_r,b}(u)\Big) = \frac{1}{4}\sum_{b' \in \mathbb{F}_4} \Lambda(b')\chi\Big(\mathrm{Tr}_1^2(bb')\Big). \tag{17.25}$$

Finally, by formula (17.24), one gets (since $\chi(\mathrm{Tr}_1^2(1)) = 1$ and $\chi(\mathrm{Tr}_1^2(\beta)) = \chi(\mathrm{Tr}_1^2(\beta^2)) = -1$)

$$\begin{aligned}
\sum_{u \in U} \chi\Big(g_{a_r,1}(u)\Big) &= \sum_{v \in V} \chi\Big(g_{a_r,0}(v)\Big) \\
&\quad - \sum_{v \in V} \chi\Big(g_{a_r,0}(\zeta v)\Big) - \sum_{v \in V} \chi\Big(g_{a_r,0}(\zeta^2 v)\Big). \\
\sum_{u \in U} \chi\Big(g_{a_r,\beta}(u)\Big) &= -\sum_{v \in V} \chi\Big(g_{a_r,0}(v)\Big) \\
&\quad - \sum_{v \in V} \chi\Big(g_{a_r,0}(\zeta v)\Big) + \sum_{v \in V} \chi\Big(g_{a_r,0}(\zeta^2 v)\Big).
\end{aligned}$$

$$\sum_{u\in U}\chi\left(g_{a_r,\beta^2}(u)\right) = -\sum_{v\in V}\chi\left(g_{a_r,0}(v)\right) + \sum_{v\in V}\chi\left(g_{a_r,0}(\zeta v)\right) - \sum_{v\in V}\chi\left(g_{a_r,0}(\zeta^2 v)\right).$$

To conclude, note that one has

$$\sum_{v\in V}\chi\left(g_{a_r,0}(\zeta v)\right) = \sum_{v\in V}\chi\left(g_{a_r,0}(\zeta^2 v)\right). \tag{17.26}$$

Indeed, since the trace function is invariant under the Frobenius automorphism $x \mapsto x^2$, we get, applying m times, the Frobenius automorphism : $\forall x \in \mathbb{F}_{2^n}$,

$$g_{a_r,0}(x) = \sum_{r\in R}\mathrm{Tr}_1^n\left(a_r^{2^m}x^{2^m r(2^m-1)}\right) = \sum_{r\in R}\mathrm{Tr}_1^n\left(a_r x^{2^m r(2^m-1)}\right) = g_{a_r,0}(x^{2^m})$$

because the a_r's are in $\mathbb{F}_{2^m}^\star$. Hence,

$$\begin{aligned}\sum_{v\in V}\chi\left(g_{a_r,0}(\zeta v)\right) &= \sum_{v\in V}\chi\left(g_{a_r,0}(\zeta^{2^m}v^{2^m})\right)\\ &= \sum_{v\in V}\chi\left(g_{a_r,0}(\zeta^2(\zeta^{2^m-2}v^{2^m}))\right).\end{aligned}$$

Now, since m is odd, 3 divides 2^m+1 and thus divides 2^m-2. Hence, ζ^{2^m-2} is a cube of U and the mapping $v \mapsto \zeta^{(2^m-2)}v^{2^m}$ is a permutation of V. The relation (17.26) follows. □

Corollary 17.5.4. *[27]] For $b \in \mathbb{F}_4^\star$ and $a_r \in \mathbb{F}_{2^m}^\star$, we denote by $g_{a_r,b}$ the function defined on $\mathbb{F}_{2^n}$ by $\sum_{r\in R}\mathrm{Tr}_1^n(a_r x^{r(2^m-1)}) + \mathrm{Tr}_1^2(bx^{\frac{2^n-1}{3}})$, and by h_{a_r} the function defined on $\mathbb{F}_{2^m}$ by $h_{a_r}(x) = \sum_{r\in R}\mathrm{Tr}_1^m(a_r D_r(x))$, where $D_r(x)$ is the Dickson polynomial of degree r. Then,*

1. $\sum_{u\in U}\chi\left(g_{a_r,\beta}(u)\right) = 1$ *if and only if* $\sum_{u\in U}\chi\left(g_{a_r,\beta^2}(u)\right) = 1$ *if and only if,*

$$\begin{aligned}&\sum_{x\in\mathbb{F}_{2^m}}\chi\left(\mathrm{Tr}_1^m(x^{-1}) + h_{a_r}(D_3(x))\right)\\ &= 2^m - 2\,\mathrm{wt}(h_{a_r}\circ D_3) + 4.\end{aligned}$$

2. $\sum_{u\in U}\chi\Big(g_{a_r,1}(u)\Big) = 1$ *if and only if*

$$\begin{aligned}
&3\sum_{x\in\mathbb{F}_{2^m}}\chi\Big(\mathrm{Tr}_1^m(x^{-1})+h_{a_r}(x)\Big)\\
&-2\sum_{x\in\mathbb{F}_{2^m}}\chi\Big(\mathrm{Tr}_1^m(x^{-1})+h_{a_r}(D_3(x))\Big)\\
&=4+2^m+4\,\mathrm{wt}(h_{a_r}\circ D_3)-6\,\mathrm{wt}(h_{a_r}).
\end{aligned}$$

To prove the corollary, we need the following useful lemma.

Lemma 17.5.5 ([27]). *Keeping the same notations as in Corollary 17.5.4, for any positive integer p, we have*

$$\sum_{u\in U}\chi\Big(g_{a_r,0}(u^p)\Big)=1+\sum_{x\in\mathbb{F}_{2^m}}\chi\Big(h_{a_r}(D_p(x))\Big)-\sum_{x\in\mathbb{F}_{2^m}}\chi\Big(\mathrm{Tr}_1^m(x^{-1})+h_{a_r}(D_p(x))\Big).$$

Proof. One has

$$\sum_{u\in U}\chi\Big(g_{a_r,0}(u^p)\Big)=1+2\sum_{x\in\mathbb{F}_{2^m}^\star,\mathrm{Tr}_1^m(x^{-1})=1}\chi\Big(h_{a_r}(D_p(x))\Big).$$

Now, note that the indicator of the set $\{x\in\mathbb{F}_{2^m}^\star \mid \mathrm{Tr}_1^m(x^{-1})=1\}$ can be written as $\frac{1}{2}\left(1-\chi(\mathrm{Tr}_1^m(x^{-1}))\right)$. Hence, $\sum_{x\in\mathbb{F}_{2^m}^\star,\mathrm{Tr}_1^m(x^{-1})=1}\chi\Big(h_{a_r}(D_p(x))\Big)$

$$\begin{aligned}
&=\frac{1}{2}\sum_{x\in\mathbb{F}_{2^m}^\star}\chi\Big(h_{a_r}(D_p(x))\Big)\\
&-\frac{1}{2}\sum_{x\in\mathbb{F}_{2^m}^\star}\chi\Big(\mathrm{Tr}_1^m(x^{-1})+h_{a_r}(D_p(x))\Big)\\
&=\frac{1}{2}\sum_{x\in\mathbb{F}_{2^m}}\chi\Big(h_{a_r}(D_p(x))\Big)\\
&-\frac{1}{2}\sum_{x\in\mathbb{F}_{2^m}}\chi\Big(\mathrm{Tr}_1^m(x^{-1})+h_{a_r}(D_p(x))\Big).
\end{aligned}$$

Therefore,

$$\begin{aligned}
\sum_{u\in U}\chi\Big(g_{a_r,0}(u^p)\Big)&=1+\sum_{x\in\mathbb{F}_{2^m}}\chi\Big(h_{a_r}(D_p(x))\Big)\\
&-\sum_{x\in\mathbb{F}_{2^m}}\chi\Big(\mathrm{Tr}_1^m(x^{-1})+h_{a_r}(D_p(x))\Big).
\end{aligned}$$

□

Now, we prove Corollary 17.5.4.

Proof. 1. According to Proposition 17.5.3, $\sum_{u\in U}\chi\Big(g_{a_r,\beta}(u)\Big)=1$ if and only if,

$$\sum_{u\in U}\chi\Big(g_{a_r,\beta^2}(u)\Big)=1$$

if and only if,

$$\sum_{v\in V}\chi\Big(g_{a_r,0}(v)\Big)=-1.$$

We have

$$\sum_{v\in V}\chi\Big(g_{a_r,0}(v)\Big)=\frac{1}{3}\sum_{u\in U}\chi\Big(g_{a_r,0}(u^3)\Big)$$

Now, take $p=3$ in Lemma 17.5.5:

$$\sum_{u\in U}\chi\Big(g_{a_r,0}(u^3)\Big)=1+\sum_{x\in\mathbb{F}_{2^m}}\chi\Big(h_{a_r}(D_3(x))\Big)-\sum_{x\in\mathbb{F}_{2^m}}\chi\Big(\mathrm{Tr}_1^m(x^{-1})+h_{a_r}(D_3(x))\Big).$$

Hence $\sum_{u\in U}\chi\Big(g_{a_r,\beta}(u)\Big)=1$ if and only if, $\sum_{u\in U}\chi\Big(g_{a_r,\beta^2}(u)\Big)=1$) if and only if,

$$\sum_{x\in\mathbb{F}_{2^m}}\chi\Big(\mathrm{Tr}_1^m(x^{-1})+h_{a_r}(D_3(x))\Big)=4+\sum_{x\in\mathbb{F}_{2^m}}\chi\Big(h_{a_r}(D_3(x))\Big).$$

Now, using the fact that, for a Boolean function f defined on $\mathbb{F}_{2^n}$, $\sum_{x\in\mathbb{F}_{2^n}}\chi\Big(f(x)\Big)=2^n-2\,\mathrm{wt}(f)$, we finally get that $\sum_{u\in U}\chi\Big(g_{a_r,\beta}(u)\Big)=1$ if and only if, $\sum_{u\in U}\chi\Big(g_{a_r,\beta^2}(u)\Big)=1$ if and only if,

$$\sum_{x\in\mathbb{F}_{2^m}}\chi\Big(\mathrm{Tr}_1^m(x^{-1})+h_{a_r}(D_3(x))\Big)=4+2^m-2\,\mathrm{wt}(h_{a_r}\circ D_3).$$

The assertion 1) follows.

2. By Proposition 17.5.3, $\sum_{u\in U}\chi\Big(g_{a_r,1}(u)\Big)=1$ if and only if,

$$\sum_{v\in V}\chi\Big(g_{a_r,0}(v)\Big)-2\sum_{v\in V}\chi\Big(g_{a_r,0}(\zeta v)\Big)=1.$$

Note that we have

$$\sum_{u\in U}\chi\Big(g_{a_r,0}(u)\Big)=\sum_{v\in V}\chi\Big(g_{a_r,0}(v)\Big)+\sum_{v\in V}\chi\Big(g_{a_r,0}(\zeta v)\Big)+\sum_{v\in V}\chi\Big(g_{a_r,0}(\zeta^2 v)\Big)$$

Using relation (17.26) and the fact that $\sum_{v\in V}\chi(g_{a_r,0}(v))=\frac{1}{3}\sum_{u\in U}\chi(g_{a_r,0}(u^3))$, one gets $\sum_{u\in U}\chi\Big(g_{a_r,1}(u)\Big)=1$ if and only if, $\frac{2}{3}\sum_{u\in U}\chi(g_{a_r,0}(u^3))-\sum_{u\in U}\chi(g_{a_r,0}(u))=1$.

Now, apply Lemma 17.5.5 for $p=3$ and $p=1$:

$$\sum_{u\in U}\chi\Big(g_{a_r,0}(u^3)\Big)=1+\sum_{x\in\mathbb{F}_{2^m}}\chi\Big(h_{a_r}(D_3(x))\Big)-\sum_{x\in\mathbb{F}_{2^m}}\chi\Big(\mathrm{Tr}_1^m(x^{-1})+h_{a_r}(D_3(x))\Big)$$

and (since $D_1(x)=x$)

$$\sum_{u\in U}\chi\Big(g_{a_r,0}(u)\Big)=1+\sum_{x\in\mathbb{F}_{2^m}}\chi\Big(h_{a_r}(x)\Big)-\sum_{x\in\mathbb{F}_{2^m}}\chi\Big(\mathrm{Tr}_1^m(x^{-1})+h_{a_r}(x)\Big).$$

The condition

$$\frac{2}{3}\sum_{u\in U}\chi(g_{a_r,0}(u^3))-\sum_{u\in U}\chi(g_{a_r,0}(u))=1$$

is then equivalent to

$$2/3+2/3\sum_{x\in\mathbb{F}_{2^m}}\chi\Big(h_{a_r}(D_3(x))\Big)-2/3\sum_{x\in\mathbb{F}_{2^m}}\chi\Big(\mathrm{Tr}_1^m(x^{-1})+h_{a_r}(D_3(x))\Big)$$
$$-1-\sum_{x\in\mathbb{F}_{2^m}}\chi\Big(h_{a_r}(x)\Big)+\sum_{x\in\mathbb{F}_{2^m}}\chi\Big(\mathrm{Tr}_1^m(x^{-1})+h_{a_r}(x)\Big)=1.$$

Now,

$$\sum_{x\in\mathbb{F}_{2^m}}\chi\Big(h_{a_r}(D_3(x))\Big)=2^m-2\,\mathrm{wt}(h_{a_r}\circ D_3)$$

and

$$\sum_{x\in\mathbb{F}_{2^m}}\chi\Big(h_{a_r}(x)\Big)=2^m-2\,\mathrm{wt}(h_{a_r}).$$

The latter condition is equivalent to

$$2/3 + 2/3(2^m - 2\,\mathrm{wt}(h_{a_r} \circ D_3)) - 2/3 \sum_{x \in \mathbb{F}_{2^m}} \chi\Big(\mathrm{Tr}_1^m(x^{-1}) + h_{a_r}(D_3(x))\Big)$$

$$-1 - (2^m - 2\,\mathrm{wt}(h_{a_r})) + \sum_{x \in \mathbb{F}_{2^m}} \chi\Big(\mathrm{Tr}_1^m(x^{-1}) + h_{a_r}(x)\Big) = 1.$$

that is,

$$3 \sum_{x \in \mathbb{F}_{2^m}} \chi\Big(\mathrm{Tr}_1^m(x^{-1}) + h_{a_r}(x)\Big) - 2 \sum_{x \in \mathbb{F}_{2^m}} \chi\Big(\mathrm{Tr}_1^m(x^{-1}) + h_{a_r}(D_3(x))\Big)$$
$$= 4 + 2^m + 4\,\mathrm{wt}(h_{a_r} \circ D_3) - 6\,\mathrm{wt}(h_{a_r}).$$

□

Using the previous results, we prove the following characterization of semi-bentness for functions in the form (17.21).

Theorem 17.5.6 ([27]). *Let $n = 2m$ with m odd. Let $b \in \mathbb{F}_4^\star$, β be a primitive element of $\mathbb{F}_4$ and $c \in \mathbb{F}_{2^m}^\star$. Let $f_{a_r,b,c}$ be the function defined on $\mathbb{F}_{2^n}$ whose expression is of the form (17.21). Let h_{a_r} be the related function defined on $\mathbb{F}_{2^m}$ by $h_{a_r}(x) = \sum_{r \in R} \mathrm{Tr}_1^m(a_r D_r(x))$, where $D_r(x)$ is the Dickson polynomial of degree r. Then*

1. $f_{a_r,\beta,c}$ is semi-bent if and only if, $f_{a_r,\beta^2,c}$ is semi-bent, if and only if,

$$\sum_{x \in \mathbb{F}_{2^m}} \chi\Big(\mathrm{Tr}_1^m(x^{-1}) + h_{a_r}(D_3(x))\Big) = 2^m - 2\,\mathrm{wt}(h_{a_r} \circ D_3) + 4.$$

2. $f_{a_r,1,c}$ is semi-bent if and only if,

$$3 \sum_{x \in \mathbb{F}_{2^m}} \chi\Big(\mathrm{Tr}_1^m(x^{-1}) + h_{a_r}(x)\Big) - 2 \sum_{x \in \mathbb{F}_{2^m}} \chi\Big(\mathrm{Tr}_1^m(x^{-1}) + h_{a_r}(D_3(x))\Big)$$
$$= 4 + 2^m + 4\,\mathrm{wt}(h_{a_r} \circ D_3) - 6\,\mathrm{wt}(h_{a_r}).$$

Proof. For $b \in \mathbb{F}_4^\star$, $a_r \in \mathbb{F}_{2^m}^\star$, denote by $g_{a_r,b}$ the function $\sum_{r \in R} \mathrm{Tr}_1^n(a_r x^{r(2^m-1)}) + \mathrm{Tr}_1^2(bx^{\frac{2^n-1}{3}})$ and by $g_{a_r,0}$ the function $\sum_{r \in R} \mathrm{Tr}_1^n(a_r x^{r(2^m-1)})$. Since m is odd, the function $g_{a_r,b}$ is constant on each (multiplicative) coset $u\mathbb{F}_{2^m}^\star$ $(u \in U)$ that is, we have:

$$\forall u \in U, \forall y \in \mathbb{F}_{2^m}^\star, g_{a_r,b}(uy) = g_{a_r,b}(u).$$

Using the polar decomposition, the Walsh transform of $f_{a_r,b,c}$ at every $\omega \in \mathbb{F}_{2^n}$ is given by

$$\widehat{\chi_{f_{a_r,b,c}}}(\omega) = \sum_{x\in\mathbb{F}_{2^n}} \chi\Big(f_{a_r,b,c}(x) + Tr_1^n(xw)\Big)$$

$$= 1 + \sum_{x\in\mathbb{F}_{2^n}^\star} \chi\Big(f_{a_r,b,c}(x) + Tr_1^n(xw)\Big)$$

$$= 1 + \sum_{u\in U} \chi\Big(g_{a_r,b}(u)\Big)\sum_{y\in\mathbb{F}_{2^m}^\star}\chi\Big(Tr_1^m(cy^{2^m+1} + Tr_m^n(wu)y)\Big)$$

$$= 1 + \sum_{u\in U} \chi\Big(g_{a_r,b}(u)\Big)\sum_{y\in\mathbb{F}_{2^m}^\star}\chi\Big(Tr_1^m(c^{\frac{1}{2}}y + Tr_m^n(wu)y)\Big)$$

$$= 1 - \sum_{u\in U} \chi\Big(g_{a_r,b}(u)\Big)$$

$$+ \sum_{u\in U} \chi\Big(g_{a_r,b}(u)\Big)\sum_{y\in\mathbb{F}_{2^m}}\chi\Big(Tr_1^m((c^{\frac{1}{2}} + Tr_m^n(wu))y)\Big)$$

$$= 1 - \sum_{u\in U} \chi\Big(g_{a_r,b}(u)\Big) + 2^m \sum_{u\in U|c^{\frac{1}{2}}+Tr_m^n(wu)=0} \chi\Big(g_{a_r,b}(u)\Big).$$

Thanks to Lemma 17.2.2, we obtain

$$\widehat{\chi_{f_{a_r,b,c}}}(\omega) \equiv 1 - \sum_{u\in U} \chi\Big(g_{a_r,b}(u)\Big) \pmod{2^{m+1}}.$$

But

$$-2^{m+1} < -2^m \le 1 - \sum_{u\in U} \chi\Big(g_{a_r,b}(u)\Big) \le 2^m + 2 < 2^{m+1}$$

therefore, $f_{a_r,b,c}$ is semi-bent if and only if, $\sum_{u\in U} \chi\Big(g_{a_r,b}(u)\Big) = 1$. We conclude thanks to Corollary 17.5.4. □

Proposition 17.5.7 ([27]). *Let $n = 2m$ with m odd. For $r \in R$, $a_r \in \mathbb{F}_{2^m}^\star$, β a primitive element of $\mathbb{F}_4$ and $c \in \mathbb{F}_{2^m}^\star$, let $f_{a_r,\beta,c}$ a function of the form (17.21).*

1. Let d be a positive integer such that $\gcd(d, \frac{2^m+1}{3}) = 1$. Let $h_{a_r,\beta,c}$ be the function

$$h_{a_r,\beta,c}(x) = \sum_{r\in R} Tr_1^n(a_r x^{dr(2^m-1)})Tr_1^2(\beta x^{\frac{2^n-1}{3}}) + Tr_1^m(cx^{2^m+1}).$$

Then, $h_{a_r,\beta,c}$ is semi-bent if and only if, $f_{a_r,\beta,c}$ is semi-bent.

2. *Suppose* $m \not\equiv 3 \pmod 6$. *Let* d *be a positive integer such that* $\gcd(d, 2^m + 1) = 3$. *Let* $h_{a_r,1,c}$ *be the function*

$$h_{a_r,1,c}(x) = \sum_{r \in R} \mathrm{Tr}_1^n(a_r x^{dr(2^m-1)}) + \mathrm{Tr}_1^2(x^{\frac{2^n-1}{3}}) + \mathrm{Tr}_1^m(cx^{2^m+1}).$$

If $f_{a_r,\beta,c}$ *is semi-bent, then* $h_{a_r,1,c}$ *is semi-bent.*

Proof. For two integers r and d, set

$$g_{a_r,0}(x) = \sum_{r \in R} \mathrm{Tr}_1^n(a_r x^{r(2^m-1)});$$

$$h_{a_r,0}(x) = \sum_{r \in R} \mathrm{Tr}_1^n(a_r x^{dr(2^m-1)}).$$

Proof of 1): according to the proof of Theorem 17.5.6 and relation (17.22), $h_{a_r,\beta,c}$ (Resp. $f_{a_r,\beta,c}$) is semi-bent if and only if $\sum_{v \in V} \chi\Big(h_{a_r,0}(v)\Big) = -1$ (Resp. $\sum_{v \in V} \chi\Big(g_{a_r,0}(v)\Big) = -1$). Now, the integers $\frac{2^m+1}{3}$ and d are co-prime thus, the mapping $v \mapsto v^d$ is then a permutation of V. Therefore,

$$\sum_{v \in V} \chi\Big(h_{a_r,0}(v)\Big) = \sum_{v \in V} \chi\Big(g_{a_r,0}(v^d)\Big) = \sum_{v \in V} \chi\Big(g_{a_r,0}(v)\Big).$$

The result follows.
Proof of 2): the function $f_{a_r,\beta,c}$ is semi-bent thus, according to the proof of Theorem 17.5.6 and relation (17.22), $\sum_{v \in V} \chi\Big(g_{a_r,0}(v)\Big) = -1$. We have to prove that $h_{a_r,1,c}$ is semi-bent, that is, $\sum_{v \in V} \chi\Big(h_{a_r,0}(v)\Big) - 2\sum_{v \in V} \chi\Big(h_{a_r,0}(\zeta v)\Big) = 1$, according to the proof of Theorem 17.5.6 and relation (17.23). But

$$\sum_{v \in V} \chi\Big(h_{a_r,0}(v)\Big) + \sum_{v \in V} \chi\Big(h_{a_r,0}(\zeta v)\Big) + \sum_{v \in V} \chi\Big(h_{a_r,0}(\zeta^2 v)\Big) = \sum_{u \in U} \chi\Big(h_{a_r,0}(u)\Big)$$

and according to relation (17.26) we have,

$$\sum_{v \in V} \chi\Big(h_{a_r,0}(\zeta v)\Big) = \sum_{v \in V} \chi\Big(h_{a_r,0}(\zeta^2 v)\Big).$$

Therefore, the condition

$$\sum_{v \in V} \chi\Big(h_{a_r,0}(v)\Big) - 2\sum_{v \in V} \chi\Big(h_{a_r,0}(\zeta v)\Big) = 1$$

is equivalent to

$$2\sum_{v\in V}\chi\Big(h_{a_r,0}(v)\Big)-\sum_{u\in U}\chi\Big(h_{a_r,0}(u)\Big)=1.$$

Now, since $\gcd(d,2^m+1)=3$ and the mapping $v\mapsto v^3$ is a permutation when $m\not\equiv 3 \pmod 6$, one has

$$\sum_{v\in V}\chi\Big(h_{a_r,0}(v)\Big)=\sum_{v\in V}\chi\Big(g_{a_r,0}(v^d)\Big)=\sum_{v\in V}\chi\Big(g_{a_r,0}(v^3)\Big)=\sum_{v\in V}\chi\Big(g_{a_r,0}(v)\Big).$$

On the other hand, note that (since $\gcd(d,2^m+1)=3$)

$$\sum_{u\in U}\chi\Big(h_{a_r,0}(u)\Big)=\sum_{u\in U}\chi\Big(g_{a_r,0}(u^d)\Big)=\sum_{u\in U}\chi\Big(g_{a_r,0}(u^3)\Big)=3\sum_{v\in V}\chi\Big(g_{a_r,0}(v)\Big).$$

Hence, $2\sum_{v\in V}\chi\Big(h_{a_r,0}(v)\Big)-\sum_{u\in U}\chi\Big(h_{a_r,0}(u)\Big)=-2-(-3)=1$, proving that $h_{a_r,1,c}$ is semi-bent. □

The reader notices that it is possible to get efficient characterizations of the semi-bentness property of several families of Boolean functions in univariate representation with multiple trace terms expressed by means of trace functions via Dillon-like exponents and Niho exponents with even number of variables.

The following statement is due to Carlet and the author [7]. An alternative direct proof has been proposed by Cohen and the author [11].

Theorem 17.5.8 ([7, 11]). *Denote by Ω_n the set of Boolean functions f defined on $\mathbb{F}_{2^n}$ by $f(x)=\sum_{i\in\Gamma_{n,m}}\mathrm{Tr}_1^{o(i)}(a_ix^i)$ where $\Gamma_{n,m}$ is the set of cyclotomic cosets $[i]$ such that $i\equiv 0 \pmod{2^m-1}$. Denote by Δ_n the set of Boolean functions f defined on $\mathbb{F}_{2^n}$ by $f(x)=\sum_{i\in\Lambda'_{n,m}}\mathrm{Tr}_1^{o(i)}(a_ix^i)$ where $\Lambda'_{n,m}$ is the set of cyclotomic cosets $[i]$ such that $i\equiv 2^j \pmod{2^m-1}$ for some j ($j<n$). Set*

$$\mathcal{D}_n:=\{f\in\Omega_n \text{ such that } f \text{ is bent with } f(0)=0\}$$

and set

$$\mathcal{N}_n:=\{f\in\Delta_n \text{ such that } f \text{ is bent with } f(0)=0\}.$$

Let $g\in\mathcal{D}_n$ and $h\in\mathcal{N}_n$. Then $g+h$ is semi-bent on $\mathbb{F}_{2^n}$.

Let us specify some infinite families of semi-bent functions in univariate form. Firstly, we give a list of infinite families containing bent functions defined on $\mathbb{F}_{2^n}$ belonging to the class $\mathcal{PS}_{ap}$; here, $K_m(a):=\sum_{x\in\mathbb{F}_{2^m}}\chi\big(\mathrm{Tr}_1^m(ax+\frac{1}{x})\big)$ denotes the binary Kloosterman sums on $\mathbb{F}_{2^m}$ and $C_m(a,a):=\sum_{x\in\mathbb{F}_{2^m}}\chi\big(\mathrm{Tr}_1^m(ax^3+ax)\big)$ denotes the cubic sums on $\mathbb{F}_{2^m}$:

- $g_1(x) = Tr_1^n(ax^{r(2^m-1)})$; $\gcd(r, 2^m+1) = 1$, $a \in \mathbb{F}_{2^m}^\star$ such that $K_m(a) = 0$ [8];
- $g_2(x) = Tr_1^n(ax^{r(2^m-1)}) + Tr_1^2(bx^{\frac{2^n-1}{3}})$; $\gcd(r, 2^m+1) = 1$, $m > 3$ odd, $b \in \mathbb{F}_4^\star$, $a \in \mathbb{F}_{2^m}^\star$ such that $K_m(a) = 4$ [25];
- $g_3(x) = Tr_1^n(a\zeta^i x^{3(2^m-1)}) + Tr_1^2(\beta^j x^{\frac{2^n-1}{3}})$; m odd and $m \not\equiv 3 \pmod 6$, β is a primitive element of $\mathbb{F}_4$, ζ is a generator of the cyclic group U of (2^m+1)-th of unity, $(i,j) \in \{0,1,2\}^2$, $a \in \mathbb{F}_{2^m}^\star$ such that $K_m(a) = 4$ and $Tr_1^m(a^{1/3}) = 0$ [24];
- $g_4(x) = Tr_1^n(a\zeta^i x^{3(2^m-1)}) + Tr_1^2(\beta^j x^{\frac{2^n-1}{3}})$; m odd and $m \not\equiv 3 \pmod 6$, β is a primitive element of $\mathbb{F}_4$, ζ is a generator of the cyclic group U of (2^m+1)-th of unity, $i \in \{1,2\}, j \in \{0,1,2\}$, $a \in \mathbb{F}_{2^m}^\star$ such that $K_m(a) + C_m(a,a) = 4$ and $Tr_1^m(a^{1/3}) = 1$ [24];
- $g_5(x) = \sum_{i=1}^{2^{m-1}-1} Tr_1^n\left(\beta x^{i(2^m-1)}\right)$; $\beta \in \mathbb{F}_{2^m} \setminus \mathbb{F}_2$ [16];
- $g_6(x) = \sum_{i=1}^{2^{m-2}-1} Tr_1^n\left(\beta x^{i(2^m-1)}\right)$; m odd and $\beta^{(2^m-4)^{-1}} \in \{x \in \mathbb{F}_{2^m}^\star; Tr_1^m(x) = 0\}$ [16];

Secondly, we give a list of known Niho bent functions in $\mathcal{N}_n$:

- $h_1(x) = Tr_1^m\left(a_1 x^{2^m+1}\right)$; $a_1 \in \mathbb{F}_{2^m}^\star$;
- $h_2(x) = Tr_1^n\left(a_1 x^{(2^m-1)\frac{1}{2}+1} + a_2 x^{(2^m-1)3+1}\right)$;
  $a_1 \in \mathbb{F}_{2^n}^\star$, $a_2^{2^m+1} = a_1 + a_1^{2^m} = \beta^5$ for some $\beta \in \mathbb{F}_{2^n}^\star$ [14];
- $h_3(x) = Tr_1^n\left(a_1 x^{(2^m-1)\frac{1}{2}+1} + a_2 x^{(2^m-1)\frac{1}{4}+1}\right)$;
  $a_1 \in \mathbb{F}_{2^n}^\star$ $a_2^{2^m+1} = a_1 + a_1^{2^m}$, m odd [14];
- $h_4(x) = Tr_1^n\left(a_1 x^{(2^m-1)\frac{1}{2}+1} + a_2 x^{(2^m-1)\frac{1}{6}+1}\right)$; $a_1 \in \mathbb{F}_{2^n}^\star$ $a_2^{2^m+1} = a_1 + a_1^{2^m}$, m even [14];
- $h_5(x) = Tr_1^n(\alpha x^{2^m+1} + \sum_{i=1}^{2^{r-1}-1} x^{s_i})$, $r > 1$ such that $gcd(r,m) = 1$, $\alpha \in \mathbb{F}_{2^n}$ such that $\alpha + \alpha^{2^m} = 1$, $s_i = (2^m-1)\frac{i}{2^r} \pmod{2^m+1} + 1$, $i \in \{1, \cdots, 2^{r-1}-1\}$ [18].

By Theorem 17.5.8, we recover the families in univariate form containing semi-bent functions derived previously by the author in [27].

A complete list of the known functions in $\mathcal{D}_n$ (which has been introduced and discussed in Chap. 14) can be found in [31] with additional functions in [19]. Now, note that $\mathcal{D}_n$ coincides with the set of Boolean functions $f : \mathbb{F}_{2^n} \to \mathbb{F}_2$ such that the restriction to $u\mathbb{F}_{2^m}^\star$ is constant for every $u \in U$ with $f(0) = 0$ while $\mathcal{L}_n$ coincides with the set of Boolean functions on $\mathbb{F}_{2^n}$ such that the restriction to $u\mathbb{F}_{2^m}^\star$ is linear for every $u \in U$ with $f(0) = 0$. Moreover, recall that $\mathcal{D}_n$ is also the set of functions belonging to $\mathcal{PS}_{ap}^\#$ which vanish at 0.

A stronger version of the previous statement has been proved in [7].

Theorem 17.5.9 ([7]). *Let $n = 2m$ with $m > 2$. Keeping the same notation as in Theorem 17.5.8. Set*

$$\mathcal{A}_n := \{f : \mathbb{F}_{2^n} \to \mathbb{F}_2 \text{ s.t the restriction to } u\mathbb{F}_{2^m}^\star \text{ is affine for every } u \in U\}.$$

Then a function f in $\mathcal{A}_n$ is semi-bent if and only if f can be written as the sum of a function in $\mathcal{D}_n$ and a function in $\mathcal{L}_n$.

Example 17.5.10. Identify the semi-bent Boolean function f over $\mathbb{F}_{64}$ of the form $f(x) = \mathrm{Tr}_1^6(ax^{36}) + \mathrm{Tr}_1^6(bx^{32}) + \mathrm{Tr}_1^6(cx^{56})$. Set $f = g + h$ where $g : x \in \mathbb{F}_{64} \mapsto \mathrm{Tr}_1^6(cx^{56})$ and $h : x \in \mathbb{F}_{64} \mapsto \mathrm{Tr}_1^6(ax^{36}) + \mathrm{Tr}_1^6(bx^{32})$. We have $36 \equiv 1 \pmod 7$, $36 \equiv 2^2 \pmod 7$ and $56 \equiv 0 \pmod 7$. So 36 and 32 are Niho exponents, while 56 is a Dillon exponent. According to the above result, f is semi-bent, if and only if its Niho part (that is, the function h) is bent and its Dillon part (that is, the function g) is bent. On one hand, the bentness of h depends only on the bentness of $x \mapsto \mathrm{Tr}_1^6(ax^{36})$ (since $x \mapsto \mathrm{Tr}_1^6(bx^{32})$ is linear). But $36 = 7 \times \frac{1}{2} + 1$ where $\frac{1}{2}$ is understood modulo 9. Thus, the function $x \mapsto \mathrm{Tr}_1^6(ax^{36})$ is bent if and only if $\mathrm{Tr}_3^6(a) = a + a^8 \neq 0$. Hence, h is bent if and only if $a+a^8 \neq 0$ ($a \in \mathbb{F}_{64}$). On the other hand, $g(x)$ is of the form $\mathrm{Tr}_1^n(cx^{2m-1})$ with $m = \frac{n}{2} = 3$ (the size of the cyclotomic class of 56 modulo $2^6 - 1 = 63$ is 6). Therefore, g is bent, if and only if $K_m(c^{2^m+1}) = K_3(c^9) = 0$ where K_m denotes the Kloosterman sums over $\mathbb{F}_{2^m}$. Let α be a primitive element of $\mathbb{F}_8$ such that $\alpha^3 + \alpha^2 + 1 = 0$. Then, it is easy to check that g is bent, if and only if $c^9 \in \{\alpha, \alpha^2, \alpha^4\}$, that is, $c^9 = \alpha^{2^j}$ for some j (since the Kloosterman sums is invariant under the Frobenius mapping). Finally, one can conclude that f is semi-bent on $\mathbb{F}_{64}$, if and only if $a + a^8 \neq 0$ and $c^9 = \alpha^{2^j}$ for some j where $\alpha \in \mathbb{F}_8$ such that $\alpha^3 + \alpha^2 + 1 = 0$.

Recall [13] that a *spread* is a collection $\{E_i,\, i = 1, \ldots, 2^m+1\}$ of vectorspaces of dimension $m = n/2$ such that $E_i \cap E_j = \{0\}$ for every i and j and $\bigcup_{i=1}^{2^m+1} E_i = \mathbb{F}_{2^n}$. The classical example of spread is $\{u\mathbb{F}_{2^m}\,;\, u \in U\}$ where U is the multiplicative group $\{u \in \mathbb{F}_{2^n}; u^{2^m+1} = 1\}$. Theorem 17.5.9 can be stated in more general setting as follows.

Theorem 17.5.11 ([7]). *Let $m \geq 2$ and $n = 2m$. Let $\{E_i,\, i = 1, \ldots, 2^m + 1\}$ be a spread in $\mathbb{F}_{2^n}$ and h a Boolean function whose restriction to every E_i is linear (possibly null). Let S be any subset of $\{1, \ldots, 2^m + 1\}$ and $g = \sum_{i \in S} 1_{E_i} \pmod 2$ where 1_{E_i} is the indicator of E_i. Then $g + h$ is semi-bent if and only if g and h are bent.*

We call g a $\mathcal{PS}_{ap}$-like bent function.

Proof. We may without loss of generality assume that $g(0) = 0$, that is, S has even size (otherwise, we replace g by $g + 1$). Let us then compute the Walsh–Hadamard transform of $g + h$. We have for all $c \in \mathbb{F}_{2^n}$:

$$\widehat{\chi_{g+h}}(c) = \sum_{x \in \mathbb{F}_{2^n}} \chi((g + h)(x) + \mathrm{Tr}_1^n(cx))$$

$$= 1 + \sum_{i=1}^{2^m+1} \sum_{e \in E_i^\star} \chi(g(e) + h(e) + \mathrm{Tr}_1^n(ce))$$

since $\bigcup_{i=1}^{2^m+1} E_i^\star = \mathbb{F}_{2^n}^\star$ and $E_i^\star \cap E_j^\star = \emptyset$. Let us denote by g_i the value of g on $E_i^\star$, by h_i the restriction of h to E_i and by $I(c)$ the set $\{i \in [1,\ldots,2^m+1]; \forall e \in E_i,\, h(e) = \mathrm{Tr}_1^n(ce)\}$. We have, for every $c \in \mathbb{F}_{2^n}$:

$$\begin{aligned}\widehat{\chi_{g+h}}(c) &= 1 + \sum_{i=1}^{2^m+1} \chi(g_i) \sum_{e\in E_i^\star} \chi(h_i(e) + \mathrm{Tr}_1^n(ce))\\ &= 1 - \sum_{i=1}^{2^m+1} \chi(g_i)\\ &\quad + \sum_{i=1}^{2^m+1} \chi(g_i) \sum_{e\in E_i} \chi(h_i(e) + \mathrm{Tr}_1^n(ce)).\end{aligned}$$

Since h_i is linear on E_i, one has $\sum_{e\in E_i} \chi(h_i(e) + \mathrm{Tr}_1^n(ce)) = 2^m$ if $i \in I(c)$ and 0 otherwise. Therefore:

$$\forall c \in \mathbb{F}_{2^n},\quad \widehat{\chi_{g+h}}(c) = 1 - \sum_{i=1}^{2^m+1} \chi(g_i) + 2^m \sum_{i\in I(c)} \chi(g_i). \tag{17.27}$$

On the other hand, the Walsh–Hadamard transform of h is (take $g = 0$ in the preceding calculation):

$$\widehat{\chi_h}(c) = 2^m(\#I(c) - 1). \tag{17.28}$$

If g is bent then we know that $\sum_{i=1}^{2^m+1} \chi(g_i) = 1$. If h is bent then, according to (17.28), $\#I(c) \in \{0, 2\}$. Hence, if g and h are bent then, $\forall c \in \mathbb{F}_{2^n}$, $\widehat{\chi_{g+h}}(c) = 2^m \sum_{i\in I(c)} \chi(g_i) \in \{0, \pm 2^{m+1}\}$, proving that $g + h$ is semi-bent.

Conversely, let us assume that $g+h$ is semi-bent and let us show that, necessarily, g and h are bent. According to (17.27), we have $\sum_{i=1}^{2^m+1} \chi(g_i) \equiv 1 \pmod{2^m}$. In other words, $\sum_{i=1}^{2^m+1} \chi(g_i) = 1 + \epsilon 2^m$ with $\epsilon \in \{0, \pm 1\}$. Suppose that $\epsilon \in \{-1, 1\}$. Then $I(c)$ is non-empty, for every c. Indeed, if $I(c) = \emptyset$, $\widehat{\chi_{g+h}}(c) = -\epsilon 2^m \notin \{0, \pm 2^{m+1}\}$; this implies that the Walsh–Hadamard transform of h is non-negative and we have seen in Sect. 2.2 (Chap. 1) that h is then linear, say $h(x) = \mathrm{Tr}_1^n(ax)$. We have then, according to (17.28): $\#I(c) = 1$ for $c \neq a$ and $\#I(c) = 2^m + 1$ for $c = a$; thus,

$$\begin{aligned}\widehat{\chi_{g+h}}(a) &= -\epsilon 2^m + 2^m \sum_{i=1}^{2^m+1} \chi(g_i)\\ &= -\epsilon 2^m + 2^m + \epsilon 2^n\\ &= (1-\epsilon)2^m + \epsilon 2^n \in \{2^n, 2^{m+1} - 2^n\}\end{aligned}$$

a contradiction with the fact that $g + h$ is semi-bent. Therefore, we have $\epsilon = 0$, $\sum_{i=1}^{2^m+1} \chi(g_i) = 1$, which implies that g is bent. Let us now prove that h is bent. One has necessarily $\sum_{i \in I(c)} \chi(g_i) \in \{-2, 0, 2\}$. Thus, $I(c)$ is of even size for every c, which implies that $\widehat{\chi_h}(c)$ is congruent to 2^m modulo 2^{m+1}, which according to Lemma 1 in [4] implies that h is bent (that is, $\#I(c) \in \{0, 2\}$ for every c). □

Given a spread $(E_i)_{i=1,\dots,2^m+1}$, the previous theorem provides a characterization of the semi-bentness for a function whose restriction to every E_i^* is affine (i.e. equal to the sum of a function whose restriction to every E_i is linear and of a function whose restriction to every E_i^* is constant).

Remark 17.5.12. One can modify the hypothesis of Theorem 17.5.11 by assuming that we have only a partial spread. There exists an example for m even due to Dillon [13] of a partial spread in $\mathbb{F}_{2^n} \approx \mathbb{F}_{2^m} \times \mathbb{F}_{2^m}$ which is not included in a spread: $E_\infty = \{0\} \times \{0\} \times \mathbb{F}_{2^{m-1}} \times \mathbb{F}_2$ and $E_a = \{(x, \epsilon, a^2x + a\mathrm{Tr}_1^{m-1}(ax) + a\epsilon, \mathrm{Tr}_1^{m-1}(ax)); (x, \epsilon) \in \mathbb{F}_{2^{m-1}} \times \mathbb{F}_2\}$ for $a \in \mathbb{F}_{2^{m-1}}$ (the corresponding function g is quadratic bent). By modifying the hypothesis, we need then to add a condition on the E_i's, and we have only a sufficient condition for $g + h$ being semi-bent:

Let g be a bent function in the $\mathcal{PS}$ class, equal to the sum modulo 2 of the indicators of $l := 2^{m-1}$ or $2^{m-1} + 1$ pairwise "disjoint" vectorspaces E_i having dimension m, and h a bent function which is linear on each E_i. Assume additionally that for every $c \in \mathbb{F}_{2^n}$ there exist at most 2 indices i such that $\forall e \in E_i$, $h(e) = \mathrm{Tr}_1^n(ce)$. Then $g + h$ is semi-bent.

Problem 17.5.13. Find semi-bent functions obtained by applying the result of Remark 17.5.12.

Problem 17.5.14. Show that some semi-bent functions obtained above in [7] are not extendable to $(n + 2)$-variable bent functions (or deduce new bent functions from them).

17.5.2.2 Primary Constructions in Bivariate Representation from the Class $\mathcal{H}$ of Bent Functions

Semi-bent functions in bivariate representation have been derived from the class $\mathcal{H}$ of bent functions introduced by Carlet and the author in [6] and from the Partial Spread class $\mathcal{PS}_{ap}$ of bent functions introduced by Dillon [13].

The functions from class $\mathcal{PS}_{ap}$ are those whose supports can be uniquely written as $\bigcup_{u \in S} u\mathbb{F}_{2^m}^\star$ where U is the set $\{u \in \mathbb{F}_{2^n}; u^{2^m+1} = 1\}$ and S is a subset of U of size 2^{m-1}. We shall also include in $\mathcal{PS}_{ap}$ the complements of these functions.

Now, functions of the class $\mathcal{H}$ are defined in bivariate form as follows.

Definition 17.5.15 ([6]). Functions h of the class $\mathcal{H}$ defined on $\mathbb{F}_{2^m} \times \mathbb{F}_{2^m}$ are of the form

$$h(x, y) = \begin{cases} \mathrm{Tr}_1^m\left(x\psi\left(\frac{y}{x}\right)\right) & \text{if } x \neq 0 \\ \mathrm{Tr}_1^m(\mu y) & \text{if } x = 0 \end{cases} \tag{17.29}$$

where $\psi : \mathbb{F}_{2^m} \to \mathbb{F}_{2^m}$ and $\mu \in \mathbb{F}_{2^m}$ and satisfying the following condition:

$$\forall \beta \in \mathbb{F}_{2^m}^{\star}, \text{the function } z \mapsto G(z) + \beta z \text{ is 2-to-1 on } \mathbb{F}_{2^m}. \tag{17.30}$$

where G is defined as : $G(z) := \psi(z) + \mu z$.

The current list of examples of functions h from the class $\mathcal{H}$ is the following:

- $h(x, y) = \mathrm{Tr}_1^m(x^{-5}y^6)$, m odd;
- $h(x, y) = \mathrm{Tr}_1^m(x^{\frac{5}{6}}y^{\frac{1}{6}})$, m odd;
- $h(x, y) = \mathrm{Tr}_1^m(x^{-3\cdot(2^k+1)}y^{3\cdot 2^k+4})$, $m = 2k - 1$;
- $h(x, y) = \mathrm{Tr}_1^m(x^{-3\cdot(2^{k-1}-1)}y^{3\cdot 2^{k-1}-2})$, $m = 2k - 1$;
- $h(x, y) = \mathrm{Tr}_1^m(x^{1-2^k-2^{2k}}y^{2^k+2^{2k}})$, $m = 4k - 1$;
- $h(x, y) = \mathrm{Tr}_1^m(x^{2^{3k-1}-2^{2k}+2^k}y^{1-2^{3k-1}+2^{2k}-2^k})$, $m = 4k - 1$;
- $h(x, y) = \mathrm{Tr}_1^m(x^{1-2^{2k+1}-2^{3k+1}}y^{2^{2k+1}+2^{3k+1}})$, $m = 4k + 1$;
- $h(x, y) = \mathrm{Tr}_1^m(x^{2^{3k+1}-2^{2k+1}+2^k}y^{1-2^{3k+1}+2^{2k+1}-2^k})$, $m = 4k + 1$;
- $h(x, y) = \mathrm{Tr}_1^m(x^{1-2^k}y^{2^k} + x^{-(2^k+1)}y^{2^k+2} + x^{-3\cdot(2^k+1)}y^{3\cdot 2^k+4})$, $m = 2k - 1$;
- $h(x, y) = \mathrm{Tr}_1^m(y(y^{2^k+1}x^{-(2^k+1)} + y^3x^{-3} + yx^{-1})^{2^{k-1}-1})$, $m = 2k - 1$;
- $h(x, y) = \mathrm{Tr}_1^m(x^{\frac{5}{6}}y^{\frac{1}{6}} + x^{\frac{1}{2}}y^{\frac{1}{2}} + x^{\frac{1}{6}}y^{\frac{5}{6}})$, m odd;
- $h(x, y) = \mathrm{Tr}_1^m(x[D_{\frac{1}{5}}(\frac{y}{x})]^6)$, m odd, where $D_{\frac{1}{5}}$ is the Dickson polynomial of index $\frac{1}{5}$.

The following result provide constructions of semi-bent functions from the classes $\mathcal{H}$ and $\mathcal{PS}_{ap}$

Theorem 17.5.16 ([7]). *The sum of a function defined on $\mathbb{F}_{2^m} \times \mathbb{F}_{2^m}$ from the class $\mathcal{PS}_{ap}$ and a function defined on $\mathbb{F}_{2^m} \times \mathbb{F}_{2^m}$ from the class $\mathcal{H}$ is semi-bent on $\mathbb{F}_{2^m} \times \mathbb{F}_{2^m}$.*

17.5.2.3 A Construction from Bent Functions via the Indirect Sum

In [5], Carlet has introduced a secondary construction (which means a construction of new functions from ones having the same properties) of bent functions. Later, such a construction was called the "*indirect sum*" because it generalizes the well-known direct sum introduced by Dillon and Rothaus [13, 32]. The indirect sum is defined as follows.

Definition 17.5.17 ([5]). Let $n = r + s$ where r and s are positive integers. Let f_1, f_2 be Boolean functions defined on $\mathbb{F}_{2^r}$ and g_2, g_2 be two Boolean functions defined on $\mathbb{F}_{2^s}$. Define h as follows (that is, h is the concatenation of the four functions f_1, $f_1 \oplus 1$, f_2 and $f_2 \oplus 1$, in an order controlled by $g_1(y)$ and $g_2(y)$):

$$\forall (x, y) \in \mathbb{F}_{2^r} \times \mathbb{F}_{2^s}, \quad h(x, y) = f_1(x) + g_1(y) + (f_1(x) + f_2(x))(g_1(y) + g_2(y)).$$

Using the indirect sum, we derive a general constructions of semi-bent functions from both bent and semi-bent functions.

Theorem 17.5.18. *Let $n = r + s$ with r and s two even integers. Let h be as in Definition 17.5.17. Assume that f_1 and f_2 are semi-bent on $\mathbb{F}_{2^r}$ and that g_1 and g_2 are bent on $\mathbb{F}_{2^s}$. Then h is semi-bent on $\mathbb{F}_{2^n}$.*

Proof. Set $r = 2\rho$ and $s = 2\sigma$. Let's compute the Walsh transform of h for every $(a, b) \in \mathbb{F}_{2^r} \times \mathbb{F}_{2^s}$. We have

$$\widehat{\chi_h}(a,b) = \sum_{x\in\mathbb{F}_{2^r}}\sum_{y\in\mathbb{F}_{2^s}} \chi(f_1(x) + g_1(y) + (f_1(x) + f_2(x))(g_1(y) + g_2(y)) + Tr_1^r(ax) + Tr_1^s(by)).$$

Now, one can split the sum depending whether $g_1(y) + g_2(y)$ is equal to 1 or not :

$$\widehat{\chi_h}(a,b) = \sum_{x\in\mathbb{F}_{2^r}}\sum_{y\in\mathbb{F}_{2^s}\mid g_1(y)+g_2(y)=1} \chi(f_2(x) + g_1(y) + Tr_1^r(ax) + Tr_1^s(by))$$
$$+ \sum_{y\in\mathbb{F}_{2^s}\mid g_1(y)+g_2(y)=0} \chi(f_1(x) + g_1(y) + Tr_1^r(ax) + Tr_1^s(by)).$$

Now, note that the indicator of the set $\{y \in \mathbb{F}_{2^s} \mid g_1(y) + g_2(y) = 1\}$ can be written as $\frac{1-\chi(g_1(y)+g_2(y))}{2}$. Similarly, one can write the indicator of the set $\{y \in \mathbb{F}_{2^s} \mid g_1(y) + g_2(y) = 0\}$ as $\frac{1+\chi(g_1(y)+g_2(y))}{2}$. Hence,

$$\widehat{\chi_h}(a,b) = \widehat{\chi_{f_1}}(a)\left(\frac{\widehat{\chi_{g_1}}(b) + \widehat{\chi_{g_2}}(b)}{2}\right) + \widehat{\chi_{f_2}}(a)\left(\frac{\widehat{\chi_{g_1}}(b) - \widehat{\chi_{g_2}}(b)}{2}\right).$$

Now, if g_1 and g_2 are bent, then

$$\left(\frac{\widehat{\chi_{g_1}}(b) - \widehat{\chi_{g_2}}(b)}{2}\right)\left(\frac{\widehat{\chi_{g_1}}(b) + \widehat{\chi_{g_2}}(b)}{2}\right) = \frac{1}{4}\left((\widehat{\chi_{g_1}}(b))^2 - (\widehat{\chi_{g_2}}(b))^2\right) = 0$$

and thus only the two following situations can occur

$$\frac{\widehat{\chi_{g_1}}(b) - \widehat{\chi_{g_2}}(b)}{2} = 0 \text{ and } \frac{\widehat{\chi_{g_1}}(b) + \widehat{\chi_{g_2}}(b)}{2} = \pm 2^{\sigma}$$

or

$$\frac{\widehat{\chi_{g_1}}(b) - \widehat{\chi_{g_2}}(b)}{2} = \pm 2^{\sigma} \text{ and } \frac{\widehat{\chi_{g_1}}(b) + \widehat{\chi_{g_2}}(b)}{2} = 0.$$

Now f_1 and f_2 being semi-bent : $\widehat{\chi_{f_1}}(a) \in \{0, \pm 2^{\rho+1}\}$ and $\widehat{\chi_{f_2}}(a) \in \{0, \pm 2^{\rho+1}\}$. Therefore $\widehat{\chi_h}(a,b) \in \{0, \pm 2^{\rho+\sigma+1}\}$ proving that h is semi-bent. □

Remark 17.5.19. Obviously, the roles of f_1 and f_2 can be exchanged with those of g_1 and g_2. This means that one can exchange the property of bentness and semi-bentness in Theorem 17.5.18.

17.5.2.4 A Simple Construction of Semi-bent Functions from Bent Functions by Field Extension

Another kind of construction of semi-bent functions from bent functions is given by the simple following statement. When we identify $\mathbb{F}_{2^n}$ with the vectorspace $\mathbb{F}_2^n$, it corresponds to a simple construction of an $(n+2)$-variable semi-bent function from an n-variable bent function. We have the following statement due to Cohen and the author.

Proposition 17.5.20 ([11]). *Let n be an even positive integer. Let f be a Boolean function over $\mathbb{F}_{2^n} \simeq \mathbb{F}_{2^n} \times \mathbb{F}_4$. For $\delta \in \mathbb{F}_4$, we define a Boolean function f_δ over $\mathbb{F}_{2^n} \times \mathbb{F}_4$ by*

$$f_\delta(y,z) = f(y) + \mathrm{Tr}_1^2(\delta z), \forall y \in \mathbb{F}_{2^n}, z \in \mathbb{F}_4.$$

If f is bent over $\mathbb{F}_{2^n}$ then f_δ is semi-bent over $\mathbb{F}_{2^{n+2}}$.

17.5.2.5 Construction of Semi-bent Functions from Bent Functions by Considering the Derivatives Functions

The following construction of semi-bent functions from bent functions under a strong condition on the derivatives functions has been shown in [34].

Theorem 17.5.21 ([34]). *Let n be an even positive integer. Let f and g be two bent functions over $\mathbb{F}_{2^n}$. Assume that there exists $a \in \mathbb{F}_{2^n}$ such that $D_af(x) = D_ag(x) + 1$ for all $x \in \mathbb{F}_{2^n}$. Then the function $h = f + g + D_af + D_a(fg)$ is semi-bent over $\mathbb{F}_{2^n}$.*

A possible construction of semi-bent functions by applying Theorem 17.5.21 is provided by Cohen and the author:

Proposition 17.5.22 ([11]). *Let f be a bent function defined over $\mathbb{F}_{2^n}$ (with n even). Define a Boolean function g by $g(x) = f(x+a) + \mathrm{Tr}_1^n(bx), \forall x \in \mathbb{F}_{2^n}$ where a and b are elements of $\mathbb{F}_{2^n}$ such that $\mathrm{Tr}_1^n(ab) = 1$. Then the function $h = f + g + D_af + D_a(fg)$ is semi-bent over $\mathbb{F}_{2^n}$.*

Proof. The bentness is invariant under the addition of linear functions. Thus g is also bent. Moreover, one has $D_ag(x) = g(x) + g(x+a) = f(x+a) + \mathrm{Tr}_1^n(bx) + f(x) + \mathrm{Tr}_1^n(bx) + \mathrm{Tr}_1^n(ab) = D_af(x) + \mathrm{Tr}_1^n(ab) = D_af(x) + 1$. The proposition follows from Theorem 17.5.21. □

Notice that quadratics semi-bent functions can be easily derived from Proposition 17.5.22.

Problem 17.5.23. Find other examples of constructions of non-quadratic semi-bent functions h starting from two bent functions f and g satisfying $D_af(x) = D_ag(x) + 1$ for some $a \in \mathbb{F}_{2^n}$.

17.5.3 A General Construction of Semi-bent Functions Based on Maiorana–McFarland's Construction

Recall that the Maiorana–McFarland's constructions are the best known primary constructions of bent functions [13, 22]. The *Maiorana–McFarland class* is the set of all the Boolean functions on $\mathbb{F}_{2^m} \times \mathbb{F}_{2^m}$ of the form : $f(x, y) = x \cdot \pi(y) + g(y)$; $x, y \in \mathbb{F}_{2^m}$ where "$\cdot$" denotes an inner product in $\mathbb{F}_{2^m}$, π is any permutation on $\mathbb{F}_{2^m}$ and g is any Boolean function on $\mathbb{F}_{2^m}$. Any such function is bent (the bijectivity of π is a necessary and sufficient condition for f being bent). By computing the Walsh transform, it is easy to see that if π is a 2-to-1 mapping from $\mathbb{F}_{2^m}$ to on $\mathbb{F}_{2^m}$, then f is semi-bent on $\mathbb{F}_{2^m} \times \mathbb{F}_{2^m}$. Consequently, the reader notices that using the Maiorana–McFarland method, any permutation leads to the construction of bent functions and any mapping 2-to-1 leads to the construction of semi-bent functions.

The following statement provides an example of construction of semi-bent functions via the Maiorana–McFarland method.

Proposition 17.5.24. *Let r be a positive integer. Set $m = 2r - 1$. Let g be any Boolean function over $\mathbb{F}_{2^m}$. Define over $\mathbb{F}_{2^m} \times \mathbb{F}_{2^m}$ a Boolean function by $f(x, y) = \mathrm{Tr}_1^m(xy^{2^r+2} + xy) + g(y)$, $\forall (x, y) \in \mathbb{F}_{2^m} \times \mathbb{F}_{2^m}$. Then f is semi-bent.*

Proof. We have to prove that f is semi-bent, that is, its Walsh transform takes only the values 0, 2^{m+1} and -2^{m+1}. Compute the Walsh transform of f. For every $(a, b) \in \mathbb{F}_{2^m} \times \mathbb{F}_{2^m}$, we have:

$$
\begin{aligned}
\widehat{\chi_f}(a, b) &= \sum_{x\in\mathbb{F}_{2^m}}\sum_{y\in\mathbb{F}_{2^m}} (-1)^{\mathrm{Tr}_1^m(xy^{2^r+2}+xy)+g(y)+\mathrm{Tr}_1^m(ax)+\mathrm{Tr}_1^m(by)} \\
&= \sum_{y\in\mathbb{F}_{2^m}}(-1)^{g(y)+\mathrm{Tr}_1^m(by)}\sum_{x\in\mathbb{F}_{2^m}}(-1)^{\mathrm{Tr}_1^m(xy^{2^r+2}+xy))+\mathrm{Tr}_1^m(ax)} \\
&= \sum_{y\in\mathbb{F}_{2^m}} (-1)^{g(y)+\mathrm{Tr}_1^m(by)} \sum_{x\in\mathbb{F}_{2^m}} (-1)^{\mathrm{Tr}_1^m((y^{2^r+2}+y)x)} \\
&= 2^m \sum_{y\in\mathbb{F}_{2^m} \mid y^{2^r+2}+y=a} (-1)^{g(y)+\mathrm{Tr}_1^m(by)}.
\end{aligned}
$$

Now, according to Cusick and Dobbertin [12], the equation $y^{2^r+2} + y = a$ has 0 or 2 solutions in $\mathbb{F}_{2^m}$. The mapping $y \in \mathbb{F}_{2^m} \mapsto y^{2^r+2} + y + a$ is 2-to-1 for every $a \in \mathbb{F}_{2^m}$. Therefore,

$$\widehat{\chi_f}(a, b) \in \{0, \pm 2^{m+1}\}$$

which completes the proof. □

17.5.4 A Construction from APN Functions

Let us recall the definition of *almost perfect nonlinear* (APN) functions.

Definition 17.5.25. Let F be a mapping from $\mathbb{F}_{2^m}$ to itself (m a positive integer). The function f is said to be APN if, $max_{a\in\mathbb{F}_{2^m}^\star} max_{b\in\mathbb{F}_{2^m}} \#\{x \in \mathbb{F}_{2^m} \mid F(x+a) + F(x) = b\} = 2$.

APN functions are important research objects in cryptography and coding theory. Given an APN function, one can derive a construction of semi-bent function in the sprit of Maiorana–McFarland's method.

Proposition 17.5.26. *Let m be a positive integer. Let $F : \mathbb{F}_{2^m} \to \mathbb{F}_{2^m}$ be an APN function, g a Boolean function over $\mathbb{F}_{2^m}$ and $\alpha \in \mathbb{F}_{2^m}^\star$. Denote by $D_\alpha F$ the derivative function of F with respect to α defined by $D_\alpha F(x) = F(x+\alpha) + F(x), \forall x \in \mathbb{F}_{2^m}$. Define over $\mathbb{F}_{2^m} \times \mathbb{F}_{2^m}$ a Boolean function by $f(x,y) = \mathrm{Tr}_1^m(xD_\alpha F(y)) + g(y), \forall (x,y) \in \mathbb{F}_{2^m} \times \mathbb{F}_{2^m}$. Then f is semi-bent.*

Proof. Let us compute the Walsh transform of f. For every $(a,b) \in \mathbb{F}_{2^m} \times \mathbb{F}_{2^m}$, we have:

$$\begin{aligned}\widehat{\chi_f}(a,b) &= \sum_{x\in\mathbb{F}_{2^m}}\sum_{y\in\mathbb{F}_{2^m}} (-1)^{\mathrm{Tr}_1^m(xD_\alpha F(y))+g(y)+\mathrm{Tr}_1^m(ax)+\mathrm{Tr}_1^m(by)} \\ &= \sum_{y\in\mathbb{F}_{2^m}}(-1)^{g(y)+\mathrm{Tr}_1^m(by)}\sum_{x\in\mathbb{F}_{2^m}}(-1)^{\mathrm{Tr}_1^m(x(D_\alpha F(y)+a))} \\ &= 2^m \sum_{y\in\mathbb{F}_{2^m} \mid D_\alpha F(y)=a} (-1)^{g(y)+\mathrm{Tr}_1^m(by)}.\end{aligned}$$

Now, since F is APN, the mapping $y \in \mathbb{F}_{2^m} \mapsto D_\alpha F(y)$ is 2-to-1 for every $\alpha \in \mathbb{F}_{2^m}^\star$. Hence, $\widehat{\chi_f}(a,b) \in \{0, \pm 2^{m+1}\}$ which completes the proof. □

17.5.5 Several Constructions from Hyperovals and Oval Polynomials

Let $PG_2(2^n)$ be the two-dimensional projective space over $\mathbb{F}_{2^n}$. The one-dimensional subspaces of $\mathbb{F}_{2^n}^3$ are then the points and the two-dimensional subspaces of $\mathbb{F}_{2^n}^3$ are called the lines. Recall that there is a close connection between the hyperovals and the o-polynomials since a hyperoval of $PG_2(2^n)$ can be represented by $D(G)$ where G is an o-polynomial on $\mathbb{F}_{2^n}$. The current list, up to equivalence, of the known o-polynomials on $\mathbb{F}_{2^m}$ in given in [6].

A simple construction of semi-bent functions from hyperovals of $PG_2(2^m)$ with $m > 2$ is given by the following statement.

Theorem 17.5.27. *Let k be a positive integer such that $2 \le k \le 2^m - 2$. Let $D(k) := \{(1, t, t^k), t \in \mathbb{F}_{2^m}\} \cup \{(0, 0, 1), (0, 1, 0)\}$ $(m > 2)$ be a hyperoval of $PG_2(2^m)$ and g be a Boolean function on $\mathbb{F}_{2^m}$. Then the function f defined over $\mathbb{F}_{2^m} \times \mathbb{F}_{2^m}$ by $f(x, y) = Tr_1^m(xy^k + xy) + g(y)$ is semi-bent.*

Proof. We have to prove that f is semi-bent, that is, its Walsh transform takes only the values 0, 2^{m+1} and -2^{m+1}. Compute the Walsh transform of f. For every $(a, b) \in \mathbb{F}_{2^m} \times \mathbb{F}_{2^m}$, we have:

$$\begin{aligned}
\widehat{\chi_f}(a, b) &= \sum_{x \in \mathbb{F}_{2^m}} \sum_{y \in \mathbb{F}_{2^m}} \chi\Big(Tr_1^m(xy^k + xy) + g(y) + Tr_1^m(ax) + Tr_1^m(by)\Big) \\
&= \sum_{y \in \mathbb{F}_{2^m}} \chi\Big(g(y) + Tr_1^m(by)\Big) \sum_{x \in \mathbb{F}_{2^m}} \chi\Big(Tr_1^m(xy^k + xy) + Tr_1^m(ax)\Big) \\
&= \sum_{y \in \mathbb{F}_{2^m}} \chi\Big(g(y) + Tr_1^m(by)\Big) \sum_{x \in \mathbb{F}_{2^m}} \chi\Big(Tr_1^m((y^k + y + a)x)\Big) \\
&= 2^m \sum_{y \in \mathbb{F}_{2^m} \mid y^k + y = a} \chi\Big(g(y) + Tr_1^m(by)\Big).
\end{aligned}$$

Now, since $D(k)$ is a hyperoval of $PG_2(2^m)$ then according to Maschietti [21], the equation $y^k + y + a = 0$ has either zero or two distinct solutions in $\mathbb{F}_{2^m}$ for every $a \in \mathbb{F}_{2^m}$ $(m > 2)$.Therefore, $\widehat{\chi_f}(a, b) \in \{0, \pm 2^{m+1}\}$ which completes the proof. □

An application of Theorem 17.5.27 is given by the next proposition.

Proposition 17.5.28. *Let m be a positive odd integer with $m > 2$. Let g be a Boolean function on $\mathbb{F}_{2^m}$. Then the function f defined over $\mathbb{F}_{2^m} \times \mathbb{F}_{2^m}$ by $f(x, y) = Tr_1^m(xy^6 + xy) + g(y)$ is semi-bent.*

Proof. According to Theorem 17.5.27, f is semi-bent if $D(6) := \{(1, t, t^6), t \in \mathbb{F}_{2^m}\} \cup \{(0, 0, 1), (0, 1, 0)\}$ $(m > 2)$ is a hyperoval of $PG_2(2^m)$. By Segre and Bartocci [33], for m odd with $m > 3$, $D(6)$ is a hyperoval of $PG_2(2^m)$. It remains to check the case $m = 3$. By Maschietti [21], it suffices to prove that the equation $y^6 + y = a$ has either zero solution or two distinct solutions in $\mathbb{F}_{2^m}$, for every $a \in \mathbb{F}_{2^m}$. The result is trivial for $a = 0$. Now, let $a \in \mathbb{F}_{2^m}^\star$. Using the fact that $y^7 = 1$ for $y \neq 0$, it is easy to see that the number of solutions of the equation $y^6 + y = a$ in $\mathbb{F}_{2^m}$ is equal to the number of solutions of $y^2 + ay + 1 = 0$ in $\mathbb{F}_{2^m}^\star$, which equals 2 (since if $y^2 + ay + 1 = 0$ has two identical solutions implies that $a = 0$, which contradicts the hypothesis). □

In the following we show how one can construct several infinite classes of semi-bent functions from o-polynomials. The first result in this direction was given in [7] which is closely related to the construction of semi-bent functions in bivariate representation from the class $\mathcal{H}$ of bent functions and the class of partial spreads $\mathcal{PS}_{ap}$ given by Theorem 17.5.16.

Theorem 17.5.29 ([7]). *Let G be an o-polynomial on $\mathbb{F}_{2^m}$, g be Boolean function on $\mathbb{F}_{2^m}$ such that $g(0) = 0$ and $wt(g) = 2^{m-1}$ (that is, g is balanced on $\mathbb{F}_{2^m}$). Let $\mu \in \mathbb{F}_{2^m}$. Define over $\mathbb{F}_{2^m} \times \mathbb{F}_{2^m}$ the Boolean function f by:*

$$f(x,y) = Tr_1^m(\mu y + xG(yx^{2^m-2})) + g(yx^{2^m-2}), \quad (x,y) \in \mathbb{F}_{2^m} \times \mathbb{F}_{2^m}.$$

Then f is semi-bent.

Very recently, several more constructions of semi-bent functions have been derived from o-polynomials [29]. An important point is that the notion of oval polynomial over $\mathbb{F}_{2^m}$ appears to be suitable to build 2-to-1 mappings on $\mathbb{F}_{2^m}$. Such a property has been used to built infinite classes of semi-bent functions.

Theorem 17.5.30 ([29]). *Let α be a primitive element of $\mathbb{F}_{2^m}$ and j a positive integer in the range $[0, 2^m - 2]$. Let G be an o-polynomial on $\mathbb{F}_{2^m}$ and g a Boolean function on $\mathbb{F}_{2^m}$. Define over $\mathbb{F}_{2^m} \times \mathbb{F}_{2^m}$ a Boolean function f by:*

$$f(x,y) = Tr_1^m(xG(y) + \alpha^j xy) + g(y), \quad (x,y) \in \mathbb{F}_{2^m} \times \mathbb{F}_{2^m}.$$

Then f is semi-bent.

Problem 17.5.31. Find other permutations G than oval polynomials having the property that $y \mapsto G(y) + \alpha^j y$ is 2-to-1 (which is the key in the proof of Theorem 17.5.30).

In the following, we emphasize the following observation.

Proposition 17.5.32 ([29]). *Any semi-bent function of Theorem 17.5.30 is the sum of two bent functions in the class of Maiorana–McFarland.*

Remark 17.5.33. Note that if we take at random two bent functions, even in the class of Maiorana–McFarland, their sum would not semi-bent in most cases (the reader should notice that semi-bent functions of Theorem 17.5.29 can also be decomposed in the sum of two bent functions).

Problem 17.5.34. Find new constructions of semi-bent functions using permutations other than oval polynomials.

Another construction of semi-bent function in bivariate representation has been derived by the author in [29].

Theorem 17.5.35 ([29]). *Let m be a positive integer. Assume $m = 2m_1 + 1$ odd. Let G be an o-polynomial on $\mathbb{F}_{2^m}$ and g be a Boolean function on $\mathbb{F}_{2^m}$. Define a Boolean function f in bivariate representation as:*

$$
\begin{aligned}
f(x,y) &= \mathrm{Tr}_1^m\Big(xG^{2^{m_1+1}+1}(y) + xyG^{2^{m_1}+1}(y) + xG^3(y) + xyG^2(y)\Big)\\
&\quad + \mathrm{Tr}_1^m\Big((xy^{2^{m_1}+1} + xy^2 + x)G(y) + xy^{2^{m_1+1}+1} + xy + xy^3\Big)\\
&\quad + g(y), (x,y) \in \mathbb{F}_{2^m} \times \mathbb{F}_{2^m}.
\end{aligned}
$$

Then f is semi-bent on $\mathbb{F}_{2^m} \times \mathbb{F}_{2^m}$.

Now, Theorem 17.5.30 and Theorem 17.5.35 can be generalized since other semi-bent functions of a more general form can be obtained from o-polynomials.

Theorem 17.5.36 ([29]). *Let π_1 and π_2 be two permutations of $\mathbb{F}_{2^m}$ whose composition $\pi_1 \circ \pi_2^{-1}$ is an o-polynomial on $\mathbb{F}_{2^m}$. Let g be a Boolean function over $\mathbb{F}_{2^m}$. Let f be the Boolean function defined on $\mathbb{F}_{2^m} \times \mathbb{F}_{2^m}$ by*

$$
(x,y) \in \mathbb{F}_{2^m} \times \mathbb{F}_{2^m}, \quad f(x,y) = \mathrm{Tr}_1^m(x(\pi_1(y) + \pi_2(y))) + g(y).
$$

Then f is semi-bent.

A first consequence of the previous theorem is the following statement which provides another primary construction of semi-bent functions.

Theorem 17.5.37 ([29]). *Let m be an odd positive integer. Define the Boolean function f on $\mathbb{F}_{2^m} \times \mathbb{F}_{2^m}$ as*

$$
(x,y) \in \mathbb{F}_{2^m} \times \mathbb{F}_{2^m}, \quad f(x,y) = \mathrm{Tr}_1^m\left(y^6x + y^5x + y^3x + yx\right) + g(y)
$$

where g is any Boolean function over $\mathbb{F}_{2^m}$. Then f is semi-bent.

A generalization of Theorem 17.5.35 is given by the following statement.

Theorem 17.5.38 ([29]). *Let π be a permutation of $\mathbb{F}_{2^m}$. Let α be a primitive element of $\mathbb{F}_{2^m}$ and j a nonnegative integer. Let G be an o-polynomial and g a Boolean function over $\mathbb{F}_{2^m}$. Define*

$$
\forall (x,y) \in \mathbb{F}_{2^m} \times \mathbb{F}_{2^m}, \quad f(x,y) = \mathrm{Tr}_1^m(\pi(G(y) + \alpha^j y)x) + g(y).
$$

Then f is semi-bent.

Let $L(x) = \sum_{s=0}^{m-1} \alpha_s x^{2^s}$ and $l(x) = \sum_{s=0}^{m-1} \alpha_s x^s$ be two polynomials over $\mathbb{F}_{2^m}$. $L(x)$ and $l(x)$ are the 2-associate of each other. More specifically, $l(x)$ is the conventional 2-associate of $L(x)$ and $L(x)$ is the linearized 2-associate of $l(x)$. It is well known that L is a linear permutation polynomial, if and only if, the determinant of the matrix $(\alpha_{i-j}^{2^i})_{0 \le i,j \le m-1}$ is not zero.

A possible construction of semi-bent functions involving linearized polynomials and oval polynomials is given by the following statement.

Proposition 17.5.39. *Let $L(x)$ and $l(x)$ two polynomials on $\mathbb{F}_{2^m}$ defined as above. Assume that $l(x)$ is co-prime with $x^m - 1$. Let $a \in \mathbb{F}_{2^m}$ such that $\mathrm{Tr}_1^m(a) = 0$ and δ be a non zero element of $\mathbb{F}_{2^m}$. Let G be an o-polynomial on $\mathbb{F}_{2^m}$ and g any Boolean function on $\mathbb{F}_{2^m}$. Then the function f defined on $\mathbb{F}_{2^m} \times \mathbb{F}_{2^m}$ as*

$$f(x,y) = \mathrm{Tr}_1^m\Big(ax\mathrm{Tr}_1^m(G(y) + \delta y) + xL(G(y) + \delta y)\Big) + g(y)$$

is semi-bent.

17.5.6 Secondary Constructions of Semi-bent Functions

In general, "secondary constructions" means constructions of new functions from ones having the same properties. Only few secondary constructions of semi-bent functions have been considered in the literature. An example of a secondary construction of semi-bent functions based on a strong condition on the derivative functions has been given in [34].

Theorem 17.5.40 ([34]). *Let f and g be two semi-bent functions over $\mathbb{F}_{2^n}$ (with n even). Assume that there exists an element a in $\mathbb{F}_{2^n}$ such that $D_a f = D_a g$. Then the function $h = f + D_a f(f + g)$ is semi-bent on $\mathbb{F}_{2^n}$.*

The reader notices that Theorem 17.5.18 shows that the indirect sum could be used to construct semi-bent functions from both bent and semi-bent functions. The construction derived from Theorem 17.5.18 can be therefore viewed as a secondary-like construction of semi-bent functions.

References

1. T. Berger, A. Canteaut, P. Charpin, and Y. Laigle-Chapuy. On almost perfect nonlinear functions. In *IEEE Trans. Inform. Theory, vol. 52, no. 9*, pages 4160–4170, 2006.
2. X. Cao, H. Chen, and S. Mesnager. Further results on semi-bent functions in polynomial form. In *Journal Advances in Mathematics of Communications (AMC). To appear.*
3. C. Carlet. Two new classes of bent functions. In *Proceedings of EUROCRYPT'93, Lecture Notes in Computer Science 765*, pages 77–101, 1994.
4. C. Carlet. Generalized partial spreads. In *IEEE Trans. Inform. Theory, vol. 41, no. 5*, pages 1482–1487, 1995.
5. C. Carlet. On the secondary constructions of resilient and bent functions. In *Proceedings of the Workshop on Coding, Cryptography and Combinatorics 2003, published by Birkhäuser Verlag*, pages 3–28, 2004.
6. C. Carlet and S. Mesnager. On Dillon's class H of bent functions, niho bent functions and o-polynomials. In *Journal of Combinatorial Theory, Series A, Vol 118, no. 8*, pages 2392–2410, 2011.
7. C. Carlet and S. Mesnager. On Semi-bent Boolean Functions. In *IEEE Transactions on Information Theory-IT, Vol 58 No 5*, pages 3287–3292, 2012.

8. P. Charpin and G. Gong. Hyperbent functions, Kloosterman sums and Dickson polynomials. In *IEEE Trans. Inform. Theory (54) 9*, pages 4230–4238, 2008.
9. P. Charpin, T. Helleseth, and V. Zinoviev. The divisibility modulo 24 of Kloosterman sums of $GF(2^m)$, m odd. *Journal of Combinatorial Theory, Series A*, 114:322–338, 2007.
10. P. Charpin, E. Pasalic, and C. Tavernier. On bent and semi-bent quadratic Boolean functions. In *IEEE Transactions on Information Theory, vol. 51, no. 12*, pages 4286–4298, 2005.
11. G. Cohen and S. Mesnager. On constructions of semi-bent functions from bent functions. In *Journal Contemporary Mathematics 625, Discrete Geometry and Algebraic Combinatorics, American Mathematical Society*, pages 141–154, 2014.
12. T. W. Cusick and H. Dobbertin. Some new three-valued crosscorrelation functions for binary m-sequences. In *IEEE Transactions on Information Theory 42(4)*, pages 1238–1240, 1996.
13. J. Dillon. Elementary Hadamard difference sets. In *PhD dissertation, University of Maryland.*
14. H. Dobbertin, G. Leander, A. Canteaut, C. Carlet, P. Felke, and P. Gaborit. Construction of bent functions via Niho Power Functions. In *Journal of Combinatorial therory, Serie A 113*, pages 779–798, 2006.
15. D. Dong, L. Qu, S. Fu, and C. Li. New constructions of semi-bent functions in polynomial forms. In *Mathematical and computer modeling (57)*, pages 1139–1147, 2013.
16. F. Gologlu. Almost Bent and Almost Perfect Nonlinear Functions, Exponential Sums, Geometries and Sequences. In *PhD dissertation, University of Magdeburg*, 2009.
17. G. Leander. Monomial Bent Functions. In *IEEE Trans. Inform. Theory (52) 2*, pages 738–743, 2006.
18. G. Leander and A. Kholosha. Bent functions with 2^r Niho exponents. In *IEEE Trans. Inform. Theory 52 (12)*, pages 5529–5532, 2006.
19. N. Li, T. Helleseth, X. Tang, and A. Kholosha. Several New Classes of Bent Functions From Dillon Exponents. In *IEEE Transactions on Information Theory 59 (3), pages 1818–1831*, 2013.
20. Petr Lisoněk. An efficient characterization of a family of hyperbent functions. *IEEE Transactions on Information Theory*, 57(9):6010–6014, 2011.
21. A. Maschietti. Difference sets and hyperovals. In *Des.CodesCryptogr.14(1)*, pages 89–98, 1998.
22. R. L. McFarland. A family of noncyclic difference sets. In *Journal of Comb. Theory, Series A, No. 15, pages 1–10*, 1973.
23. S. Mesnager. A new class of bent Boolean functions in polynomial forms. In *Proceedings of international Workshop on Coding and Cryptography, WCC 2009*, pages 5–18, 2009.
24. S. Mesnager. A new family of hyper-bent Boolean functions in polynomial form. In *Proceedings of Twelfth International Conference on Cryptography and Coding, Cirencester, United Kingdom. M. G. Parker (Ed.): IMACC 2009, LNCS 5921, Springer, Heidelberg*, pages 402–417, 2009.
25. S. Mesnager. A new class of bent and hyper-bent Boolean functions in polynomial forms. In *journal Design, Codes and Cryptography, 59(1–3)*, pages 265–279, 2011.
26. S. Mesnager. Bent and hyper-bent functions in polynomial form and their link with some exponential sums and Dickson polynomials. *IEEE Transactions on Information Theory*, 57(9):5996–6009, 2011.
27. S. Mesnager. Semi-bent functions from Dillon and Niho exponents, Kloosterman sums and Dickson polynomials. In *IEEE Transactions on Information Theory-IT, Vol 57, No 11*, pages 7443–7458, 2011.
28. S. Mesnager. Semi-bent functions with multiple trace terms and hyperelliptic curves. In *Proceeding of International Conference on Cryptology and Information Security in Latin America (IACR), Latincrypt 2012, LNCS 7533, Springer*, pages 18–36, 2012.
29. S. Mesnager. Semi-bent functions from oval polynomials. In *Proceedings of Fourteenth International Conference on Cryptography and Coding, Oxford, United Kingdom, IMACC 2013, LNCS 8308, Springer, Heidelberg*, pages 1–15, 2013.

30. S. Mesnager and G. Cohen. On the link of some semi-bent functions with Kloosterman sums. In *Proceeding of International Workshop on Coding and Cryptology, Y.M. Chee et al . (Eds.): IWCC 2011, LNCS 6639, Springer*, pages 263–272, 2011.
31. S. Mesnager and J-P Flori. Hyper-bent functions via Dillon-like exponents. In *IEEE Transactions on Information Theory-IT. Vol. 59 No. 5*, pages 3215–3232, 2013.
32. O.S. Rothaus. On "bent" functions. In *J. Combin.Theory Ser A 20*, pages 300–305, 1976.
33. B. Segre and U. Bartocci. Ovali ed altre curve nei piani di Galois di caratteristica due. In *Acta Arith.18 (1)*, pages 423–449, 1971.
34. G. Sun and C.Wu. Construction of Semi-Bent Boolean Functions in Even Number of Variables. In *Chinese Journal of Electronics, vol 18, No 2*, 2009.
35. J. Yuan and C. Ding. Secret sharing schemes from three classes of linear codes. In *IEEE Trans. Inf. Theory, vol. 52, no. 1*, pages 206–212, 2006.

Chapter 18
Linear Codes from Bent, Semi-bent and Almost Bent Functions

Error correcting codes are widely studied by researchers and employed by engineers and they have long been known to have applications in computer and communication systems, data storage devices (starting from the use of Reed Solomon codes in CDs) and consumer electronics. Certain special types of functions over finite fields and vector spaces over finite fields are closely related to linear or nonlinear codes. For instance, Kerdock codes (see [25]) are constructed from bent functions. In the past decade, much progress on interplays between special functions and codes has been made. In particular, APN functions, planar functions, Dickson polynomials and q-polynomials were employed to construct linear codes with optimal or almost optimal parameters. Recently, several new approaches to constructing linear codes, with special types of functions were proposed, and a lot of linear codes with excellent parameters were obtained; in particular, the constructions with few weights from special functions. Such codes have applications in secret sharing [1, 6, 16, 17, 31] authentication codes [14], association schemes [2], and strongly regular graphs [3]. Interesting two-weight and three-weight codes have been obtained in several papers. A non-exhaustive list is [7, 8, 12, 13, 15–18, 21, 24, 27, 29, 30, 32, 33] and [11].

In this chapter, we focus on the main results concerning the constructions of linear codes from bent functions. In particular, it has been shown in several recent papers [10, 26, 27, 34] that bent functions play an important role in the construction of linear codes with few weights. Note that an interesting reference dealing with bent functions in coding theory is [28] due to J. Wolfmann.

18.1 Two Generic Constructions of Linear Codes from Functions

Boolean functions or more generally p-ary functions have important applications in cryptography and coding theory. In coding theory, they have been used to construct linear codes. Historically, the Reed–Muller codes and Kerdock codes have been for

S. Mesnager, *Bent Functions*, DOI 10.1007/978-3-319-32595-8_18

a long time the two famous classes of binary codes derived from Boolean functions. Reed-Muller codes have been recalled in Chap. 4. Let us recall the Kerdock codes.

Definition 18.1.1. $\mathcal{RM}(r,m)$ stands for the Reed-Muller code of order r and length 2^n. The Kerdock codes of length 2^m consist of $\mathcal{RM}(1,m)$ together with $2^{m-1}-1$ cosets of $\mathcal{RM}(1,m)$ in $\mathcal{RM}(2,m)$

The Boolean functions associated to these cosets are quadratic bent functions with the property that the sum of any two of them is a bent function. Next, a lot of progress has been made in this direction and further codes have been derived from more general and complex functions. Nevertheless, as highlighted by Ding in his very recent survey [11], despite the advances in the past two decades, one can isolate essentially only two generic constructions of linear codes from functions. Linear codes obtained from the first generic construction are defined as the trace function of some functions involving polynomials which vanish at zero while those from the second generic construction are obtained via their defining set.

- **The first generic construction:**

  The first generic construction of linear codes from functions considers a code $\mathcal{C}(f)$ over $\mathbb{F}_p$ involving a polynomial f from $\mathbb{F}_q$ to $\mathbb{F}_q$ (where $q=p^m$) defined by

$$\mathcal{C}(f)=\{\mathbf{c}=(Tr_{q/p}(af(x)+bx))_{x\in\mathbb{F}_q^*};a\in\mathbb{F}_q,b\in\mathbb{F}_q\}.$$

  The resulting code $\mathcal{C}(f)$ from f is a linear code of length $q-1$ and its dimension is upper bounded by $2m$ which is reached in many cases. As explained by Ding, this generic construction has a long history and its importance is supported by Delsarte's Theorem [9]. In the binary case (that is, when $p=2$), the first generic construction provides a kind of a coding-theory characterization of special cryptographic functions such as APN functions, almost bent functions and semi-bent functions (see for instance [4, 5] and [22]).
- **The second generic construction:**

  The second generic construction of linear codes from functions fixes a set $D=\{d_1,d_2,\cdots,d_s\}$ in $\mathbb{F}_q$ (where $q=p^m$) and defines a linear code involving D as follows:

$$\mathcal{C}_D=\{\mathbf{c}_x=(Tr_{q/p}(xd_1),Tr_{q/p}(xd_2),\cdots,Tr_{q/p}(xd_n)),x\in\mathbb{F}_q\}.$$

  The set D is usually called the *defining set* of the code $\mathcal{C}_D$. The resulting code $\mathcal{C}_D$ is of length n and of dimension at most m. This construction is generic in the sense that many classes of known codes could be produced by selecting the defining set $D\subseteq\mathbb{F}_q$. The code quality in terms of parameters is closely related to the choice of the set D.

Define for each $x \in \mathbb{F}_q$, $\mathbf{c}_x = (Tr_{q/p}(xd_1), Tr_{q/p}(xd_1), \cdots, Tr_{q/p}(xd_n))$. The Hamming weight $wt(\mathbf{c}_x)$ of $\mathbf{c}_x$ is $n - N_x(0)$, where

$$N_x(0) = \#\{1 \le i \le n \mid Tr_{q/p}(xd_i) = 0\}, \forall x \in \mathbb{F}_q.$$

Note that

$$pN_x(0) = \sum_{i=1}^{n} \sum_{y \in \mathbb{F}_p} e^{\frac{2\pi\sqrt{-1}}{p} y Tr_{q/p}(xd_i)}$$

$$= \sum_{i=1}^{n} \sum_{y \in \mathbb{F}_p} \chi_1(yxd_i) = n + \sum_{y \in \mathbb{F}_p^*} \chi_1(yxD),$$

where χ_1 is the canonical additive character of $\mathbb{F}_q$, aD denotes the set $\{ad \mid d \in D\}$ and $\chi_1(S) := \sum_{x \in S} \chi_1(x)$ for any subset S of $\mathbb{F}_q$. Therefore,

$$wt(\mathbf{c}_x) = \frac{(p-1)}{p} n - \frac{1}{p} \sum_{y \in \mathbb{F}_p^*} \chi_1(yxD).$$

18.2 Linear Codes from Bent Functions Based on the Second Generic Construction

Recall that p-ary function $f : \mathbb{F}_{p^m} \longrightarrow \mathbb{F}_p$ is called *bent* if all its Walsh–Hadamard coefficients satisfy $|\widehat{\chi_f}(b)|^2 = p^m$. A bent function f is called *regular bent* if for every $b \in \mathbb{F}_{p^m}$, $p^{-\frac{m}{2}} \widehat{\chi_f}(b) = \xi_p^{f^\star(b)}$ for some p-ary function $f^\star : \mathbb{F}_{p^m} \to \mathbb{F}_p$ [23, Definition 3] where $\xi_p = e^{\frac{2\pi\sqrt{-1}}{p}}$ is the primitive p-th root of unity. The bent function f is called *weakly regular bent* if there exists a complex number u with $|u| = 1$ and a p-ary function $f^\star$ such that $up^{-\frac{m}{2}} \widehat{\chi_f}(b) = \xi_p^{f^\star(b)}$ for all $b \in \mathbb{F}_{p^m}$. Such function $f^*(x)$ is called the *dual* of $f(x)$. From [19, 20], a weakly regular bent function $f(x)$ satisfies that

$$\widehat{\chi_f}(b) = \epsilon \sqrt{p^*}^m \xi_p^{f^*(b)}, \tag{18.1}$$

where $\epsilon = \pm 1$ is called the sign of the Walsh transform of $f(x)$ and p^* equals $(\frac{-1}{p})p$ where $(\frac{a}{b})$ is the Legendre symbol for $1 \le a \le p-1$.

The most famous codes constructed from bent functions are the Kerdock codes. The extension to the odd characteristic of the concept of bentness in the past decade has produced several types of codes constructed from the two genetic models presented above but using new functions. In particular, some new codes have been obtained by considering weakly regular bent functions. We summarize in Table 18.1 all known weakly regular bent functions over $\mathbb{F}_{p^m}$ with odd characteristic p.

Table 18.1 Known weakly regular bent functions over $\mathbb{F}_{p^m}$, p odd

Bent functions	m	p
$\sum_{i=0}^{\lfloor m/2 \rfloor} Tr_{p^m/p}(a_i x^{p^i+1})$	arbitrary	arbitrary
$\sum_{i=0}^{p^k-1} Tr_{p^m/p}(a_i x^{i(p^k-1)}) + Tr_{p^l/p}(\delta x^{\frac{p^m-1}{e}})$, $e \mid p^k+1$	$m = 2k$	arbitrary
$Tr_{p^m/p}(ax^{\frac{3^m-1}{4}+3^k+1})$	$m = 2k$	$p = 3$
$Tr_{p^m/p}(x^{p^{3k}+p^{2k}-p^k+1} + x^2)$	$m = 4k$	arbitrary
$Tr_{p^m/p}(ax^{\frac{3^i+1}{2}})$; i odd, $gcd(i,m) = 1$	arbitrary	$p = 3$

Table 18.2 Weight distribution of the code given in Theorem 18.2.1

Weight	Multiplicity
0	1
$\frac{n_f}{2} - 2^{\frac{m-4}{2}}$	$\frac{2^m - 1 + n_f 2^{-\frac{m-2}{2}}}{2}$
$\frac{n_f}{2} + 2^{\frac{m-4}{2}}$	$\frac{2^m - 1 + n_f 2^{-\frac{m-2}{2}}}{2}$

In 2015, bent functions have been used to construct linear codes from the second generic construction [10, 27, 34]. Ding was the initiator of these works. In the following, we present linear codes obtained from the second generic construction.

1. Let $p = 2$. We know that a function f from $\mathbb{F}_{2^m}$ to $\mathbb{F}_2$ is bent if and only if its support $D_f := \{x \in \mathbb{F}_{2^m} \mid f(x) = 1\}$ is a difference set in $(\mathbb{F}_{2^m}, +)$ with parameters

$$(2^m, 2^{m-1} \pm 2^{\frac{(m-2)}{2}}, 2^{m-2} \pm 2^{\frac{(m-2)}{2}}).$$

When f is bent, we have

$$n_f := |D_f| = 2^{m-1} \pm 2^{\frac{(m-2)}{2}}.$$

It has been shown by Ding [10] the following result.

Theorem 18.2.1 ([10]). *Let f be a Boolean function from $\mathbb{F}_{2^m}$ to $\mathbb{F}_2$ with $f(0) = 0$ where m even and $m \geq 4$. Then the code $\mathcal{C}_{D_f}$ is an $[n_f, m, (n_f - 2^{\frac{(m-2)}{2}})/2]$ two-weight binary code with weight distribution given by Table 18.2*

Consequently, any bent function can be plugged into Theorem 18.2.1 to obtain a two-weight binary linear code.

2. Using Ding's approach [10, 17], Zhou et al. [34] have derived several classes of p-ary linear codes with two or three weights constructed from quadratic bent functions over $\mathbb{F}_p$ where p is an odd prime. More precisely, let Q be a quadratic function from $\mathbb{F}_{p^m}$ to $\mathbb{F}_p$. Recall that any quadratic form $Q(x)$ over $\mathbb{F}_p$ of rank r in n variables is equivalent (under a change of coordinates) to one of the following three standard types (recall that we write $\bar{x}$ when an element is thought of as a vector in $\mathbb{F}_p^n$ and x when the same vector is thought of as an element of $\mathbb{F}_p^n$):

Table 18.3 Weight distribution of the code $\mathcal{C}_{D_Q}$ given in Theorem 18.2.2 when m is odd

Weight	Multiplicity
0	1
$(p-1)(p^{m-2}-p^{\frac{m-3}{2}})$	$\frac{p-1}{2}(p^{m-1}+p^{\frac{m-1}{2}})$
$(p-1)p^{m-2}$	$p^{m-1}-1$
$(p-1)(p^{m-2}+p^{\frac{m-3}{2}})$	$\frac{p-1}{2}(p^{m-1}-p^{\frac{m-1}{2}})$

Table 18.4 Weight distribution of the code $\mathcal{C}_{D_Q}$ given in Theorem 18.2.3 when m is even

Weight	Multiplicity
0	1
$(p-1)p^{m-2}$	$p^{m-1}+\epsilon(p-1)p^{\frac{m-2}{2}}-1$
$(p-1)(p^{m-2}+\epsilon p^{\frac{m-2}{2}})$	$(p-1)(p^{m-1}-\epsilon p^{\frac{m-2}{2}})$

(a) **Type I**: $\tilde{Q}_r(\bar{x})$, r even;
(b) **Type II**: $\tilde{Q}_{r-1}(\bar{x})+\mu x_n^2$, r odd;
(c) **Type III**: $\tilde{Q}_{r-2}(\bar{x})+x_{r-1}^2-\eta x_r^2$, r even;

where $\mu\in\{1,\eta\}$ and η is a fixed nonsquare in $\mathbb{F}_p$.

Define $D_Q=\{x\in\mathbb{F}_{p^m}^*\mid Q(x)=0\}$. Then if m is odd, we have $\#D_Q=p^{m-1}-1$ and if m is even, we have $\#D_Q=p^{m-1}+\epsilon(p-1)p^{\frac{m-2}{2}}$ where $\epsilon\in\{-1,1\}$ (more precisely, $\epsilon=1$ if Q is equivalent to type I and $\epsilon=-1$ if Q is equivalent to type III.). Zhou et al have proved the following results on the weight distribution of the associate code $\mathcal{C}_{D_Q}$ from the second generic construction.

Theorem 18.2.2 ([34]). *If m is odd, then $\mathcal{C}_{D_Q}$ is a three-weight linear code with parameters $[p^{m-1}-1,m]$ whose weight distribution is listed in Table 18.3.*

Theorem 18.2.3 ([34]). *If m is even, then $\mathcal{C}_{D_Q}$ is a two- weight linear code with parameters $[p^{m-1}+\epsilon(p-1)p^{\frac{m-2}{2}}-1,m]$ whose weight distribution is listed in Table 18.4 where $\epsilon=1$ if Q is equivalent to type I and $\epsilon=-1$ if Q is equivalent to type II.*

3. Inspired by the work of C. Ding [10] and K. Ding and C. Ding [17], Tang et al. [27] have generalized their approach to weakly regular bent functions. More precisely, they derived linear codes with two or three weights from a sub-class of p-ary weakly regular bent functions $\mathcal{RF}$. Functions of the set $\mathcal{RF}$ (p odd) vanish at 0 and satisfying the following condition (18.2):

$$\exists h\in\mathbb{N}\mid gcd(h-1,p-1)=1 \text{ and } f(ax)=a^hf(x),\forall(a,x)\in\mathbb{F}_{p^m}^\star\times\mathbb{F}_{p^m}. \tag{18.2}$$

Given a p-ary function $f:\mathbb{F}_{p^m}\to\mathbb{F}_p$. Define $D_f:=\{x\in\mathbb{F}_{p^m}\mid f(x)=0\}$. They proved the two following results.

Theorem 18.2.4 ([27]). *Let m be an even integer and f be a function in $\mathcal{RF}$. Then $\mathcal{C}_{D_f}$ is a two-weight linear code with parameters $[p^{m-1}-1+\epsilon(p-1)p^{(m-2)/2},m]$ (where ϵ denotes sign of the Walsh transform of f) whose weight distribution is listed in Table 18.5.*

Table 18.5 Weight distribution of the code $\mathcal{C}_{D_f}$ given in Theorem 18.2.4 when m is even

Weight	Multiplicity
0	1
$(p-1)p^{m-2}$	$p^{m-1} - 1 + \epsilon(\frac{-1}{p})^{m/2}(p-1)p^{(m-2)/2}$
$(p-1)(p^{m-2} + \epsilon(\frac{-1}{p})^{m/2}p^{(m-2)/2}$	$(p-1)(p^{m-1} - \epsilon(\frac{-1}{p})^{m/2}p^{(m-2)/2}$

Table 18.6 Weight distribution of the code $\mathcal{C}_{D_f}$ given in Theorem 18.2.5 when m is odd

Weight	Multiplicity
0	1
$(p-1)p^{m-2}$	$p^{m-1} - 1$
$(p-1)(p^{m-2} - p^{(m-3)/2})$	$\frac{p-1}{2}(p^{m-1} + p^{(m-1)/2})$
$(p-1)(p^{m-2} + p^{(m-3)/2})$	$\frac{p-1}{2}(p^{m-1} - p^{(m-1)/2})$

Theorem 18.2.5 ([27]). *Let m be a odd integer and f be a function in $\mathcal{RF}$. Then $\mathcal{C}_{D_f}$ is a three-weight linear code with parameters $[p^{m-1} - 1, m]$ whose weight distribution is listed in Table 18.6.*

18.3 Linear Codes from Bent Functions Based on the First Generic Construction

In 2015, weakly bent functions have been used by the author [26] to construct new class of linear codes based on the first generic construction. In the following, we present such a class. The general construction of linear codes from mappings over finite fields is given as follow.

For any $\alpha, \beta \in \mathbb{F}_{p^m}$, defined as:

$$\begin{aligned} f_{\alpha,\beta} : \mathbb{F}_{q^r} &\longrightarrow \mathbb{F}_q \\ x &\longmapsto f_{\alpha,\beta}(x) := Tr_{q^r/q}(\alpha\Psi(x) - \beta x) \end{aligned}$$

where Ψ is a mapping from $\mathbb{F}_{q^r}$ to $\mathbb{F}_{q^r}$ such that $\Psi(0) = 0$.

We now define a linear code $\mathcal{C}_\Psi$ over $\mathbb{F}_q$ as :

$$\mathcal{C}_\Psi := \{\bar{c}_{\alpha,\beta} = (f_{\alpha,\beta}(\zeta_1), f_{\alpha,\beta}(\zeta_2), \cdots, f_{\alpha,\beta}(\zeta_{q^r-1})), \alpha, \beta \in \mathbb{F}_{q^r}\}$$

where $\zeta_1, \cdots, \zeta_{q^r-1}$ denote the nonzero elements of $\mathbb{F}_{q^r}$.

One can show the following statement.

Proposition 18.3.1 ([26]). *The linear code $\mathcal{C}_\Psi$ is of length $q^r - 1$. If the mapping Ψ has no linear component then $\mathcal{C}_\Psi$ is of dimension $k = \frac{2m}{h} = 2r$. Otherwise, k is less than $2r$.*

Moreover, the following statement shows that the weight distribution of the code $\mathcal{C}_\Psi$ of length $q^r - 1$ can be expressed by means of the Walsh transform of some absolute trace functions over $\mathbb{F}_{p^m}$ involving the map Ψ.

Proposition 18.3.2 ([26]). *We keep the notation above. Let $a \in \mathbb{F}_{p^m}$. Let us denote by ψ a mapping from $\mathbb{F}_{p^m}$ to $\mathbb{F}_p$ defined as:*

$$\psi_a(x) = Tr_{p^m/p}(a\Psi(x)).$$

For $\bar{c}_{\alpha,\beta} \in \mathcal{C}_\Psi$, we have:

$$wt(\bar{c}_{\alpha,\beta}) = p^m - \frac{1}{q}\sum_{\omega\in\mathbb{F}_q} \widehat{\chi_{\psi_{\omega\alpha}}}(\omega\beta).$$

Now, let us present a particular subclass of the previous family of linear codes. We shall fix $h = 1$ and assume $\alpha \in \mathbb{F}_p$. Let then $g_{\alpha,\beta}$ be the p-ary function from $\mathbb{F}_{p^m}$ to $\mathbb{F}_p$ given by $g_{\alpha,\beta}(x) = \alpha Tr_{p^m/p}(\Psi(x)) - Tr_{p^m/p}(\beta x)$. Note that $g_{\alpha,\beta}(x) = \alpha\psi_1(x) - Tr_{p^m/p}(\beta x)$ where ψ_a is defined as above by $\psi_a(x) = Tr_{p^m/p}(a\Psi(x))$.

Now let us defined a subcode $\mathcal{C}$ of $\mathcal{C}_\Psi$ as follows:

$$\mathcal{C} := \{\tilde{c}_{\alpha,\beta} = (g_{\alpha,\beta}(\zeta_1), g_{\alpha,\beta}(\zeta_2), \cdots, g_{\alpha,\beta}(\zeta_{p^m-1})), \alpha \in F_p, \beta \in \mathbb{F}_{p^m}\}. \tag{18.3}$$

where $\zeta_1, \cdots, \zeta_{p^m-1}$ denote the nonzero elements of $\mathbb{F}_{p^m}$.

According to Proposition 18.3.2 (where $Tr_{q/p}$ is the identity function),

$$\begin{aligned}
wt(\tilde{c}_{\alpha,\beta}) &= p^m - \frac{1}{p}\sum_{\omega\in\mathbb{F}_p} \widehat{\chi_{\psi_{\omega\alpha}}}(\omega\beta)\\
&= p^m - p^{m-1} - \frac{1}{p}\sum_{\omega\in\mathbb{F}_p^\star} \widehat{\chi_{\psi_{\omega\alpha}}}(\omega\beta)\\
&= p^m - p^{m-1} - \frac{1}{p}\sum_{\omega\in\mathbb{F}_p^\star}\sum_{x\in\mathbb{F}_{p^m}} \xi_p^{\psi_{\omega\alpha}(x) - Tr_{p^m/p}(\omega\beta x)}\\
&= p^m - p^{m-1} - \frac{1}{p}\sum_{\omega\in\mathbb{F}_p^\star}\sum_{x\in\mathbb{F}_{p^m}} \xi_p^{Tr_{p^m/p}(\omega\alpha\Psi(x) - \omega\beta x)}\\
&= p^m - p^{m-1} - \frac{1}{p}\sum_{\omega\in\mathbb{F}_p^\star}\sum_{x\in\mathbb{F}_{p^m}} \xi_p^{\omega Tr_{p^m/p}(\alpha\Psi(x) - \beta x)}\\
&= p^m - p^{m-1} - \frac{1}{p}\sum_{\omega\in\mathbb{F}_p^\star} \sigma_\omega(\widehat{\chi_{\psi_\alpha}}(\beta)).
\end{aligned}$$

But $\widehat{\chi_{\psi_\alpha}}(\beta) = \sigma_\alpha(\widehat{\chi_{\psi_1}}(\bar{\alpha}\beta))$ where $\bar{\alpha}$ satisfies $\bar{\alpha}\alpha = 1$ in $\mathbb{F}_p$. Indeed,

$$\begin{aligned}\sigma_\alpha(\widehat{\chi_{\psi_1}}(\bar{\alpha}\beta)) &= \sigma_\alpha\Big(\sum_{x\in\mathbb{F}_{p^m}} \xi_p^{\psi_1(x)-Tr_{p^m/p}(\bar{\alpha}\beta x)}\Big)\\ &= \sum_{x\in\mathbb{F}_{p^m}} \xi_p^{\alpha\psi_1(x)-Tr_{p^m/p}(\beta x)}\\ &= \widehat{\chi_{\alpha\psi_1}}(\beta) = \widehat{\chi_{\psi_\alpha}}(\beta).\end{aligned}$$

Consequently,

$$wt(\tilde{c}_{\alpha,\beta}) = p^m - p^{m-1} - \frac{1}{p}\sum_{\omega\in\mathbb{F}_p^\star} \sigma_\omega(\sigma_\alpha(\widehat{\chi_{\psi_1}}(\bar{\alpha}\beta))).$$

By computing exponentials sums, we get (where $\epsilon = \pm 1$)

$$\begin{aligned}wt(\tilde{c}_{\alpha,\beta}) &= p^m - p^{m-1} - \frac{1}{p}\sum_{\omega\in\mathbb{F}_p^\star} \sigma_\omega\Big(\epsilon(\frac{\alpha}{p})^m\sqrt{p^*}^m\xi_p^{\alpha\psi_1^*(\bar{\alpha}\beta)}\Big)\\ &= p^m - p^{m-1} - \frac{1}{p}\epsilon(\frac{\alpha}{p})^m\sum_{\omega\in\mathbb{F}_p^\star}\sigma_\omega(\sqrt{p^*}^m)\xi_p^{\omega\alpha\psi_1^*(\bar{\alpha}\beta)}\\ &= p^m - p^{m-1} - \frac{1}{p}\epsilon(\frac{\alpha}{p})^m\sum_{\omega\in\mathbb{F}_p^\star}(\frac{\omega}{p})^m\sqrt{p^*}^m\xi_p^{\omega\alpha\psi_1^*(\bar{\alpha}\beta)}\\ &= \begin{cases} p^m - p^{m-1} - \frac{1}{p}\epsilon(\frac{\alpha}{p})\sqrt{p^*}^m\sum_{\omega\in\mathbb{F}_p^\star}(\frac{\omega}{p})\xi_p^{\omega\alpha\psi_1^*(\bar{\alpha}\beta)} & \text{if } m \text{ odd}\\ p^m - p^{m-1} - p^{\frac{m}{2}-1}\epsilon\sum_{\omega\in\mathbb{F}_p^\star}\xi_p^{\omega\alpha\psi_1^*(\bar{\alpha}\beta)} & \text{if } m \text{ even.}\end{cases}\end{aligned}$$

After calculations, we obtain the following result.

Theorem 18.3.3 ([26]). *We keep the notation above let C be the linear code defined by (18.3) whose codewords are denoted by $\tilde{c}_{\alpha,\beta}$. Assume that the function $\psi_1 := Tr_{p^m/p}(\Psi)$ is bent or weakly regular bent if $p = 2$ or p odd, respectively. We denote by $\psi_1^\star$ its dual function. Then the weight distribution of C is given as follows. In any characteristic, $wt(\tilde{c}_{0,0}) = 0$ and for $\beta \neq 0$, $wt(\tilde{c}_{0,\beta}) = p^m - p^{m-1}$. Moreover,*

1. *if $p = 2$ then the Hamming weight of $\tilde{c}_{1,\beta}$ ($\beta \in \mathbb{F}_{2^m}^\star$) is given by $wt(\tilde{c}_{1,\beta}) = 2^{m-1} - (-1)^{\psi_1^*(\beta)}2^{\frac{m}{2}-1}$.*
2. *if p is odd then ($\bar{\alpha}$ denotes the inverse of α in $\mathbb{F}_p^\star$*
   - *if m is odd then the Hamming weight of $\tilde{c}_{\alpha,\beta}$ is given by*

$$\begin{cases} p^m - p^{m-1} \text{ if } \alpha\in\mathbb{F}_p^\star \text{ and } \psi_1^*(\bar{\alpha}\beta) = 0;\\ p^m - p^{m-1} - \epsilon(\frac{-1}{p})^{\frac{m+1}{2}}p^{\frac{m-1}{2}}\Big(\frac{\psi_1^*(\bar{\alpha}\beta)}{p}\Big) \text{ if } \alpha\in\mathbb{F}_p^\star \text{ and } \psi_1^*(\bar{\alpha}\beta)\in\mathbb{F}_{p^m}^\star.\end{cases}$$

Table 18.7 Weight distribution of C when m is even, p odd

Hamming weight	Multiplicity
0	1
$p^m - p^{m-1}$	$p^m - 1$
$p^m - p^{m-1} - \epsilon p^{\frac{m}{2}-1}(p-1)$	$p^m - p^{m-1} + \epsilon p^{\frac{m}{2}-1}(p-1)^2$
$p^m - p^{m-1} + \epsilon p^{\frac{m}{2}-1}$	$(p^m - p^{m-1})(p-1) - \epsilon p^{\frac{m}{2}-1}(p-1)^2$

Table 18.8 Weight distribution of C when m is even, $p = 2$

Hamming weight	Multiplicity
0	1
2^{m-1}	$2^m - 1$
$2^{m-1} - 2^{\frac{m}{2}-1}$	$2^{m-1} + 2^{\frac{m}{2}-1}$
$2^{m-1} + 2^{\frac{m}{2}-1}$	$2^{m-1} - 2^{\frac{m}{2}-1}$

Table 18.9 Weight distribution of C when m is odd

Hamming weight	Multiplicity
0	1
$p^m - p^{m-1}$	$2p^m - p^{m-1} - 1$
$p^m - p^{m-1} - \epsilon(\frac{-1}{p})^{\frac{m+1}{2}} p^{\frac{m-1}{2}}$	$(p^{m-1} + \epsilon p^{\frac{m-1}{2}})\frac{(p-1)^2}{2}$
$p^m - p^{m-1} + \epsilon(\frac{-1}{p})^{\frac{m+1}{2}} p^{\frac{m-1}{2}}$	$(p^{m-1} - \epsilon p^{\frac{m-1}{2}})\frac{(p-1)^2}{2}$

- *if m is even then the Hamming weight of $\tilde{c}_{\alpha,\beta}$ is given by*

$$\begin{cases} p^m - p^{m-1} - p^{\frac{m}{2}-1}\epsilon(p-1) \text{ if } \alpha \in \mathbb{F}_p^\star \text{ and } \psi_1^*(\bar{\alpha}\beta) = 0; \\ p^m - p^{m-1} + p^{\frac{m}{2}-1}\epsilon \text{ if } \alpha \in \mathbb{F}_p^\star \text{ and } \psi_1^*(\bar{\alpha}\beta) \in \mathbb{F}_{p^m}^\star. \end{cases}$$

Moreover, the weight distribution of C in the case where p is an odd prime is given by the next statements in cases m odd and even, separately.

Theorem 18.3.4 ([26]). *Let $p = 2$ or p an odd prime. If m is even, then the weight distribution of C (which is of dimension $m + 1$) is given by Tables 18.7 and 18.8.*

Theorem 18.3.5 ([26]). *Let p be an odd prime. If m is odd, then the weight distribution of C (which is of dimension $m + 1$) is given by Table 18.9.*

The general idea of the construction above is a classical one but the specific choice of the function employed is new. The former codes are different from those studied in the literature [15, 16, 27] due to the differences in the lengths and dimensions. Moreover, we notice that the codes considered here have dimension $m + 1$, they have the same weight distribution as a subcode of some of the codes in [4] when $p = 2$ and the same weight distribution as a subcode of some of the codes in [6, 30] when p is odd.

18.4 Linear Codes from Semi-bent Functions Based on the Second Generic Construction

Let f be a semi-bent Boolean function on $\mathbb{F}_{2^m}$ with m odd. Denote by D_f its support ans by n_f the cardinality of D_f. Recall that in this case the Walsh spectrum $\widehat{\chi_f}$ of f is equal to $\{0, \pm 2^{\frac{m+1}{2}}\}$. It therefore follows from the definition of semi-bent functions that

$$n_f = \#D_f = \begin{cases} 2^{m-1} - 2^{\frac{(m-1)}{2}} & \text{if } \widehat{\chi_f}(0) = -2^{\frac{(m+1)}{2}} \\ 2^{m-1} + 2^{\frac{(m-1)}{2}} & \text{if } \widehat{\chi_f}(0) = -2^{\frac{(m+1)}{2}} \\ 2^{m-1} & \text{if } \widehat{\chi_f}(0) = 0. \end{cases}$$

Let $\mathcal{C}_{D_f}$ be the linear code obtained via the second generic construction whose defined set D_f with f semi-bent. Ding [10] has proved the following result on $\mathcal{C}_{D_f}$.

Theorem 18.4.1 ([10]). *Let f be a Boolean function on $\mathbb{F}_{2^m}$ with $f(0) = 0$ where m is odd. Then $\mathcal{C}_{D_f}$ is an $[n_f, m, (n_f - 2^{(m-1)/2})/2]$ three-weight binary code with the weight distribution is given in Table 18.10 where n_f is defined by*

$$n_f = \#D_f = \begin{cases} 2^{m-1} - 2^{\frac{(m-1)}{2}} & \textit{if } \widehat{\chi_f}(0) = -2^{\frac{(m+1)}{2}} \\ 2^{m-1} + 2^{\frac{(m-1)}{2}} & \textit{if } \widehat{\chi_f}(0) = -2^{\frac{(m+1)}{2}} \\ 2^{m-1} & \textit{if } \widehat{\chi_f}(0) = 0. \end{cases}$$

if and only if the function f is semi-bent.

All semi-bent functions can be plugged into Theorem 18.4.1 to obtain three-weight binary linear codes. The known semi-bent functions are given in Chap. 17.

18.5 Linear Codes from Almost Bent Functions Based on the Second Generic Construction

Recall that a function F from $\mathbb{F}_{2^m}$ to $\mathbb{F}_{2^m}$ is said to be almost bent (AB) if $W_F(a,b) := \sum_{x\in\mathbb{F}_{2^m}}(-1)^{\mathrm{Tr}_1^m(aF(x)+bx)}$ takes the values 0 or $\pm 2^{(m+1)/2}$ for every pair $(a,b) \in \mathbb{F}_{2^m}^{\star} \times \mathbb{F}_{2^m}$. By definition, almost bent functions over $\mathbb{F}_{2^m}$ exist only

Table 18.10 Weight distribution of the code $\mathcal{C}_{D_f}$ given in Theorem 18.4.1

Hamming weight	Multiplicity
0	1
$\frac{n_f - 2^{(m-1)/2}}{2}$	$n_f(2^m - n_f)2^{-m} - n_f 2^{-(m+1)/2}$
$\frac{n_f}{2}$	$2^m - 1 - n_f(2^m - n_f)2^{-(m-1)}$
$\frac{n_f + 2^{(m-1)/2}}{2}$	$n_f(2^m - n_f)2^{-m} + n_f 2^{-(m+1)/2}$

for m odd. Moreover, by definition, for every almost bent function F, the sum $W_F(1,0) \in \{0, \pm 2^{(m+1)/2}\}$. Hence, if F is an almost bent function from $\mathbb{F}_{2^m}$ to $\mathbb{F}_{2^m}$, then by defining $f = \mathrm{Tr}_1^m(F)$, one straightforwardly get

$$n_f = \#D_{\mathrm{Tr}_1^m(F)} = \begin{cases} 2^{m-1} + 2^{\frac{(m-1)}{2}} & \text{if } W_F(1,0) = -2^{\frac{(m+1)}{2}} \\ 2^{m-1} - 2^{\frac{(m-1)}{2}} & \text{if } W_F(1,0) = 2^{\frac{(m+1)}{2}} \\ 2^{m-1} & \text{if } W_F(1,0) = 0. \end{cases}$$

Ding [10] has deduced the following result.

Theorem 18.5.1. *Let F be an almost bent function from $\mathbb{F}_{2^m}$ to $\mathbb{F}_{2^m}$ where m is odd. Define $f := \mathrm{Tr}_1^m(F)$. Assume $f(0) = 0$. Then $\mathcal{C}_{D_f}$ is an $[n_f, m, (n_f - 2^{(m-1)/2})/2]$ three-weight binary code with the weight distribution is given by Table 18.10, where n_f is defined as follows.*

$$n_f = \#D_{\mathrm{Tr}_1^m(F)} = \begin{cases} 2^{m-1} + 2^{\frac{(m-1)}{2}} & \textit{if } W_F(1,0) = -2^{\frac{(m+1)}{2}} \\ 2^{m-1} - 2^{\frac{(m-1)}{2}} & \textit{if } W_F(1,0) = 2^{\frac{(m+1)}{2}} \\ 2^{m-1} & \textit{if } W_F(1,0) = 0. \end{cases}$$

The list of the five known (up to now !) almost bent functions F of the form $F(x) = x^d$ is given in Chap. 12. The length of the code $\mathcal{C}_{\mathrm{Tr}_1^m(F)}$ is equal to 2^{m-1}.

References

1. R. Anderson, C. Ding, T. Helleseth, and T. Kløve. How to build robust shared control systems. In *Designs, Codes Cryptography, vol. 15, No. 2*, pages 111–124, 1998.
2. A. R. Calderbank and J. M. Goethals. Three-weight codes and association schemes. In *Philips J. Res., vol. 39*, pages 143–152, 1984.
3. A. R. Calderbank and W. M. Kantor. The geometry of two-weight codes. In *Bull. London Math. Soc.,vol. 18*, pages 97–122, 1986.
4. A. Canteaut, P. Charpin, and H. Dobbertin. Weight divisibility of cyclic codes, highly nonlinear functions on $\mathbf{F}_{2^m}$, and crosscorrelation of maximum-length sequences. *SIAM J. Discrete Math.*, 13(1):105–138 (electronic), 2000.
5. C. Carlet, P. Charpin, , and V. Zinoviev. Codes, bent functions and permutations suitable for DES-like cryptosystems. In *Designs, Codes and Cryptography, 15(2), pp. 125–156*, 1998.
6. C. Carlet, C. Ding, and J. Yuan. Linear codes from perfect nonlinear mappings and their secret sharing schemes. In *IEEE Trans. Inform. Theory, vol. 51, No.6*, pages 2089–2102, 2005.
7. S.-T. Choi, J.-Y. Kim, J.-S. No, and H. Chung. Weight distribution of some cyclic codes. In *IEEE Int. Symp. Inf. Theory*, pages 2901–2903, 2012.
8. B. Courteau and J. Wolfmann. On triple-sum-sets and two or three weights codes. In *Discrete Math. vol. 50*, pages 179–191, 1984.
9. P. Delsarte. On subfield subcodes of modified Reed-Solomon codes. In *IEEE Trans. Inf. Theory 21(5)*, pages 575–576, 1975.
10. C. Ding. Linear codes from some 2-Designs. In *IEEE Transactions on Information Theory 61(6)*, pages 3265–3275, 2015.
11. C. Ding. A construction of binary linear codes from Boolean functions. In *arXiv:1511.00321*, 2015.

12. C. Ding, C. Li, N. Li, and Z. Zhou. Three-weight cyclic codes and their weight distributions. In *Preprint*.
13. C. Ding, J. Luo, and H. Niederreiter. Two weight codes punctured from irreducible cyclic codes. In *Proc. 1st Int. Workshop Coding Theory Cryp- tography, Y. Li, S. Ling, H. Niederreiter, H. Wang, C. Xing, and S. Zhang, Eds., Singapore*, pages 119–124, 2008.
14. C. Ding and X. Wang. A coding theory construction of new systematic authentication codes. In *Theoretical Comput. Sci. vol. 330, no. 1*, pages 81–99, 2005.
15. K. Ding and C. Ding. Binary linear codes with three Weights. In *IEEE Communications Letters 18(11)*, pages 1879–1882, 2014.
16. K. Ding and C. Ding. A Class of two-weight and three-weight codes and their applications in secret sharing. In *IEEE Transactions on Information Theory 61(11)*, pages 5835–5842, 2015.
17. K. Ding and C. Ding. A Class of two-weight and three-weight codes and their applications in secret sharing. In *CoRR abs/1503.06512*, 2015.
18. K. Feng and J. Luo. Value distribution of exponential sums from perfect nonlinear functions and their applications. In *IEEE Trans. Inf. Theory, vol. 53, no. 9*, pages 3035–3041, 2007.
19. T. Helleseth and A. Kholosha. Monomial and quadratic bent functions over the finite fields of odd characteristic. In *IEEE Transactions on Information Theory, 52(5), pages 2018–2032*, 2006.
20. T. Helleseth and A. Kholosha. New binomial bent functions over the finite fields of odd characteristic. In *IEEE Transactions on Information Theory, 56(9), pages 4646–4652*, 2010.
21. Z. Heng and Q. Yue. Several class of cyclic codes with either optimal three weights or a few weights. In *arxiv.org/pdf/1510.05355*, 2015.
22. H. D. L. Hollmann and Q. Xiang. A proof of the Welch and Niho conjectures on cross-correlations of binary m-sequences. In *Finite Fields Appl. 12(1)*, pages 253–286, 2001.
23. P. V. Kumar, R. A. Scholtz, and L.R. Welch. Generalized bent functions. In *J. Combin. Theory. Ser. A. Vol. 40, No 1, pages 90–107*, 1985.
24. C. Li, Q. Yue, and F. Li. Hamming weights of the duals of cyclic codes with two zeros. In *IEEE Trans. Inf. Theory, vol. 60, no. 7*, pages 3895–3902, 2014.
25. F. J. MacWilliams and N. J. Sloane. *The theory of error-correcting codes*, Amsterdam, North Holland, 1977.
26. S. Mesnager. Linear codes with few weights from weakly regular bent functions based on a generic construction. In *Preprint*, 2015.
27. C. Tang, N. Li, Y. Qi, Z. Zhou, and T. Helleseth. Linear codes with two or three weights from weakly regular bent functions. In *IEEE Transaction Information Theory Vol. 62, No. 3*, pages 1166–1176, 2016.
28. J. Wolfmann. Bent functions and coding theory. *Difference Sets, Sequences and their Correlation Properties,* A. Pott, P. V. Kumar, T. Helleseth and D. Jungnickel, eds., pp. 393–417. Amsterdam: Kluwer 1999.
29. G. Xu and X. Cao. Linear codes with two or three weights from some functions with low Walsh spectrum in odd characteristic. In *arXiv:1510.01031*, 2015.
30. J. Yuan, C. Carlet, and C. Ding. The weight distribution of a class of linear codes from perfect nonlinear functions. In *IEEE Transaction Information Theory Vol. 52, No. 2*, pages 712–717, 2006.
31. J. Yuan and C. Ding. Secret sharing schemes from three classes of linear codes. In *IEEE Trans. Inf. Theory, vol. 52, no. 1*, pages 206–212, 2006.
32. X. Zeng, L. Hu, W. Jiang, Q. Yue, and X. Cao. The weight distribution of a class of p-ary cyclic codes. In *Finite Fields Appl., vol. 16, no. 1*, pages 56–73, 2010.
33. Z. Zhou and C. Ding. A class of three-weight codes. In *Finite Fields Appl., vol. 25*, pages 79–93, 2014.
34. Z. Zhou, N. Li, C. Fan, and T. Helleseth. Linear Codes with Two or Three Weights From Quadratic Bent Functions. In *ArXiv: 1506.06830v1, 2015.*

Index

S. Mesnager, *Bent Functions*, DOI 10.1007/978-3-319-32595-8